Recent Advances in Processing Technologies for Substance Extraction, Separation, and Enrichment

Recent Advances in Processing Technologies for Substance Extraction, Separation, and Enrichment

Editors

Yanlin Zhang
Prashant K. Sarswat

MDPI

Basel • Beijing • Wuhan • Barcelona • Belgrade • Novi Sad • Cluj • Manchester

Editors
Yanlin Zhang
School of Chemistry
The University of Melbourne
Melbourne
Australia

Prashant K. Sarswat
Materials Science and Engineering
The University of Utah
Salt Lake City, UT
USA

Editorial Office
MDPI AG
Grosspeteranlage 5
4052 Basel, Switzerland

This is a reprint of articles from the Special Issue published online in the open access journal *Processes* (ISSN 2227-9717) (available at: https://www.mdpi.com/journal/processes/special_issues/Extraction_hydrometallurgy).

For citation purposes, cite each article independently as indicated on the article page online and as indicated below:

Lastname, A.A.; Lastname, B.B. Article Title. *Journal Name* **Year**, *Volume Number*, Page Range.

ISBN 978-3-7258-1909-6 (Hbk)
ISBN 978-3-7258-1910-2 (PDF)
doi.org/10.3390/books978-3-7258-1910-2

Cover image courtesy of Yanlin Zhang and Prashant K. Sarswat

Contents

About the Editors

Yanlin Zhang

Yanlin Zhang received his PhD degree from Monash University. He has 15 years' work experience in the petrochemical industry as an engineer and 10 years' experience as a Researcher at Melbourne University. He is now a Senior Research Fellow at the Institute of Process Engineering, the Chinese Academy of Sciences. He has expertise in analytical science, separation technology, and materials science. He is the author of over 20 publications in many leading journals.

Prashant K. Sarswat

Prashant K. Sarswat is a Research Associate Professor in the Department of Materials Science and Engineering at the University of Utah. He has over 18 years of experience in Materials Science and Nanotechnology. He earned his PhD from the University of Utah and holds a B.Tech degree from the Indian Institute of Technology. He has authored over 100 publications in prestigious journals.

processes

Editorial

Special Issue "Recent Advances in Processing Technologies for Substance Extraction, Separation, and Enrichment"

Yanlin Zhang [1,*] and Prashank K. Sarswat [2,*]

1 School of Chemistry, The University of Melbourne, Parkville, VIC 3010, Australia
2 Department of Materials Science and Engineering, University of Utah, Salt Lake City, UT 84112, USA
* Correspondence: yanlinz@unimelb.edu.au (Y.Z.); prashant.sarswat@utah.edu (P.K.S.); Tel.: +86-15053328709 (Y.Z.); +1-8015206919 (P.K.S.)

Substance separation has always been the foundation of production processes in a variety of industrial sectors such as hydrometallurgy, natural compounds extraction, food industry, pharmaceutical industry, and environmental engineering. Separation processes can be performed for different purposes, e.g., extraction of valuable components from less valuable matrices, removal of impurities, and enrichment of trace components. Conventional techniques, including distillation, liquid-phase extraction (LPE), solid-phase extraction (SPE), precipitation, and electrodeposition, are the dominant technologies in the separation of various types of substances and have found a broad application in both lab-scale research and industrial production. With the development of new theories, materials, and technologies, significant advancements have occurred in recent decades that have demonstrated advantages with regards to efficiency, energy saving, environment friendliness, costs, and value addition. In particular, new modeling techniques for process optimization, novel extraction and separation systems, green solvents, and novel solid-phase materials (e.g., mesoporous and nanostructured materials) for extraction have been developed and adopted by engineers for various applications. Typical examples include the modeling and simulation of electrochemical processes (deposition and separation), liquid–liquid extraction process for the optimization of mass transfer performance, minimization of energy consumption, application of green solvents and new ionic liquids, tailored materials for highly selective separation, membranes for SPE, extraction of natural products for nutritional, and medical and health care applications. The use of these new materials and technologies can effectively improve the mass transfer performance and separation efficiency, facilitate environmental and/or user friendliness, and improve people's quality of life.

This Special Issue on the "Recent Advances in Processing Technologies for Substance Extraction, Separation and Enrichment" aimed to provide a forum for scientists to present their research findings in new solvents, extractants, solid-phase materials, techniques, as well as modeling and optimization methodologies, to address some of the most challenging problems faced by industries today and to demonstrate the potential value of a variety of resources which may previously have been overlooked. Thanks to the tremendous efforts of our esteemed authors, this Special Issue is brimming with excitement, featuring twenty-one cutting-edge research papers and five comprehensive review articles. Dive into the forefront of innovation where we explore diverse extraction systems—from the depths of hydrometallurgy to the extraction of nutritious natural compounds. Discover how novel green solvents and advanced technologies are revolutionizing the utilization of secondary resources. This collection promises to ignite your curiosity and inspire breakthroughs in extraction science.

Separation technologies play a crucial role in hydrometallurgy. Several research papers and review articles have presented the use of different extraction techniques for the separation of different elements [1–7]. Their applications cover the separation of base metals,

Citation: Zhang, Y.; Sarswat, P.K. Special Issue "Recent Advances in Processing Technologies for Substance Extraction, Separation, and Enrichment". *Processes* **2024**, *12*, 1447. https://doi.org/10.3390/pr12071447

Received: 20 June 2024
Accepted: 8 July 2024
Published: 11 July 2024

recovery of precious metals, and separation of rare earth metals. Waste water treatment in this industry via separation technology has also been presented in a research paper [8]. The utilization of natural organic substances (such as bioactive compounds, proteins, and nutritional components in natural resources) and food industry residues (such as nutrients, health care products, and skin care components) has been a fascinating research topic in recent decades. This Special Issue involves a number of research and review articles that present the use of different technologies for the extraction of such components from different resources [9–22]. In many of these contributions, various novel reagents and methods have been utilized by the authors to demonstrate the value of new technologies, including three-phase partitioning [10], reactive extraction [12], supercritical fluid extraction [14,15], deep eutectic solvents [17,20], enzymatic processing [11,23], ultrasound assisted extraction [11,13,15,19,22], microwave-assisted extraction [13], centrifugal extractor [20], pulsed electric field technology [19,21], and high-pressure homogenization [13,22]. Many of the papers dealing with natural compounds are accompanied by an assessment of their properties, such as antioxidant activity, providing an indication of the extraction's value [13,15,17,19,24].

Different forms of membrane separation technologies that feature a reduced footprint, low-energy requirements, and that are environmentally benign are attractive areas for research and applications, as indicated by the papers in this Special Issue [25,26].

As the guest editors, we would like to express our gratitude to all the authors who generously contributed their excellent work to this Special Issue. We also acknowledge the colleagues and assistant editors involved in this Issue, especially Lily Liu and Scott Pan, for their collaboration and cooperation in the whole editing process.

Conflicts of Interest: The authors declare no conflicts of interest.

References

1. Baimbetov, B.; Moldabayeva, G.; Yeleuliyeva, A.; Jumankulova, S.; Taimassova, A.; Adilzhan, Z.; Baisultanov, R.; Yakob, E.; Serikbayev, V. Prospects of Processing Tungsten Ores from the Akchatau Deposit. *Processes* **2024**, *12*, 77. [CrossRef]
2. Domańska, U.; Wiśniewska, A.; Dąbrowski, Z. Recovery of Strategic Metals from Waste Printed Circuit Boards with Deep Eutectic Solvents and Ionic Liquids. *Processes* **2024**, *12*, 530. [CrossRef]
3. Kostanyan, A.E.; Belova, V.V.; Tsareva, Y.V.; Petyaeva, M.M. Separation of Rare Earth Elements in Multistage Extraction Columns in Chromatography Mode: Experimental Study and Mathematical Simulation. *Processes* **2023**, *11*, 1757. [CrossRef]
4. Mwase, J.M.; Safarian, J. Desilication of Sodium Aluminate Solutions from the Alkaline Leaching of Calcium-Aluminate Slags. *Processes* **2022**, *10*, 1769. [CrossRef]
5. Pathapati, S.V.S.H.; Free, M.L.; Sarswat, P.K. A Comparative Study on Recent Developments for Individual Rare Earth Elements Separation. *Processes* **2023**, *11*, 2070. [CrossRef]
6. Zhang, Y.; Duan, X.; Hu, C.; Du, G.; Wang, Y. Green and Cost-Effective Separation of Cadmium from Base Metals in Chloride Medium with Halide-Loaded Anion Exchanger. *Processes* **2023**, *11*, 1051. [CrossRef]
7. Free, M.L.; Ilunga, J.K.; Podder, P.; Sarswat, P.K. Assessing and Improving Biooxidation for Acid Generation and Rare Earth Element Extraction. *Processes* **2023**, *11*, 1005. [CrossRef]
8. Ding, X.; Jiang, J.; Zhao, Y.; Dong, Z.; Wang, L.; Liu, Y. The Recovery of Sulfuric Acid in the Presence of Zr(IV) and Hf(IV) by Solvent Extraction with TEHA and Its Mixtures. *Processes* **2024**, *12*, 940. [CrossRef]
9. Kim, S.J. Experimental Study on Enrichment of Indole in Wash Oil by a Combined Process of Extraction and Re-Extraction. *Processes* **2023**, *11*, 3097. [CrossRef]
10. Liu, B.; Li, X.; Liu, Q.; Tan, Z. Extraction and Purification of Kiwifruit Seed Oil Using Three-Phase Partitioning: An Efficient and Value-Adding Method for Agro-Industrial Residue Utilization. *Processes* **2023**, *11*, 2581. [CrossRef]
11. Anita Šalić, A.; Lina Šepić, L.; Iva Turkalj, I.; Bruno Zelić, B.; Dunja Šamec, D. Comparative Analysis of Enzyme-, Ultrasound-, Mechanical-, and Chemical-Assisted Extraction of Biflavonoids from Ginkgo Leaves. *Processes* **2024**, *12*, 982. [CrossRef]
12. Blaga, A.C.; Kloetzer, L.; Cascaval, D.; Galaction, A.-I.; Tucaliuc, A. Studies on Reactive Extraction of Itaconic Acid from Fermentation Broths. *Processes* **2024**, *12*, 725. [CrossRef]
13. Cegledi, E.; Dobroslavić, E.; Zorić, Z.; Repajić, M.; Garofulić, I.E. Antioxidant Activity of Carob Tree (*Ceratonia siliqua* L.) Leaf Extracts Obtained by Advanced Extraction Techniques. *Processes* **2024**, *12*, 658. [CrossRef]
14. Cestonaro, G.; Gonzalez-Ortega, R.; Grosso, A.L.; Morozova, K.; Ferrentino, G.; Scampicchio, M.; Costanzo, E. Comparing Quality and Functional Properties of Protein Isolates from Soybean Cakes: Effect of De-Oiling Technologies. *Processes* **2024**, *12*, 600. [CrossRef]

15. Čulina, P.; Balbino, S.; Jokić, S.; Dragović-Uzelac, V.; Pedisić, S. Efficiency of Supercritical CO_2 and Ultrasound-Assisted Extraction Techniques for Isolation of Bioactive Molecules from Sea Buckthorn (*Elaeagnus rhamnoides* (L.) A. Nelson) Berry Oils and Cakes. *Processes* **2024**, *12*, 698. [CrossRef]
16. Ispiryan, A.; Kraujutiene, I.; Viskelis, J. Retaining Resveratrol Content in Berries and Berry Products with Agricultural and Processing Techniques: Review. *Processes* **2024**, *12*, 1216. [CrossRef]
17. Ivkovic, D.; Cvijetic, I.; Radoicic, A.; Stojkovic-Filipovic, J.; Trifkovic, J.; Ristivojevic, M.K.; Ristivojevic, P. NADES-Based Extracts of Selected Medicinal Herbs as Promising Formulations for Cosmetic Usage. *Processes* **2024**, *12*, 992. [CrossRef]
18. Kaliniewicz, Z.; Markowska-Mendik, A.; Warechowska, M.; Lipiński, S.; Gasparis, S. Correlations between a Friabilin Content Indicator and Selected Physicochemical and Mechanical Properties of Wheat Grain for Processing Suitability Assessment. *Processes* **2024**, *12*, 398. [CrossRef]
19. Kourelatou, A.; Chatzimitakos, T.; Athanasiadis, V.; Kotsou, K.; Makrygiannis, I.; Bozinou, E.; Lalas, S.I. Seeking Optimal Extraction Method for Augmenting Hibiscus sabdariffa Bioactive Compounds and Antioxidant Activity. *Processes* **2024**, *12*, 581. [CrossRef]
20. Lobovich, D.V.; Zinov'eva, I.V.; Milevskii, N.A.; Kostanyan, A.E.; Zakhodyaeva, Y.A.; Voshkin, A.A. Extraction Kinetics of Pyridine, Quinoline, and Indole from the Organic Phase with Natural Deep Eutectic Solvents and Separation Study Using a Centrifugal Extractor. *Processes* **2024**, *12*, 488. [CrossRef]
21. Ramaswamy, R.; Krishnan, S.B.; Leong, S.S.J. Pulsed Electric Field Technology for Recovery of Proteins from Waste Plant Resources and Deformed Mushrooms: A Review. *Processes* **2024**, *12*, 342. [CrossRef]
22. Rida, H.; Peydecastaing, J.; Takache, H.; Ismail, A.; Pontalier, P.-Y. Comparison of Tetraselmis Suecica Cell Disruption Techniques: Kinetic Study and Extraction of Hydrosoluble Compounds. *Processes* **2024**, *12*, 1139. [CrossRef]
23. Alzahrani, F.; Akanbi, T.O.; Scarlett, C.J.; Aryee, A.N.A. The Use of Immobilised Enzymes for Lipid and Dairy Processing and Their Waste Products: A Review of Current Progress. *Processes* **2024**, *12*, 634. [CrossRef]
24. de Meneses Costa Ferreira, L.M.; de Souza, P.D.Q.; Pereira, R.R.; da Silva, E.O.; Barbosa, W.L.R.; Silva-Júnior, J.O.C.; Converti, A.; Ribeiro-Costa, R.M. Preliminary Study on the Chemical and Biological Properties of Propolis Extract from Stingless Bees from the Northern Region of Brazil. *Processes* **2024**, *12*, 700. [CrossRef]
25. Alguacil, F.J.; Robla, J.I. Solvent Extraction of Gold(III) by 2-Ethylhexanol and Modeling of Facilitated Transport across a Supported Liquid Membrane. *Processes* **2024**, *12*, 771. [CrossRef]
26. Alammar, A.Y.; Choi, S.-H.; Buonomenna, M.G. Hollow Fiber Membrane Modification by Interfacial Polymerization for Organic Solvent Nanofiltration. *Processes* **2024**, *12*, 563. [CrossRef]

processes MDPI

Article

Desilication of Sodium Aluminate Solutions from the Alkaline Leaching of Calcium-Aluminate Slags

James Malumbo Mwase * and Jafar Safarian

Department of Materials Science and Engineering, Norwegian University of Science and Technology (NTNU), 7491 Trondheim, Norway
* Correspondence: james.mwase@ntnu.no

Abstract: The desilication of sodium aluminate solutions prior to precipitation of aluminum tri-hydroxides is an essential step in the production of high purity alumina for aluminum production. This study evaluates the desilication of sodium aluminate solutions derived from the leaching of calcium-aluminate slags with sodium carbonate, using CaO, $Ca(OH)_2$, and MgO fine particles. The influence of the amount of CaO used, temperature, and comparisons with $Ca(OH)_2$ and MgO were explored. Laboratory scale test work showed that the optimal conditions for this process were using 6 g/L of CaO at 90 °C for 90 min. This resulted in 92% of the Si being removed with as little as 7% co-precipitation of Al. The other desilicating agents, namely $Ca(OH)_2$ and MgO, also proved effective in removing Si but at slower rates and higher amounts of Al co-precipitated. The characteristics of solid residue obtained after the process indicated that the desilication is via the formation of hydrogarnet, Grossular, and hydrotalcite dominant phases for CaO, $Ca(OH)_2$ and MgO agents, respectively.

Keywords: desilication; silica; pedersen process; CaO

Citation: Mwase, J.M.; Safarian, J. Desilication of Sodium Aluminate Solutions from the Alkaline Leaching of Calcium-Aluminate Slags. *Processes* **2022**, *10*, 1769. https://doi.org/10.3390/pr10091769

Academic Editors: Yanlin Zhang and Prashant K. Sarswat

Received: 10 August 2022
Accepted: 1 September 2022
Published: 3 September 2022

1. Introduction

Desilication of sodium aluminate solutions is an essential step in the production of alumina through the Bayer process. In this process, bauxite ores containing silicon are leached in an alkaline media, with the primary purpose of extracting aluminum. However, silicon is often co-extracted due to a reaction with sodium hydroxide (Equation (1)), which can contaminate the final alumina product. To prevent this, a desilication process to reduce the amount of silicon in solution is conducted prior to precipitating hydrated alumina. In the Bayer process, bauxite ores are pressure leached at a high temperature (100–250 °C) using sodium hydroxide solution. The leachate solution is then cooled and seeded to precipitate alumina hydrates. Desilication of this leachate prior to precipitation is achieved through the addition of CaO solid particles in the leaching phase. This also aids in the regulation of carbonates and phosphates, which in high concentrations are detrimental to the precipitation process. Further, the presence of CaO accelerates the leaching of aluminum when it is in the mineral phase diaspore, which is the most difficult alumina mineral to leach. The chemistry of Si during the desilication has been described by a few studies [1–3] as follows.

$$SiO_{2(s)} + 2NaOH = Na_2SiO_{3(aq)} + H_2O \quad (1)$$

The soluble products formed in leaching, namely $NaAlO_2$ and Na_2SiO_3, react to form non-soluble aluminosilicate precipitates with zeolite structures and are termed desilication products (DSP) of $Na_2O.Al_2O_3.2SiO_2$ or $Na_8Al_6Si_6O_{24}(OH)_2$. These DSPs further react with sodium hydroxide and carbonates in the solution to form sodalite ($Na_8Al_6Si_6O_{24}(CO_3).2H_2O$). The whole process can be considered a 'self-desilication'. The addition of CaO results in the rest of the Si reacting to form cancrinite ($Na_6Ca_2Al_6Si_6O_{24}(CO_3)_2.2H_2O$), which is a slightly more soluble phase.

A recent study investigated the use of the Pedersen process as an alternative to the Bayer process [4]. The study also evaluated the process as a means to valorize solid waste residues from the Bayer process such as bauxite residue (red mud). In the investigation calcium aluminate slags obtained via a prior smelting-reduction process, were leached with sodium carbonate solution to extract alumina. The solution was then sparged with CO_2 gas to precipitate hydrated alumina (carbonation). Unlike the Bayer process where CaO is added during leaching, the study [4] added it after leaching, adapting the method used in a similar investigation [5]. It was observed that without the desilication step, the alumina precipitated had a high silica content exceeding the standard for metallurgical grade alumina [4]. Both studies [4,5] used this method at laboratory testing stage. It is speculated that if added during the leaching phase of the Pedersen process, the CaO may be consumed to form $CaCO_3$ (grey mud) rather than desilication products. Before carbonating with CO_2 gas to precipitate hydrated alumina the leachate solution was desilicated using up to 28 g of powdered CaO per 1 L of leachate solution at a temperature of 70 °C. The process successfully reduced the amount of silicon in solution from concentrations of 0.4 g/L to 0.02 g/L [4]. Considerable Al co-precipitation, ranging from 18 to 50%, was also observed. The process yielded a product that was mostly calcium monocarboaluminate (50–70%), followed by calcite (11–36%), portlandite (6–10%), and hydrogarnite (1–18%). It was concluded that the desilication mostly occurred via the formation of hydrogarnite ($Ca_3Al_2(SiO_4)_{3-x}(OH)_{4x}$). Prior to using this method, the chemistry and mechanisms of this process were not known and are still not fully understood. Based on the results the authors [4] proposed three possible mechanisms for the process based on the previous studies [6–10].

In the first mechanism Si in the form of $H_7SiO_6^-$ reacts with CaO to form hydrogarnet by the reaction:

$$3Ca^{2+} + 2AlO_2^- + (3-x)H_7SiO_6^- + (x+1)OH^- = Ca_3Al_2(SiO_4)_{3-x}(OH)_{4x} + (11-5x)H_2O \quad (2)$$

In a second mechanism Si is in the form of $HSiO_3^-$ reacting to form hydrogarnet by the reaction:

$$3Ca^{2+} + 2AlO_2^- + (3-x)HSiO_3^- + (x+1)OH^- + (2x-1)H_2O = Ca_3Al_2(SiO_4)_{3-x}(OH)_{4x} \quad (3)$$

It was further proposed that Si was removed from the solution by the formation of calcium monocarboaluminate, which in turn reacted with the Si in solution forming hydrogarnite. In some tests there was no formation of hydrogarnet, despite up to 27% of the Si being removed. In these cases, it is proposed that the monocarboaluminate may have adsorbed the Si ions and not necessarily reacted with them. Calcium monocarboaluminate is most likely the result of a complex reaction between the CaO, unreacted sodium carbonate and the newly formed sodium aluminate complex from leaching.

The other products of note, namely calcite and portlandite were possibly formed by reactions identical or similar to the causticisation reactions described by [11]. Calcium oxide first reacts with water to give portlandite (Equation (4)).

$$CaO_{(s)} + H_2O \rightarrow Ca(OH)_{2(s)} \quad (4)$$

Some of the portlandite reacts with unreacted sodium carbonate to give calcite (Equation (5)).

$$Ca(OH)_{2(s)} + CO^{2-}{}_3 \rightarrow CaCO_{3(s)} + 2OH^- \quad (5)$$

The present study aims to optimize the use of the desilication process as part of the Pedersen process. The influence of the amount of CaO used, temperature, and comparisons with other desilication agents (MgO and $Ca(OH)_2$) will be explored.

2. Materials and Methods

2.1. Solution Preparation

The solutions for the desilication tests were obtained by leaching a 100 g calcium aluminate slag sample (Table 1) with 1 L of 60 g/L of sodium carbonate solution at a temperature of 90 °C for a duration of 90 min. It is worth noting that the slag was produced via the smelting-reduction of bauxite with lime and carbon. On completion of each leaching test the slurry was vacuum filtered, recovering around 950 mL of the solution, which was allowed to cool overnight before the desilication tests.

Table 1. XRF analysis of the calcium aluminate slag.

Al_2O_3	CaO	SiO_2	TiO_2	MgO	Fe	SO_3	P_2O_5
43.90	52.12	2.00	1.30	0.44	0.25	0.07	0.01

2.2. Desilication and Analyses

Before each test commenced, a sample of solution was taken to determine the amount of Si, Al, and Ca at the start of each test. The experiments were conducted in a jacketed glass reactor. The mixture was heated using silicon oil circulated using a VWR thermobath circulator. The temperature of the mixture was measured using a temperature probe connected to the circulator. The mixture was agitated using an overhead stirrer with a shaft and paddle impeller. Figure 1 shows a drawing of the experimental set-up. All the tests were conducted by adding the leachate solution to analytical reagent grade powdered CaO in the glass reactor. In each test, the mixture was heated to the desired temperature while being agitated at 400 rpm for 90 min. The time it took to reach the desired temperature was included in the entire duration of each experiment (90 min), and it varied according to the set temperature. Samples of solution were withdrawn at 15 min intervals for the duration of the test and vacuum filtered with 0.22 μm membrane for ICP-MS analysis of Si, Al, and Ca. At the end of each test, the reactor was emptied via a drainage valve at the bottom of the vessel. The contents were vacuum filtered with general purpose Whatman filter paper. The filter cake was dried at 60 °C for 2 days and analyzed via XRD. The amount of CaO used varied between 1 and 12 g. Three different temperatures were used, namely 40, 70, and 90 °C.

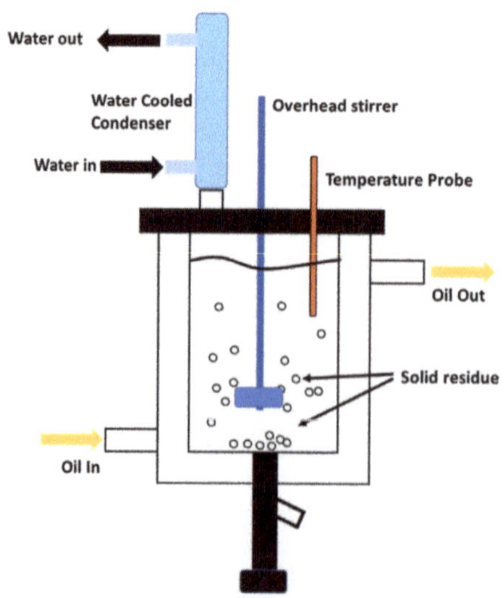

Figure 1. Schematic drawing of experimental equipment.

Further, comparison was made with the results from using $Ca(OH)_2$ and MgO as the desilication agent as opposed to CaO.

3. Results

This section may be divided by subheadings. It should provide a concise and precise description of the experimental results, their interpretation, as well as the experimental conclusions that can be drawn.

3.1. ICP-MS Analysis of Desilicated Solutions

The results from the analyses of solution samples and the solid residues are detailed below and discussed in the next section. The results for the tests using 1 and 2 g of CaO are excluded, as those amounts were ineffective and did not remove any Si. Table 2 shows the percentage reduction of Si with time and in total how much Si was removed. Moreover, Table 3 shows the reduction of Al concentration in g/L with time and in total how much Al was co-precipitated or lost at the end of the test.

Table 2. The desilication conditions and percentage reduction of Si with time.

Test No.	**1**	**2**	**3**	**4**	**5**	**6**	**7**	**8**	**9**
CaO added (g)	12	6	4	6	6	4	4		
$Ca(OH)_2$ added (g)								6	
MgO added (g)									6
Temp (°C)	90	90	70	40	70	40	90	90	90
Initial Si concn. (g/L)	0.3	0.2	0.2	0.2	0.2	0.2	0.2	0.2	0.2
Time (mins)									
0	100	100	100	100	100	100	100	100	100
15	21	77	74	66	81	71	68	74	58
30	19	59	67	64	68	65	69	45	52
45	21	31	51	61	59	63	61	26	40
60	17	16	44	53	52	62	40	15	29
75	15	9	33	47	49	58	58	10	27
90	13	9	25	34	49	52	61	8	25
% of Si removed	87	91	75	66	51	48	39	92	75

Table 3. Reduction in Al concentration (g/L) with time.

Test No. **Time (mins)**	**1**	**2**	**3**	**4**	**5**	**6**	**7**	**8**	**9**
0	16.4	15.1	13.7	13.3	12.6	14.4	13.3	14.8	13.4
15	14.3	15.0	12.1	10.9	12.4	12.8	12.3	13.3	10.6
30	12.6	14.6	12.7	11.9	12.5	12.8	12.4	13.0	12.0
45	13.5	13.5	13.1	12.6	12.5	12.8	11.9	13.3	11.4
60	13.0	13.6	13.1	12.6	12.3	13.3	8.9	12.9	9.9
75	13.5	14.1	13.0	12.5	12.0	13.5	12.0	12.8	10.7
90	13.4	14.1	12.8	11.1	11.6	12.6	12.8	12.5	10.3
% of Al co-precipitated	18.3	6.6	3.2	16.1	8.5	12.5	3.2	15.4	22.9

3.2. XRD Analysis of Solid Residue

Table 4 shows the various phases present in the solid residue recovered after each desilication test. The samples were analyzed using a Bruker D8 A25 DaVinci™ X-ray Diffractometer with CuKα radiation. The identification and quantification of phases in all residue samples were carried out by the Rietveld method using EVA and TOPAS software respectively. Obviously, silicon removal has occurred through the formation of some Si-containing phases, while Al has been partially lost to the precipitate. Figure 2 shows a typical XRD spectrum for test 2 solid residue with the identified phases. A few peaks, minor phases, such as the one at 11.5 degrees, were not unfortunately identified for this sample, while for the others they were most identified.

Table 4. XRD analysis quantifying mineral phases by wt% in solid residue.

Test No. **Mineral Phase**	**1**	**2**	**3**	**4**	**5**	**6**	**7**	**8**	**9**
Hydrogarnet [1]	71.4	80.0	47.3	47.8	1.0	22.0	7.0	28.0	
Calcium monocarboaluminate [2]	24.4		29.7	30.8	63.0	60.0	21.0		
Portlandite ($Ca(OH)_2$)	4.0	8.0	5.2	4.9	30.0	11.0	5.0	4.0	
Calcite ($CaCO_3$)	0.2	10.0	2.5	2.7	6.1	0.5	53.0	1.0	
Hydrotalcite [3]		2.0						1.0	64.0
Strätlingite [4]			15.3	13.8		6.5	14.0		
Grossular [5]								64.0	
MgO									36.0

[1] $Ca_3Al_2(SiO4)_{3-x}(OH)_{4x}$ [x = 1.5–3]. [2] $Ca_4Al_2(OH)12CO_3.5H_2O$. [3] $Mg_4Al_2(CO_3)(OH)_{12}.3H_2O$. [4] $Ca_2Al(AlSi)_2O_2(OH)_{10}$. [5] Less crystalline hydrogarnet.

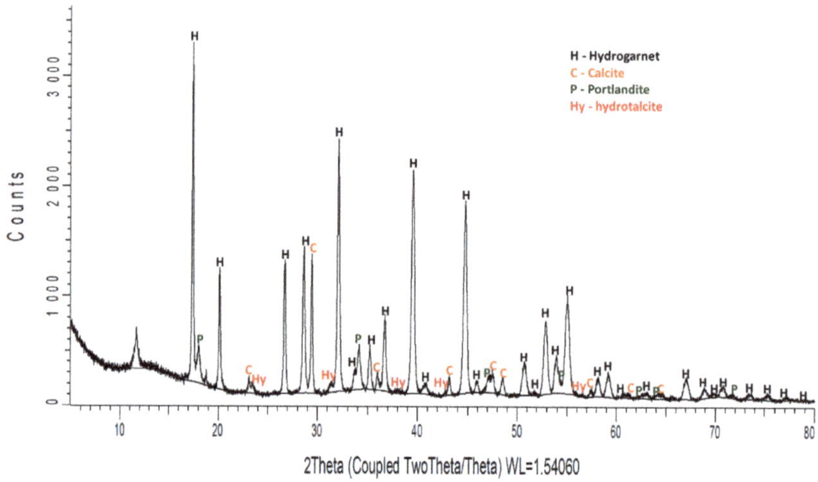

Figure 2. XRD spectrum for test 2 residue.

4. Discussion

4.1. Effect of Added CaO Mass

An initial comparison is made between tests 1, 2, and 7. These three tests were conducted at 90 °C using 12, 6, and 4 g of CaO. The results in percentage reduction of Si over time can be seen in Figure 3. It should be noted that Figures 3–6 show the reduction of silica in solution as a percentage of the original amount before desilication commenced. Tests 1 and 2 had similar amounts of Si removed, with the test 2 removing a bit more than test 1. In test 2, the amount of Si was reduced to 9% of the starting amount. This means CaO effectively removed 91% of the Si. Test 7 lagged behind the other two, only removing 39% of the Si. Further, tests 1 and 2 removed comparable amounts of Si to the 28 g tests conducted by [12]. This shows that the desilication can be successfully conducted using significantly lower CaO per volume of leachate solution that we experienced before. Comparing the two tests shows that the amount of CaO present clearly influenced the kinetics of the reaction. In test 1, the Si is mostly removed in a period of 20 min, while test 2 required 60 min to remove an equal amount of Si. The XRD analysis results of the solid residue (Table 4) indicate a larger presence of hydrogarnet in tests 1 and 2 than was observed in the tests conducted by [12] where 28 g of CaO was used. In those tests only up to 12% of the residue was hydrogarnet with the majority being calcium monocarboaluminate followed by portlandite and calcite. This means that the hydrogarnet is the more thermodynamically stable and favored product, reactions (2) and (3), and once equilibrium is reached the remaining or excess CaO reacts with unreacted $NaCO_3$ to form calcium monocarboaluminate. As before, it is proposed that the Si removal occurred due to the formation of hydrogarnet and possibly by adsorption of Si ions to the multi-layered double hydroxides in the calcium monocarboaluminate also present in tests 1 and 7. The mineral phase profile of the solid residue from test 7 is curious in that it contains Strätlingte. This phase has not been seen in previous results. From its formula it is clear it is responsible for some Si removal as well as some co-precipitation of Al. As expected, co-precipitation of Al occurred, and Table 3 shows the reduction in concentration with time and the final amount of Al lost as a percentage of the starting amount at time 0.

Similar to test 7 (4g), there were several tests whose results had data points that appeared to be Si gains. These cases are most likely Si being re-dissolved back into solution, the reason for which is unknown. This may be the Si that is not chemically bonded and part of the hydrogarnet compound, but that which has been adsorbed by the monocarboaluminate.

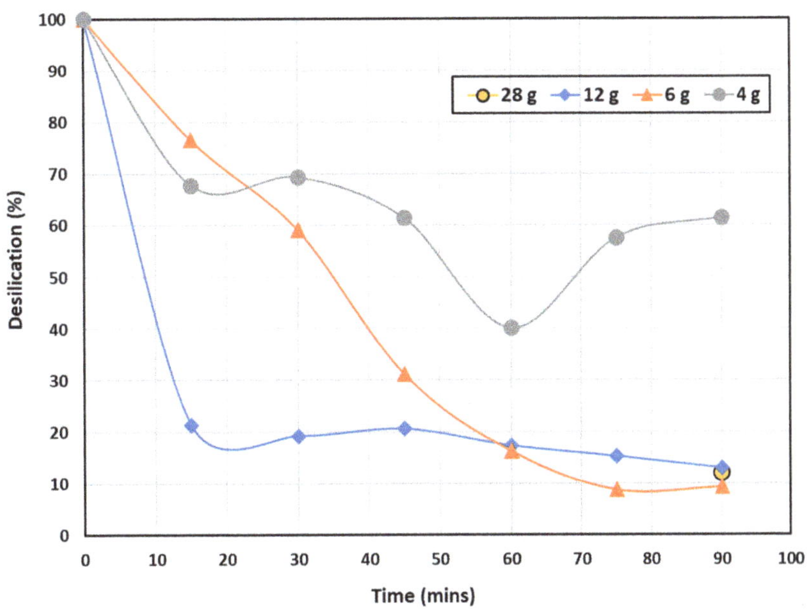

Figure 3. Percentage reduction of Si with time using 28 g [12], and tests 1, 2, and 7 with 12, 6, and 4 g of CaO at 90 °C for 90 min.

4.2. *Effect of Temperature*

Further tests were conducted with 6 and 4 g of CaO added and the influence of temperatures of 70 and 40 °C were noted and compared with the 90 °C tests. Figure 4 shows that the most effective temperature for desilication using 6 g of CaO is 90 °C (test 2). Test 4 and 5 performed similarly, although given test 5 had an operating temperature closer to test 2, it was expected it would remove an amount of Si closer to this test and not less than test 4. Table 4 shows that the solid residue of test 2 was significantly different from test 4 and 5. The higher temperature of 90 °C favored the formation of hydrogarnet over calcium monocarboaluminate, which formed more readily at 70 and 40 °C. This perhaps also explains why test 2 had markedly better Si removal that test 4 and 5. The removal of about 50% of Si in test 5 and the formation of calcium monocarboaluminate and portlandite as the dominant phases in the residue may indicate that these phases have Si impurity in their structure.

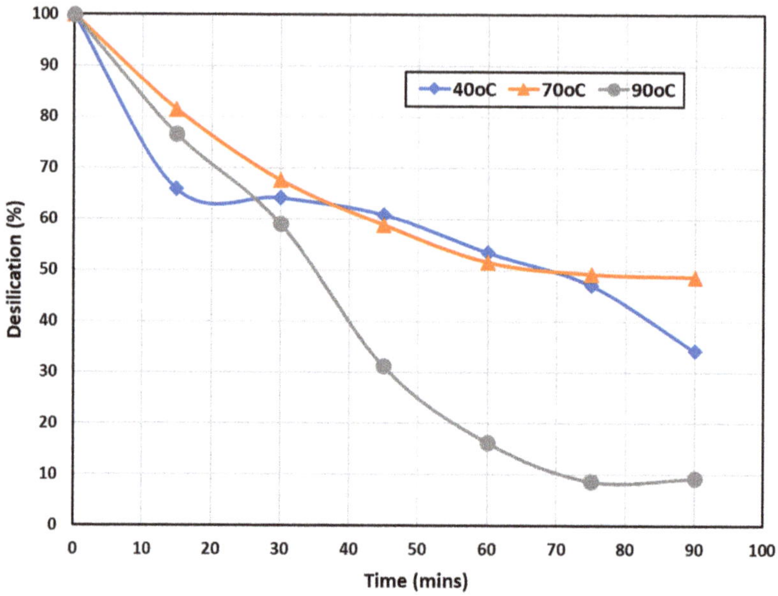

Figure 4. Percentage reduction of Si with time using 6 g CaO at temperatures 40 °C (Test 4), 70 °C (Test 5), and 90 °C (Test 2).

In the case of the tests using 4 g of CaO, Figure 5, the best performing test, was test 3 conducted at 70 °C. This contrasts with the tests using 6 g, where the best performing test was the one conducted at 90 °C. In this case the test 7 at 90 °C performed poorly, and extracted even less Si than test 6 at 40 °C. Table 4 shows that the mineral phase composition was qualitatively similar across the three tests with variation in the amounts. As already seen the test with the highest amount of hydrogarnet (test 3) has the highest Si removal. Test 7 with the lowest amount of hydrogarnet and highest amount of calcium monocarboaluminate. The formation of strätlingite phase in these three tests and formation of almost the same amounts of this phase in tests 3 and 7 may strengthen the idea that calcium-monocarboaluminate can potentially contain some Si impurity.

Considering the different results at various temperatures with different amounts of CaO added may be related to the equilibrium in the system and the sequence of reactions. For the best result obtained, test 2 at 90 °C, we observe hydrogarnet as the dominant Si-containing phase. However, for the test 7 at the same conditions with less CaO addition, hydrogarnet is not the dominant phase, while calcite, calcium monocarboaluminate, and strätlingite are dominant phases, respectively. It may be possible that more CaO addition into the system at 90 °C is accompanied with the conversion of strätlingite to hydrogarnet. Hence, with CaO addition at 90 °C, we may have the first calcite formation via chemical reactions (4) and (5), and simultaneously strätlingite precipitation. If CaO addition is added more, more Ca is available and hydrogarnet is formed.

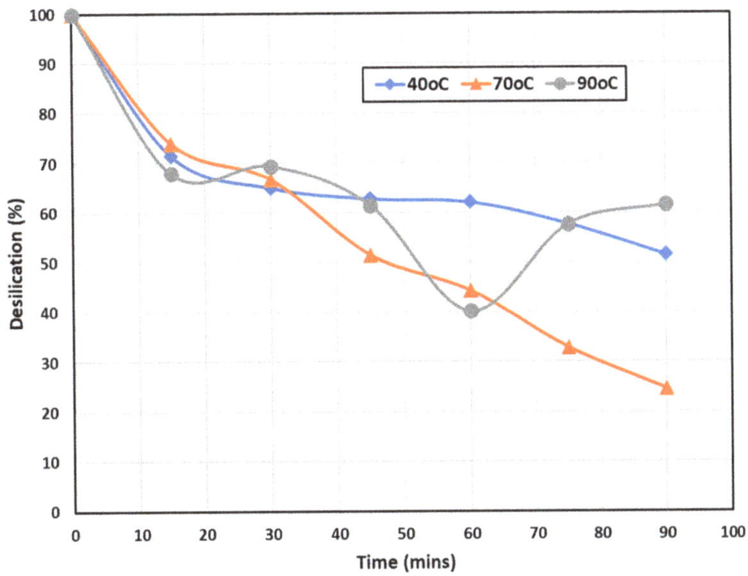

Figure 5. Percentage reduction of Si with time using 4 g CaO at temperatures 40 °C (Test 6), 70 °C (Test 3), and 90 °C (Test 7).

4.3. Effect of Type of Desilication Agent

Figure 6 shows a comparison and results when using different desilicating agents under identical conditions. Overall, $Ca(OH)_2$ removed the most Si, equaling 93% (Table 2). However, this is only 1% more than CaO. Table 4 shows the product from $Ca(OH)_2$ desilication was mostly hydrogarnet. Specifically, two types of hydrogarnet differ in the degree of crystallinity. The mechanism of hydrogarnet formation using $Ca(OH)_2$ is most likely identical to the ones proposed for CaO in Section 1 via chemical reactions (2) and (3). Although MgO removed the least amount (75%), it is worth noting it removed a considerable amount of Si. The major product in this case was hydrotalcite. Based on the chemical formula, it is assumed that this formed due to a complex reaction between the MgO, $NaAlO_2$, and unreacted sodium carbonate. However, an attempt to balance this equation and rewrite it in different ionic forms using HSC Chemistry software showed that the reaction was not possible. It is therefore assumed that the reaction takes place in a series of reactions forming intermediate products before the hydrotalcite is formed. Hydrotalcite is a layered double hydroxide [13]. Hence, the Si removal in the case of the MgO is probably due to the Si ions being adsorbed in between these double layers. This accounts for the Si removal in this test despite none of the products having Si in their chemical formulas.

The different rates of reaction are also quite noticeable among the different desilicating agents. CaO, as noted earlier, has a very short reaction time to achieve maximum Si removal, while $Ca(OH)_2$ is more gradual taking three times the amount of time to achieve the same result. Deciding the best desilicating agent will also take into consideration the amount of Al co-precipitated. It can be seen in Table 3 that CaO experiences the least amount of co-precipitation (6.7%). Although $Ca(OH)_2$ is just as effective as CaO in removing Si, it co-precipitates more than double the amount (15%) of Al as CaO. MgO is completely discounted, as it co-precipitates 23% of the Al.

Comparing the characteristics of the resulting residues from the three different desilication agents, it is observed that when CaO and $Ca(OH)_2$ have been used for 90 min, they were mostly converted to the other products, while for MgO there is a significant amount of MgO unreacted (Table 3), and Figure 6 clearly shows that it may be due to the slower desilication rate and need for longer reaction time for this agent. If MgO has not completely reacted due to the passivation via the formation of hydrotalcite over MgO particle, the particle size of this agent must be optimized, which is probably not a case for CaO and $Ca(OH)_2$ agents.

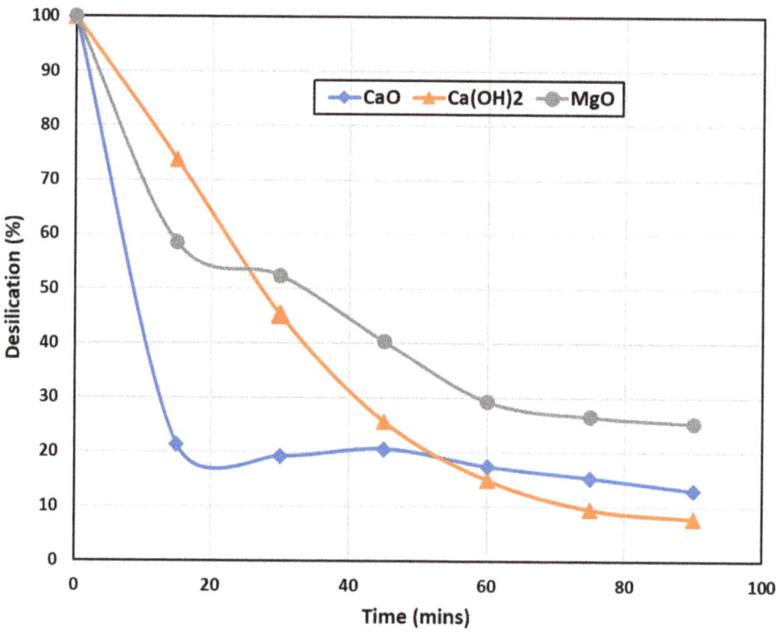

Figure 6. Percentage reduction of Si with time using 6 g of CaO, $Ca(OH)_2$, and MgO at 90 °C for 90 min.

4.4. Characteristics of the Desilicated Solutions

A comparison is made between the desilication results achieved in the current study and one which investigated adapting the Pedersen process for ferruginous bauxites of the Pacific Northwest in USA [5]. In that study, after several attempts, a leachate with 0.19 g/L of Si was obtained, using comparable leaching parameters to this study. Desilication including a first step using 28 g/L CaO at 90 °C for 1 h, then a second step using 2g/L CaO, resulted in 95% Si removal and up to 27% co-precipitation of Al. This is similar to the results achieved in this study, but with more CaO used.

A comparison with the desilication in the Bayer process is not easily conducted, as the amounts of CaO vary widely depending on the type of ore processed and the leaching temperature. Karstic type bauxites require high processing temperatures and addition of CaO, while Tropical type bauxites require lower leaching temperatures and sometimes no addition of CaO. What this study has established is that in comparison to the other works in this field [4,5], desilication can be achieved with a lower addition of CaO and correspondingly lower Al loss from the solution in this step. It has also shed some light on the mechanisms of the process.

5. Conclusions

The desilication of a sodium aluminate solution obtained from calcium aluminate slags was carried out using CaO, $Ca(OH)_2$, and MgO. The concluding remarks are summarized as:

- Calcium oxide has shown to be an effective desilicating agent of leachate solutions, depending on the amount added and applied temperature desilication varies.
- CaO works faster than other desilicating agents tested, namely calcium hydroxide and magnesium oxide, and co-precipitates less alumina.

- Desilication by CaO is mainly via the formation of hydrogarnet, while it is via the formation of both hydrogarnet and grossular by $Ca(OH)_2$. When MgO is used, silicon is removed via hydrotalcite formation.
- The desilication occurs largely due to the formation of hydrogarnet, and to a smaller extent, the adsorption of Si in between the layers of calcium monocarboaluminate, which is a layered double hydroxide. The optimal conditions for this process are 6 g of CaO for 1 L of leachate solution, with 0.3 g/L Si at 90 °C for 90 min.

Author Contributions: Conceptualization, J.M.M. and J.S.; methodology, J.M.M. and J.S.; software, J.M.M. and J.S.; validation, J.M.M. and J.S.; formal analysis, J.M.M. and J.S.; investigation, J.M.M.; resources, J.S.; data curation, J.M.M.; writing—original draft preparation, J.M.M.; writing—review and editing, J.M.M. and J.S.; visualization, J.M.M. and J.S.; supervision, J.S.; project administration, J.M.M. and J.S.; funding acquisition, J.S. All authors have read and agreed to the published version of the manuscript.

Funding: This project has received funding from the European Union's Horizon 2020 research and innovation programme under grant agreement 767533.

Institutional Review Board Statement: Not applicable.

Informed Consent Statement: Not applicable.

Data Availability Statement: Not applicable.

Acknowledgments: We also thank the partners in the ENSUREAL consortium for their continued support.

Conflicts of Interest: The authors declare no conflict of interest.

References

1. George, D.; Kwat, I. Dessilication of Bayer Process Solution. U.S. Patent 5118484, 2 June 1992.
2. Metson, J. Production of alumina. In *Fundamentals of Aluminium Metallurgy: Production, Processing and Applications*; Lumley, R., Ed.; Woodhead Publishing: Sawston, UK, 2011; pp. 23–48.
3. Balomenos, E.; Panias, D. *Application of Hydrometallurgy in Alumina Production: The Bayer Process*; Online Lecture in the HydroMetEC Introduction to Hydrometallurgy Course; 2021.
4. Mwase, J.M.; Vafeias, M.; Marinos, D.; Dimitrios, P.; Safarian, J. Investigating Aluminum Tri-Hydroxide Production from Sodium Aluminate Solutions in the Pedersen Process. *Processes* **2022**, *10*, 1370. [CrossRef]
5. Blake, H.E.; Fursman, O.C.; Fugate, A.D.; Banning, L.H. *Adaptation of the Pedersen Process to the Ferruginous Bauxites of the Pacific Northwest*; Report of Investigations 6939; United States Department of Interior, Bureau of Mines: Albany, OR, USA, 1966.
6. Yuan, J.; Zhang, Y. Desiliconization reaction in sodium aluminate solution by adding tricalcium hydroaluminate. *Hydrometallurgy* **2009**, *95*, 166–169. [CrossRef]
7. Nikolaychuk, P.A. The revised pourbaix diagram for silicon. *Silicon* **2014**, *6*, 109–116. [CrossRef]
8. Dilnesa, B.Z.; Lothenbach, B.; Renaudin, G.; Wichser, A.; Kulik, D. Synthesis and characterization of hydrogarnet. *Cem. Concr. Res.* **2014**, *59*, 96–111. [CrossRef]
9. Roach, G.I.D. The Equilibrium Approach to Causticisation for Optimising Liquor Causticisity. In *Essential Readings in Light Metals: Volume 1 Alumina and Bauxite*; Donaldson, D., Raahauge, B.E., Eds.; Springer International Publishing: Cham, Switzerland, 2016; pp. 228–234.
10. Rosenberg, S.P.; Wilson, D.J.; Heath, C.A. Some Aspects of Calcium Chemistry in the Bayer Process. In *Essential Readings in Light Metals: Volume 1 Alumina and Bauxite*; Donaldson, D., Raahauge, B.E., Eds.; Springer International Publishing: Cham, Switzerland, 2016; pp. 210–216.
11. Xu, B.A.; Giles, D.E.; Ritchie, I.M. Reactions of lime with carbonate-containing solutions. *Hydrometallurgy* **1998**, *48*, 205–224. [CrossRef]
12. Mwase, J.M.; Safarian, J. Production of Aluminium Tri-hydroxides from Secondary Bauxite Materials. In Proceedings of the EMC 2021 European Metallurgical Conference, Virtual Conference, 27–30 June 2021; pp. 547–558.
13. Labuschagné, F.J.W.J.; Wiid, A.; Venter, H.P.; Gevers, B.R.; Leuteritz, A. Green synthesis of hydrotalcite from untreated magnesium oxide and aluminum hydroxide. *Green Chem. Lett. Rev.* **2018**, *11*, 18–28. [CrossRef]

processes

MDPI

Article

Assessing and Improving Biooxidation for Acid Generation and Rare Earth Element Extraction

Michael L. Free, Joel K. Ilunga, Prasenjit Podder and Prashant K. Sarswat *

Department of Materials Science and Engineering, University of Utah, Salt Lake City, UT 84112, USA
* Correspondence: prashant.sarswat@utah.edu; Tel.: +1-801-820-6919

Abstract: Microorganisms (*Acidithiobacillus ferrooxidanns*) are effective in oxidizing ferrous ions that can be used to oxidize pyrite and produce sulfuric acid. Many coal waste resources contain significant concentrations of rare earth elements (REE) and critical materials (CM) that can be extracted using sulfuric acid. These coal waste resources often contain significant concentrations of pyrite, which if not utilized or removed present a future environmental liability for potential acid mine drainage. Thus, the combination of pyrite and REE/CM in coal waste provides a significant resource opportunity for sulfuric acid generation that can be utilized using biooxidation. In addition, the pyrite concentrate used for acid generation also contains REE/CM content that is released during biooxidation of the pyrite concentrate that augments the REE/CM release from the main ore being leached with the acid generated from the pyrite. Thus, this approach provides a very significant environmental advantage as well as augmented REE/CM recovery. Although there are many studies associated with biooxidation in relation to mineral oxidation, there is a lack of information regarding the effects of operating parameters on biooxidation performance and optimization for practical applications. In this study, findings from research in assessing and improving biooxidation for acid generation for REE/CM extraction are presented. Results show that bacteria can very effectively and efficiently oxidize ferrous ions to ferric ions, which oxidize pyrite to produce acid for REE/CM extraction. The factors that showed significant impact on biooxidation performance include air flow rate, stirring speed, residence time, solids concentration, and temperature. The dominance of *Leptospirillum ferriphilum* species was noted in the bioreactor after a prolonged period of operation, although *Acidithiobacillus ferrooxidanns* was used in the beginning.

Keywords: biooxidation; rare earth elements extraction; critical materials extraction; acid generation; biohydrometallurgy; operating parameters

Citation: Free, M.L.; Ilunga, J.K.; Podder, P.; Sarswat, P.K. Assessing and Improving Biooxidation for Acid Generation and Rare Earth Element Extraction. *Processes* **2023**, *11*, 1005. https://doi.org/10.3390/pr11041005

Academic Editors: Bipro R. Dhar and Sebastien Farnaud

Received: 5 January 2023
Revised: 18 March 2023
Accepted: 20 March 2023
Published: 26 March 2023

1. Introduction

Microorganisms have been utilized to assist in metal extraction from minerals and waste for many years [1–6]. Through metabolic procedures such as metal ion reduction, acidogenesis, and biotransformation, these bacteria can release metal ions from minerals. Bacteria have been used in connection with mineral oxidation since around 1900 [7]. However, it was not until 1947 when bacteria associated with metal extraction from minerals, were isolated from a coal mine and identified by Colmer and Hinkle [8] as *Thiobacillus ferrooxidans*, which is now recognized as *Acidithiobacillus ferrooxidanns*. In order to produce sulfuric acid, *Acidithiobacillus* bacteria have been used in the mining industry. By oxidizing iron sulfide minerals like pyrite, these acid-producing bacteria create sulfuric acid as a byproduct. It is possible to extract valuable metals from ore by using this acid to dissolve metal sulfides. In order to remove heavy metals from contaminated soils and water, *Acidithiobacillus* bacteria can also be utilized in bioremediation procedures. They are chemolithotrophic, -proteobacterium microorganisms [9] that obtain their growth energy from the oxidation of minerals that include sulfur and iron. It fixes both nitrogen and carbon from the environment and grows at pH levels as low as 1–2.

It was not until the 1950s that the role of bacteria in dump leaching of some minerals was demonstrated. In 1984, the Fairview plant in South Africa applied biooxidation in a pilot plant with a capacity of ~750 kg/day that was expanded to 10,000 kg/day in 1991 [10]. Most of the bacteria that have been identified for their role in mineral oxidation are acidophilic and need acid to access dissolved ions such as ferrous ions for oxidation. The bacteria are also mesophilic and tend to perform best between 20 and 40 °C. These bacteria are also autotrophic and utilize the carbon from carbon dioxide to supply their carbon needs. Thus, the conditions needed for successful biooxidation for metal extraction from minerals are generally acidic with temperatures controlled between 20 °C and 40 °C, and significant air flow to provide both oxygen and carbon dioxide.

In the context of this research, acid is needed for REE dissolution. The REEs in the ore matrix are released into a solution and become extractable in acidic media. The kind of acid employed is determined by the ore's mineral makeup and the particular REEs being extracted. Due to its ability to dissolve a variety of minerals containing REEs, sulfuric acid is frequently employed for the leaching of REEs. Depending on the exact mineral and REE being leached, other acids such as nitric acid or hydrochloric acid may also be utilized in some instances. To maximize REE extraction efficiency and reduce environmental effects, leaching conditions such as temperature, duration, and agitation must be properly managed.

Sulfuric acid needed for REE extraction from coal waste can be generated from pyrite found in coal waste using bacteria to facilitate rapid, low-cost acid generation. Furthermore, the use of the coal waste pyrite for acid generation prevents future acid mine drainage from it. Thus, the authors have utilized this biooxidation route for acid production for REE and critical materials (CM) extraction. Some of the recent literature discussed the details about industrial scale production of sulfuric acid using microorganisms as well as work from the authors [11,12], the present manuscript mainly deals with fine tuning of processing parameters.

2. Experimental Procedures

2.1. Size Distribution Analysis

Size analysis experiments were conducted using a set of 6 different United States (US) mesh sieves, and the sequence was maintained by following the USA standard mesh chart. After putting the samples on the top of the sieves, the entire set was vibrated by an automatic sieve vibrator for 20 min.

2.2. SEM Imaging and EDS Mapping Technique

SEM imaging and EDS mapping of different pyrite concentrations have been conducted by "HITACHI S-4800" high-resolution field emission SEM using 1.0 kV for all samples during the imaging. All the images have been collected at the same magnification. The applied voltage was increased to 20 kV during EDS mapping analysis to investigate the particles. EDS mapping was also done at the same magnification.

2.3. XRD Analysis Procedure

XRD analysis has been done using a RIGAKU MINIFLEX 600 XRAY DIFFRACTOMETER (XRD). MATCH 30 has been used to analyze the data collected from the XRD analyzer. The same contact angle and voltage have been maintained.

2.4. ICP-MS Analysis Information

The ICP-MS analysis was performed by American West Analytical Laboratory (AWAL) using standard equipment and procedures.

2.5. Acid Consumption Titration Procedure

Acid titration was conducted to determine the effective $CaCO_3$ concentration in the feed. The procedures include preparation of 20 mL of 0.5 M HCl in a 50 mL beaker, grinding

1 g of sample which is added to the solution for 24 h, followed by titrating with 0.5 M NaOH solution to pH 7.0 and recording the volume of NaOH solution consumed.

2.6. Eh, and pH Measurements

pH and Eh values were measured using Thermo Scientific™ Orion Star™ A211 Benchtop pH Meter, and a Mettler Toledo™ InLab™ Redox ORP Electrode was used to measure the ORP value, and a Thermo Scientific™ Orion™ Triode™ 3-in-1 pH/ATC Probe was used to measure the pH. The acidity of the back titration solution was checked after adding NaOH to the titrate solution based on measured values from the pH electrode/meter. The pH of the ferric sulfate solution was also measured during the experiments to see the changes in acidity. Acidity and ORP data were collected at least two times a day every day at the same time right before feeding was started.

2.7. Bio Reactor Experimental Setup

Two 3-liter Chemglass jacketed bioreactors have been utilized using appropriate safety protocols for the project. The bioreactor was purchased from the Chemglass Lifesciences INC and assembled in the lab. Previously preserved bacterial cultures were used to inoculate the reactors for bioleaching experiments in two different buildings. Reactors were run in separate facilities to avoid discontinuity in case of shutdowns during the COVID-19 response. Necessary installation processes such as cleaning steps, installing all parts, re-filling new solutions, maintaining temperature (inside the reactor ~35 °C), pH control (~maintenance of pH 1.4 using H_2SO_4 if needed), and providing the supply of 9K medium also has been initiated in the laboratory. The pH of 1.4 was set to keep the pH in range of preferred bacterial activity and performance. The reactors generally operated below this pH, but in cases where there were upsets in flows that caused the pH to rise above 1.7, acid was added to bring the pH to approximately 1.4. Figure 1 shows the experimental setup of one reactor in the lab. The authors have used standard *Acidithiobacillus ferrooxidanns* strain (ATCC 23,270) as well as the indigenous bacteria in the feed to prepare bioleaching solutions. The microorganisms were grown in updated ATCC 2436 9k medium where iron sulfate was eliminated (~0 g/mL). Later, it was found that and only ammonium sulfate and dipotassium phosphate were needed from the 9 k medium due to the presence of trace amounts of other elements present in the feed. An initial residence time of 8 days was maintained with an air flow rate of 1 L/M. Pyrite concentrate was used as a feedstock with 5% solid feeding.

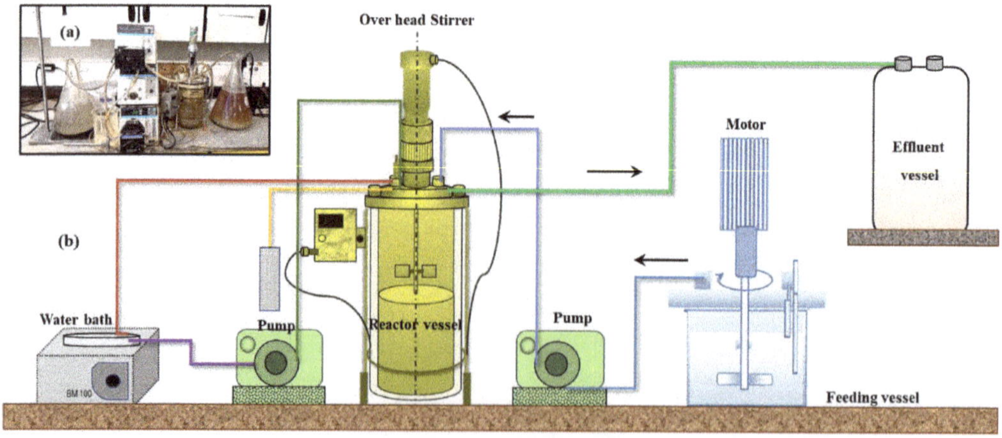

Figure 1. (**a**) A digital photograph of bioreactor/related support equipment and tubing. (**b**) a schematic diagram of bioreactor setup. The red line is the heated water, the green line is the effluent, the yellow line is the air input, and the blue line is the feed solution.

2.8. Feeding Solution Making and Flow Rate Setup Procedure

The feeding solution was made with 50% DI water and 50% of the 9 k medium without iron sulfate. Then, feed solids were added by following the parameter requirement. The automatic feeding system was activated after every 6 hours to perform the automatic feeding and effluent extraction using a timer. The solution's flow rate volume has been controlled using two multichannel peristaltic cartridge pumps. The feeding solution was stirred for at least ten minutes before each feeding to make a homogenous feeding solution.

2.9. Ferrous Oxidation Rate Test

The activity of the bacteria has been evaluated by doing a ferrous oxidation test. Every day, this test was performed by adding 1 g of $FeSO_4{\cdot}7H_2O$ into the system. Before adding the $FeSO_4$ to reactor, Eh of the solution was measured. After applying 1 g ferrous sulfate, Eh was measured every 30 s to see the changes in Eh until it reached its initial Eh value. Then the data was analyzed using the ferrous oxidation plotter spreadsheet developed in our group to determine the ferrous oxidation rate.

By adding ferrous sulfate to the solution and then monitoring the change in oxidation reduction potential (ORP) as a function of time, it was possible to determine the rate of ferrous ion oxidation by bacteria. The associated Nernst Equation for half animal reaction $Fe^{3+} + e^- \leftrightarrow Fe^{2+}$ is [13]:

$$E = E_o + \frac{2.303RT}{nF} ln \frac{\gamma_{Fe^{3+}} m_{Fe^{3+}}}{\gamma_{Fe^{2+}} m_{Fe^{2+}}} \quad (1)$$

where E_o is the standard potential, E is the electrochemical potential, T is the absolute temperature, R is the gas constant, n is the number of electrons, a is the activity, F is the Faraday constant, and γ is the activity coefficient, which can be calculated using the Davies Equation (13):

$$-log\gamma = \frac{Az^2\sqrt{I}}{1+\sqrt{I}} - 0.2I \quad (2)$$

If z is the ion's charge, and I, the ionic strength, can be calculated by [13]:

$$I = 0.5\sum_{i=1}^{n} m_i z_i^2 \quad (3)$$

The concentration of ferrous ions initially present can be estimated using the preceding equations by knowing the starting potential from an ORP probe and monitoring the new ORP value after adding a small amount of ferrous or ferric sulfate:

$$m_{Fe^{2+}_{init}} = \frac{m_{Fe^{2+}_{added}}}{exp\frac{(E_{init} - E_{afterFe2+addition})nF}{RT} - 1} \quad (4)$$

The equation that results from the foregoing calculations can be used to determine the quantity of free (i.e., uncomplexed) ferric ions that were initially present [13]:

$$m_{Fe^{3+}_{init}} = exp\frac{(E_{init} - E_o)nF}{RT} \frac{m_{Fe^{2+}_{init}} \gamma_{Fe^{2+}}}{\gamma_{Fe^{3+}}} \quad (5)$$

The ferrous ion concentration as a function of time can be calculated using these values as substitutes in the Nernst Equation. E_o is equal to 0.77 V in the ferrous/ferric half-cell reaction. By measuring the rate of ferrous ion oxidation, we can determine important information about the bacterial population and activity.

Assuming that all the basic needs are met, the oxidation rate is associated with ferrous ions. The ferrous biooxidation rate can be determined using traditional Michaelis–Menten or Monod kinetics [14]:

$$R_{Fe+2ox.} = \frac{C_{cells}\mu_{max}C_{Fe+2}}{Y_c(C_{Fe+2} + K_m)} \quad (6)$$

2.10. Acid Generation Rate Tracking Procedure

A titration test was conducted to determine the acid concentration in the leaching solution. The test procedures consisted of preparing a 0.1 M NaOH solution by dissolving NaOH pellets into DI water, taking 20 mL of the leaching solution from the bioreactor and filtering it to remove solids particles, preparing a burette filled with the 0.1 M NaOH solution, recording the initial reading on the burette, placing 10 mL of the filtered leaching solution into an Erlenmeyer flask and put it under the burette, titrating until the pH reading is 2.5, and closing the valve, and recording the final reading on the burette.

3. Results and Discussion

3.1. Feed Material Analysis

Inductively coupled plasma-mass spectroscopy (ICP-MS) and X-ray diffraction analysis of typical feed materials result in estimates of 68% pyrite, 16% calcite, and 16% other minerals such as aluminum-bearing silicate minerals. A representative scanning electron microscope (SEM) view of the feed material with corresponding electron dispersive X-ray spectrometer (EDS) mapping for specific elements is shown in Figure S1 (S represents the figures in supporting information section). The d80 size of the feed material is 75 μm. The feed material is a pyrite concentrate produced from coal waste and provided by Rick Honaker's research group at the University of Kentucky. Figure 2 shows the SEM imaging/EDS mapping of Fine Pyrite at different magnifications. Fine pyrite particles have a uniform and sharp crystal structure and smooth surface. It can be seen that a significant number of FeS_2 particles can be observed because Fe and S can be traced in identical points. Silica, clay materials, and calcite can be determined by identifying their positions in the map images. From the point analysis, it can be clearly seen that there are traces for Fe, S, C, Ca, Al, Si, and K.

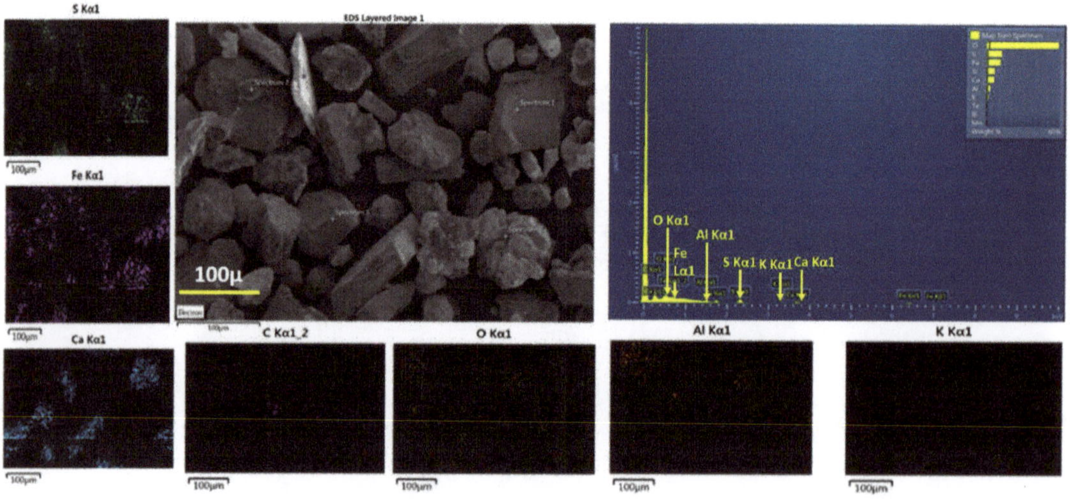

Figure 2. SEM imaging and EDS mapping of fine pyrite.

In this study, key operating parameters that impact key performance indicators for biooxidation, which are acid production and the biooxidation rate, were evaluated. The key parameters that were evaluated are residence time, solids concentration, air flow rate, stirring speed, temperature, and nutrient availability. The evaluations and assessments were performed using bioreactors such as those shown in Figure 1.

As general background information, the bacteria catalyze reactions that can produce acid from sulfide minerals. A simplified set of reactions that describes what can happen generally is given as [12]:

$$2Fe^{2+} + 0.5O_2 + 2H^+ \leftrightarrow 2Fe^{3+} + H_2O \text{ (Biotic-ferrous ion oxidation)} \quad (7)$$

$$FeS_2 + 2Fe^{3+} \leftrightarrow 3Fe^{2+} + 2S_o \text{ (Abiotic-indirect bioleaching)} \quad (8)$$

$$2S_o + 3O_2 + 2H_2O \leftrightarrow 2SO_4{}^{2-} + 4H^+ \text{ (Biotic-direct bioleaching)} \quad (9)$$

Or

$$15Fe^{2+} + 3.75O_2 + 15H^+ \leftrightarrow 15Fe^{3+} + 7.5H_2O \text{ (Biotic-ferrous ion oxidation)} \quad (10)$$

$$FeS_2 + 6Fe^{3+} + 3H_2O \leftrightarrow 7Fe^{2+} + 6H^+ + S_2O_4{}^{2-} \text{ (Abiotic-indirect)} \quad (11)$$

$$S_2O_4{}^{2-} + 9Fe^{3+} + 5H_2O \leftrightarrow 8Fe^{2+} + 9H^+ + HSO_4{}^- + FeSO_4{}^+ \text{ (Abiotic-indirect)} \quad (12)$$

Or

$$15Fe^{2+} + 3.75O_2 + 15H^+ \leftrightarrow 15Fe^{3+} + 7.5H_2O \text{ (Biotic-ferrous ion oxidation)} \quad (13)$$

$$FeS_2 + 15Fe^{3+} + 8H_2O \leftrightarrow 15Fe^{2+} + 15H^+ + HSO_4{}^- + FeSO_4{}^+ \text{ (Abiotic-indirect)} \quad (14)$$

Note that the last two sets of reactions sum up to the same ratio of ferric ions per pyrite (15 to 1) even though one pathway has thiosulfate ions as an intermediate species. The top set of reactions involves some direct leaching in which the bacteria are attached to the elemental sulfur substrate at which the reaction takes place, whereas in the reactions involving ferric ions as the oxidant, the bacteria can facilitate pyrite leaching indirectly through the solution with the need for attachment. In the case of direct leaching, the ratio of ferric ions needed per pyrite molecule is 2 to 1. Correspondingly, the bacteria can cause direct or indirect leaching, and they can produce ferric ions from ferrous ions using oxygen and acid, then produce acid from pyrite [15–18]. Testing with pure pyrite leaching in ferric sulfate media indicates that the ratio of Fe^{3+} consumed per FeS_2 is 14.8. This indicates the primary chemical oxidation of pyrite results primarily in sulfate production rather than elemental sulfur. This also indicates that the demand for ferric ions in pyrite oxidation is high.

Chemical analyses, performed using carbon disulfide dissolution of the elemental sulfur after heat treating the residual bioleached solids sample at 108 °C, indicated that the residual solids contained only 0.23% elemental sulfur, thereby suggesting that the pyrite dissolution pathway is not dominated by elemental sulfur formation. This provides corroboratory evidence that the main pathway for pyrite dissolution involves indirect leaching with ferric ions providing the primary oxidant. A schematic diagram of biooxidation of pyrite is shown in Figure 3.

3.2. *Biooxidation Rates and Data Analysis*

The rate of biooxidation was found to be as high as 4 g of ferrous ions oxidized per liter per hour based on testing with additions of ferrous sulfate, followed by corresponding electrochemical potential measurements and conversion to ferrous ion oxidation rate, and example of which is presented in Figure S2. The calculation of the ferrous ion concentration from the Eh measurement is as follows:

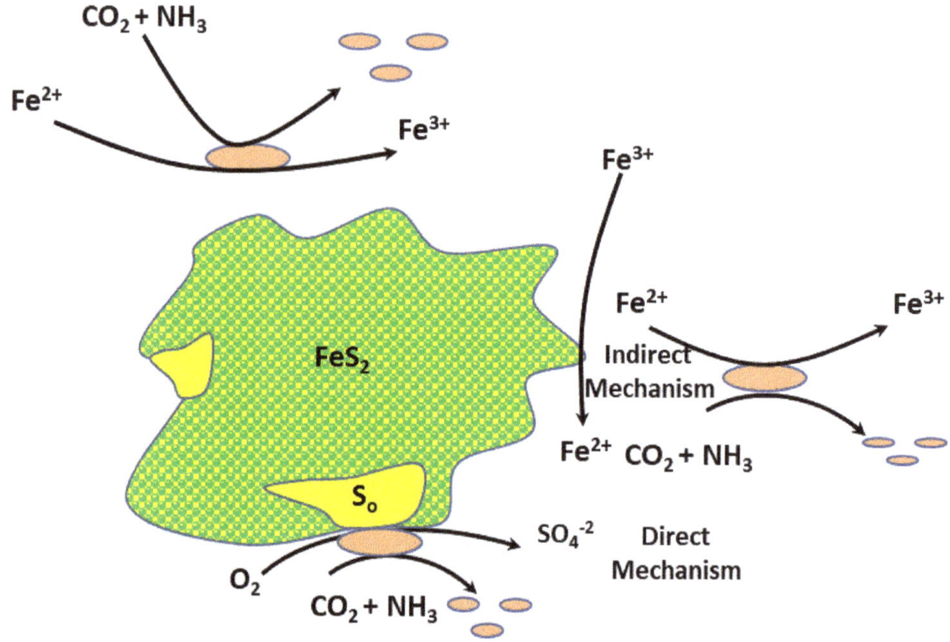

Figure 3. Schematic diagram of biooxidation of pyrite.

Based on a sustained rate of 3.1 g of ferric ions produced/liter/hour. If 15 moles of ferric ions are needed per mole of pyrite, a reactor with a residence time of 96 h would theoretically produce:

$$[H^+] = \frac{\frac{\frac{3.1gFe^{3+}}{liter}}{hr}}{\frac{55.85gFe^{3+}}{moleFe}} \frac{mole\ H^+}{15moles\ Fe^{3+}} 96hr = 0.35MH^+$$

A reactor with 6% feed concentrate containing 70% pyrite with enough time to oxidize nearly all it (usually around 96 h) can produce:

$$[H^+] = \frac{60\ g\ concentrate}{l} \frac{0.70\ g\ pyrite}{g\ concentrate} \frac{1\ mole\ pyrite}{119.98\ g\ pyrite} \frac{1\ mole\ H^+}{mole\ pyrite} = 0.35MH^+$$

This is even matching of bacteria capacity with optimal concentrate feed if no acid consuming minerals are present.

A significant portion of the acid is consumed by the calcite and acid-consuming minerals in the feed concentrate. For the reaction of $CaCO_3 + 2H^+ \rightarrow Ca^{2+} + H_2O + CO_2$, the potential consumption of acid is:

$$[H^+] = \frac{60\ g\ concentrate}{l} \frac{0.17\ g\ calcite}{g\ concentrate} \frac{1\ mole\ calcite}{100\ g\ calcite} \frac{2\ mole\ H^+}{mole\ calcite} = 0.20\ MH^+$$

The observed apparent acid consumption is generally about 0.2 MH^+, so about half of the acid the bacteria produce is consumed by calcite and other acid-consuming minerals in the feed. Note that the resulting amount of carbon dioxide released for bacterial use is equivalent to 0.0336 g of carbon dioxide per liter per hour, which is 2.7 times greater than the amount of carbon dioxide supplied by the typical air flow rate of 1 L/min in 2.5 L of reactor solution, thereby providing additional carbon dioxide for enhanced bacterial growth rates.

The corresponding amount of oxygen consumption was determined using an oxygen analyzer (Cerion OA-1 oxygen analyzer with auto ranging facility). The biooxidation consumed approximately 6% of the incoming oxygen (1.2% out of 20.2%) with an air flow rate of 1.5 L/min or 0.005 g O_2/L/min.

This is similar to the rate of dissolved oxygen depletion of 0.003 g of O_2/L/min measured using a dissolved oxygen (DO) probe and meter under similar conditions. The dissolved oxygen concentration was measured using an Oakton DO Six+ Portable dissolved oxygen meter. The corresponding probe was submerged inside the bioreactor until the reading was stable then the air supply is shut off and a DO value is collected every 30 s until close to complete depletion then the air supply is turned back on until equilibrium was reached again. The amount of acid that can be produced from 0.004 g of O_2/L/min is theoretically 0.384 M assuming 3.75 moles of O_2 per mole of pyrite and per mole of acid. This is consistent with observations.

Approximately 90% of the feed pyrite was oxidized by the bacteria in most tests. A notable exception was 70% oxidation of pyrite in the 2.5-day residence time test in which there was insufficient time for additional oxidation.

These Figures S3 and S4 show that solids concentration has a significant influence on performance. Although 9 wt% solids had a slight improvement in acid production rate, there is also a decrease in pyrite utilization efficiency and in the biooxidation rate. Previous calculations also suggest that 6 wt% pyrite concentrate in the feed is an appropriate level for the solids concentration that matches the ability of the bacteria to utilize the available pyrite in a typical 4-day residence time.

Representative plots of pH data along with operator notes for one of the bioreactors during most of 2022 are presented in Figure 4a,b. Representative plots of Eh data for one of the bioreactors during most of 2022 are presented in Figure 5a,b. Representative plots of ferrous ion oxidation data for one of the bioreactors during most of 2022 are presented in Figure S3a,b.

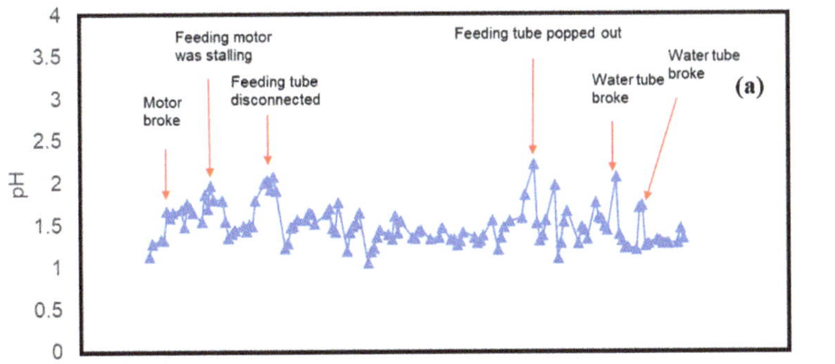

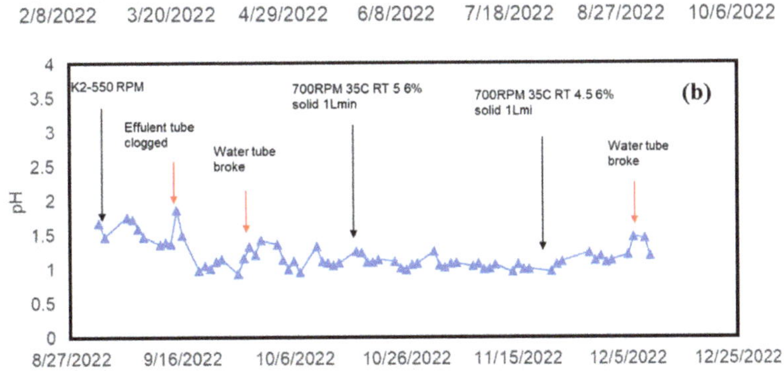

Figure 4. (**a**) Comparison of pH data in the WBB 419 bioreactor from 3 March 2022 to 31 August 2022. (**b**) Comparison of pH data in the WBB 419 bioreactor from 9 January 2022 to 12 September 2022.

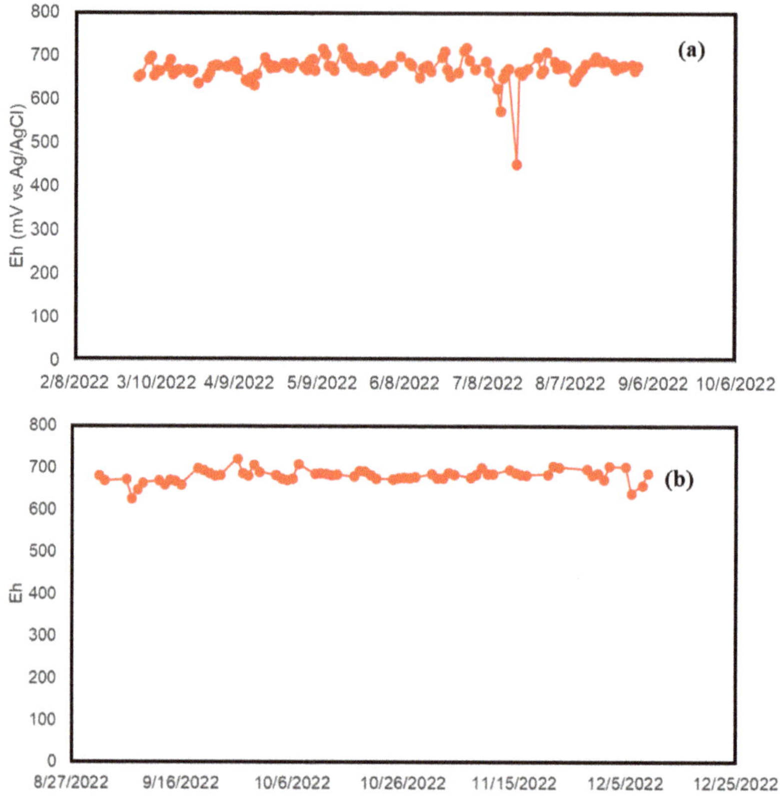

Figure 5. Comparison of Eh data in the WBB 419 bioreactor: (**a**) from 3 March 2022 to 31 August 2022. (**b**) from 9 January 2022 to 12 September 2022.

These bioreactor data plots illustrate the variations associated with parameter changes as well as laboratory-related issues that result while running these reactors 24 h per day, seven days per week. The data also shows that steady-state conditions can be achieved for sustained periods of time, provided that there are no mechanical or laboratory facility related issues that impact the operation of the bioreactors. The effects of changing residence time, solids concentration, air flow, stirring speed, temperature, and nutrient availability on biooxidation performance were assessed. Figure S4a shows the effect of solids concentration on biooxidation rate and Figure S4b shows the related net acid production rate.

The effects of air flow rate on performance are shown in Figure S5a,b. These figures show the related biooxidation rate and acid production rate performance levels, respectively. These result show that a flow rate of 1 L/min is best for the associated rotational speed of 500 rpm.

The effects of changing nutrient concentrations are illustrated in Figure S6. Note that the general nutrient formulation for 9 K medium, which is the general nutrient addition standard is:

9 K Medium (without ferrous sulfate heptahydrate)

$(NH_4)_2SO_4$: 6.0 g/L
KCl: 0.2 g/L
K_2HPO_4: 1 g/L
$MgSO_4.7H_2O$: 1 g/L
$Ca(NO_3)_2$: 0.02 g/L

Note that because pyrite is FeS_2 there is more than enough iron in the feed solids, so the iron is missing from the 9 K medium in this application.

The results of the nutrient evaluation indicate the nutrient addition can be reduced substantially to include only 0.8 g of K_2HPO_4 per liter and 1.6 g of $(NH_4)_2SO_4$ per liter. The reduction in the use of nutrients provides a significant cost savings relative to initial projections. The effect of temperature on biooxidation performance is shown in Figure 6a. The data show that increasing temperature is beneficial up to 35 °C. However, the data also show that increasing the temperature to 40 °C is very detrimental to performance. Thus, temperature is a very critical parameter to control well. Note that temperature has a significant impact on how effectively bioleaching occurs. The pace of bacterial metabolic activities and, thus, the rate of metal extraction can both be increased by raising the temperature. High temperatures can, however, potentially harm microorganisms, resulting in diminished viability and diminished effectiveness.

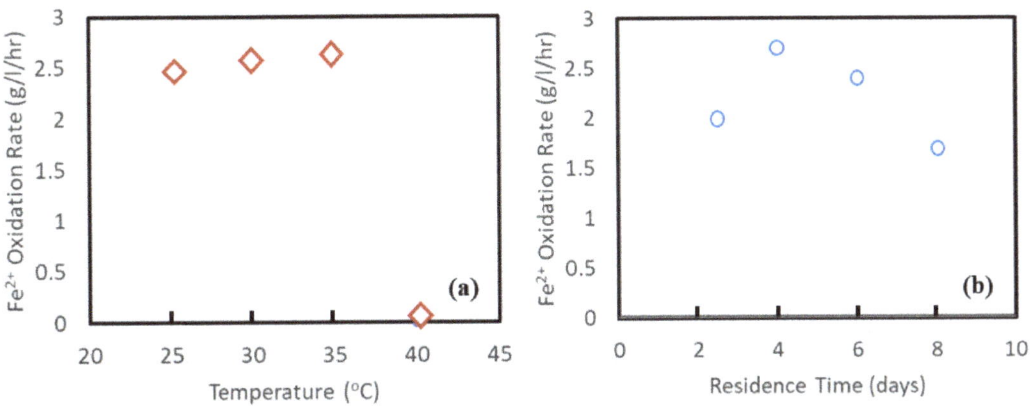

Figure 6. (**a**) Effect of temperature on ferrous ion production rate. (**b**) Comparison of residence time and ferrous ion oxidation rate.

The effects of residence time on biooxidation performance are presented in Figures 6b and 7a. These data show that a residence time of 4 days resulted in the best performance. The lower performance for 2.5 days of residence is attributed to the fast rate of effluent removal of bacteria relative to the time required for the bacteria to reproduce. The short residence time may also have reduced the overall utilization of pyrite, thereby decreasing performance. The longer residence times above 4 days can result in depletion of resources as pyrite is completely dissolved at longer residence times, making the available feed in limited supply as residence times become long, thereby reducing performance. The effect of stirring speed on biooxidation performance is presented in Figure 7b. These data indicate that performance increases as stirring speed increases. In general, stirring encourages close contact between the microorganisms and the substrate by distributing the substrate (ore or waste material) and bacteria throughout the leaching reactor uniformly. This can speed up the extraction of metal and help to guarantee reliable outcomes. The effectiveness of bioleaching, however, might also be negatively impacted by excessive stirring. Very high stirring rates can create high shear forces at impeller tips that can harm the bacteria mechanically and decrease their viability.

The data from these tests indicated there were some clear optimum levels of performance for parameters such as air flow rate and residence time. The data also showed potential for increasing performance by increasing stirring rate. Therefore, additional testing was performed to determine optimum levels near observed performance peaks presented in previous plots. The optimization testing focused on stirring speed and residence time. The corresponding results are presented in Figure 8a,b. Note that the optimum performance is achieved at a stirring speed of 650 RPM and 5 days of residence time. Apart from those factors, appropriate adaptation conditions are also essential for the health of microorganisms and the associated performance [19]. It was noted in the case of Zn^{2+}, where *Acidithiobacillus* bacteria used for sphalerite concentrate, leaching was significantly improved when the appropriate protocol was followed [19].

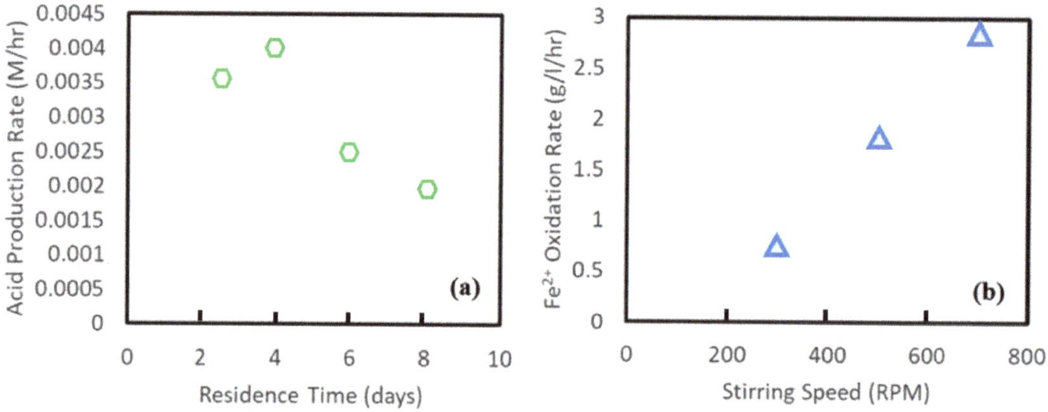

Figure 7. (**a**) Comparison of residence time and acid production rate. (**b**) Initial comparison of stirring speed and ferrous ion oxidation rate.

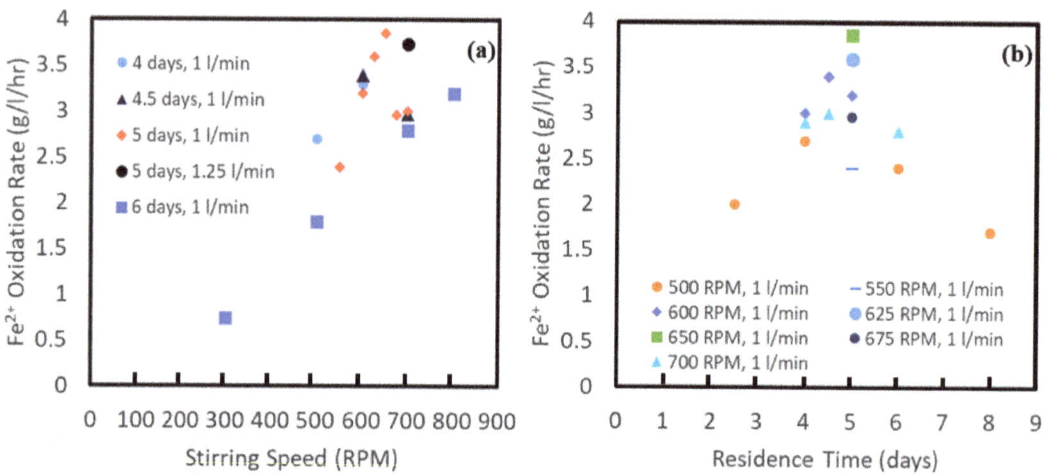

Figure 8. Detailed, more comprehensive comparison of: (**a**) stirring speed and ferrous ion oxidation rate; (**b**) residence time and ferrous ion oxidation rate.

3.3. Analysis of Bacteria

Genome sequencing was performed to understand the species of microorganisms present in different operating conditions. Bacterial concentrations were also analyzed through these sequencing methods. An HPC test was conducted to see how the bacterial colonies can be formed under other conditions. Figure 9 shows the presence of *Acidithiobacillus*, *Leptospirillum*, and *Sulfolobacillus* bacteria.

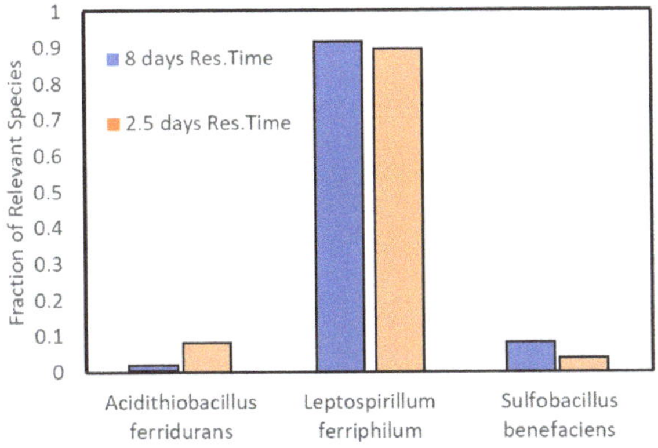

Figure 9. Genome sequencing data from samples originating from bioreactor solutions from 2.5- and 8-days residence time tests at 500 RPM, 1 L/min air flow, 35 °C, and 6 wt% solids conditions showing the relative fraction of species that are generally identified with biooxidation of pyrite.

The dominance of *Leptospirillum ferriphilum* in the reactors is consistent with the related literature [20,21]. Gram-negative, vibrio- or spiral-shaped, and obligately chemolithotrophic Leptospirillum species are connected to the deep branching class Nitrospira phylogenetically [22].

They are widely known to be the crucial biological catalysts in various metal sulfide biooxidation processes and can be found in a wide range of acidophilic microbial communities [22]. Leptospirillum species play important roles in the biogeochemical cycle of iron, as they are the most common iron-oxidizing bacteria in metal-tolerant, acidophilic microbial consortia that cause ferric iron [Fe(III)]-mediated oxidative dissolution of sulfide minerals. Leptospirilla have been identified to be the main contributors [22] responsible for the development of acid mine drainage under the environmental circumstances characterized by temperature above 40 °C and pH value below 1.0. Apart from genome sequencing, oxidized pyrite solids were investigated through a high magnification SEM imaging system to view potential bacterial attachment to the pyrite surface. These SEM images are presented in Figure 10a–f. More detailed analyses have been presented in previous report [12].

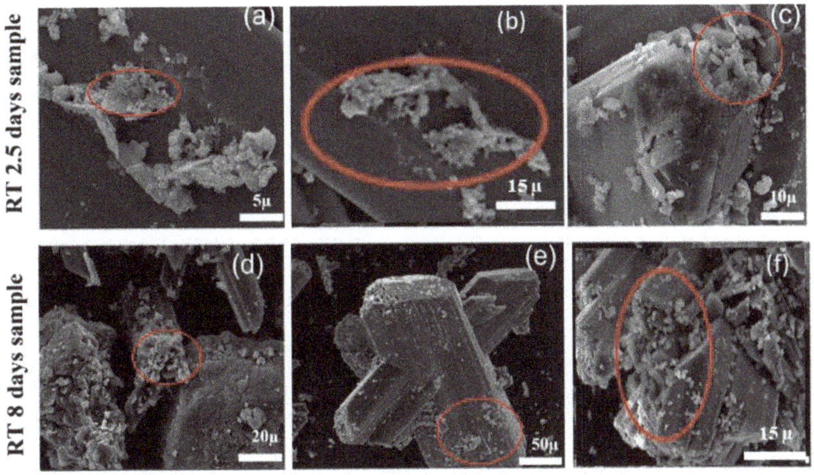

Figure 10. (a–f). View of magnified pyrite surface from samples originating from bioreactor solutions from 2.5- and 8-days residence time tests at 500 RPM, 1 L/min air flow, 35 °C, and 6 wt% solids conditions showing the relative fraction of species that are generally identified with biooxidation of pyrite.

3.4. Assessment of Leaching of REE/CM from Pyrite Concentrate

The pyrite feed concentrate contains some REE/CM content that is extracted during the pyrite biooxidation. Table 1 provides the feed concentration of selected REE/CM for which significant extraction occurs during the pyrite concentrate biooxidation. The data show that although the concentrations are not high; the extraction for many of these elements is significant and would augment the subsequent REE/CM extraction from the coal waste ore for which the acid is generated.

Table 1. Pyrite concentrate feed material REE/CM content for selected elements for which the extraction exceeds 10% during biooxidation.

Element	Feed (ppm)	Extraction (%)
Al	1383	14.71
Co	32.86	76.11
Dy	1.91	41.36
Er	0.98	56.12
Gd	6.65	20.75
Ho	0.3	56.67
Lu	0.14	50.00
Mg	1991	30.84
Sc	3.86	47.67
Tb	0.31	38.71
Tm	0.09	55.56
Y	7.31	53.63
Yb	1.94	30.41
Zn	1482	11.76

4. Conclusions

Overall, microorganism-assisted sulfuric acid production processing parameters were assessed in this research. The acid produced from biooxidation of pyrite can be utilized for REE dissolution from the ore matrix as well as from the pyrite as shown in this paper. The bacteria are adapted to the bioleaching solution containing pyrite concentrate, with a ferrous oxidation rate of up to 4 g/L/h. Pyrite utilization was nearly complete based on analyses of residual solids as well as general mass balancing. The elemental sulfur analysis confirmed that most of the sulfur was converted to sulfate for acid generation. Air flow rate, stirring speed, residence time, feed solids concentration, and temperature were all shown to have significant effects on biooxidation performance. It was observed that a typical residence time of 4–5 days is most effective for biooxidation performance. Genome sequencing tests confirm the dominance of Leptospirillum species.

Supplementary Materials: The following supporting information can be downloaded at: https://www.mdpi.com/article/10.3390/pr11041005/s1.

Author Contributions: Conceptualization, M.L.F. and P.K.S.; methodology, P.K.S.; software, P.P. and P.K.S.; validation, P.P., J.K.I. and M.L.F.; formal analysis, P.P. and J.K.I.; investigation, M.L.F. and P.K.S.; resources, M.L.F.; data curation, J.K.I. and P.P.; writing—original draft preparation, M.L.F.; writing—review and editing, P.K.S.; visualization, M.L.F.; supervision, M.L.F. and P.K.S.; project administration, M.L.F. and P.K.S.; funding acquisition, M.L.F. All authors have read and agreed to the published version of the manuscript.

Funding: This research was funded by the U.S. Department of Energy, grant number DE-FE0031827. This report was prepared as an account of work sponsored by an agency of the United States government. Neither the United States government nor any agency thereof, nor any of their employees, make any warranty, express or implied, or assumes any legal liability or responsibility for the accuracy, completeness, or usefulness of any information, apparatus, product, or process disclosed, or represents that its use would not infringe privately owned rights. Reference to any specific commercial product, process, or service by trade name, trademark, manufacturer, or otherwise does not necessarily constitute or imply its endorsement, recommendation, or favoring by the United States government or any agency there. The views and opinions of author's expressed herein do not necessarily state or knows of the United States government or any agency thereof.

Data Availability Statement: The data presented in this study are available on request from the authors.

Acknowledgments: The pyrite concentrate feed material used in this study was provided by the research group of Rick Honaker at the University of Kentucky.

Conflicts of Interest: The authors declare no conflict of interest.

References

1. Vera, M.; Schippers, A.; Sand, W. Progress in bioleaching: Fundamentals and mechanisms of bacterial metal sulfide oxidation—Part A. *Appl. Microbiol. Biotechnol.* **2013**, *97*, 7529–7541. [CrossRef]
2. Rawlings, D.E. Heavy metal mining using microbes. *Annu. Rev. Microbiol.* **2002**, *56*, 65–91. [CrossRef]
3. Rawlings, D.E.; Dew, D.; du Plessis, C. Biomineralization of metal-containing ores and concentrates. *TRENDS Biotechnol.* **2003**, *21*, 38–44. [CrossRef]
4. Olson, G.; Brierley, J.; Brierley, C. Bioleaching review part B. *Appl. Microbiol. Biotechnol.* **2003**, *63*, 249–257. [CrossRef] [PubMed]
5. Brierley, J.A. A perspective on developments in biohydrometallurgy. *Hydrometallurgy* **2008**, *94*, 2–7. [CrossRef]
6. Johnson, D.B. Development and application of biotechnologies in the metal mining industry. *Environ. Sci. Pollut. Res.* **2013**, *20*, 7768–7776. [CrossRef]
7. Brierley, C.L.; Le Roux, N.W. Bacterial leaching. *CRC Crit. Rev. Microbiol.* **1978**, *6*, 207–262. [CrossRef] [PubMed]
8. Colmer, A.R.; Hinkle, M. The role of microorganisms in acid mine drainage: A preliminary report. *Science* **1947**, *106*, 253–256. [CrossRef]
9. Valdés, J.; Pedroso, I.; Quatrini, R.; Dodson, R.J.; Tettelin, H.; Blake, R.; Eisen, J.A.; Holmes, D.S. *Acidithiobacillus ferrooxidans* metabolism: From genome sequence to industrial applications. *BMC Genom.* **2008**, *9*, 597. [CrossRef]
10. Van Aswegen, P.; Godfrey, M.; Miller, D.; Haines, A. Developments and innovations in bacterial oxidation of refractory ores. *Min. Metall. Explor.* **1991**, *8*, 188–191. [CrossRef]
11. Yang, Y.; Qian, C.; Shi, X.; Tian, B.; Chu, H.; Wang, J.; Xin, B. Efficient and industrial production of H_2SO_4 from sulfur sludge by acidophilic cells in a membrane bioreactor via optimizing process. *J. Clean. Prod.* **2020**, *250*, 119444. [CrossRef]
12. Sarswat, P.K.; Podder, P.; Zhang, Z.; Free, M.L. Autogenous acid production using a regulated bio-oxidation method for economical recovery of REEs and critical metals from coal-based resources. *Appl. Surf. Sci. Adv.* **2022**, *11*, 100283. [CrossRef]
13. Sarswat, P.K.; Leake, M.; Allen, L.; Free, M.L.; Hu, X.; Kim, D.; Noble, A.; Luttrell, G.H. Efficient recovery of rare earth elements from coal based resources: A bioleaching approach. *Mater. Today Chem.* **2020**, *16*, 100246. [CrossRef]
14. Bailey, J.; Ollis, D. Applied enzyme catalysis. *Biochem. Eng. Fundam.* **1986**, *2*, 157–227.
15. Podder, P.; Free, M.L.; Sarswat, P.K. Extraction and Recovery of Rare-Earth Elements and Critical Materials from Coal Waste Using Low Cost Processing Methods: Acid Generation Using Acidithiobacillus Ferrooxidans Mediated Bio-Oxidation of Pyrite. In *Rare Metal Technology*; Ouchi, T., Azimi, G., Forsberg, K., Kim, H., Alam, S., Neelameggham, N.R., Baba, A.A., Peng, H., Eds.; Springer International Publishing: Cham, Switzerland, 2022; pp. 51–62.
16. Podder, P.; Zhang, Z.; Honaker, R.Q.; Free, M.L.; Sarswat, P.K. Evaluating and Enhancing Iron Removal via Filterable Iron Precipitates Formation during Coal-Waste Bioleaching. *Eng* **2021**, *2*, 632–642. [CrossRef]
17. Zhang, Z.; Allen, L.; Podder, P.; Free, M.L.; Sarswat, P.K. Recovery and Enhanced Upgrading of Rare Earth Elements from Coal-Based Resources: Bioleaching and Precipitation. *Minerals* **2021**, *11*, 484. [CrossRef]
18. Murali, A.; Plummer, M.J.; Shine, A.E.; Free, M.L.; Sarswat, P.K. Optimized bioengineered copper recovery from electronic wastes to increase recycling and reduce environmental impact. *J. Hazard. Mater. Adv.* **2022**, *5*, 100031. [CrossRef]
19. Haghshenas, D.F.; Alamdari, E.K.; Torkmahalleh, M.A.; Bonakdarpour, B.; Nasernejad, B. Adaptation of Acidithiobacillus ferrooxidans to high grade sphalerite concentrate. *Miner. Eng.* **2009**, *22*, 1299–1306. [CrossRef]
20. Rawlings, D.E.; Johnson, D.B. The microbiology of biomining: Development and optimization of mineral-oxidizing microbial consortia. *Microbiology* **2007**, *153*, 315–324. [CrossRef]

21. Johnson, D.B. Biomining—Biotechnologies for extracting and recovering metals from ores and waste materials. *Curr. Opin. Biotechnol.* **2014**, *30*, 24–31. [CrossRef]
22. Zhang, X.; Liu, X.; Yang, F.; Chen, L. Pan-Genome Analysis Links the Hereditary Variation of Leptospirillum ferriphilum With Its Evolutionary Adaptation. *Front. Microbiol.* **2018**, *9*, 577. [CrossRef] [PubMed]

processes

Article

Green and Cost-Effective Separation of Cadmium from Base Metals in Chloride Medium with Halide-Loaded Anion Exchanger

Yanlin Zhang [1,2], Xiaofei Duan [2], Chaoquan Hu [1,3], Guanshang Du [1] and Yong Wang [3,4,5,*]

1 Zhongke Nanjing Institute of Green Manufacturing Industry, Nanjing 211135, China
2 School of Chemistry, The University of Melbourne, Parkville, Melbourne, VIC 3010, Australia
3 Institute of Process Engineering, Chinese Academy of Sciences, Beijing 100190, China
4 National Engineering Research Centre of Green Recycling for Strategic Metal Resources, Beijing 100190, China
5 The University of Chinese Academy of Sciences, Beijing 101400, China
* Correspondence: wangyong@ipe.ac.cn

Abstract: A novel strategy for the separation of cadmium from base metals in chloride medium by anion exchange based on the formation of cadmium halo-complexes is presented in this article. Instead of using excess Br^- or I^- in the feed solution, the anion exchange resin or fibre was first preloaded with Br^- or I^- for Cd^{2+} adsorption from Cl^- solution. Thus, the consumption of iodide or bromide was minimized while promising high adsorption stability of Cd^{2+}. The adsorption thermodynamics and kinetics of Cd^{2+} were studied. The adsorption thermodynamics results were processed with Langmuir and Freundlich models and adsorption kinetics results were treated with Pseudo first order model, Pseudo second order model and intraparticle diffusion model. The adsorption mechanism was studied with the aid of FTIR and elemental analysis. Three strategies were applied for the stripping of Cd^{2+}, including the use of water for disequilibrating its halo-complex formation, the formation of positively charged species with ammonia, and the formation of positively charged chelate with multidentate ligand ethylenediamine. On the bases of these results, separation performance was investigated with strongly basic anion exchange resin and fibre, and it was shown that Cd^{2+} can be efficiently separated from base metals in a wide range of acidity, and that pure Cd^{2+} solution can be obtained, thus providing a robust, environmentally friendly, and economic technology for its separation.

Keywords: cadmium separation; base metals; halide-loaded anion exchanger; chloride medium

Citation: Zhang, Y.; Duan, X.; Hu, C.; Du, G.; Wang, Y. Green and Cost-Effective Separation of Cadmium from Base Metals in Chloride Medium with Halide-Loaded Anion Exchanger. *Processes* **2023**, *11*, 1051. https://doi.org/10.3390/pr11041051

Academic Editor: Anna Wołowicz

Received: 19 January 2023
Revised: 7 March 2023
Accepted: 23 March 2023
Published: 31 March 2023

1. Introduction

Cadmium commonly exists in natural minerals. It is a valuable resource, but also an unwanted impurity in industrial products, especially in some metals, and hence, its removal and/or recovery are common issues in the production of these products [1–3]. Typical examples with industrial significance include the removal of cadmium from phosphoric acid used in fertilizer production [4], cadmium removal from effluents and process streams for environmental protection [5], and recycling of spent batteries for recovery of the valuable metals [6–10]. One of the largest areas for the separation of cadmium, however, is in the production of some metals from their ores, particularly lead and zinc, where cadmium commonly occurs with these metals and must be removed during the refining processes to obtain pure metal products [2,11]. Separation of Cd and other co-existed elements from their ores often generates large amounts of solid residues containing different metal elements such as Cd, Cu, Pb, Ni, etc. [12,13]. Great efforts have been made to make use of these secondary resources, and cost-effective extraction techniques for these elements have been a hot topic in metallurgical study [11,14].

Although various techniques have been studied for the separation and/or recovery of Cd from mineral resources and much progress has been made in industrial application, many of the techniques still have some shortcomings in terms of complexity, efficiency, facility investment, operational parameters, cost-effectiveness, and pollution. Pyrometallurgical processes were conventionally used for the separation and recovery of Cd from various resources and are still applied in many industries. For instance, in the recycling of spent Ni-Cd battery, 99.92% Cd is separated from the matrix at 900 °C with coal (anthracite) as a carbonaceous material, whereas Ni and Co form an alloy [6]. Distillation has been used for the recovery of Cd under high temperature with three processes, i.e., SNAM—SAVAM, SAB—NIFE, and INMETCO [6,15]. The use of vacuum can facilitate metallurgical separation with high efficiency and low pollution [16]. Although the pyrometallurgical techniques have the advantages of high reaction rate, high overall efficiency and productivity and need fewer chemicals than hydrometallurgical techniques, the co-existed metals with high boiling point usually form new slag after Cd separation and the recovery of valuable metals from the slag still needs hydrometallurgical refining [6]. In addition, such pyrometallurgical processes often need waste gas treatment because harmful gas is released, and such techniques are not profitable for low-grade resources considering the high energy consumption and low valuable metal productivity [6].

Compared with pyrometallurgical processes, hydrometallurgical technologies are more appreciated from economic and environmental perspectives in the separation of cadmium from many resources because less energy is consumed and less dust collecting and gas cleaning are needed [3]. Hydrometallurgical technologies developed for this purpose include precipitation [17,18], cementation [14,19], electrodeposition [8,20], solvent extraction [2,11,21–27], membrane technology [28–31], imprinted sorbent separation [32–34], water-based two-phase system extraction [35], selective leaching [36,37], and solid phase extraction [38,39]. Some of these techniques, such as imprinted sorbent separation, have only been practiced in lab-scale research due to their technical limitations. In comparison, several technologies including cementation, solvent extraction, and solid phase extraction have found wide industrial application. Cementation is often employed in industry to remove Cd from Zn concentrate [14]. However, recovery of metal values including Cd from secondary resources such as Cd-bearing purification filter cake requires a series of dissolution and re-cementation stages, usually culminating in the electrowinning of Cd metal to obtain a product of sufficient purity for commercial demands [40]. In addition, this process is not applicable to Cd removal from moderately acidic medium (pH 1.5–2) because unacceptable quantities of zinc dust would be consumed for neutralizing the excess acid [41].

Solvent extraction has been extensively studied and applied to the separation of Cd from base metals [42]. The separation efficiency is basically determined by the properties of the coexisted elements and the performance of the extractants. In particular, due to their close physical and chemical properties, separation of Cd from Zn is highly challenging [43], especially when a single extractant is used [2,11,23,25], despite the fact that high separation efficiency has been achieved by careful optimization in some examples [3,23]. As a consequence, synergistic extraction has received more attention in the recent decades and better selectivity has been achieved [2,21–23]. For example, Preston and co-workers [21] found that the addition of 0.50 M triisobutylphosphine sulphide (TIBPS or Cyanex 471X) to 0.50 M 3,5-diisopropylsalicylic acid (DIPSA) resulted in a strong synergistic extraction effect for Cd^{2+}, as indicated by a shift of pH_{50} (the pH at which 50% of metal is extracted) by 1.9 pH units, while pH_{50} shifts of less than 0.6 pH unit for Zn^{2+}, Pb^{2+}, Mn^{2+}, Fe^{3+}, Co^{2+}, Ni^{2+}, Ca^{2+}, and Mg^{2+} were observed, which thus enabled the selective extraction of Cd^{2+} from these metals. In a study by Moradkhani and co-workers [2], pH_{50} and ΔpH_{50} values were determined and compared for a number of organic systems for synergistic extraction of Cd^{2+} from Zn^{2+}. The largest $\Delta pH_{50(Zn-Cd)}$ value was 1.53 pH units obtained with nonylsalicylic acid (HRJ-4277)/TIBPS system, implying excellent separation of Cd^{2+} from Zn^{2+} with this system. However, the pH_{50} values for Cd^{2+} were still not low enough for acidic industrial materials and hence, neutralization is often necessary before the extraction is carried out [2,21–23]. Moreover, the $\Delta pH_{50(Cu\text{-}Cd)}$ and $\Delta pH_{50(Pb\text{-}Cd)}$ values are small [11,21] and multi-stage extraction is often needed to realize satisfactory Cd separation [11]. In addition, many of the extractants, especially organophosphorus and organosulfur compounds, are highly toxic.

Solid phase extraction (SPE), especially ion exchange, has exhibited many advantages over other technologies, including simpler operational processes without the use of volatile and toxic solvents, lower pollution than solvent involved extraction processes, avoidance of sludge generation, milder operating conditions, and tolerance for wider pH range [44]. A number of studies on both anion and cation exchange methods based on halo-complex formation have been reported for the separation of Cd from other metals [1,38,39,41,45]. In anion exchange methods, Cd^{2+} was usually adsorbed by strong base anion exchange resin in the form of $[CdX_4]^{2-}$ in the presence of excess of X^- and subsequently stripped with highly concentrated $NaNO_3$ or HNO_3 [38,45]. In cation exchange methods, Cd^{2+} and other metal cations were adsorbed onto a strongly acidic cation exchanger and Cd^{2+} was selectively eluted with halide solution [46–48]. Alternatively, Cd^{2+} passes through the cation exchange resin in the form of $[CdI_4]^{2-}$ while other cations are retained by the resin due to their different stability of iodo-complexes [49]. In the anion exchange procedures, moderate to high concentrations of ligand (X^-) were used in the feed solutions and this often means high operating costs when using bromide or iodide, and the striping of Cd^{2+} by $NaNO_3$ or HNO_3 is usually very slow because of the strong retention of $[CdX_4]^{2-}$ [45,50,51]. In addition, highly concentrated HNO_3 as the stripping agent can oxide I^- adsorbed on the resin [50]. Halide solution with high concentration has also been used for the elution of Cd^{2+}, but the stripping was not fast and efficient [38].

Considering the above-mentioned shortcomings of the conventional techniques, it is of significance to develop a more efficient, environmentally benign and economic technique for the separation and recovery of Cd for industrial application. Therefore, a new strategy for ion exchange separation of Cd^{2+} is presented in this article. The stability of Cd^{2+} halo-complexes increases in the order of $[CdCl_4]^{2-} < [CdBr_4]^{2-} < [CdI_4]^{2-}$ [52]. As such, the favorable order of using halide as the ligand for Cd^{2+} complexation is in the order of $I^- > Br^- > Cl^-$. In this new method, anion exchange resin is first loaded with Br^- or I^-, and such modified resins are used for the adsorption of Cd^{2+} from Cl^- solution, thus consumption of bromide and iodide is minimized while achieving high adsorption performance and selectivity because of the higher affinity of Br^- and I^- to Cd^{2+} than Cl^-. Such a method has been applied to lab-scale in-line separation of Cd^{2+} from base metals in the digest of solid environment samples for Cd determination in conjunction with atomic absorption spectrometry with a detection limit of 3 ng L^{-1} and precision of around 5% achieved [53]. However, some fundamental issues relating to the adsorption, desorption and separation processes, including thermodynamics, kinetics, and mechanism, are to be further clarified. The present work aims at elucidating these issues with experimental data. It will investigate the performance of halide preloaded strong base anion exchangers in the separation of Cd from other base metals including Mn, Fe, Co, Ni, Cu, Zn, and Pb. The adsorption isotherms and kinetics of Cd^{2+} on Cl^--, Br^--, or I^--loaded anion exchange resin and the elution strength of water, ammonia, and ethylenediamine solutions from the loaded resin and fiber will be examined. The separation performance of Cd from the above-mentioned base metals will be presented. An Amberlite IRA 900 macroporous strongly basic anion exchange resin will be tested as the adsorbent for the thermodynamics and kinetics study. At the same time, a ZB-2 strongly basic anion exchange fiber will be used for comparison in desorption and separation performance experiments. This research aims at achieving a scalable technique for industrial application.

2. Materials and Methods

2.1. Reagents and Apparatus

All chemicals are reagent grade unless otherwise stated. Amberlite IRA 900 strongly basic anion exchange resin (Sigma-Aldrich Chemie GmbH, Buchs, France) and ZB-2 strongly basic anion exchange fibre (Guilin Zhenghan Science and Technology Developing Co. Ltd., Guilin, China) were used for the study. The Amberlite IRA 900 resin has particle size of 297–841 micron, exchange capacity of 1.0 meq g^{-1}, and pore size of 100 nm. The ZB-2 fibre has linear density of 1.7–2.3 dtex and exchange capacity of ≥3 meq g^{-1}.

Individual Cd^{2+}, Ni^{2+}, and Pb^{2+} stock solutions of 1% (m/v) and Mn^{2+}, Fe^{3+}, Co^{2+}, Cu^{2+} and Zn^{2+} stock solutions of 5% (m/v) were prepared by first dissolving their chloride or oxide in 2% hydrochloric acid. The final HCl concentration in the solutions was 0.15 M. Standard solutions of Cd^{2+}, Zn^{2+}, Mn^{2+}, Fe^{3+}, Co^{2+}, Ni^{2+}, Cu^{2+} and Pb^{2+} of 1000 mg L^{-1} in 1 M HNO_3 (National non-ferrous metal and electronic materials analytical and testing center, Beijing, China) were used for calibration in spectroscopic analysis of the metals after adequate dilution. Potassium bromide and potassium iodide were purchased from Shanghai Macklin Biochemical Co., Ltd. (Shanghai, China). Ammonia solution (25–28%, Shanghai Lingfeng Chemical Reagent Co., Ltd., Shanghai, China) and ethylenediamine (Nanjing Chemical Reagent Co., Ltd., Nanjing, China) were used for Cd^{2+} desorption after adequate dilution with water. Milli-Q water was used in all experiments.

An Agilent 7900 inductively coupled plasma mass spectrometer (ICP-MS) (Agilent Technologies, Inc., Tokyo, Japan) was used for the detection of the metal elements. In the separation performance experiments, a multi-channel peristaltic pump (Model ISM944, ISMATEC SA, Glattbrugg, Switzerland) was used for the percolation of sample solution, washing agent and eluent with Tygon tubing of 1.6 mm inner diameter (id). A Nicolet Summit FTIR spectrometer (Thermo Fisher Scientific Inc., Waltham, MA, USA) was used for IR spectrum measurement.

2.2. Preparation of Halide-Loaded Anion Exchangers

To prepare Br^-- or I^--loaded anion exchange resin or fibre, 5 g of the exchangers were placed in 100 mL of 0.5 M KBr or KI solution for two hours under agitation with magnetic stirrer followed by filtration and thorough washing with water until no Br^- or I^- was detectable by reaction with $AgNO_3$. Cl^- preloading was not necessary because all the commercial exchanger products were supplied in Cl-form. Once immobilized on the resin or fibre, Br^- and I^- were quite stable, and considerable replacement by Cl^- was not observed under the experimental conditions in this study due to their higher affinity to the quaternary ammonium groups than Cl^-.

2.3. Adsorption Isotherm Study

For the preparation of adsorption isotherms, different volumes of Cd^{2+} stock solution of 4.39 mg Cd mL^{-1} in 0.15 M HCl were added into seven 150 mL conical glass flasks containing 0.5 g of X^--loaded Amberlite IRA 900 resin each and made up to 50 mL with 0.15 M HCl solution and agitated on a HY-2A platform mixer (Guohua Changzhou Instruments Pty. Ltd., Changzhou, China) at 150 rpm for 2 h at 25 ± 2 °C. For Cl^--loaded resin (Cl-resin) and Br^--loaded resin (Br-resin), the quantities of Cd^{2+} added to the seven flasks were 2.2, 4.4, 8.8, 17.6, 26.3, 35.1 and 43.9 mg, respectively. For I^--loaded resin (I-resin), the quantities of Cd^{2+} added to the seven flasks were 4.4, 8.8, 17.6, 26.3, 35.1, 43.9 and 65.9 mg, respectively. Cd^{2+} in the equilibrium solutions was determined with ICP-MS after adequate dilution and calculated through standard calibration and the quantity of Cd^{2+} adsorbed by the resin was calculated by mass balance.

2.4. Adsorption Kinetics Study

All the kinetic experiments were conducted using 100 mL of Cd^{2+} solution in 0.15 M HCl. Duplicate tests were conducted for each experiment and Cd^{2+} concentrations for the two parallel experiments were 40 mg L^{-1} and 80 mg L^{-1}, respectively. The solutions were added to 150-mL Pyrex glass conical flasks. 0.5 g (dry) of X^--loaded Amberlite IRA 900 resin was added into each flask and the mixture was agitated with a magnetic stirrer at 600 rpm. Aliquots of 200 μL of the 40 mg L^{-1} or 100 μL of the 80 mg L^{-1} solution were withdrawn at certain time intervals for Cd^{2+} determination by ICP-MS after dilution with water and the mass of adsorbed Cd versus time relationship was constructed.

2.5. *Effect of Acidity on Cd^{2+} Adsorption*

To investigate the effect of acidity on the adsorption efficiency of Cd^{2+}, 0.5 g of dry X-preloaded anion exchange resin was packed into a plastic syringe of 9 mm in diameter, producing 1 mL volume after saturation with water. 20 mL feed solution of 0.1% Cd^{2+} was loaded into the column and drained dropwise at a flow rate of 0.5 mL min^{-1}. Feed solutions with HCl concentrations of 0.1 M, 0.5 M, 1.0 M, 3.0 M, and 6.0 M were compared. The Cd^{2+} retention efficiency at different acidities was measured by analyzing the residual Cd^{2+} concentration in the effluents by ICP-MS.

2.6. *Stripping Agent*

Different stripping agents were compared with online method using an ion exchange column and a peristaltic pump. 0.5 g dry X^--loaded Amberlite IRA 900 resin or ZB-2 fibre was packed into a plastic syringe of 9 mm in diameter, producing 1 mL volume after saturation with water. The column was loaded with Cd^{2+} by percolating 20 mL solution containing 0.2% Cd^{2+} in 0.15 M HCl through it at 1 mL min^{-1}, followed by washing with 20 mL 0.10 M HCl to remove the un-adsorbed Cd^{2+}. Adsorbed Cd^{2+} was stripped with different agents including water, 0.5 M ammonia and 0.3 M ethylenediamine (En) solutions at 1 mL min^{-1}. The eluate was collected in fractions of 2 mL (water and NH_3) or 1 mL (En). Cd^{2+} in the effluents and eluates was determined by ICP-MS and the retention efficiency of Cd^{2+} can be calculated by mass balance according to the Cd^{2+} concentration in the effluents, while the desorption profile can be established with the Cd^{2+} concentrations in the eluate fractions.

2.7. *Separation Performance*

Separation performance was investigated in the same way as in the stripping agent comparison with a synthetic feed solution containing 0.1% Cd^{2+}, 1% Zn^{2+}, 1% Cu^{2+}, 0.1% Pb^{2+}, 0.1% Mn^{2+}, 0.5% Fe^{3+}, 0.1% Co^{2+}, and 0.1% Ni^{2+} (w/v) in 0.15 M HCl. Cd and other elements in the effluents and eluates were determined with ICP-MS so that the retention efficiency and recovery of each element and the separation efficiency of Cd from each co-existed element can also be calculated. The above concentration values were chosen just for fundamental study, rather than for any special raw material from industry. Cd usually exists in mineral resources as a minor component. As such, it was added into the synthetic feed solution with a lower concentration than some major elements such as Zn, Cu and Fe in minerals.

3. Results and Discussion

3.1. *Selection of Ligand and Concentration in the Feed Solution*

The presence of X^- in the feed solution is important for the adsorption of Cd^{2+} on an anion exchanger due to its complexation with Cd^{2+} forming the negatively charged species $[CdX_4]^{2-}$. In hydrometallurgical processing of minerals, base metals are often leached with HCl or H_2SO_4 [3,8–10,18,22,54]. After leaching with HCl, Cd^{2+} and Cl^- ions can form stable $[CdCl_4]^{2-}$ for anion exchange separation. When leached with H_2SO_4, halide ions (Cl^-, Br^- or I^-) is to be added as ligand to form complex with Cd^{2+} for its adsorption by anion exchanger. Although Br^- and I^- form more stable complexes with Cd^{2+} which is beneficial for the adsorption of Cd^{2+} by anion exchanger [45,50], this advantage is compromised by the high cost of bromide and iodide salts or acids. Therefore, HCl was chosen as the sample medium while Br^- or I^- was pre-loaded onto the anion exchangers in this study, thus improving the adsorption performance of commercial Cl^--formed anion exchanger while minimizing consumption of Br^- or I^-. In the adsorption of Cd^{2+} onto the X^--loaded exchanger, the exchanger may offer two X atoms for coordinating Cd^{2+} while the feed solution provides two Cl^- ions for the formation of $[CdX_4]$ as detailed below. Therefore, the concentration of Cl^- in the feed solution should be a determinant for the adsorption. This has been verified by our experiment which showed that the retention efficiency and adsorption rate of Cd^{2+} on the X^- loaded resins increased as increasing HCl concentration. The retention efficiency culminated with 0.15 M HCl and levelled off with higher HCl concentration. Based on these facts, the feed solution was prepared in a medium containing 0.15 M HCl in the subsequent experiments.

3.2. Adsorption Isotherms and Mechanism

Several models can be used to describe the variations of the adsorption data. The two most frequently used ones for dilute solutions are the Langmuir and Freundlich isotherms [55]. Langmuir treatment is based on three assumptions that: (a) maximum adsorption corresponds to a saturated monolayer of solute molecules or ions on the adsorbent surface, (b) the energy of adsorption is constant, and (c) there is no transmigration of adsorbate in the plane of the surface [56], which is represented by:

$$q_e = K_a q_m C_e/(1 + K_a C_e) \tag{1}$$

or, in the linear form of

$$C_e/q_e = 1/q_m K_a + C_e/q_m \tag{2}$$

where q_e is the amount of adsorbate adsorbed per unit mass of adsorbent (mg/g) at equilibrium state and C_e is the liquid phase concentration of the adsorbate (mg/L) at equilibrium. q_m and K_a are Langmuir constants representing adsorption capacity and equilibrium constant, respectively.

The Freundlich equation is derived to model multilayer adsorption on heterogeneous surfaces, which is expressed as [56]:

$$q_e = K_F\, C_e^{1/n} \tag{3}$$

or, in the linear form of

$$\log q_e = \log K_F + (1/n)\log C_e \tag{4}$$

where q_e and C_e have the same meaning as in the Langmuir model, while K_F and n are Freundlich constants related to adsorption capacity and adsorption intensity, respectively.

The equilibrium data of Cd^{2+} adsorption on X^--loaded Amberlite IRA 900 resin and the Langmuir and Freundlich isotherms are illustrated in Figure 1, and the isotherm parameters obtained from linear analysis of Cd^{2+} adsorption by the two models are listed in Table 1.

The K_a and K_F data in Table 1 show that the adsorption performance of Cd^{2+} on the halide-loaded resins rose in the order of Cl-resin < Br-resin << I-resin. This is in accordance with the stability of $[CdX_4]^{2-}$ complexes [52]. The K_a and n values show that all the adsorptions are favorable [57].

The graphs in Figure 1 show that the adsorption of Cd^{2+} on Cl-resin and Br-resin can be described by Langmuir model while the data for I-resin can be described by Freundlich model. This may indicate that different adsorption mechanisms are followed. On the Cl-resin and Br-resin, monolayer adsorption of Cd^{2+} halo-complex should happened which can be expressed as $[Cd(Cl_S)_2(Cl_L)_2]$ and $[CdBr_2(Cl_L)_2]$, where Cl_S and Cl_L refer to Cl provided by the preloaded Cl on the resin and the liquid phase, respectively. In comparison, multilayer adsorption may have occurred on I-resin. Another possibility in I-resin adsorption is the formation of mixed-ligand complexes of different compositions on the resin surface, such as $[CdI(Cl)_3]$ and $[CdI_2(Cl)_2]$ as discussed below. The former one may augment the adsorption capacity of the I-resin. The adsorption capacity data (q_m) in Table 1 also indicate the different adsorption performance of the three resins with the I-resin having significantly higher capacity. It is worth mentioning that the difference should be more profound by taking the use of the same mass of halide modified resins in the experiment and the different atomic weights of Cl, Br, and I into account.

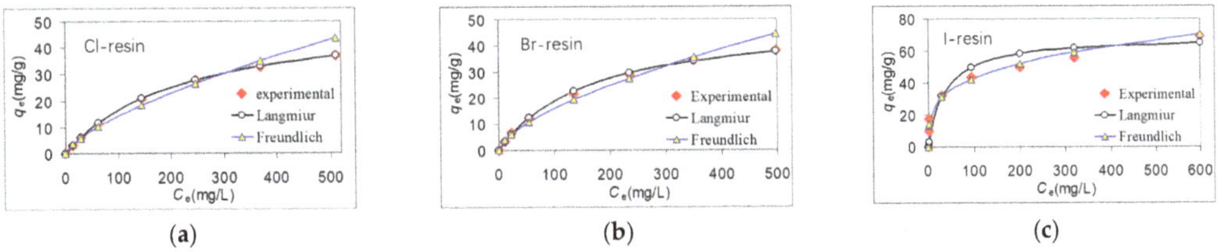

Figure 1. Theoretical and experimental non-linear isotherm plots of Cd^{2+} adsorption on halide-loaded Amberlite IRA 900 resins. (**a**) Cl-resin; (**b**) Br-resin; (**c**) I-resin.

Table 1. Isotherm parameters obtained from linear analysis of Cd^{2+} adsorption on X^--loaded resins.

		K_a	q_m(mg/g)	R^2
Langmuir model	Cl-resin	0.0045	53.19	0.9991
	Br-resin	0.0063	54.75	0.9972
	I-resin	0.0277	68.97	0.9777
		K_F	n	R^2
Freundlich model	Cl-resin	0.5867	1.446	0.9842
	Br-resin	0.8734	1.581	0.9835
	I-resin	12.325	3.671	0.9882

It has been established that Cd^{2+} complexes with mixed halide ligands are formed with high stability in multiple-halide aqueous solution [58,59]. For instance, in the presence of Cl^- and I^-, Cd^{2+} could form different species such as $[CdClI_3]^{2-}$, $[CdCl_2I_2]^{2-}$ and $[CdCl_3I]^{2-}$, depending on the relative concentrations of Cl^- and I^-. As such, we can assume that in the ion exchange process in this study, the simultaneous presence of different halide anions could also result in the formation of different $[CdX_4]^{2-}$ species with mixed halide ligands. To validate these assumptions, anion exchange resin with different forms (C-N$(CH_3)_3$-OH, C-N$(CH_3)_3$-Cl, C-N$(CH_3)_3$-I), and Cd-loaded X-resins were characterized with FTIR and the results are presented in Figure S1. When the Cl^- ions on the resin were replaced with OH^-, an intense peak at 1610 cm^{-1} was observed in the IR spectrum (Figure S1b). It is due to the deformation vibration of C-N$(CH_3)_3$ and O-H bonds [60]. In -Cl form, the peak is much weaker with a shoulder peak at about 1618 cm^{-1} (Figure S1a). Similarly, when the C-N$(CH_3)_3$ was associated with I^-, the peak was also much weaker than that with -OH groups but without the shoulder peak and the peak is more intense than that of the Cl-resin (Figure S1c). When the resin was loaded with $[CdCl_4]^{2-}$, the peak was also weaker than the -OH formed resin with the shoulder peak observed (Figure S1d), and when the resin was loaded with $[CdI_4]^{2-}$ by adsorbing Cd^{2-} from 0.1 M KI solution with I-resin, the shoulder peak disappeared but the intensity is higher than that of the $[CdCl_4]$ loaded resin (Figure S1g). In comparison, when the resin was loaded with Cd complex with mixed halide ligands (Cl^- and I^-), the peak profile and relative intensity are something between those of $[CdCl_4]^{2-}$ and $[CdI_4]^{2-}$ loaded resins (Figure S1e,f)). These results could be interpreted by the formation of mixed-ligand complexes of Cd^{2+} under different conditions. By using different Cl^- concentrations in the feed solution for Cd^{2+} adsorption by the I-resin, different species of Cd complexes, such as $[CdCl_2I_2]^{2-}$ and $[CdCl_3I]^{2-}$ could be formed. This is in line with the situation in solution chemistry [61]. It should be noted that Cd^{2+} ions were loaded from highly acidic solutions onto the resins and hence, the residual OH^- groups should be negligible. As such, the peak at 1610 cm^{-1} after loading with $[CdX_4]^{2-}$ should not be considered as a result of the presence of OH^- group associated with the quaternary ammonium groups, instead, it could reflect the change of the deformation vibration of C-N(CH_3)$_3$ group. The spectroscopic results should therefore indicate the effect of the $[CdX_4]^{2-}$ species on the C-N bond.

The change of the intense peak at 3374 cm^{-1}, corresponding to the stretching vibration of the O-H bond of the -OH group in the resin (Figure S1b), after loading the resin with different halide or Cd^{2+} halo-complex species, also indicates the influence of the Cd^{2+} halo-complex species. In particular, by loading with $[CdCl_4]^{2-}$ or $[CdI_4]^{2-}$, the peak shifted to 3390 cm^{-1} and 3500 cm^{-1}, respectively. By loading with $[CdCl_xI_y]^{2-}$ with increased x value and decreased y value (adsorbed from 0.1 M and 1 M HCl, respectively), the peak shifted towards lower wavenumber from 3572 cm^{-1} to 3500 cm^{-1}. This trend of wavenumber shifting is in accordance with the possible change in the complex composition.

However, the above spectroscopic results are not sufficient as a justification for the assumed complex species. For this reason, elemental analysis was performed by ICP-MS. Br-resin and I-resin were first saturated with Cd^{2+} by adsorption from HCl solution. After removing the excess Cd^{2+} in the resin micropores and voids with 0.1 M HCl, the resins were washed with water followed by complete stripping of Cd^{2+} with 0.3 M ethylenediamine (En). Br^- and I^- in the resins were completely stripped with 0.1 M KOH solution. Br, I and Cd in the respective water, En eluates and KOH solutions were determined by ICP-MS and the molar ratios of Br/Cd and I/Cd were calculated. The results showed that the Br/Cd ratio was 2.04 ± 0.03, and the I/Cd ratio was 1.89 ± 0.05, indicating that $[CdBr_2Cl_2]^{2-}$ and $[CdI_2Cl_2]^{2-}$ should be the dominant species on the Br-resin and I-resin, respectively. The I/Cd ratio lower than 2 may imply that species corresponding to $[CdICl_3]^{2-}$ exists on the resin, no matter if monolayer or multi-layer adsorption took place.

3.3. *Adsorption Kinetics*

3.3.1. Effect of Contact Time

Figure 2 shows the time courses of Cd^{2+} adsorption on X^--loaded Amberlite IRA 900 resins by using 40 mg L^{-1} and 80 mg L^{-1} as the initial Cd^{2+} concentrations. Adsorption processes could follow different mechanisms and be controlled by various factors, such as film diffusion, intra-particle diffusion or chemical reaction. In order to elucidate the adsorption process, several adsorption models are applicable for evaluation of the experimental data. In this study, Lagergren's pseudo-first order kinetic model, the pseudo-second order kinetic model, and intra-particle diffusion model were applied for the data treatment.

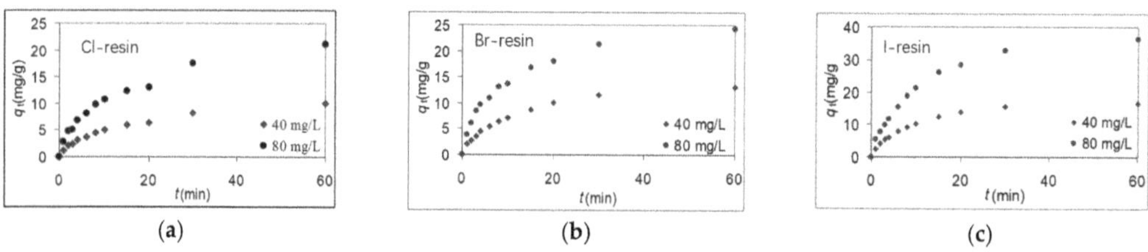

Figure 2. Mass-time profiles of Cd^{2+} adsorption on X^--loaded resins. (**a**) Cl-resin, (**b**) Br-resin, (**c**) I-resin.

3.3.2. Pseudo-First Order Kinetics

The linearized form of the pseudo-first order rate equation by Lagergren [55] is given as

$$\ln(q_e - q_t) = \ln q_e - k_1 \times t \quad (5)$$

where q_e and q_t are the amounts of the metal ion adsorbed (mg g^{-1}) at equilibrium and at time t (min), respectively and k_1 is the adsorption rate constant (min^{-1}). The linear regression of $\ln(q_e - q_t)$ versus t gives a straight line and the rate constant (k_1) can be calculated from its slope. The Lagergren's plots are given in Figure 3. The k_1, q_e and R^2 obtained from this model are listed in Table 2. It shows that the Lagergren's pseudo-first order model fits the experimental data well and the adsorption rate constants increase in the order of Cl-resin < Br-resin < I-resin. The theoretical and experimental q_e values are also in good agreement, implying the adsorption process can be assumed as a first order reaction. Obviously, the concentration of $[Cd(Cl)_4]^{2-}$ at the surface of the resin is the major controlling factor of the adsorption rate. The data with different initial Cd^{2+} concentrations also support the fitting of the pseudo-first order model. However, the difference in the rate constant with different resins implies that the adsorption process is not a pure first order reaction. Ligand exchange could also play a significant role, i.e., ligand exchange rate between X_S and Cl_L increases in the order of Cl < Br < I. X_S and Cl_L denote the pre-immobilized X^- on the resin and Cl^- ions in the liquid phase, respectively.

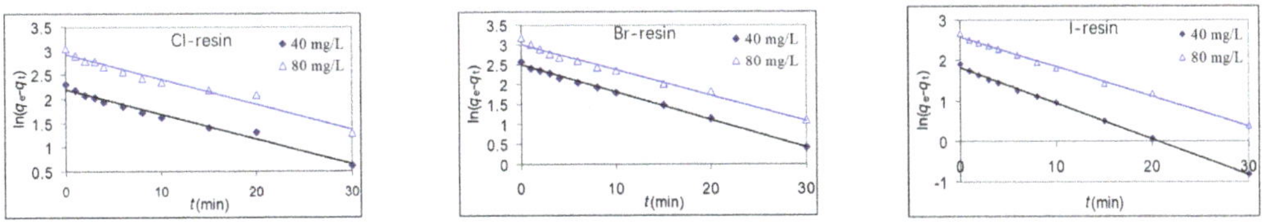

(a) (b) (c)

Figure 3. Pseudo-first order plots of Cd^{2+} adsorption on X^--loaded resins. (**a**) Cl-resin, (**b**) Br-resin, (**c**) I-resin.

Table 2. Kinetic parameters with pseudo-first order, pseudo-second order and intra-particle diffusion model analysis.

Resin	C_0 (mg/L)	Pseudo-First Order Parameters				Pseudo-Second Order Parameters			Intra-Particle Diffusion Rate Constants	
		q_e (exp., mg/g)	q_e (Theor. mg/g)	k_1 (min^{-1})	R^2	$k_2 \times 10^3$ (g/mg^{-1} min^{-1})	q_e (Theor., mg/g)	R^2	k_{id} (μg g^{-1} min$^{-1/2}$)	R^2
Cl-resin	40	10.1	8.98	0.0513	0.9792	7.3	11.7	0.9808	1.36	0.9826
	80	21.2	18.8	0.0508	0.9465	3.6	24.2	0.9754	2.83	0.9798
Br-resin	40	13.2	12.1	0.0687	0.9968	6.7	15.2	0.9938	1.85	0.9553
	80	24.3	20.6	0.0642	0.9817	4.6	27.0	0.9941	3.27	0.9448
I-resin	40	16.8	15.5	0.0878	0.9984	6.6	19.2	0.9983	2.40	0.9144
	80	36.6	33.6	0.0731	0.9951	2.5	42.4	0.9965	5.21	0.9386

3.3.3. Pseudo-Second Order Kinetics

The pseudo-second order kinetic model was also applied to the experimental data. It is given by the equation below [55]:

$$t/q_t = 1/k_2 q_e^2 + (1/q_e) \times t \qquad (6)$$

where k_2 (g mg^{-1} min^{-1}) is the rate constant of a pseudo-second order adsorption reaction. Second order kinetics is said to be applicable if the plot of t/q_t versus t relationship shows linearity. Second order kinetics is more likely to predict the behavior over whole range of adsorption and is in agreement with chemical sorption being the rate-controlling step [62]. The graphical interpretation of the data by the pseudo-second order kinetic model is given in Figure 4 and the rate constants (k_2), correlation coefficients of the plots together with the experimental q_e values are given in Table 2. The q_e values obtained from the pseudo-second order model are also reasonably consistent with the experimental ones and the correlation coefficients are very high, indicating that chemisorption is also a determinant for the adsorption process [63].

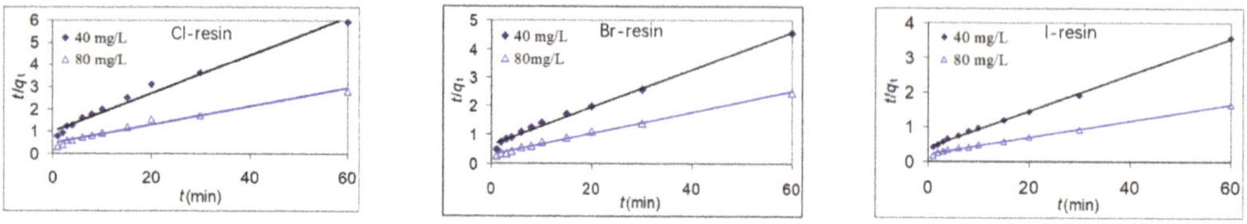

(a) (b) (c)

Figure 4. Pseudo-second order plots of Cd^{2+} adsorption on X^--loaded resins. (**a**) Cl-resin, (**b**) Br-resin, (**c**) I-resin.

3.3.4. Intra-Particle Diffusion Kinetics

The overall rate of ion exchange is often controlled by the intra-particle diffusion step with porous adsorbent and an intra-particle diffusion model is expressed in the equation given by Weber and Morris [64]:

$$q_t = k_{id}t^{1/2} \tag{7}$$

where q_t is the quantity of metal ions adsorbed at time t (mg g^{-1}) and k_{id} is the intra-particle diffusion rate constant (mg g^{-1} $min^{-1/2}$). Plots of q_t versus $t^{1/2}$ are shown in Figure 5 and the k_{id} and R^2 data are listed in Table 2. The correlation co-efficient data suggest that intra-particle diffusion plays a considerable part in the adsorption process but not the only rate controlling factor because the plots do not pass through the origin [65]. As pointed out above, ligand exchange may also play a significant role in determining the overall adsorption rate.

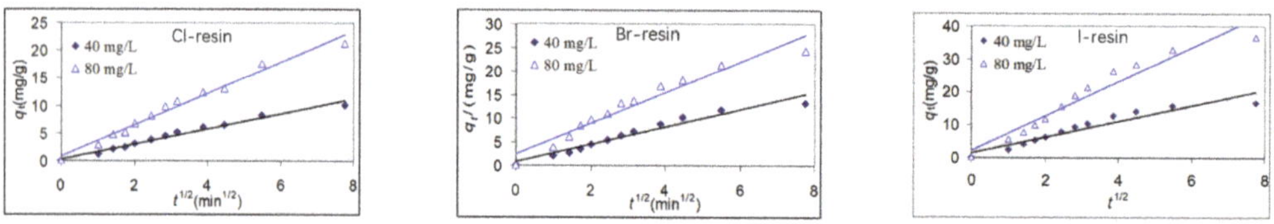

(a) (b) (c)

Figure 5. Intra-particle diffusion plots of Cd^{2+} adsorption on X^--loaded resins. (**a**) Cl-resin, (**b**) Br-resin, (**c**) I-resin.

3.4. Comparison of Dynamic Retention Efficiency of Cd^{2+}

The separation of Cd^{2+} from other metal ions in this study is based on their different retention efficiency by the anion exchangers. The performance of the three modified resins were compared experimentally in the presence of some major metal ions including Zn^{2+}, Cu^{2+} and Pb^{2+} which are competing ions against Cd^{2+} in the ion exchange process. Because ion exchange can be conveniently performed with a column, the comparative retention efficiency of Cd^{2+} on X^--loaded Amberlite IRA 900 resin or ZB-2 fibre in a dynamic process was studied by percolating a 20 mL feed solution containing 0.2% Cd^{2+}, 1% Zn^{2+}, 1% Cu^{2+} and 0.1% Pb^{2+} (w/v) at 1 mL min^{-1} through an ion exchange column packed with 0.5 g of dry resin or fibre, which gives a wet volume of 1 mL after saturation with water. Cd^{2+} in the effluent was analyzed and its retention efficiency was calculated by mass balance and the results are given in Table 3. The retention of Cd^{2+} increased in the order of Cl-resin < Br-resin < I-resin. The fibres have much higher retention efficiency than the corresponding resins under the same operational conditions. This could be attributed to their different properties. The fibre has much higher adsorption capacity than the resin, i.e., 3 meq g^{-1} versus 1 meq g^{-1}. It also has the functional groups on the surface, thus intra-particle diffusion was not involved in the adsorption process.

Although the retention efficiency obtained in most of the experiments is not sufficient for complete separation of Cd^{2+} from other metal ions under the operational conditions, this can be compensated by using different strategies. Potential ways of improving the efficiency include using larger column volume, optimization of the feeding flow rate, re-percolating the effluent through the same column before washing and elution, and use of multi-stage adsorption system. The data in the third row of Table 3 show that after re-percolating the effluent, the total retention efficiency improved considerably. We can assume from these results that quantitative Cd^{2+} retention could be attained by using a lower flow rate or multi-stage adsorption. In addition, when Cd^{2+} concentration in the feed solution was 200 mg L^{-1}, retention efficiency was all over 99%, implying that efficient adsorption can also be achieved when the initial Cd^{2+} concentration is low.

It is worth mentioning that Zn^{2+} could have adverse impact on the retention of Cd^{2+} and this should be especially profound for feed solutions with high Zn^{2+} concentration and when using Cl^-- and Br^--loaded exchangers due to the competition of Zn^{2+} against Cd^{2+} for ion exchange sites. Moreover, Cu^{2+} has significant impact on Cd^{2+} adsorption by I^--loaded exchangers as further discussed in the following section. Therefore, the retention efficiency of Cd^{2+} for real samples should be examined according to the concentrations of these co-existed elements.

Table 3. Retention efficiency of Cd^{2+} by X^--loaded resin and fibre.

Ion Exchanger	Cl-Resin	Br-Resin	I-Resin	Cl-Fibre	Br-Fibre	I-Fibre
Retention efficiency (%)	66.3	67.1	94.5	81.6	86.8	99.5
Retention efficiency after re-percolating the effluent (%)	84.3	89.3	99.1			

3.5. Effect of Acidity

The effect of acidity on the Cd^{2+} adsorption was studied by using different HCl concentrations in the feed solution, and the results are presented in Figure 6. It is evident that the HCl concentration had significant effect on the Cd^{2+} adsorption efficiency. We can assume that the Cl^- concentration is the major factor determining the Cd^{2+} adsorption. It is clear that when the HCl concentration was under 0.5 M, the adsorption had constant efficiency. With the increase in HCl concentration, the adsorption became less efficient. In 6 M HCl, the adsorption was very low. This can be attributed to the equilibrium between the solid resin phase and the liquid feed solution. In addition, high HCl concentration (>1 M) leads to oxidation of I^- by air, but 0.1% ascorbic acid dissolved in the feed solution effectively eliminated the oxidation. Although the HCl concentration has a profound influence on the adsorption, this method tolerated much higher acidity than solvent extraction and cementation.

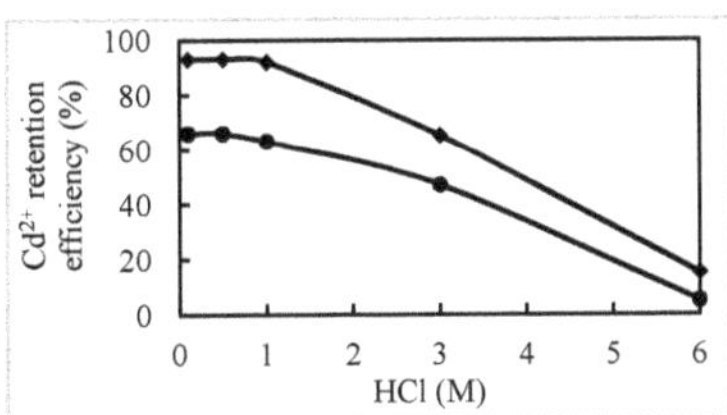

Figure 6. Effect of HCl concentration on Cd^{2+} adsorption efficiency on Br-resin (•) and I-resin (♦).

3.6. Removal of Retained Foreign Metal Ions

To separate the co-existed elements from Cd^{2+} as completely as possible, chemically and physically retained ions of these elements are to be removed prior to the elution of Cd^{2+}. Mn^{2+}, Fe^{3+}, Co^{2+}, Ni^{2+}, Cu^{2+}, Zn^{2+}, and Pb^{2+} were examined in this study as they are the most commonly encountered elements with Cd in minerals and mineral processing residues. Some of these elements form halo-complexes with appreciable stability and could be chemically retained by the exchangers. At the same time, the residual feed solution left in the pores and voids of the resin phase also contains considerable quantities of these elements. To remove them prior to the Cd^{2+} stripping stage, a washing step was applied following the sample loading by using 0.10 M HCl. It was found that Mn^{2+}, Fe^{3+}, Co^{2+}, and Ni^{2+} were readily removed by washing with 20 mL of 0.1 M HCl from the 1 mL resin or fiber column. Cu^{2+} and Pb^{2+} were also removed easily from Cl^-- or Br^--loaded resins. This is due to the weak complexation of these ions with the ligands. However, Cu^{2+} can be easily reduced to Cu^+ by I^-, resulting in consumption of the latter and strong retention of Cu^+ via formation of stable complex with I^- and/or CuI precipitate. In addition, the transformation of I^- to I_2 could deteriorate the performance of I^--loaded resin and fibre. Therefore, the separation of Cu^{2+} by I^--loaded resin or fibre was not considered. Coordination between I^- and Pb^{2+} has considerable strength [52] and Pb^{2+} was partially retained by I^--loaded resin or fibre and resulted in incomplete separation from Cd^{2+}. Stability of Zn^{2+} complexes decreases in the order of $[ZnCl_4]^{2-}$ > $[ZnBr_4]^{2-}$ > $[ZnI_4]^{2-}$ [50] and its retention efficiency declined in the same order. Compared with the other co-existed elements, halo-complexes of Zn^{2+} have higher stability and competitiveness against Cd^{2+} in the ion exchange process. Nevertheless, experimental results have shown that 40 mL of the washing solution completely removed Zn^{2+}. Under these conditions, Cd^{2+} was not lost from the exchangers. The use of HCl solution higher than 0.1 M was less effective in the removal of Zn^{2+} whereas a HCl lower than 0.1 M caused the loss of Cd^{2+} in the washing stage.

3.7. Stripping of Cd^{2+}

A common strategy for stripping Cd^{2+} from anion exchanger is using highly concentrated anions such as NO_3^- (HNO_3 or $NaNO_3$), but consumption of such eluent is massive due to the strong retention of Cd^{2+} on the anion exchanger [38,45,50,51,61]. In this study, we used water to disequilibrate the formation of Cd^{2+} halo-complex on the exchanger by diluting the Cl^- concentration in the system. This showed that 50–68% of the adsorbed Cd^{2+} was eluted with 20 bed-volumes (20 mL) of water (Figure S2). The stripping efficiency was defined as the stripped fraction as a percentage of the total retained amount. More water did not increase the stripping efficiency considerably. In general, the stripping efficiency with water was in reverse order with the stability of the halo-complexes. Although the stripping is not sufficient under the experimental conditions, this technique enjoys the benefit of environmental-benign asset.

Another stripping strategy is to use a ligand which forms positively charged species with Cd^{2+} so that its halo-complex can be decomposed. Ammonia is a typical candidate for this purpose [38,45] and has the potential to accelerate the stripping compared with water by forming $[Cd(NH_3)_4]^{2+}$ which spontaneously dissociates from the C-N$(CH_3)^+$ groups of the exchanger. As shown in Figure S3, 10 bed-volumes (10 mL) of 0.5 M NH_3 can strip 50% to 60% of the adsorbed Cd^{2+}, much more efficient than water. However, a further increase in the volume of ammonia solution did not make appreciable improvement in the stripping efficiency. This could be due to the formation of $Cd(OH)_2$ precipitate on the surface of the exchanger which hindered further stripping of Cd^{2+}. It also shows that the different stability of the halo-complexes and exchangers (i.e., resin vs. fibre) did not cause significant difference in stripping efficiency and this could be attributed to the high stability of $[Cd(NH_3)_4]^{2+}$ [52].

Compared to ammonia, amines should have higher stripping ability due to the formation of more stable complex with Cd^{2+} [52] and this should be more promising with multidentate ligands such as ethylenediamine. Stripping results with En are shown in Figure S4. Compared with H_2O and NH_3, En is a much more efficient eluent because only 5 bed-volumes (5 mL) of 0.3 M En were consumed to strip over 90% of Cd^{2+} from I^--loaded fibre. This is attributed to the formation of highly stable chelate $[CdEn_2]^{2+}$ and its spontaneous dissociation from the exchanger.

3.8. Separation Performance

The separation performance was investigated by percolating 20 mL of multielement feed solution at 1 mL min^{-1} through a column of 1 mL wet volume filled with halide loaded Amberlite IRA 900 resin or ZB-2 fibre followed by washing out the residual co-existed elements with 20 mL of 0.10 M HCl and eluting the adsorbed Cd^{2+}. Elements in the effluent, wash-out, and eluate were determined by ICP-MS and the recovery of the elements were calculated by taking the percentage of the recovered amount in the stripped solution over the total input amount in the feed solution. The concentrations of the elements in the feed solution were 0.1% Cd, 1% Zn, 1% Cu, 1% Pb, 0.1% Mn, 0.5% Fe, 0.1% Co, and 0.1% Ni (w/v), respectively. The results are given in Tables 4 and 5. It is clear that Zn^{2+} is the most difficult element for separation from Cd^{2+}, followed by Cu^{2+} and Pb^{2+}. In comparison, Mn^{2+}, Fe^{3+}, Co^{2+} and Ni^{2+} were easily separated from Cd^{2+} due to their low halo-complex stability. The separation of Cu^{2+} by I^--loaded resin was not examined due to its reduction by I^-.

Considering the difficulty in complete Zn^{2+} separation from Cd^{2+}, a separate experiment was conducted which involved the use of a feed solution containing 0.1% Cd^{2+} and 1% Zn^{2+}, 1 mL of wet I^--loaded anion exchange fibre, 40 mL of 0.1 M HCl as the washing agent and En solutions as the stripping agents. The results given in Table 6 show that by using these parameters, Zn^{2+} was separated from Cd^{2+} with high efficiency.

Table 4. Metal recovery by water elution.

Element	Element Recovery in Eluate (% of Total Input)					
	Cl-Resin	Br-Resin	I-Resin	Cl-Fibre	Br-Fibre	I-Fibre
Cd^{2+}	24.18	24.39	24.20	63.86	61.63	49.28
Zn^{2+}	1.85	1.67	0.98	0.24	0.08	0.06
Cu^{2+}	ND [a]	ND	NE [b]	ND	0.02	NE
Pb^{2+}	ND	ND	2.5	ND	ND	7.5
Mn^{2+}	ND	ND	ND	ND	ND	ND
Fe^{3+}	ND	ND	ND	ND	ND	ND
Co^{2+}	ND	ND	ND	ND	ND	ND
Ni^{2+}	ND	ND	ND	ND	ND	ND

Colum size 1 mL, 40 mL H_2O for elution, [a] Not detectable, [b] Not examined.

Table 5. Metal recovery by NH_3 elution.

Element	Element Recovery in Eluate (% of Total Input)					
	Cl-Resin	Br-Resin	I-Resin	Cl-Fibre	Br-Fibre	I-Fibre
Cd^{2+}	35.45	37.30	67.15	60.45	54.40	68.20
Zn^{2+}	2.12	1.79	1.19	0.30	0.24	0.20
Cu^{2+}	0.023	0.029	NE	0.036	0.066	NE
Pb^{2+}	0.06	0.04	0.13	<0.01	<0.01	0.17
Mn^{2+}	ND	ND	ND	ND	ND	ND
Fe^{3+}	ND	ND	ND	ND	ND	ND
Co^{2+}	ND	ND	ND	ND	ND	ND
Ni^{2+}	ND	ND	ND	ND	ND	ND

* 1 mL wet column volume, 10 mL 0.5 M NH_3 for elution.** Cu^{2+} was not examined by I^--loaded columns due to its reduction.

Table 6. Cd and Zn recovery (% of total input) by En elution from I^--loaded ZB-2 fibre.

En Concentration (M)	Cd^{2+} Recovery (%)	Zn^{2+} Recovery (%)	Separation Factor ($f_{Cd/Zn}$)
0.1	71.73	0.13	5518
0.2	88.45	0.04	22,113
0.3	97.03	0.02	48,515

* 1 mL column volume and 6 mL En for elution.

It is worth-noting that higher concentration of En led to higher Cd^{2+} recovery from the I^--loaded fibre but lower Zn^{2+} recovery although the $[ZnEn_2]^{2+}$ complex has higher stability than $[CdEn_2]^{2+}$ ($\log K_f$ 14.11 for $[ZnEn_2]^{2+}$ and 12.09 for $[CdEn_2]^{2+}$, respectively) [52]. This was possibly caused by the much higher stability of Zn-OH complex than that of Cd-OH complex ($\log K_f$ 17.66 for Zn-OH complex and 8.62 for Cd-OH complex, respectively) [52] which led to the relatively stable retention of Zn^{2+} by ion exchange or precipitation due to the basic property of the En solution and lower stripping efficiency. This is desirable for complete separation of the two elements.

3.9. *Comparison with Conventional and Reported Techniques*

Compared with some typical hydrometallurgical techniques applied in industry and reported in the literature, several advantages can be recognized in the present method. Solvent extraction and cementation are the two typical techniques for Cd separation applied in industry. In comparison with solvent extraction, the present technique eliminated the use of organic solvents, and it's also much more tolerant to high acidity. The benefit of the latter feature underlies its value for industrial application because mineral raw materials are often leached with acid. Compared with cementation, the present technique is also much less affected by high acidity, and there is no zinc dust consumption and Cd is separated from the matrix as a pure element in this method, thus no slag is produced in its removal process and no further separation from other metals is needed for its recovery. Many other techniques, such as imprinted sorbent separation, polymer inclusion membrane extraction, and water-based two-phase system extraction, have been only practiced in lab-scale study due to their limitations, whereas the present method is highly scalable for industrial application given the availability of commercial ion exchanger products.

4. Conclusions

By pre-loading strongly basic anion exchange resin and fibre with halide (Br^- and I^-), higher retention efficiency for Cd^{2+} from Cl^- solution was achieved by forming mixed ligand complexes than using Cl-formed exchanger. Thus, no excess Br^- or I^- is needed in the feed solution, and the operating cost can be minimized. Adsorption isotherms showed that the adsorption capacity increases in the order of Cl-resin < Br-resin < I-resin. The Langmuir model fitted well for Cl-resin and Br-resin while the Freundlich model fitted the I-resin better than Langmuir model. $[CdX_2Cl_2]^{2-}$ (X=Cl, Br or I) was assumed to be the dominant species retained by the exchangers, but different species could be formed on I^--loaded resin and fibre. Kinetics data of all the resins were well described by both Langergren pseudo-first order model and pseudo-second order model, implying that both chemical and physical factors are important for the sorption. Intra-particle diffusion also plays a significant role in the adsorption process with the resin. Under the optimized conditions, Mn^{2+}, Fe^{3+}, Co^{2+}, and Ni^{2+} were not retained by any of the ion exchangers and thus separated from Cd^{2+} completely. I^--loaded exchangers exhibited superior performance for the separation of Cd^{2+} from Zn^{2+} over Cl^-- or Br^--loaded exchangers, whereas the latter ones are more ideal for separating Cd^{2+} from Cu^{2+} and Pb^{2+}. Water exhibited appreciable stripping capability for Cd^{2+} ions from the resin or fibre and ammonia accelerated the elution considerably, but the overall stripping efficiency of Cd^{2+} by water or ammonia was not sufficient. Ethylenediamine showed much higher stripping capability with elution efficiency of over 90% under adequate conditions. This study represents a potentially scalable, environmentally friendly, and cost-effective technology for the separation of cadmium from base metals compared with conventional techniques.

Supplementary Materials: The following supporting information can be downloaded at: https://www.mdpi.com/article/10.3390/pr11041051/s1, Figure S1: IR spectra of halide and $[CdX_4]$ loaded anion exchange resins; Figure S2. Stripping efficiency of Cd^{2+} with 20 bed volume of H_2O from X^--loaded Amberlite IRA 900 resin; Figure S3. Stripping efficiency of Cd^{2+} with 10 bed volume of

0.5 M NH_3 from X^- loaded Amberlite IRA 900 resins; Figure S4. Stripping efficiency of Cd^{2+} with 5 bed volume of ethylenediamine from I^--loaded ZB-2 fibre.

Author Contributions: Conceptualization: Y.Z., C.H., X.D. and Y.W.; methodology: Y.Z., C.H. and Y.W.; resources: C.H. and Y.W.; Investigation: Y.Z., X.D. and G.D.; data processing: Y.Z. and G.D.; writing, manuscript drat preparation: Y.Z.; writing—review and editing: Y.W. and C.H.; funding acquisition: Y.W. All authors have read and agreed to the published version of the manuscript.

Funding: This research was funded by National Natural Science Foundation of China (No. 22178350).

Data Availability Statement: The research created experimental data that can be found in the tables and figures presented in this manuscript.

Acknowledgments: The authors would like to acknowledge the funding provided by National Natural Science Foundation of China (No. 22178350).

Conflicts of Interest: The authors declare no conflict of interest.

References

1. Kekesi, T.; Isshiki, M. Ultra high purification of copper chloride solutions by anion exchange. *Hydrometallurgy* **1997**, *45*, 345–361. [CrossRef]
2. Moradkhani, D.; Urbani, M.; Cheng, C.Y.; Askari, M.; Bastani, D. The separation of cadmium from zinc with a synergistic mixture of nonylsalicyclic acid and triisobutylphosphine sulphide. *Hydrometallurgy* **2005**, *78*, 129–136. [CrossRef]
3. Safarzadeh, M.S.; Bafghi, M.S.; Moradkhani, D.; Ilkhchi, M.K. A review on hydrometallurgical extraction and recovery of cadmium from various resources. *Miner. Eng.* **2007**, *20*, 211–220. [CrossRef]
4. Kouzbour, S.; Gourich, B.; Gros, F.; Vial, C.; Allam, F.; Stiriba, Y. Comparative analysis of industrial processes for cadmium removal from phosphoric acid: A review. *Hydrometallurgy* **2019**, *188*, 222–247. [CrossRef]
5. Mashhadikhan, S.; Amooghin, A.E.; Sanaeepur, H.; Shirazi, M.M.A. A critical review on cadmium recovery from wastewater towards environmental sustainability. *Desalination* **2022**, *535*, 115815. [CrossRef]
6. Blumbergs, E.; Serga, V.; Platacis, E.; Maiorov, M.; Shishkin, A. Cadmium recovery from spent Ni-Cd batteries: A brief review. *Metals* **2021**, *11*, 1714. [CrossRef]
7. Łukomska, A.; Wiśniewska, A.; Dąbrowski, Z.; Kolasa, D.; Lach, J.; Wróbel, K.; Domańska, U. New method for recovery of nickel and cadmium from the "black mass" of spent Ni-Cd batteries by solvent extraction. *J. Mol. Liq.* **2022**, *357*, 119087. [CrossRef]
8. Freitas, M.B.J.G.; Rosalém, S.F. Electrochemical recovery of cadmium from spent Ni-Cd batteries. *J. Power Sources* **2005**, *139*, 366–370. [CrossRef]
9. Reddy, B.R.; Priya, D.N. Chloride leaching and solvent extraction of cadmium, cobalt and nickel from spent nickel-cadmium batteries using Cyanex 923 and 272. *J. Power Sources* **2006**, *161*, 1428–1434. [CrossRef]
10. Rudnik, E.; Nikiel, M. Hydrometallurgical recovery of cadmium and nickel from spent Ni–Cd batteries. *Hydrometallurgy* **2007**, *89*, 61–71. [CrossRef]
11. Gotfryd, L.; Cox, M. The selective recovery of cadmium(II) from sulfate solutions by a counter-current extraction-stripping process using a mixture of diisopropylsalicyclic acid and Cyanex 471X. *Hydrometallurgy* **2006**, *81*, 226–233. [CrossRef]
12. Liu, H.; Wang, S.; Fu, L.; Zhang, G.; Zuo, Y.; Zhang, L. Mechanism and kinetics analysis of valuable metals leaching from copper-cadmium slag assisted by ultrasound cavitation. *J. Clean. Prod.* **2022**, *379*, 134775. [CrossRef]
13. Han, G.; Wang, J.; Sun, H.; Liu, B.; Huang, Y. A critical review on the removal and recovery of hazardous Cd from Cd-containing secondary resources in Cu-Pb-Zn smelting processes. *Metals* **2022**, *12*, 1846. [CrossRef]
14. Gouvea, L.R.; Morais, C.A. Recovery of zinc and cadmium from industrial waste by leaching/cementation. *Miner. Eng.* **2007**, *20*, 956–958. [CrossRef]
15. Espinosa, D.C.R.; Bernardes, A.M.; Tenório, J.A.S. An overview on the current processes for the recycling of batteries. *J. Power Sources* **2004**, *135*, 311–319. [CrossRef]
16. Espinosa, D.C.R.; Tenório, J.A.S. Fundamental aspects of recycling of nickel–cadmium batteries through vacuum distillation. *J. Power Sources* **2004**, *135*, 320–326. [CrossRef]
17. Mauchauffée, S.; Meux, E.; Schneider, M. Selective precipitation of cadmium from nickel cadmium sulphate solutions using sodium decanoate. *Sep. Purif. Technol.* **2008**, *62*, 394–400. [CrossRef]
18. Freitas, M.B.J.G.; Penha, T.R.; Sirtoli, S. Chemical and electrochemical recycling of the negative electrodes from spent Ni-Cd batteries. *J. Power Sources* **2007**, *163*, 1114–1119. [CrossRef]
19. Amin, N.K.; El-Ashtoukhy, E.-S.Z.; Abdelwahab, O. Rate of cadmium ions removal from dilute solutions by cementation on zinc using a rotating fixed bed reactor. *Hydrometallurgy* **2007**, *89*, 224–232. [CrossRef]
20. Safarzadeh, M.S.; Moradkhani, D. The electrowinning of cadmium in the presence of zinc. *Hydrometallurgy* **2010**, *105*, 168–171. [CrossRef]

21. Preston, J.S.; Patrick, J.H.; Steinbach, G. The selective solvent extraction of cadmium by mixtures of carboxylic acids and trialkylphosphine sulphides. Part 2. Practical applications in the separation of cadmium of cadmium from zinc and nickel. *Hydrometallurgy* **1994**, *36*, 143–160. [CrossRef]
22. Gupta, B.; Deep, A.; Malik, P. Extraction and recovery of cadmium using Cyanex 923. *Hydrometallurgy* **2001**, *61*, 65–67. [CrossRef]
23. Itabashi, H.; Yoshida, M.; Kawamoto, H. Kinetically controlled separation of cadmium(II) from zinc(II) with dithizone in the presence of nitrilotriacetic acid. *Anal. Sci.* **2001**, *17*, 1301–1304. [CrossRef] [PubMed]
24. Reddy, B.R.; Priya, D.N.; Kumar, J.R. Solvent extraction of cadmium (II) from sulphate solutions using TOPS 99, PC 88A, Cyanex 272 and their mixtures. *Hydrometallurgy* **2004**, *74*, 277–283. [CrossRef]
25. Yaftian, M.R.; Zamani, A.A.; Parinejad, M.; Shams, E. Ion-pair extraction of cadmium complex anions from hydrochloric acid media using oxonium ion-dicyclohexyl-18-crown-6 complex. *Sep. Purif. Technol.* **2005**, *42*, 175–180. [CrossRef]
26. Babakhani, A.; Rashchi, F.; Zakeri, A.; Vahidi, E. Selective separation of nickel and cadmium from sulfate solutions of spent nickel–cadmium batteries using mixtures of D2EHPA and Cyanex 302. *J. Power Sources* **2014**, *247*, 127–133. [CrossRef]
27. Lee, L.Y.; Morad, N.; Ismail, N.; Talebi, A.; Rafatullah, M. Optimization for liquid-liquid extraction of Cd(II) over Cu(II) ions from aueous slutions uing inic lquid Aliquat 336 with tibutyl posphate. *Int. J. Mol. Sci.* **2020**, *21*, 6860. [CrossRef]
28. Li, Q.; Liu, Q.; Zhang, Q.; Wei, X.; Guo, J. Separation study of cadmium through an emulsion liquid membrane using triisooctylamine as mobile carrier. *Talanta* **1998**, *46*, 927–932. [CrossRef]
29. Choi, D.W.; Kim, Y.H. Cadmium removal using hollow fiber membrane with organic extractant. *Korean J. Chem. Eng.* **2003**, *20*, 768–771. [CrossRef]
30. Swain, B.; Sarangi, K.; Das, R.P. Effect of different anions on the separation of cadmium and zinc by supported liquid membrane using TOPS-99 as mobile carrier. *J. Membr. Sci.* **2006**, *277*, 240–248. [CrossRef]
31. Adelung, S.; Lohrengel, B.; Nghiem, L.D. Selective transport of cadmium by PVC/Aliquat 336 polymer inclusion membranes (PIMs): The role of membrane composition and solution chemistry. *Membr. Water Treat.* **2012**, *3*, 123–131. [CrossRef]
32. Lu, Y.; Yan, X. An imprinted organic-inorganic hybrid sorbent for selective separation of cadmium from aqueous solution. *Anal. Chem.* **2004**, *76*, 453–457. [CrossRef]
33. Zhai, Y.; Liu, Y.; Chang, X.; Chen, S.; Huang, X. Selective solid-phase extraction of trace cadmium(II) with an ionic imprinted polymer prepared from a dual-ligand monomer. *Anal. Chim. Acta* **2007**, *593*, 123–128. [CrossRef]
34. Li, F.; Jiang, H.; Zhang, S. An ion-imprinted silica-supported organic-inorganic hybrid sorbent prepared by a surface imprinting technique combined with a polysaccharide incorporated sol-gel process for selective separation of cadmium(II) from aqueous solution. *Talanta* **2007**, *71*, 1487–1493. [CrossRef]
35. Akama, Y.; Ito, M.; Tanaka, S. Selective separation of cadmium from cobalt, copper, iron(III) and zinc by water-based two-phase system of tetrabutylammonium bromide. *Talanta* **2000**, *53*, 645–650. [CrossRef]
36. Aparajith, B.; Kumar, A.; Hodder, D.; Gupta, M.L. Recovery of cadmium from hydrometallurgical zinc smelter by selective leaching. *Hydrometallurgy* **2010**, *102*, 31–36. [CrossRef]
37. Turan, M.D.; Safarzadeh, M.S. Separation of zinc, cadmium and nickel from ZnO–CdO–NiO mixture through baking with ammonium chloride and leaching. *Hydrometallurgy* **2012**, *119–120*, 1–7. [CrossRef]
38. Dybczyński, R.; Samczyński, Z. The use of amphoteric ion exchange resin for the selective separation of cadmium from other radioelements present in neutron irradiated biological materials. *J. Radioanal. Nucl. Chem.* **1991**, *150*, 143–153. [CrossRef]
39. Fröhlich, S.; Sewing, D. The BATENUS process for recycling mixed battery waste. *J. Power Sources* **1995**, *57*, 27–30. [CrossRef]
40. Ghatas, N.E. Purification of Zinc Sulfate Solutions. Canadian Patent 1,046,288, 25 September 1979.
41. Saeed, M.M.; Ahmed, M. Retention, kinetics and thermodynamics profile of cadmium and adsorption from iodide medium onto polyurethane foam and its separation from zinc bulk. *Anal. Chim. Acta* **2004**, *525*, 289–297. [CrossRef]
42. Jha, M.K.; Kumar, V.; Jeong, J.; Lee, J.-C. Review on solvent extraction of cadmium from various solutions. *Hydrometallurgy* **2012**, *111–112*, 1–9. [CrossRef]
43. Hadioui, M.; Mecherri, M.O.; Harrache, Z. Predictive simulation for separation of toxic metal ions by ion-exchange resins in complex media. *J. Chem. Biochem. Mol. Biol.* **2008**, *2*, 1–15.
44. Barakat, M.A. New trends in removing heavy metals from industrial wastewater. *Arab. J. Chem.* **2011**, *4*, 361–377. [CrossRef]
45. Strelow, F.W.E. Improved separation of cadmium from indium, zinc, gallium and other elements by anion-exchange chromatography in hydrochloric-nitric acid mixtures. *Anal. Chim. Acta* **1978**, *100*, 577–588. [CrossRef]
46. Kallmann, S.; Oberthin, H.; Liu, R. Determination of cadmium in zinc concentrates and other zinc-rich materials: A cation exchange procedure. *Anal. Chem.* **1960**, *34*, 58–60. [CrossRef]
47. Korkisch, J.; Sorio, A. Determination of cadmium, copper and lead in natural waters after anion-exchange separation. *Anal. Chim. Acta* **1975**, *76*, 393–399. [CrossRef]
48. Fritz, J.S.; Garralda, B.B. Cation exchange separation of metal ions with hydrochloric acid. *Anal. Chem.* **1962**, *34*, 102–106. [CrossRef]
49. Strelow, F.W.E. Separation of cadmium from uranium, cobalt, nickel, manganese, zinc, coppor, titanium, and other elements by cation exchange chromatography. *Anal. Chem.* **1960**, *32*, 363–365. [CrossRef]
50. Kallmann, S.; Oberthin, H.; Lin, R. Determination of cadmium in zinc concentrates and other zinc-rich materials: Anion exchange procedure. *Anal. Chem.* **1958**, *30*, 1846–1848. [CrossRef]

51. Korkisch, J.; Feik, F. Anion-exchange separation of cadmium in organic solvent-nitric acid media. *Anal. Chim. Acta* **1965**, *32*, 110–116. [CrossRef]
52. Patnaik, P. *Dean's Analytical Chemistry Handbook*; McGraw-Hill Companies, Inc.: New York, NY, USA, 2004.
53. Adeloju, S.B.; Zhang, Y. Vapor generation atomic absorption spectrometric determination of cadmium in environmental samples with in-line anion exchange separation. *Anal. Chem.* **2009**, *81*, 4249–4255. [CrossRef]
54. Bartolozzi, M.; Braccini, G.; Bonvini, S.; Marconi, P.F. Hydrometallurgical recovery process for nickel-cadmium spent batteries. *J. Power Sources* **1995**, *55*, 247–250. [CrossRef]
55. Sari, A.; Tuzen, M.; Citak, D.; Soylak, M. Equilibrium, kinetic and thermodynamic studies of adsorption of Pb(II) from aqueous solution onto Turkish kaolinite clay. *J. Hazard. Mater.* **2007**, *149*, 283–291. [CrossRef]
56. Jain, C.K. Adsorption of zinc onto bed sediments of the River Ganga: Adsorption models and kinetics. *Hydrol. Sci.* **2001**, *46*, 419–434. [CrossRef]
57. Dada, A.O.; Olalekan, A.P.; Olatunya, A.M.; Dada, O. Langmuir, Freundlich, Temkin and Dubinin–Radushkevich isotherms studies of equilibrium sorption of Zn^{2+} onto phosphoric acid modified rice husk. *IOSR-J. Appl. Chem.* **2012**, *3*, 38–45. [CrossRef]
58. Eliezer, I.; Moreno, A. Potentiometric study of mixed cadmium halide complexes in aqueous solution. *J. Chem. Eng. Data* **1974**, *19*, 226–228. [CrossRef]
59. Zelić, M.; Branica, M. Voltammetric study of cadmium complexation in halogenide mixtures of high ionic strength. *Electroanalysis* **1992**, *4*, 701–706. [CrossRef]
60. Wołowicz, A.; Wawrzkiewicz, M. Screening of Ion Exchange Resins for Hazardous Ni(II) Removal from Aqueous Solutions: Kinetic and Equilibrium Batch Adsorption Method. *Processes* **2021**, *9*, 285. [CrossRef]
61. Marcus, Y.; Eliezer, I. The anion exchange of metal complexes-XII. The cadmium and mercury-halide systems. *J. Inorg. Nucl. Chem.* **1963**, *25*, 867–874. [CrossRef]
62. Ünlü, N.; Ersoz, M. Adsorption characteristics of heavy metal ions onto a low cost biopolymeric sorbent from aqueous solutions. *J. Hazard. Mater.* **2006**, *136*, 272–280. [CrossRef]
63. Farouq, R.; Yousef, N.S. Equilibrium and kinetics studies of adsorption of copper(II) ions on natural biosorbent. *Int. J. Chem. Eng. Appl.* **2015**, *6*, 319–324. [CrossRef]
64. Weber, W.J.; Morris, J.C. Kinetics of adsorption on carbon from solution. *J. Sanit. Eng. Div.* **1963**, *89*, 31–60. [CrossRef]
65. Zhang, X.; Liu, H.; Yang, J.; Zhang, L.; Cao, B.; Liu, L.; Gong, W. Removal of cadmium and lead from aqueous solutions using iron phosphate-modified pollen microspheres as adsorbents. *Rev. Adv. Mater. Sci.* **2021**, *60*, 365–376. [CrossRef]

processes

MDPI

Article

Separation of Rare Earth Elements in Multistage Extraction Columns in Chromatography Mode: Experimental Study and Mathematical Simulation

Artak E. Kostanyan, Vera V. Belova *, Yulia V. Tsareva and Maria M. Petyaeva

Kurnakov Institute of General and Inorganic Chemistry, Russian Academy of Sciences, 31 Leninskii pr., 119991 Moscow, Russia; kost@igic.ras.ru (A.E.K.); tsareva.juliana@ya.ru (Y.V.T.); maria.mikhailovna1994@yandex.ru (M.M.P.)
* Correspondence: belova@igic.ras.ru

Abstract: The application of liquid–liquid chromatography principles to solvent extraction processes in hydrometallurgy can greatly simplify rare earth metal separation technologies by separating multicomponent mixtures in one technological operation. In this study, the chromatographic separation of rare earth elements (REEs) in multistage extraction columns was experimentally studied under conditions of impulse sample injection—single and multiple loading of large volumes of metal salt solution into the installation. The results obtained showed the feasibility of operating sieve plate extraction columns in the liquid–liquid chromatography mode. A closed-loop recycling technology is proposed for the separation of rare earth elements in multistage extraction columns operating in the liquid–liquid chromatography mode. For further development and industrial implementation of this technology, experimental studies should be conducted on intensified multistage extraction columns, such as sectioned columns with agitators and vibrating plate columns. Computer simulation of the chromatographic separation of rare earth elements by closed-loop recycling liquid–liquid chromatography was carried out.

Keywords: separation of rare earth elements; liquid–liquid chromatography; solvent extraction; multistage extraction columns; closed-loop recycling chromatography

Citation: Kostanyan, A.E.; Belova, V.V.; Tsareva, Y.V.; Petyaeva, M.M. Separation of Rare Earth Elements in Multistage Extraction Columns in Chromatography Mode: Experimental Study and Mathematical Simulation. *Processes* **2023**, *11*, 1757. https://doi.org/10.3390/pr11061757

Academic Editors: Yanlin Zhang and Prashant K. Sarswat

Received: 22 March 2023
Revised: 6 June 2023
Accepted: 7 June 2023
Published: 9 June 2023

1. Introduction

Solvent extraction methods are widely used in hydrometallurgy to separate metals with similar properties, in particular for the separation and purification of rare earth elements [1–11]. Rare earth elements are widely applied in different fields: electronics, medicine, permanent magnet manufacturing, and catalysis. The separation and purification of individual rare earth elements is challenging due to their remarkably similar chemical properties. At present, in addition to extraction methods, various rare earth separation and recovery methods have been developed, which include chemical precipitation, ion exchange, crystallization, adsorption, chromatography, membrane separation, etc. However, solvent extraction processes are most commonly used to separate rare earth elements in industry. Currently, 2-ethylhexyl phosphinic acid mono-2-ethylhexyl ester (P507) is most often used for the REE separation. For this extraction system, the separation factors of individual REEs have been established, and operational parameters have been developed for the equipment and chemical reagents used [5]. Compared with other extractants, systems with P507 exhibit relatively high separation factors between REEs and good physical characteristics. However, this process has several disadvantages. The separation factors for some elemental pairs (for example, Nd/Pr, Gd/Eu, Lu/Yb) are very small, requiring many equilibrium extraction stages for achieving high purities in the separated products. In addition, the stripping acidity required for the heavier elements is quite high. Cyanex

272 (bis-2,4,4-trimethylpentylphosphinic acid) has for a long time been considered as an alternative extractant to P507 [6]. This phosphinic acid has a lower acid dissociation constant (Ka) than P507, i.e., it is a weaker extractant, of which the stripping processes proceed at a lower aqueous acidity. However, the use of Cyanex 272 is limited due to the poor physical characteristics of this extractant. Subsequently, it was found that in systems involving mixtures of P507 and Cyanex 272, stripping processes proceeded more easily than with P507 alone [9,10]. An improvement in separation factors was noted for some pairs of heavy REEs. Later, a new extractant was released under the name Cyanex 572 (Cytec Industries), specially designed for the separation of rare earth elements, which is a mixture of phosphinic and phosphonic acids (P507 and Cyanex 272) [11]. This extractant provides efficient extraction of heavy rare earth elements simultaneously, with the stripping of these metals using mineral acid solutions with a lower concentration than in the systems with P507. Commercially available extractants such as Cyanex 572 are immediately available for industrial use and have shown significant potential for optimizing and improving the separation and concentration of individual rare earth elements [11].

The application of chromatography principles to solvent extraction processes can greatly improve industrial separation technologies. In solvent extraction and in liquid–liquid chromatography, referred to as "counter-current chromatography" (CCC) [12–16] and "centrifugal partition chromatography" (CPC) [17–19], the separation of substances is based on their different solubility in two liquid phases. Countercurrent chromatography differs from traditional liquid chromatography in that the so-called stationary phase in a CCC device is not fixed on a solid carrier but is kept mobile in the device due to centrifugal forces. The undoubted advantage of CCC separation methods over the methods of solvent extraction is the possibility of separating multicomponent mixtures in one technological operation (within a single process stage), which greatly simplifies technological schemes. In addition, since the organic phase containing the extractant is not removed from the chromatographic device, the consumption of chemicals is significantly reduced, which allows the use of more selective, but more expensive, chemicals.

Currently, CCC separations are mainly used for analytical and preparative purposes for the separation and purification of herbal pharmaceutical products [12–19]. CCC devices in the form of a tube wound in one or several layers around the drum of a planetary centrifuge (hydrodynamic devices), or in the form of a cascade of chambers on the surface of a cylinder, or on a pack of discs mounted on the shaft of a conventional centrifuge (hydrostatic devices), are difficult to scale up. In addition to the complexity of the equipment, the existing CCC devices are characterized by low productivity. Thus, the disadvantage of CCC processes is the complexity of instrumentation and the associated low productivity, which prevents their use in large-capacity industrial production, for example, in hydrometallurgy. The productivity of extraction apparatuses used in industry (extraction columns and cascades of series-connected mixer–settlers) exceeds the productivity of chromatographic apparatuses by several orders of magnitude. Therefore, for creating high-performance devices that can be used in industry for the isolation of chemical compounds from solutions, their separation and purification by methods of liquid–liquid chromatography is an urgent task. To overcome the above disadvantages of CCC apparatuses, high-performance industrial-scale CCC separations can be developed using conventional liquid extraction equipment [20–26]. In [21], the chromatographic behavior of europium, gadolinium, dysprosium, samarium, terbium, and yttrium in the cascade consisting of 70 centrifugal mixer–settler extractors operating in the isocratic elution mode was experimentally studied. A mixture of 30 vol.% Cyanex 572 + 10 vol.% tributylphosphate in a hydrocarbon diluent was used as the stationary phase, and aqueous nitric acid was used as the mobile phase. It was experimentally shown that modified centrifugal mixer–settlers (the stationary organic phase recirculates between the mixing and separating chambers) can operate in chromatography mode and be used for large-scale separation of rare earth elements [21].

Multicomponent two-phase solvent systems used in liquid–liquid chromatography can be an alternative to conventional extraction systems [27–29]. At present, several multicomponent two-phase solvent systems are applied in countercurrent chromatography, for example, ARIZONA solvent systems (ethyl acetate–water–heptane–methanol) [30], HEMWat systems (hexane–ethyl acetate–methanol–water) [31], and others. Multicomponent solvent systems differ from extraction systems in the possibility of varying the number of solvent components and their ratio, which allows achieving appropriate distribution coefficients of substances. However, countercurrent chromatography is mainly used to separate natural products in physical extraction. The addition of extractants to multicomponent two-phase solvent systems makes it possible to create a new class of extraction systems for separating metals [28,29]. Previously, we studied the distribution of REEs in ternary solvent systems containing di(2-ethylhexyl)phosphoric acid [32,33], binary extractants of various compositions and their mixtures [28,34,35], and Cyanex 572 [29,36]. It was established that the hexane–isopropanol–water (2:1:1.5) ternary system with Cyanex 572 is promising for the separation of metals of heavy-light and medium groups. The extraction of REEs with Cyanex 572 proceeds by the cation-exchange mechanism at a relatively low acidity of the aqueous phase, and the stripping of metals can be performed using weak solutions of mineral acids.

In this work, using multicomponent two-phase solvent systems, an experimental study on the extraction separation of REEs in columns with sieve plates operating in the liquid–liquid chromatography mode was conducted to assess the potential of this extraction equipment for large-scale chromatographic separation of rare earth elements. Compared with the CCC devices, these apparatuses are extremely simple and, as mentioned above, have a significantly higher productivity. In this study, for the first time, the possibility of operating a cascade of sieve plate extraction columns in the chromatography mode has been shown. For the first time, the chromatographic separation of REEs in the cascade of sieve plate extraction columns was experimentally studied under conditions of injections of large volumes of metal salt solution into the installation; it has been shown that this high-performance industrial equipment can be used for large-scale separation of rare earth elements. The main advantages of the separation of metals in a cascade of multi-stage extraction columns or in a cascade of mixer–settlers operating in the chromatographic mode over extraction methods are the separation of multicomponent mixtures within a single process stage (in one technological operation), which simplifies technological schemes; higher purity of the separated products; and low consumption of reagents and organic solvents, since they remain in the system. Thus, the application of chromatography mode can lead to significant cost savings.

Multistage extraction columns operating in liquid–liquid chromatography mode

There are two types of multistage extraction columns: (1) intensified columns, in which for the dispersion of one of the phases kinetic energy is supplied with the help of mixers (sectioned columns with various types of stirrers and pulsation columns); (2) spray columns with perforated trays, in which one of the phases is dispersed due to gravitational forces as it flows out of the tray holes. When operating in the countercurrent extraction mode, the columns have settling zones at the top and at the bottom for separation and withdrawal of the light and heavy phases. To operate in the conventional eluent chromatography mode, for the separation and withdrawal of the mobile phase it is sufficient to equip the columns with only one settling zone. Obviously, to carry out real countercurrent chromatography separations, the columns must be equipped with two settling zones.

Columns with sieve plates are the simplest type of extraction apparatus widely used in large-scale production of chemical products and petrochemistry. When using a cascade of these columns as an installation for liquid–liquid chromatography, depending on the design of the sieve plates and the physicochemical properties of the liquids, two hydrodynamic modes of separation processes are possible. In the first case, after crushing in the volume of the continuous stationary phase between adjacent trays, drops of the mobile phase coalesce with the formation of a continuous layer on the lower tray (when the mobile phase is the heavy phase) or under the upper tray (when the mobile phase is the light phase). In the second case, the mobile phase is crushed at the top of the column (when the mobile phase is the heavy phase) or at the bottom of the column (when the mobile phase is the light phase); its drops pass through the entire column, and their coalescence occurs in the settling zone at the bottom or at the top of the column.

2. Materials and Methods

The experimental setup consisted of four sieve tray columns connected in series along the heavy mobile phase flow. Each column (from a 6.4 × 9.6 mm FEP tube) contained 25 PTFE plates 3 mm thick with 13 holes 0.25 mm in diameter spaced every 35 mm. The total volume of the setup was 119 mL; the volume of each stage in the columns was 1.13 mL; the number of stages in each column was 26; and the total number of stages in the setup was 104. The scheme of the experimental setup is shown in Figure 1.

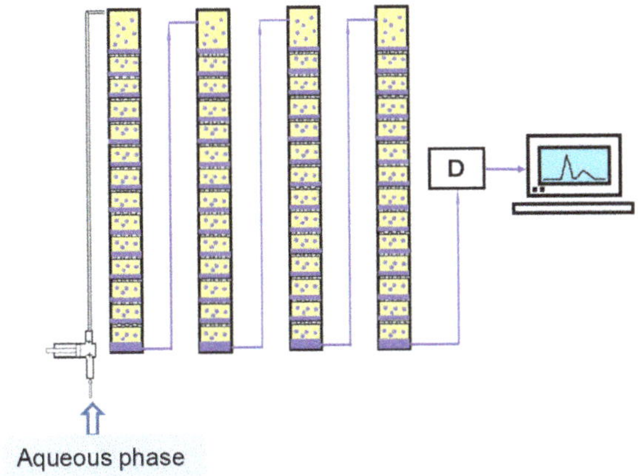

Figure 1. Mixing–settling mode of operation of a liquid–liquid chromatography unit, consisting of series-connected sieve plate columns.

The experimental setup operates as follows: the columns are filled with the light stationary extractant phase, and the heavy mobile phase is pumped through them with a piston or peristaltic pump.

The stock solutions of lanthanum(III) and neodymium(III) chlorides were prepared with dissolution of $LaCl_3{\cdot}7H_2O$ and $NdCl_3{\cdot}6H_2O$ salts (all chemically pure) in distilled water. The stock solutions of Sm(III), Eu(III), Er(III), and Tb(III) chlorides were obtained by dissolving weights of oxides of corresponding metals (all chemically pure) in concentrated HCl followed by evaporating the solutions in a water bath for removing excess acid.

The stock solutions of REE chlorides were used to prepare aqueous solutions of metal chlorides with addition of NaCl (0.1 M) and HCl (0.01 M). Cyanex 572 (a mixture of phosphonic and phosphinic acids) produced in Cytec was applied as an extractant. Hexane and isopropyl alcohol (all reagent grade) were applied as solvents. The feasibility of extraction–chromatographic separation of REEs in a cascade of extraction columns was experimentally tested. For this purpose, the chromatographic behavior of REE chlorides was studied in a 2-phase liquid system: 0.1 M Cyanex 572 in hexane–isopropyl alcohol—0.02 M aqueous solution of Ln(III) (0.01 M HCl and 0.1 M NaCl). The volume ratio of all components was 2:1:1.5.

After the establishment of a stable hydrodynamic regime, in which only the mobile phase leaves the system, an aqueous solution of metal salts was introduced into the upper stage of the first column using a syringe, and the relative content of metals in the mobile phase was continuously measured and recorded by the conductometric method at the outlet of the last column using a flow-through conductivity electrode connected to the PC.

3. Results and Discussion

Preliminary experiments established that in the selected multicomponent 2-phase solvent system, the retention of the stationary phase, depending on the mode of pumping through the columns of the mobile phase, varied by 80–90%. First, experiments were carried out using a piston pump, which provided a discrete supply of the mobile phase to the columns and the pulsating movement of liquids in them.

Individual chromatographic peaks of La(III) and Sm(III) chlorides and the chromatogram of a mixture of REE chlorides are shown in Figure 2.

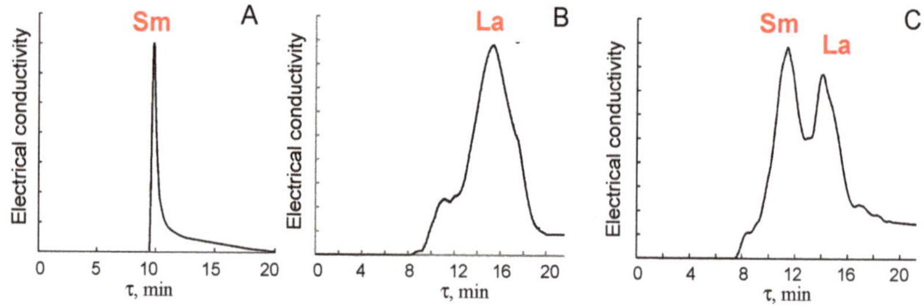

Figure 2. Experimental chromatographic peaks of $SmCl_3$ (**A**) and $LaCl_3$ (**B**) and the chromatogram of a mixture of $LaCl_3$ and $SmCl_3$ (**C**).

The unusual elution sequence of Sm and La may be explained by the following: in the multicomponent system, which involves hydrophilic and hydrophobic components, as well as a component with hydrophilic and hydrophobic properties (water–hexane–isopropyl alcohol), in addition to the chemical reaction between lanthanide ions and the Cyanex 572 extractant, the physical distribution of the extracted species between the phases proceeds. This difference between the distribution mechanisms in the multicomponent and extraction systems can affect the order of REE distribution in the multicomponent system.

The experimental chromatograms in Figure 2 confirm the possibility of separating metals in a cascade of series-connected sieve plate columns operating in the elution chromatography mode. The process of extracting metals is usually associated with a chemical reaction at the interface. Therefore, intensified multistage columns are to be used for the chromatographic separation of metals. The efficiency of metal separation in such devices can also be significantly increased by carrying out the process of extraction–chromatographic separation in a closed circulation loop. The sample components are repeatedly pumped through the columns when organizing the separation processes in the circulation loop; therefore, for the separation of metals according to the recirculation scheme, significantly fewer extraction stages will be required than in the scheme without recirculation. As an example, Figure 3 shows the experimental chromatogram of samarium obtained in the above column cascade after two sample passes.

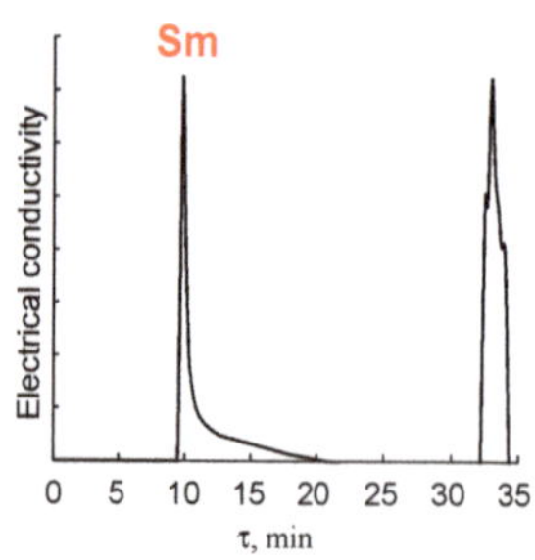

Figure 3. Experimental chromatographic peaks of $SmCl_3$ after two sample passes.

When certain conditions are met for introducing a solution of components into the cascade, it is possible to isolate concentrated fractions of the target components in the recycling process [24]. To model the considered methods of REE separation, the analytical dependences published in [22,24,25] can be used.

High-throughput chromatographic separation can be achieved by long and/or multiple sample loading. Figure 4 shows experimental chromatograms of binary mixtures of REE salts obtained in the mode of the continuous flow of the dispersed mobile phase (by using the peristaltic pump) under conditions of prolonged (A) and repeated (B) sample loading. In the first case (Figure 4A), due to the greater difference in the partition coefficients K_D, a more efficient separation of metals occurs. According to the obtained experimental results, the practical complete separation of lanthanum and erbium occurs within 60 min (Figure 4A), while the incomplete separation of lanthanum and samarium takes 20 min (Figure 2C). As above, this can be explained by differences in the partition coefficients K_D of REEs. The values of K_D calculated directly from the chromatographic peaks are K_{DSm} = 0.05; K_{DLa} = 0.2; K_{DEr} = 0.55; and K_{DTb} = 0.42.

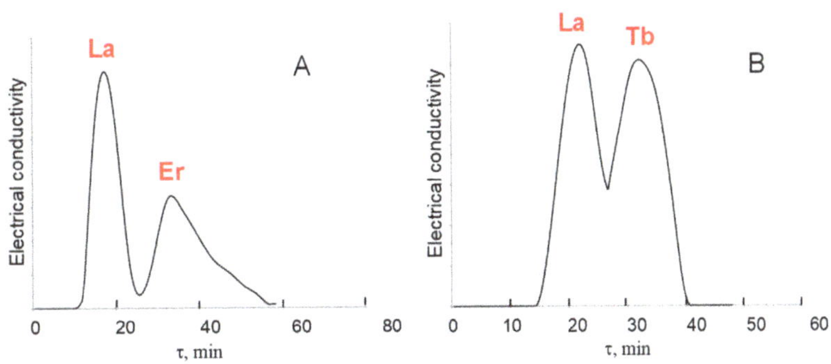

Figure 4. Experimental chromatograms of mixtures of REE chlorides at (**A**) long (sample loaded for 6 min) and (**B**) multiple sample loading (sample loaded 6 times within a minute with a 30 s interval for feeding the mobile phase).

Based on the above, we can propose the following technology for the chromatographic separation of rare earth elements: a continuous separation of REEs is conducted in a closed-loop recycling liquid–liquid chromatography mode, using a cascade of intensified multistage extraction columns; the recycling system includes a pump and connecting pipelines. At certain time intervals, the loop is opened and the solution of REEs is fed into the first column with the flow rate equal to the rate of the mobile phase circulation in the loop. After the loading of the REE solution, the loop is closed, and during a certain number of cycles the metals are separated. Then, the loop is opened again; an appropriate mobile phase is fed into the system with the flow rate equal to the rate of the mobile phase circulation in the loop, and the separated metal fractions are removed from the cascade. The supply of the feed solution and the removal of the metal fractions can be carried out at different times.

To evaluate the applicability of the proposed technology in the separation of rare earth elements, computer simulation of these processes was carried out.

4. Simulation of Chromatographic Separation of Rare Earth Elements by Closed-Loop Recycling Liquid–Liquid Chromatography

In liquid–liquid chromatography, as in solvent extraction, the separation of compounds is influenced by interphase mass transfer and longitudinal dispersion of the compounds in a chromatographic column caused by axial mixing in stationary and mobile phases. To describe the elution profile of compounds in chromatographic devices and to interpret the experimental results, two models were used: the continuous diffusion model and staged cell model [37–39]. Both models take into account two band-broadening mechanisms:

interfacial mass transfer and longitudinal mixing due to the uneven flow velocity profile of the mobile phase, turbulence, and molecular diffusion. For low degrees of longitudinal mixing, as is the case in chromatographic devices, simulations by continuous and staged models give practically identical results. However, the mathematical description of chromatography by the cell model, which represents a chain of ideally mixed, equally sized tanks (cells), is much simpler. In the cell model, the effects of interfacial mass transfer and longitudinal mixing is characterized by one parameter—the number of cells N:

$$N = \frac{nT(1+K')^2}{T(1+K')^2 + 2nK'^2} \tag{1}$$

Equation (1) determines the contribution of longitudinal mixing, characterized by the number of ideally mixed cells n, and interfacial mass transfer, characterized by the number of transfer units $T = a_c\, k_x\, V_c/F$, to the efficiency of a separation process in liquid–liquid chromatography.

In Equation (1), $K' = \frac{K_D S_f}{1-S_f}$ is the dimensionless capacity factor; $K_D = y/x$ is the distribution coefficient; x is the concentration in the mobile phase; y is the concentration in the stationary phase; S_f is the fraction of column volume (V_c) occupied by the stationary phase; a_c is the specific contact surface of the phases; k_x is the mass transfer coefficient; and F is the volumetric flow rate of the mobile phase.

Replacing the number of theoretical plates in the known equilibrium chromatography model by the number of theoretical cells defined by Equation (1) makes it possible to use a simple model of theoretical plates (or equilibrium cells) in the analysis of non-equilibrium processes of liquid–liquid chromatography.

Based on the cell model and the results of our previous studies [16,25,39–42], the following equations can be developed to simulate the separation of a binary mixture of rare earth elements by liquid–liquid chromatography in the conventional elution mode:

$$X(t) = \frac{a\sqrt{6N}}{\sqrt{\pi(Na^2t_s^2+12)}} exp\left[-\frac{3N(2+at_s-2at)^2}{2(Na^2t_s^2+12)}\right] \tag{2}$$

$$X_{1,2}(t) = q_1X_1(t) + q_2X_2(t) \tag{3}$$

Equation (2) describes the elution profile of individual rare earth elements, and Equation (3) describes the chromatogram of a mixture of two rare earth elements after the loading of a solution, containing q_1 and $q_2 = 1 - q_1$ proportions of the first and second rare earth elements. When developing Equation (3), it was assumed that there is no significant interaction between the components of the mixture.

In Equations (2) and (3), $a = \frac{1}{1-S_f+S_fK_D}$; $q_1 = Q_1/Q$, $q_2 = Q_2/Q$: Q_1 and Q_2 are the amounts of the first and the second rare earth elements in a loaded solution of rare earth elements; $Q = Q_1 + Q_2$ is the total amount of rare earth elements in the loaded solution; $X = x/\bar{x}$ is the dimensionless (normalized) concentration of individual rare earth elements; $\bar{x} = x_sF\tau_s/V_c$ is the mean concentration in the chromatographic device after the loading of the rare earth element solution; $X_{1,2}$ is the normalized total concentration of the metals; $t_s = \tau_sF/V_c$ is the normalized rare earth element solution loading time; x_s is the concentration of a rare earth element in the feed; and τ_s is the rare earth element solution loading time.

The results of the simulation of the separation of the La/Er mixture by Equations (2) and (3) are shown in Figure 5. The process parameters $N = 40$; $K_{DLa} = 0.2$; and $K_{DEr} = 0.55$ were determined from the experimental chromatograms in Figure 4. The results in Figures 4 and 5 demonstrate an acceptable agreement between the experiment and the mathematical model presented above.

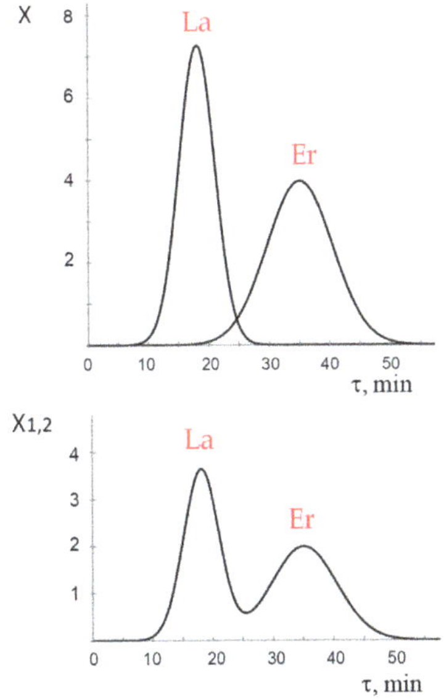

Figure 5. Simulation of the separation of La/Er mixture by Equations (2) and (3). Process parameters: $N = 40$; $K_{DLa} = 0.2$; $K_{DEr} = 0.55$; $S_f = 0.9$; $q_1 = 0.5$.

In liquid–liquid chromatography devices, the high-volume fraction of the stationary phase enhances the separation of components of a mixture. However, the separation efficiency of liquid–liquid chromatography (LLC) is much lower than that of HPLC. The efficiency of LLC separations can be significantly improved by using elution schemes and modes that simulate the increase in the number of theoretical cells (plates) of the LLC system, such as dual mode, multiple dual mode [26,39], and closed-loop recycling mode. The closed-loop recycling LLC method is the simplest way to increase the number of theoretical plates of LLC plants. It provides significant solvent savings and requires no additional equipment. In this method, the sample or certain parts of the elution profile are recycled several times in the LLC system until the required degree of separation of the components is reached. With an increase in the number of sample passages through the plant (in the number of cycles), the quality of the separation improves due to the increase in the number of theoretical plates. However, at the same time, the chromatograms of neighboring cycles begin to converge: components with low distribution coefficients of the current cycle catch up with components with high distribution coefficients from the previous cycle, and, after a certain number of cycles, the chromatograms of neighboring cycles begin to overlap. To take into account these counteracting phenomena, two approaches are used [40–42]: (1) When the volume of the recycling system consisting of connecting lines, valves, a pump, a detector, etc., does not exceed one percent of the volume of the LLC plant, the effects of extra-column dispersion on the separation can be neglected [40]. (2) The recycling system, which allows for the recycling of the mobile phase through the LLC plant, is included in the mathematical model, replacing it with a cascade of ideally mixed cells. Thus, there are two options to design an LLC installation operating in closed-loop recycling mode: (1) with short recycling tubing (small volume of the recycling system compared with the LLC device); (2) with long recycling tubing (with a certain volume of the recycling system). In a recycling system with long, small-diameter tubing, where a large volume of the recycling system can be provided due to the length of the connecting pipes, the flow of the mobile phase approaches the ideal displacement mode, which allows an increase in the number of sample passages through the LLC device and therefore improves separation. In this case, the following equation can be used to simulate the separation of rare earth elements by closed-loop recycling liquid–liquid chromatography:

$$X_m(t) = \sum_{i=1}^{m} \frac{a\sqrt{12N}}{\sqrt{2\pi(Na^2t_s^2 + 12i)}} exp\left[-\frac{3N(2i + at_s + 2ab(i-1) - 2at)^2}{2(Na^2t_s^2 + 12i)}\right] \quad (4)$$

In Equation (4), b is the ratio of the recycling system volume and the LLC device volume; m is the number of cycles. Equation (4) describes the change in the concentration profiles of individual rare earth elements at the outlet of the installation for the entire circulation time from the first to the last cycle m. Figure 6 shows the results of the simulation of the separation of samarium (K_{DSm} = 0.05), lanthanum (K_{DLa} = 0.2), and erbium (K_{DEr} = 0.55) in the liquid–liquid chromatography plant with N = 50 and b = 0.8 operating in the closed-loop recycling mode. Figure 6 shows the chromatograms after the first and second cycles. As seen, an acceptable separation of the metals can be reached after two cycles.

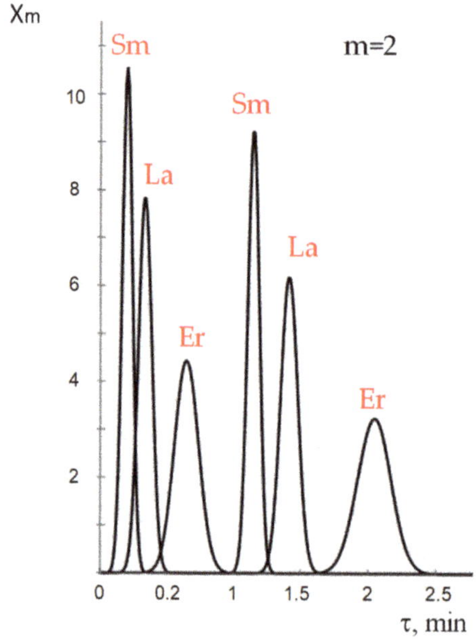

Figure 6. Simulation of the separation of samarium (K_{DSm} = 0.05), lanthanum (K_{DLa} = 0.2), and erbium (K_{DEr} = 0.55) in a liquid–liquid chromatography plant operating in the closed-loop recycling mode. Process parameters: N = 50; S_f = 0.9; and b = 0.8.

5. Conclusions

In this study, based on a cascade of multistage extraction columns, a chromatographic technology is proposed for the separation of REEs.

The chromatographic separation of REEs in the cascade of sieve plate extraction columns was experimentally studied under conditions of single and multiple injections of large volumes of metal salt solution into the installation. The feasibility of operating the cascade of sieve plate extraction columns in the chromatography mode was demonstrated.

For the separation of rare earth elements, a method of closed-loop recycling liquid–liquid chromatography is proposed. Computer simulation of the chromatographic separation of rare earth elements by closed-loop recycling liquid–liquid chromatography showed the feasibility and potential of this method. Industrial separation of rare earth elements by closed-loop liquid–liquid chromatography can be implemented on the basis of a cascade of multistage extraction columns or a cascade of mixer–settler extractors with semi-continuous loading of the solution of the metals to be separated.

For the further development and industrial implementation of the considered technology, experimental studies should be conducted using intensified multistage extraction columns, such as sectioned columns with agitators and vibrating plate columns.

Author Contributions: Conceptualization, A.E.K.; methodology, A.E.K. and V.V.B.; validation, V.V.B. and Y.V.T.; formal analysis, A.E.K. and V.V.B.; writing, review, and editing, A.E.K. and V.V.B.; visualization, V.V.B., Y.V.T. and M.M.P.; project administration, V.V.B. All authors have read and agreed to the published version of the manuscript.

Funding: This research was funded by the Russian Science Foundation, grant number 23-29-00162.

Institutional Review Board Statement: Not applicable.

Informed Consent Statement: Not applicable.

Data Availability Statement: Not applicable.

Conflicts of Interest: The authors declare no conflict of interest.

References

1. Fan, S.J.; Zhao, X.W.; Song, N.Z.; Shi, Y.F.; Jia, Q.; Liao, W.P. Studies on the synergistic extraction of rare earths from nitrate medium with mixtures of secnonylphenoxy acetic acid and 1.10-phenanthroline. *Sep. Purif. Technol.* **2010**, *71*, 241–245. [CrossRef]
2. Jha, M.K.; Kumar, A.; Panda, R.; Kumar, J.R.; Yoo, K.; Lee, J.Y. Review on hydrometallurgical recovery of rare earth metals. *Hydrometallurgy* **2016**, *165*, 2–26. [CrossRef]
3. Chen, L.; Wu, Y.; Dong, H.; Meng, M.; Li, C.; Yan, Y.; Chen, J. An overview on membrane strategies for rare earths extraction and separation. *Sep. Purif. Technol.* **2018**, *197*, 70–85. [CrossRef]
4. Kostanyan, A.E. Staged versions of liquid membrane extraction processes. *Solvent Extr. Ion Exch.* **2013**, *31*, 297–305. [CrossRef]
5. Fontana, D.; Pietrelli, L. Separation of middle rare earths by solvent extraction using 2-ethylhexylphosphonic acid mono-2-ethylhexyl ester as an extractant. *J. Rare Earth* **2009**, *27*, 830–833. [CrossRef]
6. Li, K.; Freiser, H. Extraction of lanthanide metals with bis (2,4,4-trimethylpentyl)phosphinic acid. *Solvent Extr. Ion Exch.* **1986**, *4*, 739–755. [CrossRef]
7. Xie, F.; Zhang, T.A.; Dreisinger, D.; Doyle, F. A critical review on solvent extraction of rare earths from aqueous solutions. *Miner. Eng.* **2014**, *56*, 10–28. [CrossRef]
8. Belova, V.V.; Voshkin, A.A.; Kholkin, A.I.; Payrtman, A.K. Solvent extraction of some lanthanides from chloride and nitrate solutions by binary extractants. *Hydrometallurgy* **2009**, *97*, 198–203. [CrossRef]
9. Kuang, S.; Zhang, Z.; Li, Y.; Wei, H.; Liao, W. Synergistic extraction and separation of rare earths from chloride medium by the mixture of HEHAPP and D2EHPA. *Hydrometallurgy* **2017**, *174*, 78–83. [CrossRef]
10. Xiong, Y.; Wang, X.; Li, D. Synergistic extraction and separation of heavy lanthanide by mixtures of bis(2,4,4-trimethylpentyl)phosphinic acid and 2-ethylhexyl phosphinic acid mono-2-ethylhexyl ester. *Sep. Sci. Technol.* **2005**, *40*, 2325–2336. [CrossRef]
11. Wang, Y. The novel extraction process based on Cyanex 572 for separating heavy rare earths from ion-adsorbed deposit. *Sep. Purif. Technol.* **2015**, *151*, 303–308. [CrossRef]
12. Friesen, J.B.; McAlpine, J.B.; Chen, S.-N.; Pauli, G.F. Countercurrent separation of natural products: An update. *J. Nat. Prod.* **2015**, *78*, 1765–1796. [CrossRef]
13. Kostanyan, A.E.; Ignatova, S.; Sutherland, I.; Hewitson, P.; Zakhodjaeva, Y.A.; Erastov, A.A. Steadystate and nonsteady state operation of counter-current chromatography devices. *J. Chromatogr. A* **2013**, *1314*, 94–105. [CrossRef] [PubMed]
14. Huang, X.-Y.; Pei, D.; Liu, J.-F.; Di, D.-L. A review on chiral separation by counter-current chromatography: Development, applications and future outlook. *J. Chromatogr. A* **2018**, *1531*, 1–12. [CrossRef] [PubMed]
15. Li, H.; Zhang, F.; Jin, Q.; Zhu, T. Preparative separation and purification of Cyclosporin D from fungus *Hypoxylon* Spp. by improved closed-loop recycling counter-current chromatography. *J. Chromatogr. A* **2021**, *1649*, 462221. [CrossRef]
16. Kostanyan, A.E. On influence of sample loading conditions on peak shape and separation efficiency in preparative isocratic counter-current chromatography. *J. Chromatogr. A* **2012**, *1254*, 71–77. [CrossRef]
17. Chami, M.C.; Bouju, E.; Lequemener, C.; Vaumas, R.; Hadji-Minaglou, F.; Fernandez, X.; Michel, T. Purification of two valepotriates from *Centranthus ruber* by centrifugal partition chromatography: From analytical to preparative scale. *J. Chromatogr. A* **2018**, *1580*, 126–133. [CrossRef]
18. Guilois-Dubois, S.; Guyot, S.; Poupard, P. Preparative isolation of apple flavan-3-ols monomers and oligomers using pH-zone-refining centrifugal partition chromatography combined with reversed-phase liquid chromatography. *J. Chromatogr. A* **2021**, *1653*, 462382. [CrossRef]

19. Sutherland, I.A.; Hewitson, P.; Siebers, R.; Van den Heuvel, R.; Arbenz, L.; Kinkel, J.; Fisher, D. Scale-up of protein purifications using aqueous two-phase systems: Comparing multilayer toroidal coil chromatography with centrifugal partition chromatography. *J. Chromatogr. A* **2011**, *1218*, 5527–5530. [CrossRef]
20. Kostanyan, A.E.; Voshkin, A.A.; Kodin, N.V. Controlled-cycle pulsed liquid–liquid chromatography: A modified version of Craig's counter-current distribution. *J. Chromatogr. A* **2011**, *1218*, 6412–6415. [CrossRef]
21. Kostanyan, A.E.; Galieva, Z.N.; Semenov, A.A.; Aldushkin, A.V. Chromatographic behavior of six lanthanides on a centrifugal mixer-settler extractor cascade. *J. Chromatogr. A* **2020**, *1634*, 461686. [CrossRef] [PubMed]
22. Kostanyan, A.E. Extraction chromatographic separation of rare-earth metals in a cascade of centrifugal extractors. *Russ. J. Inorg. Chem.* **2018**, *63*, 287–292. [CrossRef]
23. Kostanyan, A.E.; Erastov, A.A. Industrial countercurrent chromatography separations based on a cascade of centrifugal mixer-settler extractors. *J. Chromatogr. A* **2018**, *1572*, 212–216. [CrossRef] [PubMed]
24. Kostanyan, A.; Martynova, M.; Erastov, A.; Belova, V. Simultaneous concentration and separation of target compounds from multicomponent mixtures by closed-loop recycling countercurrent chromatography. *J. Chromatogr. A* **2018**, *1560*, 26–34. [CrossRef] [PubMed]
25. Kostanyan, A.E.; Erastov, A. Theoretical study of closed loop recycling liquid–liquid chromatography and experimental verification of the theory. *J. Chromatogr. A* **2016**, *1462*, 55–62. [CrossRef]
26. Kostanyan, A.E.; Voshkin, A.A.; Belova, V.V. Analytical, preparative, and industrial-scale separation of substances by methods of countercurrent liquid-liquid chromatography. *Molecules* **2020**, *25*, 6020. [CrossRef]
27. Zhang, L.; Wu, S. Hydrophobic and hydrophilic interactions in countercurrent chromatography. *J. Chromatogr. A* **2020**, *1611*, 460576. [CrossRef]
28. Belova, V.V.; Petyaeva, M.M.; Tsareva, J.V.; Kostanyan, A.E. Solvent extraction of lanthanides(III) from chloride and nitrate media with di(2-ethylhexyl)phosphoric acid and ionic liquids in three-component two-phase systems. *Hydrometallurgy* **2021**, *199*, 105526. [CrossRef]
29. Belova, V.V.; Tsareva, Y.V.; Kostanyan, A.E. Extraction of rare-earth elements from chloride solutions in multicomponent systems using CYANEX 572. *Theor. Found. Chem. Eng.* **2022**, *56*, 915–919. [CrossRef]
30. Berthod, A.; Hassoun, M.; Ruiz-Angel, M.J. Alkane effect in the Arizona liquid systems used in countercurrent chromatography. *Anal. Bioanal. Chem.* **2005**, *383*, 327–340. [CrossRef]
31. Liu, Y.; Friesen, J.B.; Klein, L.L.; McAlpine, J.B.; Lankin, D.C.; Pauli, G.F.; Chen, S.-N. The generally useful estimate of solvent systems (GUESS) method enables the rapid purification of methylpyridoxine regioisomers by countercurrent chromatography. *J. Chromatogr. A* **2015**, *1426*, 248–251. [CrossRef] [PubMed]
32. Belova, V.V.; Martynova, M.M. Interphase distribution of lanthanide chlorides in multicomponent aqueous-organic two-phase systems containing DEHPA. *Theor. Found. Chem. Eng.* **2019**, *53*, 961–966. [CrossRef]
33. Belova, V.V.; Martynova, M.M.; Erastov, A.A. Extraction of lanthanide nitrates in multicomponent aqueous-organic two-phase systems with D2EHPA. *Russ. J. Inorg. Chem.* **2018**, *63*, 1651–1655. [CrossRef]
34. Belova, V.V.; Tsareva, Y.V. Interphase distribution of lanthanide nitrates in aqueous organic two-phase systems with amine and organic acid salts. *Theor. Found. Chem. Eng.* **2020**, *54*, 769–774. [CrossRef]
35. Belova, V.V.; Tsareva, Y.V. Extraction of lanthanide chlorides in aqueous-organic two-phase systems with salts of tertiary and secondary amines. *Theor. Found. Chem. Eng.* **2020**, *54*, 1108–1113. [CrossRef]
36. Belova, V.V.; Petyaeva, M.M.; Kostanyan, A.E. Extraction of lanthanides from chloride solutions in hexane–isopropanol–water systems using Cyanex 572. *Theor. Found. Chem. Eng.* **2022**, *56*, 595–599. [CrossRef]
37. Kostanian, A.E. Modelling counter-current chromatography: A chemical engineering perspective. *J. Chromatogr. A* **2002**, *937*, 39–46. [CrossRef] [PubMed]
38. Hewitson, P.; Sutherland, I.; Kostanyan, A.; Voshkin, A.; Ignatova, S. Intermittent counter-current extraction-Equilibrium cell model, scaling and an improved bobbin design. *J. Chromatogr. A* **2013**, *1303*, 18–27. [CrossRef]
39. Kostanyan, A.E. Increasing efficiency of the separation of substance mixtures by methods of liquid–liquid chromatography. *J. Anal. Chem.* **2020**, *75*, 1384–1398. [CrossRef]
40. Kostanyan, A.E. Modeling of preparative closed-loop recycling liquid–liquid chromatography with specified duration of sample loading. *J. Chromatogr. A* **2016**, *1471*, 94–101. [CrossRef]
41. Kostanyan, A.; Martynova, M. Modeling of two semi-continuous methods in liquid–liquid chromatography: Comparing conventional and closed-loop recycling modes. *J. Chromatogr. A* **2020**, *1614*, 460735. [CrossRef] [PubMed]
42. Kostanyan, A.E. Simple equations to simulate closed-loop recycling liquid-liquid chromatography: Ideal and non-ideal recycling models. *J. Chromatogr. A* **2015**, *1423*, 71–78. [CrossRef] [PubMed]

processes

MDPI

Article

Prospects of Processing Tungsten Ores from the Akchatau Deposit

Bolotpay Baimbetov [1], Gulnara Moldabayeva [1,*], Akmaral Yeleuliyeva [1], Saltanat Jumankulova [1], Akbota Taimassova [1], Zhanargul Adilzhan [1], Rinat Baisultanov [1], Elena Yakob [2] and Valery Serikbayev [2]

1 Department of Metallurgy and Mineral Processing, Non-Profit Joint Stock Company (NJSC) Kazakh National Research Technical University Named K.I. Satbayev, Almaty 050013, Kazakhstan; b.baimbetov@satbayev.university (B.B.); yeleuli.a@gmail.com (A.Y.); s.jumankulova@satbayev.university (S.J.); a.taimassova@satbayev.university (A.T.); z.adilzhan@satbayev.university (Z.A.); rinat.baisultanov@gmail.com (R.B.)

2 "KSPSteel" LLP, Pavlodar 140000, Kazakhstan; yakob_ev@ksp-steel.kz (E.Y.); valery.serikbayev@gmail.com (V.S.)

* Correspondence: gulnara.moldabayeva@satbayev.university or g.moldabaeva@bk.ru

Abstract: An overview of the composition of wolframite ores of the Akchatau deposit and the technologies for processing concentrates using NaOH and Na_2CO_3 by hydro- and pyrometallurgical methods is given, and the disadvantages associated with both the technology and the equipment are noted. To develop a technology for processing Akchatau wolframite concentrates, samples of ore materials were taken, the chemical and mineralogical composition of the samples was studied, and enrichment was carried out to obtain rich concentrates. The kinetics of the sintering of the wolframite concentrate with soda was investigated, the dependences of the degree of transformation of the tungsten minerals into sodium tungstate were obtained, and the rate constants, the order of the reaction, and the values of the apparent activation energy were calculated. The results of sintering an enlarged sample of wolframite concentrate with soda in a muffle furnace are presented. After the subsequent leaching, studies were carried out to purify the obtained solutions of sodium tungstate from the impurities while eliminating the operations of the neutralizing solutions through the use of electrodialysis with an MK-40 cation-exchange membrane. The scheme of processing the wolframite ores of Akchatau is proposed.

Keywords: wolframite; scheelite; soda; sodium carbonate; roasting; sintering; leaching; electrodialysis

Citation: Baimbetov, B.; Moldabayeva, G.; Yeleuliyeva, A.; Jumankulova, S.; Taimassova, A.; Adilzhan, Z.; Baisultanov, R.; Yakob, E.; Serikbayev, V. Prospects of Processing Tungsten Ores from the Akchatau Deposit. *Processes* **2024**, *12*, 77. https://doi.org/10.3390/pr12010077

Academic Editors: Yanlin Zhang, Prashant K. Sarswat and Liming Lu

Received: 8 November 2023
Revised: 14 December 2023
Accepted: 22 December 2023
Published: 28 December 2023

1. Introduction

As noted in the article [1] devoted to tungsten resources, the world production of tungsten in terms of WO_3 was estimated at 106 thousand tons, and proven world reserves were 3400 thousand tons in 2020. The main tungsten producers are China, Vietnam, Russia, Mongolia, Bolivia, Austria, and Rwanda, where the main tungsten deposits are concentrated [2–7]. At the present stage, tungsten is widely demanded in applications requiring hardness, high density, wear resistance, and high-temperature resistance [8]. The strategic importance of this refractory metal is related to the ever-increasing application of both tungsten itself and its alloys and chemical compounds in the electronics, military engineering, aerospace, and nuclear industries [3,9–12]. However, the authors [1,9] draw attention to the limited sources; therefore, many researchers work towards the processing of old deposits [13], tungsten ore dressing tailings [14], as well as other tungsten wastes [5,15–17], which have great potential for the exploration, mining, and additional extraction, not only of tungsten, but also other related valuable components [5,18].

The Akchatau molybdenum–tungsten deposit [19,20] is located in the Shet district of the Karaganda region, the Republic of Kazakhstan. This high-temperature hydrothermal tungsten–molybdenum deposit was developed until the 1990s of the 20th century to produce gravity tungsten concentrates. According to the standards, the content of WO_3 in the concentrates of the first grade was not lower than 65%, and in the second grade, not lower than 60%; the content of such impurities as P, S, As, Sn, Cu, Pb, Sb, and Bi were within the range from hundredths of a percent to 1.0%. Currently, the deposit is not mined, and the average tungsten grade in the deposit ore is tenths of a percent (0.1–0.3%).

The balance reserves of the major minerals approved by the State Committee of Mineral Resources include:

- Tungsten: 37,239 tons are under the C1 category, 8884 tons are under the C2 category, 18,228 tons are off-balance ones, 22,131 tons were mined since the beginning of development, and 17,279 tons are residuals of the C1 category;
- Beryllium: 6987 tons are under the C1 category, 1989 tons are under the C2 category, 5630 tons are off-balance ones, 3820 tons were produced since the beginning of development, and 4746 tons are residuals of the C1 category;
- Molybdenum: 8647 tons are under the C1 category, 1807 tons are under the C2 category, 7610 tons are off-balance ones, 3811 tons were mined since the beginning of development, and 4485 tons are C1 category residues.

The mineral composition of the ores includes mineral groups according to geological exploration data:

- Quartz, muscovite, topaz, and pyrite are the most common;
- Wolframite, molybdenite, and fluorite are moderately widespread;
- Tourmaline, biotite, feldspars, scheelite, bismuthite, sphalerite, chalcopyrite, cassiterite, and bismite are not very common.

Ore processing is profitable at a WO_3 content of 0.14% and higher [3]. Wolframite ores are beneficiated with the use of the following methods—gravity, magnetic, and electrostatic separation [4,21,22]. Beneficiation produces standard concentrates containing 55–65% WO_3. Concentrates are used to smelt ferrotungsten or to produce technical oxide WO_3, used as a feedstock for the production of tungsten carbide, or metallic tungsten by pyro- and hydrometallurgical technologies [8].

All of the used methods of WO_3 production include the following technological stages: the decomposition of concentrates; the leaching of alloy or sinter; the purification of sodium tungstate solutions from impurities; obtaining technical tungstic acid; obtaining tungsten trioxide [8,23–29].

The autoclave–soda method is used to process scheelite concentrates and intermediate products. Currently, this method is used at a number of plants [8,23,24]; however, the technology is associated with a large consumption of Na_2CO_3. A high concentration of excess soda in leaching solutions (80–120 g/dm^3) requires significant consumption of acids for the neutralization of sodium tungstate solutions during the purification from impurities that, accordingly, increases the cost of disposal of the discharge solutions.

Rich tungsten concentrates (65–70% WO_3) are decomposed by a 25–40% sodium hydroxide solution [8,23,24] at 110–120 °C according to the reaction:

$$MeWO_4 + 2NaOH = Na_2WO_4 + Me(OH)_2$$

where, Me—Fe, Mn.

Complete decomposition (98–99%) is achieved by the treatment of a finely ground concentrate with a 25–40% sodium hydroxide solution at 110–120 °C. The required excess of alkali is 50% and higher.

The most widely used method to process tungsten concentrates is the technology that includes sintering with soda, leaching with water, the purification of solutions from impurities, and the separation of tungstic acid [8,23,24,30,31].

Wolframite concentrate is sintered with soda in the presence of oxygen at 800–900 °C [23,24]. The main reactions occurring during sintering include:

$$2FeWO_4 + 2Na_2CO_3 + 0.5O_2 = 2Na_2WO_4 + Fe_2O_3 + 2CO_2$$
$$3MnWO_4 + 3Na_2CO_3 + 0.5O_2 = 3Na_2WO_4 + Mn_3O_4 + 3CO_2$$

The complete decomposition of the concentrate is achieved at an excess of Na_2CO_3 in the charge of 10–15% over the stoichiometric amount. A 1–4% nitrate is sometimes added to the charge to accelerate the oxidation of iron and manganese. Impurities of silicon, phosphorus, arsenic, and molybdenum minerals contained in the wolframite concentrate form soluble salts when they are sintered with Na_2CO_3: Na_2SiO_3, Na_3PO_4, Na_3AsO_4, and Na_2MoO_4.

Wolframite is sintered with Na_2CO_3 in tubular rotary furnaces lined with fireclay bricks. The furnace, with a length of 20 m and an external diameter of 2.2 m, at a rotation speed of 0.4 rpm and an inclination angle of 3°, has a charge capacity of 25 tons per day. To avoid the melting of the charge and the formation of crust, tailings from sintered mass leaching containing iron and manganese oxides are added to the charge, thus reducing the content of WO_3 in the charge to 20–22%.

There is also known technology to produce liquid–fluid melts in a tubular rotary furnace from rich tungsten concentrates, where the initial charge is not diluted, with poor turnovers [24]. However, the main disadvantage of this technology is the rapid wear of the furnace lining, and, if an unlined drum is used, there is rapid wear of the body, which is made of alloy steel, by the liquid alkali melt.

Sodium tungstate solutions containing 80–150 g/dm^3 WO_3 are obtained during the leaching of alkaline melts and sintered mass, as well as the autoclave leaching of concentrates with soda [32]. The solutions are processed according to a diagram that includes the purification from impurities (Si, P, As, F, Mo); the precipitation of calcium tungstate (artificial scheelite), with its subsequent decomposition by acids and the obtainment of technical tungstic acid; the dissolution of tungstic acid in ammonia water, with the subsequent evaporation of the solution and the crystallization of ammonium paravolphramate (PVA); the calcination of PVA to obtain pure tungsten trioxide, so to produce tungsten trioxide [32,33]. The disadvantages of this technology are related to the presence of several operations in the process, the performance of the operations in a periodic mode, and their duration.

Extraction and ion-exchange technology to convert Na_2WO_4 solutions into $(NH_4)_2WO_4$ solutions have been developed and are used in some enterprises. However, the use of ammonia solutions and gaseous ammonia in these technologies requires the careful sealing of the equipment and the regeneration of the reagent that were used [8].

The technology for sintering with soda, the subsequent leaching of the sintered mass with water, and the extraction of tungsten acid with the subsequent production of tungsten trioxide and tungsten carbide is the most acceptable for the wolframite concentrates in the Akchatau deposit. Some experiments with ore materials from the Akchatau deposit were performed to choose the tungsten concentrate processing technology. The pelletization of the charge with soda, as well as the thermodynamics and kinetics of the roasting process, were studied to improve the performance of the main process of the concentrate sintering. Exploratory technological experiments for the sintering of the charge, the leaching of the obtained sintered mass with water, and the purification of the sodium tungstate solutions from impurities, with the subsequent separation of tungstic acid, were performed to assess the possibility to reduce the consumption of inorganic acids through the use of the membrane processes of electrodialysis.

2. Materials and Methods

Ore beneficiation. Representatives of the company holding the license for the deposit took samples of the rich ore materials. The composition of the ore samples and the resulting products from the gravity concentration and pyritic flotation are summarized in Tables 1–4.

Table 1. Chemical analysis of the ore and beneficiation products.

Material	The Content of Chemical Elements, %											
	WO_3	Fe_2O_3	SiO_2	Al_2O_3	P_2O_5	S	CaO	MnO	MgO	MoO_3	Re	C
Ore	20.1	10.7	12.3	2.34	0.046	30.7	6.54	4.07	2.39	0.57	0.1	8.08
Concentrate	68.8	6.52	2.96	2.51	0.037	1.94	5.51	7.31	2.65	0.35	0.06	0.73
Tailings	2.4	8.51	44.7	3.33	0.09	24.5	6.06	0.83	2.5	0.2	0.08	6.76

Table 2. Phase (structural) ore analysis.

Material	Quantum Distribution, %										
	$FeWO_4$	WO_3	FeS_2	S	SiO_2	MoO_3	MgO	MnO	Al_2O_3	CaO	C
Ore	24	5	14	22	12	-	2	4	2	7	8

Table 3. Phase (structural) analysis of the tungsten concentrate.

Material	Quantum Distribution, %									
	$FeWO_4$	WO_3	$MnWO_4$	SiO_2	CaO	Al_2O_3	MgO	MoO_3	S	C
Concentrate	27	25	31	3	6	3	3	-	2	-

Table 4. Phase (structural) analysis of the beneficiation tailings.

Material	Quantum Distribution, %												
	WO_3	$FeWO_4$	$MnWO_4$	SiO_2	CaO	Al_2O_3	MgO	FeS_2	MnO	S	C	Fe_2O_3	MoO_3
Tailings	-	4	-	45	7	3	3	16	1	14	7	-	-

The chemical analysis and determination of the tungsten in the samples and beneficiation products were performed by calorimetry using an FEC instrument.

X-ray diffractometric analysis was performed with the help of an automated diffractometer, the DRON-3, with $Cu_{K\alpha}$ radiation and a β-filter. The conditions for obtaining the diffractograms include: U = 35 kV; I = 20 mA; imaging, θ-2θ; detector, 2 deg/min. The diffractograms were interpreted with the use of data from the ICDD card catalog: the PDF2 (Powder Diffraction File) database and diffractograms of impurity-free minerals.

The beneficiation results provided data on the distribution of the tungsten and other components.

A mixture of three products with different contents of components was used for the experiments on the sintering kinetics. Its composition is shown in Table 5 and Figure 1.

It follows from the data of the X-ray phase analysis that tungsten is present mainly in the form of iron wolframate—$FeWO_4$ (ferberite), manganese wolframate—$MnWO_4$ (manganese wolframate), water, and anhydrous oxides—$WO_2 \cdot H_2O$, $WO_3 \cdot H_2O$, WO_3, and WO_2.

Calculations of the thermodynamic characteristics of the main sintering reactions were performed using the licensed program—HSChemisry 8.1.5.

The sintering kinetics of the wolframite concentrate with soda was studied in a laboratory unit (Figure 2) with two rotary electric furnaces and nichrome heaters. The temperature was measured and controlled by a Pt-Pt-Rh thermocouple and an IRT-2 electronic unit, with an accuracy of ±5 °C. The O_2, CO_2, and CO и SO_2 content was analyzed with a Gasochrom-1 chromatograph and a MAG-6T gas analyzer. The charge was placed in the crucible, and air was supplied under the charge layer through an inner tube.

Table 5. Composition of the samples of the tungsten concentrates.

The Indicator Being Determined	Test			
	13,463	213,464	13,465	Average
Chemical composition, %				
SiO_2	30.65	10.45	50.4	30.5
Al_2O_3	0.27	0.6	0.66	0.5
FeO	5.58	7.19	3.2	5.3
P_2O_5	0.041	0.036	0.1	0.0
S	1.19	1.36	1.14	1.2
CaO	2.2	4.22	1.7	2.7
MoO_2	0.60	0.60	0.7	0.6
TiO_2	0.44	0.58	0.54	0.5
Cu	0.27	0.25	0.2	0.3
WO_3	54.57	70.31	37.0	54.0
V	0.41	0.44	0.50	0.5
C	0.64	0.48	0.47	0.5
MnO	2.46	2.34	2.4	2.4
Re	0.35	0.41	0.5	0.4
Phase analysis, quantum distribution, %				
SiO_2	34.0	24.0	53.0	37.0
$FeWO_4$ (Ferberite)	29.0	35.0	15.0	26.3
WO_2	5.0	5.0	4.0	4.7
WO_3	4.0	5.0	3.0	4.0
$WO_2{\cdot}H_2O$ ($WO_3{\cdot}H_2O$)	4.0	6.0	6.0	5.3
$MnWO_4$ (Manganese Wolframate)	12.0	11.0	11.0	11.3
$ReMnO_4$	0.5	0.5	0.5	0.5
Re_4Si_7 (Rhenium Silicide)	1.0	1.0	1.0	1.0
$Mo_{0.97}Re_{0.03}S_2$ (Molybdenite 2H (Rebearing))	0.5	0.5	0.5	0.5
TiO_2 (Titanium Oxide)	0.5	0.5	0.5	0.5
$Cu_2V_4O_{11}$ (Copper Vanadium Oxide)	0.5	0.5	0.5	0.5
CaO (Calcium Oxide)	5.0	7.0	2.0	4.7
C_8O_5Al (Calcite)	4.0	4.0	3.0	3.7

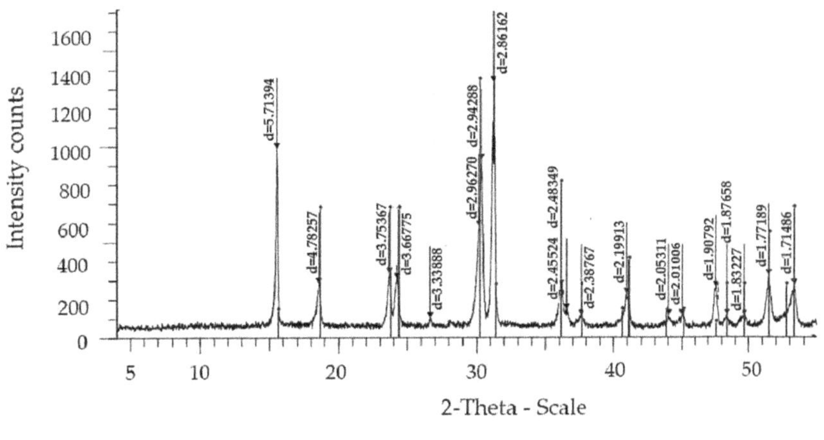

♦ —wolframite $(Fe,Mn)WO_4$

Figure 1. Diffractogram of the wolframite concentrate.

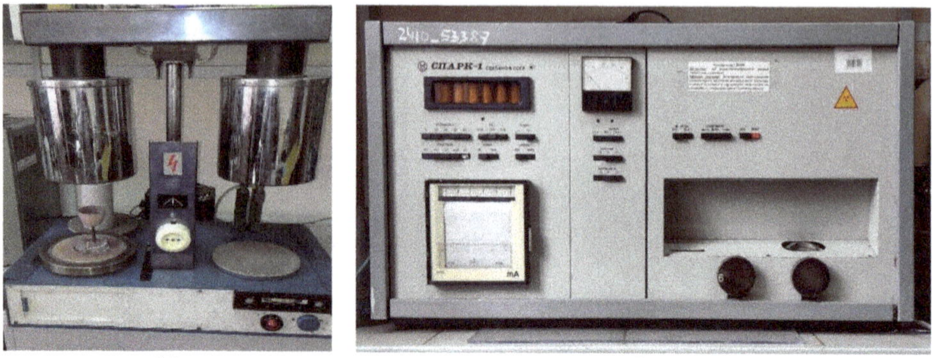

(a) (b)

Figure 2. Firing unit with the IRT-2 temperature controller (**a**) and the SPARK-1 X-ray fluorescence analysis device (**b**).

The starting materials and products were analyzed for W, Fe, and Mn on the X-ray fluorescence analysis unit, the SPARK-1, with a Cu cathode at a voltage on the tube of U = 25 kV and an exposure of 40 s. The samples for the analysis were pressed into tablets at a pressure of P = 4.0 tons.

Enlarged suspensions of the wolframite concentrate charge in a mixture with soda in a SNOL muffle furnace were sintered to leach and obtain sodium tungstate solutions. A 6.0 kg charge was sintered in alundum crucibles with a 110 mm diameter and 150 mm height. The sintered mass was leached in porcelain beakers with the use of laboratory stirrers with thermostats. The electrodialysis of the solutions was performed in a 2-section electrodialyzer with a platinized anode and steel cathode using the MK-40 cation-exchange membrane (Figure 3).

Figure 3. Laboratory 2-section electrodialyzer.

3. Results and Discussion

3.1. Thermodynamic Assessment of the Main Reactions

The calculations of the enthalpy, entropy, and Gibbs energy of the reactions of the components of the wolframite concentrate with soda were performed using the licensed program HSChemisry 8.1.5 by Outotec Technologies, taking into account the X-ray phase analysis. The results of the calculations for the temperatures of 500 °C and 800 °C, as provided in Table 6, confirm the thermodynamic feasibility of the process. All reactions proceeded with the release of carbon dioxide and, therefore, the process could be controlled by analyzing the gases leaving the reactor for the CO_2 content. The reactions of the tungsten minerals with soda (WO_2, $MnWO_4$, $FeWO_4$) occurred in the presence of oxygen.

Table 6. Calculation of the thermodynamic characteristics of the concentrate reactions with soda.

№	Reactions	T °C	ΔH kJ	ΔS J/K	ΔG kJ	K	Log K
1	$2CaWO_4 + 2Na_2CO_3 + SiO_2 = 2Na_2WO_4 + Ca_2SiO_4 + 2CO_2(g)$	500.0	434,181	590,405	−22,291	3207×10^1	1506
		800.0	643,177	820,640	−237,494	3637×10^{11}	11,561
2	$CaWO_4 + Na_2CO_3 = Na_2WO_4 + CaO + CO_2(g)$	500.0	257,742	260,910	56,020	1640×10^{-4}	−3785
		800.0	358,462	371,950	−40,696	9572×10^1	1981
3	$2WO_2 + 2Na_2CO_3 + O_2(g) = 2Na_2WO_4 + 2CO_2(g)$	500.0	−451,033	143,889	−562,280	9802×10^{37}	37,991
		800.0	−336,803	270,752	−627,360	3458×10^{30}	30,539
4	$6MnWO_4 + 6Na_2CO_3 + O_2(g) = 6Na_2WO_4 + 2Mn_3O_4 + 6CO_2(g)$	500.0	161,625	597,391	−300,247	1935×10^{20}	20,287
		800.0	499,140	972,439	−544,433	3177×10^{26}	26,502
5	$4FeWO_4 + 4Na_2CO_3 + O_2(g) = 4Na_2WO_4 + 2Fe_2O_3 + 4CO_2(g)$	500.0	−22,009	872,171	−696,329	1118×10^{47}	47,048
		800.0	404,064	1342,023	−1,036,128	2735×10^{50}	50,437
6	$WO_3 + Na_2CO_3 = Na_2WO_4 + CO_2(g)$	500.0	25,135	144,731	−86,764	7283×10^5	5862
		800.0	79,044	204,864	−140,805	7148×10^6	6854
7	$SiO_2 + Na_2CO_3 = Na_2SiO_3 + CO_2(g)$	500.0	82,701	132,429	−19,687	2139×10^1	1330
		800.0	75,904	125,063	−58,308	6892×10^2	2838

3.2. Kinetic Studies of the Wolframite Concentrate Sintering with Soda

The charge was pelletized on a plate pelletizer in order to improve the contact of the tungsten concentrate particles of a 0.1 mm size with soda. The character of the sintering kinetic curves changed, and a high transformation degree was achieved in a shorter process duration during the pellet firing.

Figure 4 shows the curves of the carbon dioxide content in the waste gas and the conversion degree for a pellet suspension of 10 g of the concentrate (54% WO_3) and 3.5 g of soda.

The CO_2 content in the exhaust gases increased and reached a maximum in 15–20 min (Figure 4a), and then gradually decreased when the transformation degree of the suspension material increased above 45–50% at a constant firing temperature (Figure 4b). This form of kinetic curve was characteristic of heterogeneous topochemical processes.

The amount of carbon dioxide due to the reactions of the tungsten compounds with Na_2CO_3 was calculated from the kinetic curves, taking into account the amount of exhaust gases. The transformation degree with the formation of Na_2WO_4 was calculated as the ratio of the amount of CO_2 in the gases to the amount in the roasting charge in the form of soda. No carbon dioxide emission was observed during heating of the pure soda suspension without the concentrate. Additionally, the formation of sodium silicate was insignificant, which was confirmed in the purification results of the sodium tungstate solutions from the silicon impurity.

The experiments were performed with the suspensions of class 2–4 pellets, weighing 10 g and containing 21.4% of soda, in the heating regime from 500 to 800 °C, based on the differential method [34,35] and dynamic method [36] applied for nonisothermal processes. The heating duration was 20 min and the average heating rate was 15 °C/min. The transformation degree in the nonisothermal regime did not exceed 50–80%. Parallel experiments were performed in the nonisothermal regime, and the dependences, as shown in Figure 5, were plotted to determine the kinetic constants.

The dependences of the CO_2 content in the gases on the temperature characterize the course of the reactions of the tungsten compounds with soda. The effect of the dissociation reactions of the calcium and magnesium carbonates can be neglected due to their low content in the charge. The curve of the CO_2 content in the gases, in the mode of continuous heating, as shown in Figure 5a, shows that the interaction reactions of the minerals with the soda occur at temperatures above 520 °C. The dependence of the CO_2 content on the duration was used for the calculations of the transformation degree.

The kinetics of the heterogeneous firing processes can be described with the use of the transformation degree according to the Roginsky–Schultz–Thodes–Mampel, Prout–Tompkins, and Kolmogorov–Yerofeyev–Avrami equations [34,35,37].

During the sintering of the wolframite concentrate, the process takes place, not in the volume, but on the surface of the interface, so the rate is defined as the change in the transformation degree in time:

$$W = da/d\tau.$$

To process the experimental data, the Yerofeyev equation was used in the form of

$$\left.\begin{array}{l} -\ln(1-\alpha) = kt^n \\ \alpha = 1 - e^{-kt^n} \end{array}\right\}.$$

The constants k and n were found by the graphical processing of the experimental data according to the linear form of the equation.

$$\ln[-\ln(1-\alpha)] = \ln k + n \cdot \ln t.$$

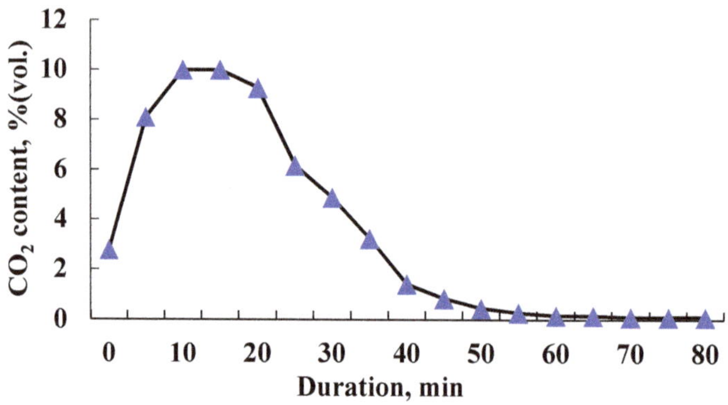

(a)

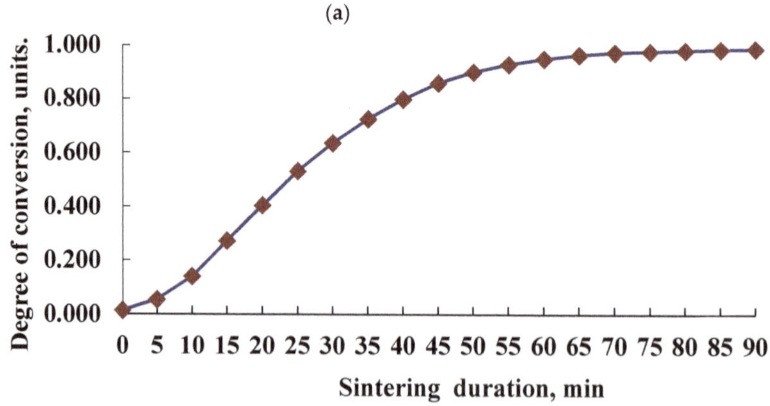

(b)

Figure 4. Dependence of the CO_2 content in the gases (**a**) during the sintering of the wolframite concentrate with Na_2CO_3 and the degree of transformation (**b**) at the roasting temperature of 800 °C.

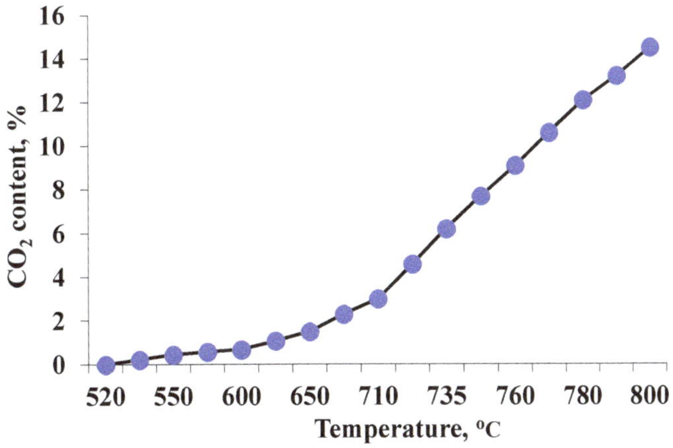

(a)

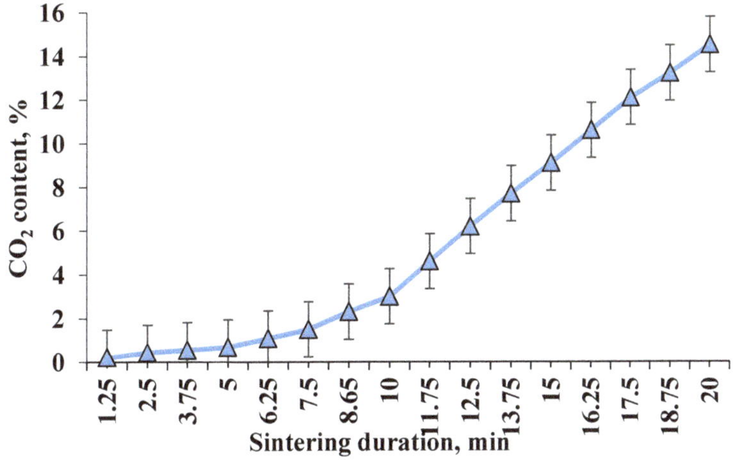

(b)

Figure 5. Dependences of the CO_2 content on temperature (**a**) and heating duration (**b**).

The free term in the approximation equations in Figure 6a,b represents $\ln k$, while the constant n is defined as the tangent of the slope angle of the straight lines

The dependence of the reaction rate constant on the temperature is described by the Arrhenius equation in the integral form:

$$K = K_0 \cdot e^{-\frac{E}{RT}},$$

where

K_0 is the pre-exponential multiplier;
E is the activation energy;
T is the temperature of the reaction.

The following is obtained after its logarithmization:

$$\ln K = \ln A - E_a / R \cdot T.$$

The study results of the roasting kinetics of the concentrate pelletized with soda are summarized in Table 7.

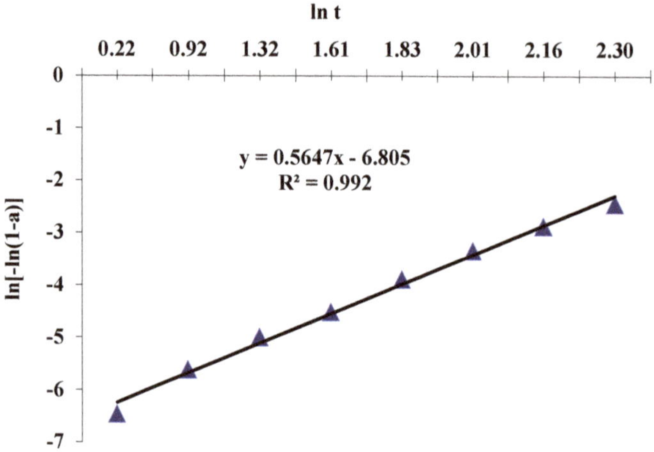

(a)

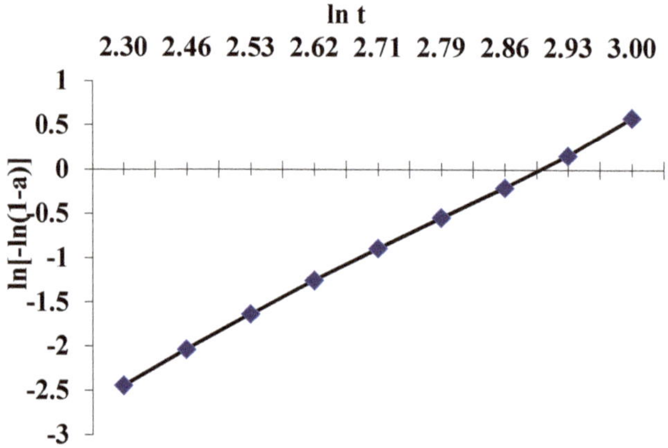

(b)

Figure 6. Dependence of the transformation degree on the duration, ln[−ln(1 − a)] = f [ln(t)] in the interval: (**a**) 520–710 °C; (**b**) 710–800 °C.

Table 7. Sintering constants of the pellets of the wolframite concentrate with soda.

Interval, °C	T, °C	T, K	$1/(T \cdot 10^{-4})$	Ln K	n	E1, J/mol	K
520–710	615	888	11.26	−6805	1933	16.07	0.0011
710–800	755	1028	9.73	−2769	4374	36.37	0.0627

The apparent activation energy value of the sintering process for the tungsten concentrate pellets with soda indicates a transient diffusion–kinetic regime within 525–710 °C and the transition to the kinetic regime within 710–800 °C. This is probably due to a liquid phase appearing due to the formed sodium tungstate melting and an increase in the oxygen diffusion rate to the reaction surface and carbon dioxide removal. It may be due to the liquid phase appearing due to the melting of the formed sodium tungstate and an increase in the rate of the oxygen diffusion to the reaction surface and the carbon dioxide removal. In general, the wolframite concentrate sintering with soda can be attributed to topochemical processes, confirmed with the kinetic curves of the transformation degree and the reaction rate.

3.3. Roasting of the Tungsten Concentrate with Soda in a Muffle Furnace

The concentrate was ground to a −100 μm and mixed with soda to assess the leaching of the sintered mass and to produce sodium tungstate solutions. The obtained charge was pelletized on a plate laboratory pelletizer with a bowl diameter of 0.3 m at a rotation speed of 45 rpm. The pellets, with a diameter of 6–8 mm, were sintered after drying under an air supply and a gas analysis for CO_2. The enlarged samples of the wolframite concentrate charge in a mixture with soda were sintered in a SNOL muffle furnace in alumina crucibles with a diameter of 110 and a height of 150 mm.

A charge of 6 kg of the concentrate with an average tungsten content of 61.45% (weight) or in terms of 77.5% WO_3 was prepared for sintering. Soda consumption was calculated from the condition of the complete binding of tungsten trioxide in Na_2WO_4. The theoretical consumption of the soda for 6 kg of the concentrate was 2.06 kg, and the practical consumption at a 10% excess, was 2.27 kg of Na_2CO_3. The charge after mixing was fired in corundum crucibles in a muffle furnace. The decomposition of the wolframite concentrate by sintering with soda in the presence of oxygen proceeds with the release of carbon dioxide.

The decarbonization degree of the soda that characterized the conversion degree of the tungsten minerals into sodium tungstate was estimated by the weight loss. The firing conditions and the results obtained for the three parallel experiments for each temperature are summarized in Table 8 below.

Table 8. Sintering of the wolframite concentrate with soda in a muffle furnace.

Experience Number	T, °C	Concentrate, g	Soda, g	Charge, P_1, g	Sintering Mass, P_2, g	H_2O	$CO_2 \cdot 0.9$	P_1–P_2	a, %
1	700	1500	567	2067	1871.1	5.85	211.83	195.6	89.58
2	710	1959.3	740.7	2700	2434.2	7.62	276.72	265.8	93.30
3	725	2033	769.5	2802.5	2521.6	7.84	287.45	280.9	94.99
4	800	500	189	689	616.8	1.95	70.61	72.2	99.49

The charge weight in the crucible was 690–900 g. The charge was stirred every 30 min during firing. As the firing progressed, the charge volume in the crucible decreased by two times, and the charge gradually changed color from gray to black. The duration of the firing was 2 h. The charge melted, forming a liquid–fluid melt with a viscosity of about 5–7 Pz at the end of the firing. The smelting was formed at 700–710 °C. It was explained by the high content of WO_3 in the concentrate and the low melting point of the Na_2WO_4 (according to different sources t_m = 696, 698, 700 °C). A black-colored melt was obtained after cooling. It was easily crushed and pulverized.

3.4. Leaching of the Sintered Mass with Distilled Water

When the sintered masses were leached with water [19], sodium tungstate and soluble impurities passed into the solution (Na_2SiO_3, Na_2HPO_4, Na_2HAsO_4, Na_2MoO_4, and Na_2SO_4), as well as an excess of Na_2CO_3. The extraction of the tungsten from the sintered mass into the solution was 98–99%. Strong solutions contained 150–250 g/dm^3 WO_3. The composition of the cake after the first stage of the leaching before the water-washing operations is shown in Table 9.

Table 9. Cake composition after the leaching of the sintering product.

№	The Name of the Indicators	Unit of Measurement	Result
1	Mass fraction of calcium oxide (CaO)	%	5.79
2	Mass fraction of lead (Pb)	%	0.12
3	Mass fraction of tin dioxide (SnO_2)	%	0.28
4	Mass fraction of sulfur (S)	%	0.95
5	Mass fraction of phosphorus oxide (P_2O_5)	%	0.010
6	Mass fraction of silicon oxide (SiO_2)	%	16.9
7	Mass fraction of manganese oxide (MnO)	%	23.5
8	Mass fraction of tungsten oxide (WO_3)	%	7.2
9	Mass fraction of molybdenum (Mo)	%	0.13
10	Mass fraction of total iron (Fe_{total})	%	19.2
11	Mass fraction of sodium oxide (Na_2O)	%	23.6
12	Calcination losses	%	2.03

The WO_3 content in the filtered and washed leach tailings ranged from 1.0 to 2%. The pulp was filtered on vacuum filters. It was relatively easy to settle and filter the leach slurry at L:S = 3:1.

The solutions of the sodium tungstate, in order to obtain the tungsten trioxide of the required purity, were processed according to the traditional scheme [8,23,24], including purification from compounds of impurity elements (Si, P, As, F, Mo), the precipitation and production of technical tungstic acid, and the obtaining tungsten trioxide. The composition of the filtrate is shown in Table 10.

Table 10. Composition of the filtrate after the leaching of the sintering product.

№	The Name of the Indicators	Unit of Measurement	Result
1	Mass fraction of sodium tungstate (Na_2WO_4)	%	51.0
2	Mass fraction of sodium hydroxide (NaOH)	%	4.0
3	Mass fraction of calcium oxide (CaO)	%	0.10
4	Mass fraction of sulfur (S)	%	0.074
5	Mass fraction of phosphorus oxide (P_2O_5)	%	0.011
6	Mass fraction of silicon oxide (SiO_2)	%	0.074
7	Mass fraction of manganese oxide (MnO)	%	0.55
8	Mass fraction of molybdenum (MoO_2)	%	0.02
9	Mass fraction of total iron (Fe_{total})	%	0.11

The purification of the solution from the silicon impurity was based on the hydrolytic decomposition of Na_2SiO_3 by boiling the solution, neutralized to a pH = 8–9 with the release of silicic acid. The solutions were neutralized with hydrochloric acid added to the heated sodium tungstate solution by a thin jet with stirring. The silicon purification process was relatively easy according to the literature data [24], and the solution was filtered well through a white filter.

Subsequent stages of the purification from impurities—P, As, F, and Mo—did not lead to precipitation due to the low content of these elements in the feedstock. Apparently, these impurities will accumulate in solutions during the circulation of solutions in continuous production, and then there will be a need for their removal in the form of precipitation by known technologies.

A part of the solution was evaporated to obtain a precipitate in order to verify the composition of the leaching and the purification of the solutions. Its XRD analysis is shown in Figure 7 and Table 11. These data indicate the content of water-soluble forms of tungsten and rhenium in the specimen taken to leaching.

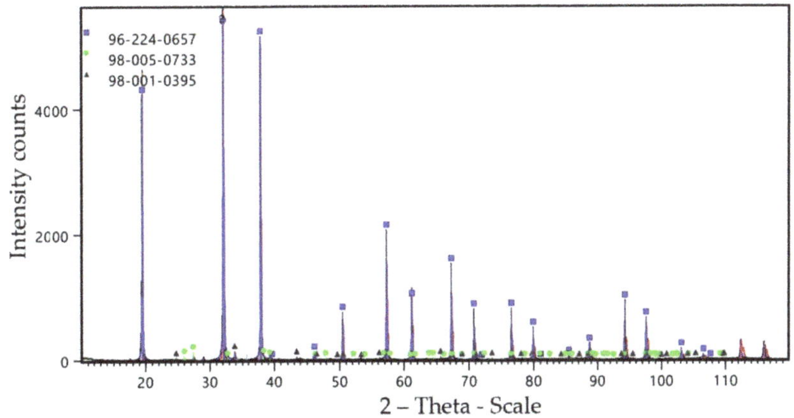

Figure 7. Diffractogram of precipitate evaporated from the sodium tungstate solution.

Table 11. Phase composition of the precipitate evaporated from a solution of sodium tungstate.

Visible	Name of the Chemical Compound	Chemical Formula	Weight Fraction
1	Disodium Tungsten (VI) Oxide	Na_2WO_4	97%
2	Tungsten Oxide—Alpha	WO_3	2%
3	Rhenium Oxide	ReO_3	1%

Tungstic acid was extracted from the sodium tungstate solution purified from the impurities by the neutralization with sulfuric or hydrochloric acid, but certain technological problems arose at this stage:

- Acid consumption for the following reactions was quite significant:

$$Na_2WO_4 + H_2SO_4 = H_2WO_4 + Na_2SO_4,$$
$$Na_2WO_4 + 2HCl = H_2WO_4 + 2NaCl;$$

- Acidity at the end of the tungsten acid precipitation had to be brought to 100 g/dm^3, and further increased the acid consumption;
- The white (yellow when heated) tungstic acid formed was poorly sedimented, and it was difficult to filter, resulting in problems with its washing from the impurities of sodium salts;
- The direct extraction of tungsten acid produced large volumes of sodium sulphate or sodium chloride solutions that require further processing.

It is possible to neutralize excess soda and remove sodium ions from sodium tungstate solutions with the use of membrane electrodialysis processes to reduce and even completely eliminate the use of mineral acids in the solution purification, as well as to prevent the formation of salt effluents. The possibility of electrodialysis of sodium tungstate solutions was reported in [38–41].

A decarbonization method of tungstate–soda solutions with the regeneration of sodium hydroxide by electrodialysis in a two-chamber electrolyzer using the MKK-4 cationic membrane based on the cationite SG-1 was proposed in the copyright certificate SU 1750719 A1 [38], and the electrodialysis process was conducted at a current density of 500–1000 A/m^2 to a pH = 2.5–3.0 in the anode chamber. This process eliminates both the stage of the treatment of the tungstate solutions with mineral acid and the discharge of the salt solutions.

The authors [39,40] propose to conduct the electrolysis process at an anodic current density of 75–85 A/m^2 and to maintain the solution concentration in terms of tungsten oxide within 0.05–0.2 wt% to obtain tungstic acid from the sodium tungstate solution. The advantage of the proposed method is that the use of electrolysis eliminates the use of mineral acids required for the deposition of tungstic acid, and reduction in the anodic current density to 75–85 A/m^2 reduces the energy consumption. The recovery of tungsten oxide into acid was 92.5%.

The membrane electrolysis process was used to develop a cleaner process for the production of ammonium paravolphramate in studies [41]. The results showed that the separation of tungsten and sodium improved with the increase in the current density and initial cathodic NaOH concentration, and the decrease in the initial anodic NaOH concentration.

We performed laboratory- and large-scale laboratory studies for the purification of sodium tungstate solutions from impurities, with the exclusion of operations of the neutralization of solutions due to the use of electrodialysis with an MK-40 cation-exchange membrane. Our studies showed the possibility of using electrodialysis with the removal of sodium ions into the catholyte solution and the subsequent regeneration of the resulting alkali to the initial reagent-caustic soda. A platinum anode and a nickel cathode were used during the electrodialysis in the laboratory unit, while bipolar platinized titanium-based electrodes were used in the large-sale laboratory unit. The duration of electrodialysis was 12–15 h at the bath voltage of 5.28–6.97 V and a current of 0.7–0.9 A. During the electrodialysis process, the acidity of the solution in the anodic area decreased within the range of pH = 2.0–0.1, and in the cathodic area, it increased within the range of pH = 11.0–12.0. After electrolysis, the resulting white precipitate was evaporated and dried at a temperature of 105 °C, and the color of the precipitate changed to yellow. The precipitate was calcined at the temperature of 800 °C for 90 min. After calcination, the precipitate acquired a dark-green color. Photographs of the yellow tungstic acid precipitate and the calcined product are shown in Figure 8, and the phase composition is presented in Table 12.

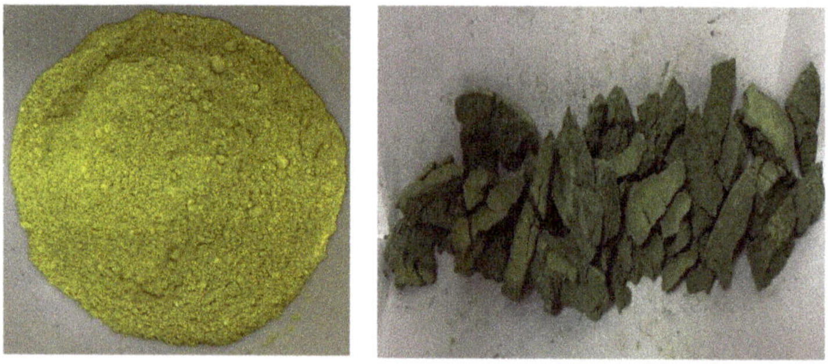

(a) (b)

Figure 8. Photos of the samples of the tungstic acid (**a**) and the product after calcination (**b**).

Table 12. Phase composition of the yellow tungstic acid and calcined product.

Name of the Chemical Compound	Chemical Formula	Weight Fraction
	Yellow Tungstic Acid	
Tungsten Oxide Hydrate	H_2WO_4	80%
Sodium Tungsten Oxide	$Na(WO_3)$	11%
Tungsten Oxide	WO_3	9%
	The product after calcination at 800 °C	
Tungsten Oxide	WO_3	95%
Sodium Tungsten Oxide	$Na(WO_3)$	5%

The purification of the sodium tungstate solutions by the known reagents [8,24], in combination with electrodialysis, excludes the consumption of mineral acids and obtaining effluents requiring utilization.

The analysis of the known processing technologies for the wolframite and scheelite raw materials, as well as the conducted studies, allowed us to justify the following diagram to process the Akchatau wolframite ores, as shown in Figure 9.

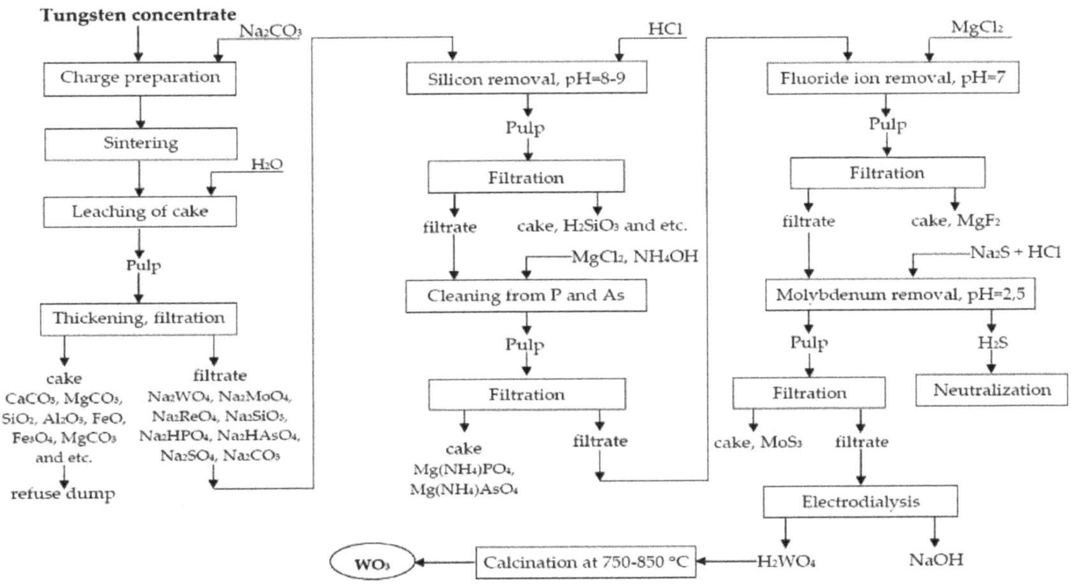

Figure 9. Process flow diagram for the wolframite concentrates.

The difference of the technology from the known ones [8,24] is the wet grinding of the gravity concentrate to 100 microns, the pelletization in a mixture with soda, the roasting of the pellets in a shaft-type reactor, as well as obtaining alkali (NaOH) and tungsten acid in an electrodialyzer. It is possible to regenerate the Na_2CO_3 soda by use of the CO_2 gases of the furnace.

4. Conclusions

A brief review of technologies for processing tungsten concentrates using NaOH and Na_2CO_3 by the autoclave–soda method and sintering method is given, and the shortcomings of the technologies and apparatus design are noted. A high concentration of excess soda in the leaching solutions (80–120 g/dm^3) requires the significant consumption of acids for the neutralization of the solutions during the cleaning from the impurities in the autoclave process, and accordingly, the costs for the disposal of the discharge solutions. The main disadvantage in the sintering–leaching technology is the rapid wear of the furnace lining, and the wear of the body made of alloy steel by the liquid alkali melt in the case of using an unlined drum.

Samples of the ore materials were taken to develop a technology intended to process the wolframite concentrates from the Akchatau deposit, their chemical and mineralogical composition was studied, and beneficiation was performed to produce rich concentrates. Kinetic studies of the sintering process of the tungsten concentrate with soda were performed with a laboratory unit with gas analysis on a Gasochrom-1 chromatograph and MAG-6T gas analyzer. The kinetic dependences of the changes in the composition of the gases and the transformation degree of the tungsten minerals into sodium tungstate were obtained. The rate constants and values of the apparent activation energy were calculated using differential and dynamic methods.

It was established that the sintering process of a mixture of rich wolframite concentrates with soda at 550–800 °C proceeded in diffusion-kinetic and kinetic modes. During pellet firing, the character of the kinetic curves of the sintering changes and the high transformation degree was achieved for a shorter duration of the process, and it was explained by the better contact of the concentrate and soda particles. The dependence of the transformation degree on the duration of the sintering of the wolframite concentrate with the soda from 15% to 90% was satisfactorily described by the kinetic equation of Yerofeyev.

The roasting results of an enlarged sample of the wolframite concentrate with soda in a muffle furnace were presented. A charge containing 6 kg of the concentrate with an average tungsten content of 61.45% (weight) was prepared for sintering. The firing duration was 2 h, and the charge melted at the end of the firing formed a liquid–fluid melt. The melt formation was also observed at 700–710 °C, and it was explained by the high content of WO_3 in the concentrate and the low melting point of Na_2WO_4.

The obtained liquor was subject to aqueous leaching and purification from impurities. The recovery of the tungsten in the solution was 98–99%, and the solutions contained 150–250 g/dm^3 WO_3 during the aqueous leaching of the sintered mass. A search of studies of the purification of the sodium tungstate solutions from the impurities, excluding the neutralization operations of the solutions due to the application of electrodialysis with the MK-40 cation-exchange membrane, were performed. A platinum anode and a nickel cathode were used in the laboratory unit. It was established that the purification of the sodium tungstate solutions by known reagents, in combination with electrodialysis, excludes the consumption of mineral acids and the formation of effluents requiring utilization.

The processing flow diagram for the wolframite ores, including the wet grinding of the concentrate to the size of 100 microns, the pelletization of the concentrate with soda, the drying and roasting of the charge in a shaft-type reactor, the aqueous leaching of the sodium tungstate, the purification of the solutions from the impurities according to the standard flow diagram, the regeneration of the alkali NaOH, and the obtainment of a solution of tungstic acid in an electrodialyzer were proposed.

The proposed technology has the following advantages:

- A significant reduction in the hardware and process flow diagram;
- The elimination of lining corrosion problems in a pit-type reactor;
- The elimination of acid consumption during neutralization of solutions through electrodialysis.

In general, the authors of the work do not claim to provide a complete review of the technologies and the completeness of the research on the issues of the sintering of the tungsten concentrates and the subsequent processing of the solutions using electrodialysis. The firing and sintering issues require further research to study the effect of such factors as the pellet size, the air blast consumption, and the degree of the oxygen beneficiation of the blast. It is necessary to study a set of issues related to the design of the units, the membrane types, the energy consumption, and the scalability in the technology of the electrodialysis of the sodium tungstate solutions. Additionally, the thermodynamic modeling of the studied multicomponent systems, as well as the mathematical and simulation modeling of the processes of the technological diagram, are also important.

Author Contributions: Conceptualization, B.B.; Data curation, G.M.; Formal analysis, S.J. and Z.A.; Investigation, A.T. and Z.A.; Methodology, G.M.; Project administration, B.B.; Resources, E.Y. and V.S.; Software, A.Y. and R.B.; Visualization, A.Y., S.J., A.T. and R.B.; Writing—original draft, B.B.; Writing—review and editing, G.M. All authors have read and agreed to the published version of the manuscript.

Funding: This article was carried out with the financial support of the Science Committee of the Ministry of Science and Higher Education of the Republic of Kazakhstan (No. AP19677216).

Data Availability Statement: All data to support the results of this study are included in this article.

Conflicts of Interest: Authors Elena Yakob and Valery Serikbayev were employed by the company "KSPSteel" LLP. The remaining authors declare that the research was conducted in the absence of any commercial or financial relationships that could be construed as a potential conflict of interest.

References

1. Han, Z.; Golev, A.; Edraki, M. A Review of Tungsten Resources and Potential Extraction from Mine Waste. *Minerals* **2021**, *11*, 701. [CrossRef]
2. United States Geological Survey. *Mineral Commodity Summaries 2022*; United States Geological Survey: Reston, VA, USA, 2022.
3. Leal-Ayala, D.R.; Allwood, J.M.; Petavratzi, E.; Brown, T.J.; Gunn, G. Mapping the global flow of tungsten to identify key material efficiency and supply security opportunities. *Resour. Conserv. Recycl.* **2015**, *103*, 19–28. [CrossRef]
4. Wang, X.; Qin, W.; Jiao, F.; Dong, L.; Guo, J.; Zhang, J.; Yang, C. Review of tungsten resource reserves, tungsten concentrate production and tungsten beneficiation technology in China. *Trans. Nonferrous Met. Soc. China* **2022**, *32*, 2318–2338. [CrossRef]
5. Liu, H.; Liu, H.; Nie, C.; Zhang, J.; Steenari, B.-M.; Ekberg, C. Comprehensive treatments of tungsten slags in China: A critical review. *J. Environ. Manag.* **2020**, *270*, 110927. [CrossRef] [PubMed]
6. Damdinova, L.B.; Damdinov, B.B. Tungsten Ores of the Dzhida W-Mo Ore Field (Southwestern Transbaikalia, Russia): Mineral Composition and Physical-Chemical Conditions of Formation. *Minerals* **2021**, *11*, 725. [CrossRef]
7. Damdinova, L.B.; Damdinov, B.B. Mineral composition and formation conditions of the Inkur tungsten deposit ores (Dzhidinsky ore field, South-Western Transbaikalia). *Earth Sci. Subsoil Use* **2020**, *43*, 290–306. [CrossRef]
8. Lassner, E.; Schubert, W.-D. *Tungsten: Properties, Chemistry, Technology of the Element, Alloys, and Chemical Compounds*; Springer Science & Business Media: Berlin/Heidelberg, Germany, 1999; p. 447. ISBN 0-306-15053-4.
9. Shen, L.; Li, X.; Lindberg, D.; Taskinen, P. Tungsten extractive metallurgy: A review of processes and their challenges for sustainability. *Miner. Eng.* **2019**, *142*, 105934. [CrossRef]
10. Keller, T.; Litnovsky, A.; Mauer, G.; Linsmeier, C.; Guillon, O. Innovative Tungsten Coatings for an Application in Modern and Future Fusion Devices. *Metals* **2023**, *13*, 531. [CrossRef]
11. Manikandan, R.; Raja Annamalai, A. Tungsten Heavy Alloys Processing via Microwave Sintering, Spark Plasma Sintering, and Additive Manufacturing: A Review. *Processes* **2022**, *10*, 2352. [CrossRef]
12. Wurster, S.; Baluc, N.; Battabyal, M.; Crosby, T.; Du, J.; García-Rosales, C.; Hasegawa, A.; Hoffmann, A.; Kimura, A.; Kurishita, H.; et al. Recent progress in R&D on tungsten alloys for divertor structural and plasma facing materials. *J. Nucl. Mater.* **2013**, *442* (Suppl. 1), S181–S189. [CrossRef]
13. Mulenshi, J.; Khavari, P.; Chelgani, S.C.; Rosenkranz, J. Characterization and Beneficiation Options for Tungsten Recovery from Yxsjöberg Historical Ore Tailings. *Processes* **2019**, *7*, 895. [CrossRef]
14. Clemente, D.; Newling, P.; de Sousa, A.B.; LeJeune, G.; Barber, S.P.; Tucker, P. Reprocessing slimes tailings from a tungsten mine. *Miner. Eng.* **1993**, *6*, 831–839. [CrossRef]
15. Wang, X.; Ma, X.D.; Su, K.; Liao, C.F.; Zhao, B.J. Fundamental studies for high temperature processing of tungsten leaching residues for alloy formation. *Tungsten* **2020**, *2*, 362–370. [CrossRef]
16. Su, K.; Ma, X.; Zhao, B. Harmless Treatment and Valuable Metals Recovery of Tungsten Leaching Residues: A Thermodynamic and Experimental Study. *JOM* **2021**, *73*, 1937–1946. [CrossRef]
17. Vadasdi, K. Effluent-free manufacture of ammonium paratungstate (APT) by recycling the byproducts. *Int. J. Refract. Met. Hard Mater.* **1995**, *13*, 45–59. [CrossRef]
18. 18 Deng, P.; Cheng, L.; Li, A.; Zeng, Z.; Liao, C. Recent Advances in the Utilization of Tungsten Residue: A Mini Review of China. *Metals* **2023**, *13*, 1481. [CrossRef]
19. Bogolepov, V.G.; Shirkunov, M.V.; Gulyayeva, N. Structure, composition and distribution of mineralization in the main greisen bodies of the Akchatau rare metal deposit. *Mineral. Geochem. Tungsten Deposits.* **1971**, 96–104.
20. Bychkov, A.; Matveeva, S. Thermodynamic model of the formation of ore bodies at the Akchatau wolframite greisenvein deposit. *Geochem. Int.* **2008**, *46*, 867–886. [CrossRef]
21. Mountain Encyclopedia. Available online: http://www.mining-enc.ru/v/volframovaya-promyshlennost (accessed on 10 January 2022).
22. Yang, X. Beneficiation studies of tungsten ores–A review. *Miner. Eng.* **2018**, *125*, 111–119. [CrossRef]
23. Belskiy, S.S. Processing of wolframite concentrate. In *Bulletin of Irkutsk State Technical University*; Metallurgy and Metal Science; Irkutsk State Technical University: Irkutsk Oblast, Russia, 2018; pp. 186–193.
24. Zelikman, A.N.; Korshunov, B.G. Metallurgy of Rare Metals. Metallurgy, 1991. 432p. Available online: https://reallib.org/reader?file=412937 (accessed on 10 January 2022).
25. Zhao, Z.; Li, J.; Wang, S.; Li, H.; Liu, M.; Sun, P.; Li, Y. Extracting tungsten from scheelite concentrate with caustic soda by autoclaving process. *Hydrometallurgy* **2011**, *108*, 152–156. [CrossRef]
26. Sun, P.M.; Li, H.G.; Li, Y.J.; Zhao, Z.W.; Huo, G.S.; Sun, Z.M.; Liu, M.S. Decomposing scheelite and scheelite-wolframite mixed concentrate by caustic soda digestion. *J. Cent. South Univ. Technol.* **2003**, *10*, 297–300. [CrossRef]
27. Queneau, P.; Huggins, D.; Beckstead, L. Autoclave Soda Digestion of Refractory Scheelite Concentrates. U.S. Patent 4320095A, 16 March 1982.

28. Premchand. Processing of low grade tungsten ore concentrates by hydrometallurgical route with particular reference to India. *Bull. Mater. Sci.* **1996**, *19*, 295–312. [CrossRef]
29. Lei, Y.; Sun, F.; Liu, X.; Zhao, Z. Understanding the wet decomposition processes of tungsten ore: Phase, thermodynamics and kinetics. *Hydrometallurgy* **2022**, *213*, 105928. [CrossRef]
30. Yang, L. Study of processing low-grade tungsten mineral and waste residue of tungsten by soda sintering. *China Molybd Ind* **2008**, *32*, 25.
31. Xu, L.; Zhao, B. Extraction of Sodium Tungstate from Tungsten Ore by Pyrometallurgical Smelting. *Metals* **2023**, *13*, 312. [CrossRef]
32. Chernyak, A. Ore chemical beneficiation. In *Mineral Wealth*; 1987; 224p, Available online: https://reallib.org/reader?file=320869 (accessed on 15 March 2022).
33. Plyushcheeva, V.E. *(Translator) Handbook of Rare Metals*; Mir: Moscow, Russia, 1965; p. 922.
34. Belousova, N.; Belousov, O. *Chemical Kinetics: An Electronic Textbook*; Electronic data (2 Mb); Institute of Continuing Education Siberian Federal University: Krasnoyarsk, Russia, 2009; pp. 25–26, 53–54. Available online: https://bik.sfu-kras.ru/elib/view?id=UMKD-UMKD544.4(075)/%D0%A5+46-932110 (accessed on 19 May 2022).
35. Belousova, N.; Belousov, O. (Eds.) *Chemical Kinetics [Electronic Resource]: Practicum*; Electronic data (2 MB); Institute of Continuing Education Siberian Federal University: Krasnoyarsk, Russia, 2009; pp. 11–12, 52–53. Available online: https://www.twirpx.com/file/820264/ (accessed on 19 May 2022).
36. Kubrakova, I.; Khamizov, R. The use of microwave heating to quickly determine the kinetic parameters of reactions. Kinetics of Decomposition of Organic Substances by Nitric Acid. *Russ. Chem. Bull.* **2005**, *6*, 1371–1375.
37. Sitnikova, V.; Ponomareva, A.; Uspenskaya, M. *Methods of Thermal Analysis*; The workshop. National Research University of Information Technologies, Mechanics and Optics: St. Petersburg, Russia, 2021; pp. 91–93. Available online: https://books.ifmo.ru/file/pdf/2722.pdf (accessed on 19 May 2022).
38. Smirnova, N.M.; Glukhova, L.P.; Rusakova, I.S. Method for Decarbonization of Tungstate-Soda Solutions and Regeneration of Sodium Hydroxide by Electrodialysis. SU 1750719 A1, 30 July 1992. Available online: https://yandex.ru/patents/doc/SU1750719A1_19920730 (accessed on 20 October 2022).
39. Butukhanov, V.L.; Lebukhova, N.V.; Ershova, T.B. Method for Producing Tungstic Acid. RF Patent 2073644, 20 February 1997. Available online: https://yandex.ru/patents/doc/RU2073644C1_19970220 (accessed on 20 October 2022).
40. Verkhoturov, A.D.; Makienko, V.M.; Butukhanov, V.L. Possibilities of Processing Tungsten Raw Materials in Environmentally Friendly Conditions. Scientific Notes of Komsomolsk-on-Amur State Technical University. No. III-1 (15). 2013, pp. 102–106. Available online: https://www.uzknastu.ru/files/pdf/2013-15-1/102-106.pdf (accessed on 20 October 2022).
41. Olayiwola, A.U.; Du, H.; Wang, S.N.; Liu, B.; Lv, Y.Q.; Pan, B. Cleaner production of ammonium paratungstate by membrane electrolysis-precipitation of sodium tungstate solution. *Tungsten* **2023**, *5*, 145–159. [CrossRef]

processes

Article

Recovery of Strategic Metals from Waste Printed Circuit Boards with Deep Eutectic Solvents and Ionic Liquids

Urszula Domańska *, Anna Wiśniewska and Zbigniew Dąbrowski

ŁUKASIEWICZ Research Network—Industrial Chemistry Institute, Rydygiera 8 Str., 01-793 Warsaw, Poland
* Correspondence: urszula.domanska-zelazna@ichp.lukasiewicz.gov.pl; Tel.: +48-517-883150

Abstract: The recycling of metals from waste printed circuit boards (WPCBs) has been presented as a solid–liquid extraction process using two deep eutectic solvents (DESs) and four ionic liquids (ILs). The extraction and separation of Cu(II), Ag(I), and other metals, such as Al(III), Fe(II), and Zn(II), from the solid WPCBs (after the physical, mechanical, and thermal pre-treatments) with different solvents are demonstrated. Two popular DESs were used to recover valuable metal ions: (1) choline chloride + malonic acid, 1:1, and (2) choline chloride + ethylene glycol, 1:2. The extraction efficiencies of DES 1 after two extraction and two stripping stages were only 15.7 wt% for Cu(II) and 17.6 wt% for Ag(I). The obtained results were compared with those obtained with four newly synthetized ILs as follows: didecyldimethylammonium propionate ($[N_{10,10,1,1}][C_2H_5COO]$), didecylmethylammonium hydrogen sulphate ($[N_{10,10,1,H}][HSO_4]$), didecyldimethylammonium dihydrogen phosphate ($[N_{10,10,1,1}][H_2PO_4]$), and tetrabutylphosphonium dihydrogen phosphate ($[P_{4,4,4,4}][H_2PO_4]$). Various additives, such as didecyldimethyl ammonium chloride surfactant, DDACl; hydrogen peroxide, H_2O_2; trichloroisocyanuric acid, TCCA; and glycine or pentapotassium bis(peroxymonosulphate) bis(sulphate), PHM, were used with ILs during the extraction process. The solvent concentration, quantity of additivities, extraction temperature, pH, and solid/liquid, as well as organic/water ratios, and the selectivity and distribution ratios were described for all of the systems. The utilization of DESs and the new ILs with different additives presented in this work can serve as potential alternative extractants. This will help to compare these extractants, additives, extraction efficiency, temperature, and time of extraction with those of others with different formulas and procedures. The metal ion content in aqueous and stripped organic solutions was determined by the ICP-MS or ICP-OES methods. The obtained results all show that solvent extraction can successfully replace traditional hydrometallurgical and pyrometallurgical methods in new technologies for the extraction of metal ions from a secondary electronic waste, WPCBs.

Keywords: spent solid WPCBs; metals extraction/recovery; DESs; ionic liquids

Citation: Domańska, U.; Wiśniewska, A.; Dąbrowski, Z. Recovery of Strategic Metals from Waste Printed Circuit Boards with Deep Eutectic Solvents and Ionic Liquids. *Processes* **2024**, *12*, 530. https://doi.org/10.3390/pr12030530

Academic Editors: Yanlin Zhang and Prashant K. Sarswat

Received: 15 January 2024
Revised: 2 February 2024
Accepted: 6 February 2024
Published: 6 March 2024

1. Introduction

Recently, the increasing amount of electronic waste (e-waste) has become a global issue, and it is reported to be the fastest-growing waste stream. It is estimated to have an annual growth rate of over 70 million tons per year. Waste printed circuit boards (WPCBs) constitute about 4–7% of the total mass of e-waste [1]. WPCBs contain metals, such as Au, Cu, Ag, Zn, Al, and Pb, as well as polymers, ceramics, and other substances [2]. Therefore, the resource recovery of metals from WPCBs with an efficient and green method is key to contemporary recycling. Presently, only about 8–12 chemical elements from the list of 60 are being studied for strategic recycling [3–8]. Currently, the consumption of metals has increased, and recycling is an important parameter for metal sustainability. The application of 21st-century solvents—ionic liquids (ILs)—in the recycling of metals has been popular for many years [3]. E-waste generated globally is a source of many metals, such as Cu, Zn, Ni, Al, Fe, Pb, and Sn, as well as precious metals (Au, Ag, Pd, and Pt), and rare earth metals (Y, Eu, Ce, Gd, and La) [8–11]. The most popular methods for recycling metals from WPCBs are the well-known hydrometallurgical and pyrometallurgical processes [12–14]. Many reviews have been published in the literature on the current status of extraction methods and perspectives [13–28]. These have been presented as physical, mechanical, and thermal pre-treatment steps, followed by physical, pyrolysis, supercritical fluid, pyrometallurgical (smelting-refining), and hydrometallurgical processes with chemical leaching. Various approaches and techniques for the selective recovery of metal elements from the leachate, including ionic liquids, DESs, solvent extraction, electrowinning, adsorption, and different precipitation methods, have been discussed in the literature in recent years ([13–28], and literature cited therein).

WPCBs contain precious metals, like Au(III) and Pd(II), which are separated by cementation with Cu powder from the leaching solutions containing many metals with a mixture of HCl and NaClO and by solvent extraction with bis(2,4,4-trimethylpentyl) phosphinic acid, Cyanex 272 [29]. A new review dedicated to porous structure-based materials for application in the recycling of Au from e-waste has also been published [30]. The reusability, high selectivity, stability, and ease of handling have been presented [30].

A sustainable green and closed-loop method has been proposed for bromine recovery from crude bromine salt by-products of WPCB smelting [31]. WPCBs contain Cu, an important metal in the future global copper cycle in different sectors and research in recent years [32,33].

The hydrometallurgical processing of Cu and Au dissolution from the feed of WPCBs and processed fractions using different lixiviants have been studied [34]. The extraction efficiency for Cu of electrodeposition is 75%. Au can be successfully precipitated from leaching solutions using urea, followed by sodium bisulfide. As a result, 4.1 g of Au and 9 kg of Cu can be recovered from 10 kg of WPCBs [34].

The recovery of gold (Au) from WPCBs derived from discarded mobile phones has been proposed with thiosulfate leaching systems that utilize a cobalt (Co(II))–glycine complex as a catalyst with low thiosulfate consumption and a Au leaching efficiency of 97.8% [35]. The other metals could be extracted using HCl and a H_2SO_4–H_2O_2 system. The recovery of Au from the leachate solution was demonstrated with N1923 [35].

Thiosulfate leaching with nickel-ammonia, cobalt-ammonia, and copper-ammonia catalysis thiosulfate processes were proposed for gold extraction from an oxide gold concentrate. The electrochemical leaching mechanism, combined with XPS and SEM-EDS analyses, has been presented [36]. A systematic review of gold extraction and other aqueous leaching or electrochemical methods has also been presented [37–39].

The two pre-treatment routes to extract (Au, Ag, Cu, etc.) from WPCBs by dimethylacetamide (DMA) at the optimized conditions of T = 423 K for 16 h with a solid–liquid ratio of 3:10 and ultrasonic treatment have been recently presented [40].

The leaching of Cu, Co, Au, and Ag from the WPCBs of cellular phones has been carried out using the imidazolium-based ionic liquids (ILs) 1-butyl-3-methylimidazolium hydrogen sulphate ([BMIM][HSO_4]) and 1-hexyl-3-methylimidazolium hydrogen sulphate ([HMIM][HSO_4]) for Cu and Co extraction, as well as 1-butyl-3-methylimidazolium chloride ([BMIM][Cl]) and 1-butyl-3-methylimidazolium bromide ([BMIM][Br]) for Au and Ag extraction within a temperature range of T = 333–353 K [41]. The extraction efficiency for each metal was as follows: 86.2% for copper, 99.5% for cobalt, 40.8% for gold, and 44.6% for silver [41].

A new solvent extraction procedure has been proposed by many authors using thermal pre-treatment, different leaching procedures, and different ILs with different additives to recover Cu, Ag, Pd, Ni, Al, Au, Fe, and Zn from the solid material of WPCBs [3,5,42–46]. In hydrometallurgical processes for recycling metals, different acids and alkalis are commonly used to extract metals such as copper, gold, silver, and palladium [5]. Copper extraction from solid e-waste without leaching using ammonium and imidazolium-based ILs, e.g., dimethyloctylammonium hydrogen sulphate, [$N_{1,1,8,H}$][HSO_4]/20 wt% H_2O_2 at a temperature of T = 348 K after one hour with a selectivity of 20–33 wt%, has been presented [43]. Copper extraction from solid material at a level of 98% was obtained using hydrogen sulphate ILs as follows:

1-propylsulfonic-3-methylimidazolium hydrogen sulphate ([BMIM][HSO_4]), 1-carboxymethyl-3-mehylimidazolium hydrogen sulphate ([PS-MIM][HSO_4]), diacethylimidazolium hydrogen sulphate ([CM-MIM][HSO_4]), [di-Ac-IM][HSO_4] with the addition of H_2O_2, after the primary extraction with 1-butyl-3-methylimidazolium tetrafluoroborate ([BMIM][BF_4]) [44]. The extraction of Cu, Ag, and Au from WPCBs was presented with 1-butyl-3-methylimidazolium bis{(trfluoromethyl)sulfonyl}amide ([BMIM][NTf_2]), 1-butyl-3-methylimidazolium hexafluorophosphate ([BMIM][PF_6]), and trihexyltetradecylphosphonium chloride, also known as Cyphos 101 ([$P_{6,6,6,14}$][Cl]), after preliminary leaching with (H_2SO_4 + H_2O_2) and (35% HCl + 55% HNO_3) [45]. The extraction efficiency for copper was at levels of 90% and 99% at the temperature of T = 343 K at pH = 6, with a S/L ratio of 1:1 [45].

In our recent study, extraction with the IL, DESs and Cyanex 272 in the recovery of metals from WPCBs after thermal pre-treatment and leaching with different acids was proposed [46]. For the ILs, the aqueous biphasic system (ABS) method and DESs were used to extract metal ions from the leachate and the solid phase after leaching to the extent of 20–30 wt% of metals [46]. Various methods of recycling valuable metals have been developed with ILs and DESs with different additives, as well as using organophosphorus-based acids, and are presented in many works [14,19,47,48].

In recent studies, the electrodeposition of copper and silver after extraction with ILs has been proposed [49]. One of the popular ILs used for metal extraction from the waste of petroleum catalysts and e-waste is [BMIM][HSO_4] [49]. This IL has been used for the extraction of Cu (82%) and Zn (99%) from brass waste, with the addition of H_2O_2 and $KHSO_5$ at room temperature at an IL/A ratio of 1:1 v/v [49]. The IL [BMIM][HSO_4], with the addition of H_2O_2, was used to extract copper, zinc, and aluminium from spent WPCBs at a temperature of T = 343 K for 2 h [50]. The extraction efficiency of copper was 100% using a solution of 25 cm^3 of 80% IL v/v and 10 cm^3 of 30% H_2O_2 with an S/L ratio of 1:25 g/cm^3 [50]. The ILs with the $[HSO_4]^-$ anion were also used in a recently published work for the extraction of Cu(II) and Ag(I) [51].

ILs and DESs have been used by us for the extraction of metals from aqueous solutions and from the solid phase, such as the "black mass" of Li-ion batteries [52,53]. The DESs {choline chloride + ethylene glycol, 1:2, 1:3, 1:4, and 1:10} with iodine species were currently used as oxidizing agents of e-waste metals [54].

The use of tributylmethylammonium chloride ([$N_{4,4,4,1}$][Cl]) with trichloroizocyanuric acid (TCCA) showed 100% extraction of the metals: Au, Pd, Cu, and Ag at a low temperature of T = 298 K [55]. TCCA is a simple substance used as a cheap oxidant for disinfecting swimming pools. Glycine and sodium cyanide were used for the extraction of metals (Au, Ag, Pd, and Pt) from WPCBs [56], as well as Cu(II) [57]. However, the very inconvenient ratio of S/L= 1:100 was applied with the addition of 10% of H_2O_2 at the temperature T = 303 K for 2 h at pH = 6–6.5 [57]. The extraction efficiency of Cu(II) was 94%. Glycine is known as an amino acid used for the creation of heteronuclear complexes with metal ions.

Pourbaix diagrams for the Cu–S–H_2O, Zn–S–H_2O, or Ni–S–H_2O systems were presented after the leaching of WPCBs with H_2SO_4 and showed the possibility of metal ion extraction [58].

The extraction of approximately 60% aluminium, 94% copper, 76% zinc, 50% nickel, and residual iron from the non-magnetic fraction of WPCBs was obtained using bis-(2-ethylhexyl) hydrogen phosphate, D2EHPA [59]. Initially, 100 wt% Zn, Fe, and Al were retrieved at pH 3.5, 2:1 A/O, 10% (v/v) D2EHPA, and then 100% of the Cu was extracted at pH 3.5, 1:1 A/O, 20% (v/v) D2EHPA in the second stage.

In this work, the extraction of Cu(II), Ag(I), Al(III), Fe(II), and Zn(II) in the solid–liquid extraction of metal ions with various solvents was evaluated. The single- or two-stage extraction of metals from WPCBs after thermal pre-treatment (T = 1023 K) for 7 h was investigated with two DESs as follows: (1) {choline chloride + malonic acid, 1:1} and (2) {choline chloride + ethylene glycol, 1:2}. The extraction ability of four synthetized ILs, such as didecyldimethylammonium propionate ([$N_{10,10,1,1}$][C_2H_5COO]), didecylmethylammonium hydrogen sulphate ([$N_{10,10,1,H}$][HSO_4]), didecyldimethylammonium dihydrogen phosphate

([$N_{10,10,1,1}$][H_2PO_4]), and tetrabutylphosphonium dihydrogen phosphate ([$P_{4,4,4,4}$][H_2PO_4]), is also presented. Different additives such as didecyldimethylammonium chloride surfactant (DDACl), diethyl phosphite ester (DFE), hydrogen peroxide (H_2O_2), trichloroisocyanuric acid (TCCA), glycine or pentapotassium bis(peroxymonosulphate) bis(sulphate), PHM were used.

These processes, consisting mainly of solid–liquid extraction, were carried out at a different pH, temperature, time, and concentration of additives, such as H_2O_2, TCCA, glycine, PHM, and the (S/L) phase ratio. The concentration of metal ions in the aqueous and stripped organic solutions was determined by the ICP-MS or ICP-OES methods.

2. Experimental Design

2.1. Preparation of the Solid Material

The WPCB samples were delivered by Elemental H2Tech Waste Management in Poland. The WPCB blend was used after the process of mechanical destruction. In the next step, the thermal pre-treatment at T = 1023 K for 7 h was carried out in a resistance chamber furnace (IZO), 16.1 kW, as described in our previous work [46] (see Figure 1). The high temperature helps to decompose the bonding force between particles of the material. The collected material was then manually shredded into small particles of a diameter range of 1–2 mm. The determination of the metal content and the apparatus used have been described earlier [46]. The composition of the solid material is presented in Table 1. WPCB samples, which were delivered by Elemental H2Tech Waste Management in Poland, did not contain any gold.

Figure 1. Solid WPCB material after thermal pre-treatment.

Table 1. Metal content in the starting WPCB material and after thermal pre-treatment at T = 1023 K for 7 h. (microwave digestion/FAAS and ICP-MS methods) [46].

Content of Metals in the Solid Phase					Mass of Sample
Cu(II)	Ag(I)	Al(III)	Fe(II)	Zn(II)	kg
g/kg	mg/kg		g/kg		
		Primary material			
224	483	81.9	79.4	18.1	1.000
		Primary material after thermal pre-treatment			
335	721	122	119	27.0	0.670

2.2. Chemicals

Basic information on the ILs is listed in Table 2, and on other chemicals, it is listed in Table 3.

Table 2. Data of the ionic liquids used for extraction: structure, name, abbreviation of name, supplier, CAS number, molar mass (M), and mass fraction purity.

Chemical Structure	Name, Abbreviation of Name, Supplier, CAS Number	Molar Mass M/ (g mol^{-1})	Purity * in Mass Percent (%)
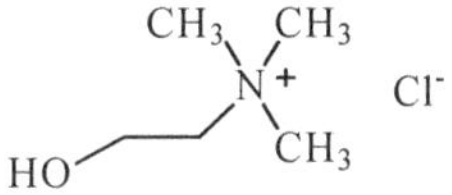	Choline chloride ([$N_{2OH,1,1,1}$][Cl]), Sigma-Aldrich, Darmstadt, Germany, CAS: 67-48-1	139.62	>98
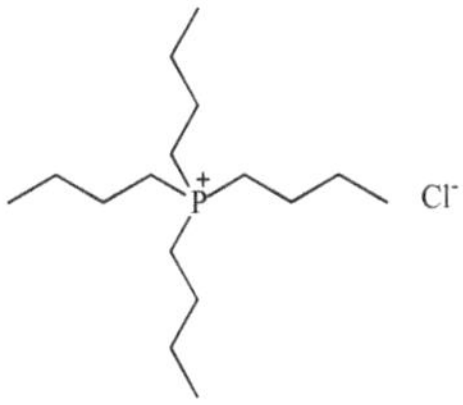	Tetrabutylphosphonium chloride ([$P_{4,4,4,4}$][Cl]), IoLiTec, Heilbronn, Germany, CAS: 2304-30-5	294.88	>95
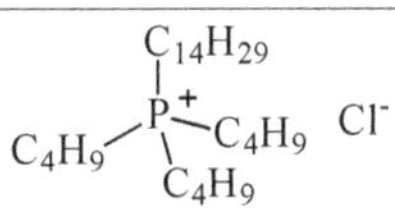	Tributyltetradecylphosphonium chloride ([$P_{4,4,4,14}$][Cl]), IoLiTec, Heilbronn, Germany CAS: 81741-28-8	435.24	>95
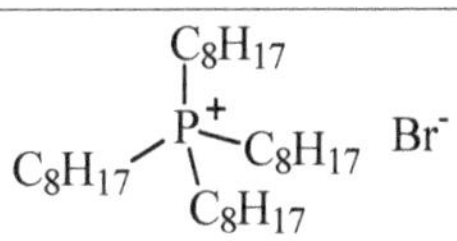	Tetraoctylphosphonium bromide ([$P_{8,8,8,8}$][Br]), IoLiTec, Heilbronn, Germany CAS: 23906-97-0	563.76	>95
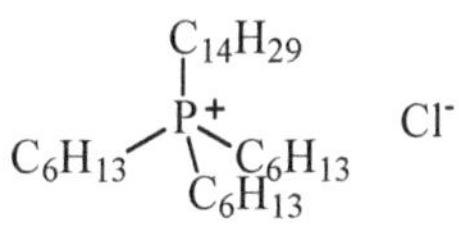	Trihexyltetradecylphosphonium chloride, known as Cyphos IL 101, ([$P_{6,6,6,14}$][Cl]), IoLiTec, Heilbronn, Germany CAS 258864-54-9	519.42	>95
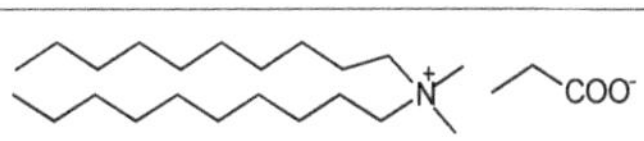	Didecyldimethylammonium propionate ([$N_{10,10,1,1}$][C_2H_5COO]), synthesized	399.79	>95
	Didecyldimethylammonium hydrogensulphate ([$N_{10,10,1,H}$][HSO_4]), synthesized	409.67	>95
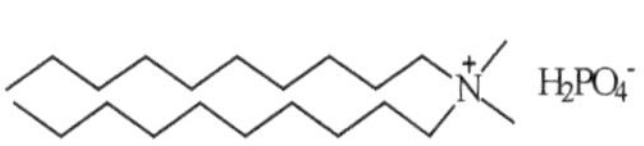	Didecyldimethylammonium bis(2-ethylhexyl) dihydrogen phosphate ([$N_{10,10,1,1}$][H_2PO_4]), synthesized	423.7	95
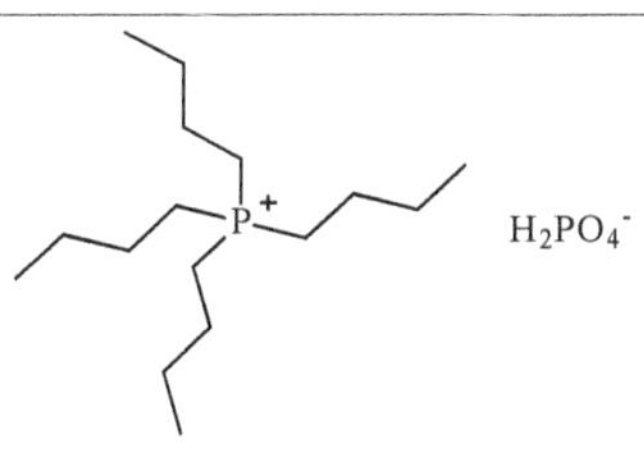	Tetrabutylphosphonium dihydrogen phosphate ([$P_{4,4,4,4}$][H_2PO_4]), synthesized	356.48	95

* As stated by the supplier.

Table 3. Basic data of the chemicals used in the extraction process: name, supplier, CAS number, molar mass (M), and purity.

Name, Molecular Formula, Supplier, CAS Number	Molar Mass ($g \cdot mol^{-1}$)	Purity * in Mass Percent (%)
Malonic acid, $C_3H_4O_4$, Sigma-Aldrich, Darmstadt, Germany, CAS 141-82-2	104.06	99.0
Ethylene glycol, $HOCH_2CH_2OH$, Riedel-de-Haën, Seelze 1, Germany, CAS 107-21-1	62.08	99.0
Didecyldimethylammonium chloride, $[N_{10,10,1,1}][Cl]$, DDACl, Alpinus Sp. z o.o., Solec Kujawski, Poland, CAS 7173-51-5	362.16	50 wt% aq. solution
Trichloroisocyanuric acid (TCCA), Thermo Fisher Scientific, Karlsruhe, Germany, CAS 87-90-1	232.41	99.0
Glycine, $C_2H_5NO_2$, Sigma-Aldrich, Heilbronn, Germany, CAS 56-40-6	75.07	99.0
Pentapotassium bis(peroxymonosulphate)bis(sulphate) PHM, ($2KHSO_5 \cdot KHSO_4 \cdot K_2SO_4$), Sigma-Aldrich, Heilbronn, Germany, CAS 70693-62-8	614.76	98.0
Sulphuric acid, H_2SO_4, Riedel-de Haën, Seelze 1, Gemany, CAS 7664-93-9	98.08	96.0
Sodium chloride, NaCl, Chempur, Karlsruhe, Germany, CAS 7647-14-5	58.44	99.9
Sodium sulphate, Na_2SO_4, Chempur, Karlsruhe, Germany, CAS 7757-82-6	142.04	99.0

* As stated by the supplier.

Taking into consideration the economic prospects for the recovery of valuable metals, two DESs were tested as follows: DES 1 (choline chloride, $[N_{2OH,1,1,1}][Cl]$ + malonic acid, 1:1 [60]) and DES 2 (choline chloride, $[N_{2OH,1,1,1}][Cl]$ + ethylene glycol, 1:2 [61]). The $[N_{2OH,1,1,1}][Cl]$ used for the synthesis of DESs was dried under reduced pressure (10 hPa) at T = 323 K for 8 h.

All the ILs used in the extraction were dried for 72 h at T = 340 K under reduced pressure, p = 6 kPa, and analyzed by Karl–Fischer titration (Metrohm, Herisau, Switzerland, 716 DMS Titrino) to obtain information about the exact mass of the added IL. The water content was less than 760×10^{-6} g with an uncertainty of u(w.c.) = 10×10^{-6} g. The uncertainty of the temperature measurements was ±0.1 K. The Mettler Toledo AB 204-S balance, with an accuracy of $\pm 1 \times 10^{-4}$ g, was used. The pH was measured by litmus bromothymol blue papers in a small reactor.

2.3. Recovery Procedure

A mixture of 15 cm^3 of DES 1 or DES 2, 1.5 g of the black powder, 8 cm^3 of DDACl (50 wt% aqueous solution), 4 cm^3 of H_2O_2 (30 wt% aqueous solution), and 3 cm^3 of water was stirred with a coated magnetic stirring bar under reflux for 2 h, 3000 rpm at T = 333 K, and pH = 5. After the sedimentation of the residual solid phase, the solid material was dried at a temperature of T = 323 K for 5 h, and in many experiments, a second extraction stage was proposed using the same proportion of the solid phase, DES, and additives as above. The volumetric ratio of the organic to aqueous phase was O/A = 1:1. The solid WPCB material to liquid (S/L) ratio was 1.5:30 g/cm^3. The solid phase to DDACl surfactant (S/DDACl) ratio was kept at 1.5:8 g/cm^3. The liquid phases (DES 1: 27 cm^3 of the aqueous lower phase and 3 cm^3 of the organic upper phase; DES 2: 44 cm^3 of the aqueous lower phase and 9 cm^3 of the organic upper phase) were mixed together after two extractions and analyzed for the content of metal ions. It may be observed that during the process, a large amount of the DES is in the aqueous phase.

Extraction with the $[N_{10,10,1,1}][C_2H_5COO]$ or $[N_{10,10,1,H}][HSO_4]$ ILs was performed as follows: 16 g of the IL was added to 1.5 g of the black powder, 8 cm^3 of DDACl (50 wt% aqueous solution), 8 cm^3 of H_2O_2 (30 wt% aqueous solution), and 10 cm^3 of water. The process was carried out for 3 h at 3000 rpm at T = 333 K at pH = 6 or 1.5-3 for $[N_{10,10,1,H}][HSO_4]$. The ratio O/A = 0.6. The (S/L) ratio was kept at 1.5:42 g/cm^3. After the sedimentation process, the solid phase was dried at the temperature T = 323 K for 5 h, and the second extraction stage was carried out as described above. The acidic aqueous phases were mixed together and analyzed as follows: For $[N_{10,10,1,1}][C_2H_5COO]$, one phase of 32 cm^3 in the single-extraction stage and 46 cm^3 in the two-extraction stage; $[N_{10,10,1,H}][HSO_4]$: 36 cm^3 of the lower phase and 47 cm^3 of the upper phase.

The extraction procedure with the $[N_{10,10,1,1}][H_2PO_4]$ and $[P_{4,4,4,4}][H_2PO_4]$ ILs was performed as described above. The solid-to-liquid (S/L) ratio was kept at 1.5:42 g/cm^3. Only a single phase was obtained after extraction. After sedimentation, the solid phase was dried at a temperature of T = 323 K for 5 h. The single- and two-extraction stages were used for these two ILs. After the two-extraction stage, the acidic aqueous phases were combined and analyzed (single phase: 36 cm^3 in the single-extraction stage and 58 cm^3 in the two-extraction stage).

Extraction with the addition of the oxidizing agent TCCA was performed using the following ILs: $[P_{4,4,4,4}][Cl]$, $[P_{4,4,4,14}][Cl]$, $[P_{8,8,8,8}][[Br]$, $[P_{6,6,6,14}][Cl]$, $[N_{10,10,1,1}][C_2H_5COO]$, $[N_{10,10,1,H}][HSO_4]$, $[N_{10,10,1,1}][H_2PO_4]$, or $[P_{4,4,4,4}][H_2PO_4]$. A total amount of 16 g of the IL was added to 1.5 g of the black powder, 10 cm^3 of water, 8 cm^3 of DDACl (50 wt% aqueous solution), and 4 g, 8 g or 12 g of TCCA and 11, 22, or 33 cm^3 of acetone, respectively. The O/A ratio = 2:1. The (S/L) ratio was kept at 1.5:55 g/cm^3. The extraction process was run for 2 h at 3000 rpm at T = 318 K, pH = 2. A single aqueous phase, or a single organic phase ($[P_{6,6,6,14}][Cl]$), or two phases ($[P_{8,8,8,8}][[Br]$ O/A phases, 29 cm^3/22 cm^3; $[N_{10,10,1,H}][HSO_4]$ O:A phases, 27 cm^3/78 cm^3) were obtained after the extraction. The acidic aqueous phase and the stripped (with sulphuric acid, 1.2 M H_2SO_4 at T = 323 K for 20 min.) organic phase were analyzed. The second extraction of metal ions from the solid material that was dried at a temperature of T = 323 K for 5 h was carried out with $[P_{4,4,4,4}][Cl]$, $[P_{4,4,4,14}][Cl]$, $[P_{8,8,8,8}][[Br]$, $[P_{6,6,6,14}][Cl]$, and $[N_{10,10,1,1}][H_2PO_4]$.

The addition of glycine and H_2O_2 was performed using the following ILs: $[P_{4,4,4,4}][Cl]$, $[P_{8,8,8,8}][Br]$, or $[N_{10,10,1,1}][C_2H_5COO]$. A total amount of 16 g of the IL was added to 1.5 g of the black powder, 8 cm^3 of DDACl (50 wt% aqueous solution), 4 cm^3 of H_2O_2 (30 wt% aqueous solution), 10 cm^3 of water, and 4 g or 12 g of glycine. The extraction process, as in the previous case, was carried out for 2 h at 3000 rpm at T = 333 K, pH = 6. The O/A ratio = 0.9–1.3. The (S/L) ratio was kept at 1.5:42–50 g/cm^3. A single phase was obtained after the extraction. The acidic aqueous phase was analyzed (one phase of 31–40 cm^3 in single-extraction stage).

Extraction with the addition of PHM was performed using the only two ILs found to be the best in previous IL treatments, such as $[P_{4,4,4,4}][Cl]$ or $[P_{4,4,4,14}][Cl]$. A total amount of 16 g of the IL was added to 1.5 g of the black powder, 8 cm^3 of DDACl (50 wt% aqueous solution), and 4 g, 8 g, or 16 g of PHM and 14 cm^3 of water at a time of 2 h, 3000 rpm at T = 333 K, pH = 3. The O/A ratio = 0.6–0.4. The (S/L) ratio was kept at 1.5:42–54 g/cm^3. Two extraction stages were carried out. The two phases obtained after extraction were analyzed as follows: ($[P_{4,4,4,4}][Cl]$, 76/6 cm^3, 98/16 cm^3, 83/49 cm^3 organic phase/aqueous phase for 4 g, 8 g, and 16 g of PHM, respectively, and $[P_{4,4,4,14}][Cl]$, 49/30 cm^3 organic phase/aqueous phase for 4 g of PHM). The O/A ratio = 1:1.

2.4. Synthesis of ILs

The synthesis of $[N_{10,10,1,1}][C_2H_5COO]$, $[N_{10,10,1,H}][HSO_4]$, $[N_{10,10,1,1}][H_2PO_4]$, and $[P_{4,4,4,4}][H_2PO_4]$ are presented together with the NMR spectra (recorded on the spectrometer at 300 MHz in the presence of tetramethyl silane (TMS) as an internal standard in Figures S1–S4 in the Supplementary Materials) [62].

3. Results and Discussion

3.1. Solid WPCBs' Content

The metal content in spent WPCB material after thermal pre-treatment, obtained with microwave digestion/FAAS and the ICP-MS method, is listed in Table 1 [46].

The starting WPCB solid material contained the following metals: aluminium (Al), iron (Fe), nickel (Ni), copper (Cu), zinc (Zn), silver (Ag), and lead (Pb). The metal content was in the range of 1–220 g/kg and only the content of silver was much lower, 483 mg/kg. The solid material reduced in weight after the thermal pre-treatment process, and the Cu and Ag contents were 335g/kg and 721 mg/kg, respectively [46].

3.2. Extraction with DESs and ILs

The metal extraction efficiency (*E*) from the solid to liquid phases (the leaching process) and the distribution ratio (*D*) were calculated according to the following equations:

$$E\ (\text{wt\%}) = 100 \times (g_{E,A} + g_{E,O})/g_0 \quad (1)$$

$$D = (g_{E,A} + g_{E,O})/g_0 \quad (2)$$

where g_0 (g) is the metal content in the black powder, $g_{E,A}$ (g) is the amount of metal ions in the aqueous phase (A), and $g_{E,O}$ (g) is the organic phase (O) after extraction.

The DES 1 or DES 2 solution was used directly in the extraction from the solid phase. The two-extraction stage with DES 1 led to an effective but simultaneous extraction of all metal ions, including Cu(II) and Ag(I) ions, at a low temperature of T = 333 K and for a short period of time, 2 h at pH = 5, E_{Cu} = 15.7 wt%, D_{Cu} = 0.2, and E_{Ag} = 17.6 wt%, D_{Ag}= 0.2. The results, presented in Table 4, also indicate a higher extraction of other metals such as aluminium, E_{Al} = 47.5 wt%, D_{Al} = 0.5, and iron E_{Fe} = 47.9 wt%, D_{Fe} = 0.5.

Table 4. Results of metals extraction with DESs at T = 333 K, extraction efficiency, *E* (wt%), distribution ratio, *D*, and pH of the aqueous phase after the two-extraction stage.

Extracting Solvent	Ion	g_0 * (mg)	g_E * (mg)	E (wt%)	D	pH
DES 1 + H_2O_2	Cu(II)	502.50	78.73	15.7	0.2	5
	Ag(I)	1.082	0.191	17.6	0.2	
	Al(III)	183.00	86.89	47.5	0.5	
	Fe(II)	178.50	85.46	47.9	0.5	
	Zn(II)	40.50	5.45	13.5	0.1	
DES 2 + H_2O_2	Cu(II)	502.50	21.13	4.2	0	5
	Ag(I)	1.082	0.09	8.3	0	
	Al(III)	183.00	60.98	33.3	0.3	
	Fe(II)	178.50	15.76	8.8	0	
	Zn(II)	40.50	3.03	7.5	0	

g_0 *—metal content in the solid phase before the extraction; g_E *—metal ion content in the aqueous phase after extraction and from the organic phase after stripping.

A high solubility of the solid material in DES 1 during the extraction processes was observed. These results confirm that DES 1 is suitable for the extraction of metals from the solid phase to the liquid phase with a high extraction efficiency in the presence of H_2O_2 and the surfactant DDACl. However, the separation of particular metal anions was unsuccessful. Unfortunately, the extraction efficiency using DES 2 was not attractive, and only the extraction of aluminium was on a level of E_{Al} = 33.3 wt%, D_{Al} = 0.3. This means that DES 2 may only be used for the extraction of Al(III) at low temperatures and for a short duration.

The hypothesized mechanisms of recovery of metal ions from the aqueous (A) to the organic (O) DES 1 phase are proposed as an "ion exchange" or/and of an "ion pairing" extraction process as follows:

$$2\,Me^{+}{}_{(A)} + [Cl]^{-}{}_{(A)} + [COO]^{-}{}_{(O)} = MeCl_{(A)} + Me[COO]_{(O)} \quad (3)$$

This mechanism has been proposed in the literature for the extraction of metal ions from the solid material to the liquid leachate acidic phase in the presence of H_2O_2 [63].

The synthetized IL $[N_{10,10,1,1}][C_2H_5COO]$ was very attractive for the extraction of silver, as follows: E_{Ag} = 101 wt%, D_{Ag} = 1 after the single-extraction stage, and E_{Ag} = 108 wt%, D_{Ag} = 1 after the two-extraction stage at the low temperature of T = 333 K. This IL also exhibited over 50 wt% extraction efficiency for Zn (67.2–71.9 wt%) after a single- or two-extraction stage. The results are listed in Table 5.

Table 5. Results of single- or two-extraction stage metal extraction with $[N_{10,10,1,1}][C_2H_5COO]$ and $[N_{10,10,1,H}][HSO_4]$ at T = 333 K, extraction efficiency, E (wt%), distribution ratio, D, and pH of the aqueous phase after extraction.

Extracting Solvent	Ion	g_0 * (mg)	g_E * (mg)	E (wt%)	D	pH
$[N_{10,10,1,1}][C_2H_5COO] + H_2O_2$ (single-extraction stage)	Cu(II)	502.50	48.45	9.6	0	6
	Ag(I)	1.082	1.096	100	1	
	Al(III)	183.00	37.12	20.3	0.2	
	Fe(II)	178.50	23.84	13.3	0.1	
	Zn(II)	40.50	27.23	67.2	0.7	
$[N_{10,10,1,1}][C_2H_5COO] + H_2O_2$ (two-extraction stage)	Cu(II)	502.50	69.45	13.8	0.1	6
	Ag(I)	1.082	1.168	100	1	
	Al(III)	183.00	54.34	29.7	0.3	
	Fe(II)	178.50	24.32	13.6	0.1	
	Zn(II)	40.50	29.11	71.9	0.7	
$[N_{10,10,1,H}][HSO_4] + H_2O_2$ (two-extraction stage)	Cu(II)	502.50	113.35	22.6	0.2	3–1.5
	Ag(I)	1.082	0.555	51.3	0.5	
	Al(III)	183.00	108.24	59.1	0.6	
	Fe(II)	178.50	21.57	12.8	0.1	
	Zn(II)	40.50	5.93	14.6	0.1	

g_0 *—metal content in the solid phase before the extraction; g_E *—metal ion content in the aqueous phase after extraction.

The synthetized ($[N_{10,10,1,H}][HSO_4] + H_2O_2$) exhibited an extraction of only a little above 50 wt% for silver, E_{Ag} = 51.3 wt%, D_{Ag} = 0.5, and aluminium, E_{Al} = 59.1 wt%, D_{Al} = 0.6, after the two-extraction stage (see Table 5). It is well-known from the literature that imidazolium-based ILs [BMIM][HSO_4], [HMIM][HSO_4], [BMIM][Cl], and [BMIM][Br] have shown an extraction efficiency for copper at 86.2 wt% and for silver at only 44.6 wt% [41]. Using [BMIM][HSO_4] + H_2O_2, the extraction of Cu was 82 wt%, Zn (99 wt%) [49], and Cu (100 wt%) [50]. It can be easily observed that changing the cation of the IL leads to a higher extraction efficiency for copper and silver.

The synthetized ILs with the $[H_2PO_4]^{-}$ anion revealed an extraction of silver at a level of E_{Ag} = 20.7 wt%, D_{Ag} = 0.2 with ($[N_{10,10,1,1}][H_2PO_4] + H_2O_2$) after the two-extraction stage at a low temperature of $T = 333$ K. The second IL with this anion, $[P_{4,4,4,4}][H_2PO_4]$, exhibited a low extraction efficiency for Cu, Al, and Zn after the two-extraction stage. An average value of the extraction efficiency was observed for copper, $E_{Cu} = 11.4$ wt%, $D_{Cu} = 0.1$ (see Table 6). Thus, the long aliphatic chains at the cation of the IL ($[N_{10,10,1,1}][H_2PO_4]$), proposed as the additional influence of the surfactant on the extraction from the solid phase, were not very attractive with this anion [63]. The extraction efficiency after the single- or two-extraction stages for DES 1, DES 2, and different ILs + H_2O_2 at $T = 318$–333 K for 2–3 h, 3000 rpm, and at different pHs is summarized in Figure 2.

Table 6. Results of single- or two-extraction stage of metal extraction with ILs with $[H_2PO_4]^-$ anion at $T = 333$ K, extraction efficiency, *E* (wt%), distribution ratio, *D*, and pH of the aqueous phase after extraction.

Extracting Solvent	Ion	g_0 * (mg)	g_E * (mg)	E (wt%)	D	pH
	Cu(II)	502.50	15.33	3.0	0	
	Ag(I)	1.082	0.214	19.8	0.2	
$[N_{10,10,1,1}][H_2PO_4] + H_2O_2$ (single-extraction stage)	Al(III)	183.00	0.58	0.32	0	5.5
	Fe(II)	178.50	0.116	0.06	0	
	Zn(II)	40.50	2.04	5.0	0	
	Cu(II)	502.50	22.56	4.5	0	
	Ag(I)	1.082	0.224	20.7	0.2	
$[N_{10,10,1,1}][H_2PO_4] + H_2O_2$ (two-extraction stage)	Al(III)	183.00	0.433	0.24	0	5.5
	Fe(II)	178.50	0.163	0.09	0	
	Zn(II)	40.50	4.25	10.5	0.1	
	Cu(II)	502.50	30.15	6.0	0	
	Ag(I)	1.082	0.089	8.2	0	
$[P_{4,4,4,4}][H_2PO_4] + H_2O_2$ (single-extraction stage)	Al(III)	183.00	17.78	9.7	0	6
	Fe(II)	178.50	5.22	2.9	0	
	Zn(II)	40.50	3.94	9.7	0	
	Cu(II)	502.50	57.17	11.4	0.1	
	Ag(I)	1.082	0.104	9.6	0	
$[P_{4,4,4,4}][H_2PO_4] + H_2O_2$ (two-extraction stage)	Al(III)	183.00	25.31	13.8	0.1	6
	Fe(II)	178.50	5.72	3.2		
	Zn(II)	40.50	4.68	11.6	0.1	

g_0 *—metal content in the solid phase before the extraction; g_E *—metal ion content in the aqueous phase after extraction.

The extraction process of silver, for example, from the solid WPCB material in the presence of the proposed ILs and H_2O_2 may be interpreted as similar to the extraction of Cu from the solid material as follows:

$$2\,Ag + 2\,[N_{10,10,1,1}][C_2H_5COO] + H_2O_2 + 2\,H^+ = 2\,Ag[C_2H_5COO] + 2\,H_2O + 2\,[N_{10,10,1,1}]^+ \quad (4)$$

According to reaction (4), the ions of metals (for example, Ag^+) enter into the solution.

The obtained results of the extraction with various ILs with the addition of the oxidizing agent TCCA at a low temperature of $T = 333$ K were very successful. The choice of a suitable solvent and a special additive is always crucial [64]. The results of the extraction of Ag(I) with ($[P_{4,4,4,4}][Cl]$ + TCCA, 4 g, single-extraction stage) and with ($[P_{4,4,4,4}][Cl]$ + TCCA, 8g, two-extraction stage) are the same, $E_{Ag} = 100$ wt%, $D_{Ag} = 1$. A lower extraction efficiency of Cu(II) from the solid material was also observed with

($[P_{4,4,4,4}][Cl]$ + TCCA, 4 g, single-extraction stage), having an efficiency of E_{Cu} = 53.2 wt%, D_{Cu} = 0.5, and with ($[P_{4,4,4,4}][Cl]$ + TCCA, 8g, two-extraction stage), E_{Cu} = 68.9 wt%, D_{Cu} = 0.7. The IL $[P_{4,4,4,4}][Cl]$ is much more effective than $[P_{8,8,8,8}][Br]$ and $[P_{6,6,6,14}][Cl]$ under the same conditions. The best result of aluminium extraction was obtained with ($[P_{4,4,4,4}][Cl]$ + TCCA, 8 g, two-extraction stage): E_{Al} = 81,3 wt%, D_{Al} = 0.8, and with ($[P_{4,4,4,14}][Cl]$ + TCCA, 8 g, two-extraction stage): E_{Al} = 82,3 wt%, D_{Al} = 0.8. The extraction efficiency of Ag(I) under the same conditions was E_{Ag} = 100 wt%, D_{Ag} = 1. While reviewing Table 7, we can observe that the extraction efficiency of silver is E_{Ag} = 100 wt%, D_{Ag} = 1 using ($[N_{10,10,1,1}][H_2PO_4]$ + TCCA, 4 g, single-extraction stage), and of the remaining metal ions, with ($[N_{10,10,1,1}][C_2H_5COO]$ + TCCA, 8g, two-extraction stage), was E_{Cu} = 50.1 wt%, D_{Ag} = 0.5; E_{Ag} = 100 wt%, D_{Ag} = 1; and of aluminum, it was E_{Al} = 52.5 wt%, D_{Al} = 0.5. Increasing the amount of TCCA to 12 g in the single-extraction stage did not lead to better results in the extraction efficiency of metal ions. The use of ($[N_{10,10,1,H}][HSO_4]$ + TCCA, 8 g, two-extraction stage) revealed a high extraction efficiency only for aluminum, E_{Al} = 83 wt%, D_{Al} = 0.8. The complete results are listed in Table 7. In comparison, the use of $[N_{4,4,4,1}][Cl]$ with the addition of TCCA showed a 100% extraction of the metals Au, Pd, Cu, and Ag at a low temperature of T = 298 K [55].

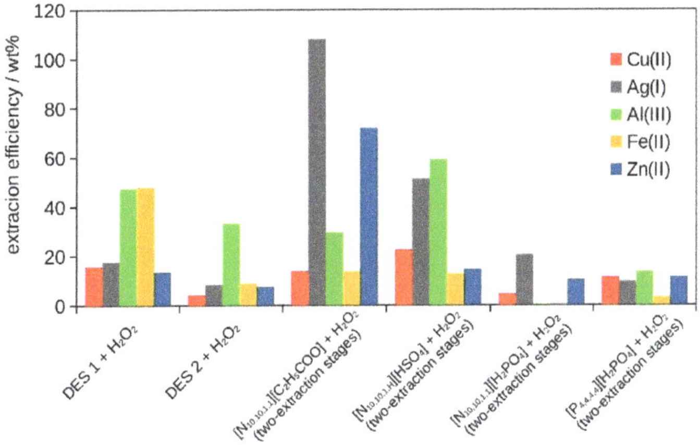

Figure 2. Extraction efficiency (wt%) after single- or two-extraction stages with DES 1, DES 2, and different ILs + H_2O_2. The extraction at each cycle was performed at T = 318-333 K for 2–3 h at 3000 rpm at different pHs.

Table 7. Results of single- or two-extraction stage of metals with ILs and TCCA at T = 333 K, extraction efficiency, E (wt%), distribution ratio, D, and pH of the aqueous phase after extraction.

Extracting Solvent	Ion	g_0 * (mg)	g_E * (mg)	E (wt%)	D	pH
$[P_{4,4,4,4}][Cl]$ + TCCA (4 g) (single-extraction stage)	Cu(II)	502.50	267.6	53.2	0.5	2
	Ag(I)	1.082	1.10	100	1	
	Al(III)	183.00	92.0	50.3	0.5	
	Fe(II)	178.50	62.4	35.0	0.3	
	Zn(II)	40.50	10.5	25.9	0.3	
$[P_{4,4,4,4}][Cl]$ + TCCA (8 g) (two-extraction stage)	Cu(II)	502.50	346.4	68.9	0.7	2
	Ag(I)	1.082	1.08	100	1	
	Al(III)	183.00	148.7	81.3	0.8	
	Fe(II)	178.50	45.4	25.4	0.2	
	Zn(II)	40.50	17.7	43.7	0.4	

Table 7. *Cont.*

Extracting Solvent	Ion	g_0 * (mg)	g_E * (mg)	E (wt%)	D	pH
[$P_{4,4,4,4}$][Cl] + TCCA (12 g) (single-extraction stage)	Cu(II)	502.50	190.1	37.8	0.4	2
	Ag(I)	1.082	1.122	100	1	
	Al(III)	183.00	80.79	44.1	0.4	
	Fe(II)	178.50	29.24	16.4	0.2	
	Zn(II)	40.50	4.69	11.6	0.1	
[$P_{4,4,4,14}$][Cl] + TCCA (8 g) (two-extraction stage)	Cu(II)	502.50	197.1	39.2	0.4	2
	Ag(I)	1.082	1.09	100	1	
	Al(III)	183.00	150.6	82.3	0.8	
	Fe(II)	178.50	29.57	16.6	0.2	
	Zn(II)	40.50	8.74	21.6	0.2	
[$P_{8,8,8,8}$][Br] + TCCA (8 g) (single-extraction stage)	Cu(II)	502.50	124.08	24.7	0.2	3
	Ag(I)	1.082	0.284	26.2	0.3	
	Al(III)	183.00	54.85	30.0	0.3	
	Fe(II)	178.50	12.40	6.9	0	
	Zn(II)	40.50	1.79	4.4	0	
[$P_{8,8,8,8}$][Br] + TCCA (8 g) (two-extraction stage)	Cu(II)	502.50	154.5	30.7	0.3	3
	Ag(I)	1.082	0.36	33.3	0.3	
	Al(III)	183.00	62.81	34.3	0.3	
	Fe(II)	178.50	13.73	7.7	0	
	Zn(II)	40.50	2.16	5.3	0	
[$P_{6,6,6,14}$][Cl] + TCCA (4 g) (organic phase after the demineralization)(single-extraction stage)	Cu(II)	502.50	100.7	20.0	0.2	2
	Ag(I)	1.082	0.30	27.7	0.3	
	Al(III)	183.00	66.18	36.2	0.4	
	Fe(II)	178.50	72.12	40.4	0.4	
	Zn(II)	40.50	7.54	18.6	0.2	
[$N_{10,10,1,1}$][C_2H_5COO] + TCCA (4 g) (two-extraction stage)	Cu(II)	502.50	147.6	29.4	0.3	2
	Ag(I)	1.082	0.12	11.1	0.1	
	Al(III)	183.00	38.11	20.8	0.2	
	Fe(II)	178.50	8.96	5.0	0	
	Zn(II)	40.50	3.76	9.3	0	
[$N_{10,10,1,1}$][C_2H_5COO] + TCCA (8 g), (two-extraction stage)	Cu(II)	502.50	251.6	50.1	0.5	2
	Ag(I)	1.082	1.08	100	1	
	Al(III)	183.00	96.05	52.5	0.5	
	Fe(II)	178.50	19.6	11.0	0.1	
	Zn(II)	40.50	13.0	32.1	0.3	
[$N_{10,10,1,1}$][C_2H_5COO] + TCCA (12 g), (single-extraction stage)	Cu(II)	502.50	133.4	26.5	0.3	2
	Ag(I)	1.082	0.347	32.1	0.3	
	Al(III)	183.00	80.58	44.0	0.4	
	Fe(II)	178.50	15.72	8.8	0	
	Zn(II)	40.50	8.40	20.7	0.2	

Table 7. *Cont.*

Extracting Solvent	Ion	g_0 * (mg)	g_E * (mg)	E (wt%)	D	pH
$[N_{10,10,1,1}][H_2PO_4]$ + TCCA(4 g) (single-extraction stage)	Cu(II)	502.50	199.8	39.8	0.4	2
	Ag(I)	1.082	1.14	100	1	
	Al(III)	183.00	98.8	54.0	0.5	
	Fe(II)	178.50	23.1	12.9	0.1	
	Zn(II)	40.50	8.3	20.5	0.2	
$[N_{10,10,1,H}][HSO_4]$ + TCCA (8 g) (two-extraction stage)	Cu(II)	502.50	118.8	23.6	0.2	1.5
	Ag(I)	1.082	0.417	38.6	0.4	
	Al(III)	183.00	151.8	83.0	0.8	
	Fe(II)	178.50	72.75	40.8	0.4	
	Zn(II)	40.50	8.28	20.5	0.2	

g_0 *—metal content in the solid phase before the extraction; g_E *—metal ion content in the aqueous phase after extraction.

The extraction efficiency after single- or two-extraction stages for different ILs + TCCA (4 g or 8 g) at T = 318 K for 2 h at 3000 rpm, pH = 1.5–3, is presented in Figure 3.

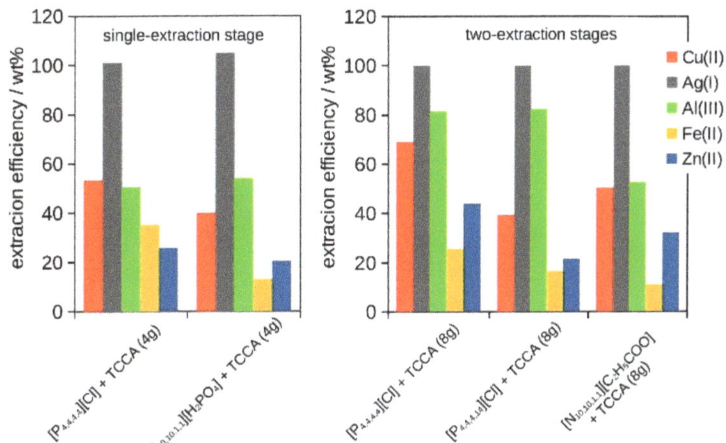

Figure 3. Extraction efficiency (wt%) after single- or two-extraction stages with different ILs + TCCA (4 g or 8 g). The extraction at each cycle was performed at T = 318 K for 2 h at 3000 rpm, pH = 1.5–3.

The single-extraction stage with the addition of glycine and H_2O_2 exhibited a hundred percent efficiency of the extraction of silver using ($[P_{8,8,8,8}][Br]$ + glycine, 4 g) and ($[P_{4,4,4,4}][Cl]$, or $[N_{10,10,1,1}][C_2H_5COO]$ + glycine, 12 g); see Table 8.

Glycine and NaCN were used for the extraction of metals (Au, Ag, Pd, and Pt) [56] and of Cu(II) [57] from the WPCBs (S/L= 1:100) using 10% of H_2O_2 at the temperature T = 303 K for 2 h, pH = 6–6.5, and exhibited an extraction efficiency of Cu(II) 94 wt% [57]. This is a much better result for the extraction of copper than in our experiments, but the use of NaCN and such a large volume of liquid phase is neither economical nor green.

The extraction of Ag(I) with ($[P_{4,4,4,4}][Cl]$ + PHM, 16 g, two-extraction stages) was E_{Ag} = 71.2 wt%, D_{Ag} = 0.7, while at the same time, the low efficiency of Cu(II) extraction was observed, E_{Cu} = 21.2 wt%, D_{Cu} = 0.2 (see Table 9).

Table 8. Results of single-extraction stages of metals with ILs and (glycine + H_2O_2) at T = 333 K, extraction efficiency, E (wt%), distribution ratio, D, and pH of the aqueous phase after the single extraction stage.

Extracting Solvent	Ion	g_0 * (mg)	g_E * (mg)	E (wt%)	D	pH
$[P_{4,4,4,4}][Cl]$ + glycine (4 g)	Cu(II)	502.50	25.88	5.1	0	6
	Ag(I)	1.082	0.086	8.0	0	
	Al(III)	183.00	1.48	0.8	0	
	Fe(II)	178.50	1.87	1.0	0	
	Zn(II)	40.50	1.40	3.4	0	
$[P_{4,4,4,4}][Cl]$ + glycine (12 g)	Cu(II)	502.50	48.40	9.6	0	6
	Ag(I)	1.082	1.09	100	1	
	Al(III)	183.00	12.80	7.0	0	
	Fe(II)	178.50	9.84	5.5	0	
	Zn(II)	40.50	14.69	36.3	0.4	
$[P_{8,8,8,8}][Br]$ + glycine (4 g)	Cu(II)	502.50	35.28	7.02	0	6
	Ag(I)	1.082	1.072	100	1	
	Al(III)	183.00	17.22	9.41	0	
	Fe(II)	178.50	8.79	4.92	0	
	Zn(II)	40.50	2.832	6.99	0	
$[N_{10,10,1,1}][C_2H_5COO]$ + glycine (12 g)	Cu(II)	502.50	25.52	5.1	0	6
	Ag(I)	1.082	1.16	100	1	
	Al(III)	183.00	54.80	29.9	0.3	
	Fe(II)	178.50	7.32	4.1	0	
	Zn(II)	40.50	13.24	32.7	0.3	

g_0 *—metal content in the solid phase before the extraction; g_E *—metal ion content in the aqueous phase after extraction.

Table 9. Results of two-stage extraction of metals with ILs and PHM at T = 333 K, extraction efficiency, E (wt%), distribution ratio, D, and pH of the aqueous phase after the two-extraction stage.

Extracting Solvent	Ion	g_0 * (mg)	g_E * (mg)	E (wt%)	D	pH
$[P_{4,4,4,4}][Cl]$ + PHM (4g) (2 phases)	Cu(II)	502.50	62.3	12.4	0.1	3
	Ag(I)	1.082	0.07	6.5	0	
	Al(III)	183.00	8.52	4.6	0	
	Fe(II)	178.50	18.28	10.2	0.1	
	Zn(II)	40.50	3.58	8.8	0	
$[P_{4,4,4,4}][Cl]$ + PHM (8g) (2 phases)	Cu(II)	502.50	120.5	24.0	0.2	3
	Ag(I)	1.082	0.136	12.6	0.1	
	Al(III)	183.00	12.47	6.8	0	
	Fe(II)	178.50	16.81	9.4	0	
	Zn(II)	40.50	5.73	14.1	0.1	

Table 9. *Cont.*

Extracting Solvent	Ion	g_0 * (mg)	g_E * (mg)	E (wt%)	D	pH
$[P_{4,4,4,4}][Cl]$ + PHM (16g) (2 phases)	Cu(II)	502.50	106.5	21.2	0.2	3
	Ag(I)	1.082	0.770	71.2	0.7	
	Al(III)	183.00	41.29	22.6	0.2	
	Fe(II)	178.50	23.29	13.0	0.1	
	Zn(II)	40.50	3.56	8.8	0	
$[P_{4,4,4,14}][Cl]$ + PHM (4g) (2 phases)	Cu(II)	502.50	19.81	3.9	0	3
	Ag(I)	1.082	0.206	19.1	0.2	
	Al(III)	183.00	28.18	15.4	0.1	
	Fe(II)	178.50	47.64	26.7	0.3	
	Zn(II)	40.50	19.19	47.4	0.5	

g_0 *—metal content in the solid phase before the extraction; g_E *—metal ion content in the aqueous phase after extraction.

The content of additives was at a level of 2–12 g or only 16 g for PHM/1.5 g of WPCB solid material. In order to increase the extraction efficiency, a much larger amount of additives is probably needed, which will generate higher costs.

The results of the extraction of metals with {Cyanex 272) + diethyl phosphite ester + H_2O_2} and with {D2EHPA + diethyl phosphite ester + H_2O_2} after the two-extraction stage at T = 333 K, pH = 5–6, have shown an extraction efficiency for comparison, E_{Cu} = 16.0 wt% and E_{Cu} = 20.1 wt%, respectively (not shown in this work).

4. Conclusions

Various methods for the extraction of metal ions from waste printed circuit board (WPCBs) powder after the process of cutting them into small pieces, crushing in a hydraulic press, and finally, thermal pre-treatment at a temperature of T = 1023 K for 7 h, have been presented. The synthesis of two DESs (1) {choline chloride + malonic acid, 1:1} and (2) {choline chloride + ethylene glycol, 1:2} and four new ILs, such as didecyldimethylammonium propionate ($[N_{10,10,1,1}][C_2H_5COO]$), didecylmethylammonium hydrogen sulphate ($[N_{10,10,1,H}][HSO_4]$), didecyldimethylammonium dihydrogen phosphate, ($[N_{10,10,1,1}][H_2PO_4]$), and tetrabutylphosphonium dihydrogen phosphate ($[P_{4,4,4,4}][H_2PO_4]$), are also presented. The efficiency of the extraction of various metals such as Cu(II), Ag(I), Al(III), Fe(II), and Zn(II) from solid material to the liquid phase with DES 1, DES 2, and ILs, with the addition of H_2O_2, TCCA, PHM, or glycine after a single- or two-extraction stage, is presented. Finally, a 100% extraction efficiency of Ag(I) ions from the solid WPCB material was achieved using the following ILs: $[P_{4,4,4,4}][Cl]$, $[P_{4,4,4,14}][Cl]$, $[P_{8,8,8,8}][Br]$, $[N_{10,10,1,1}][C_2H_5COO]$, and $[N_{10,10,1,1}][H_2PO_4]$, with the addition of H_2O_2, TCCA, or (glycine + H_2O_2) in a single- or two-extraction stage. The best extraction of Cu(II), E_{Cu} = 68.9 wt%, was with $[P_{4,4,4,4}][Cl]$ + TCCA (8 g, two-extraction stage), and for E_{Cu} = 50.1 wt%, it was with $[N_{10,10,1,1}][C_2H_5COO]$ + TCCA (8 g, two-extraction stage). The best extraction of Zn(II) was E_{Zn} = 67.2–71.9 wt% with $[N_{10,10,1,1}][C_2H_5COO]$ in a single- or two-extraction stage. The best extraction of Al(III), E_{Al} = 81.2 wt%, was with $[P_{4,4,4,4}][Cl]$ + TCCA (8 g, two-extraction stage), and for E_{Al} = 52.5 wt%, it was with $[N_{10,10,1,1}][C_2H_5COO]$ + TCCA (8g, two-extraction stage). It has been demonstrated in a series of experiments that it is possible to use ILs or DESs for the extraction of metal ions from the solid WPCB e-waste after thermal pre-treatment to the liquid phase with a 60–100 wt% efficiency. However, the next step—the separation of copper, silver, or other metals from the final liquid solution—still has to be resolved.

Supplementary Materials: The following supporting information can be downloaded at: https://www.mdpi.com/article/10.3390/pr12030530/s1, 1. Synthesis of ILs; Figure S1: NMR of didecyldimethylammonium propionate, $[N_{10,10,1,1}][C_2H_5COO]$; Figure S2: NMR of didecylmethylammonium hydrogen sulphate, $[N_{10,10,1,H}][HSO_4]$; Figure S3: NMR of didecyldimethylammonium dihydrogen phosphate, $[N_{10,10,1,1}][H_2PO_4]$: Figure S4: NMR of tetrabutylphosphonium dihydrogen phosphate $[P_{4,4,4,4}][H_2PO_4]$.

Author Contributions: U.D.: Conceptualization, supervision, funding acquisition, methodology, formal analysis, and writing—original draft; A.W.: Data curation and investigation; Z.D.: Data curation and investigation. All authors have read and agreed to the published version of the manuscript.

Funding: This research was funded by the ŁUKASIEWICZ Research Network—Industrial Chemistry Research Institute, Rydygiera 8, 01-793 Warsaw, Poland, grant number 841321A, 2022.

Data Availability Statement: Data are contained within the article.

Conflicts of Interest: The authors declare that they have no known competing financial interests or personal relationships that could have appeared to influence the work reported in this paper.

References

1. Zhu, X.-N.; Nie, C.C.; Zhang, H.; Lyu, X.-J.; Tao, Y.-J.; Quin, J.; Li, L.; Zhang, G.-W. Recovery of high-grade copper from waste printed circuit boards by mechanical grinding assisted flotation. *J. Clean. Prod.* **2019**, *232*, 1251–1256. [CrossRef]
2. Zhu, X.-N.; Nie, C.-C.; Wang, S.-S.; Xie, Y.; Zhang, H.; Lyu, X.-J.; Qiu, J.; Li, L. Cleaner approach to the recycling of metals in waste printed circuit boards by magnetic and gravity separation. *J. Clean. Prod.* **2020**, *248*, 119235. [CrossRef]
3. Park, Y.J.; Fray, D.J. Recovery of high purity precious metals from printed circuit boards. *J. Hazard. Mater.* **2009**, *164*, 1152–1158. [CrossRef] [PubMed]
4. Chen, Z.; Yang, M.; Shi, M.Q.; Kuang, X.; Qi, H.J.; Wang, T. Recycling waste circuit board efficiently and environmentally friendly through small-molecule assisted dissolution. *Sci. Rep.* **2019**, *9*, 17902. [CrossRef] [PubMed]
5. Debnath, B.; Chowdhury, R.; Ghosh, S.K. Sustainability of metal recovery from E-waste. *Front. Environ. Sci. Eng.* **2018**, *12*, 2. [CrossRef]
6. Gorewoda, T.; Eschen, M.; Charasińska, J.; Knapik, M.; Kozłowicz, S.; Anyszkiewicz, J.; Jadwiński, M.; Potempa, M.; Gawliczek, M.; Chmielarz, A.; et al. Determination of metals' content in components mounted on printed circuit boards from end-of-life mobile phones. *Recycling* **2020**, *5*, 20. [CrossRef]
7. Maurice, A.A.; Ngoc Dinh, K.; Charpentier, N.M.; Brambilla, A.; Gabriel, J.-C.P. Dismantling of printed circuit boards enabling electronic components sorting and their subsequent treatment open improved elemental sustainability opportunities. *Sustainability* **2021**, *13*, 10357. [CrossRef]
8. Panda, R.; Mishra, S.; Pant, K.K.; Bhaskar, T.; Naik, S.N. A closed loop recycling strategy for sustainable recovery of group 11 metals (Cu, Au, and Ag) from waste PCBs: An amalgamation of low-temperature NH_4Cl roasting, HCl leaching and cementation. *Sustain. Mater. Technol.* **2023**, *37*, e00652. [CrossRef]
9. Pavón, S.; Lorenz, T.; Fortuny, A.; Sastre, A.M.; Bertau, M. Rare earth elements recovery from secondary wastes by solid-state chlorination and selective organic leaching. *Waste Manag.* **2021**, *122*, 55–63. [CrossRef]
10. Bilesan, M.R.; Makarova, I.; Wickman, B.; Repo, E. Efficient separation of precious metals from computer waste printed circuit boards by hydrocyclone and dilution gravity methods. *J. Clean. Prod.* **2021**, *286*, 125505. [CrossRef]
11. Beiki, V.; Naseri, T.; Mousavi, S.M. Comprehensive characterization and environmental implications of spent telecommunication printed circuit boards: Towards a cleaner and sustainable environment. *J. Environ. Manag.* **2023**, *325*, 116482. [CrossRef] [PubMed]
12. Kamberovic, Z.; Korac, M.; Ranitovic, M. Hydrometallurgical Process for extraction of metals from electronic waste—Part II: Development of the processes for the recovery of copper from Printed Circuit Boards (PCB). *Metalurgija* **2011**, *17*, 139–149. Available online: https://www.researchgate.net/publication/267985628 (accessed on 5 August 2023).
13. Wang, H.; Zhang, S.; Li, B.; Pan, D.; Wu, Y.; Zuo, T. Recovery of waste printed circuit boards through pyrometallurgical processing: A review. *Resour. Conserv. Recycl.* **2017**, *126*, 209–218. [CrossRef]
14. Li, H.; Eksteen, J.; Oraby, E. Hydrometallurgical recovery of metals from waste printed circuit boards (WPCBs): Current status and perspectives—A review. *Resour. Conserv. Recycl.* **2018**, *139*, 122–139. [CrossRef]
15. Akcil, A.; Erust, C.; Gahan, C.S.; Ozgun, M.; Sahin, M.; Tuncuk, A. Precious metal recovery from waste printed circuit boards using cyanide and non-cyanide lixiviants—A review. *Waste Manag.* **2015**, *45*, 258–271. [CrossRef]
16. Lu, Y.; Xu, Z. Precious metals recovery from waste printed circuit boards: A review for current status and perspective. *Resour. Conserv. Recycl.* **2016**, *113*, 28–39. [CrossRef]
17. Rocchetti, L.; Amato, A.; Beolchini, F. Printed circuit board recycling: A patent review. *J. Clean. Prod.* **2018**, *178*, 814–832. [CrossRef]
18. Baniasadi, M.; Vakilchap, F.; Bahaloo-Horeh, N.; Mousavi, S.M.; Farnaud, S. Advances in bioleaching as a sustainable method for metal recovery from e-waste: A review. *J. Ind. Eng. Chem.* **2019**, *76*, 75–90. [CrossRef]

19. Rai, V.; Liu, D.; Xia, D.; Jayaraman, Y.; Gabriel, J.-C.P. Electrochemical approaches for the recovery of metals from electronic waste: A critical review. *Recycling* **2021**, *6*, 53. [CrossRef]
20. Korf, N.; Løvik, A.N.; Figi, R.; Schreiner, C.; Kuntz, C.; Mählitz, P.M.; Rösslein, M.; Wäger, P.; Rotte, V.S. Multi-element chemical analysis of printed circuit boards—Challenges and pitfalls. *Waste Manag.* **2019**, *92*, 124–136. [CrossRef]
21. Qiu, R.; Lin, M.; Ruan, J.; Fu, Y.; Hu, J.; Deng, M.; Tang, Y.; Qiu, R. Recovering full metallic resources from waste printed circuit boards: A refined review. *J. Clean. Prod.* **2020**, *244*, 118690. [CrossRef]
22. Krishnan, S.; Zulkapli, N.S.; Kamyab, H.; Mat Taib, S. Current technologies for recovery of metals from industrial wastes: An overview. *Env. Technol. Innov.* **2021**, *22*, 101525. [CrossRef]
23. Mishra, G.; Jha, R.; Rao, M.D.; Meshram, A.; Singh, K.K. Recovery of silver from waste printed circuit boards (WPCBs) through hydrometallurgical route: A review. *Environ. Chall.* **2021**, *4*, 100073. [CrossRef]
24. Mir, S.; Dhawan, N. A comprehensive review on the recycling of discarded printed circuit boards for resource recovery. *Resour. Conserv. Recyc.* **2022**, *178*, 106027. [CrossRef]
25. Wu, C.; Awasthi, A.K.; Qin, W.; Liu, W.; Yang, C. Recycling value materials from waste PCBs focus on electronic components: Technologies, obstruction and prospects. *J. Environ. Chem. Eng.* **2022**, *10*, 108516. [CrossRef]
26. Zhao, W.; Xu, J.; Fei, W.; Liu, Z.; He, W.; Li, G. The reuse of electronic components from waste printed circuit boards: A critical review. *Environ. Sci. Adv.* **2023**, *2*, 196–214. [CrossRef]
27. Do, M.H.; Nguyen, G.T.; Thach, U.D.; Lee, Y.; Bui, T.H. Advances in hydrometallurgical approaches for gold recovery from E-waste: A comprehensive review and perspectives. *Miner. Eng.* **2023**, *191*, 107977. [CrossRef]
28. Niu, B.; Shanshan, E.; Xu, Z.; Guo, J. How to efficient and high-value recycling of electronic components mounted on waste printed circuit boards: Recent progress, challenge, and future perspectives. *J. Clean. Prod.* **2023**, *415*, 137815. [CrossRef]
29. Hau Nguyen, T.N.; Lee, M.S. Recovery of Au and Pd from the letching solution of printed circuit boards by cementation, solvent extraction, reduction, and precipitation. *J. Ind. Eng. Chem.* **2023**, *126*, 214–223. [CrossRef]
30. Ly, N.H.; Joo, S.-W.; Choo, J.; Vasseghian, Y.R.; Cho, J.; Rezania, S. Sustainable cutting-edge techniques for gold valorization from electronic wastes. *Chem. Eng. J.* **2023**, *471*, 44324. [CrossRef]
31. Zhao, H.; Zhang, Q.; Li, G.; Zhang, X.; Pan, D. Bromine recovery from crude bromine salts by-products of waste printed circuit boards smelting. *Sep. Purif. Technol.* **2023**, *320*, 124190. [CrossRef]
32. Klose, S.; Pauliuk, S. Sector-level estimates for global future copper demand and the potential for resource efficiency. *Resour. Conserv. Recycl.* **2023**, *193*, 106941. [CrossRef]
33. Li, Y.; An, H.; Gao, X.; Liu, S.; Sun, Q.; Zhao, Y. The influence of copper trade relation structure on copper price: From the perspective of industrial chain. *Resour. Conserv. Recycl.* **2023**, *192*, 106933. [CrossRef]
34. Barnwal, A.; Dhawan, N. Recycling of discarded mobile printed circuit boards for extraction of gold and copper. *Sustain. Mat. Technol.* **2020**, *25*, e00164. [CrossRef]
35. Hao, J.; Wang, X.; Wang, Y.; Guo, F.; Wu, Y. Study of gold leaching from pre-treated waste printed circuit boards by thiosulfate-cobalt-glycine system and separation by solvent extraction. *Hydrometallurgy* **2023**, *221*, 106141. [CrossRef]
36. Dong, Z.; Jiang, T.; Xu, B.; Zhang, B.; Liu, G.; Li, Q.; Yang, Y. A systematic and comparative study of copper, nickel and cobalt-ammonia catalysed thiosulfate processes for eco-friendly and efficient gold extraction from an oxide gold concentrate. *Sep. Purif. Technol.* **2021**, *272*, 118929. [CrossRef]
37. Liu, Z.; Guo, X.; Tian, Q.; Zhang, L. A systematic review of gold extraction: Fundamentals, advancements, and challenges toward alternative lixiviants. *J. Hazardous Mater.* **2022**, *440*, 129778. [CrossRef] [PubMed]
38. Wang, R.; Liu, C.; Yang, Y.; Zhou, J.; Guo, Y.; Xie, B. Pretreatment of refractory gold concentrate calcine using ammonium bifluoride and sulfuric acid solution. *Miner. Eng.* **2022**, *187*, 107778. [CrossRef]
39. Cotty, S.R.; Kim, N.; Su, X. Electrochemically Mediated Recovery and Purification of Gold for Sustainable Mining and Electronic Waste Recycling. *ACS Sustain. Chem. Eng.* **2023**, *11*, 3975–3986. [CrossRef]
40. Jha, R.; Sharma, R.; Agrawal, M.; Rao, M.D.; Singh, K.K. Exploring the pretreatment routes of waste printed circuit boards for enhanced metal recovery. *Mater. Today Proc.* 2023, *in press*. [CrossRef]
41. Barrueto, Y.; Hernández, P.; Jiménez, Y.; Morales, J. Leaching of metals from printed circuit boards using ionic liquids. *J. Mat. Cyc. Waste Manag.* **2021**, *23*, 2028–2036. [CrossRef]
42. Kilicarslan, A.; Nezihi Saridede, M.; Stopic, S.; Friedrich, B. Use of ionic liquid in leaching process of brass wastes for copper and zinc recovery. *Int. J. Miner. Metall. Mater.* **2014**, *21*, 138–143. [CrossRef]
43. Wstawski, S.; Emmons-Burzyńska, M.; Rzelewska-Piekut, M.; Skrzypczak, A.; Regel-Rosocka, M. Studies on copper (II) leaching from e-waste with hydrogen sulfate ionic liquids: Effect of hydrogen peroxide. *Hydrometallurgy* **2021**, *205*, 105730. [CrossRef]
44. Zhang, D.-J.; Dong, L.; Li, Y.-T.; Wu, Y.; Ma, Y.-X.; Yang, B. Copper leaching from waste printed circuit boards using typical acidic ionic liquids recovery of e-wastes' surplus value. *Waste Manag.* **2018**, *78*, 191–197. [CrossRef]
45. Masilela, M.; Ndlovu, S. Extraction of Ag and Au from chloride electronic waste leach solutions using ionic liquids. *J. Environ. Chem. Eng.* **2019**, *7*, 102810. [CrossRef]
46. Łukomska, A.; Wiśniewska, A.; Dąbrowski, Z.; Lach, J.; Wróbel, K.; Kolasa, D.; Domańska, U. Recovery of metals from electronic waste—Printed circuit boards by ionic liquids, DESs and organophosphorous-based acid extraction. *Molecules* **2022**, *27*, 4984. [CrossRef]

47. Schaeffer, N.; Passos, H.; Billard, I.; Papaiconomou, N.; Coutinho, J.A.P. Recovery of metals from waste electrical and electronic equipment (WEEE) using unconventional solvents based on ionic liquids. *Environ. Sci. Technol.* **2018**, *48*, 859–922. [CrossRef]
48. Aliakbari, R.; Marfavi, Y.; Kowsari, E.; Ramakrishna, S. Recent studies on ionic liquids in metal recovery from e-waste and secondary sources by liquid-liquid extraction and electrodeposition: A Review. *Mater. Circ. Econ.* **2020**, *2*, 10–18. [CrossRef]
49. Niknam, K.; Damya, M. 1-Butyl-3-methylimidazolium hydrogen sulfate [Bmim]HSO_4: An efficient reusable acidic ionic liquid for the synthesis of 1,8-dioxo-octahydroxanthenes. *J. Chin. Chem. Soc.* **2009**, *56*, 659–665. [CrossRef]
50. Huang, J.; Chen, M.; Chen, H.; Chen, S.; Sun, Q. Leaching behavior of copper from waste printed circuit boards with Brønsted acidic ionic liquid. *Waste Manag.* **2014**, *34*, 483–488. [CrossRef] [PubMed]
51. Barruetoc, Y.; Hernández, P.; Jiménezeb, Y.P.; Moralesc, J. Properties and application of ionic liquids in leaching base/precious metals from e-waste. A review. *Hydrometallurgy* **2022**, *212*, 105895. [CrossRef]
52. Łukomska, A.; Wiśniewska, A.; Dąbrowski, Z.; Domańska, U. Liquid-liquid extraction of cobalt(II) and zinc(II) from aqueous solutions using novel ionic liquids as extractants. *J. Mol. Liq.* **2020**, *307*, 112955. [CrossRef]
53. Łukomska, A.; Wiśniewska, A.; Dąbrowski, Z.; Kolasa, D.; Luchcińska, S.; Domańska, U. Separation of cobalt, lithium and nickel from the "black mass" of waste Li-ion batteries by ionic liquids, DESs and organophosphorous-based acids extraction. *J. Mol. Liq.* **2021**, *343*, 117694. [CrossRef]
54. Hartley, J.M.; Scott, S.; Rivera, R.M.; Hunt, P.; Lucio, A.J.; Bird, P.; Harris, R.; Jenkin, G.R.T. Tailoring lixiviant properties to optimise selectivity in E-waste recycling. *RSC Sustain.* **2023**, *1*, 107–116. [CrossRef]
55. Feng, F.; Sun, Y.; Rui, J.; Yu, L.; Liu, J.; Zhang, N.; Zhao, M.; Wei, L.; Lu, C.; Zhao, J.; et al. Study of the "Oxidation-Complexation" coordination composite ionic liquid system for dissolving precious metal. *Appl. Sci.* **2020**, *10*, 3625. [CrossRef]
56. Li, H.; Oraby, E.; Eksteen, J. Extraction of precious metals from waste printed circuit boards using cyanide-free alkaline glycine solution in the presence of an oxidant. *Miner. Eng.* **2022**, *181*, 107501. [CrossRef]
57. Han, Y.; Yi, X.; Wang, R.; Huang, J.; Chen, M.; SUN, Z.; Sun, S.; Shu, J. Copper extraction from waste printed circuit boards by glicyne. *Sep. Purif. Technol.* **2020**, *253*, 117463. [CrossRef]
58. Martinez-Ballesteros, G.; Valenzuela-Garci, J.L.; Gomez-Alvarez, A.; Encinas-Romero, M.A.; Mejia-Zamudio, F.A.; de Jesus Rosas-Durazo, A. Base Metals Extraction from Printed Circuit Boards by Pressure Acid Leaching. *Minerals* **2023**, *13*, 98. [CrossRef]
59. Correa, M.M.J.; Silvas, F.P.C.; Aliprandini, P.; Tavares de Moraes, V.; Dreisinger, D.; Espinosa, D.C.R. Separation of copper from a leaching solution of printed circuit boards by using solvent extraction with D2EHPA. *Braz. J. Chem. Eng.* **2018**, *35*, 919–930. [CrossRef]
60. Abbott, A.P.; Boothby, D.; Capper, G.; Davies, D.L.; Rasheed, R.K. Deep eutectic solvents formed between choline chloride and carboxylic acids: Versatile alternatives to ionic liquids. *J. Am. Chem. Soc.* **2004**, *126*, 9142–9147. [CrossRef]
61. Łukomska, A.; Wiśniewska, A.; Dąbrowski, Z.; Kolasa, D.; Luchcińska, S.; Lach, J.; Wróbel, K.; Domańska, U. Recovery of zinc and manganese from "black mass" of waste Zn-MnO_2 alkaline batteries by solvent extraction technique with ionic liquids, DESs and organophosphorous-based acids. *J. Mol. Liq.* **2021**, *338*, 116590. [CrossRef]
62. Domańska-Żelazna, U.; Wiśniewska, A.; Dąbrowski, Z. New ammonium ionic liquids, methods of synthesis and their application. *Pol. Pat. Appl.* **2023**, *198*, 444.
63. Boudesocque, S.; Mohamadou, A.; Dupont, L.; Martinez, A.; Dechamps, I. Use of dicyanamide ionic liquids for extraction of metal ions. *RSC Adv.* **2016**, *6*, 107894. [CrossRef]
64. Domańska-Żelazna, U.; Wiśniewska, A.; Dąbrowski, Z. The method of extraction of silver, copper, aluminium, iron and zinc from electronic waste with ionic liquids. *Pol. Pat. Appl.* **2023**, *199*, 444.

processes

MDPI

Article

Solvent Extraction of Gold(III) by 2-Ethylhexanol and Modeling of Facilitated Transport across a Supported Liquid Membrane

Francisco Jose Alguacil * and Jose Ignacio Robla *

Centro Nacional de Investigaciones Metalurgicas (CSIC), Avda. Gregorio del Amo 8, 28040 Madrid, Spain
* Correspondence: fjalgua@cenim.csic.es (F.J.A.); jrobla@cenim.csic.es (J.I.R.)

Abstract: The solvent extraction of gold(III) by undiluted 2-ethylhexanol or dissolved in toluene from a HCl solution has been investigated in this paper. The numerical analysis of gold distribution data suggests the formation of $HAuCl_4 \cdot L$ and $HAuCl_4 \cdot 2L$ (L = 2-ethylhexanol) species in the organic phase, with formation constants K_{11} = 38 and K_{12} = 309, respectively. The results derived from gold(III) distribution have been implemented in a solid-supported liquid membrane system. The influence of several variables on gold transport has been considered: feed and receiving phases' stirring speeds, HCl and gold concentrations in the feed phase, and carrier concentration in the membrane phase as well as the presence of base metals (Fe, Cu, Ni) and platinum-group metals (PGMs) in the feed phase. Gold transport is influenced by the stirring speed of the feed phase and the variation in HCl and gold (feed phase) and carrier (membrane phase) concentrations. Also, diffusional resistances to mass transfer are estimated, and the contribution of each resistance to gold transport is estimated. Gold is recovered as zero-valent nanoparticles.

Keywords: gold(III); 2-ethylhexanol; extraction; membrane transport; gold nanoparticles

Citation: Alguacil, F.J.; Robla, J.I. Solvent Extraction of Gold(III) by 2-Ethylhexanol and Modeling of Facilitated Transport across a Supported Liquid Membrane. *Processes* **2024**, *12*, 771. https://doi.org/10.3390/pr12040771

Academic Editors: Yanlin Zhang and Prashant K. Sarswat

Received: 29 February 2024
Revised: 9 April 2024
Accepted: 9 April 2024
Published: 11 April 2024

1. Introduction

Over the last few years, there has been an increasing interest in the recovery and utilization of so-called strategic metals (rare earth elements, lithium, neodymium, tantalum, etc.); less mentioned was the recovery, from secondary materials or waste, of gold and platinum-group metals, which have always been of interest.

The recovery of gold has two main tendencies: (i) recovery from raw materials, which is dominated by cyanidation and gold adsorption, and (ii) recovery from secondary materials or waste (printed circuit boards, jewelry scraps, e-waste, etc.), in which the precious metal is usually dissolved in a non-cyanide medium, i.e., aqua regia, thiourea, thiosulphate, etc. From the leachates, gold is purified and recovered via different options.

Though yet with no apparent or definitive practical applications, the scientific community has considered the recovery of gold from these various non-cyanide media important, an importance which is reflected by the series of publications which, every year, appear in journals. With respect to gold adsorption onto different materials, several recent approaches have been considered, including the use of metal organic frameworks [1], metal sulfide microspheres [2], or Aliquat 336 (quaternary ammonium salt) impregnating a Nitrolite adsorbent [3].

The recovery of metals by solvent extraction processing is well known; this is also demonstrated by the number of publications on the subject over the years. Gold(III) solvent extraction has attracted interest, with recent publications discussing the use of ionic liquids [4] or pseudo-protic ionic liquids [5] as extractants for this precious metal.

Due to some environmental and operational concerns (i.e., solvent extraction is not adequate in the treatment of dilute solutions), liquid membranes have emerged as an alternative to solvent extraction; however, it must be said that, at the present time, there is still no plant using any of the different liquid membrane operational variations. Indeed, the interest of scientists in the utilization of these membranes technologies on the recovery of metals or other pollutants is still alive, as some recent publications have also demonstrated. Several reviews about the general use of liquid membranes included the transport of actinide elements using diglycolamides as carriers [6] and the application of membranes in the recovery of various elements [7–10], with a mention of the use of polymer inclusion membranes (PIMs) [11]. Experimental papers have been devoted to investigating the transport of both inorganic and organic solutes or harmful species potentially appearing in liquid waste. As organic compounds, the transport of diclofenac and ibuprofen by supported liquid membrane technology and Cyanex 923 (mixture of tri-alkylphosphine oxides) dissolved in kerosene as the carrier phase has been investigated [12,13]. Using a flat-sheet, supported liquid membrane operational mode, the transport of chromium(VI), indium(III), and the separation of Nd(III) and Er(III) have been investigated [14–16]; a variation of the above operation has been found in the investigation of the transport process (copper, zinc, or cobalt) using electromembranes [17–19].

The utilization of a more dynamic, supported liquid membrane procedure, such as the use of hollow fiber modules, has found applications in the simultaneous removal of arsenic and mercury [20] or only mercury [21]. A yet-more-advanced hollow fiber application is the so-called (though other names have been used) pseudo-emulsion-based hollow fiber membrane with strip-dispersion technology (PEHFMSD), in which, differently to conventional hollow fiber use, employing two modules (one for extraction and another for stripping), just one module is used for extraction and stripping. Moreover, in this PEHFMSD process, the organic and the receiving or stripping phase are mixed in one tank and continuously pumped into the module to improve the transport and removal of the solute from the feed to the organic and from the organic to the receiving phases. This PEHFMSD process was recently used in the removal of iron(III) [22] and in the treatment of stainless steel rinse waters [23]. It is worth noting that, against all the literature favoring the use of liquid membranes in the treatment of wastewaters, these technologies seem unpopular in the case of acid mine wastewater treatment [24].

In the application of these membranes, it is of the utmost importance to gain knowledge on the interfacial phenomena associated with metal carrier's (extractant) chemical reactions occurring in the feed phase and the membrane interfaces when the metal (solute) is transported against its concentration gradient. Before scaling up the use of flat-sheet, supported liquid membranes (the simplest form of supported liquid membranes) to spiral wound or hollow fiber module operational modes, a theoretical model of the membrane system is welcomed to design an efficient and stable process.

The present work implemented solvent extraction investigations in a solid, supported liquid membrane configuration. Firstly, a quantitative characterization of the solvent extraction of gold(III) by 2-ethylhexanol was undertaken. Further, a liquid membrane system was designed using the extraction process mentioned above, and the parameters affecting the liquid membrane—stirring speeds applied in both feed and receiving phases, composition of the feed and organic phases, etc.—were investigated. A model was developed to estimate the mass transfer coefficients relative to the transport of gold across the 2-ethylhexanol-immobilized supported liquid membrane.

2. Materials and Methods

2.1. Materials

A stock solution (1 g/L) of gold(III) was prepared by dissolving $HAuCl_4$ (Fluka, Buchs, Switzerland) in HCl. The aqueous solutions for solvent extraction or the membrane transport experiments were prepared by the dilution of this stock solution. The extractant 2-ethylhexanol (Merck, Rahway, NJ, USA) was used without further purification. The PGM stock solution was a commercially available standard solution for ICP analysis. All the other reagents used in this investigation were of AR grade.

The support for the liquid membrane was a polydifluoroethylene film (Millipore GVHP4700, Darmstadt, Germany) with a thickness of 125×10^{-4} cm, a porosity of 75%, a tortuosity of 1.67, and a 0.22 μm average pore size.

2.2. Methods

2.2.1. Liquid–Liquid Extraction Experiments

Before experimentation, undiluted 2-ethylhexanol was pre-saturated with 6 M HCl in order to avoid a change in the organic and aqueous phases' volume after gold extraction, and, from this pre-saturated alcoholic solution, all the organic phases used in the investigation were prepared. Distribution ratio experiments were performed at 20 °C by shaking (800 min^{-1}) equal volumes (25 cm^3) of the aqueous and organic phases in glass separatory funnels for the required time. After phase separation, the metal remaining in the raffinate was analyzed by AAS (±2.5 reproducibility) using a Perkin Elmer 1100B (Waltham, MA, USA) spectrophotometer. The amount of metals extracted in the organic phase was calculated by the difference with the initial concentration in the aqueous phase. The percentage of extraction was calculated as follows:

$$\%E = \frac{[M]_{aq,0} - [M]_{aq,t}}{[M]_{aq,0}} \times 100 \quad (1)$$

where $[M]_{aq,0}$ and $[M]_{aq,t}$ are the metal concentrations in the initial aqueous phase and in the raffinate, respectively.

2.2.2. Liquid Membrane Experiments

The batch transport experiments were carried out in a methacrylate permeation cell having two (200 cm^3 each) compartments separated by a solid support (Figure 1). One compartment was for the feed phase and the other for the receiving or stripping phase. The liquid membrane was prepared by impregnation of the laminar microporous support by the carrier phase over 24 h and then by leaving it to drip for twenty seconds before placing it in the cell. The metal contents in the feed and receiving phases were periodically analyzed by AAS or ICP (platinum-group metals) (±1.5 reproducibility), whereas the overall mass transfer coefficient (K_0) was calculated as follows:

$$\ln\frac{[M]_{f,t}}{[M]_{f,0}} = -\frac{A}{V}K_0 t \quad (2)$$

where $[M]_{f,0}$ and $[M]_{f,t}$ are the metal concentrations in the feed at time zero and at an elapsed time, respectively; A is the effective membrane area 11.3 cm^2; V is the volume of the feed phase (200 cm^3); and t is the elapsed time.

Figure 1. Permeation cell. A: feed phase; B: membrane support; and C: stripping phase.

3. Results and Discussion

3.1. Solvent Extraction Experiments

Firstly, the influence of the equilibration time on the attachment gold(III) extraction equilibrium was studied. In these series of experiments, the aqueous phase contained 35 g/L Au(III) in 6 M HCl, whereas the organic phase was of 75% *v/v* 2-ethylhexanol in toluene. From the results shown in Table 1, it can be concluded that equilibrium was almost reached within 2 min of contact between both phases, and the same conclusion was reached when the aqueous phase contained other (15–100 g/L) gold concentrations.

Table 1. Influence of equilibration time on the fractional attachment to the equilibrium.

Time, min	F
1	0.71
2	0.98
5	1
10	1
15	1

Temperature: 20 °C. Org/Aq volume ratio: 1.

In Table 1, the fractional attachment to the equilibrium (F) was calculated as follows:

$$F = \frac{[Au]_{org,t}}{[Au]_{org,e}} \quad (3)$$

where $[Au]_{org,t}$ and $[Au]_{org,e}$ are the gold concentrations in the organic phase at an elapsed time and at the equilibrium, respectively. According to these experimental data, all the subsequent experiments were carried out over 5 min of contact between the respective organic and aqueous phases.

The variation in the initial gold concentration from the extraction of this precious metal from a 6 M HCl medium was also investigated. In these cases, the initial aqueous solutions contained varying gold concentrations in the 5–200 g/L range, and the organic phases were of undiluted 2-ethylhexanol. The results from this set of experiments are shown in Table 2.

Table 2. Percentage of gold extraction at various initial metal concentrations in the aqueous phase.

$[Au]_0$, g/L	% Extraction
5	98
15	98
25	98
35	98
50	98
75	97
100	96
150	93
200	87

Temperature: 20 °C. Org/Aq volume ratio: 1.

These results concluded that, in the 5–100 g/L gold concentration range, the variation in the percentage of gold extraction during the organic phase was almost negligible (96–98%), whereas, at higher initial gold concentrations, in the aqueous solutions, these percentages decreased from 93% at a 100 g/L gold concentration to 87% at a 200 g/L gold concentration. This decrease in the percentage of gold extraction with the increase in the initial metal concentration is not rare [25–28] and can be due to the polymerization of the metal in the aqueous phase and also to the common ion effect, according to which the organic phase is dominated by the metal anion ($AuCl_4^-$ in this work) and the aqueous phase by the halide ions (Cl^- in this case) [29].

The influence of the variation in the initial extractant concentration on gold extraction was also investigated using organic phases containing 25–90% v/v 2-ethylhexanol in toluene or an undiluted extractant. The aqueous phase was 35 g/L Au(III) in a 6 M HCl medium; the results from these experiments are shown in Table 3.

Table 3. Percentage of gold extraction at various extractant concentrations in the organic phase.

[2-Ethylhexanol], % v/v	% Extraction
25	57
35	75
50	88
60	92
70	94
75	95
80	96
90	98
undiluted	98

Temperature: 20 °C. Org/Aq volume ratio: 1.

These results indicated that there was an increase in the percentage of gold extraction with the increase in the extractant concentration in the organic phase from 25 to 60% v/v, though, for extractant concentrations higher than 70% v/v, the variation in gold extraction did not exceed 5%.

The stoichiometry of the extracted species and their equilibrium constants were determined through the treatment of the experimental data using a numerical program, which minimized the U function, defined as follows:

$$U = \sum \log(D_{cal} - D_{exp})^2 \tag{4}$$

where D_{cal} and D_{exp} are the respective distribution ratios calculated from the mass balance equations for the various models, experimentally calculated as

$$D_{exp} = \frac{[Au]_{org}}{[Au]_{aq}}. \tag{5}$$

The results of the numerical treatment indicated that the extraction of gold(III) can be explained (U = 0.009) by the formation of $HAuCl_4 \cdot L$ ($\log K_{ext} = 1.58 \pm 0.09$) and $HAuCl_4 \cdot 2L$ ($\log K_{ext} = 2.49 \pm 0.26$) species in the gold-loaded organic phases. In the above formulations, L represents 2-ethylhexanol molecules.

Once gold had been extracted in the organic phase, a stripping operation was performed using water as the stripping agent. In this case, from an organic phase of undiluted 2-ethylhexanol containing 34 g/L gold, using the same volume of water (Org/Aq = 1), near 89% of gold was recovered at 20 °C in one step. From these solutions, gold can be recovered by precipitation with sodium borohydride, oxalic acid, etc.; however, this last step was not investigated in our study, unlike in the case of solutions derived from supported liquid membrane experimentation (see further).

The selectivity of the present system against the presence of Cu(II), Zn(II), and Ni(II) in the aqueous phase was also investigated. These experiments used organic phases of undiluted 2-ethylhexanol and aqueous phases containing, separately, 0.18 M of the element in 6 M HCl. The results from this set of experiments are summarized in Table 4.

The separation factor (SF) values showed the extreme selectivity of the system; these separation factors were calculated using the following equation:

$$SF = \frac{D_{Au}}{D_M}. \tag{6}$$

Table 4. Extraction of Au(III), Cu(II), Fe(III), and Ni(II).

Element	% Extraction	D	SF
Au(III)	98	69	
Cu(II)	21	0.25	276
Zn(II)	11	0.12	575
Ni(II)	12	0.13	530

Temperature: 20 °C. Org/Aq volume ratio: 1.

Further experiments were performed to explain the separation of Au(III)/Cu(II) using this same organic phase, an aqueous phase containing both elements in 6 M HCl, and various Au(III):Cu(II) molar concentration ratios. Table 5 summarizes these results.

Table 5. Extraction and separation of Au(III) and Cu(II) at various molar concentration ratios.

$[Au]_0$:$[Cu]_0$	D_{Au}	D_{Cu}	SF
2:1	69	0.045	1533
3:1	69	0.047	1468
6:1	69	0.044	1568
18:1	69	0.049	1408

$[Au]_0$: 0.18 M; temperature: 20 °C; and Org/Aq volume ratio: 1.

These results show that, in mixed Au(III)-Cu(II) solutions, the extraction of this last element decreased with respect to the results obtained from the use of single-metal-bearing solutions. This is why the SF values resulted to be in the 1400–1500 order, though the variation in the $[Au(III)]_0$:$[Cu(II)]_0$ ratio did not seem to greatly influence this separation.

3.2. Supported Liquid Membrane Experiments

In these membrane systems, the extraction (transport) of a given metal species depends not only on the equilibrium parameters but also on the kinetics parameters.

The influence of the variation (600–1600 rpm) in the stirring speed of the feed phase on gold transport was investigated. The results from these experiments are shown in Table 6, and it can be observed that the overall mass transfer coefficient value (K_O) increased with the increase in the stirring speed from 600 to 1000 rpm, and, beyond that, no increase in gold transport was observed. In fact, there was a decrease in transport from 1400 rpm due to membrane instability, probably attributable to the displacement of the organic phase from the membrane pores caused by the turbulence from these higher stirring speeds.

Table 6. Influence of the feed-phase stirring speed on gold transport.

Stirring Speed, rpm	$K_O \times 10^3$, cm/s
600	0.95
800	2.6
1000	5.7
1200	5.7
1400	5.5
1600	4.9

Feed phase: 0.01 g/L Au(III) in 6 M HCl. Membrane phase: 50% v/v 2-ethylhexanol in toluene immobilized on GVHP4700 support. Receiving phase: water. Receiving-phase stirring speed: 1200 rpm. Temperature: 20 °C.

In supported liquid membrane experimentation, two types of diffusional resistance are usually found: (i) one due to the feed-phase boundary layer, and (ii) another in relation to the membrane support. Often, the magnitudes of the values of both resistances compete between them [30]. These experimental results show that, in the 1000–1200 rpm range, the feed-phase boundary layer reaches a minimum, the aqueous resistance to mass transfer should be minimized, and, thus, the diffusion contribution of the aqueous species to mass transfer phenomena is constant [31].

The influence of the variation (800–1200 rpm) in the stirring speed of the receiving phase on gold transport was also investigated using the same experimental conditions as in Table 6 but with a stirring speed of 1000 rpm in the feed phase. The results from these experiments showed that the variation in the stirring speed applied to the receiving phase had a negligible influence on gold transport. In the case of the receiving phase and if the stirrer in the receiving compartment was close to the support, the thickness of the boundary layer was minimized, and the resistance in the receiving side was neglected [32]. Thus, subsequent experiments were performed using stirring speeds of 1000 rpm for both feed and receiving phases.

Another variable investigated was the variation (0.5–6 M) in the HCl concentration in the feed phase on gold transport, keeping the carrier concentration in the membrane support constant. The results are shown in Table 7. It can be seen that gold permeation increased when the acid concentration in the feed solution was increased up to 1 M, with metal permeation being independent of the HCl concentration in the 3–6 M range due to the following equilibrium:

$$HAuCl_4 \leftrightarrow AuCl_4^- + H^+ \quad (7)$$

where the equilibrium is shifted completely to the left, and $HAuCl_4$ is the predominant species in the feed solution.

Table 7. Gold(III) permeation at various HCl concentrations in the feed phase.

[HCl], M	$K_O \times 10^3$, cm/s
0.5	3.8
1	4.7
3	5.7
6	5.7

Feed phase: 0.01 g/L Au(III) and HCl. Membrane phase: 50% v/v 2-ethylhexanol in toluene immobilized on GVHP4700 support. Receiving phase: water. Temperature: 20 °C.

The influence of the variation in the carrier concentration on gold transport was investigated next. In these experiments, the feed phase was 0.01 g/l Au(III) in 6 M HCl, whereas the organic phase contained 10–75% v/v 2-ethylhexanol in toluene or undiluted extractant. Table 8 shows the variation in the overall mass transfer coefficient with the variation in the carrier concentration in the membrane phase.

Table 8. Gold transport at various carrier concentrations.

[Carrier], % v/v	$K_O \times 10^3$, cm/s	[a] % Gold Recovery
10	1.2	95
25	2.7	95
35	4.0	94
45	5.6	93
50	5.7	92
65	5.5	93
75	3.0	90
undiluted	1.2	89

Membrane support: GVHP4700. Receiving phase: water. [a] In the receiving phase, after 3 h.

These results demonstrated that gold transport increased from 10 to 50% v/v carrier concentration in the membrane phase, then leveled off, and, after, decreased at the highest carrier concentrations. These phenomena could be attributed to the fact that, at low carrier concentrations, metal transport was dominated by membrane diffusion, whereas, in the 50% v/v range, the contribution of membrane diffusion was negligible and gold transport was controlled by the diffusion in the stagnant film of the feed phase. At this maximum K_O value [33],

$$K_O = \frac{D_{aq}}{d_{aq}} \tag{8}$$

where D_{aq} represents the average aqueous diffusion coefficient (10^{-5} cm^2/s) of the metal species in the feed phase [34], and d_{aq} is the minimum thickness of the feed phase boundary layer. Thus, d_{aq}, for the present system, was estimated at 1.8×10^{-3} cm. The decrease in gold transport at the highest carrier concentration was due to the increase in the organic-phase viscosity, which decreased the Au(III)-2-ethylhexanol complexes' diffusion coefficients values [35]. Moreover, the diffusion coefficients of the gold carrier species in the membrane phase were calculated as follows [36]:

$$D_{org,Au} = \frac{J \cdot d_{org}}{[carrier]} \tag{9}$$

The results are summarized in Table 9. In this equation, J is the metal flux (Equation (10)), and d_{org} is the thickness of the membrane support (125×10^{-4} cm). It can be seen that the increase in the carrier concentration was accompanied by a decrease in the diffusion coefficient and, thus, a decrease in the corresponding value of the overall permeation coefficient (Table 8).

Table 9. Variation in the diffusion coefficient in the membrane phase with the carrier concentration.

Carrier Concentration, %*v*/*v* (M)	$D_{org,Au} \times 10^9$, cm^2/s
50 (3.2)	1.1
65 (4.2)	0.83
75 (4.8)	0.39
Undiluted (6.4)	0.12

During our investigation on the effect of the initial concentration (0.01–0.1 g/L) of gold, in the feed phase, on metal transport (Table 10), it was observed that the increase in the initial gold concentration was accompanied by a continuous decrease in metal transport, whereas the initial metal flux (J) was defined as follows:

$$J = K_O[Au]_0 \tag{10}$$

It initially increased from 0.005 to 0.04 g/L in the initial gold concentrations, and, beyond this concentrations range, J became almost independent of the initial gold concentration in the feed phase.

Table 10. Variation in gold transport and initial flux at various metal concentrations.

$[Au]_0$, g/L	$K_O \times 10^3$, cm/s	J, mol/cm²s	[a] % Gold Recovery
0.01	5.7	2.9	95
0.02	4.5	4.6	93
0.04	3.0	6.0	94
0.06	2.2	6.6	93
0.08	1.7	6.8	94
0.1	1.4	7.0	95

Feed phase: gold(III) in 6 M HCl. Membrane phase: 50% *v*/*v* 2-ethylhexanol in toluene immobilized on GVHP4700 support. Receiving phase: water. [a] In the receiving phase, after 3 h.

According to these results, at low metal concentrations (0.01–0.04 g/L), the transport process was controlled by the diffusion of gold species, whereas, in the highest metal concentrations range, the near-constant metal flux values were due to a change in the rate-determining step for the transport process. At these higher gold concentrations, the membrane became saturated by metal carrier species on the feed membrane interface, resulting in a lower effective membrane area, which led to a near-constant initial flux value.

Following the same considerations described in the literature [37], it can be concluded that the gold transport rate was determined by the rate of diffusion of the $HAuCl_4$ species across the feed-phase diffusion layer and the rate of diffusion of the gold carrier species through the liquid membrane. Thus, making the same assumptions as [38,39], a final expression for the overall mass transfer coefficient can be derived as follows:

$$K_O = \frac{K_{11}\left[ROH\right]_{org} + K_{12}\left[ROH\right]^2_{org}}{\Delta_{org} + \Delta_f\left(K_{11}\left[ROH\right]_{org} + K_{12}\left[ROH\right]^2_{org}\right)} \quad (11)$$

where Δ_{org} and Δ_f are the transport resistances due to diffusion across the membrane and the aqueous feed boundary layer, respectively. The above equation combined, in one expression, both the equilibrium and diffusional parameters involved in the gold(III) transport process from the feed phase, across supported liquid membranes, using 2-ethylhexanol as the carrier.

To estimate the values of the resistance to the mass transfer, the above equation was linearized, resulting in the following:

$$\frac{1}{K_O} = \Delta_f + \Delta_{org}\frac{1}{K_{11}\left[ROH\right]_{org} + K_{12}\left[ROH\right]^2_{org}} \quad (12)$$

A plot of $1/K_O$ versus $1/(K_{11}\left[ROH\right]_{org} + K_{12}\left[ROH\right]^2_{org})$ might result in a straight line, with slope Δ_{org} and intercept Δ_f. From the plot, it resulted that Δ_{org} and Δ_f were 2580 s/cm and 147 s/cm, respectively, with r^2 = 0.9895. The utility of Equation (12) in describing gold transport across the supported liquid membrane by 2-ethylhexanol is shown in Figure 2, where the experimental values and those calculated using this equation have been represented against the initial carrier concentration.

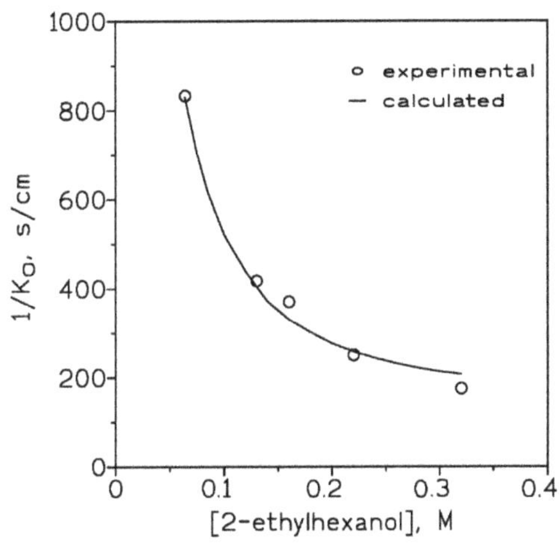

Figure 2. Plot of $1/K_O$ versus [2-ethylexanol]. Feed phase: 0.01 g/L Au(III) in 6 M HCl. Membrane phase: 10–50% *v*/*v* 2-ethylhexanol in toluene immobilized on GVHP4700. Receiving phase: water.

As the first term ($1/K_O$) of Equation (12) represents the value of the total resistance (R_T) and this resistance is the sum of the mass transfer resistances due to the feed and membrane phases, Equation (12) can be expressed as follows:

$$R_T = R_f + R_m. \tag{13}$$

The total resistance calculated from the experiments in Table 6 presented values in the 175–1052 s/cm range. In comparison, the total resistance calculated by the model is 206 s/cm, which indicates that the resistance due to the membrane is dominant at low stirring speeds.

The contribution of the fractional resistances due to each step of the total transport process, R_f^o and R_m^o, can be expressed using the following equations:

$$\%R_f^o = \frac{R_f}{R_T} \times 100 \tag{14}$$

$$\%R_m^o = \frac{R_T - R_f}{R_T} \times 100. \tag{15}$$

The values of $\%R_f^o$ and $\%R_m^o$ under various experimental conditions are summarized in Table 11.

Table 11. Contribution of the mass transfer resistances to the gold(III) transport process.

Experimental Condition	R_T, s/cm	[a] R_T, s/cm	$\%R_f^o$	$\%R_m^o$
Carrier 10–50% v/v	370–175	147	40–84	60–16
Gold 0.01–0.1 g/L	175–714	147	84–21	16–79
HCl 6 M	175	147	84	16

[a] Value from the theoretical model (Equation (12)).

Taking into account that

$$D_{org} = \frac{d_m}{\Delta_{org}}, \tag{16}$$

where D_{org} represents the diffusion coefficient of the gold-containing species in the organic phase immobilized on the solid support and d_m is the membrane thickness, then, D_{org} is found to be 4.8×10^{-6} cm^2/s. The diffusion coefficient of the gold transported species in the bulk organic phase can be estimated using the following equation [40]:

$$D_{org,b} = \frac{D_{org}\tau^2}{\epsilon} \tag{17}$$

where τ and ε are the membrane tortuosity and porosity values, respectively. The above resulted in $D_{org,b}$ being calculated as 1.8×10^{-5} cm^2/s. A comparison of the D_{org} and $D_{org,b}$ values showed that D_{org} was lower than $D_{org,b}$, which can be attributed to the diffusional resistance caused by the solid support separating the feed and receiving phases.

The selectivity of the present system against the presence of other metals in the feed phase was investigated. In this case, the feed phase contained 0.01 g/L each of Au(III), Ni(II), Cu(II), Fe(III), and PGMs (Ir, Os, Pd, Pt, Re, Rh, and Ru) in a 6 M HCl medium, with the organic phase being 50% v/v 2-ethylhexanol in toluene immobilized on a Durapore GVHP4700 support. Like in all the previous experimentations, water was used as the receiving phase. The results indicated that PGMs were not transported under the present experimental conditions, whereas the transport of gold and base metals followed the Au(III) > Fe(III) > Cu(II) = Ni(III) order, with separation factors Au/Fe, Au/Ni, and Au/Cu of 1.3, 45, and 47, respectively. In these transport experiments, the separation factors were calculated as follows:

$$SF = \frac{K_{O,Au}}{K_{O,M}} \tag{18}$$

However, the presence of all these elements in the feed phase produced a decrease in the overall gold mass transfer coefficient, from 5.7×10^{-3} cm/s to 2.3×10^{-3} cm/s. This behavior is not rare in supported liquid membrane processes and is attributable to multi-ion competition or the crowding effect [41].

Once gold is transported from the feed solution to the membrane phase and, finally, to the receiving phase, it can be recovered from this last solution by precipitation as zero-valent gold nanoparticles through the use of sodium borohydride. This procedure was used by one of the authors years ago [42]. The importance of the recovery of this precious metal in the form of nanoparticles has also been demonstrated [43], and the use of these nanoparticles in different fields (catalysis, medicine, sensors, etc.) has regularly appeared in the literature [44–47]. In the present work, the receiving solution containing 0.009 g/L gold was precipitated by the direct action of solid sodium borohydride, and, after the reaction had stopped, the dark solid was filtered. From this dry solid, the following images (Figures 3 and 4) were obtained.

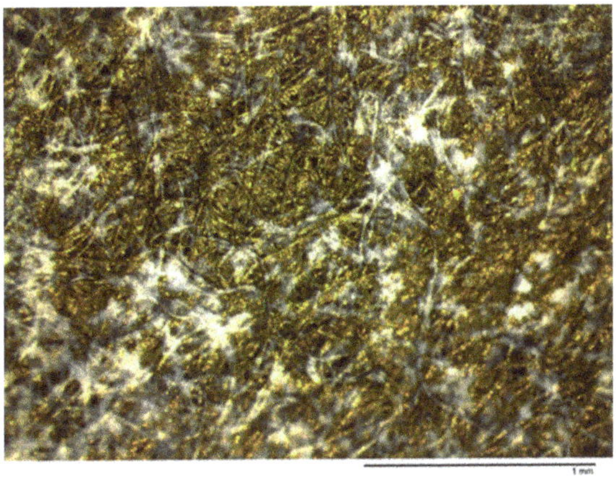

Figure 3. Gold nanoparticles under a magnifier.

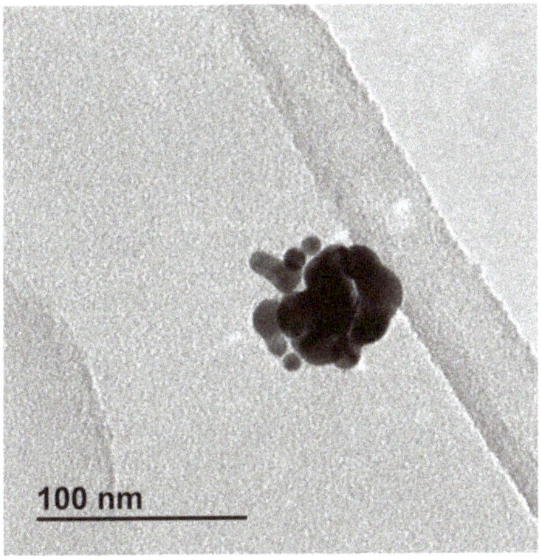

Figure 4. TEM image of gold nanoparticles. Some degree of agglomeration can be observed.

4. Conclusions

2-ethylhexanol was used to extract gold(III) from a 6 M HCl medium. The extraction process reached equilibrium within a few minutes, with metal extraction being dependent on gold and extractant concentrations. On the basis of numerical simulation, gold(III) extraction by 2-ethylhexanol can be explained by the formation of $HAuCl_4{\cdot}L$ and $HAuCl_4{\cdot}2L$, with extraction constants of 38 and 309, respectively, and L being represented as the extractant molecule. The extraction process is highly selective with respect to gold(III) in the presence of copper(II), nickel(II), and zinc(II). Gold can be stripped from the metal-loaded organic phase with water.

The extraction system was implemented in a solid, supported liquid membrane system in which the metal flux was dependent on the initial gold concentration (up to 0.04 g/L) but almost independent from this variable above this concentration. The metal's transport was also dependent on the extractant's concentration, but, for 2-ethylhexanol concentrations in the 50% v/v range in toluene, a limiting value for the overall mass transfer coefficient was obtained. The mass transfer coefficients in the feed and membrane phases were calculated as 6.8×10^{-3} cm/s and 3.9×10^{-4} cm/s, respectively. The minimum thickness of the feed boundary layer was calculated to be 1.8×10^{-3} cm. The transport process was controlled by the mixed diffusion of gold across the feed phase and the diffusion of gold–2-ethylhexanol complexes in the liquid membrane phase. The system was selective with respect to Au(III) in the presence of platinum-group metals Cu(II) and Ni(II) in the aqueous feed phase and, to a lesser extent, with respect to the presence of Fe(III) in this phase as well (Au-Fe separation factor of 1.3). From the water-receiving phase, gold was recovered as zero-valent gold nanoparticles using sodium borohydride.

Author Contributions: Conceptualization, F.J.A.; methodology, F.J.A.; formal analysis, F.J.A.; investigation, F.J.A. and J.I.R.; writing—original draft preparation, F.J.A.; writing—review and editing, F.J.A. and J.I.R.; funding acquisition, J.I.R. All authors have read and agreed to the published version of the manuscript.

Funding: This research was funded by the CSIC-Project 202250E019.

Data Availability Statement: Data are contained within the article.

Acknowledgments: We would like to thank the CSIC (Spain) for the support and M.I. Maher for checking the English of this paper.

Conflicts of Interest: The authors declare no conflicts of interest.

References

1. Hu, G.; Wang, Z.; Zhang, W.; He, H.; Zhang, Y.; Deng, X.; Li, W. MIL-161 metal–organic framework for efficient Au(III) recovery from secondary resources: Performance, mechanism, and DFT calculations. *Molecules* **2023**, *28*, 5459. [CrossRef] [PubMed]
2. Zhao, L.; Zhou, Q.; Yang, Y.; Zhang, Y.; Qiu, Y.; Chen, Y.; Jin, X.; Yang, X.; Wang, S. Ultra-high adsorption capacity and selectivity of photo-enhanced sulfur-rich M2S3 (M[dbnd]Bi and Sb) for gold recovery from electronic wastewater. *J. Water Proc. Eng.* **2024**, *57*, 104572. [CrossRef]
3. Hubicki, Z.; Zinkowska, K.; Wójcik, G. A new impregnated adsorbent for noble metal ion sorption. *Molecules* **2023**, *28*, 6040. [CrossRef]
4. Thombre, A.V.; Kundu, D. Ionic liquid promoted extraction of gold(III) from electronic waste: A modeling study. *Sep. Sci. Technol.* **2023**, *58*, 2641–2654. [CrossRef]
5. Alguacil, F.J.; Robla, J.I. On the use of pseudo-protic ionic liquids to extract gold(III) from HCl solutions. *Int. J. Mol. Sci.* **2023**, *24*, 6305. [CrossRef] [PubMed]
6. Ansari, S.A.; Mohapatra, P.K. Diglycolamides as highly efficient carrier ligands for actinide transport in supported liquid membranes. *Desalin. Water Treat.* **2023**, *285*, 47–66. [CrossRef]
7. Botelho Junior, A.B.; Tenório, J.A.S.; Espinosa, D.C.R. Separation of critical metals by membrane technology under a circular economy framework: A review of the state-of-the-art. *Processes* **2023**, *11*, 1256. [CrossRef]
8. Kaczorowska, M.A. The latest achievements of liquid membranes for rare earth elements recovery from aqueous solutions—A mini review. *Membranes* **2023**, *13*, 839. [CrossRef]
9. Kostanyan, A.E.; Voshkin, A.A.; Belova, V.V.; Zakhodyaeva, Y.A. Modelling and comparative analysis of different methods of liquid membrane separations. *Membranes* **2023**, *13*, 554. [CrossRef]
10. Rzelewska-Piekut, M.; Regel-Rosocka, M. Liquid membranes for separation of metal ions from wastewaters. *Phys. Sci. Rev.* **2023**, *8*, 937–982. [CrossRef]
11. Wang, B.; Wang, H.; Yan, J.; Wang, Y.; Xu, T. Research progress of polymer inclusion membrane in metal separation and recovery. *Huagong Jinzhan Chem. Ind. Eng. Prog.* **2023**, *42*, 3990–4004. [CrossRef]

12. Farah, M.; Giralt, J.; Stüber, F.; Font, J.; Fabregat, A.; Fortuny, A. Hollow fiber liquid membrane: A promising approach for elimination of pharmaceutical compounds from wastewater. *J. Environ. Chem. Eng.* **2023**, *11*, 111544. [CrossRef]
13. Farah, M.; Giralt, J.; Stüber, F.; Font, J.; Fabregat, A.; Fortuny, A. Intensification of diclofenac removal through supported liquid membrane and ozonation. *Environ. Technol. Innov.* **2024**, *33*, 103469. [CrossRef]
14. Alguacil, F.J.; Robla, J.I. Transport of chromium(VI) across a aupported liquid membrane containing Cyanex 921 or Cyanex 923 dissolved in Solvesso 100 as carrier phase: Estimation of diffusional parameters. *Membranes* **2023**, *13*, 177. [CrossRef]
15. Kumar, R.; Dhiman, S.; Gupta, H. Indium extraction from nitrate medium using Cyphos ionic liquid 104 and its mathematical modeling. *Environ. Sci. Pollut. Res.* **2023**, *30*, 107341–107349. [CrossRef] [PubMed]
16. Middleton, A.; Hsu-Kim, H. Separation of rare-earth elements by supported liquid membranes: Impacts of soluble iron, aluminum, and pH in low-grade feedstocks. *ACS EST Eng.* **2023**, *3*, 1197–1204. [CrossRef]
17. Kadhim, N.R.; Abbar, A.H.; Flayeh, H.M. Removal of copper from a simulated wastewater by electromembrane extraction technique using a novel electrolytic cell provided with a flat polypropylene membrane infused with 1-octanol and DEHP as a carrier. *Case Stud. Chem. Environ. Eng.* **2023**, *8*, 100430. [CrossRef]
18. Kadhim, N.R.; Flayeh, H.M.; Abbar, A.H. Zinc (II) removal from simulated wastewater by electro-membrane extraction approach: Adopting an electrolysis cell with a flat sheet supported liquid membrane. *J. Electrochem. Sci. Eng.* **2023**, *13*, 1097–1112. [CrossRef]
19. Kadhim, N.R.; Flayeh, H.M.; Abbar, A.H. A new approach for cobalt (II) removal from simulated wastewater using electro membrane extraction with a flat sheet supported liquid membrane. *Heliyon* **2023**, *9*, e22343. [CrossRef] [PubMed]
20. Suren, S.; Punyain, W.; Maneeintr, K.; Nootong, K.; Pancharoen, U. The simultaneous elimination of arsenic and mercury ions via hollow fiber supported liquid membrane and their reaction mechanisms: Experimental and modeling based on DFT and generating function. *Arab. J. Chem.* **2023**, *16*, 104501. [CrossRef]
21. Traiwongsa, N.; Suren, S.; Pancharoen, U.; Nootong, K.; Maneeintr, K.; Punyain, W.; Lothongkum, A.W. Mechanisms of mercury ions separation by non-toxic organic liquid membrane via DFT, thermodynamics, kinetics and mass transfer model. *J. Ind. Eng. Chem.* **2023**, *117*, 522–537. [CrossRef]
22. Alguacil, F.J.; Robla, J.I. Iron control in liquid effluents: Pseudo-emulsion based hollow fiber membrane with strip dispersion technology with pseudo-protic ionic liquid ($RNH_3^+HSO_4^-$) as mobile carrier. *Membranes* **2023**, *13*, 723. [CrossRef] [PubMed]
23. Alguacil, F.J.; Robla, J.I. Treatment of stainless steel rinse waters using non-dispersive extraction and strip dispersion membrane technology. *Membranes* **2023**, *13*, 902. [CrossRef] [PubMed]
24. Wang, Y.; Cao, J.; Biswas, A.; Fang, W.; Chen, L. Acid mine wastewater treatment: A scientometrics review. *J. Water Proc. Eng.* **2024**, *57*, 104713. [CrossRef]
25. Pacary, V.; Burdet, F.; Duchesne, M.T. Experimental and modeling of extraction of lanthanides in system HNO_3-TEDGA-{DMDOHEMA-HDEHP}. *Proc. Chem.* **2012**, *7*, 328–333. [CrossRef]
26. Zhang, J.; Liu, R.; Liu, X.; Li, J.; Wu, R.; Yang, Y. Study on extraction of Rh(III) by DABCO-based ionic liquid from hydrochloric acid medium. *Sep. Purif. Technol.* **2023**, *324*, 124578. [CrossRef]
27. Li, H.; Mabhiza, T.; Ning, Y.; Yu, Z.; Lv, C.; Shen, X.; Wei, G.; Qu, J. Solvent extraction and recovery of beryllium from hydrochloric acid solution with naphthenic acid. *Sep. Purif. Technol.* **2024**, *331*, 125594. [CrossRef]
28. Dong, H.; Ci, E.; Zhao, T.; Chen, P.; Liu, F.; Hu, G.; Yang, L. Hydrophobic deep eutectic solvents as the green media for highly efficient extraction of Cr(VI) over a broad pH range and low oil-water ratio. *Sep. Purif. Technol.* **2024**, *334*, 126104. [CrossRef]
29. Diamond, R.M. The solvent extraction behavior of inorganic compounds. III. Variation of the distribution quotient with metal ion concentration. *J. Phys. Chem.* **1957**, *61*, 75–81. [CrossRef]
30. Bohrer, M.P. Diffusional boundary layer resistance for membrane transport. *Ind. Eng. Chem. Fundam.* **1983**, *22*, 72–78. [CrossRef]
31. Alguacil, F.J.; Martinez, S. Permeation of iron(III) by an immobilised liquid membrane using Cyanex 923 as mobile carrier. *J. Membr. Sci.* **2000**, *176*, 249–255. [CrossRef]
32. Pavón, S.; Fortuny, A.; Coll, M.T.; Bertau, M.; Sastre, A.M. Permeability dependencies on the carrier concentration and membrane viscosity for Y(III) and Eu(III) transport by using liquid membranes. *Sep. Purif. Technol.* **2020**, *239*, 116573. [CrossRef]
33. Alguacil, F.J.; Alonso, M.; Lopez, F.; Lopez-Delgado, A. Uphill permeation of Cr(VI) using Hostarex A327 as ionophore by membrane-solvent extraction processing. *Chemosphere* **2008**, *72*, 684–689. [CrossRef] [PubMed]
34. El Aamrani, F.Z.; Kumar, A.; Beyer, L.; Cortina, J.L.; Sastre, A.M. Uphill permeation model of gold(III) and its separation from base metals using thioures derivatives as ionophores across a liquid membrane. *Hydrometallurgy* **1998**, *50*, 315–330. [CrossRef]
35. Gupta, S.K.; Rathore, N.S.; Sonawane, J.V.; Pabby, A.K.; Janardan, P.; Changrani, R.D.; Dey, P.K. Dispersion-free solvent extraction of U(VI) in macro amount from nitric acid solutions using hollow fiber contactors. *J. Membr. Sci.* **2007**, *300*, 131–136. [CrossRef]
36. Bromberg, L.; Levin, G.; Libman, J.; Shanzer, A. A novel tetradentate hydroxamate as ion carrier in liquid membranes. *J. Membr. Sci.* **1992**, *69*, 143–153. [CrossRef]
37. Sastre, A.M.; Madi, A.; Cortina, J.L.; Miralles, N. Modelling of mass transfer in facilitated supported liquid membrane transport of gold(III) using phospholene derivatives as carriers. *J. Membr. Sci.* **1998**, *139*, 57–65. [CrossRef]
38. Haghighi, H.K.; Irannajad, M.; Fortuny, A.; Sastre, A.M. Mathematical modeling on non-dispersive extraction of germanium from aqueous solutions using Aliquat 336. *Water Sci. Technol.* **2018**, *78*, 2489–2499. [CrossRef] [PubMed]
39. Alguacil, F.J. Mechanistic investigation of facilitated transport of gold(III) from HCl media using ionic liquid Cyphos IL102 as carrier across a supported liquid membrane. *Gold Bull.* **2019**, *52*, 145–151. [CrossRef]

40. Huang, T.-C.; Juang, R.-S. Rate and mechanism of divalent metal transport through supported liquid membrane containing di(2-ethylhexyl) phosphoric acid as a mobile carrier. *J. Chem. Technol. Biotechnol.* **1998**, *42*, 3–17. [CrossRef]
41. De Gyves, J.; De San Miguel, E.R. Metals separations by supported liquid membranes. *Ind. Eng. Chem. Res.* **1999**, *38*, 2182–2202. [CrossRef]
42. Alguacil, F.J.; Adeva, P.; Alonso, M. Processing of residual gold(III) solutions via ion exchange. *Gold. Bull.* **2005**, *38*, 9–13. [CrossRef]
43. Oestreicher, V.; García, C.S.; Soler-Illia, G.J.A.A.; Angelomé, P.C. Gold recycling at laboratory scale: From nanowaste to nanospheres. *ChemSusChem* **2019**, *12*, 4882–4888. [CrossRef] [PubMed]
44. Layek, K. Hydroxyapatite supported gold nanoparticles catalyzed efficient catalysis for the reduction of nitroarenes and degradation of azo dyes. *Catal. Surv. Asia* **2024**, *27*, 349–362. [CrossRef]
45. Cheng, C.-W.; Lee, S.-Y.; Zhan, S.-Q.; Huang, C.-L.; Chen, T.-Y.; Yuan, J.-M.P.; Huang, S.-T.; Chiu, C.-M.; Liang, J.-Y. The effect of photolysis of sodium citrate treated with gold chloride using coloured light on the generation of gold nanoparticles and the repression of WiDr colon cancer cells. *J. Photochem. Photobiol. B Biol.* **2024**, *251*, 112844. [CrossRef] [PubMed]
46. Hiral, V.; Kokila, P.; Jyotindra, M. Biogenic synthesis of gold nanoparticles using bark extract of Bauhinia variegata: Antibacterial and in vitro anticancer study. *Res. J. Chem. Environ.* **2024**, *28*, 48–56. [CrossRef]
47. Ajayi, R.F.; Nqunqa, S.; Ngema, N.P.P.; Barry, S.C.L.; Feleni, U.; Mulaudzi, T. UV–Vis detection of *E. coli* 0157:H7 using *Vitis vinifera* and *Musa paradaisica* modified Au-NPs. *MethodsX* **2024**, *12*, 102522. [CrossRef]

processes MDPI

Article

The Recovery of Sulfuric Acid in the Presence of Zr(IV) and Hf(IV) by Solvent Extraction with TEHA and Its Mixtures

Xiaoxi Ding [1], Jiaxin Jiang [1], Yafang Zhao [1], Zaichao Dong [1], Lingyun Wang [1,2,*] and Yang Liu [3,*]

1 College of Chemical Engineering, Qingdao University of Science and Technology, Qingdao 266042, China; dingxiaoxi1998@163.com (X.D.); jiangjiaxin1998@163.com (J.J.); zhaoyafang2023@163.com (Y.Z.); dongzaichao0571@163.com (Z.D.)
2 Sino-German Engineering College, Qingdao University of Science and Technology, Qingdao 266061, China
3 College of Environment and Safety Engineering, Qingdao University of Science and Technology, Qingdao 266042, China
* Correspondence: wangly412@qust.edu.cn (L.W.); liuyang@qust.edu.cn (Y.L.)

Abstract: The recovery of sulfuric acid in the presence of Zr(IV) and Hf(IV) was studied via solvent extraction using TEHA (tri-2-ethylhexyl amine) and its mixtures. A solidification phenomenon occurred in the loaded organic phase when a single TEHA was employed in the extraction of 1 to 5 M H_2SO_4. Octanol, decanol and TBP (tri butyl phosphate) were mixed with TEHA, separately, to prevent the solidification of sulfuric-acid-loaded organic. Due to the relatively high aqueous solubility of octanol and decanol, the mixture of TEHA + TBP was selected as the optimal system for the extraction of H_2SO_4. Simulated counter-current extraction and stripping experiments were performed on the basis of the McCabe–Thiele diagrams, indicating that sulfuric acid could be reduced by TEHA + TBP from 4.2 to around 0.5 M without Zr(IV) and Hf(IV) extraction and recovered by its complete stripping with water. The proposed sulfuric acid recovery step would contribute to the completion of the closed-circuit of the Zr(IV) and Hf(IV) separation process in our previous work and help to re-separate the remaining Zr(IV) and Hf(IV) in the sulfuric acid stripping solution.

Keywords: sulfuric acid; recovery; TEHA; mixtures; closed-circuit process

Citation: Ding, X.; Jiang, J.; Zhao, Y.; Dong, Z.; Wang, L.; Liu, Y. The Recovery of Sulfuric Acid in the Presence of Zr(IV) and Hf(IV) by Solvent Extraction with TEHA and Its Mixtures. *Processes* **2024**, *12*, 940. https://doi.org/10.3390/pr12050940

Academic Editor: Blaž Likozar

Received: 22 March 2024
Revised: 17 April 2024
Accepted: 27 April 2024
Published: 6 May 2024

1. Introduction

It is well known that sulfuric acid (H_2SO_4) is extensively used in hydrometallurgical processes for various treatments, such as electrowinning, leaching, ion exchange and solvent extraction. The removal and recovery of sulfuric acid from the acidic effluent are considered to be valuable from both the economic and environmentally friendly points of view. To solve the sulfuric acid discharge issue, solutions including direct disposal, precipitation, electrohydrolysis, diffusion dialysis and solvent extraction have been proposed [1–6]. Nevertheless, only a few of these methods are recommended due to the sulfuric acid recovery concern.

Recent efforts to recover and regenerate sulfuric acid on a commercial scale have focused on solvent extraction due to its easy operation and extractant regenerability. Organophosphorus solvents containing P=O, such as Cyanex 923 (hexyl/octyl phosphine oxides) [7,8] and TOPO (tri-n-octylphosphine oxide) [9] were proposed as possible extractants for the extraction of sulfuric acid. Alguacil and Lopez [7] tested the sulfuric acid extraction equilibrium of Cyanex 923 with different diluents. It was found that diluents showed a negligible influence on sulfuric acid extraction. And, the formation of a third phase was observed above a H_2SO_4 concentration of 3 M when Cyanex 923 was diluted in aliphatic diluents. Liao et al. studied the phase behavior of a Cyanex 923-heptane/H_2SO_4 system [8]. The third phase was observed at different concentrations of sulfuric acid with a change in the Cyanex 923 concentration due to the transportation of both H_2SO_4 and H_2O by their coordination with Cyanex 923. The removal of sulfuric acid from electrowin solutions by reactive extraction using an Alamine 336 (tri-octyl/decyl amines), isodecanol and shell sol 2325 solvent mixture was studied to evaluate the effectiveness of this system as carriers of the acid [10]. The extraction of sulfuric acid using binary extractants was reported by Brian [11]. The binary extractants studied consisted of mixtures of the acidic extractants and amine-type extractant. The best binary extractant in terms of sulfuric acid extraction efficiency turned out to be the Alamine + D2EHPA system. TBP in kerosene was also employed in the extraction of sulfuric acid by Xun et al., and the reported predominant extracted complex was $(TBP)_2{\cdot}H_2SO_4{\cdot}(H_2O)_3$ or $TBP{\cdot}H_2SO_4{\cdot}H_2O$ [12].

TEHA has been extensively used for the extraction of inorganic acids. The recovery of sulfuric acid from electrolyte bleed streams was developed by using TEHA as the extractant, shellsol 2046 as the diluent and octanol as the modifier [13,14]. The equilibrium data and simulation results were also compared with Cyanex 923 in these studies. Furthermore, a comparison between TEHA and Cyanex 923 by Haghshenas et al. [15] demonstrated that TEHA can extract more H_2SO_4 with fewer extraction stages than Cyanex 923 and that the stripping of sulfuric acid from TEHA is more readily. Kesieme et al. [16] demonstrated that the TEHA system was the optimization for the extraction of sulfuric acid among TEHA, Alamine 336 and Cyanex 923. An optimal organic composition consisting of 50% TEHA, 40% octanol and 10% shellsol A150 was determined. A slope analysis showed that the extracted species consisted of one acid molecule, one TEHA (A) molecule and two octanol (O) molecules with a formula of $(H_2SO_4)AO_2$. Literature surveys indicated that TEHA seems to be an optimal extractant for the extraction of sulfuric acid owing to its outstanding acid extraction capacity [5,16] and ease of stripping [15,17]. Meanwhile, modifiers such as octanol and decanol should be presented in the organic phase with TEHA to prevent the formation of a third phase [13,15–18].

Zirconium was invariably accompanied by hafnium in its major sources such as zircon ($ZrSiO_4$) [19]. Due to the great difference in the transmission of thermal neutrons, zirconium and hafnium play opposite roles in the nuclear industry. Zirconium is generally used as raw material for nuclear neutron insulator construction, while hafnium is a powerful nuclear neutron absorber [20]. The demand for zirconium and hafnium with high purity for application in the nuclear industry was thus highlighted. However, the separation of zirconium and hafnium is quite difficult because of the great similarities in their chemical and physical properties [20,21]. The separation of Zr(IV) and Hf(IV) from various media has thus been widely studied with different extractants [20–24]. Previous work revealed that zirconium was extracted into the organic phase prior to hafnium by most extraction systems [22]. Wang and Lee [23] reported that Zr(IV) was separated over Hf(IV) with amine extractants in the 0.1–3 M H_2SO_4 concentration range due to its stronger complexing tendency with sulfate ions. The proposed extraction mechanism was anion exchange. However, the selective extraction of Hf(IV) over Zr(IV) was only seen in strong sulfuric acid solutions using acidic organophosphorus extractants, such as D2EHPA [22]. The extraction mechanism was demonstrated to be cation exchange. And thus, the extraction efficiency was greatly influenced by the sulfuric acid concentration in the aqueous phase. In order to separate Hf(IV) over Zr(IV) using D2EHPA, a separation process in which metals extracted at 0.5 M H_2SO_4 and then selectively stripped by 4 M H_2SO_4 and oxalic acid solutions was proposed. It has been verified by simulated counter-current extraction and stripping experiments that around 20% of Zr(IV) and a portion of Hf(IV) are detected in the first stripping step together with sulfuric acid with a concentration of 4 M [20]. With the purpose of re-separating these amounts of metals, a reduction in H_2SO_4 from 4 to 0.5 M in the presence of Zr(IV) and Hf(IV) is thus a required step.

In this work, the extraction and recovery of sulfuric acid in the presence of Zr(IV) and Hf(IV) were investigated using TEHA and its mixtures with octanol, decanol or TBP to prevent the solidification of the loaded organic phase. The mixture of TEHA and TBP in kerosene was found to be the best extractant system for sulfuric acid extraction in terms of the extraction efficiency. Complete stripping of sulfuric acid from the loaded organic was achieved with water. Moreover, extraction and stripping isotherms were constructed and verified by batch simulation experiments for the counter-current extraction and stripping.

2. Materials and Methods

2.1. Reagents

The commercial solvent TEHA (Tri-2-ethylhexyl amine, purity 99%) was supplied by BASF Chem Co., Ltd. (Beijing, China), while TBP (Tri butyl phosphate, purity 98%) was purchased from Yakuri pure chemical Co., Ltd. (Kyoto, Japan) The molecular structures of TEHA and TBP are presented in Table 1. Octanol (Acros Organics, purity 99%) and decanol (Acros Organics, purity 98%) were utilized as the modifiers and kerosene (Daejung Chemicals and Metals Co., Ltd., Siheung, Republic of Korea, purity 95%) as a diluent. All of the employed reagents in this work were of analytical grade and without any further purification.

Table 1. Molecular structures of TEHA and TBP.

Extractant	Structure	R
TEHA Tris 2-ethylhexyl amine	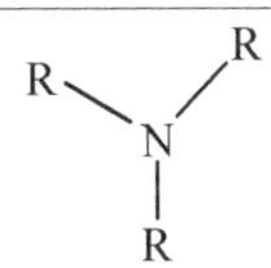	C_4H_9—CH(C_2H_5)—CH_2—
TBP Tri-butyl phosphate	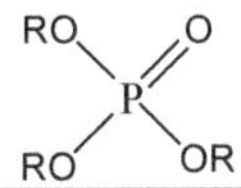	CH_3-CH_2-CH_2-CH_2-

The mixed feed solutions containing sulfuric acid and metals were prepared by dissolving the necessary amounts of $Zr(SO_4)_2{\cdot}4H_2O$ (Alfa Aesar, Johnson Matthey Company (Ward Hill, MA, USA), purity 98+%), $Hf(SO_4)_2$ (Alfa Aesar, Johnson Matthey Company, purity 99.9%) and analytical grade sulfuric acid in double deionized water. The concentrations of both zirconium(IV) and hafnium(IV) were kept at 0.2 g/L throughout the experimental work, while the sulfuric acid concentration was varied from 1 to 5 M.

2.2. Experimental Procedure

The general batch experiments for extraction and stripping were conducted by adding equal volumes of the aqueous and organic phases in a 100 mL sealed glass container and by mechanically shaking the two phases for 30 min using a wrist action shaker (Burrell model 75, Burrell Scientific, Zelienople, PA, USA). After shaking, separation funnels were used for disengagement of the two phases. To determine the number of extraction and stripping stages in a continuous operation, a McCabe–Thiele plot was constructed based on the acid extraction/stripping isotherms. During the extraction and stripping equilibrium isotherm experiments, the aqueous and organic phases were mixed with different phase ratios at a constant total volume of the solution. Subsequently, counter-current simulation experiments were performed in a batch-wise scale. All of the experiments were performed at ambient temperature (25 ± 1 °C).

2.3. Analytical Techniques and Data Analysis

The concentration of zirconium and hafnium in the aqueous phase were measured by ICP-OES (Optima 4300DV, PerkinElmer Ltd., Shelton, CT, USA), and H_2SO_4 concentration was analyzed volumetrically using the acid–base titration method with a standard Na_2CO_3 solution [25]. The amount of metal or sulfuric acid extracted by the organic solvent was estimated with the mass balance. D (distribution coefficient) was defined as the amount of H_2SO_4 in the extracted organic phase compared to that in the raffinate when the extraction reached equilibrium. The extraction percentage was calculated as the difference in H_2SO_4 amount at equilibrium compared with the initial amount in the aqueous phase.

3. Results

TEHA was reported to be the most suitable extractant for H_2SO_4 recovery via the solvent extraction method due to its good extraction efficiency and ease of stripping [15–18]. In this study, the extraction of sulfuric acid was investigated by using TEHA and its mixtures when Zr(IV) and Hf(IV) metals co-exist in aqueous solutions. The concentrations of both zirconium and hafnium were fixed at 0.2 g/L, and the concentration of H_2SO_4 was varied from 1 to 5 M. According to the research work on the study of the complexing formation of Zr(IV) and Hf(IV), anionic zirconium/hafnium complexes ($Zr(SO_4)_3^{2-}$ and $Hf(SO_4)_3^{2-}$) are the prevailing species in an aqueous solution when the concentration of H_2SO_4 is higher than 0.25 M. Kerosene was used as a diluent in the organic phase.

3.1. Extraction of H_2SO_4 by TEHA in Kerosene

Firstly, the effect of the sulfuric acid concentration in a feed solution was investigated using 1 and 1.5 M TEHA in kerosene, as shown in Figure 1. The extraction percentage of H_2SO_4 by both concentrations of TEHA was remarkably affected by the initial H_2SO_4 concentration. Below a 2 M sulfuric acid concentration in the initial aqueous phase, the extraction percentage increased rapidly with the increasing of H_2SO_4 concentration. It was reported by Nguyen and Lee [18] that there was a minimum sulfuric acid concentration remaining in the raffinate due to the hydrolysis and complexing effect of sulfate ion and other ions. This could be used to interpret the low extraction percentage presented when the initial concentration of sulfuric acid was low. A maximum extraction percentage of 34.3% and 49.3% was observed when the concentration of H_2SO_4 was 2 M, with 1 and 1.5 M TEHA, respectively. After that, the extraction of H_2SO_4 decreases steadily from 2 to 5 M H_2SO_4 concentrations. The decrease at a high aqueous phase concentration of H_2SO_4 is supposed to be related to an overload of sulfuric acid in the organic phase.

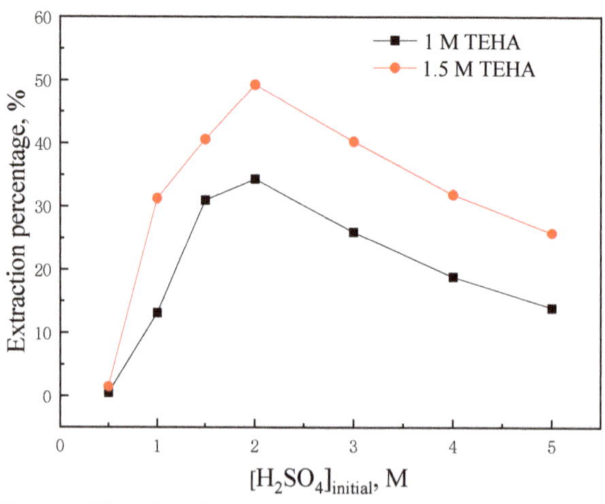

Figure 1. Effect of H_2SO_4 concentration on the extraction of sulfuric acid by 1 and 1.5 M TEHA in the presence of Zr(IV) and Hf(IV).

To illustrate the extraction behavior of H_2SO_4 in Figure 1, Figure 2 was constructed to show the variation in sulfuric acid concentration in the loaded organic phase and raffinate after extraction by 1 and 1.5 M TEHA. As reflected in Figure 2, the organic loaded H_2SO_4 increased up to 3 M initial H_2SO_4 and then leveled off. This phenomenon demonstrates that the newly added sulfuric acid can no longer be extracted by TEHA into the organic phase. Thus, the decreased in sulfuric acid extraction percentage in Figure 1 in the range of 3 to 5 M for the initial H_2SO_4 concentration was determined. In addition, the dashed curves in Figure 2 also proved that the minimum sulfuric acid in raffinate after extraction is around 0.47 M.

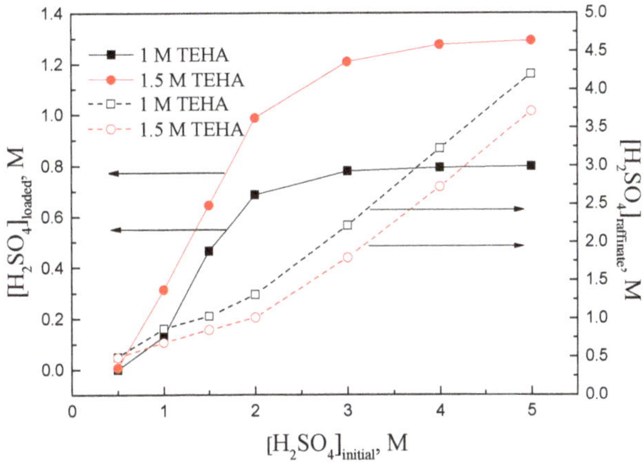

Figure 2. Variation in H_2SO_4 concentration in the loaded organic and raffinate after extraction by 1 and 1.5 M TEHA.

3.2. Extraction of H_2SO_4 by TEHA in Kerosene

To study the effect of TEHA concentration on the extraction of sulfuric acid, solvent extraction experiments were performed by varying the concentration of TEHA from 0.5 to 1.5 M at the initial sulfuric acid concentrations of 1.5, 3 and 5 M, respectively. As shown in Figure 3, the extraction tendency was similar with different concentrations of the initial sulfuric acid while the TEHA concentration increased from 0.5 to 1.5 M. From the initial sulfuric acid concentration of 3 and 5 M, it was observed that the extraction percentage of H_2SO_4 was quite low when the TEHA concentration was lower than 1 M. In order to identify the extraction reaction, the log D against log[TEHA] was plotted for the extraction of H_2SO_4 at the initial concentration of 1.5 M, and the slope analysis method was applied. In consideration of the effect of organic phase overloading, the extraction results obtained from the sulfuric acid initial concentrations of 3 and 5 M are not presented. Figure 4 indicates that the line has a value equal to that of the slope, implying that 1 mol of TEHA takes part in the reaction of one mole of sulfuric acid. In other words, the extracted species was proposed to be H_2SO_4•TEHA. This result matched well with the reported extraction reaction by Haghshenas et al. [15].

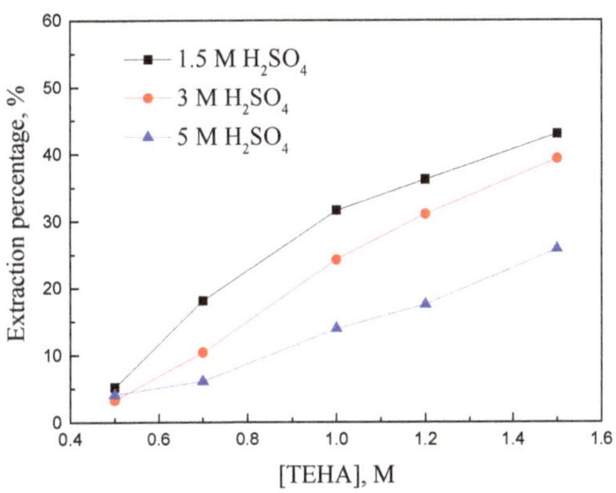

Figure 3. Effect of TEHA concentration on the extraction of 3 M H_2SO_4 in the presence of Zr(IV) and Hf(IV).

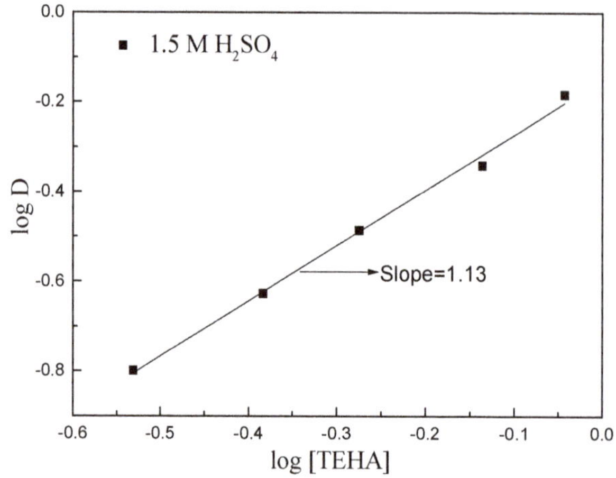

Figure 4. Plot of log D vs. log[TEHA] for the extraction of sulfuric acid at its initial concentration of 1.5 M.

3.3. Extraction of H_2SO_4 by TEHA Mixed with Octanol, Decanol or TBP

Although the extraction of sulfuric acid by a single TEHA in kerosene displayed a fine extraction ability, it has to be pointed out that the solidification of the organic phase occurred when TEHA was loaded by H_2SO_4. To prevent TEHA from being solidified after being loaded by H_2SO_4, octanol, decanol or TBP was added into the organic phase. Figure 5 exhibited the extraction behaviors of sulfuric acid when using TEHA mixed with octanol or decanol in kerosene. The concentration of sulfuric acid in the feed solution was constant at 4 M. As can be seen, the extraction percentage slightly decreased with the addition of various volumes of octanol or decanol to 1 M TEHA. The solidification of the loaded organic phase could be prevented when the volume percentage of octanol or decanol was higher than 20 vol%. However, the high volume percentage of octanol or decanol in the organic phase is not operable or commercially available due to their high aqueous solubility.

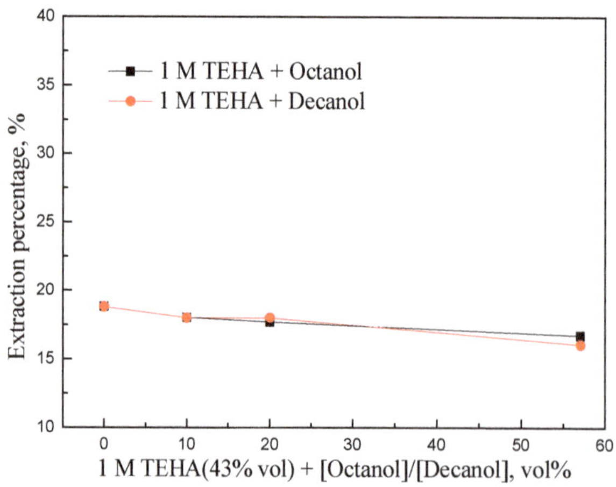

Figure 5. Effect of adding octanol and decanol to 1 M TEHA on the extraction of sulfuric acid in the presence of Zr(IV) and Hf(IV).

In the following experiments, TBP was added to TEHA to improve the solidification phenomenon of the loaded organic phase. The concentration of TEHA was kept at 1 M, while that of TBP was varied from 0.1 to 1 M. As shown in Figure 6, a slight increase in the sulfuric acid extraction from 18.8 to 23.0% was observed with the increasing concentration of the added TBP. This increase is explained by the fact that TBP also showed extraction abilities on acids, as reported by Sarangi et al. [26]. When 1 M TBP was added to 1 M TEHA, solidification was prevented. With lower concentrations of TBP in the mixtures, the solidification of the organic phase after loading by sulfuric acid cannot be prevented.

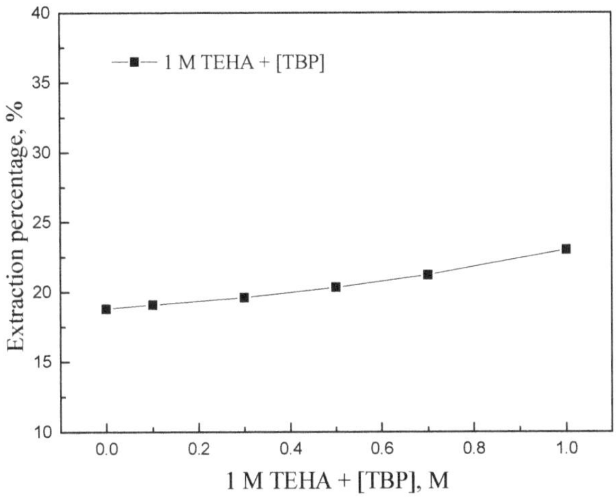

Figure 6. Effect of adding TBP to 1 M TEHA on the extraction of sulfuric acid in the presence of Zr(IV) and Hf(IV).

In general, the polymerization of the extracted species or its low dissolution rate in organic diluents was considered the main reason for the formation of a third phase or the loaded organic solidification phenomenon. TEHA is one kind of branched tertiary amine. The extraction of acid (HX) by a tertiary amine, denoted as R_3N, is generally regarded as $R_3NH^+{\bullet}X^-$. Octanol and decanol containing oxhydryl (R-OH) are considered to be electronegativity polar molecular and thus have good affinity for R_3NH^+. They may replace the hydrogen-surrounded R_3NH^+ and thereby promote the dissolution of loaded TEHA in kerosene. In terms of the addition of TBP to TEHA, Xun et al. [12] investigated the three-phase behavior of the TBP-kerosene/H_2SO_4-H_2O extraction system and the composition of the extracted complexes in the middle phase. They reported that H_2SO_4 and H_2O could be transferred into the middle phase mainly by their coordination with TBP and the outer layer of water composed by P=O---H_3O^+. Based on this, it is assumed that the addition of TBP to TEHA may promote the generation of $[R_3NH{\bullet}(H_2O)_n$---$O{=}P(OC_4H_9)_3]^+$ and HSO_4^- [8,12,27]. Due to the steric hindrance, both $[R_3NH{\bullet}(H_2O)_n$---$O{=}P(OC_4H_9)_3]^+$ and HSO_4^- would be maintained in an organic as cations and anions, hence facilitating dissolution in a diluent and preventing the generation of polymerization. Therefore, the solidification of a loaded organic phase could also be prevented by adding TBP.

In comparison with octanol or decanol, TBP has been commercially used in solvent extraction processes. Furthermore, a mixture of TEHA with TBP revealed a relatively higher extraction efficiency of sulfuric acid than that mixed with octanol or decanol. A mixture of 1 M TEHA and 1 M TBP was expected to be an effective and suitable system for sulfuric acid extraction and recovery. Furthermore, an analysis of the raffinate after the extraction of H_2SO_4 indicated that no Zr(IV) or Hf(IV) were extracted by TEHA and its mixtures.

3.4. Stripping of H_2SO_4 from Loaded TEHA + TBP

To regenerate sulfuric acid, it is of great importance to select a suitable reagent for its stripping from the loaded organic phase. Water has been demonstrated to be an effective strippant for sulfuric acid stripping from a single loaded TEHA [18,28]. To study the stripping efficiency of sulfuric acid from mixtures of TEHA and TBP by water, stripping experiments were performed by using water with different aqueous to organic phase volume ratios (A/O ratios). The stripping results are presented in Table 2. The loaded organic phases used were obtained when 1 M TEHA and 1 M TEHA +1 M TBP made contact with 4 M sulfuric acid, separately. The effective stripping of sulfuric acid from loaded TEHA + TBP by water is demonstrated in Table 2. Compared to the stripping results of TEHA + TBP with a single TEHA, the stripping efficiency was improved by using TEHA + TBP when the A/O ratio exceeded the unit. Additionally, 99.3% of sulfuric acid was stripped by water from loaded TEHA when the A/O ratio reached 3/1.

Table 2. Stripping of sulfuric acid from loaded TEHA and loaded TEHA + TBP with water under different A/O ratios.

A/O Ratio	**Stripping Percentage, %**	
	Loaded TEHA	**Loaded TEHA + TBP**
1/3	30.1	27.3
1/2	42.0	35.4
1/1	60.1	58.7
2/1	75.2	91.9
3/1	82.7	99.3

3.5. *Extraction Isotherm of Sulfuric Acid by TEHA + TBP in Kerosene*

A mixture of 1 M TEHA + 1 M TBP was employed to construct the extraction isotherm of H_2SO_4 from the solution containing 4 M H_2SO_4. The organic to aqueous phase volume ratio (O/A ratio) was varied from 1/3 to 9/1, while the other experimental parameters were kept constant. The resultant extraction isotherm is shown in Figure 7. McCabe–Thiele plot diagram indicates that four extraction stages are needed to extract most of the H_2SO_4 from the aqueous phase by 1 M TEHA + 1 M TBP at an O/A ratio of 4. However, the extraction of sulfuric acid differs from the solvent extraction of metals in the fact that there is a minimum sulfuric acid concentration remaining in raffinate of about 0.47 M, as mentioned in Section 3.1. In other words, the complete extraction of H_2SO_4 is infeasible. According to Figure 7, to reduce the sulfuric acid concentration from 4.2 M to its minimum concentration, three extraction stages are sufficient at an O/A ratio of 4.

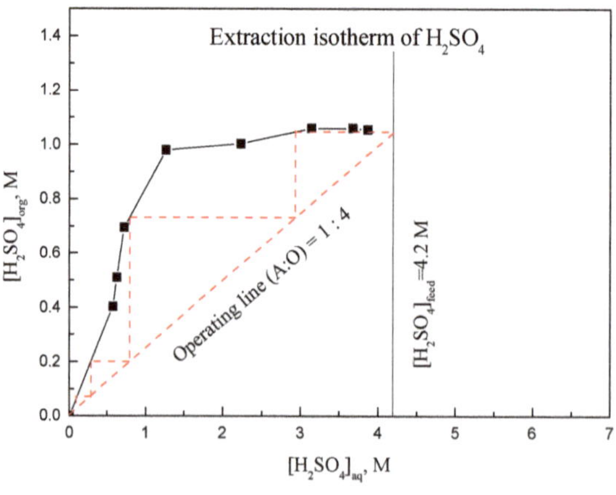

Figure 7. McCabe–Thiele plot for H_2SO_4 extraction with TEHA + TBP. Experimental condition: $[H_2SO_4]$ = 4.2 M, V_{org}/V_{aq} = 1/3–9, [Extractant] = 1 M TEHA + 1 M TBP, Diluent = Kerosene.

To verify this, a three-stage counter-current batch simulation experiment was conducted at an O/A ratio of 4 using 1 M TEHA + 1 M TBP. As shown in Figure 8, a feed solution in the aqueous phase contained 4.2 M sulfuric acid together with 0.2 g/L Zr(IV) and 0.2 g/L Hf(IV). After a three-stage counter-current extraction, the sulfuric acid concentration was reduced from 4.2 to 0.48 M, while all of Zr(IV) and Hf(IV) were maintained in raffinate. A loaded TEHA + TBP containing 0.83 M sulfuric acid was prepared for the subsequent counter-current stripping study.

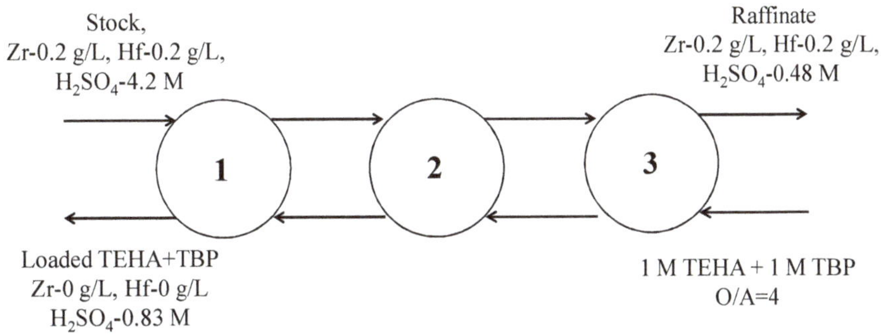

Figure 8. The extraction of 4.2 M sulfuric acid in the presence of Zr(IV) and Hf(IV) by 1 M TEHA + 1 M TBP in kerosene.

3.6. Stripping Isotherm of Sulfuric Acid by TEHA + TBP in Kerosene

A McCabe–Thiele plot of the sulfuric acid stripping isotherm was constructed to estimate the number of required stages and the flow rate of the two phases to achieve the maximum stripping of sulfuric acid from loaded TEHA + TBP by water. Stripping experiments were carried out at different A/O ratios from 1/5 to 3/1. Figure 9 suggests that complete stripping of sulfuric acid by water is possible in two counter-current stripping stages at an A/O ratio of 2/1. Based on the results obtained from the McCabe–Thiele plot, a two-stage batch simulation experiment on the counter-current stripping of H_2SO_4 was carried out. The loaded organic was obtained from the extraction batch simulation experiments in Section 3.5, and the concentration of H_2SO_4 in the loaded organic was 0.83 M. Figure 10 indicates that H_2SO_4 could be quantitatively stripped by water after 2 counter-current stripping steps at an A/O ratio of 2. Finally, a sulfuric acid solution with the concentration of 0.5 M was obtained in the stripping solutions.

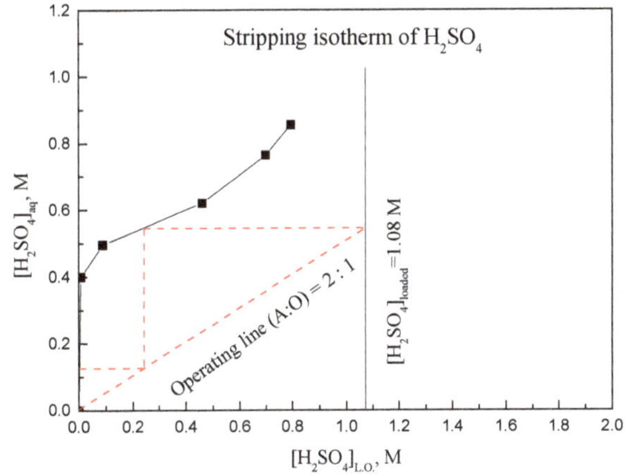

Figure 9. McCabe–Thiele plot for H_2SO_4 stripping from loaded TEHA + TBP with water. Stripping condition: [Extractant] = 1 M TEHA + 1 M TBP, Diluent = Kerosene, $[H_2SO_4]_{loaded}$ = 1.08 M, [Strippant] = water, V_{org}/V_{aq} = 1/3–5.

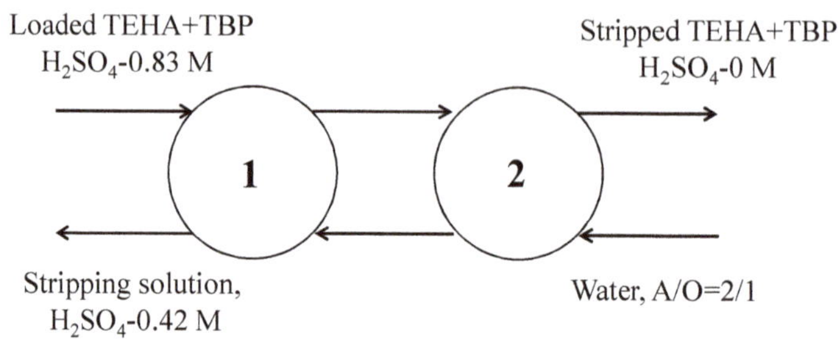

Figure 10. Counter-current simulation of the stripping of sulfuric acid from loaded TEHA + TBP by water.

3.7. Proposed Closed-Circuit Process

Based on the aforementioned data, the process developed for the separation of Zr(IV) and Hf(IV) by inserting a H_2SO_4 recovery step is presented in Figure 11. As denoted by the green box in Figure 11, the regeneration of sulfuric acid was achieved after stripping the H_2SO_4-loaded organic with water. Moreover, the concentration of H_2SO_4 reduced from 4 M to 0.5 M after extraction by 1 M TEHA + 1 M TBP, without loss of Zr(IV) and Hf(IV). This indicates that the raffinate obtained after H_2SO_4 extraction can be directed back toward the Zr(IV) and Hf(IV) starting solution for metal re-separation. A good closed-circuit could be created by inserting a sulfuric acid recovery step into the whole flowsheet, making the process economic and eco-friendly. Furthermore, the recovery and regeneration of sulfuric acid in the effluent is also necessary to address environmental issues.

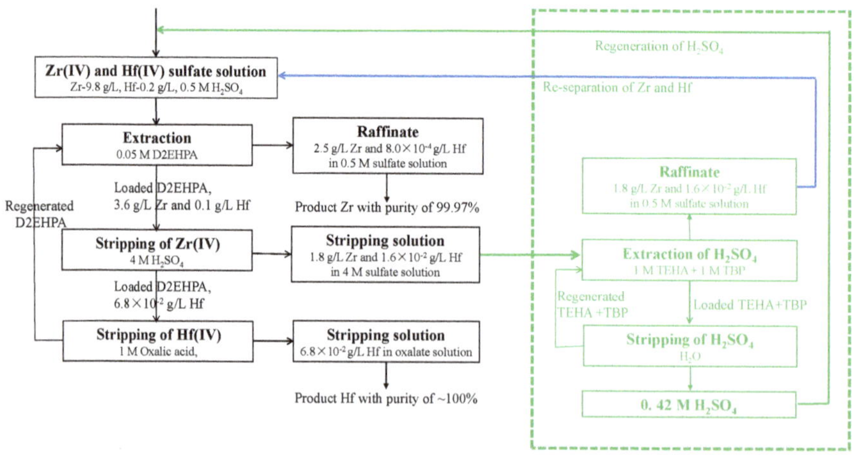

Figure 11. Flowsheet of the proposed closed-circuit process for the separation of Zr(IV) and Hf(IV) by inserting a H_2SO_4 recovery step.

4. Conclusions

TEHA and its mixtures with octanol, decanol and TBP were employed to recover sulfuric acid in the presence of Zr(IV) and Hf(IV). The extraction of H_2SO_4 by TEHA in kerosene was found to be influenced by the minimum sulfuric acid concentration in raffinate when the feed concentration of H_2SO_4 was low and to be limited by the loading capacity of TEHA in a high initial sulfuric acid range. The extracted species was determined to be H_2SO_4•TEHA based on the slope analysis method. The solidification phenomenon could be prevented when more than 20% vol of octanol or decanol, or 1 M TBP were added. The possible roles played by the three additives on the prevention of solidification was interpreted by analyzing the characteristics of their functional group. The mixture of TEHA and TBP in kerosene was found to be the best system for the extraction of H_2SO_4. The effective stripping of sulfuric acid from the loaded TEHA + TBP was achieved by using water. In the whole experimental procedure, Zr(IV) and Hf(IV) could not be extracted into TEHA or its mixtures. McCabe–Thiele diagrams for the extraction and stripping isotherms were also constructed. Batch-type simulation experiments for the multi-stage counter-current extraction and stripping demonstrated that 4.2 M sulfuric acid could be reduced to a minimum concentration of 0.5 M in raffinate by 1 M TEHA + 1 M TBP without the co-extraction of Zr(IV) and Hf(IV) and could be easily stripped by water.

Author Contributions: Conceptualization, L.W. and Y.L.; methodology, L.W.; validation, X.D., J.J. and Y.Z.; formal analysis, Y.Z.; investigation, L.W.; resources, Z.D.; data curation, X.D.; writing—original draft preparation, L.W.; writing—review and editing, Y.L.; visualization, J.J.; supervision, Y.L.; project administration, Y.L. All authors have read and agreed to the published version of the manuscript.

Funding: This research was funded by the Shandong Provincial Natural Science Foundation (ZR2022MB036, ZR2022MB111).

Data Availability Statement: The data are contained within the article.

Conflicts of Interest: The authors declare no conflicts of interest.

References

1. Cifuentes, L.; García, I.; Ortiz, R.; Casas, J.M. The use of electrohydrolysis for the recovery of sulphuric acid from copper-containing solutions. *Sep. Purif. Technol.* **2006**, *50*, 167–174. [CrossRef]
2. Jeong, J.; Kim, M.-S.; Kim, B.-S.; Kim, S.-K.; Kim, W.-B.; Lee, J.-C. Recovery of h2so4 from waste acid solution by a diffusion dialysis method. *J. Hazard. Mater.* **2005**, *124*, 230–235. [CrossRef]
3. Moraes, L.d.S.; Kronemberger, F.d.A.; Ferraz, H.C.; Habert, A.C. Liquid–liquid extraction of succinic acid using a hollow fiber membrane contactor. *J. Ind. Eng. Chem.* **2015**, *21*, 206–211. [CrossRef]
4. Dufour, J.; Marrón, J.O.; Negro, C.; Latorre, R.; Formoso, A.; López-Mateos, F. Mechanism and kinetic control of the oxyprecipitation of sulphuric liquors from steel pickling. *Chem. Eng. J.* **1997**, *68*, 173–187. [CrossRef]
5. Agrawal, A.; Kumari, S.; Sahu, K.K. Liquid–liquid extraction of sulphuric acid from zinc bleed stream. *Hydrometallurgy* **2008**, *92*, 42–47. [CrossRef]
6. Palatý, Z.; Žáková, A. Separation of H_2SO_4 + $CuSO_4$ mixture by diffusion dialysis. *J. Hazard. Mater.* **2004**, *114*, 69–74. [CrossRef]
7. Alguacil, F.J.; López, F.A. The extraction of mineral acids by the phosphine oxide Cyanex 923. *Hydrometallurgy* **1996**, *42*, 245–255. [CrossRef]
8. Liao, W.; Shang, Q.; Yu, G.; Li, D. Three-phase extraction study of Cyanex 923–n-heptane/H_2SO_4 system. *Talanta* **2002**, *57*, 1085–1092. [CrossRef] [PubMed]
9. Petkwl, C.D.M.; Kopdnl, M.M.; Mltrovl, C.A.A. Solvent extraction of nitric, hydrochloric and sulfuric acid and their uranyl salts with tri-n-octylphosphine oxide. *Solvent Extr. Ion Exc.* **1992**, *10*, 685–696.
10. Ntengwe, F.W. Removal of sulphuric acid from electrowin solutions by reactive extraction using Alamine 336 in isodecanol-shell sol 2325 solvent mixture. *Int. J. ChemTech Res.* **2010**, *2*, 2131–2139.
11. Tait, B.K. The extraction of sulphuric acid by some binary extractants. *Hydrometallurgy* **1993**, *33*, 245–251. [CrossRef]
12. Fu, X.; Xiaopeng, H.; Zuhua, Z.; Zhengshui, H.; Debao, W. Three phase extraction study. I. Tri-butyl phosphate-kerosene/ H_2SO_4–H_2O extraction system. *Colloids Surf. A* **1999**, *152*, 335–343. [CrossRef]
13. Gottliebsen, K.; Grinbaum, B.; Chen, D.; Stevens, G.W. Recovery of sulfuric acid from copper tank house electrolyte bleeds. *Hydrometallurgy* **2000**, *56*, 293–307. [CrossRef]
14. Gottliebsen, K.; Grinbaum, B.; Chen, D.; Stevens, G.W. The use of pulsed perforated plate extraction column for recovery of sulphuric acid from copper tank house electrolyte bleeds. *Hydrometallurgy* **2000**, *58*, 203–213. [CrossRef]
15. Haghshenas, D.F.; Darvishi, D.; Rafieipour, H.; Alamdari, E.K.; Salardini, A.A. A comparison between teha and Cyanex 923 on the separation and the recovery of sulfuric acid from aqueous solutions. *Hydrometallurgy* **2009**, *97*, 173–179. [CrossRef]
16. Kesieme, U.K.; Aral, H.; Duke, M.; Milne, N.; Cheng, C.Y. Recovery of sulphuric acid from waste and process solutions using solvent extraction. *Hydrometallurgy* **2013**, *138*, 14–20. [CrossRef]
17. Banda, R.; Nguyen, T.H.; Lee, M.S. Recovery of hcl from chloride leach solution of spent hds catalyst by solvent extraction. *Chem. Process Eng.* **2013**, *34*, 153–163. [CrossRef]
18. Nguyen, T.H.; Lee, M.S. Extraction and stripping of inorganic acids by tris 2-ethylhexyl amine. *Korean J. Met. Mater.* **2014**, *52*, 799–803.
19. Taghizadeh, M.; Ghanadi, M.; Zolfonoun, E. Separation of zirconiumand hafniumby solvent extraction usingmixture of TBP and Cyanex 923. *J. Nucl. Mater.* **2011**, *412*, 334–337. [CrossRef]

20. Wang, L.Y.; Lee, M.S. A review on the aqueous chemistry of Zr(IV) and Hf(IV) and their separation by solvent extraction. *J. Ind. Eng. Chem.* **2016**, *39*, 1–9. [CrossRef]
21. Wang, L.Y.; Lee, H.Y.; Lee, M.S. Solvent extraction of zirconium and hafnium from hydrochloric acid solutions using acidic organophosphorus extractants and their mixtures with TOPO. *Mater. Trans.* **2013**, *54*, 1460–1466. [CrossRef]
22. Wang, L.Y.; Lee, M.S. Development of a separation process for the selective extraction of hafnium(IV) over zirconium(IV) from sulfuric acid solutions by using D2EHPA. *Hydrometallurgy* **2016**, *160*, 12–17. [CrossRef]
23. Wang, L.Y.; Lee, M.S. Separation of Zr and Hf from sulfuric acid solutions with amine-based extractants by solvent extraction. *Sep. Purif. Technol.* **2015**, *142*, 83–89. [CrossRef]
24. Amaral, J.C.B.S.; Morais, C.A. Equilibrium of zirconium and hafnium in the process of extraction with TBP in nitric medium—Influence in the Zr/Hf separation. *Miner. Eng.* **2020**, *146*, 106138. [CrossRef]
25. Basset, J.; Denney, R.; Jeffery, G.; Mendham, J. *Vogel's Textbook of Quantitative Inorganic Analysis*; John Wiley & Sons, Inc.: New York, NY, USA, 1978.
26. Sarangi, K.; Padhan, E.; Sarma, P.V.R.B.; Park, K.H.; Das, R.P. Removal/recovery of hydrochloric acid using Alamine 336, Aliquat 336, TBP and Cyanex 923. *Hydrometallurgy* **2006**, *84*, 125–129. [CrossRef]
27. Margulis-Goshen, K.; Magdassi, S. Organic nanoparticles from microemulsions: Formation and applications. *Curr. Opin. Colloid Interface Sci.* **2012**, *17*, 290–296. [CrossRef]
28. Agrawal, A.; Sahu, K.K. An overview of the recovery of acid from spent acidic solutions from steel and electroplating industries. *J. Hazard. Mater.* **2009**, *171*, 61–75. [CrossRef]

processes

MDPI

Article

Extraction and Purification of Kiwifruit Seed Oil Using Three-Phase Partitioning: An Efficient and Value-Adding Method for Agro-Industrial Residue Utilization

Bangfu Liu [1], Xiufang Li [2,*], Qingchun Liu [3] and Zhijian Tan [4,*]

1 Hunan Electronic Information Industry Institute, Changsha 410012, China; liubangfu2006@163.com
2 College of Pharmacy, Heze University, Heze 274015, China
3 Xiangxi Laodie Biotechnology Co., Ltd., Xiangxi 416714, China; laodieshengwu@126.com
4 Institute of Bast Fiber Crops & Center of Southern Economic Crops, Chinese Academy of Agricultural Sciences, Changsha 410205, China
* Correspondence: lixiufang_2030@163.com (X.L.); tanzhijian@caas.cn (Z.T.)

Abstract: Kiwifruit seeds are usually discarded as waste in the kiwifruit processing industry. However, kiwifruit seeds are rich in oil, which implies their value as a potential resource. Therefore, three-phase partitioning (TPP) using ammonium sulfate and n-butanol has been developed to extract kiwifruit seed oil (KSO) from discarded kiwifruit seeds. The factors influencing the extraction process have been investigated. The highest extraction yield of 36.06% for KSO was obtained under the following conditions: an extraction time of 18 h, a liquid-solid ratio of 4:1, an amount of ammonium sulfate of 28.5 wt%, and a phase ratio (top phase/bottom phase) of 1/1. The composition of obtained KSO was analyzed by gas chromatography-mass spectrometry (GC-MS), and the results indicated the high content of α-linolenic acid. This simple and low-cost method can be used as an efficient approach for utilizing the value of kiwifruit seeds.

Keywords: kiwifruit seed oil; extraction and separation; three-phase partitioning; waste utilization

Citation: Liu, B.; Li, X.; Liu, Q.; Tan, Z. Extraction and Purification of Kiwifruit Seed Oil Using Three-Phase Partitioning: An Efficient and Value-Adding Method for Agro-Industrial Residue Utilization. *Processes* **2023**, *11*, 2581. https://doi.org/10.3390/pr11092581

Academic Editors: Prashant K. Sarswat and Yanlin Zhang

Received: 17 July 2023
Revised: 19 August 2023
Accepted: 22 August 2023
Published: 29 August 2023

1. Introduction

Kiwifruit (*Actinidia chinensis* Planch.) is a valuable agricultural produce that grows in the central and southern regions of China [1]. Kiwifruit contains a variety of essential nutrients, such as amino acids, proteins, polysaccharides, vitamin C, and minerals [2,3]. Kiwifruits are abundant in vitamin C, which can strengthen the immune system and promote wound healing and absorption of iron. The content of vitamin C in fresh kiwifruit is 3–14 times that of citrus, and 30–140 times that of pears. Kiwifruits are also abundant in inositol and amino acids, which can inhibit depression and supplement the nutrients consumed by mental work. The perfect ratio of low sodium and high potassium in kiwifruits can supplement the physical energy lost by staying up late and working overtime, and is beneficial for maintaining cardiovascular health [4]. In recent decades, there has been rapid progress in the industrial processing of kiwifruit. Kiwifruit has been processed to obtain various food products, including juice, jam, jelly, syrup, and candy [5–7]. Kiwifruit seeds, as a by-product of kiwifruit processing, are usually discarded as agro-industrial residue [8]. Kiwifruit seeds are abundant in a variety of unsaturated fat acids, lipids, flavonoids, phenols, vitamins, trace element selenium, and other bioactive substances. Linoleic acid, linolenic acid, and other main unsaturated fat acids account for more than 75% of the total quantity of unsaturated fat acids, especially linolenic acid content of 64.1% of the total quantity of unsaturated fat acids, which is currently the highest in natural vegetable oils except for perilla oil, and is also the premium resource for high-quality natural polyenoic acids. Modern research and clinical experiments have demonstrated that kiwifruit seed oil (KSO) possesses the ability to regulate blood lipids, soften blood vessels, improve obesity, and delay aging [1,9]. Berry et al. conducted intervention tests on 399 male healthy subjects, and found that systolic pressure, diastolic pressure, and mean arterial pressure decreased by 5 mmHg [10], accompanied by α-linolenic acid increase of 1% in human adipose tissue. Ghafoorunissa et al. fed newly lactating rats with saturated fat acid, monounsaturated fatty acid, and polyunsaturated fatty acid, respectively, for 3 months, and induced insulin resistance with sucrose instead of starch [11]. The results showed that using α-linolenic acid instead of linoleic acid in food could improve insulin sensitivity of sucrose-induced rats. For these reasons, KSO has broad applications in medicine, health food, and beauty products [12–14]. Therefore, it is worthy to recover the KSO from discarded kiwifruit seeds.

In previous studies, organic solvent extraction, supercritical extraction, and subcritical extraction were commonly used to extract other oil and KSO [15–17]. Renata et al. used compressed propane as part of a liquid–liquid extraction system for separation of kiwi fruit seed oil and compared the total extraction rate with the Soxhlet extraction using n-hexane [2]. Most extraction conditions using compressed propane have high extraction rates and a total extraction rate of 31.7 ± 1.4%. Nedasadat et al. used supercritical extraction and mathematical modeling for Elaeagnus angustifolia oil [11]. The experimental design and optimization were completed through Box Behnken design (BBD) to analyze the impact of main operating parameters, including pressure, temperature, and particle size. The highest extraction rate (84.9% w/w) was achieved under the optimal conditions. However, these methods have some drawbacks. For example, the organic solvent extraction method is simple; however, additional purification steps are required to eliminate impurities such as pigments and proteins. Supercritical extraction and subcritical extraction are efficient, but require expensive equipment, which can be a significant cost factor. Therefore, it is necessary to find a simple, efficient, and low-cost method for the extraction and purification of KSO.

Three-phase partitioning (TPP) is a cheap and efficient separation technology first developed in 1972 [18]. Usually, TPP was performed by adding ammonium sulfate and t-butanol to the crude extract [19,20]. TPP has been widely used for the extraction and purification of various bioactive components, including polysaccharides, proteins, enzymes, oils, and other substances [21–27]. TPP had been previously applied in the separation and purification of proteins and enzymes by the formation of a solid middle phase to effectively preserve their biological activity. Subsequently, it was found that TPP also exhibited excellent performance in the extraction and purification of plant oils. For example, TPP based on t-butanol and ammonium sulfate was used to extract oil from *Crotalaria juncea* seeds to obtain a yield of 37%, which was much higher than that (13%) obtained by Soxhlet extraction [28]. In our previous study, enzyme-assisted TPP based on t-butanol and ammonium sulfate was also developed for the extraction of flaxseed oil [29], and 71.68% extraction yield was obtained under the optimal conditions. Through TPP was used for oil extraction, a preliminary purification can be achieved. In this process, impurities such as proteins or pigments are either dissolved in the salt phase or deposited in the intermediate phase, leaving the oil enriched and separated in the top phase.

In this study, TPP was initially performed using n-butanol and ammonium sulfate. This system was then employed for the extraction, separation, and purification of oil from discarded kiwifruit seeds. Single-factor experiments were conducted to investigate the influencing factors involved in the preparation of crude extract. These factors include the type of solvent, the extraction time, and the solid-liquid ratio. Then TPP was developed for the separation and purification of KSO by adding salts to the crude extract. The types of salts added to the system, as well as the effects of salt concentration and phase ratio (top n-butanol phase/bottom salt phase), were investigated. With these single factor experiments, the optimal conditions for separating kiwifruit seed oil through TPP were determined, and the final extraction yield of oil was 79.23% under the optimal conditions of the total oil content. Finally, the main components of KSO extracted by TPP were analyzed by GC-MS. The highest content of oil obtained is α-linolenic acid (51.26%), followed by Squalene

(15.07%), and then Linoleic acid (14.49%), which are more than 10% of all other components. This TPP method can provide some guidance for kiwifruit processing industry.

2. Materials and Methods

2.1. Materials and Reagents

The kiwifruit seeds were provided by Xiangxi Laodie Biotechnology Co., Ltd. (Jishou City, China). Kiwifruit seeds were crushed and passed through a 40-mesh sieve to obtain fine and uniform particles. Intermittent operation was employed during the crushing process to prevent excessive heat buildup. Kiwifruit seeds were dried to constant weight at 60 °C in an electric thermostatic drying oven. Ethanol (>98%), acetone (>98%), n-propanol (>98%), isopropanol (>98%), t-butanol(>98%), n-butanol (>98%), ammonium sulfate ($(NH_4)_2SO_4$) (>98%), sodium chloride (NaCl) (>98%), anhydrous sodium carbonate (Na_2CO_3) (>98%), anhydrous sodium sulfate (Na_2SO_4) (>98%), dipotassium hydrogen phosphate trihydrate ($K_2HPO_4{\cdot}3H_2O$) (>98%), and ammonium citrate ($C_6H_5O_7(NH_4)_3$) (>98%) were purchased from Shanghai Titan Technol. Co., Ltd. (Shanghai, China). All the reagents are of analytical reagent grade and used without further treatment.

2.2. Preparation of the Crude Extract

Kwifruit seeds of 2.0 g and an organic solvent (ethanol, acetone, n-propanol, isopropanol, t-butanol, or n-butanol) of 4.0 mL were added to a tube. The amount of kiwifruit seeds added was accurately weighed. For the extraction of KSO, various intensification methods were employed, including ultrasonic treatment, microwave irradiation, high-temperature conditions, or stirring. These techniques were used to enhance the extraction process and improve the yield of kiwifruit seed oil. After extraction, the supernatant was separated by centrifugation at 4000 r/min for 5 min. After determining the optimal conditions for the extraction stage, the crude extract of KSO was obtained under the optimal conditions, then stored in a refrigerator for further use.

2.3. Extraction of KSO Using TPP

TPP was performed by adding certain concentration of ammonium sulfate solution to the crude extract of KSO. This led to the formation of a three-phase system comprising a solvent top phase, a solid middle phase, and a salt bottom phase. The solvent top phase, which contained the desired kiwifruit seed oil, was collected from the TPP system. Subsequently, the solvent was removed using rotary evaporation and then dried in a vacuum oven until reaching a constant weight, resulting in the isolated oil as the final product. The oil weight was measured, and the extraction yield (Y, %) of KSO was calculated using Equation (1).

$$Y = \frac{Mass\ of\ obtained\ KSO}{Mass\ of\ Kiwifruit\ seeds} \times 100\% \quad (1)$$

2.4. Low Field Nuclear Magnetic Resonance Relaxation Measurements

KSO is measured using low field nuclear magnetic resonance instrument (MesoMR23-060H-I, Suzhou Niumag Analytical Instrument Co., Ltd., Suzhou, China) [30]. The analyzer operates at 32 °C and has a proton resonance of 20 MHz, with a sample tube diameter of 25 mm. The sequence measurement of spin lateral relaxation of the Carr Purcell Meiboom Gill (CPMG) sequence was used. This sequence has the following parameters: 90° pulse width (P1) = 5.52 µs,180 ° pulse width (P2) = 12.48 µs, waiting time (TW) = 2000 ms, number of scans (NS) = 32. The T2 distribution curve is fitted with multiple exponents using the Simultaneous Iterative Reconstruction Technique (SIRT) algorithm. The relaxation time, peak area, and ratio of peak area were recorded.

2.5. GC-MS Analysis of KSO

The components of KSO were analyzed by using a GC-MS (an Agilent 7890B GC coupled with a 5977A MS). For this purpose, the KSO was first esterified to fatty acid methyl ester according to previous reports [31]. The specific details are as follows:100 μL KSO were taken on centrifuge tube and 2 mL of n-hexane were added into the centrifuge tube, then 2 mL of 0.4 mol/L NaOH methanol solution were added to the centrifuge tube. The centrifuge tube was sealed and vibrated violently. Then it was placed at 70 °C water bath for heating and refluxing for 10 min. These solutions were then transferred to another graduated centrifuge tube, and saturated salt water was added to 10 mL, shaken and centrifuged. The top layer of solution was taken and subjected to 0.22 μm membrane filtration. The fatty acid methyl ester was analyzed by GC-MS. The detailed analysis conditions are listed in Table 1.

Table 1. The GC-MS conditions for the analysis of KSO.

GC Conditions		MS Conditions	
Chromatographic column	HP-5MS (30.0 m × 250 μm, 0.25 μm)	Tune	Autotune
Carrier gas	He	Acquisition mode	scan mode
Carrier gas flow rate	1.0 mL/min	Electron energy	70 eV
Oven program	Initial 50 °C (hold 5 min); 150 °C (hold 0 min) at 15 °C/min; 180 °C (hold 0 min) at 5 °C/min; 325 °C (hold 10 min) at 8 °C/min; max 350 °C (1 min equilibrium time)	Solvent delay	3.5 min
Inlet temperature	320 °C	MS temperature	230 °C (Source); 150 °C (Quad)
Transmission line temperature	320 °C	Scan range	20–600 amu
Injection mode	Split (2:1)		
Injection volume	0.2 μL		

2.6. Statistical Analysis

Determination was repeated three times in this study. The statistical analysis was done using the SPSS 26.1 software. The one-way analysis of variance (ANOVA) was used for statistical significance analysis. A "*p*-value" of less than 0.05 ($p < 0.05$) was regarded as statistical significance. Different letters in all the figures represent the statistically significant difference.

3. Results and Discussion

3.1. Factors Influencing the Preparation of Crude Extract

3.1.1. Screening of the Extractant

The choice of organic solvent plays a crucial role in the extraction of KSO using TPP. The type of solvent used can significantly influence the efficiency and effectiveness of the extraction process. Different solvents may exhibit varying abilities to dissolve and separate the components of kiwifruit seed oil, thereby affecting the overall extraction yield and purity. Therefore, six common solvents were chosen as the extractants with stirring assisted extraction. According to the data presented in Figure 1, it is evident that n-butanol used as the extractant resulted in the highest extraction yield of 19.6% for KSO, surpassing the commonly used t-butanol by a significant margin. This difference in extraction efficiency can be attributed to the non-polar nature of n-butanol, which appears to be better suited for extracting KSO. While t-butanol has demonstrated efficient performance in TPP, the similarity in structure and properties between n-butanol and t-butanol suggests that n-butanol is the more appropriate choice for subsequent studies. n-propanol and isopropanol have the lowest extraction efficiency for KSO. This is because the polarity of propanol series solvents is greater than that of butanol; their viscosity is higher than that of methanol and ethanol at the same time. After careful consideration of these factors, n-butanol was selected for use in the following experiments.

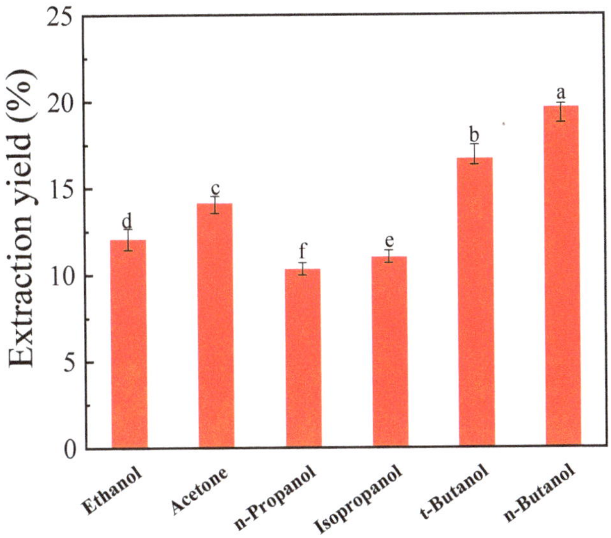

Figure 1. Extraction yields of KSO based on TPP using different organic solvents with stirring assisted extraction. The extraction conditions were as follows: 3:1 liquid–solid ratio, 6 h extraction time, 4/3 phase ratio (top/bottom phase), and 25 wt% ammonium sulfate. Different letters represent the statistically significant difference.

3.1.2. Effect of the Intensification Methods for Extraction

In this study, four intensification methods, namely, ultrasonic treatment, microwave irradiation, high-temperature, and stirring, were utilized to aid the extraction of KSO. The static extraction was used for comparison. The conditions for ultrasound-assisted extraction in this study were as follows: 200 W power ultrasound was applied for 10 min, followed by a resting period of 10 min. This process was repeated six times to complete the extraction procedure. Then, it was allowed to rest for 6 h. The conditions for microwave assisted extraction were as follows: 200 W power microwave was used for 5 min and then stood for 2 min; the procedure was repeated 6 times. The total duration of ultrasound assisted extraction was about 6 h. The high-temperature assisted extraction was performed by placing a tube in a water bath at 60 °C for 6 h using n-butanol as the extractant. The stirring assisted extraction was performed for 6 h at a stirring rate of 200 r/min using n-butanol as the extractant. The static extraction was performed for 6 h using n-butanol as the extractant. As shown in Figure 2, the stirring method has the highest extraction yield of 26.22%. This is because that stirring can make n-butanol efficiently contact with kiwifruit seeds, which is beneficial to the extraction. Therefore, the stirring assisted method was chosen for further experiments.

3.1.3. Effect of Extraction Time

The extraction time was a crucial factor in the solid-liquid extraction process. In this study, a range of 1 to 24 h was selected to investigate its impact on the efficiency and yield of the extraction of KSO from kiwifruit seeds. As shown in Figure 3, the extraction yield of KSO significantly increased in the interval 1–6 h. After that, there was an increase of 0.70% (for 12 h), for 0.74% (for 18 h) and 0.46% (for 24 h), the extraction yield did not significantly increase with a further increase of time. At the early extraction stage, oil can more easily diffuse into the solvent, resulting in the rapid increase of extraction yield [32]. As time increases, oil in the solid phase is more difficult to be extracted. Therefore, considering both extraction efficiency and time cost, an extraction time of 18 h was selected as the extraction condition for subsequent experiments.

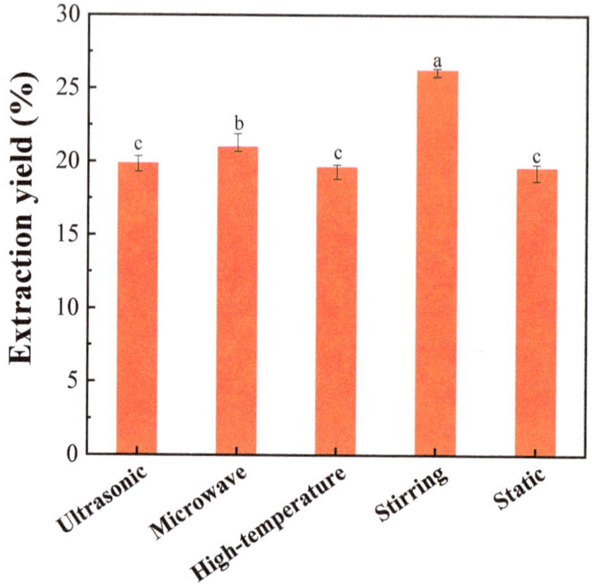

Figure 2. Extraction yields of KSO using n-butanol as the extractant and different intensification methods. The extraction conditions were as follows: 3:1 liquid–solid ratio, 6 h extraction time, 4/3 phase ratio (top/bottom phase), and 25 wt% ammonium sulfate. Different letters represent the statistically significant difference.

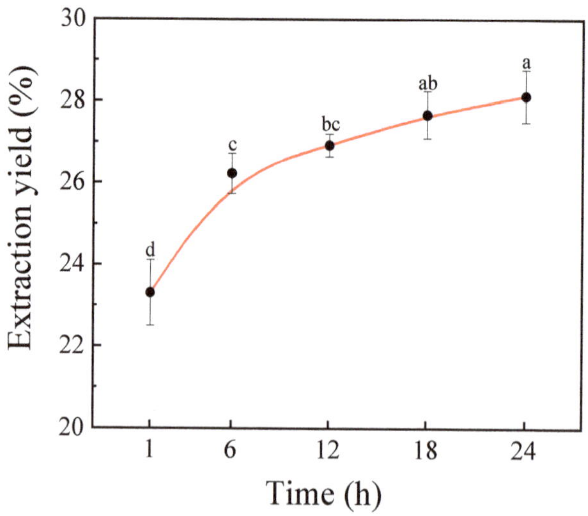

Figure 3. Extraction yields of KSO at different time lengths. The extraction conditions were as follows: n-butanol used as the extractant with stirring assisted extraction, 3:1 liquid-solid ratio, 4/3 phase ratio, and 25 wt% ammonium sulfate. Different letters represent the statistically significant difference.

3.1.4. Effect of Liquid-Solid Ratio

The liquid-solid ratios of 1:1 to 6:1 were compared. As shown in Figure 4, the extraction yield increased when the liquid-solid ratio increased from 1:1 to 4:1, but when the liquid-solid ratio was further increased, the extraction yield decreased. These results may be attributed to the hydraulic effect formed by the narrow and long tube at high liquid levels, which hindered the internal and external oil diffusion. Therefore, a liquid-solid ratio of 4:1 was chosen.

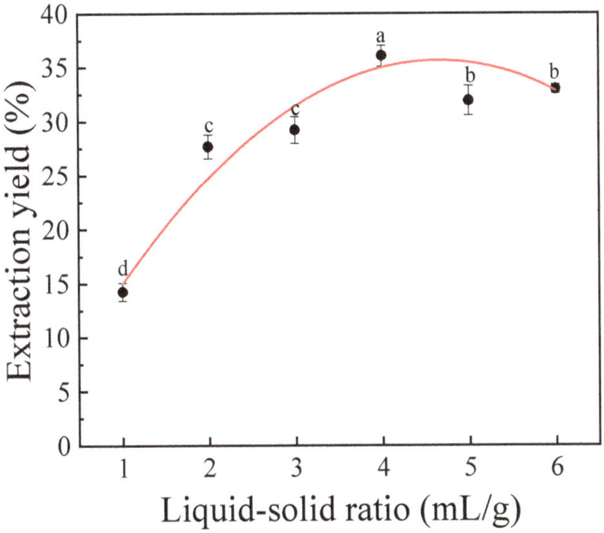

Figure 4. Extraction yields of KSO at different liquid-solid ratios. The extraction conditions were as follows: n-butanol used as the extractant with stirring assisted extraction, 18 h extraction time, 4/3 phase ratio, and 25 wt% ammonium sulfate. Different letters represent the statistically significant difference.

3.2. Factors Influencing the Purification of KSO Using TPP

3.2.1. Effect of Salt Type

As an essential component of TPP, salt plays an important role in the separation process, especially in floating proteins and enzymes through the salting-out effect. The addition of salt induces changes in the solubility of proteins and enzymes in the solvent, leading to their precipitation and separation from the desired KSO, thereby contributing to the overall separation effectiveness [33,34]. Six common types of salt have been compared in this study, including NaCl, Na_2CO_3, Na_2SO_4, K_2HPO_4, $C_6H_5O_7(NH_4)_3$, and $(NH_4)_2SO_4$. As shown in Figure 5, the salt type has a relatively small impact on KSO extraction. The TPP formed by $C_6H_5O_7(NH_4)_3$ results in the lowest extraction efficiency of KSO, as $C_6H_5O_7(NH_4)_3$ has lower phase forming ability than other salts. This is mainly caused by its anions. Conversely, ammonium sulfate has the optimal extraction yield. This is because ammonium sulfate is more suitable for salting out the proteins, enzymes, and other components. Therefore, ammonium sulfate was chosen for subsequent experiments.

3.2.2. Effect of Ammonium Sulfate Concentration

The ammonium sulfate concentration observably affects the extraction efficiency of TPP, so its concentration range of 20–35 wt% was studied in this study. As shown in Figure 6, the maximum extraction yield was obtained at 28.5 wt% ammonium sulfate concentration. When the salt concentration is lower, it is difficult to form three-phase [35]. The salting out effect strengthens with the increase of salt concentration, resulting in an increase in extraction yield. However, when the ammonium sulfate concentration is further increased, the extraction yield decreases, which can be attributed to that the stronger salting out effect influences the dissolving capacity of n-butanol. Therefore, the ammonium sulfate concentration of 28.5 wt% was chosen.

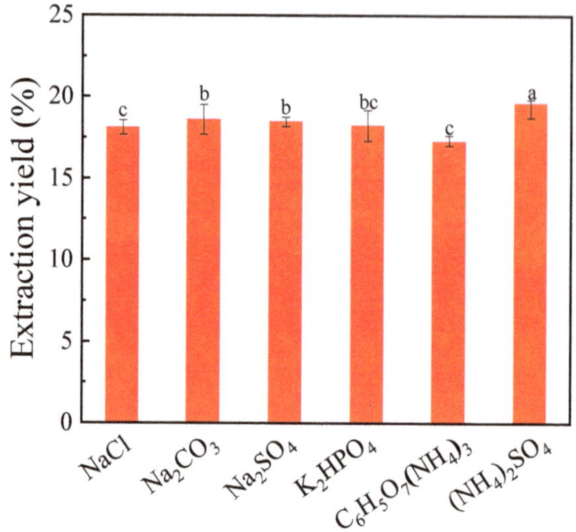

Figure 5. Extraction yields of KSO using different types of salt. The extraction conditions were as follows: n-butanol used as the extractant with stirring assisted extraction, 18 h extraction time, 3:1 liquid–solid ratio, 4/3 phase ratio, and 25 wt% salt concentration. Different letters represent the statistically significant difference.

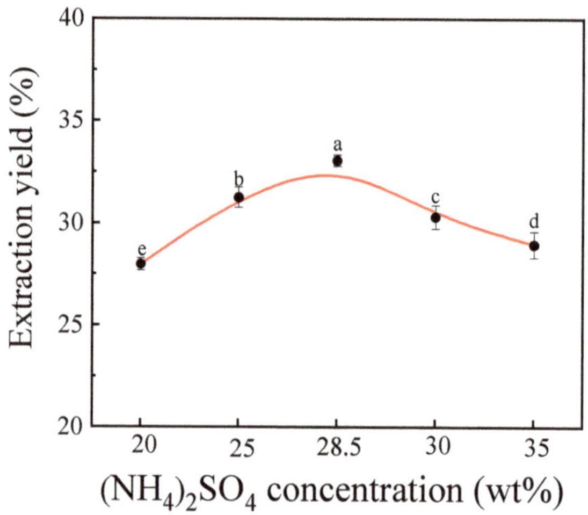

Figure 6. Extraction yields of KSO at different ammonium sulfate concentrations. The extraction conditions were as follows: n-butanol was used as the extractant with stirring assisted extraction, 18 h extraction time, 4:1 liquid–solid ratio, and 4/3 phase ratio. Different letters represent the statistically significant difference.

3.2.3. Effect of Phase Ratio

In the TPP system, the phase ratio (n-butanol top phase/ammonium sulfate bottom phase) was studied. As shown in Figure 7, the maximum extraction yield was obtained at 1/1 phase ratio. When the phase ratio is 4/1, there is no TPP formation. The middle phase rich in protein combines with the salt to form a homogeneous gel phase. Oil is stuck in the gel, resulting in low extraction yield. When the phase volume of salt phase further exceeds the n-butanol phase, the extraction yield decreases. This can be possibly attributed to the insufficient dissolution of oil in the middle phase. Therefore, a phase ratio (n-butanol phase/ammonium sulfate phase of 1/1) was chosen.

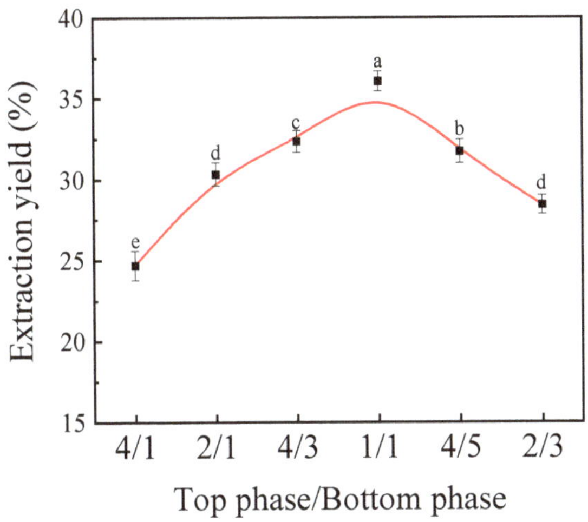

Figure 7. Extraction yields of KSO at different phase ratios. The extraction conditions were as follows: n-butanol used as the extractant with stirring assisted extraction, 18 h extraction time, 4:1 liquid–solid ratio, and 28.5 wt% ammonium sulfate. Different letters represent the statistically significant difference.

3.3. *Repeated Extraction of KSO from Kiwifruit Seeds*

To maximize the recovery of KSO, repeated extraction was performed using n-butanol. The total oil content was 45.51% when using low field nuclear magnetic resonance for kiwifruit seeds used in this study. The kiwifruit seeds were repeatedly extracted using TPP. As shown in Figure 8, the total extraction yield was 43.70% after five runs of extraction, which was 96.02% of the total KSO content. The first extraction resulted in an extraction yield of 36.06% oil (79.23% of the total oil content), while the second extraction was 5.47% oil. The further repeated extraction resulted in only a low recovery of the oil. Therefore, only one run of extraction was performed for the samples.

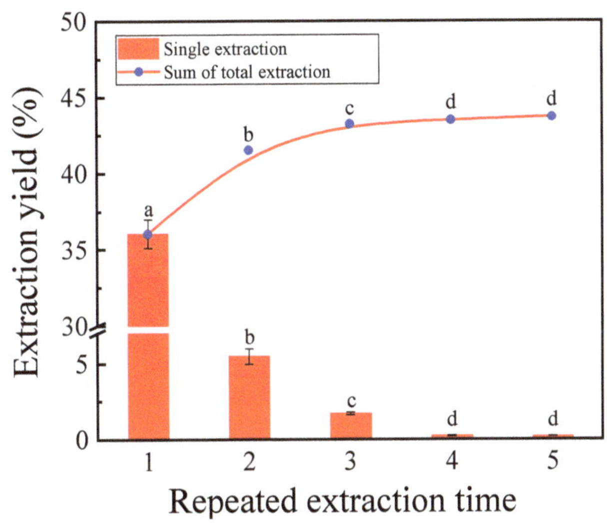

Figure 8. The repeat extraction of KSO. Different letters represent the statistically significant difference.

3.4. Analysis of KSO

GC-MS was used to analyze KSO, and the results are shown in Table 2. The content of α-linolenic acid reaches 51.26%, which is in agreement with previous reports [36]. α-Linolenic acid is a core substance for life and an important component for human brain cell development and histiocyte [37]. Therefore, α-linolenic acid is considered an essential nutrient required for human life. The content in mass of linoleic acid is 14.49%. Linoleic acid can regulate blood lipid levels and reduce cholesterol, which can effectively reduce the occurrence of hypercholesterolemia and reduce the risk of developing coronary heart disease [38]. The proportion of squalene reaches 15.07% for KSO obtained through this method. Squalene is a unique active nutrient, tricosahexaene, which has many physiological effects such as improving the activity of superoxide dismutase in the body, enhancing the immunity of the body, anti-fatigue, anti-aging, anti-tumor, et cetera. Smith et al. found through the A/J mouse experiment that dietary olive oil and squalene can effectively inhibit the occurrence of lung tumors induced by 4-(methylnitrosamine)-1-(3-pyridyl)-1-butanone (NNK) [39]. Hernández-Diaz et al. found that linoleic acid diet management significantly reduced blood pressure, blood sugar, cholesterol, and triglyceride levels in spontaneously hypertensive rats, as well as decreased the body fat index of adipose tissue in the heart, abdomen, and epididymis [40]. These effects were accompanied by tumor necrosis factor. The decrease in resistance in secretion suggests that linoleic acid may be an effective inhibitor of preadipocyte proliferation and differentiation. These results indicate that KSO can become a good supplementary nutrient for human health.

Table 2. Composition of KSO from TPP extraction.

Constituents	Content (%)
α-Linolenic acid	51.26 ± 1.04
Squalene	15.07 ± 1.11
Linoleic acid	14.49 ± 0.97
Conjugated linoleic acid	7.53 ± 0.46
Palmitic acid	5.18 ± 1.08
Oleic acid	4.39 ± 0.13
γ- linolenic acid	1.42 ± 0.39
Linolenic alcohol	0.66 ± 0.15

4. Conclusions

In this study, TPP formed by ammonium sulfate and n-butanol was used to extract KSO from kiwifruit seeds. The highest extraction yield of 36.06% was obtained for KSO under the following conditions: n-butanol as the extractant with stirring assisted extraction, 28.5 wt% ammonium sulfate, 18 h extraction time, 4:1 liquid–solid ratio, and 1/1 phase ratio. The one-run extraction recovered 79.23% of the total oil content. The composition of the KSO was obtained by GC-MS, indicating the highest content of 51.26% α-linolenic acid in the KSO obtained by this method. This method is simple, low-cost, and efficient, which can provide a guidance for agro-industrial residue utilization like kiwifruit seeds and other plant resources.

Author Contributions: Investigation, methodology, data curation, validation B.L. and Z.T.; writing-original draft preparation, B.L. and X.L.; resources, Q.L.; conceptualization, B.L. and Z.T.; supervision, Z.T. and X.L.; funding acquisition, Z.T.; project administration, B.L. and Z.T.; writing-review and editing, X.L. and Z.T. All authors have read and agreed to the published version of the manuscript.

Funding: This work was financially supported by the Agricultural Science and Technology Innovation Program (ASTIP-IBFC08), Central Public-interest Scientific Institution Basal Research Fund (1610242023001), Natural Science Foundation of Shandong Province (ZR2022MB037), Heze University Pharmaceutical Research Institute Cultivation Fund Project (2021YXYPY01), and Heze University Research Fund Science and Technology Project (XY21BS38).

Data Availability Statement: Not applicable.

Conflicts of Interest: The authors declare that they have no known competing financial interests or personal relationships in this paper.

References

1. Deng, J.; Liu, Q.; Zhang, Q.; Zhang, C.; Liu, D.; Fan, D.; Yang, H. Comparative study on composition, physicochemical and antioxidant characteristics of different varieties of kiwifruit seed oil in China. *Food Chem.* **2018**, *264*, 411–418. [CrossRef] [PubMed]
2. Coelho, R.; Kanda, L.R.S.; Hamerski, F.; Masson, M.L.; Corazza, M.L. Extraction of kiwifruit seed (*Actinidia deliciosa*) oil using compressed propane. *J. Food Process Eng.* **2016**, *39*, 335–344. [CrossRef]
3. Sivakumaran, S.; Huffman, L.; Sivakumaran, S.; Drummond, L. The nutritional composition of Zespri® SunGold Kiwifruit and Zespri® Sweet Green Kiwifruit. *Food Chem.* **2018**, *238*, 195–202. [CrossRef]
4. Schröder, R.; Nicolas, P.; Vincent, S.J.F.; Fischer, M.; Reymond, S.; Redgwell, R.J. Purification and characterisation of a galactoglucomannan from kiwifruit (*Actinidia deliciosa*). *Carbohydr. Res.* **2001**, *331*, 291–306. [CrossRef]
5. Wang, J.; Wang, J.; Vanga, S.K.; Raghavan, V. High-intensity ultrasound processing of kiwifruit juice: Effects on the microstructure, pectin, carbohydrates and rheological properties. *Food Chem.* **2020**, *313*, 126121. [CrossRef]
6. Sanz, V.; López-Hortas, L.; Torres, M.D.; Domínguez, H. Trends in kiwifruit and byproducts valorization. *Trend Food Sci. Technol.* **2021**, *107*, 401–414. [CrossRef]
7. Wang, J.; Zhang, L.; Shi, J.; Vanga, S.K.; Raghavan, V. Effect of microwave processing on the nutritional properties and allergenic potential of kiwifruit. *Food Chem.* **2023**, *401*, 134189. [CrossRef]
8. Windauer, L.B.; Insausti, P.; Biganzoli, F.; Benech-Arnold, R.; Izaguirre, M.M. Dormancy and germination responses of kiwifruit (*Actinidia deliciosa*) seeds to environmental cues. *Seed Sci. Res.* **2016**, *26*, 342–350. [CrossRef]
9. Qu, L.; Liu, Q.; Zhang, Q.; Tuo, X.; Fan, D.; Deng, J.; Yang, H. Kiwifruit seed oil prevents obesity by regulating inflammation, thermogenesis, and gut microbiota in high-fat diet-induced obese C57BL/6 mice. *Food Chem. Toxicol.* **2019**, *125*, 85–94. [CrossRef]
10. Berry, E.M.; Hirsch, J. Does dietary linolenic acid influence blood pressure? *Am. J. Clinic. Nutrit.* **1986**, *44*, 336–340. [CrossRef]
11. Ghafoorunissa; Ibrahim, A.; Natarajan, S. Substituting dietary linoleic acid with α-linolenic acid improves insulin sensitivity in sucrose fed rats. *BBA—Mol. Cell Biol. Lipids* **2005**, *1733*, 67–75. [CrossRef] [PubMed]
12. Maojun, Y.A.O.; Fei, L.I.U. Study on supercritical CO_2 extraction and microcapsulation of kiwi fruit seed oil. *Food Sci.* **2007**, *28*, 162–165.
13. Santos, T.R.J.; de Aquino Santana, L.C.L. Antimicrobial potential of exotic fruits residues. *S. Afr. J. Bot.* **2019**, *124*, 338–344. [CrossRef]
14. Wu Cai, E.; Xu, K.; Li, Y.; Li, T.; Li, W. Study on the solubility of Kiwi fruit seed oil in supercritical carbon dioxide. *Sci. Agr. Sin.* **2007**, *40*, 1242–1247.
15. Krist, S. Kiwi Seed Oil. In *Vegetable Fats and Oils*; Krist, S., Ed.; Springer International Publishing: Cham, Switzerland, 2020; pp. 395–398.
16. Thilakarathna, R.C.N.; Siow, L.F.; Tang, T.K.; Lee, Y.Y. A review on application of ultrasound and ultrasound assisted technology for seed oil extraction. *J. Food Sci. Technol.* **2023**, *60*, 1222–1236. [CrossRef]
17. Cravotto, G.; Bicchi, C.; Mantegna, S.; Binello, A.; Tomao, V.; Chemat, F. Extraction of kiwi seed oil: Soxhlet versus four different non-conventional techniques. *Nat. Prod. Res.* **2011**, *25*, 974–981. [CrossRef]
18. Tan, K.H.; Lovrien, R. Enzymology in aqueous-organic cosolvent binary mixtures. *J. Biol. Chem.* **1972**, *247*, 3278–3285. [CrossRef]
19. Panadare, D.C.; Rathod, V.K. Three phase partitioning for extraction of oil: A review. *Trends Food Sci. Technol.* **2017**, *68*, 145–151. [CrossRef]
20. Rajendran, D.S.; Chidambaram, A.; Kumar, P.S.; Venkataraman, S.; Muthusamy, S.; Vo, D.V.N.; Rangasamy, G.; Vaithyanathan, V.K.; Vaidyanathan, V.K. Three-phase partitioning for the separation of proteins, enzymes, biopolymers, oils and pigments: A review. *Environ. Chem. Lett.* **2023**, *21*, 911–934. [CrossRef]
21. Carpaneto, E.; Chicco, G.; Akilimali, J.S. Loss partitioning and loss allocation in three-phase radial distribution systems with distributed generation. *IEEE T. Power Syst.* **2008**, *23*, 1039–1049. [CrossRef]
22. Dembczynski, R.; Dolata, A.M.; Bombczynska, K.; Bialas, W.; Jankowski, T. Separation of lactoferrin from whey in three-phase extraction systems (TPP method). *Zywn-Nauka Technol. Jakość* **2013**, *20*, 119–136.
23. Liu, Y.F.; Garcia, A.A. Three phase countercurrent extraction. *Chem. Eng. Commun.* **2000**, *182*, 239–259. [CrossRef]
24. Sharma, A.; Gupta, M.N. Macroaffinity ligand-facilitated three-phase partitioning (MLFTPP) for purification of xylanase. *Biotechnol. Bioeng.* **2002**, *80*, 228–232. [CrossRef]
25. Varakumar, S.; Umesh, K.V.; Singhal, R.S. Enhanced extraction of oleoresin from ginger (*Zingiber officinale*) rhizome powder using enzyme-assisted three phase partitioning. *Food Chem.* **2017**, *216*, 27–36. [CrossRef]
26. Wang, Q.; Huo, Y.; Wang, J.; Zhang, Q.; Ren, Y.; Ge, F. On Determination of serum total cholesterol cholesterol oxidase by three-phase partitioning and its clinic application. *J. Southwest China Norm. Univ. Nat. Sci. Ed.* **2014**, *39*, 47–55.
27. Wei, S.; Zhou, Q.; Qian, W.; Zhang, C. Preparation of papain by three-phase partitioning method. *Modern Chem. Ind.* **2017**, *37*, 114–117.
28. Dutta, R.; Sarkar, U.; Mukherjee, A. Process optimization for the extraction of oil from *Crotalaria juncea* using three phase partitioning. *Ind. Crop Prod.* **2015**, *71*, 89–96. [CrossRef]

29. Tan, Z.; Yang, Z.; Yi, Y.; Wang, H.; Zhou, W.; Li, F.; Wang, C. Extraction of oil from flaxseed (*Linum usitatissimum* L.) using enzyme-assisted three-phase partitioning. *Appl. Biochem. Biotechnol.* **2016**, *179*, 1325–1335. [CrossRef]
30. Tang, S.; Zhang, Y.; Li, W.; Tang, X.; Huang, X. Rapid and simultaneous measurement of fat and moisture contents in pork by low-field nuclear magnetic resonance. *Foods* **2023**, *12*, 147. [CrossRef] [PubMed]
31. Cai, C.; Chen, X.; Li, F.; Tan, Z. Three-phase partitioning based on CO_2-responsive deep eutectic solvents for the green and sustainable extraction of lipid from *Nannochloropsis* sp. *Sep. Purif. Technol.* **2021**, *279*, 119685. [CrossRef]
32. Zeb, L.; Shafiq, M.; Chi, Z.-Y.; Xiu, Z.L. Separation of microalgal docosahexaenoic acid-rich oils using a microwave-assisted three-phase partitioning system. *Sep. Purif. Technol.* **2020**, *252*, 117441. [CrossRef]
33. Chew, K.W.; Ling, T.C.; Show, P.L. Recent Developments and applications of three-phase partitioning for the recovery of proteins. *Sep. Purif. Rev.* **2019**, *48*, 52–64. [CrossRef]
34. Yan, J.-K.; Wang, Y.-Y.; Qiu, W.-Y.; Ma, H.; Wang, Z.-B.; Wu, J.-Y. Three-phase partitioning as an elegant and versatile platform applied to nonchromatographic bioseparation processes. *Crit. Rev. Food Sci. Nutr.* **2018**, *58*, 2416–2431. [CrossRef]
35. Yan, J.-K.; Wang, Y.-Y.; Qiu, W.-Y.; Wang, Z.-B.; Ma, H. Ultrasound synergized with three-phase partitioning for extraction and separation of *Corbicula fluminea* polysaccharides and possible relevant mechanisms. *Ultrason. Sonochem.* **2018**, *40*, 128–134. [CrossRef] [PubMed]
36. Qu, L.; Liu, Q.; Zhang, Q.; Liu, D.; Zhang, C.; Fan, D.; Deng, J.; Yang, H. Kiwifruit seed oil ameliorates inflammation and hepatic fat metabolism in high-fat diet-induced obese mice. *J. Funct. Foods* **2019**, *52*, 715–723. [CrossRef]
37. Yue, H.; Qiu, B.; Jia, M.; Liu, W.; Guo, X.; Li, N.; Xu, Z.; Du, F.; Xu, T.; Li, D. Effects of α-linolenic acid intake on blood lipid profiles: A systematic review and meta-analysis of randomized controlled trials. *Crit. Rev. Food Sci.* **2021**, *61*, 2894–2910. [CrossRef] [PubMed]
38. den Hartigh, L.J. Conjugated linoleic acid effects on cancer, obesity, and atherosclerosis: A review of pre-clinical and human trials with current perspectives. *Nutrients* **2019**, *11*, 370. [CrossRef]
39. Smith, T.J.; Yang, G.Y.; Seril, D.N.; Liao, J.; Kim, S. Inhibition of 4-(methylnitrosamino)-1-(3-pyridyl)-1-butanone-induced lung tumorigenesis by dietary olive oil and squalene. *Carcinogenesis* **1998**, *19*, 703–706. [CrossRef]
40. Hernández-Díaz, G.; Alexander-Aguilera, A.; Arzaba-Villalba, A.; Soto-Rodríguez, I.; García, H.S. Effect of conjugated linoleic acid on body fat, tumor necrosis factor alpha and resistin secretion in spontaneously hypertensive rats. *Prostag. Leukotr. Ess.* **2010**, *82*, 105–109. [CrossRef]

processes

MDPI

Article

Experimental Study on Enrichment of Indole in Wash Oil by a Combined Process of Extraction and Re-Extraction

Su Jin Kim

Department of Chemical & Biological Engineering, Chungwoon University, Incheon 22100, Republic of Korea; sujkim@chungwoon.ac.kr

Abstract: Wash oil, one of the by-products of coal tar distillation, contains useful nitrogen heterocyclic compounds (*NHC*) such as indole (*IN*), which is used as a raw material for perfume, essential amino acid, and pharmaceuticals. In this study, as a pre-treatment step for purification of *IN*, which is about 5.75% contained in wash oil, the enrichment of *IN* in wash oil was examined by combining formamide extraction (FE) and n-hexane re-extraction (HRE). The raw material for the re-extraction was a mixture of the preceding formamide extract phase acquired through batchwise 5-stage co-current extraction trial under fixed FE condition. As the 5-stage FE and HRE progressed, the *IN* yield based on *IN* in the wash oil was approximately 73.3%, and the *IN* concentration in the raffinate oil acquired from the re-extraction run was highly enriched to about 92.4%. In addition, the *IN* enrichment efficiency of the FE-HRE combined process (FHC) employed in this study was compared with methanol extraction (ME)-hexane re-extraction (HRE) combination (MHC).

Keywords: indole; wash oil; extraction; re-extraction; enrichment combination

Citation: Kim, S.J. Experimental Study on Enrichment of Indole in Wash Oil by a Combined Process of Extraction and Re-Extraction. *Processes* **2023**, *11*, 3097. https://doi.org/10.3390/pr11113097

Academic Editors: Raymond Cecil Everson, Prashant K. Sarswat and Yanlin Zhang

Received: 4 September 2023
Revised: 23 October 2023
Accepted: 25 October 2023
Published: 27 October 2023

1. Introduction

Wash oil, one of the by-products of coal tar distillation, contains useful nitrogen heterocyclic compound (*NHC*) such as indole (*IN*), which is used as a raw material for perfume, essential amino acid, and pharmaceuticals [1–7]. Therefore, if it is possible to separate and purify *IN* contained in wash oil used only as fuel, it is very meaningful in terms of resource utilization.

Although it depends on the manufacturer from which the wash oil is produced, wash oil used in this study contained approximately *IN* of 5.75%. From Table 1, we see that the boiling points of the various compounds present in the wash oil are almost the same, so separation between components through distillation operation is very difficult. From this fact, we think that the following multi-step procedure is required for purification of *IN* in wash oil: (i) The recovery of *IN* contained in wash oil by the solvent extraction method etc.; (ii) Enrichment of *IN* presenting in extract phase through distillation [6] or re-extraction method [3] etc.; (iii) The purification of *IN* using as a raw material, which are *IN*-enriched distillate acquired through distillation operation or *IN*-enriched raffinate recovered through re-extraction operation.

Until now, the separation of *NHC* contained in coal tar distillation by-products has been mainly achieved by reaction extraction using acid and base substances, but this separation method uses both a strong alkaline solution (e.g., NaOH) and an acidic solution (e.g., H_2SO_4) [8,9]. It has the disadvantage of being expensive, corroding the metal separation apparatus and seriously polluting the environment, and also complicating the separation process because the separation agent cannot be regenerated. Therefore, it is of great significance to find a novel alternative separation method that is simpler and more efficient than reactive extraction.

Table 1. Physical property and composition of each compound in feed.

Compound	Physical Property		Composition (wt%)	
	b.p. * [K]	m.p. ** [K]	Extraction	Re-Extraction
Naphthalene (*NA*, $C_{10}H_8$)	491	351–353	0.960	0.013
Quinoline (*QU*, C_9H_7N)	511	257	1.379	0.157
Indole (*IN*, C_8H_7N)	526	325	5.753	0.993
2-Methylnaphthalene (2*MNA*, $C_{11}H_{10}$)	514–515	307–309	53.251	0.391
1-Methylnaphthalene (1*MNA*, $C_{11}H_{10}$)	513–516	251	21.854	0.165
Biphenyl (*BP*, $C_{12}H_{10}$)	528	342	11.339	0.074
Others			5.464	98.207 (with solvent)

*: boiling point, **: melting point.

For the development of a novel separation method, until now the recovery and enrichment of *IN* contained in coal tar fraction was reviewed through separation operations, such as traditional extraction methods [1,3–6,10] using organic solvents, supercritical extraction [9,11,12], extraction using ionic liquids [2,7,13–18], and azeotropic distillation etc. The purification of *IN* using the *IN*-enriched distillate acquired by distillation operation of the extract oil was examined by purification operations such as crystallization [19], adsorption [20], and inclusion complexation [21,22].

In the previous study [3], we examined the recovery and enriching of *IN* in wash oil through methanol extraction (ME)-hexane re-extraction (HRE) combination (MHC) as a pre-treatment operation to purify 5.75% *IN* contained in the wash oil. As a result of the review, the solubility of *IN* is really excellent, but the selectivity is very low, so the concentration of *IN* (about 12.4 wt%) in the extract phase recovered by ME run was very low. In addition, when the extract phase finally acquired through ME run is distilled, methanol flows out as an overhead product, and water and extracted oil remain as bottom products. Therefore, in order to recover the extracted oil, a phase separation process for the bottom product is required. As the phase separation is added, the process for *IN* enrichment becomes more complicated. To solve the disadvantages of ME, that is, the low selectivity of *IN* and the complexity of the *IN* enrichment process, this study first examined the recovery of *IN* in wash oil by using formamide as the solvent, which shows relatively high selectivity of *IN* [1,5] and does not cause phase separation during distillation operation for solvent recovery. Next, the enrichment of *IN* by re-extraction was studied using extract phase which is acquired from formamide extraction (FE) as a raw material and n-hexane as a solvent, respectively. For this, we examined the effects of operation factors and conditions of FE and n-hexane re-extraction (HRE) on the enrichment of *IN*. Furthermore, we compared the experimental results of this study obtained by FE-HRE combination (FHC) with those of the previous study [3] obtained by MHC in terms of the enrichment efficiency of *IN*.

2. Experimental Method and Conditions

2.1. Equilibrium Extraction

In this study, equilibrium extraction of batch 1-stage and co-current 5-stage was performed by the same method as MHC in the previous study [3]. The formamide was mixed with tap water (electrical conductivity: 86 μS/cm, pH: 7.3) to make a solvent of a certain concentration. In order to reach liquid-liquid equilibrium, the 1 L Erlenmeyer flask containing a constant volume of feed and fresh solvent was placed in a shaking water bath maintained at a constant temperature and shaken for a constant time. After equilibrium was reached, it was allowed to settle for a certain time, and the extract phase and the raffinate phase were separated using a 1 L separatory funnel, and then the masses of the two phases were measured. The raffinate phase separated from each stage and a fresh solvent were used as the feed and a solvent of the next stage, respectively. The analysis conditions of two phases by the gas chromatography (GC) are shown in Table 2. The concentration of each compound contained in the two phases was determined using a calibration curve for each compound obtained using six standard reagents, which were selected considering the constituent compounds of wash oil [3].

Table 2. Equipment and conditions of analysis by gas chromatography (GC).

Gas Chromatograph	**HP 6890 (Hewlett Packard Co., Palo Alto, CA, USA)**
Column	Capillary column HP-1
length [m] inner diameter [mm] film [μm]	60 0.32 0.25
Carrier gas	N_2
split ratio [-]	0.025
flow rate in column [mL/min]	2
Sample volume [μL]	1
Injection temperature [K]	523
Column temperature [K]	383–593
temperature program [K/min]	0 (for 0–3 min) 5 (383–523 K) 14 (523–593 K)
Detector (FID) temperature [K]	593

Table 3 shows the material systems and experimental conditions. Wash oil (provided by OCI Company Ltd., Seoul, Republic of Korea), and the raffinate phases which recovered from each stage run, were used as feed for extraction operation, and aqueous formamide solution was used as extraction solvent. The volume fraction of water in solvent in initial ($y_{w,0}$), the volume ratio of solvent to feed in initial (E_0/R_0, E_0 and R_0: volume of fresh solvent and feed put into n = 1), the stage number of batch co-current equilibrium extraction (n), and the shaking time (t) were changed, the temperature (T) and the volume of solvent added to each stage (E_0–E_4) were fixed.

Table 3. Material systems and experimental conditions.

Equilibrium Extraction		**Equilibrium Re-Extraction**
Feed: wash oil or raffinate phase *	*MFEP* ** or raffinate phase ***	
Solvent: aqueous solution of formamide	n-hexane	
Shaking time, t (h)	72	72
Stage number of batch co-current Equilibrium extraction, n (-)	1–5	1–5
Temperature, T (K)	303	303
Volume of fresh solvent (E_0–E_4), (mL)	320	320
Volume fraction of water in solvent in initial, $y_{w,0}$ (-)	0.1–0.3	–
Volume ratio of solvent to feed in initial, E_0/R_0 (-)	0.5–3	0.5–3

* raffinate phase acquired from FE run of each stage, ** mixed formamide extract phase ($MFEP$, $\sum_{n=1}^{5} E_n$), which is a mixture of each formamide extract phase acquired from FE run of each stage, *** raffinate phase acquired from HRE run of each stage.

All experimental data acquired from extraction and re-extraction operation of this study were obtained by performing the experiment only once, rather than repeating it 2–3 times under the same conditions.

2.2. Equilibrium Re-Extraction

Liquid-liquid equilibrium was reached in the same method as the above-mentioned FE using a certain amount of raw materials and a fresh solvent, and then separated into a raffinate phase and an extraction phase. The analysis conditions of two phases by GC were the same as the FE described above (see Table 2).

Material system and experimental conditions are shown in Table 3. A mixed formamide extract phase ($MFEP$, $\sum_{n=1}^{5} E_n$), which is a mixture of each extract phase acquired from FE run of each stage (n = 1–5), and n-hexane were used as feed and solvent, respectively. T, t, and E_0–E_4 were fixed, E_0/R_0 and n were varied in this study.

3. Results and Discussion

3.1. Calculation Formula

In extraction and re-extraction, the distribution coefficient of compound i ($m_{i,n}$) was calculated using Equation (1).

$$m_{i,n} = y_{i,n}/x_{i,n} \text{ (n = 1-5)} \quad (1)$$

$x_{i,n}$ and $y_{i,n}$ in Equation (1), respectively, represent the mass fraction of component i in a raffinate phase and an extract phase which acquired at the nth stage.

The yield of IN ($Y_{IN,5}$) in FE run of n = 5 was calculated using the Equation (2).

$$Y_{IN,5} = \sum_{n=1}^{5} E_n \times y_{IN,n})/(R_0 \times x_{IN,0}) \times 100\% \text{ (n = 1–5)} \quad (2)$$

$x_{IN,0}$ and R_0 represent the mass fraction of IN in wash oil, and the mass of wash oil, respectively. Additionally, E_n and $y_{IN,n}$ denote the mass of the extract phase and the mass fraction of IN in the extract phase acquired from the nth stage extraction, respectively.

$Y_{IN,5}$, the yield of IN obtained after HRE run of n = 5, was calculated using the Equation (3).

$$Y_{IN,5} = (R_5 \times x_{IN,5})/(R_0 \times x_{IN,0}) \times 100\% \text{ (n = 1–5)} \quad (3)$$

in the denominator of Equation (3), R_0 and $x_{IN,0}$ represent the mass of the wash oil and the mass fraction of IN in the wash oil, respectively. In addition, R_5 and $x_{IN,5}$ in the numerator, respectively, represent the mass of the raffinate phase acquired from the 5th re-extraction run and the mass fraction of IN presenting in the raffinate phase.

The selectivity of compound i in reference to 2-methylnaphthalene (2MNA) or IN acquired from each operation of the nth stage, $(\beta_{i,2MNA})_n$ or $(\beta_{i,IN})_n$, were calculated using the following equation, respectively.

$$(\beta_{i,2MNA})_n = m_{i,n}/m_{2MNA,n},\ (\beta_{i,IN})_n = m_{i,n}/m_{IN,n} \text{ (n = 1–5)} \quad (4)$$

where $m_{i,n}$, $m_{2MNA,n}$, and $m_{IN,n}$ represent the distribution coefficients of compound i, 2MNA, and IN obtained from each operation of the nth stage, respectively.

3.2. Equilibrium Extraction

3.2.1. GC Chromatogram of Wash Oil (Extraction Feed)

Figure 1a shows the GC chromatogram of wash oil and the names of the identified compounds [3]. We have added a small amount of six standard reagents purchased from Sigma Aldrich Korea, Seoul, Republic of Korea, one compound each, to the wash oil and accurately identified each compound presented in the wash oil based on the change in height of the GC chromatogram peak. From identification by GC, we found that wash oil contained 2MNA, 1-methylnaphthalene (1MNA), and biphenyl (BP) in large amounts compared to naphthalene (*NA*), quinoline (*QU*), and *IN*. Table 1 shows the composition of the quantified compounds. The composition of *IN* was about 5.75%.

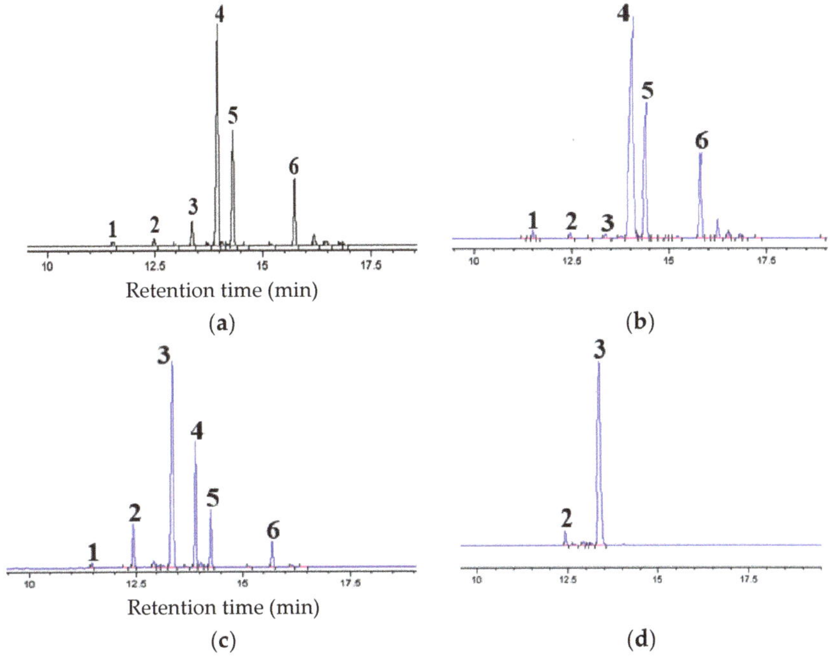

Figure 1. GC chromatograms of (**a**) wash oil [3], (**b**) raffinate phase (R_5) acquired from FE run of n = 5, (**c**) *MFEP* ($\sum_{n=1}^{5} E_n$, re-extraction feed), (**d**) raffinate phase (R_5) acquired from HRE run of n = 5 using *MFEP* as feed. Extraction conditions: n = 1–5, $y_{w,0} = 0.1$, $E_0/R_0 = 1$, T = 303 K, t = 72 h; Re-extraction conditions: n = 1–5, $E_0/R_0 = 0.5$, T = 303 K, t = 72 h. Peak number 1: *NA*, 2: *QU*, 3: *IN*, 4: 2*MNA*, 5: 1*MNA*, and 6: *BP*.

3.2.2. Confirmation of Equilibration Time

Although the figure was omitted, the two phases obtained from contact between the feed and fresh solvent (shaking times 24, 48, 72, and 96 h) were analyzed to confirm the time taken to reach equilibrium. Regardless of shaking time and operation, the compositions of the two phases acquired contacting 72 and 96 h were almost identical. For this reason, all experiments in this study were performed with the shaking time maintained at 72 h.

3.2.3. Recovery Performance of IN

Figure 2a,b, respectively, show the solvent comparison results for the effect of $y_{w,0}$ on the distribution coefficient of compound *i* ($m_{i,1}$), the selectivity of compound *i* in reference to 2*MNA* ($(\beta_{i,2MNA})_1$, the ratio of $m_{i,1}$ in ME run to FE run $\left(\frac{(m_{i,1})_{\mathrm{ME}}}{(m_{i,1})_{\mathrm{FE}}}\right)$ and the ratio of $(\beta_{i,2MNA})_1$ in FE run to ME run $\left(\frac{(\beta_{i,2MNA})_{1,\mathrm{FE}}}{(\beta_{i,2MNA})_{1,\mathrm{ME}}}\right)$ obtained in n = 1. The abbreviations FE and ME refer to the formamide extraction and methanol extraction, respectively. For comparison of solvents, the recovering performance data of *IN* obtained from a previous study [3] using methanol aqueous solution as an extraction solvent are added to this figure. For two solvents used, the $m_{i,1}$ and $(\beta_{i,2MNA})_1$ values of 2*MNA* and 1*MNA*, which are structural isomers, are almost identical in Figure 2a, so the keys of 2*MNA* and 1*MNA* overlap, the key of 2*MNA* is not visible. Under the same $y_{w,0}$, the $m_{i,1}$ of two kinds of *NHC* was much higher when methanol, which is a greater polar than formamide, was used as the extraction solvent, but on the contrary, the $(\beta_{i,2MNA})_1$ when formamide was used was much higher than that of methanol. Regardless of the extraction solvent, compared to *NA*, *1MNA*, *2MNA* and *BP*, $m_{i,1}$ of *NHC* with high polarity was much larger. Additionally, an increase in $y_{w,0}$ increased the polarity of the extraction phase, so $m_{i,1}$ of each compound decreased. Conversely, $(\beta_{i,2MNA})_1$ of two kinds of *NHC* decreased as $y_{w,0}$ decreased in the two solvents used. These results for each compound suggest that the selectivity of entire *NHC* based on entire bicyclic aromatic compound (*BAC*) contained in the wash oil may be very high. Consistent with the previously published methanol results [3], for each compound using formamide, $m_{i,1}$ and $(\beta_{i,2MNA})_1$ showed the order *IN* > *QU* > *NA* > *1MNA* = *2MNA* > *BP*. As predicted from the results of Figure 2a, we can see in Figure 2b that $\frac{(m_{i,1})_{ME}}{(m_{i,1})_{FE}}$ and $\frac{(\beta_{i,2MNA})_{1,FE}}{(\beta_{i,2MNA})_{1,ME}}$ rapidly increase as $y_{w,0}$ decreases. In the range of $0.1 \le y_{w,0} \le 0.3$, we found that $(m_{i,1})_{ME}$ of all compounds was about 2 to 30.7 times greater than $(m_{i,1})_{FE}$, whereas $(\beta_{i,2MNA})_{1,FE}$ was about 1 to 9.8 times greater than $(\beta_{i,2MNA})_{1,ME}$. The $\frac{(m_{i,1})_{ME}}{(m_{i,1})_{FE}}$ and $\frac{(\beta_{i,2MNA})_{1,FE}}{(\beta_{i,2MNA})_{1,ME}}$ of *i* = *IN*, which is the compound for recovery and enriching in this study, showed about 3.1 and 9.8 at $y_{w,0} = 0.1$ and about 2 and 3.9 at $y_{w,0} = 0.3$. Considering the balance between $m_{i,1}$ and $(\beta_{i,2MNA})_1$ of *IN*, the FE run of $y_{w,0} = 0.1$ was thought to be the most appropriate condition.

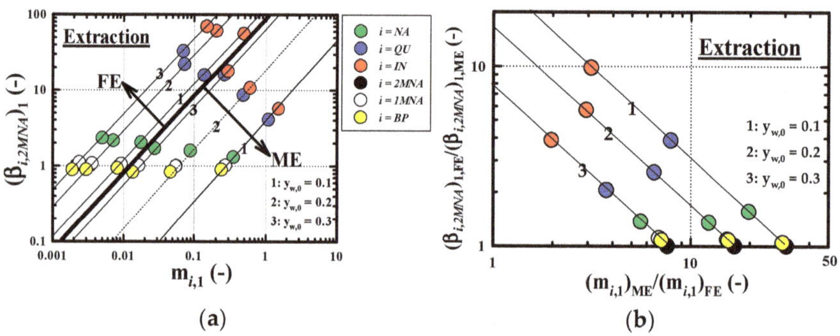

Figure 2. Solvent comparison for effect of $y_{w,0}$ on (**a**) $m_{i,1}$ and $(\beta_{i,2MNA})_1$, (**b**) $\frac{(m_{i,1})_{ME}}{(m_{i,1})_{FE}}$ and $\frac{(\beta_{i,2MNA})_{1,FE}}{(\beta_{i,2MNA})_{1,ME}}$. FE: formamide extraction, ME: methanol extraction [3]. Experimental conditions: n = 1, $y_{w,0}$ = 0.1–0.3, $E_0/R_0 = 0.1$, T = 303 K, t = 72 h. Keys: *NA* naphthalene, *QU* quinoline, *IN* indole, *2MNA* 2-methylnaphthalene, *1MNA* 1-methylnaphthalene, *BP* biphenyl.

Figure 3a,b, respectively, present the solvent comparison results for the effect of E_0/R_0 on $m_{i,1}$ and $(\beta_{i,2MNA})_1$, $\frac{(m_{i,1})_{ME}}{(m_{i,1})_{FE}}$ and $\frac{(\beta_{i,2MNA})_{1,FE}}{(\beta_{i,2MNA})_{1,ME}}$ obtained in n = 1. Irrespective of the solvent used, the effects of E_0/R_0 on $m_{i,1}$ and $(\beta_{i,2MNA})_1$ are not significant for all compounds. In addition, in $0.5 \le E_0/R_0 \le 3$, we found that $(m_{i,1})_{ME}$ for all compounds was about 2.5 to 30.7 times greater than $(m_{i,1})_{FE}$, whereas $(\beta_{i,2MNA})_{1,FE}$ was about 1 to 9.8 times greater than $(\beta_{i,2MNA})_{1,ME}$. The $\frac{(m_{i,1})_{ME}}{(m_{i,1})_{FE}}$ and $\frac{(\beta_{i,2MNA})_{1,FE}}{(\beta_{i,2MNA})_{1,ME}}$ of *i* = *IN*, which is the compound for recovery and enriching in this study, showed about 2.5 and 9.3 at $E_0/R_0 = 0.5$ and about 3.2 and 9.8 at $E_0/R_0 = 3$. $E_0/R_0 = 1$ was thought the optimal condition for FE run when considering the recovery efficiency of *IN* per unit solvent volume.

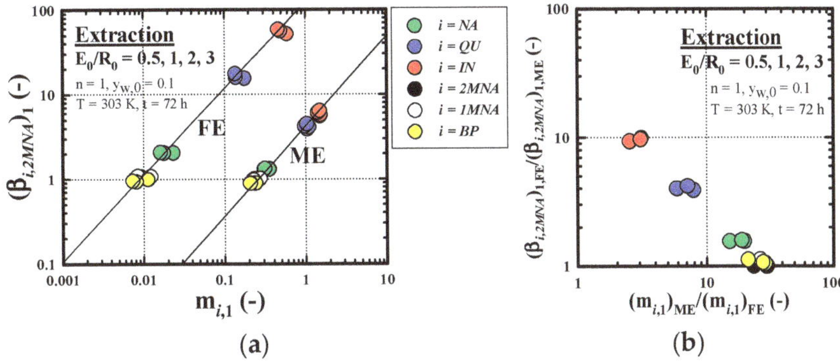

Figure 3. Solvent comparison for effect of E_0/R_0 on (**a**) $m_{i,1}$ and $(\beta_{i,2MNA})_1$, (**b**) $\frac{(m_{i,1})_{ME}}{(m_{i,1})_{FE}}$ and $\frac{(\beta_{i,2MNA})_{1,FE}}{(\beta_{i,2MNA})_{1,ME}}$. FE: formamide extraction, ME: methanol extraction [3].

Figure 4a–c, respectively, show the solvent comparison results for the effect of n on $x_{i,n}$ (without solvent), $m_{i,n}$ and $(\beta_{i,2MNA})_n$, $\frac{(m_{i,n})_{ME}}{(m_{i,n})_{FE}}$ and $\frac{(\beta_{i,2MNA})_{n,FE}}{(\beta_{i,2MNA})_{n,ME}}$ obtained in the nth FE run. The n = 0 shown in Figure 4a represents wash oil. Regardless of the solvent used, as n increases, large amounts of highly polar *NHC* transferred into the extract phase, resulting in a sharp decrease in the $x_{i,n}$ of *NHC* in the raffinate phase. In ME run of n = 5, the key did not appear because *IN* was not contained within R_5. Conversely, except for *NHC*, *BAC* (*NA*, 2*MNA*, 1*MNA*) and *BP* showed almost constant $x_{i,\,n}$ values regardless of n and solvent. It is considered that this is the result of reflecting the content of each compound included in the wash oil [3]. In the range of n = 1–5, $x_{IN,n}$ ($i = IN$) showed the ranges of about 0.73–3.93% in the case of FE and about 0–3.3% in the case of ME [3], respectively. Irrespective of the solvent used, improvement in wash oil quality is expected from $x_{IN,5}$. In addition, from Figure 4b,c, for all the compounds in this study, $m_{i,n}$ and $(\beta_{i,2MNA})_n$, $\frac{(m_{i,n})_{ME}}{(m_{i,n})_{FE}}$ and $\frac{(\beta_{i,2MNA})_{n,FE}}{(\beta_{i,2MNA})_{n,ME}}$ show almost the same value regardless of used solvent and n, therefore, there is no considerable difference by using different solvents for the extraction. Furthermore, $\frac{(m_{i,n})_{ME}}{(m_{i,n})_{FE}}$ and $\frac{(\beta_{i,2MNA})_{n,FE}}{(\beta_{i,2MNA})_{n,ME}}$ of *IN*, showed about 3 and 11 at the ranges of n = 1–5.

The GC chromatogram of R_5 recovered through FE is presented in Figure 1b. As a result of comparing R_5 and wash oil, we found that the peaks of *QU* and *IN* were very small in R_5, but the peak heights of other compounds were almost the same as those of wash oil. From this fact, we were able to reconfirm that the quality of wash oil (raffinate oil) was improved through FE run by confirming the reduction of *NHC* (*QU*, *IN*), undesirable impurities in wash oil due to their contribution to air pollution and unpleasant odors [12,14]. In addition, when the GC chromatogram of the *MFEP* in Figure 1c was compared with those of wash oil or R_5, the peak heights of *NHC* in Figure 1c were much higher than that of wash oil or R_5. From this result, it can be seen that FE is a very good solvent in terms of selective separation between the *NHC* (particularly *IN*) and the *BAC* and *BP* contained in the wash oil.

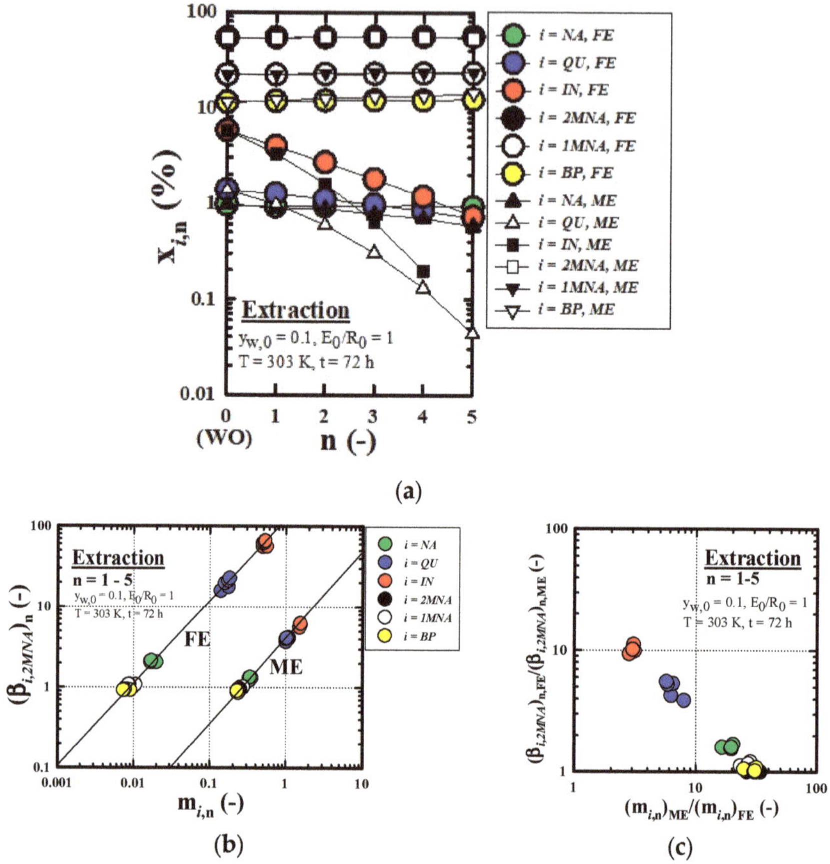

Figure 4. Solvent comparison for effect of n on (**a**) $x_{i,n}$ (without solvent), (**b**) $m_{i,1}$ and $(\beta_{i,2MNA})_1$, (**c**) $\frac{(m_{i,1})_{ME}}{(m_{i,1})_{FE}}$ and $\frac{(\beta_{i,2MNA})_{1,FE}}{(\beta_{i,2MNA})_{1,ME}}$. WO: wash oil, FE: formamide extraction, ME: methanol extraction [3].

3.3. *Equilibrium Re-Extraction*

3.3.1. GC Chromatogram of MFEP (Re-Extraction Feed)

A *MFEP* of above-mentioned Figure 1c was used as the feed. The composition of the *MFEP* including the extraction solvent is shown in Table 1. The concentration in feed of *IN*, the target compound for enrichment of this study, was approximately 1%.

3.3.2. Enrichment Performance of *IN*

Figure 5a,b, respectively, present the effect of E_0/R_0 and the re-extraction feed on $m_{i,1}$ and $(\beta_{i,2MNA})_1$, $\frac{(m_{i,n})_{MFEP}}{(m_{i,n})_{MMEP}}$ and $\frac{(\beta_{i,IN})_{1,MFEP}}{(\beta_{i,IN})_{1,MMEP}}$ obtained in n = 1. The abbreviations *MFEP* and *MMEP* refer to the mixed formamide and methanol extract phases used as feed, respectively. In Figure 5a, when *MFEP* was used as the feed, $x_{i,n}$ for four compounds except *NHC* did not appear in above $E_0/R_0 = 1$. This is because the concentration of the four compounds contained in the raffinate phase was too low to analyze under conditions of $E_0/R_0 = 1$ or higher. Irrespective of feed used from Figure 5a, contrary to the above-mentioned FE run, $m_{i,1}$ and $(\beta_{i,IN})_1$ of *NHC* showed much smaller values compared to *BAC* and *BP*. From this fact, it can be seen that compounds easily acquired from wash oil by FE run are enriched in raffinate phase by re-extraction. Therefore, contrary to the results of FE run, $m_{i,1}$ and $(\beta_{i,IN})_1$ of each compound showed the order *2MNA* = *1MNA* > *BP* > *NA* > *QU* > *IN*. The effect of raw material type and E_0/R_0 on $m_{i,1}$ and $(\beta_{i,IN})_1$ was not differentiable for all compounds.

Through Figure 5b, for all compounds of the range of $0.5 \leq E_0/R_0 \leq 3$, $(m_{i,1})_{MFEP}$ was about 0.6 to 22 times greater than $(m_{i,1})_{MMEP}$ and $(\beta_{i,IN})_{1,MFEP}$ was 1 to about 21 times greater than $(\beta_{i,IN})_{n,MMEP}$. Since $m_{IN,1}$ (*i* = *IN*) showed the smallest value among the six components, it was found that *IN* was highly enriched in the raffinate phase. Considering the enrichment efficiency of *IN* per unit solvent volume, $E_0/R_0 = 0.5$ was thought to be the optimal condition for HRE run.

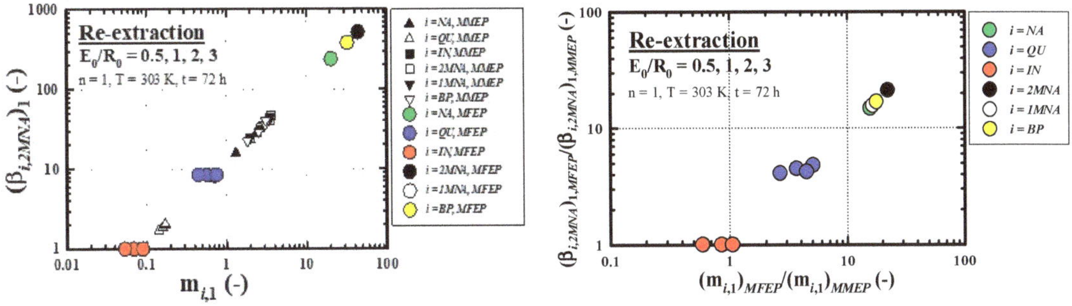

(a) **(b)**

Figure 5. Effects of E_0/R_0 and re-extraction feed on (**a**) $m_{i,1}$ and $(\beta_{i,2MNA})_1$, (**b**) $\frac{(m_{i,1})_{MFEP}}{(m_{i,1})_{MMEP}}$ and $\frac{(\beta_{i,2MNA})_{1,MFEP}}{(\beta_{i,2MNA})_{1,MMEP}}$. *MFEP*: mixed formamide extract phase, *MMEP*: mixed methanol extract phase [3].

Figure 6a–c, respectively, show the effects of n and the re-extraction feed on $x_{i,n}$ (without solvent), $m_{i,n}$ and $(\beta_{i,IN})_n$, $\frac{(m_{i,n})_{MFEP}}{(m_{i,n})_{MMEP}}$ and $\frac{(\beta_{i,IN})_{n,MFEP}}{(\beta_{i,IN})_{n,MMEP}}$ obtained in the nth HRE run. The n = 0 shown in Figure 6a represents the re-extraction feed (REF). In the case of using *MFEP*, *IN* with strong polarity compared to the other five compounds are enriched in the raffinate phase, and it can be seen that the $x_{IN,n}$ (*i* = *IN*) increases as n increases. Conversely, the $x_{i,n}$ of the four compounds with strong hydrophobic properties, except *NHC*, decreased rapidly as n increased. In *MFEP*, $x_{i,n}$ for four compounds except for *NHC* did not appear after n = 1. This is the result of four compounds not being detected in the GC analysis of the raffinate phase recovered after n = 1. At n = 5, $x_{IN,n}$ (*i* = *IN*)was about 92.4% in *MFEP* and about 73.3% in *MMEP*, respectively. In Figure 6b,c, for all compounds in this study, $m_{i,n}$ and $(\beta_{i,IN})_n$, $\frac{(m_{i,n})_{MFEP}}{(m_{i,n})_{MMEP}}$ and $\frac{(\beta_{i,IN})_{n,MFEP}}{(\beta_{i,IN})_{n,MMEP}}$ show almost the same values regardless of the re-extraction feed used and n, so the effect of re-extraction feed and n is not considerable.

Figure 1d shows the GC chromatograms of R_5 recovered through the 5th re-extraction run with *MFEP*. Comparing Figure 1d with Figure 1c, which is the re-extraction feed, it can be seen that the four compounds except *NHC* (*IN*, *QU*) are almost absent in R_5. Through this, we were able to reaffirm the feasibility of HRE run regarding the enrichment of *IN* present in the *MFEP*.

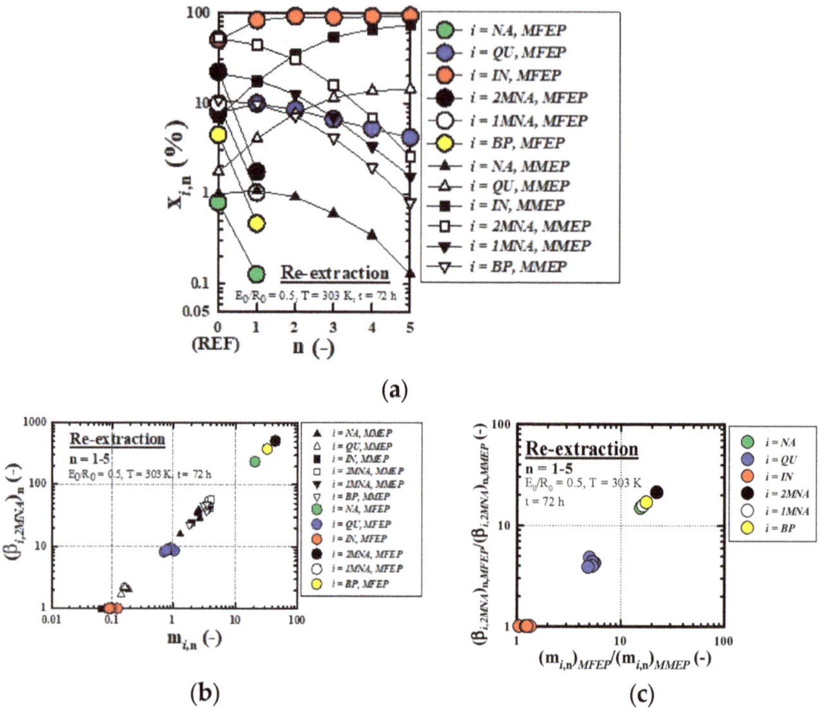

Figure 6. Effects of n and re-extraction feed on (**a**) $x_{i,n}$ (solvent-free), (**b**) $m_{i,n}$ and $(\beta_{i,2MNA})_n$, (**c**) $\frac{(m_{i,1})_{MFEP}}{(m_{i,1})_{MMEP}}$ and $\frac{(\beta_{i,2MNA})_{1,MFEP}}{(\beta_{i,2MNA})_{1,MMEP}}$. REF: re-extraction feed, *MFEP*: mixed formamide extract phase, *MMEP*: mixed methanol extract phase [3].

3.4. Comparison of Combined Enrichment Efficiency

Figure 7a,b respectively, present the yield ($Y_{IN,5}$) and concentration ($y_{IN,5}$) changes of *IN* by two types (FHC and MHC (3)) of enrichment combination after extraction and re-extraction run of n = 5. $Y_{IN,5}$ in each combination is based on the mass of *IN* contained in the wash oil, and this was calculated by Equations (2) and (3). $Y_{IN,5}$ by FHC and MHC decreased as enrichment run progressed, showing almost the same value (about 78%) regardless of the combination, but $y_{IN,5}$ conversely increased rapidly as enrichment run progressed by the two combinations. The concentration of *IN* in the wash oil was about 5.75%, but the $y_{IN,5}$ of the extract phase, which recovered by FE and ME of n = 5, was 50.1% and 12.4%, respectively. And the $y_{IN,5}$ of the raffinate phase (without solvent), which recovered by HRE of n = 5 was it showed high enrichment efficiency of 92.4% in the case of FHC and 73.3% in the case of MHC, respectively. Therefore, we evaluated FHC as a much more useful combination than MHC in terms of the enrichment efficiency of *IN* as well as the simplicity of enrichment process of *IN*. The enrichment efficiency of *IN* by FHC of this study showed really excellent results compared to the combination of FE and distillation [1], the combination of ME and distillation [6], and reaction extraction using acid and base [8,9].

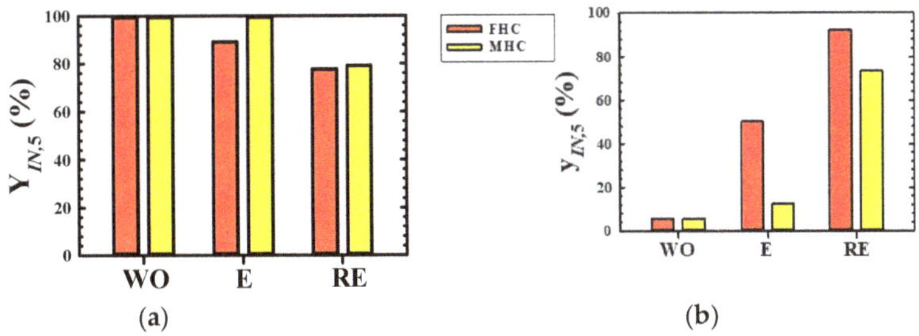

Figure 7. Comparison of enrichment combination for (**a**) $Y_{IN,5}$ and (**b**) y_{IN} of *IN*. WO: wash oil, E: extraction, RE: re-extraction, FHC: formamide-hexane combination, MHC: methanol-hexane combination [3]. Extraction conditions: n = 1–5, $y_{w,0}$ = 0.1, E_0/R_0 = 1, T = 303 K, t = 72 h; Re-extraction conditions: n = 1–5, E_0/R_0 = 0.5, T = 303 K, t = 72 h.

4. Conclusions

To develop an efficient new process for high-purity purification of IN contained in wash oil, this study first experimentally investigated the enrichment performance of IN contained in wash oil by FHC, and then compared the results with those obtained by MHC previously reported. Yield of *IN* using FHC and MHC showed almost the same value, but the concentration of *IN* is significantly higher with FHC than with MHC. FHC adopted in this study was expected to be a more effective combination than MHC not only in terms of the enrichment efficiency of *IN* included in wash oil but also in terms of simplicity of the enrichment process. For high-purity purification of *IN* in wash oil, it is necessary to remove *QU*, etc. presenting as impurity in the 92.4% *IN* concentrate recovered from FHC of this study. Since the melting point difference between *IN* and *QU* is 68 K, the purification method using this melting point difference is expected to be highly feasible for further purification of *IN*. Therefore, we would like to report the research results on the high purity purification of *IN* in a future paper by adopting crystallization as a separation method using the melting point difference.

Funding: This research received no external funding.

Data Availability Statement: The data presented in this study are available within the article (tables and figures).

Conflicts of Interest: The author declares no conflict of interest.

References

1. Kim, S.J. Separation and purification of indole in model coal tar fraction of 9 compounds system. *Polycycl. Aromat. Compd.* **2019**, *39*, 60–72. [CrossRef]
2. Gao, P.; Zhang, J.; Guo, Z.; Gao, J.; Xu, D.; Ma, Y.; Zhang, L.; Wang, Y. Experimental and quantum chemical calculations investigations of morpholine-based ionic liquids as extractants for efficient extraction of nitrogen heterocyclic neutral compounds. *Fuel* **2023**, *333*, 126446. [CrossRef]
3. Kim, S.J. Purification of indole contained in wash oil by combination of extraction and crystallization (Part 1: Recovery and concentration of indole contained in wash oil by solvent extraction). *Molecules* **2022**, *27*, 5331. [CrossRef] [PubMed]
4. Ukegawa, K.; Matsumura, A.; Kodera, Y.; Kondo, T.; Nakayama, T.; Tanabe, H.; Yoshida, S.; Mito, Y. Solvent extraction of nitrogen compounds from a coal tar fraction. (Part 1). Effect of extraction conditions on the extraction rate and the selectivities of nitrogen compounds. *J. Jpn. Pet. Inst.* **1990**, *33*, 250–254. [CrossRef]
5. Kim, S.J.; Chun, Y.J. Separation of nitrogen heterocyclic compounds from model coal tar fraction by solvent extraction. *Sep. Sci. Technol.* **2005**, *40*, 2095–2109.
6. Salim, C.; Egashira, R. Separation of coal tar distillate by solvent extraction—Separation of extract phase using distillation. *J. Jpn. Pet. Inst.* **2006**, *49*, 326–334. [CrossRef]
7. Yang, Y.; Liu, Q.; Chen, A.; Ji, Y. Extraction behavior of indole from simulated wash oil using halogen-free ionic liquids. *ACS Omega* **2021**, *6*, 16623–16630. [CrossRef]
8. Xiao, R.H.; Gao, W.H. Study on the recovery of indole from coal tar wash oil. *Coal Convers.* **1998**, *21*, 59–61.

9. Mamoru, Y.; Tomonori, K. Separation and purification of indole from coal tar by supercritical fluid extraction. *J. Chem. Eng. Jpn.* **1993**, *26*, 153–158.
10. Feng, Y.; Yang, E.; Dang, L.; Wei, H. Liquid–liquid phase equilibrium for ternary mixtures of formamide (or ethylene glycol, or monoethanolamine) + indole + 2-methylnaphthalene at 308.15 K. *Fluid Phase Equilibria* **2015**, *398*, 10–14. [CrossRef]
11. Sakanishi, K.; Obata, H.; Mochida, I.; Sakaki, T. Capture and recovery of indole from methylnaphthalene oil in a continuous supercritical CO_2 extraction apparatus over a fixed bed of anion-exchange resin. *Ind. Eng. Chem. Res.* **1996**, *35*, 335–337. [CrossRef]
12. Sakanishi, K.; Obata, H.; Mochida, I.; Sakaki, T. Removal and recovery of quinoline bases from methylnaphthalene oil in a semicontinuous supercritical CO_2 separation apparatus with a fixed bed of supported aluminum sulfate. *Ind. Eng. Chem. Res.* **1995**, *34*, 4118–4124. [CrossRef]
13. Liu, Q.; Zhang, T.; Gao, P.; Gao, J.; Xu, D.; Zhao, P.; Zhang, L.; Wang, Y. Separation of indole by designed ionic liquids with dual functional chemical sites: Mechanism exploration and experimental validation. *J. Environ. Chem. Eng.* **2021**, *9*, 105971. [CrossRef]
14. Ji, Y.; Hou, Y.; Ren, S.; Wu, W. Highly efficient separation of indole from model wash oil using tetraethyl ammonium amino acid ionic liquids. *Sep. Purif. Technol.* **2021**, *258*, 117997. [CrossRef]
15. Zhang, L.; Xu, D.; Gao, J.; Zhou, S.; Zhao, L.; Zhang, Z. Extraction and mechanism for the separation of neutral N-compounds from coal tar by ionic liquids. *Fuel* **2017**, *194*, 27–35. [CrossRef]
16. Xu, D.; Zhang, M.; Gao, J.; Zhang, L.; Zhou, S.; Wang, Y. Separation of heterocyclic nitrogen compounds from coal tar fractions via ionic liquids: COSMO-SAC screening and experimental study. *Chem. Eng. Commun.* **2019**, *206*, 1199–1217. [CrossRef]
17. Jiao, T.; Zhuang, X.; He, H.; Zhao, L.; Li, C.; Chen, H.; Zhang, S. An ionic liquid extraction process for the separation of indole from wash oil. *Green Chem.* **2015**, *17*, 3783–3790. [CrossRef]
18. Ji, Y.; Hou, Y.; Ren, S.; Niu, M.; Yao, C.; Wu, W. Efficient extraction of indole from wash oil by quaternary ammonium salts via forming deep eutectic solvents. *Fuel* **2018**, *215*, 330–338. [CrossRef]
19. Yamamoto, Y.; Sato, Y.; Ebina, T.; Yokoyama, C.; Takahasi, S.; Mito, Y.; Tanabe, H.; Nishiguchi, N.; Nagaoka, K. Separation of high purity indole from coal tar by high pressure crystallization. *Fuel* **1991**, *70*, 565–566. [CrossRef]
20. Mochida, I.; Fei, Y.Q.; Sakanishi, K. Capture and recovery of basic nitrogen species in coal tar pitch, using nickel sulfate as adsorbent. *Chem. Lett.* **1990**, *19*, 515–518. [CrossRef]
21. Uemasu, I.; Nakayama, T. Concentration of indole in coal tar using α-cyclodextrin as the host for inclusion complexation. *J. Inclus. Phenom. Molec. Recogn. Chem.* **1989**, *7*, 327–331. [CrossRef]
22. Uemasu, I. Effect of methanol-water mixture solvent on concentration of indole in coal tar using β-cyclodextrin as complexing agent. *J. Jpn. Pet. Inst.* **1991**, *34*, 371–374. [CrossRef]

processes

MDPI

Article

Correlations between a Friabilin Content Indicator and Selected Physicochemical and Mechanical Properties of Wheat Grain for Processing Suitability Assessment

Zdzisław Kaliniewicz [1,*], Agnieszka Markowska-Mendik [1], Małgorzata Warechowska [1], Seweryn Lipiński [1] and Sebastian Gasparis [2]

[1] Faculty of Technical Sciences, University of Warmia and Mazury in Olsztyn, Oczapowskiego 11, 10-719 Olsztyn, Poland; agnieszka.markowska@uwm.edu.pl (A.M.-M.); gosiaw@uwm.edu.pl (M.W.); seweryn.lipinski@uwm.edu.pl (S.L.)
[2] Department of Functional Genomics, Plant Breeding and Acclimatization Institute—National Research Institute, 05-870 Błonie, Poland; s.gasparis@ihar.edu.pl
* Correspondence: zdzislaw.kaliniewicz@uwm.edu.pl

Abstract: A new approach to determining the friabilin content of wheat grain was proposed. Electropherograms were taken, and the intensity of the friabilin bands was compared in the analyzed wheat cultivars and the cv. Chinese Spring. The friabilin content indicator was calculated in the grain of 17 common wheat cultivars, which differed mostly in their crude protein content and hardness index (HI). The basic properties of the kernels were measured in each wheat cultivar, and the correlations between the measured parameters and the friabilin content indicator were determined. In the analyzed wheat cultivars, the friabilin content indicator ranged from around 0.21 to around 0.77. This indicator was significantly correlated with the kernel length, thickness, mass, vitreousness, HI, and rupture force. The strongest correlation was observed between the friabilin content indicator and kernel length. An increase in the mean kernel length from around 5.4 mm to around 8.0 mm decreased the friabilin content indicator by approximately 51%. After the mean kernel length had been calculated in a given wheat cultivar, a certain value of the friabilin content indicator could be ascribed to this cultivar, and the energy consumption during grain grinding or milling could be partly predicted. In the group of analyzed wheat cultivars, the process of grain grinding would be the most energy-intensive in the cvs. Ceres, SMH200, and SMH214 and the least energy-intensive in the cvs. Chinese Spring, Julius, and Askalon.

Keywords: wheat grain; seed properties; friabilin content; image analysis; puroindoline

Citation: Kaliniewicz, Z.; Markowska-Mendik, A.; Warechowska, M.; Lipiński, S.; Gasparis, S. Correlations between a Friabilin Content Indicator and Selected Physicochemical and Mechanical Properties of Wheat Grain for Processing Suitability Assessment. *Processes* **2024**, *12*, 398. https://doi.org/10.3390/pr12020398

Academic Editors: Yanlin Zhang and Prashant K. Sarswat

Received: 11 January 2024
Revised: 15 February 2024
Accepted: 15 February 2024
Published: 17 February 2024

1. Introduction

Common wheat (*Triticum aestivum* L.) is one of the four most widely grown cereals in the world despite its high soil, environmental, and agronomic requirements. Wheat grain has numerous applications, mainly in the milling, food processing, cosmetic, and pharmaceutical industries. Wheat grain is also used in the production of animal feed and planting materials. Wheat grain contains 60–68% starch, 7–18% protein, 1.5–2.0% lipids, 2.0–2.5% cellulose, and 1.5–2.2% ash [1,2]. The endosperm is the most important part of each kernel, as it contains nutrients that are essential for seed germination. From a botanical point of view, the endosperm is composed of the aleurone layer and the starchy endosperm. However, from a technological point of view, the aleurone layer is regarded part of the seed coat that is removed together with bran during milling. The endosperm accounts for up to 83% of the kernel mass [2].

The processing suitability of wheat grain is determined mainly on the basis of its protein content. Grain for the bakery industry should contain a minimum of 11.5% protein on a dry matter (DM) basis, whereas grain for enhancing mixtures of medium- and low-quality grain should contain more than 14% protein [3]. Numerous authors have analyzed the correlations between the crude protein content and the mechanical properties of grain that determine the energy consumption during milling, as well as the sustainable use of this raw material in food processing. However, the reported results are often contradictory [4–9]. According to Hogg et al. [10], the kernel hardness is not determined by the total crude protein content but by the content of friabilin, a protein complex responsible for grain softness. Turnbull and Rahman [11], Laskowski and Różyło [12], and Pasha et al. [6] observed that the friabilin content was very high in kernels of soft wheat varieties and that its content decreased with a rise in the kernel hardness. Hard wheat, such as durum wheat (*Triticum turgidum* ssp. *durum*), do not contain friabilin due to the absence of genes encoding this protein. In grain, the endosperm texture is regulated by the *Pina* and *Pinb* puroindoline genes, which, together with the grain softness protein-1 gene (*Gsp-1*), are localized in the *Ha* locus on chromosome 5DS. These genes encode the puroindoline proteins PINA and PINB, which are subunits of friabilin, a protein with a molecular weight of 13 kDa [6,13–16]. The wild-type alleles of the *Pin* genes, i.e., *Pina-D1a* and *Pinb-D1a*, are responsible for grain softness, and mutations in one or both of these genes lead to the formation of hard grains [4,15,17–19]. Different types of mutations are responsible for different levels of grain hardness, such as intermediate, between soft wheat and durum wheat [6,14,20–22]. Immunodetection assays have revealed that the PIN proteins accumulate mainly in the endosperm on the surface of starch granules and in the aleurone layer [6,23–26]. According to Beecher et al. [27] and Chen et al. [28], the PIN proteins prevent the protein matrix from adhering to the starch granules, which decreases the endosperm density and leads to the formation of softer grains. In turn, hard wheat is characterized by a more compact endosperm and denser kernels.

Qualitative and quantitative analyses of the friabilin content can be performed with the use of reversed-phase high-performance liquid chromatography (RP-HPLC), which is a complex and expensive analytical technique [29–31]. In this method, peaks are registered on chromatograms and compared with a reference sample. However, according to Hogg et al. [10] and Martin et al. [32], SDS-PAGE is one of the most popular methods for identifying friabilin proteins. In this analytical technique, the proteins are separated using electrophoresis on polyacrylamide gel with the addition of sodium dodecyl sulfate (SDS) as the denaturing agent. In SDS-PAGE, the wheat cv. Chinese Spring is most often used as a reference to determine the friabilin content of other grain. Chinese Spring harbors the wild-type alleles of the puroindoline genes, *Pina-D1a* and *Pinb-D1a*, and it is characterized by a high content of PIN proteins [13,17,21]. When the proteins are separated using electrophoresis on polyacrylamide gel, the friabilin fraction produces bands with an estimated size of 13 kDa. The band intensity is determined by the content of puroindoline proteins. Li et al. [33] postulated the presence of associations between the protein content and the intensity of the corresponding bands, but they did not propose any indicators for measuring this trait. Hogg et al. [10] used an eight-point scale to quantify friabilin relative to a protein reference with a known friabilin content. However, visual inspection of the band intensity is not a highly accurate method due to observer bias. Therefore, a novel comparative approach was proposed in this study to determine the friabilin content of wheat grain. In the developed technique, the band intensity in the electropherograms was evaluated using computer-assisted image analysis. The friabilin content of the examined wheat cultivars was determined by comparing the band intensities against Chinese Spring as the reference cultivar.

An analysis of the correlations between the physical properties and the friabilin content of wheat grain may contribute to breeding wheat varieties with desirable traits in order to respond to changing consumer preferences and market demand and reduce food waste. The quality of the raw materials should be improved to streamline production processes, thus reducing energy and resource consumption. In turn, the physical properties and friabilin content of the grain of local wheat cultivars should be optimized to reduce dependency on grain imports, which translates into lower CO_2 emissions, lower fuel consumption, and more efficient utilization of local natural resources such as soil and climate. This may also reduce the use of fertilizers and pesticides, thus contributing to minimizing the adverse environmental impacts of agricultural production [15,34,35].

The friabilin content of wheat grain was determined using a friabilin content indicator, which is a novel parameter introduced by the authors. Therefore, the aim of this study was to determine changes in the value of the friabilin content indicator in the grain of selected wheat cultivars and to analyze the correlations between this parameter and the other physicochemical properties of wheat grain.

2. Materials and Methods

2.1. Materials

The experimental material comprised the grain of 18 wheat cultivars that differed considerably in the mean values of their crude protein content and hardness index (HI) (Table 1). The grain was obtained from the IHAR Group Plant Breeding Company in Strzelce (cvs. Bamberka, Chinese Spring, Fregata, Nutka, Parabola, Radunia, and Tonacja), Smolice (cvs. Ceres, SMH87, SMH196, SMH200, and SMH214), and Agro-Plon, Ostróda (cvs. Askalon, Astoria, Cytra, Julius, Magic, and Türkis), in 2013. These companies belong to the Polish Seed Trade Association in Poznań, and they produce and distribute planting materials in Poland according to the applicable national and international standards. Initial grain samples of approximately 2 kg each were obtained from each wheat cultivar and were stored in a refrigerator at a temperature of around 4 °C. Before analysis, all the grain samples were brought to moisture content equilibrium by conditioning a thin layer of the grain at a temperature of 22 ± 1 °C and 54 ± 3% humidity for five days. Each layer of conditioned grain had an estimated thickness of 1 cm.

Table 1. Approximate composition, vitreousness, and hardness index (mean value ± standard deviation) of wheat grain.

Cultivar	Moisture [% DB]	Crude Protein [% DM]	Total Ash [% DM]	Vitreousness [%]	Hardness Index HI [–]
Chinese Spring	12.63 ± 0.32 ^{de}	12.69 ± 0.03 ^{bc}	2.01 ± 0.01 ^{j}	0 ^{a}	39.74 ± 11.56 ^{a}
Türkis	12.75 ± 0.39 ^{fg}	14.17 ± 0.01 ^{g}	1.55 ± 0.01 ^{a}	2.60 ± 0.55 ^{a}	48.16 ± 16.36 ^{b}
Astoria	12.83 ± 0.39 ^{g}	14.82 ± 0.02 ^{k}	1.68 ± 0.02 ^{d}	54.00 ± 4.74 ^{c}	53.22 ± 13.79 ^{c}
Julius	12.99 ± 0.40 ^{h}	13.66 ± 0.03 ^{f}	1.64 ± 0.02 ^{cd}	10.20 ± 2.39 ^{b}	57.81 ± 16.83 ^{d}
Fregata	12.57 ± 0.36 ^{cd}	15.36 ± 0.10 ^{l}	1.68 ± 0.02 ^{d}	72.20 ± 1.92 ^{de}	66.08 ± 12.82 ^{e}
Radunia	12.60 ± 0.41 ^{de}	14.52 ± 0.08 ^{hi}	2.02 ± 0.01 ^{j}	54.60 ± 1.95 ^{c}	66.64 ± 14.09 ^{e}
Parabola	12.62 ± 0.40 ^{de}	14.78 ± 0.05 ^{jk}	2.01 ± 0.01 ^{j}	72.80 ± 3.08 ^{de}	67.26 ± 13.80 ^{e}
Cytra	12.80 ± 0.43 ^{fg}	14.59 ± 0.06 ^{ij}	1.85 ± 0.03 ^{fg}	49.00 ± 1.23 ^{c}	68.30 ± 14.66 ^{e}
Tonacja	12.58 ± 0.43 ^{de}	13.42 ± 0.12 ^{e}	1.73 ± 0.02 ^{e}	80.20 ± 1.48 ^{ef}	70.35 ± 12.93 ^{ef}
Askalon	13.08 ± 0.35 ^{h}	14.11 ± 0.14 ^{g}	1.62 ± 0.01 ^{bc}	86.80 ± 2.77 ^{gh}	73.89 ± 15.16 ^{fg}
Magic	12.81 ± 0.43 ^{fg}	14.58 ± 0.03 ^{ij}	1.57 ± 0.01 ^{ab}	61.60 ± 2.19 ^{d}	76.76 ± 14.99 ^{gh}
Nutka	12.70 ± 0.37 ^{ef}	13.14 ± 0.05 ^{d}	1.82 ± 0.01 ^{f}	82.60 ± 2.41 ^{fg}	77.87 ± 13.08 $^{g–i}$
Bamberka	13.07 ± 0.33 ^{h}	14.31 ± 0.08 ^{gh}	1.65 ± 0.01 ^{cd}	89.40 ± 0.89 ^{h}	80.56 ± 12.36 $^{h–j}$
SMH196	12.52 ± 0.48 ^{cd}	12.72 ± 0.08 ^{bc}	1.87 ± 0.01 ^{gh}	77.20 ± 1.48 ^{e}	81.89 ± 18.49 ^{ij}
SMH87	12.34 ± 0.47 ^{b}	13.52 ± 0.09 ^{ef}	1.96 ± 0.01 ^{i}	84.80 ± 1.64 $^{f–h}$	83.66 ± 16.84 ^{j}
Ceres	12.16 ± 0.43 ^{a}	12.85 ± 0.05 ^{c}	1.93 ± 0.02 ^{hi}	87.80 ± 2.27 ^{h}	85.96 ± 17.40 ^{k}
SMH200	12.46 ± 0.41 ^{c}	12.52 ± 0.05 ^{b}	1.84 ± 0.01 ^{fg}	85.40 ± 1.67 ^{gh}	88.17 ± 17.88 ^{k}
SMH214	12.81 ± 0.44 ^{g}	12.20 ± 0.02 ^{a}	1.90 ± 0.01 ^{hi}	81.20 ± 2.09 $^{e–g}$	88.62 ± 20.05 ^{k}

$^{a,\ldots,l}$—superscript letters denote significant differences between the examined properties at $p < 0.05$.

2.2. Approximate Composition, Vitreousness, and Hardness

The crude protein content of the wheat grain was determined using the Kjeldahl method (Nx5.7) and the KjelFlex K-360 apparatus (BÜCHI Labortechnik AG, Flawil, Switzerland) according to ISO standard 20483:2013 [36]. The total ash content was determined using the procedure described in ISO standard 2171:2007 [37]. Both analyses were performed in triplicate. The grain vitreousness was measured using a grain cutter (Zakład Badawczy Przemysłu Piekarskiego sp. z o.o., Bydgoszcz, Poland). A total of 50 kernels from each wheat cultivar were cut perpendicular to the long axis, and their vitreousness was visually inspected. Kernels with completely opaque surfaces were classified as starchy, and kernels with uniform, glassy surfaces were classified as vitreous [38]. Kernels were classified as semi-vitreous when both opaque and glassy areas were observed on their surface. In each grain batch, the vitreousness v (%) was calculated using the following formula:

$$v = 2a + b \tag{1}$$

where a = number of vitreous kernels and b = number of semi-vitreous kernels.

The vitreousness analysis was conducted in five replicates for each wheat cultivar.

The grain HI and relative moisture content were measured using the SKCS 4100 system (Perten Instruments, New York, NY, USA) according to AACC procedure 55-31.01 [39]. The values of only two out of the four measured parameters were included in this analysis because it was assumed that the thickness and mass of the kernels could be more accurately determined with the use of a thickness gauge and a laboratory weighing scale, respectively. In the SKCS 4100 system, the HI is calculated for each kernel by measuring its resistance to fracture and comparing the result with the reference cultivar. In addition, the results were adjusted to 12% moisture content and a kernel mass of 35 mg [40]. In each test, a grain sample of around 30 g was placed in the hopper. Random kernels were automatically fed to the crushing unit. A total of 300 kernels from each sample were analyzed in the test.

2.3. Geometric Properties, Mass, and Mechanical Properties

The grain, for analyses of its geometric properties and kernel mass, was sampled according to ISO standard 24333:2009 [41]. Each sample comprised 60 kernels from each analyzed wheat cultivar, and the geometric properties and mass of the wheat kernels were measured according to the procedure proposed by Kaliniewicz et al. [9]. The basic seed dimensions were determined using a MWM 2325 workshop microscope (PZO, Warsaw, Poland) and a self-designed instrument for measuring the seed thickness. The length (L) and width (W) of each kernel were determined with an accuracy of 0.02 mm, and the kernel thickness (T) was determined with an accuracy of 0.01 mm. The mass (m) of the wheat kernels was determined using a WAA 100/C/2 digital weighing scale (Radwag, Radom, Poland) with an accuracy of 0.1 mg.

The measured parameters were used to calculate the aspect ratio (R) and the sphericity index (Φ) of each kernel:

$$R = \frac{W}{L} \cdot 100 \tag{2}$$

$$\Phi = \frac{\sqrt[3]{T \cdot W \cdot L}}{L} \cdot 100 \tag{3}$$

The same samples were used to determine the mechanical properties of the wheat grain. The mechanical properties were analyzed at room temperature (22 ± 1 °C) according to the procedure described by Kaliniewicz et al. [9]. The measurements were performed using the Instron Mecmesin MultiTest-i tensile and compression test system (Mecmesin Ltd., Slinfold, UK) and the Emperor™ Force testing system software (Version 1.18-408). The crosshead speed was set at 10 mm·min^{-1} (within a loading range of 0–1 kN) according to the recommendations made by Różyło and Laskowski [5] and Dziki et al. [42]. The compression test was interrupted when the rupture force decreased by a minimum of 10% and the first fracture was observed on the kernel. The rupture force (F), longitudinal strain (d), and rupture energy (E) were registered using the software installed in the test system. The relative strain (Δ) was calculated for each kernel with the use of formula (4) based on the registered values of the longitudinal strain (d) and kernel thickness (T) [9,43]:

$$\Delta = 100 \cdot \frac{d}{T} \tag{4}$$

2.4. SDS-PAGE Analysis

The puroindoline proteins in the endosperm of the analyzed wheat cultivars were separated using electrophoresis on polyacrylamide gel [32,44–46]. In the first stage of the analysis, the starch fraction of the wheat endosperm was purified according to the method described by Bettge et al. [47] with the modifications proposed by Chang et al. [48]. Kernel samples of around 200 mg each were ground in a mortar; a total of 0.6 mL of 0.1 M NaCl solution was added, and the mixture was shaken for 30 min. The aqueous suspension was removed with a pipette, placed in a new vial, and centrifuged for 3 min at a RCF pf 13,000× *g* (the same parameters were applied in successive centrifugation steps). The supernatant was discarded, 1 mL of H_2O was added to the sediment, and the mixture was stirred and centrifuged. The water was removed, and 1 mL of 80% CsCl solution was added to the purified sediment. The mixture was stirred to obtain a homogeneous suspension and centrifuged, and the supernatant was discarded. The sediment was rinsed three times with 1 mL of H_2O according to the above procedure. The purified sediment was suspended in 0.5 mL of acetone and centrifuged. The acetone was removed, and the purified starch sediment was left to dry. Approximately 50 mg of purified starch was collected from each sample; then, 150 µL of 1 M NaCl solution and 150 µL of isopropanol were added; and the mixture was shaken at a temperature of 45 °C for 60 min and centrifuged to separate the suspension. The aqueous phase was removed, combined with 150 µL of cooled acetone, and left overnight at a temperature of −20 °C. Next, 450 µL of acetone cooled to the same temperature was added, and the sample was left overnight at a temperature of −20 °C. After sediment precipitation, each sample was centrifuged, the supernatant was discarded, and the sediment was suspended in 50 µL of SDS-PAGE buffer.

The puroindoline proteins were separated via electrophoresis on polyacrylamide gel with SDS as the denaturing agent, according to the procedure described by Laemmli [49]. Before loading, each sample was incubated at a temperature of 75 °C for 10 min. The proteins were separated on two gel layers, including a layer of condensing gel (5% T, 2.6% C) and a layer of separating gel (15% T, 2.6% C), on plates measuring 180 mm × 160 mm × 1.5 mm. The separation was performed using a Hoefer SE660 electrophoresis unit (Hoefer, Inc., Holliston, MA, USA) at a constant current of 25 mA. The samples were loaded into wells, and the separation was run until the dye front had reached the bottom of the gel.

The separated proteins were stained according to the method described by Gromova and Celis [50] with the modifications proposed by Gasparis [51]. The proteins were stained with 0.2% aqueous solution of silver nitrate with a 0.076% addition of formaldehyde. Protein bands were visualized according to incubation in an aqueous solution of 6% sodium carbonate with a 0.0004% addition of sodium thiosulfate and a 0.05% addition of formaldehyde for 2–5 min. Each sample was rinsed with distilled water and fixed via immersion in 12% acetic acid solution with a 20% addition of methanol for 10 min. The samples were stored in 20% methanol solution.

Images of the separated protein fractions (Figure 1) were acquired using the Kodak Gel Logic 200 Imaging System (Kodak, Molecular Imaging Systems, New Haven, CT, USA). The acquired images were analyzed using MATLAB v. R2011b (MathWorks, Natick, MA, USA). In the first step, the gel region was cropped from the image, and its brightness was adjusted by increasing the contrast to the maximum value (from 0 to 256). In the next step, the region corresponding to the friabilin bands was cropped from the image, and the pixel values in this region were averaged and normalized in the range of 0–1 relative to the intensity of the bands obtained in the reference cultivar (Chinese Spring). The value of the friabilin content indicator (c_f) was determined by comparing the intensity of the friabilin bands in the examined wheat cultivars with the intensity of the corresponding bands in the wheat cv. Chinese Spring.

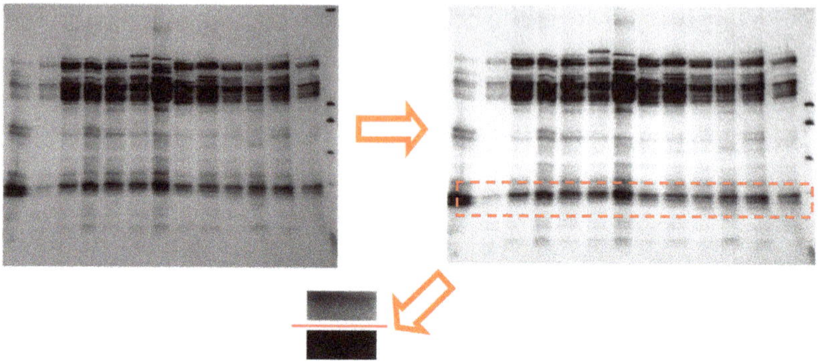

Figure 1. Determination of the friabilin content indicator in the analyzed wheat cultivars based on band intensity in SDS-PAGE electropherograms. Friabilin bands are outlined with a dashed line. A marker with a molecular weight of 15 kDa is marked with a red arrow.

2.5. Statistical Analysis

The results were processed using Statistica for Windows v. 13.3 (TIBCO, Palo Alto, CA, USA) at a significance level of $\alpha = 0.05$. The following procedures were used in the statistical analyses: descriptive statistics, one-way analysis of variance (ANOVA), correlation analysis, and regression analysis. Differences between the grain parameters in the analyzed wheat cultivars were evaluated using Tukey's test. Pearson's correlation coefficients were calculated to determine the strength of the linear relationships between the grain parameters. Regression equations were plotted for highly correlated parameters.

3. Results and Discussion

3.1. Chemical Properties, Vitreousness, and the Hardness Index

Various indicators are used in analytical practice to characterize the grain of different wheat cultivars, including moisture, protein, and ash content; vitreousness; and the HI. The values of the parameters determined in the analyzed wheat cultivars are presented in Table 1. These parameters are considerably influenced by external factors, including environmental conditions and weather, and they can differ even within the same cultivar [52–56]. In the present study, the average dry basis (DB) moisture content of the wheat grain was determined within a relatively narrow range, from around 12.3% DB (SMH 87) to around 13.1% DB (Askalon and Bamberka), which indicates that the examined grain batches can be stored over prolonged periods of time and used in the production process. The average crude protein content was highest in the wheat cv. Fregata (approx. 15.4% DB) and lowest in the cv. SMH214 (approx. 12.2% DB). Other authors have reported a similar range of values [2,8,57,58]. It should also be noted that the grain of the wheat cvs. Türkis, Astoria, Fregata, Radunia, Parabola, Cytra, Askalon, Magic, and Bamberka contained a minimum of 14% protein; therefore, it can be used to enhance mixtures of medium- and low-quality grain. The total ash content ranged from 1.55% DB (Türkis) to 2.02% DB (Radunia), and these values were highly consistent with the data from the literature [2,8,53,58]. The lower the ash content of the grain, the lighter the color of the obtained flour. Grain with a high total ash content is particularly suitable for making brown and wholegrain bread [2,42]. Chinese Spring, Radunia, Parabola, SMH 87, and Ceres were characterized by the highest total ash content.

The sample of the reference wheat cv. Chinese Spring did not contain vitreous or semi-vitreous kernels, and the vitreousness was determined at 0% in this wheat cultivar. In the remaining cultivars, the vitreousness ranged from 2.6% (Türkis) to 89.4% (Bamberka). This parameter is mostly genetically conditioned but it can be modified by changes in weather and environmental conditions, in particular during the growing season, which can contribute to the grain's apparent vitreousness [52,59]. Vitreous kernels are generally denser than starchy kernels, and they have to be soaked for a longer time before milling [60,61]. According to Woźniak and Gontarz [62], vitreous kernels are characterized by a higher processing suitability because flour with a high extraction rate can be obtained during milling. The grain of the cvs. Bamberka, Ceres, Askalon, and SMH200 was characterized by the highest vitreousness.

The mean values of the HI ranged from 39.7 (Chinese Spring) to 88.6 (SMH214). Based on this parameter, the analyzed wheats were divided into five hardness classes [63]: soft (Chinese Spring), medium soft (Türkis), medium hard (Astoria and Julius), hard (Fregata, Radunia, Parabola, Cytra, Tonacja, Askalon, Magic, and Nutka), and very hard (Bamberka, SMH 196, SMH87, Ceres, SMH200, and SMH214). The HI determined using the SKCS system is one of the most important parameters for evaluating the processing suitability of grain [15,26,64–66]. Wheat with a soft endosperm is generally processed into pastry flour; wheat with a medium hard endosperm is processed into baking flour; whereas that with a hard endosperm is used in the production of semolina, couscous, and expanded wheat. In soft wheat, the starch granules are more loosely embedded into the protein matrix than in hard wheat, and the produced flour is finer-grained [6,43,53,61,67].

Analysis of the grain parameters (Table 1) revealed the greatest similarities between the cvs. Askalon and Barmberka. Parabola and SMH296 were most similar whereas Julius was the least similar to the remaining wheat cultivars.

3.2. Geometric Properties and Mass

The mean values of the kernels' geometric properties and mass are presented in Table 2. On average, the kernel mass was lowest in the cv. Chinese Spring (26.96 mg) and highest in the cv. Ceres (52.02 mg). Chinese Spring was also characterized by the smallest kernels. The longest kernels were noted in the cv. Ceres (7.99 mm), the widest kernels in the cv. Nutka (3.63 mm), and the thickest kernels in the cv. Bamberka (3.05 mm). The geometric properties and mass of the kernels play an important role during grain processing, sorting, separation, and milling [43]. According to Dziki and Laskowski [68], larger grains, in particular grains with a high sphericity index, are more suitable for producing flours with a high extraction rate because large kernels have a more desirable endosperm-to-bran ratio. The sphericity index was highest in the grain of the cv. Bamberka (65.35%) and lowest in the cv. Ceres (53.36%). The processing suitability of grain is evaluated based not only on the size of kernels but also their homogeneity [68], which was highest in the cv. Nutka and lowest in the cv. Parabola. In addition to genetic factors, the geometric properties and mass of the kernels are also affected by environmental conditions, and grain of the same wheat cultivar can differ considerably when grown in various regions. Nonetheless, the mean values of the geometric properties and mass were determined within the range reported in the literature [7,8,69–71].

The following pairs of wheat cultivars did not differ significantly in terms of their analyzed properties: Türkis and Fregata, Türkis and Bamberka, Astoria and Julius, Julius and Fregata, Julius and Bamberka, Fregata and Bamberka, Parabola and SMH214, Cytra and Askalon, Cytra and Magic, Tonacja and Nutka, Askalon and Magic, SMH196 and SMH87, SMH196 and Ceres, and SMH200 and SMH214. The grain of the cv. Julius was most similar whereas the grain of the cv. Chinese Spring was most different from the remaining wheat cultivars.

3.3. Mechanical Properties

The mechanical properties of the grain of the evaluated wheat cultivars are presented in Table 3. Its mechanical properties significantly influence the grain milling efficiency [5,8,42]. In this study, the average rupture force required to damage the kernels ranged from around 76 N (Chinese Spring) to around 155 N (Ceres). Similar rupture force values were given by Ibrahim et al. [55] in the grain of Hungarian wheat fertilized with different nitrogen rates. The longitudinal strain was determined in the range of 0.45 mm (Julius) to 0.64 mm (Nutka and Ceres), and the relative strain ranged from around

15.3% (Julius) to around 21.8% (Nutka). Similar rupture force values in wheat grain were noted by Markowski et al. [72], Dziki et al. [42], and Kalkan and Kara [71], but the cited authors reported a higher longitudinal strain and different values for the rupture energy. In the current study, the rupture energy ranged from 11.55 mJ (Chinese Spring) to 28.89 mJ (SMH214). Similar rupture force and rupture energy values in wheat grain were noted by Gorji et al. [73], Başlar et al. [7], and Voicu et al. [74]. The process of grain grinding would be the most energy-intensive in the cvs. Ceres, SMH200, and SMH214 and the least energy-intensive in the cvs. Chinese Spring, Julius, and Askalon.

Table 2. Geometric properties and mass (mean value ± standard deviation) of wheat grain.

Cultivar	Length [mm]	Width [mm]	Thickness [mm]	Mass [mg]	Aspect Ratio [%]	Sphericity [%]
Chinese Spring	5.41 ± 0.23^{a}	2.76 ± 0.27^{a}	2.46 ± 0.17^{a}	26.960 ± 5.58^{a}	50.91 ± 3.90^{cd}	61.33 ± 2.56^{gh}
Türkis	$6.22 \pm 0.39^{c–e}$	3.61 ± 0.39^{ef}	$2.96 \pm 0.24^{d–f}$	$47.38 \pm 9.90^{d–f}$	58.03 ± 5.26^{h}	65.05 ± 2.95^{k}
Astoria	6.33 ± 0.37^{e}	$3.48 \pm 0.32^{c–f}$	$2.87 \pm 0.29^{b–e}$	$45.04 \pm 10.33^{c–e}$	$55.09 \pm 4.15^{e–g}$	$62.97 \pm 3.04^{h–j}$
Julius	6.24 ± 0.40^{de}	$3.43 \pm 0.33^{b–f}$	$2.94 \pm 0.23^{d–f}$	$46.07 \pm 9.49^{c–f}$	$55.07 \pm 4.96^{e–g}$	$63.70 \pm 2.76^{i–k}$
Fregata	6.23 ± 0.37^{de}	$3.52 \pm 0.25^{d-f}$	3.04 ± 0.27^{f}	$47.97 \pm 8.54^{d–f}$	$56.55 \pm 2.83^{f–h}$	65.10 ± 1.91^{k}
Radunia	6.34 ± 0.38^{e}	3.31 ± 0.35^{bc}	$2.95 \pm 0.29^{d–f}$	$42.86 \pm 9.99^{b–d}$	52.20 ± 4.71^{cd}	$62.31 \pm 3.16^{g–i}$
Parabola	6.96 ± 0.56^{fg}	$3.36 \pm 0.35^{b–d}$	3.02 ± 0.32^{ef}	49.06 ± 10.96^{ef}	48.38 ± 5.20^{bc}	59.36 ± 3.44^{ef}
Cytra	5.78 ± 0.41^{b}	3.29 ± 0.36^{bc}	2.75 ± 0.24^{bc}	37.46 ± 8.68^{b}	$56.97 \pm 5.45^{f–h}$	64.68 ± 3.24^{jk}
Tonacja	6.74 ± 0.40^{f}	3.62 ± 0.25^{f}	$2.97 \pm 0.23^{d–f}$	51.62 ± 8.16^{f}	53.73 ± 3.16^{de}	61.81 ± 2.12^{gh}
Askalon	$5.99 \pm 0.26^{b–d}$	3.30 ± 0.31^{bc}	2.73 ± 0.21^{b}	40.51 ± 8.35^{bc}	$55.02 \pm 4.06^{e–g}$	$62.98 \pm 2.51^{h–j}$
Magic	5.96 ± 0.38^{bc}	3.26 ± 0.29^{b}	$2.82 \pm 0.22^{b–d}$	40.89 ± 8.54^{bc}	$54.63 \pm 3.61^{d–f}$	$63.67 \pm 2.18^{i–k}$
Nutka	6.93 ± 0.37^{f}	3.63 ± 0.23^{f}	$2.94 \pm 0.16^{d–f}$	51.50 ± 6.54^{f}	$52.41 \pm 2.85^{c–e}$	60.57 ± 1.89^{fg}
Bamberka	6.26 ± 0.31^{e}	3.59 ± 0.30^{ef}	3.05 ± 0.27^{f}	49.77 ± 9.03^{ef}	57.41 ± 3.94^{gh}	65.35 ± 3.07^{k}
SMH196	7.78 ± 0.50^{jk}	3.27 ± 0.29^{b}	3.03 ± 0.22^{ef}	51.54 ± 10.10^{f}	42.13 ± 3.84^{a}	54.71 ± 2.80^{ab}
SMH87	7.54 ± 0.41^{ij}	3.24 ± 0.24^{b}	3.01 ± 0.26^{ef}	50.49 ± 9.01^{ef}	43.06 ± 2.79^{a}	55.58 ± 2.39^{bc}
Ceres	7.99 ± 0.40^{k}	3.24 ± 0.27^{b}	3.00 ± 0.24^{ef}	52.02 ± 10.40^{f}	40.57 ± 3.30^{a}	53.36 ± 2.65^{a}
SMH200	7.32 ± 0.47^{hi}	$3.37 \pm 0.28^{b–d}$	$2.90 \pm 0.20^{c–f}$	$48.10 \pm 8.38^{d–f}$	46.22 ± 4.16^{b}	56.76 ± 2.83^{cd}
SMH214	7.20 ± 0.46^{gh}	$3.42 \pm 0.26^{b–e}$	3.01 ± 0.21^{ef}	48.88 ± 9.11^{ef}	47.60 ± 3.57^{b}	58.37 ± 2.74^{de}

$^{a,...,k}$—superscript letters denote significant differences between the examined properties at $p < 0.05$.

Table 3. Mechanical properties (mean value ± standard deviation) of wheat grain.

Cultivar	Rupture Force [N]	Longitudinal Strain [mm]	Relative Strain [%]	Rupture Energy [mJ]
Chinese Spring	76.10 ± 13.22^{a}	0.46 ± 0.05^{a}	18.57 ± 1.75^{fg}	11.55 ± 3.57^{a}
Türkis	$100.26 \pm 18.70^{b–d}$	0.46 ± 0.04^{a}	15.58 ± 1.63^{ab}	$15.51 \pm 4.50^{a–d}$
Astoria	$105.00 \pm 18.93^{b–e}$	0.50 ± 0.05^{bc}	$17.50 \pm 1.93^{c–f}$	$18.69 \pm 5.93^{c–f}$
Julius	92.43 ± 22.10^{ab}	0.45 ± 0.05^{a}	15.28 ± 1.83^{a}	13.83 ± 5.07^{ab}
Fregata	$113.05 \pm 22.06^{c–f}$	$0.51 \pm 0.06^{b–e}$	$16.93 \pm 2.16^{b–d}$	20.49 ± 7.60^{ef}
Radunia	$117.50 \pm 27.90^{d–f}$	$0.50 \pm 0.06^{b–d}$	$17.10 \pm 1.73^{c–e}$	$19.74 \pm 6.83^{d–f}$
Parabola	$111.96 \pm 27.79^{c–f}$	$0.52 \pm 0.10^{c–e}$	$17.40 \pm 3.56^{c–f}$	20.49 ± 8.25^{ef}
Cytra	$102.76 \pm 17.69^{b–d}$	0.46 ± 0.04^{a}	16.79 ± 1.49^{bc}	$15.93 \pm 4.04^{a–d}$
Tonacja	140.75 ± 19.93^{gh}	$0.54 \pm 0.06^{d–f}$	18.29 ± 2.03^{ef}	25.08 ± 7.14^{gh}
Askalon	98.51 ± 21.39^{bc}	0.46 ± 0.04^{a}	$16.93 \pm 1.40^{b–d}$	$14.81 \pm 4.62^{a–c}$
Magic	$112.32 \pm 23.57^{c–f}$	0.48 ± 0.04^{ab}	$16.93 \pm 1.51^{b–d}$	$17.73 \pm 5.47^{b–e}$
Nutka	123.70 ± 26.47^{fg}	0.64 ± 0.07^{h}	21.84 ± 2.13^{j}	22.65 ± 7.63^{fg}
Bamberka	126.10 ± 17.83^{fg}	$0.52 \pm 0.05^{c–e}$	$16.98 \pm 1.17^{c–e}$	$20.95 \pm 4.49^{e–g}$
SMH196	122.41 ± 36.36^{ef}	0.55 ± 0.07^{ef}	$18.22 \pm 2.72^{d–f}$	$21.32 \pm 9.60^{e–g}$
SMH87	124.80 ± 28.39^{fg}	0.60 ± 0.06^{f}	19.91 ± 2.30^{gh}	$21.95 \pm 7.59^{e–g}$
Ceres	154.87 ± 40.96^{h}	0.64 ± 0.06^{h}	21.27 ± 2.09^{ij}	28.43 ± 10.34^{h}
SMH200	154.35 ± 28.40^{h}	0.58 ± 0.05^{fg}	19.98 ± 1.85^{hi}	27.69 ± 7.76^{h}
SMH214	152.77 ± 37.99^{h}	0.59 ± 0.06^{g}	19.82 ± 1.97^{gh}	28.89 ± 9.97^{h}

$^{a,...,j}$—superscript letters denote significant differences between the examined properties at $p < 0.05$.

No significant differences in the mechanical properties of the grain were observed in the following pairs of wheat cultivars: Türkis and Julius, Türkis and Cytra, Türkis and Askalon, Türkis and Magic, Astoria and Fregata, Astoria and Radunia, Astoria and Parabola, Astoria and Magic, Fregata and Radunia, Fregata and Parabola, Fregata and Magic, Fregata and Bamberka, Fregata and SMH196, Radunia and Parabola, Radunia and Magic, Radunia and Bamberka, Parabola and Bamberka, Parabola and SMH196, Cytra and Askalon, Cytra and Magic, Tonacja and Bamberka, Askalon and Magic, Bamberka and SMH196, and SMH200 and SMH214. The grain of the cv. Magic was most similar whereas the grain of the wheat cv. Ceres was most different from the remaining cultivars.

The cultivar pairs Cytra and Magic, Askalon and Magic, and SMH200 and SMH214 were most similar in terms of all the examined grain parameters. In turn, the greatest differences were observed between the cvs. Chinese Spring and Bamberka, and these pairs were always classified into different homogeneous groups. The grain of the cvs. Fregata and Parabola can be used to determine the physicochemical parameters of wheat kernels because these cultivars were most similar to the other wheat cultivars. In this respect, Chinese Spring differed most considerably from the remaining wheat cultivars.

3.4. Friabilin Content Indicator

According to Martin et al. [32], Chinese Spring should be used as the reference cultivar for determining the friabilin content of wheat grain in SDS-PAGE analysis. In the present study, the friabilin content indicator was set at 1 for the reference cultivar (Chinese Spring) (Figure 2). In the remaining wheat cultivars, this parameter ranged from 0.213 (SMH87) to 0.769 (Parabola). The wide practical application of the proposed approach to determining the friabilin content of wheat grain as a measurement method would require validation, which could not be conducted in this study. The results obtained with the use of this approach should also be compared with the results of measurements performed with other methods, such as RP-HPLC.

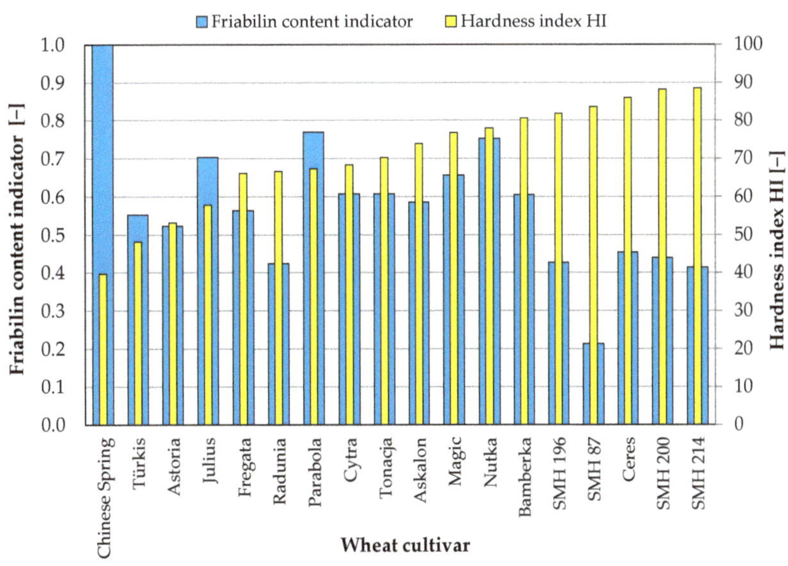

Figure 2. Values of the friabilin content indicator and the grain hardness index (HI) in the analyzed wheat cultivars.

Contrary to the observations made by other authors [6,12,75], the friabilin content did not always decrease with an increase in the HI. According to Bhave and Morris [15], the kernel hardness is not affected by the total content of friabilin proteins but only by the protein fractions that are directly associated with starch. For this reason, kernels with the same content of friabilin proteins often differ in their HI values. Similar observations were made by Anjum and Walker [76], Geneix et al. [77], and Tu and Li [26], who concluded that significant differences in the hardness of the wheat kernels with a similar content of puroindoline proteins may suggest that this parameter is also influenced by other factors. Therefore, the value of the friabilin content indicator determined for the grain of a given wheat cultivar is not always correlated with its mechanical properties.

3.5. Correlations between the Fiabilin Content Indicator and the Physicochemical Properties of Grain

The correlation coefficients describing the strength of the relationships between the friabilin content indicator and selected physicochemical properties of wheat grain are presented in Table 4. The friabilin content indicator was bound by the strongest correlation with the kernel length (0.621) and by the weakest correlation with the total ash content (0.083). In addition to the mean kernel length, the friabilin content indicator was also significantly influenced by the mean kernel thickness, mass, rupture force, rupture energy, vitreousness, and HI. The correlation coefficients noted for these pairs were negative, which testifies to their inversely proportional relationship, where an increase in the value of one parameter induces a decrease in the value of the other parameter. The presence of correlations between the friabilin content of grain, its geometric properties, and the mass of its kernels could be a unique feature of wheat cultivars grown in Polish regions because no such relationships have been reported in the literature. The observed correlations could result from changes in the puroindoline profile with an increase in kernel size, but more detailed chemical analyses are needed to confirm this hypothesis. However, the friabilin content was found to be associated with the rupture force, rupture energy and HI in other studies, although these correlations were relatively weak. Geneix et al. [77] reported higher values of coefficients of correlation between the content of puroindoline proteins, the particle size index (PSI) (0.94), and the grain hardness measured using near-infrared spectroscopy (NIRS) (0.70). According to Hogg et al. [10], the wheat grain hardness is not correlated with the total content of puroindoline proteins but with the content of the PINA and PINB proteins. This observation could explain the findings of Mikulíková [14] and Tu and Li [26], who concluded that only approximately 60% of the variation in kernel hardness can be attributed to the friabilin content.

Table 4. Correlation coefficients denoting the strength of linear relationships between friabilin content and other properties of wheat grain.

Property	Correlation with Friabilin Content
Moisture	0.368
Crude protein	0.109
Total ash	−0.083
Vitreousness	**−0.492**
Hardness index, HI	**−0.600**
Length	**−0.621**
Width	−0.153
Thickness	**−0.584**
Mass	**−0.552**
Aspect ratio	0.436
Sphericity	0.445
Rupture force	**−0.592**
Longitudinal strain	−0.435
Relative strain	−0.213
Rupture energy	**−0.552**

Values in bold denote significant correlations at 0.05. Critical value of the correlation coefficient—0.468. Sample size n = 18.

Due to the relatively weak correlations between the compared parameters, the derived regression equations were characterized by relatively low values of coefficients of determination (Figure 3). These relationships can be described with linear functions, and the highest value of the coefficient of determination was noted for the relationship between the kernel length and the friabilin content indicator ($R^2 = 0.39$).

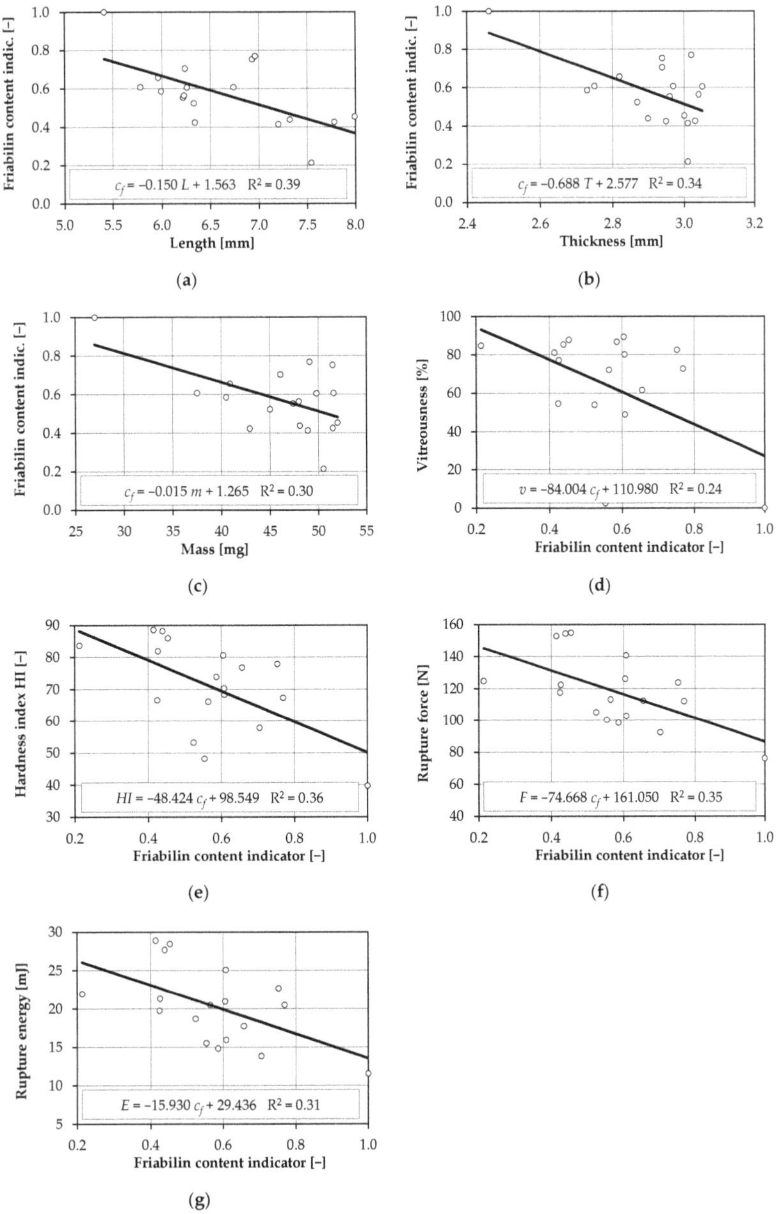

Figure 3. Relationships between the friabilin content indicator and the physical properties of wheat grain: (**a**) length and friabilin content indicator; (**b**) thickness and friabilin content indicator; (**c**) mass and friabilin content indicator; (**d**) friabilin content indicator and vitreousness; (**e**) friabilin content indicator and hardness index HI; (**f**) friabilin content indicator and rupture force; (**g**) friabilin content indicator and rupture energy.

The friabilin content indicator decreased with a rise in the mean kernel length, thickness, and mass, and the noted decrease exceeded 50% in extreme cases. An increase in the friabilin content indicator also led to a decrease in the vitreousness, HI, rupture force, and rupture energy. The vitreousness decreased by around 70%, the HI decreased by around 40%, the rupture force decreased by around 40%, and the rupture energy decreased by around 50% when the friabilin content indicator increased from 0.21 to 1. This indicates that in a given wheat cultivar, the mean kernel length is associated with a certain value of the friabilin content indicator, and this information can be used to estimate the energy consumption during grain griding and milling.

4. Conclusions

In the proposed approach, the friabilin content of wheat grain was determined by comparing the intensity of the friabilin bands in electropherograms of the analyzed wheat cultivars and the reference cultivar. Image analysis software can be used to perform accurate measurements because visual inspection of the band intensity is not highly reliable.

In the grain of the analyzed wheat cultivars, the friabilin content ranged from around 0.21 (SMH87) to around 0.77 (Parabola) relative to the Chinese Spring reference.

The friabilin content indicator was significantly correlated with the mean kernel length, thickness, mass, vitreousness, HI, rupture force, and rupture energy. Among these parameters, this indicator was bound by the strongest correlation with the kernel length and by the weakest correlation with vitreousness.

The presence of a relatively low coefficient of correlation between the friabilin content indicator and grain hardness validates the hypothesis that wheat grain's resistance to mechanical damage is conditioned not only by the content of puroindoline proteins but also by other factors.

Author Contributions: Conceptualization, A.M.-M. and M.W.; methodology, Z.K., A.M.-M., M.W. and S.G.; software, Z.K. and S.L.; validation, Z.K.; formal analysis, Z.K.; investigation, Z.K.; resources, A.M.-M.; data curation, A.M.-M. and S.G.; writing—original draft preparation, Z.K., A.M.-M. and M.W.; writing—review and editing, Z.K. and M.W.; visualization, Z.K.; supervision, Z.K.; project administration, Z.K. and M.W.; funding acquisition, Z.K. and A.M.-M. All authors have read and agreed to the published version of the manuscript.

Funding: This research received no external funding.

Institutional Review Board Statement: Not applicable.

Informed Consent Statement: Not applicable.

Data Availability Statement: The raw data supporting the conclusions of this article will be made available by the authors on request.

Conflicts of Interest: The authors declare no conflicts of interest.

References

1. Kim, K.-H.; Kim, J.-Y. Understanding wheat starch metabolism in properties, environmental stress condition, and molecular approaches for value-added utilization. *Plants* **2021**, *10*, 2282. [CrossRef]
2. Gooding, M.J.; Shewry, P.R. *Wheat: Environment, Food and Health*; Wiley Online Library: John Wiley & Sons Ltd.: Hoboken, NJ, USA, 2022. [CrossRef]
3. Biel, W.; Maciorowski, L. Assessing nutritional value of grains of selected wheat cultivars. *Zywn.-Nauk. Technol. Ja.* **2012**, *2*, 45–55. [CrossRef]
4. Groos, C.; Bervas, E.; Charmet, G. Genetic analysis of grain protein content, grain hardness and dough rheology in a hard × hard bread wheat progeny. *J. Cereal Sci.* **2004**, *40*, 93–100. [CrossRef]
5. Różyło, R.; Laskowski, J. Study of relationships between physical ad technological properties of spring wheat grain. *Acta Agroph.* **2007**, *9*, 459–470.
6. Pasha, I.; Anjum, F.M.; Morris, C.F. Grain hardness: A major determinant of wheat quality. *Food Sci. Technol. Int.* **2010**, *16*, 511–522. [CrossRef] [PubMed]
7. Başlar, M.; Kalkan, F.; Kara, M.; Ertugay, M.F. Correlation between the protein content and mechanical properties of wheat. *Turk. J. Agric. For.* **2012**, *36*, 601–607. [CrossRef]

8. Kasraei, M.; Nejadi, J.; Shafiei, S. Relationship between grain physicochemical and mechanical properties of some Iranian wheat cultivars. *J. Agric. Sci. Technol.* **2015**, *17*, 635–647.
9. Kaliniewicz, Z.; Markowska-Mendik, A.; Warechowska, M. An evaluation of selected engineering properties of polish durum wheat grain. *J. Cereal Sci.* **2022**, *104*, 103401. [CrossRef]
10. Hogg, A.C.; Sripo, T.; Beecher, B.; Martin, J.M.; Giroux, M.J. Wheat puroindolines interact to form friabilin and control wheat grain hardness. *Theor. Appl. Genet.* **2004**, *108*, 1089–1097. [CrossRef]
11. Turnbull, K.-M.; Rahman, S. Endosperm texture in wheat. *J. Cereal Sci.* **2002**, *36*, 327–337. [CrossRef]
12. Laskowski, J.; Różyło, R. Influence of gluten content and wheat glassiness on the wheat grain grinding energy. *Acta Agroph.* **2003**, *2*, 589–596.
13. Morris, C.F. Puroindolines: The molecular genetic basis of wheat grain hardness. *Plant Mol. Biol.* **2002**, *48*, 633–647. [CrossRef] [PubMed]
14. Mikulíková, D. The effect of friabilin on wheat grain hardness. *Czech J. Genet. Plant Breed.* **2007**, *43*, 35–43. [CrossRef]
15. Bhave, M.; Morris, C.F. Molecular genetics of puroindolines and related genes: Regulation of expression, membrane binding properties and applications. *Plant Mol. Biol.* **2008**, *66*, 221–231. [CrossRef] [PubMed]
16. Lullien-Pellerin, V.; Haraszi, R.; Anderssen, R.S.; Morris, C.F. Understanding the mechanics of wheat grain fractionation and the impact of puroindolines on milling and product quality. In *Wheat Quality for Improving Processing and Human Health*; Igrejas, G., Ikeda, T., Guzmán, C., Eds.; Springer: Cham, Switzerland, 2020; pp. 369–385. [CrossRef]
17. Giroux, M.J.; Morris, C.F. Wheat grain hardness results from highly conserved mutations in the friabilin components puroindoline a and b. *Proc. Natl. Acad. Sci. USA* **1998**, *95*, 6262–6266. [CrossRef]
18. Gazza, L.; Nocente, F.; Ng, P.K.W.; Pogna, N.E. Genetic and biochemical analysis of common wheat cultivars lacking puroindoline a. *Theor. Appl. Genet.* **2005**, *110*, 470–478. [CrossRef]
19. Ni, D.; Yang, F.; Lin, L.; Sun, C.; Ye, X.; Wang, L.; Kong, X. Interrelating grain hardness index of wheat with physicochemical and structural properties of starch extracted therefrom. *Foods* **2022**, *11*, 1087. [CrossRef] [PubMed]
20. Tranquilli, G.; Heaton, J.; Chicaiza, O.; Dubcovsky, J. Substitutions and deletions of genes related to grain hardness in wheat and their effect on grain texture. *Crop Sci.* **2002**, *42*, 1812–1817. [CrossRef]
21. Gasparis, S.; Nadolska-Orczyk, A. Genetic background of wheat and triticale grain hardness. *Biul. Inst. Hod. i Aklim. Roślin* **2013**, *267*, 17–29. (In Polish)
22. Nucia, A.; Okoń, S.; Tomczyńska-Mleko, M.; Nawrocka, A. Molecular and physical characterization of grain hardness in European spring common wheat (*Triticum aestivum* L.). *3 Biotech* **2021**, *11*, 345. [CrossRef]
23. Capparelli, R.; Amoroso, M.G.; Palumbo, D.; Iannaccone, M.; Faleri, C.; Cresti, M. Two plant puroindolines colocalize in wheat seed and in vitro synergistically fight against pathogens. *Plant Mol. Biol.* **2005**, *58*, 857–867. [CrossRef] [PubMed]
24. Wiley, P.R.; Tosi, P.; Evrard, A.; Lovergrove, A.; Jones, H.D.; Shewry, P.R. Promoter analysis and immunolocalisation show that puroindoline genes are exclusively expressed in starchy endosperm cells of wheat grain. *Plant Mol. Biol.* **2007**, *64*, 125–136. [CrossRef] [PubMed]
25. Feiz, L.; Wanjugi, H.W.; Melnyk, C.W.; Altossar, I.; Martin, J.M.; Giroux, M.J. Puroindolines co-localize to the starch granule surface and increase seed bound polar lipid content. *J. Cereal Sci.* **2009**, *50*, 91–98. [CrossRef]
26. Tu, M.; Li, Y. Toward the genetic basis and multiple QTLs of kernel hardness in wheat. *Plants* **2020**, *9*, 1631. [CrossRef] [PubMed]
27. Beecher, B.; Bettge, A.; Smidansky, E.; Giroux, M.J. Expression of wild-type PinB sequence in transgenic wheat complements a hard phenotype. *Theor. Appl. Genet.* **2002**, *105*, 870–877. [CrossRef] [PubMed]
28. Chen, M.; Wilkinson, M.; Tosi, P.; He, G.; Shewry, P. Novel puroindoline and grain softness protein alleles in *Aegilops* species with the C, D, S, M and U genomes. *Theor. Appl. Genet.* **2005**, *111*, 1159–1166. [CrossRef] [PubMed]
29. Qian, Y.; Preston, K.; Krokhin, O.; Mellish, J.; Ens, W. Characterization of wheat gluten proteins by HPLC and MALDI TOF mass spectrometry. *J. Am. Soc. Mass Spectrom.* **2008**, *19*, 1542–1550. [CrossRef]
30. Obuchowski, W.; Salmanowicz, B.; Banaszak, Z.; Adamski, T.; Surma, M.; Kaczmarek, Z.; Majcher, M.; Ługowska, B.; Kuczyńska, A.; Krystkowiak, K. Grain hardness of wheat bred in Poland and its relationship to starch damage during milling. *Int. Agrophys.* **2010**, *24*, 69–74.
31. Zhang, Y.; Hu, X.; Juhasz, A.; Islam, S.; Yu, Z.; Zhao, Y.; Li, G.; Ding, W.; Ma, W. Characterising avenin-like proteins (ALPs) from albumin/globulin fraction of wheat grains by RP-HPLC, SDS-PAGE, and MS/MS peptides sequencing. *BMC Plant Biol.* **2020**, *20*, 45. [CrossRef]
32. Martin, J.M.; Meyer, F.D.; Smidansky, E.D.; Wanjugi, H.; Blechl, A.E.; Giroux, M.J. Complementation of the pina (null) allele with the wild type *Pina* sequence restores a soft phenotype in transgenic wheat. *Theor. Appl. Genet.* **2006**, *113*, 1563–1570. [CrossRef]
33. Li, G.; He, Z.; Peña, R.J.; Xia, X.; Lillemo, M.; Qixin, S. Identification of novel secaloindoline-a and secaloindoline-b alleles in CIMMYT hexaploid triticale lines. *J. Cereal Sci.* **2006**, *43*, 378–386. [CrossRef]
34. Chantret, N.; Salse, J.; Sabot, F.; Rahman, S.; Bellec, A.; Laubin, B.; Dubois, I.; Dossat, C.; Sourdille, P.; Joudrier, P.; et al. Molecular basis of evolutionary events that shaped the *Hardness* locus in diploid and polyploid wheat species (*Triticum* and *Aegilops*). *Plant Cell* **2005**, *17*, 1033–1045. [CrossRef]
35. Wanjugi, H.W.; Hogg, A.C.; Martin, J.M.; Giroux, M.J. The role of puroindoline A and B individually and in combination on grain hardness and starch association. *Crop Sci.* **2007**, *47*, 67–76. [CrossRef]

36. *ISO 20483:2013*; Cereals and Pulses—Determination of the Nitrogen Content and Calculation of the Crude Protein Content—Kjeldahl Method. International Organization for Standardization: Geneva, Switzerland, 2013. Available online: https://www.iso.org/standard/59162.html (accessed on 16 January 2023).
37. *ISO 2171:2007*; Cereals, Pulses and By-Products—Determination of Ash Yield by Incineration. International Organization for Standardization: Geneva, Switzerland, 2007. Available online: https://www.iso.org/standard/37264.html (accessed on 16 January 2023).
38. Symons, S.J.; Van Schepdael, L.; Dexter, J.E. Measurement of hard vitreous kernels in durum wheat by machine vision. *Cereal Chem.* **2003**, *80*, 511–517. [CrossRef]
39. AACC method 55-31.01. Single-kernel characterization system for wheat kernel texture. In *AACC Approved Methods of Analysis*, 11th ed.; Cereals & Grains Association: Eagan, MN, USA, 2010. [CrossRef]
40. Martin, C.R.; Rousser, R.; Brabec, D.L. Development of a single-kernel wheat characterization system. *Trans. ASAE* **1993**, *36*, 1399–1404. [CrossRef]
41. *ISO 24333:2009*; Cereals and Cereal Products—Sampling. International Organization for Standardization: Geneva, Switzerland, 2009. Available online: https://www.iso.org/standard/42165.html (accessed on 16 January 2023).
42. Dziki, D.; Cacak-Pietrzak, G.; Miś, A.; Jończyk, K.; Gawlik-Dziki, U. Influence of wheat kernel physical properties on the pulverizing process. *J. Food Sci. Technol.* **2014**, *51*, 2648–2655. [CrossRef] [PubMed]
43. Babić, L.; Babić, M.; Turan, J.; Matić-Kekić, S.; Radojčin, M.; Mehandžić-Stanišić, S.; Pavkova, I.; Zoranović, M. Physical and stress-strain properties of wheat (*Triticum aestivum*) kernel. *J. Sci. Food Agric.* **2011**, *91*, 1236–1243. [CrossRef] [PubMed]
44. Morris, C.F.; Bhave, M. Reconciliation of D-genome puroindoline allele designations with current DNA sequence data. *J. Cereal Sci.* **2008**, *48*, 277–287. [CrossRef]
45. Wang, J.; Sun, J.; Liu, D.; Yang, W.; Wang, D.; Tong, Y.; Zhang, A. Analysis of *Pina* and *Pinb* alleles in the micro-core collections of Chinese wheat germplasm by ecotilling and identification of a novel *Pinb* allele. *J. Cereal Sci.* **2008**, *48*, 836–842. [CrossRef]
46. Feiz, L.; Martin, J.M.; Giroux, M.J. Creation and functional analysis of new puroindoline alleles in *Triticum aestivum*. *Theor. Appl. Genet.* **2009**, *118*, 247–257. [CrossRef] [PubMed]
47. Bettge, A.D.; Morris, C.F.; Greenblatt, G.A. Assessing genotypic softness in single wheat kernels using starch granule-associated friabilin as a biochemical marker. *Euphytica* **1995**, *86*, 65–72. [CrossRef]
48. Chang, C.; Zhang, H.P.; Xu, J.; Li, W.H.; Liu, G.T.; You, M.S.; Li, B.Y. Identification of allelic variations of puroindoline genes controlling grain hardness in wheat using a modified denaturing PAGE. *Euphytica* **2006**, *152*, 225–234. [CrossRef]
49. Laemmli, U.K. Cleavage of structural proteins during the assembly of the head of bacteriophage T4. *Nature* **1970**, *227*, 680–685. [CrossRef] [PubMed]
50. Gromova, I.; Celis, J.E. Chapter 27—Protein detection in gels by silver staining: A procedure compatible with mass spectrometry. In *Cell Biology. A Laboratory Handbook*, 3rd ed.; Celis, J.E., Carter, N., Simons, K., Eds.; Elsevier: Amsterdam, The Netherlands, 2006; Volume 4, pp. 219–223. [CrossRef]
51. Gasparis, S. *Analysis of Pina and Pinb Genes Determining Grain Hardness in Hexaploid Wheat and Their Orthologs in Triticale*; Instytut Hodowli i Aklimatyzacji Roślin—Państwowy Instytut Badawczy: Radzików, Poland, 2012. (In Polish)
52. Greffeuille, V.; Abecassis, J.; Rousset, M.; Oury, F.-X.; Faye, A.; L'Helgouac'h, C.B.; Lullien-Pellerin, V. Grain characterization and milling behaviour of near-isogenic lines differing by hardness. *Theor. Appl. Genet.* **2006**, *114*, 1–12. [CrossRef] [PubMed]
53. Hrušková, M.; Švec, I. Wheat hardness in relation to other quality factors. *Czech J. Food Sci.* **2009**, *27*, 240–248. [CrossRef]
54. Oury, F.X.; Lasme, P.; Michelet, C.; Rousset, M.; Abecassis, J.; Lullien-Pellerin, V. Relationships between wheat grain physical characteristics studied through near-isogenic lines with distinct puroindoline-b allele. *Theor. Appl. Genet.* **2015**, *128*, 913–929. [CrossRef] [PubMed]
55. Ibrahim, A.; Varga, A.C.; Jolánkai, M.; Safranyik, F. Applying infrared technique as a nondestructive method to assess wheat grain hardness. *Int. J. Sci. Qual. Anal.* **2018**, *4*, 100–107. Available online: https://article.sciencepublishinggroup.com/pdf/10.11648.j.ijsqa.20180403.16.pdf (accessed on 16 January 2023).
56. Manai-Djebali, H.; Oueslati, I.; Nouairi, I.; Taamalli, A.; Nait-Mohamed, S.; Mliki, A.; Ghorbel, A. Chemical composition of durum wheat kernels: Impact of the growing location. *Euro-Mediterr. J. Environ. Integr.* **2021**, *6*, 26. [CrossRef]
57. El-Khayat, G.H.; Samaan, J.; Manthey, F.A.; Fuller, M.P.; Brennan, C.S. Durum wheat quality I: Some physical and chemical characteristics of Syrian durum wheat genotypes. *Int. J. Food Sci. Technol.* **2006**, *41*, 22–29. [CrossRef]
58. Boz, H.; Gerçekaslan, K.E.; Karaoğlu, M.M.; Kotancilar, H.G. Differences in some physical and chemical properties of wheat grains from different parts within the spike. *Turk. J. Agric. For.* **2012**, *36*, 309–316. [CrossRef]
59. Korkut, K.Z.; Bílgín, O.; Başer, Í.; Sağlam, N. Stability of grain vitreousness in durum wheat (*Triticum durum* L. Desf.) genotypes in the north-western region of Turkey. *Turk. J. Agric. For.* **2007**, *31*, 313–318.
60. Manley, M.; du Toit, G.; Geladi, P. Tracking diffusion of conditioning water in single wheat kernels of different hardnesses by near infrared hyperspectral imaging. *Anal. Chim. Acta* **2011**, *686*, 64–75. [CrossRef] [PubMed]
61. Fu, B.X.; Wang, K.; Dupuis, B.; Taylor, D.; Nam, S. Kernel vitreousness and protein content: Relationship, interaction and synergistic effects on durum wheat quality. *J. Cereal Sci.* **2017**, *78*, 2–9. [CrossRef]
62. Woźniak, A.; Gontarz, D. Evaluation of selected quality features of grain of durum wheat cv. Floradur depending on tillage and nitrogen fertilisation. *Acta Agroph.* **2011**, *18*, 481–489.

63. Mahesh, S.; Jayas, D.S.; Paliwal, J.; White, N.D.G. Comparison of partial least squares regression (PLSR) and principal components regression (PCR) methods for protein and hardness predictions using the near-infrared (NIR) hyperspectral images of bulk samples of Canadian wheat. *Food Bioproc. Tech.* **2015**, *8*, 31–40. [CrossRef]
64. Dobraszczyk, B.J.; Withworth, M.B.; Vincent, J.F.V.; Khan, A.A. Single kernel wheat hardness and fracture properties in relation to density and the modeling of fracture in wheat endosperm. *J. Cereal Sci.* **2002**, *35*, 245–263. [CrossRef]
65. Anderssen, R.; Haraszi, R. Characterizing and exploiting the rheology of wheat hardness. *Eur. Food Res. Technol.* **2009**, *229*, 159–174. [CrossRef]
66. Erkinbaev, C.; Derksen, K.; Paliwal, J. Single kernel wheat hardness estimation using near infrared hyperspectral imaging. *Infrared Phys. Technol.* **2019**, *98*, 250–255. [CrossRef]
67. Misailidis, N.; Campbell, G. Interpreting crush response profiles from the single kernel characterization system. *J. Cereal Sci.* **2013**, *57*, 222–229. [CrossRef]
68. Dziki, D.; Laskowski, J. Wheat kernel physical properties and milling process. *Acta Agroph.* **2005**, *6*, 59–71.
69. Yücel, C.; Baloch, F.S.; Özkan, H. Genetic analysis of some physical properties of bread wheat grain (*Triticum aestivum* L. em Thell). *Turk. J. Agric. For.* **2009**, *33*, 525–535. [CrossRef]
70. Boac, J.M.; Casada, M.E.; Maghirang, R.G.; Harner, J.P., III. Material and interaction properties of selected grains and oilseeds for modeling discrete particles. *Trans. ASABE* **2010**, *53*, 1201–1216. [CrossRef]
71. Kalkan, F.; Kara, M. Handling, frictional and technological properties of wheat as affected by moisture content and cultivar. *Powder Technol.* **2011**, *213*, 116–122. [CrossRef]
72. Markowski, M.; Żuk-Gołaszewska, K.; Kwiatkowski, D. Influence of variety on selected physical and mechanical properties of wheat. *Ind. Crops Prod.* **2013**, *47*, 113–117. [CrossRef]
73. Gorji, A.; Rajabipour, A.; Tawakoli, H. Fracture resistance of wheat grain as a function of moisture content, loading rate and grain orientation. *Aust. J. Crop Sci.* **2010**, *4*, 448–452.
74. Voicu, G.; Tudosie (Stefan), E.M.; Ungureanu, N.; Constantin, G.A. Some mechanical characteristics of wheat seeds obtained by uniaxial compression tests. *UPB Sci. Bull. Ser. D* **2013**, *75*, 265–278.
75. Langer, M.; Salmanowicz, B.P. Identification of puroindoline alleles in Polish wheat cultivars by molecular marker. *Biul. Inst. Hod. Aklim. Rośl.* **2009**, *253*, 93–101. (In Polish)
76. Anjum, F.M.; Walker, C.E. Review on the significance of starch and protein to wheat kernel hardness. *J. Sci. Food Agric.* **2006**, *56*, 1–13. [CrossRef]
77. Geneix, N.; Dalgalarrondo, M.; Tassy, C.; Nadaud, I.; Barret, P.; Bakan, B.; Elmorjani, K.; Marion, D. Relationships between puroindoline A-prolamin interactions and wheat grain hardness. *PLoS ONE* **2020**, *15*, e0225293. [CrossRef]

processes MDPI

Article

Extraction Kinetics of Pyridine, Quinoline, and Indole from the Organic Phase with Natural Deep Eutectic Solvents and Separation Study Using a Centrifugal Extractor

Dmitriy V. Lobovich, Inna V. Zinov'eva, Nikita A. Milevskii, Artak E. Kostanyan *, Yulia A. Zakhodyaeva and Andrey A. Voshkin

Kurnakov Institute of General and Inorganic Chemistry, Russian Academy of Sciences, 31 Leninskii Pr., 119991 Moscow, Russia; lobovichdv@yandex.ru (D.V.L.); iz@igic.ras.ru (I.V.Z.); mna@igic.ras.ru (N.A.M.); yz@igic.ras.ru (Y.A.Z.); aav@igic.ras.ru (A.A.V.)
* Correspondence: kost@igic.ras.ru

Abstract: In this study, the kinetics of extracting pyridine, quinoline, and indole from model fuels using natural deep eutectic solvents (NaDES) composed of carboxylic acids, xylitol, and water were investigated under static conditions. This research marks the first examination of extraction kinetics in this context. The key kinetic parameters of the extraction process were identified. Notably, it was observed that the mass transfer coefficient for indole was in the range of 3.4×10^{-6} to 1.2×10^{-6}, depending on NaDES. That is significantly lower, by an order of magnitude, than for pyridine and quinoline under identical experimental conditions. The study revealed that, under specific conditions, where thermodynamic equilibrium for indole cannot be reached, it becomes possible to achieve kinetic separation of the components. The presented experimental data obtained on a centrifugal extractor showed a decrease in the degree of indole extraction with increasing flow: Extraction efficiency decreased from 63% at a flow rate of 0.05 L/h to 18% at 0.8 L/h. Moreover, the research indicated that, during indole extraction, the mass transfer coefficient in a centrifugal extractor was 1.3×10^{-4}, which is two orders of magnitude higher than under static conditions. The study underscores the potential utility of the proposed extraction systems based on environmentally friendly NaDES, comprised of carboxylic acids and xylitol, for the kinetic separation of various classes of heterocyclic compounds. Overall, the research provides valuable insights into the kinetics of extraction and the potential applications of 'green' NaDES in the separation of heterocyclic compounds from organic liquids.

Keywords: extraction kinetics; mass transfer; pyridine; quinoline; indole; natural deep eutectic solvents; centrifugal extractors

Citation: Lobovich, D.V.; Zinov'eva, I.V.; Milevskii, N.A.; Kostanyan, A.E.; Zakhodyaeva, Y.A.; Voshkin, A.A. Extraction Kinetics of Pyridine, Quinoline, and Indole from the Organic Phase with Natural Deep Eutectic Solvents and Separation Study Using a Centrifugal Extractor. *Processes* **2024**, *12*, 488. https://doi.org/10.3390/pr12030488

Academic Editors: Yanlin Zhang, Prashant K. Sarswat and Fausto Gallucci

Received: 12 January 2024
Revised: 22 February 2024
Accepted: 26 February 2024
Published: 28 February 2024

1. Introduction

Nitrogen-containing heterocyclic compounds exhibit diverse pharmacological effects, having antibacterial, antiviral, antitumor, anticonvulsant, anti-inflammatory, and analgesic properties, rendering them indispensable in pharmaceuticals and perfumery [1]. Pyridine, quinoline, isoquinoline, indole, imidazole, and other bases form the foundation of various drugs [2]. Notably, pyridine, quinoline, and indole serve as precursors for synthesizing essential compounds such as nicotinic acid, serotonin, tryptophan, and reserpine. Indole, in particular, plays a pivotal role in synthesizing biologically active compounds, such as sumatriptan, tadalafil, risatriptan, and fluvastatin [3,4]. Quinoline derivatives have antimalarial, antibacterial, antifungal, anthelmintic, cardiotonic, anticonvulsant, anti-inflammatory, and analgesic effects, so they are widely used in the pharmaceutical industry. Important commercially available alkylpyridine compounds are α, β, γ-picoline, lutidine, 5-ethyl-2-methylpyridine and 2,4,6-collidine. Pyridine has applications in the paint and pharmaceutical industries, as well as in the synthesis of herbicides, insecticides, fungicides, and plant growth regulators [5]. The rationale behind their inclusion in medicinal formulations lies in the stability and efficacy of N-heterocycles within the human body [6].

The predominant methods for acquiring heterocyclic nitrogenous compounds involve chemical synthesis and extraction from coal tar or petroleum fractions, with the latter addressing environmental concerns related to harmful NOx gas emissions during fuel combustion. Therefore, the removal of these compounds from petroleum products such as gasoline, diesel fuel, and lubricants is imperative for enhancing oil quality and safeguarding the environment.

Traditionally, refineries use hydrogenitrogenation technology, which currently does not fully meet the increasing demand for nitrogen content in fuel. This is due to the inability to remove some N-containing compounds due to steric factors, which affect the performance of the catalyst. In addition, it has significant disadvantages in terms of operating conditions, requiring the use of high temperature and pressure and expensive catalysts. In this regard, other methods of fuel purification from nitrogen impurities are being developed today: extraction, adsorption, oxidation methods, and biodenitrogenation [7–10].

Recent research has concentrated on developing extraction methods for separating nitrogen-containing compounds from liquid petroleum fractions [11–18]. This dual-purpose approach involves creating an affordable and safe method for obtaining heterocyclic nitrogenous compounds for medical and pharmaceutical applications while simultaneously purifying fuel and reducing environmental impact.

While effective extractants such as sulfolane, N-formylmorpholine, N-methylpyrrolidone, acetonitrile, dimethyl sulfoxide, sulfuric acid, and trifluoroacetic acid were proposed in the middle of the 20th century and are still in use today, their toxicity renders them unsuitable for further use in industries such as pharmaceuticals and perfumery [19–22]. These substances have been used for decades, but their use results in heavy contamination, difficulty in their recovery, loss due to volatility and emissions to the environment, fire hazard, and toxicity of the process in general. Ionic liquids have long attracted great attention from researchers because they are non-volatile, non-flammable, and have high thermal stability and high extraction capacity with respect to nitrogen-containing compounds. However, the introduction of ionic liquids into industrial processes has never occurred due to their high mutual solubility with fuels and the risk of environmental pollution. Consequently, the urgent task is to develop environmentally safe yet equally effective extraction systems for separating nitrogen-containing compounds from petroleum fractions.

The emergence of deep eutectic solvents (DES) has provided a promising avenue for effective extraction processes in denitrification and desulfurization of hydrocarbon fractions [23–26]. Based on the hydrogen bond donor (HBD) structure, the solvents are classified into glycol, glycerol, organic acid, metal, aromatic, and imidazole-based DESs. As shown by previous studies, the extraction efficiency of sulfur- and nitrogen-containing substances largely depends on the components in the DES composition, which contain functional groups capable of interacting with the extracted components through the formation of intermolecular interactions (hydrogen bonds and/or p-p stacking) [27–29].

DES offers advantages such as tunable physico-chemical properties, accessibility, and simplicity of synthesis. DES often contains salts of ammonium bases (e.g., tetrabutylammonium bromide), which act as hydrogen bond acceptors that form when DES components interact with each other [30,31]. However, their use does not allow us to speak about the "green" nature of deep eutectic solvents. To address this, natural deep eutectic solvents (NaDES) have been introduced, consisting of liquid mixtures formed by combining natural substances such as sugars, polyalcohols, amino acids, organic acids, and bases at specific molar ratios [32–34]. In [35], the authors studied the efficiency of betaine and levulinic acid-based NaDES for the extraction of toluene, pyridine, and thiophene from a model diesel fuel (n-decane). The study showed that the extraction efficiency of pyridine was higher using NaDES betaine-levulinic acid at a molar ratio of 1:7 than using sulfolane under the same conditions. NaDES choline chloride-propylene glycol (1:3) was used for the extraction of sulfur- and nitrogen-containing compounds from a real gasoline fuel sample; the nitrogen content in the fuel decreased almost 2-fold, corresponding to an extraction efficiency of 45% [36].

Studies have demonstrated the effectiveness of NaDES in extracting compounds from model diesel fuel and real gasoline fuel, showcasing their potential in desulfurization, denitrification, and dearomatization of oil refining products [2,37].

This study specifically aimed to investigate the kinetics of extracting pyridine, quinoline, and indole using NaDES based on citric (CA) and malic (MA) acids, xylitol (Xyl) and water, (MA-CA-H_2O, CA-Xyl-H_2O and MA-Xyl-H_2O at a molar ratio of 1:1:10, respectively [38]), from a model fuel (n-hexane) under static conditions. The NaDES selected in this work for the extraction of heterocyclic compounds are stable over a wide temperature range and have low viscosity. Cheap and environmentally friendly components are also used for their preparation, which is important for scaling the process [39]. The primary objectives included establishing the main kinetic parameters of the extraction process and validating the proposed extraction separation of heterocycles using a centrifugal extractor.

2. Materials and Methods

Natural deep eutectic solvents were prepared according to the procedure described in the literature [38]. The water and solid NaDES components weighed on analytical scales were placed in a beaker and stirred on a magnetic stirrer US-4150D (ULAB, Hong Kong, China) at 60 °C for 30 min until a homogeneous mixture was formed. After cooling to room temperature, all obtained mixtures remained stable.

The experiments investigating the kinetics of pyridine, quinoline, and indole extraction were conducted using a Lewis cell with a constant phase contact surface area (5.31 cm^2). A schematic representation of the setup is shown in Figure 1. Initial solutions of nitrogen-containing heterocycles and NaDES, in a volume ratio of 1:1, at room temperature (20 °C), were agitated using an agitator Nanostar 7.5 (IKA, Staufen, Germany) equipped with an overhead drive running at a speed of 600 rpm. The concentration of pyridine, quinoline, and indole in n-hexane was detected using an optico-optical probe with an optical path of 2 nm every 15 s, connected to a Cary 60 UV-VIS spectrophotometer (Agilent Tech., Santa Clara, CA, USA). The wavelengths corresponding to the absorption maximum for pyridine, indole, and quinoline were 251, 266, and 271 nm, respectively. Control of the spectrophotometer and data collection were carried out using Cary WinUV 5.1.1 software (Agilent Tech., Santa Clara, CA, USA).

The degree of extraction of the i-th component (E_i, %) was determined by the formula:

$$E_i = \frac{n^i_{in} - n^i_{org}}{n^i_{in}} \quad (1)$$

when n^i_{in}—is the initial amount of substance of the i-th component in the organic phase, mol, n^i_{org}—the amount of substance of the i-th component in the organic phase after extraction, mol.

A number of experiments were performed with a centrifugal extractor (SOLVEX, Moscow, Russia) with a mixing chamber volume of 2.5 mL. The principle scheme of the unit is presented in Figure 2 and works in the following way: The phases are supplied by two peristaltic pumps BT-100M (Shenchen Pump, Baoding, China) with pump head YZ2515X (Shenchen Pump, Baoding, China) and Shenchen silicone tubing 15# (Shenchen Pump, Baoding, China) with varying flow rates given in Table 1.

Table 1. Flow rates of the organic phase and the NaDES phase in the experiment on centrifugal extractors.

Experiment	1	2	3	4	5	6	7	8	9	10	11
NaDES flowrate, L/h	0.1	0.2	0.252	0.3	0.4	0.504	0.78	1.008	1.2	1.38	1.6
Organic phase flowrate, L/h	0.05	0.1	0.126	0.15	0.2	0.252	0.39	0.504	0.6	0.69	0.8

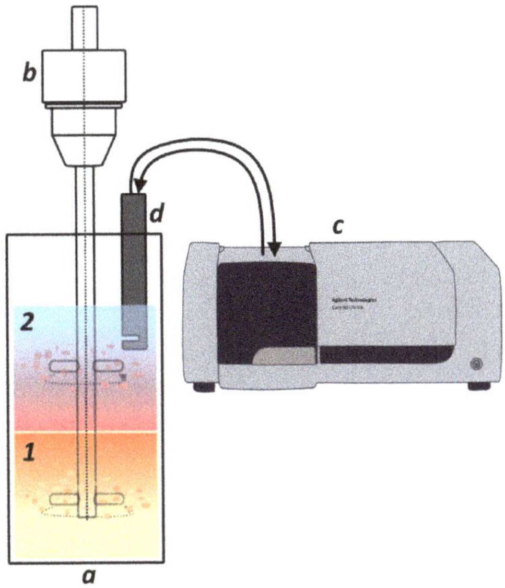

Figure 1. Scheme of an installation for studying the kinetics of extraction of nitrogen-containing heterocyclic compounds. a—Lewis cell, b—an agitator, c—spectrophotometer, d—optico-optical probe, 1—NaDES phase, 2—organic phase.

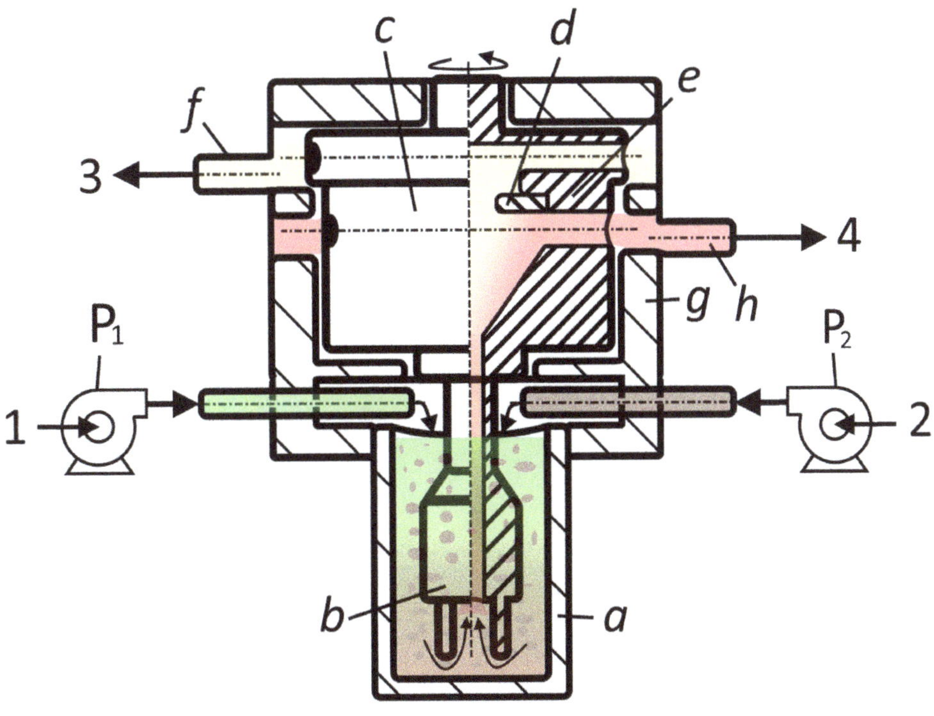

Figure 2. The extraction scheme implemented on a centrifugal apparatus: a—mixing chamber, b—agitator, c—general view of separation chamber and agitator, d—PTFE ring, e—split view of separation chamber and agitator, f—organic phase drain, g—split view of the external body of apparatus, h—NaDES phase drain, P—peristaltic pump, 1—initial NaDES phase, 2—light hydrocarbon fractions model solution, 3—light hydrocarbon fractions model solution raffinate, 4—NaDES extract.

We prepared 1 L NaDES CA-Xyl-H_2O and a solution of three components, quinoline, pyridine, and indole, in n-hexane with concentrations of 0.008, 0.005, and 0.003 mol/L, respectively. This solution was a model of a real solution of light hydrocarbon fractions. To ensure the phase ratio V_{NaDES}/V_{org} = 2/1, the NaDES phase flow rate was set 2 times higher in relation to the flow rate of the model solution. In Figure 2, yellow is the phase flow of pure extractant NaDES CA-Xyl-H_2O, and pink is the phase flow of the model solution of light hydrocarbon fractions—in this case, a solution of quinoline, pyridine, and indole in n-hexane. Both streams flow through the connections into the interior of the extractor, where they are mixed in the mixing chamber with the help of a stirrer. Figure 2 shows how the formed emulsion enters the upper part of the shaft, which is the drum of the separation chamber, where the emulsion is stratified due to the centrifugal force generated by rotating the shaft of the unit at a speed of 3000 rpm. It is worth noting that the 3000 rpm rotation speed of this unit allows it to achieve a mixing efficiency of more than 99.9%, which has been determined on test systems recommended by [40]. Centrifugal force creates a concentration of heavy phase on the periphery of the drum, after which it is discharged through drain holes leading to the peripheral channel; the light phase is concentrated in the middle part of the drum and overflows through the polytetrafluoroethylene ring into the channel, where the light phase is concentrated, after which it is discharged through drain holes. After extraction and stratification, a stream of extract contaminated with heterocyclic nitrogenous compounds—quinoline, pyridine, and indole, colored orange, and a stream of raffinate, colored blue in the figure, are formed, which are discharged from the apparatus.

All components of the device are made of stainless steel. The mixing chamber has a volume of 2.5 mL. High productivity of the extractor was achieved due to the combined action of intensive stirring created by the stirrer rotating coaxially with the centrifugal drum at a speed of 3000 rpm, which leads to high dispersion, i.e., to the formation of dispersed phase droplets with a small diameter, which creates a high value of the ratio of the contact surface area of the phases to the volume and a high mass transfer rate. The emulsion obtained in the process was sucked into the central hollow channel of the agitator due to the vortex flow created by a special design of the device and from the mixing chamber was transferred through the inner channel to the centrifugal drum, which acted as a separation chamber. Due to the rapid rotation of the system from the agitator and drum, the emulsion fell into the field of centrifugal force, which led to the separation of phases. The main volume of the emulsion was concentrated and separated in the lower part of the drum, where the heavy phase was formed along the periphery of the device and the light phase in the center. Thus, the NaDES phase was discharged through the channels of the outer ring of the drum, while the organic phase overflowed through the polytetrafluoroethylene ring through the center into the cavity of the inner ring of the drum, from where it was exited from the agitator/drum system through the channels of the inner ring to the drain shelves of the apparatus body, connected to welded fittings. Phases were introduced into the cells through inlet fittings "a" (see Figure 3). After the separation in the drum, NaDES phases were extracted from the cell through outlet fittings "b", while the organic phase was extracted through outlet fittings "c" (see Figure 3).

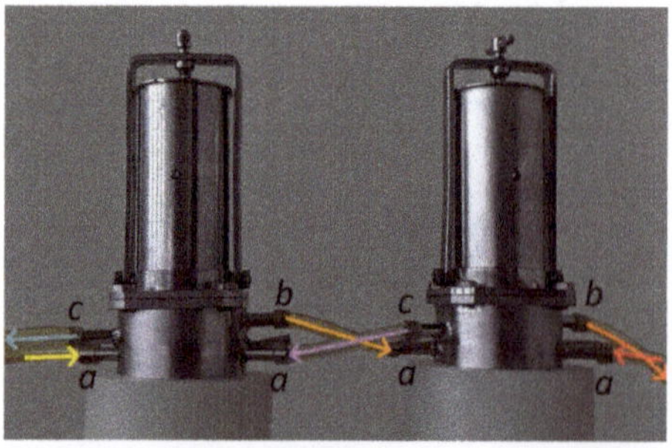

Figure 3. Outlet and inlet fittings for the centrifugal extraction equipment cell: a—inlets for the organic/NaDES phase, b—outlets for the NaDES phase, and c—outlets for the organic phase.

3. Results and Discussion

3.1. Extraction Kinetics Study

This study focused on investigating the kinetics of heterocyclic compound extraction in systems using NaDES. The examination of extraction kinetics under static conditions is crucial for the subsequent application of the process in extraction equipment. An experiment was conducted to explore the kinetics of extraction using Lewis cells. This experiment facilitated the determination of the kinetic parameters of the extraction process by using spectrophotometric detection to monitor the concentration of nitrogen-containing compounds in real time within the raffinate (n-hexane) phase.

Figure 4 shows data that illustrate the kinetics of nitrogen-containing compound extraction from n-hexane using deep eutectic solvents MA-CA-H_2O, CA-Xyl-H_2O, and MA-Xyl-H_2O. In the case of pyridine and quinoline extraction, there was virtually no significant difference in equilibrium time based on the NaDES selected. Given that pyridine is a stronger base, with a lower pKa, than quinoline (4.8 vs. 5.17, for a bound base) [41], pyridine extraction occurred more rapidly, taking only 4 min, compared with 5 to 7 min for quinoline. These observations, coupled with an understanding of the acid-base interaction extraction mechanism, suggest that extraction is likely to occur in a diffusion-limited regime.

In contrast, indole extraction required a longer equilibrium time for all NaDES, specifically 30 min. The formation of an extractable compound between indole and the extractant involves the creation of weak hydrogen bonds, resulting in a significantly slower reaction rate compared with acid-base interactions. Consequently, extraction, in this case, proceeds in a mixed regime, combining both diffusion-limited and chemical reaction-limited aspects. This distinct kinetic behavior of indole can be strategically utilized for its kinetic isolation from a mixture of heterocyclic compounds.

The mass transfer across the boundary between two liquids in a liquid-liquid system can be explained using the two-film theory, which is based on Fick's First Law [42]. The process is mathematically described by Equation (2) for the respective phase j (j—NaDES) [43]:

$$V_j \times \frac{d_{c_j}}{dt} = k_j \times a \times \left(c_{j,\infty} - c_j\right) \quad (2)$$

V_j—volume of the corresponding phase, k_j—mass transfer coefficient, a—interfacial surface, $c_{j,\infty}$ equilibrium concentration of the component in phase j, c_j—concentration of the component in phase j at time t.

Solving Equation (2) at t = 0, $c_j = c_{j,0}$, where $c_{j,0}$ is the initial concentration, yields Equation (3):

$$\ln \frac{\left(c_{j,\infty} - c_{j,0}\right)}{\left(c_{j,\infty} - c_j\right)} \times \frac{V_j}{a} = k_j \times t \quad (3)$$

Figure 5 shows graphs depicting the dependence of $\ln \frac{\left(c_{j,\infty} - c_{j,0}\right)}{\left(c_{j,\infty} - c_j\right)}$ on time for the extraction of pyridine, quinoline, and indole by NaDES. The mass transfer coefficients for the compounds investigated were calculated based on the slope of the established dependencies, and the results are summarized in Table 2.

Table 2. Mass transfer coefficients (cm/s) for pyridine, quinoline, and indole during extraction by the proposed NaDES.

	CA-Xyl-H_2O	MA-Xyl-H_2O	MA-CA-H_2O
Pyridine	3.8×10^{-5}	2.8×10^{-5}	2.7×10^{-5}
Quinoline	2.0×10^{-5}	1.9×10^{-5}	2.5×10^{-5}
Indole	3.4×10^{-6}	2.9×10^{-6}	1.2×10^{-6}

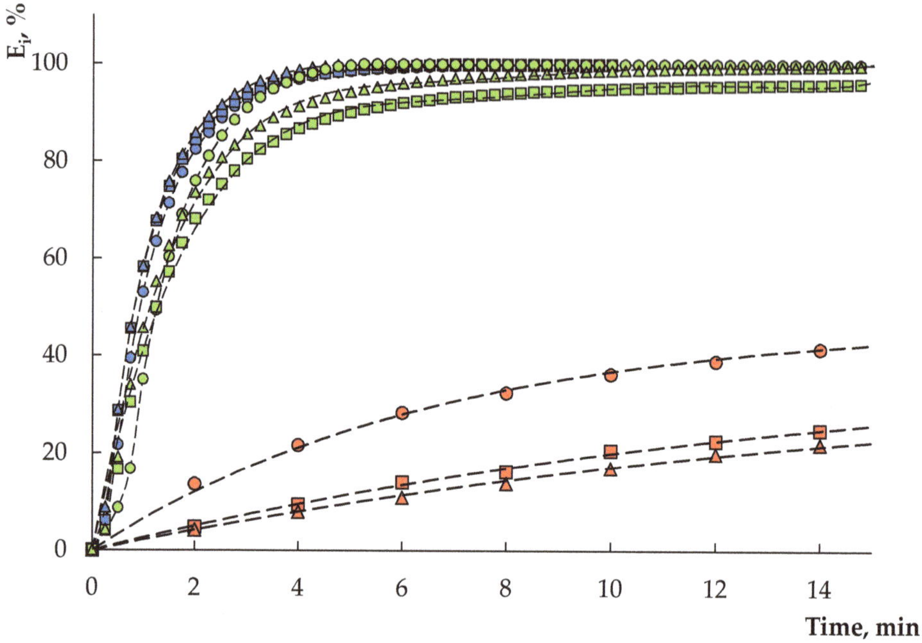

Figure 4. Dependence of the degree of extraction of pyridine (blue), quinoline (green), and indole (red) NaDES MA-CA-H_2O (circle), MA-Xyl-H_2O (square), CA-Xyl-H_2O (triangle) on the time of phase contact.

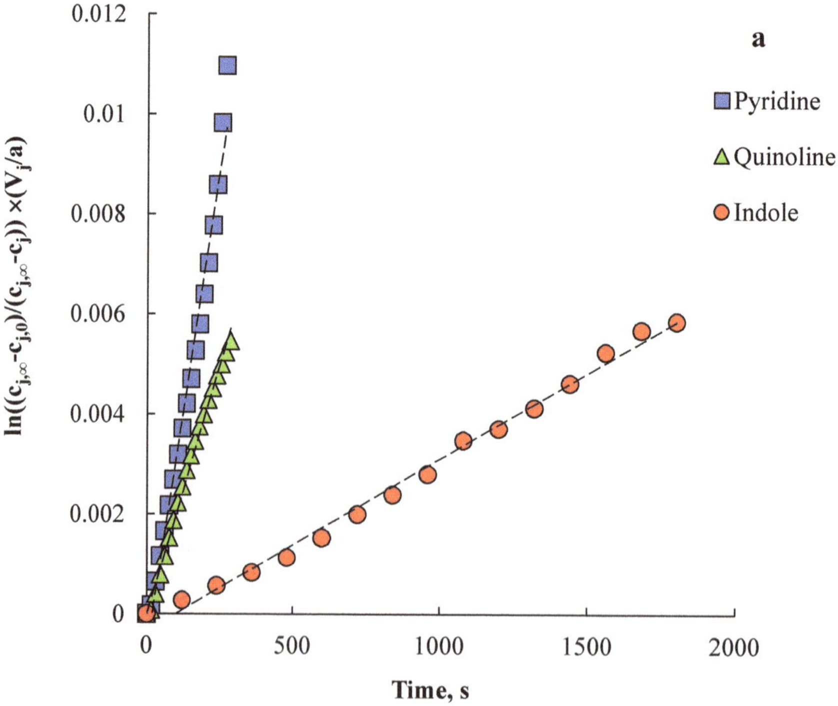

Figure 5. *Cont.*

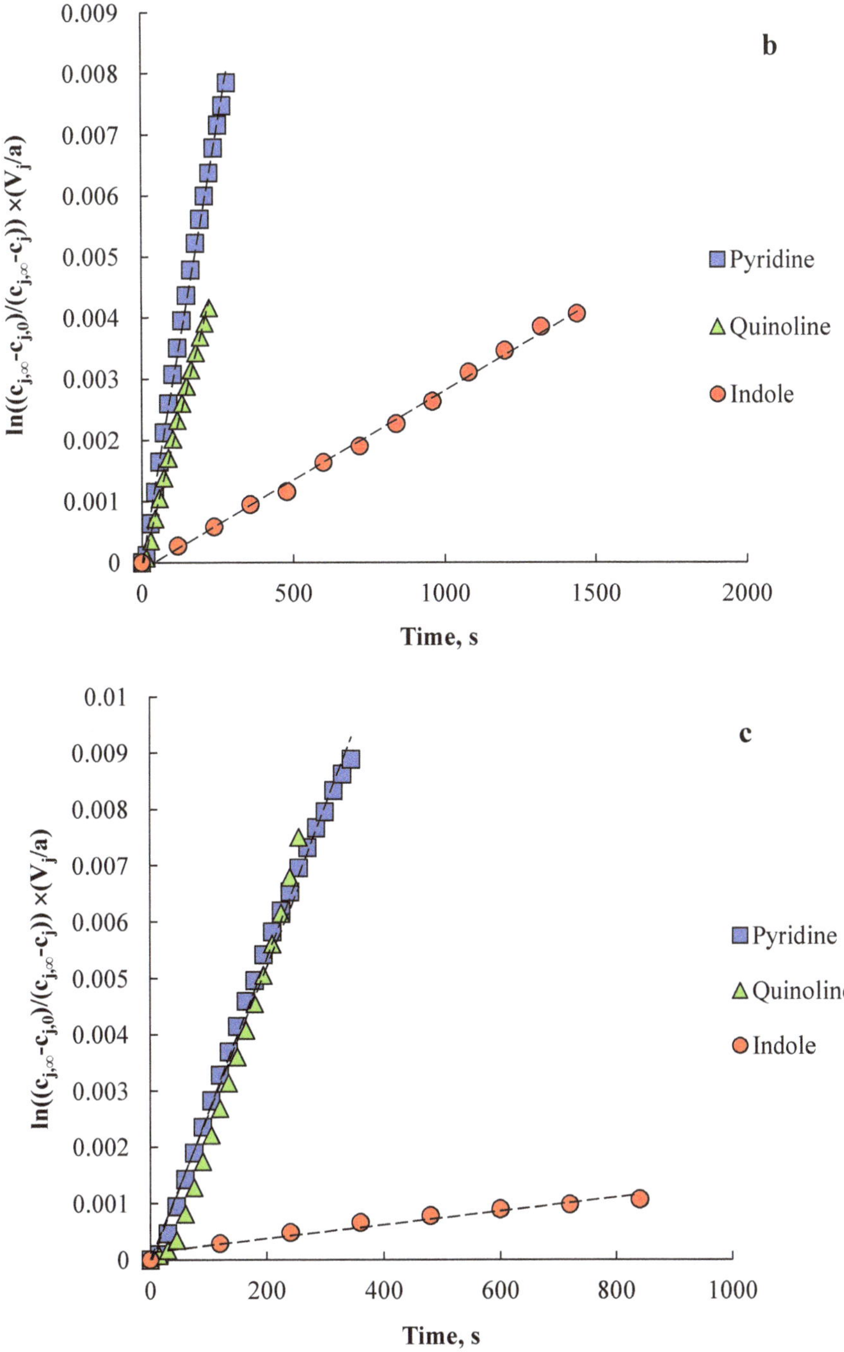

Figure 5. Dependence of $\ln \frac{(c_{j,\infty} - c_{j,0})}{(c_{j,\infty} - c_j)} \times \frac{V_j}{a}$ on time for pyridine, quinoline, and indole by NaDES extraction CA-Xyl-H_2O (**a**), MA-Xyl-H_2O (**b**) and MA-CA-H_2O (**c**).

It can be seen that the mass transfer coefficient for indole is an order of magnitude smaller than for pyridine and quinoline under the same experimental conditions. Thus, it was determined that when reaching a regime in which the thermodynamic equilibrium is not reached for indole, it is possible to realize the kinetic separation of components since pyridine and quinoline will still reach equilibrium under these conditions.

3.2. Separation Study on Centrifugal Extractors

With a constant volume of the mixing chamber of classical types of extractors such as sump mixer, centrifugal extractor, and extraction column, the phase contact time can be varied by varying the volume flow rate. Increasing the volume flow rate leads to a decrease in the phase contact time, but rapid emulsion formation under such conditions leads to improper operation of the equipment. Due to the resulting obstacle to achieving a good stratification, penetration of one phase into the other can occur, and a persistent emulsion can form, resulting in an upper limit of the volume flow rate. The way out of this situation can be a reduction in the speed of rotation of the agitators, reducing the diameter of the blades, or simplifying its shape. This technique does lead to a reduction in emulsification but reduces the mixing efficiency, which ultimately does not solve the problem. The advantage of centrifugal extractors is the combination of intensive mixing with high productivity. The high shaft speed of the apparatus not only leads to powerful mixing and emulsification but also creates a centrifugal force in the drum, which enables quality stratification in the same system. Thus, even with small mixing chamber volumes, it is possible to increase the volume flow rate many times over. Consequently, this type of extraction equipment is well-suited for solving the problem of kinetic separation.

Experiments on the extraction of heterocyclic nitrogenous compounds quinoline, pyridine, and indole simultaneously from n-hexane by the proposed NaDES on a centrifugal extractor were carried out. It is worth noting the fact that at the used flow rates, the system worked stably, no persistent emulsions were formed, and no penetration of one phase into the other occurred. After the establishment of a stationary hydrodynamic regime, the organic phase was taken for analysis through the appropriate drain connection. The residual concentration of components in n-hexane in each sample was determined by the spectrophotometric method. As can be seen from Figure 6, the degree of extraction of quinoline and pyridine remains unchanged at any value of the volume flow rate, while the degree of extraction of indole monotonically decreases as the organic phase flow rate increases above 0.125 L/h, indicating that at higher flow rates the system does not reach thermodynamic equilibrium. At the maximum possible organic phase flow rate of 0.8 L/h, the degree of indole extraction reached 18%, which indicates that under such extraction conditions, it is possible to achieve separation of components and extraction of 82% of indole in pure form. Further increase in the flow rate on these apparatuses led to mutual entrainment of phases, so the study stopped at this value. However, it should be considered that further scaling of the process is possible, and further studies with larger volume apparatuses may lead to even higher flow rates and separation values. The process of re-extraction of substances and regeneration of NaDES after use remains poorly understood. However, in a number of works, the authors used effective methods for re-extracting pure compounds from the NaDES phase after extraction, which can also be used for pyridine, quinoline, and indole [44].

Based on the findings from the kinetics study of nitrogen-containing NaDES CA-Xyl-H_2O heterocycles, the distribution coefficients ($D = y/x$, where y is the concentration in the extract and x is in the raffinate) were: indole—0.44, quinoline—approximately 300, and pyridine—418.

Using the data acquired and phase flow ratio $v_x{:}v_y = 0.5$, the extraction factor $f = v_x/(D \cdot v_y)$ could be calculated, resulting in the following values: indole—1.14, quinoline—0.0017, and pyridine—0.0012.

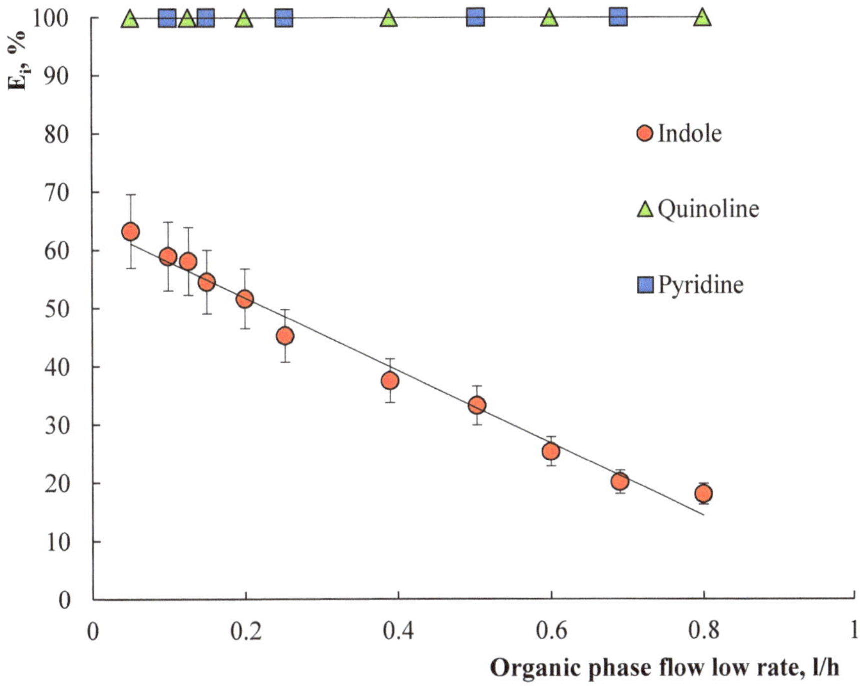

Figure 6. Dependence of the degree of extraction of indole, quinoline, and pyridine from n-hexane CA-Xyl-H_2O on a centrifugal extractor on the flow rate of the organic phase.

If equilibrium is achieved in the extractor (one theoretical stage), the output concentrations in the raffinate can be determined by Equation (4):

$$X = \frac{x_{yield}}{x_0} = \frac{f}{1+f} \quad (4)$$

The calculations yielded the following results: during indole extraction, X = 0.53, during quinoline extraction, X = 0.0017, and during pyridine extraction, X = 0.0012.

Experimental results during indole extraction at organic phase consumptions of v_x = 0.1–0.25 L/h showed X = 0.34–0.59. As the consumption of the organic phase increased (v_x = 0.39–0.8 L/h), the concentration in the raffinate also increased (X = 0.65–0.82), indicating that equilibrium was not reached in the extractor (X > 0.53). This observation aligns with the earlier graphical representation (see Figure 6).

In contrast, during the extraction of quinoline and pyridine, at all organic phase flow rates, equilibrium was achieved in the extractor. The concentrations of quinoline and pyridine in the raffinate were X = 0.0014–0.0016 and 0.0012–0.0013, respectively, consistent with the calculated results.

It is important to note that, in extractors, the ratio of phase volumes in the mixer is equal to the ratio of phase flow rates. In experiments conducted on a centrifugal extractor with a flow rate ratio of extract phases (water phase) to raffinate (organic phase) = 2:1, the ratio of phase flow rates (volumes of phases in the mixer) at 1:1 under conditions of the equilibrium distribution of components between phases resulted in the following calculated concentrations in the raffinate, using Equation (4): indole—(X = 0.7), quinoline—(X = 0.0034), pyridine—(X = 0.0024). Therefore, under equilibrium conditions, practically pure indole can be isolated by adjusting the ratio of phase flow rates, but this would lead to a more dilute solution of the product, resulting in a lower yield than with kinetic separation.

During the extraction of indole in experiments with high phase consumption (v_X = 0.39–0.8 L/h), equilibrium was not achieved in the extractor due to insufficient phase contact time. The output concentrations in the raffinate under these conditions can be determined by Equation (5):

$$X = \frac{1 + tf}{1 + t(1 + f)} \quad (5)$$

where t = a_c k_X V_c/v_X—the number of transfer units, a_c—the specific contact surface of the phases in the extractor mixer, k_X—surface mass transfer coefficient, V_c—the volume of the mixer. At high mass transfer rates (t→∞), Equation (5) converges to Equation (4).

From Equation (5), the number of transfer units can be determined from the experimental values X and f:

$$t = \frac{1 - X}{X(1 + f) - f} \quad (6)$$

Knowing the number of transfer units, t = a_c k_X V_c/v_X, it is possible to calculate the volumetric mass transfer coefficient $a_c k_X$. To calculate the surface mass transfer coefficient k_X, it is necessary to determine the value of a_c. The specific contact surface of the phases in the extractor mixer a_c depends on the average droplet size d and the fraction of the volume of the dispersed phase in the emulsion S: $a_c = 6S/d$.

The ratio of the flow rates of the extract/raffinate phases is 2:1 *S* = 0.333. The average size of droplets formed in the extractor mixer can be assumed to be 0.1 mm in the first approximation: d = 0.01 cm and $a_c \approx 200$ cm^{-1}.

Table 3 shows the values of mass transfer parameters calculated from the results of experiments during indole extraction.

Table 3. Calculated values of mass transfer parameters during indole extraction in a centrifugal extractor.

v_X (L/h)	X	t	V_c/v_X (s)	k_X a_c (1/s)	k_X (cm/s)
0.252	0.58558	3.66255	35.71429	0.10255	0.00051
0.390	0.65309	1.34664	23.07692	0.05835	0.00029
0.504	0.66744	1.15341	17.85714	0.06459	0.00032
0.600	0.74677	0.55282	15.00000	0.03685	0.00018
0.69	0.79911	0.35237	13.04348	0.02702	0.00014
0.800	0.82069	0.29095	11.25000	0.02586	0.00013

Comparing the results presented in Tables 2 and 3, it can be inferred that the mass transfer coefficient in a centrifugal extractor is two orders of magnitude higher than in experiments studying the kinetics of extraction in a Lewis cell. This discrepancy arises from variations in hydrodynamic conditions within the apparatus. Specifically, the intensity of phase movement relative to the interface in the extractor surpasses that in the Lewis cell. Additionally, in the extractor, the phase contact surface is continually refreshed due to the coalescence and redispersion of droplets. Consequently, the proportion of convective mass transfer, and hence the mass transfer rate, is notably higher in the extractor than in the Lewis cell.

In summary, the study demonstrates that the proposed extraction systems using 'green' NaDES comprised of carboxylic acids and xylitol have promise for the kinetic separation of various classes of heterocyclic compounds. The commendable performance metrics of these systems position them as being highly promising as an environmentally friendly, cost-effective, and accessible approach to producing heterocyclic nitrogenous compounds for medical and pharmaceutical purposes. Furthermore, their application contributes to a reduction in environmental impact through deep denitrification of light hydrocarbon fractions.

4. Conclusions

This study investigated the kinetics of extracting pyridine, quinoline, and indole within a system using natural deep eutectic solvents. The time required to attain equilibrium for each substance has been precisely determined, with pyridine reaching equilibrium within 4 min, quinoline within 5 to 7 min, and indole within 30 min. These disparate equilibrium times can be attributed to distinct extraction mechanisms inherent to each compound. Importantly, the study has demonstrated the potential for selective extraction of pyridine and quinoline over indole using kinetic separation.

The findings indicate a significant difference in mass transfer coefficients among the compounds studied. Specifically, the mass transfer coefficient for indole was observed to be an order of magnitude lower than that for pyridine and quinoline under identical experimental conditions. Furthermore, the experimental results have demonstrated that, during indole extraction, the mass transfer coefficient in a centrifugal extractor surpasses that in a Lewis cell by two orders of magnitude. This discovery underlines the superior efficiency of the centrifugal extraction process for indole.

This promising result opens up new possibilities for the practical application of the proposed extraction systems, which are based on carboxylic acids, xylitol, and water. The study emphasizes that these systems not only excel in equilibrium processes but also harbor the potential for efficient kinetic separation of diverse classes of heterocyclic compounds. In summary, the research contributes valuable insights into the versatile application of these extraction systems in both equilibrium and kinetic separation processes for nitrogen-containing heterocyclic compounds.

Author Contributions: Conceptualization, D.V.L. and I.V.Z.; methodology, N.A.M. and A.E.K.; software, D.V.L. and A.E.K.; validation, I.V.Z. and N.A.M.; formal analysis, Y.A.Z.; investigation, D.V.L.; data curation, Y.A.Z.; writing—original draft preparation, I.V.Z. and N.A.M.; writing—review and editing, Y.A.Z. and A.A.V.; visualization, I.V.Z.; supervision, A.E.K. and Y.A.Z.; project administration, A.A.V.; funding acquisition, A.A.V. All authors have read and agreed to the published version of the manuscript.

Funding: This work was supported by the Ministry of Science and Higher Education of the Russian Federation as part of the State Assignment of the Kurnakov Institute of General and Inorganic Chemistry of the Russian Academy of Sciences.

Data Availability Statement: Data are contained within the article.

Conflicts of Interest: The authors declare no conflicts of interest.

References

1. Marella, A.; Tanwar, O.P.; Saha, R.; Ali, M.R.; Srivastava, S.; Akhter, M.; Shaquiquzzaman, M.; Alam, M.M. Quinoline: A Versatile Heterocyclic. *Saudi Pharm. J.* **2013**, *21*, 1–12. [CrossRef]
2. Rogošić, M.; Kučan, K.Z. Deep Eutectic Solvents Based on Choline Chloride and Ethylene Glycol as Media for Extractive Denitrification/Desulfurization/Dearomatization of Motor Fuels. *J. Ind. Eng. Chem.* **2019**, *72*, 87–99. [CrossRef]
3. Inman, M.; Moody, C.J. Indole Synthesis–Something Old, Something New. *Chem. Sci.* **2013**, *4*, 29–41. [CrossRef]
4. Mindt, M.; Beyraghdar Kashkooli, A.; Suarez-Diez, M.; Ferrer, L.; Jilg, T.; Bosch, D.; Martins dos Santos, V.; Wendisch, V.F.; Cankar, K. Production of Indole by Corynebacterium Glutamicum Microbial Cell Factories for Flavor and Fragrance Applications. *Microb. Cell Factories* **2022**, *21*, 45. [CrossRef]
5. Scriven, E.F.V.; Murugan, R. Pyridine and Pyridine Derivatives. In *Kirk-Othmer Encyclopedia of Chemical Technology*; John Wiley & Sons, Inc.: Hoboken, NJ, USA, 2005.
6. Portero, C.E.; Han, Y.; Marchán-Rivadeneira, M.R. Advances on the Biosynthesis of Pyridine Rings. *Eng. Microb.* **2023**, *3*, 100064. [CrossRef]
7. Akbari, A.; Andevary, H.H. Catalytic Oxidative-Extractive Deep Desulfurization of Diesel Fuel by N-Methyl-2-Pyrrolidone-Based Protic Acidic Ionic Liquids (PAILs). *Iran. J. Chem. Chem. Eng.* **2014**, *20*, 315–323.
8. Feng, X.; Ma, X.; Li, N.; Shang, C.; Yang, X.; Chen, X.D. Adsorption of Quinoline from Liquid Hydrocarbons on Graphite Oxide and Activated Carbons. *RSC Adv.* **2015**, *5*, 74684–74691. [CrossRef]
9. Le Borgne, S.; Quintero, R. Biotechnological Processes for the Refining of Petroleum. *Fuel Process. Technol.* **2003**, *81*, 155–169. [CrossRef]

10. Koriakin, A.; Ponvel, K.M.; Lee, C.-H. Denitrogenation of Raw Diesel Fuel by Lithium-Modified Mesoporous Silica. *Chem. Eng. J.* **2010**, *162*, 649–655. [CrossRef]
11. Gao, P.; Yang, L.; Wang, J.; Gao, J.; Xu, D.; Ma, L.; Zhang, L.; Wang, Y. Integrated Investigation for Extractive Denitrogenation of Fuel Oils with Eco-Friendly Piperazine-Based Ionic Liquids. *Fuel* **2023**, *337*, 127187. [CrossRef]
12. Solov'ev, V.O.; Kostenko, M.O.; Solov'eva, S.V.; Zakhodyaeva, Y.A.; Parenago, O.O.; Sobolev, N.A.; Voshkin, A.A. A Green Hybrid Extraction Process for Thiophene, Quinoline and Indole Recovery from Light Hydrocarbon Fractions. *Chem. Eng. Res. Des.* **2023**, *191*, 1–13. [CrossRef]
13. Ci, F.; Zhang, T.; Zhang, L. Efficient Separation of Indole from Fossil Fuel Pyrolysis Products by Carboxylic Acid Non-Aromatic Ring Ionic Liquids: Experiment and Mechanism Exploration. *J. Mol. Liq.* **2023**, *391*, 123429. [CrossRef]
14. Hansmeier, A.R.; Meindersma, G.W.; de Haan, A.B. Desulfurization and Denitrogenation of Gasoline and Diesel Fuels by Means of Ionic Liquids. *Green Chem.* **2011**, *13*, 1907. [CrossRef]
15. Bertleff, B.; Haider, M.S.; Claußnitzer, J.; Korth, W.; Wasserscheid, P.; Jess, A.; Albert, J. Extractive Catalytic Oxidative Denitrogenation of Fuels and Their Promoting Effect for Desulfurization Catalyzed by Vanadium Substituted Heteropolyacids and Molecular Oxygen. *Energy Fuels* **2020**, *34*, 8099–8109. [CrossRef]
16. Solov'ev, V.O.; Solov'eva, S.V.; Zakhodyaeva, Y.A.; Voshkin, A.A. Extractive Denitrogenization of Liquid Model Fuel Using Polyethylene Glycol Methyl Ether 350. *Can. J. Chem. Eng.* **2024**, *102*, 703–712. [CrossRef]
17. Solov'ev, V.O.; Zakhodyaeva, Y.A.; Voshkin, A.A. On the Influence of Additives of Polymer, Sodium Nitrate, and 1-Methyl-2-Pyrrolidone on the Extraction of Thiophene in an n-Hexan–Water System. *Theor. Found. Chem. Eng.* **2020**, *54*, 894–899. [CrossRef]
18. Zakhodyaeva, Y.A.; Solov'ev, V.O.; Zinov'eva, I.V.; Rudakov, D.G.; Timoshenko, A.V.; Voshkin, A.A. Interphase Distribution of Thiophene, Toluene, and o-Xylene in the Hexane–Polymer–Water Extraction System. *Theor. Found. Chem. Eng.* **2019**, *53*, 550–555. [CrossRef]
19. Gaile, A.A.; Kostenko, A.V.; Semenov, L.V.; Koldobskaya, L.L. Extraction of 1-Methylnaphthalene, Benzothiophene, and Indole with N-Methylpyrrolidone from Their Mixtures with Alkanes. *Russ. J. Appl. Chem.* **2005**, *78*, 1403–1407. [CrossRef]
20. Lemaoui, T.; Benguerba, Y.; Darwish, A.S.; Hatab, F.A.; Warrag, S.E.E.; Kroon, M.C.; Alnashef, I.M. Simultaneous Dearomatization, Desulfurization, and Denitrogenation of Diesel Fuels Using Acidic Deep Eutectic Solvents as Extractive Agents: A Parametric Study. *Sep. Purif. Technol.* **2021**, *256*, 117861. [CrossRef]
21. Mokhtar, W.N.A.W.; Bakar, W.A.W.A.; Ali, R.; Kadir, A.A.A. Deep Desulfurization of Model Diesel by Extraction with N,N-Dimethylformamide: Optimization by Box–Behnken Design. *J. Taiwan Inst. Chem. Eng.* **2014**, *45*, 1542–1548. [CrossRef]
22. Mohsen-Nia, M.; Modarress, H.; Doulabi, F.; Bagheri, H. Liquid+liquid Equilibria for Ternary Mixtures of (Solvent+aromatic Hydrocarbon+alkane). *J. Chem. Thermodyn.* **2005**, *37*, 1111–1118. [CrossRef]
23. Kumar, K.; Bharti, A.; Kumar, A.; Ghosh, S.K.; Kumar, A. Choline Based Deep Eutectic Solvent for Denitrogenation of Liquid Fuel: A Molecular Dynamics Study. *J. Mol. Liq.* **2023**, *382*, 121862. [CrossRef]
24. Lima, F.; Dave, M.; Silvestre, A.J.D.; Branco, L.C.; Marrucho, I.M. Concurrent Desulfurization and Denitrogenation of Fuels Using Deep Eutectic Solvents. *ACS Sustain. Chem. Eng.* **2019**, *7*, 11341–11349. [CrossRef]
25. Hizaddin, H.F.; Hadj-Kali, M.K.; Ramalingam, A.; Ali Hashim, M. Extractive Denitrogenation of Diesel Fuel Using Ammonium- and Phosphonium-Based Deep Eutectic Solvents. *J. Chem. Thermodyn.* **2016**, *95*, 164–173. [CrossRef]
26. Ali, M.C.; Yang, Q.; Fine, A.A.; Jin, W.; Zhang, Z.; Xing, H.; Ren, Q. Efficient Removal of Both Basic and Non-Basic Nitrogen Compounds from Fuels by Deep Eutectic Solvents. *Green Chem.* **2016**, *18*, 157–164. [CrossRef]
27. Yin, J.; Wang, J.; Li, Z.; Li, D.; Yang, G.; Cui, Y.; Wang, A.; Li, C. Deep Desulfurization of Fuels Based on an Oxidation/Extraction Process with Acidic Deep Eutectic Solvents. *Green Chem.* **2015**, *17*, 4552–4559. [CrossRef]
28. Li, C.; Zhang, J.; Li, Z.; Yin, J.; Cui, Y.; Liu, Y.; Yang, G. Extraction Desulfurization of Fuels with 'Metal Ions' Based Deep Eutectic Solvents (MDESs). *Green Chem.* **2016**, *18*, 3789–3795. [CrossRef]
29. Wang, X.; Li, Z.; Wang, X.; Wu, C.; Gates, I.D.; Guo, S.; Wu, B.; Zhu, W.; Gu, M.; Gao, M.; et al. Intermolecular Interactions Induced Desulfurization/Denitrification of Oil with Deep Eutectic Solvents. *J. Mol. Liq.* **2022**, *366*, 120159. [CrossRef]
30. Alli, R.D.; Kroon, M.C. Extraction of Benzothiazole and Thiophene from Their Mixtures with N-Heptane Using Tetrahexylammonium Bromide-Based Deep Eutectic Solvents as Extractive Denitrogenation and Desulfurization Agents. *Fluid Phase Equilib.* **2018**, *477*, 1–11. [CrossRef]
31. Lee, H.; Kang, S.; Jin, Y.; Jung, D.; Park, K.; Li, K.; Lee, J. Systematic Investigation of the Extractive Desulfurization of Fuel Using Deep Eutectic Solvents from Multifarious Aspects. *Fuel* **2020**, *264*, 116848. [CrossRef]
32. Fortunato, L.; Al Fuhaid, L.; Murgolo, S.; De Ceglie, C.; Mascolo, G.; Falivene, L.; Vrouwenvelder, J.S.; Witkamp, G.-J.; Farinha, A. Removal of Polyfluoroalkyl Substances (PFAS) from Water Using Hydrophobic Natural Deep Eutectic Solvents (NADES): A Proof of Concept Study. *J. Water Proc. Eng.* **2023**, *56*, 104401. [CrossRef]
33. Zinov'eva, I.V.; Fedorov, A.Y.; Milevskii, N.A.; Zakhodyaeva, Y.A.; Voshkin, A.A. A Deep Eutectic Solvent Based on Choline Chloride and Sulfosalicylic Acid: Properties and Applications. *Theor. Found. Chem. Eng.* **2021**, *55*, 371–379. [CrossRef]
34. Ul Haq, H.; Bibi, R.; Balal Arain, M.; Safi, F.; Ullah, S.; Castro-Muñoz, R.; Boczkaj, G. Deep Eutectic Solvent (DES) with Silver Nanoparticles (Ag-NPs) Based Assay for Analysis of Lead (II) in Edible Oils. *Food Chem.* **2022**, *379*, 132085. [CrossRef] [PubMed]

35. Hatab, F.A.; Darwish, A.S.; Lemaoui, T.; Warrag, S.E.E.; Benguerba, Y.; Kroon, M.C.; AlNashef, I.M. Extraction of Thiophene, Pyridine, and Toluene from *n*-Decane as a Diesel Model Using Betaine-Based Natural Deep Eutectic Solvents. *J. Chem. Eng. Data* **2020**, *65*, 5443–5457. [CrossRef]
36. Rogošić, M.; Zagajski Kučan, K. Deep Eutectic Solvent Based on Choline Chloride and Propylene Glycol as a Potential Medium for Extraction Denitrification of Hydrocarbon Fuels. *Chem. Eng. Res. Des.* **2020**, *161*, 45–57. [CrossRef]
37. Tahir, S.; Qazi, U.Y.; Naseem, Z.; Tahir, N.; Zahid, M.; Javaid, R.; Shahid, I. Deep Eutectic Solvents as Alternative Green Solvents for the Efficient Desulfurization of Liquid Fuel: A Comprehensive Review. *Fuel* **2021**, *305*, 121502. [CrossRef]
38. Santana, A.P.R.; Mora-Vargas, J.A.; Guimarães, T.G.S.; Amaral, C.D.B.; Oliveira, A.; Gonzalez, M.H. Sustainable Synthesis of Natural Deep Eutectic Solvents (NADES) by Different Methods. *J. Mol. Liq.* **2019**, *293*, 111452. [CrossRef]
39. Lobovich, D.V.; Solov'eva, S.V.; Milevskii, N.A.; Zakhodyaeva, Y.A.; Voshkin, A.A. Denitrogenation of Light Hydrocarbon Fractions with Natural Deep Eutectic Solvents Using Commercial Extraction Equipment. *Theor. Found. Chem. Eng.* **2023**, *57*, 1276–1291.
40. Misek, T.; Berger, R.; Schröter, J. *Standart Test Systems for Liquid Extraction*, 2nd ed.; Misek, T., Ed.; European Federation of Chemical Engineering by Institution of Chemical Engineers: Rugby, UK, 1985; Volume 46.
41. Dean, J.A. *Lange's Handbook of Chemistry*; McGraw Hill Inc.: New York, NY, USA, 1992.
42. Pratt, H.R.C.; Lo, T.C. *Handbook of Solvent Extraction*; Lo, T.C., Braid, M.H.I., Hanson, C., Eds.; Wiley: New York, NY, USA, 1983.
43. Kaplanow, I.; Goerzgen, F.; Merz, J.; Schembecker, G. Mass Transfer of Proteins in Aqueous Two-Phase Systems. *Sci. Rep.* **2019**, *9*, 3692. [CrossRef]
44. Santana-Mayor, Á.; Socas-Rodríguez, B.; Rodríguez-Ramos, R.; Herrera-Herrera, A.V.; Rodríguez-Delgado, M.Á. Application of Natural Deep Eutectic Solvents for the Extraction of Plasticizers in Laying Hen and Goat Feed. *Microchem. J.* **2023**, *193*, 109072. [CrossRef]

processes

Article

Seeking Optimal Extraction Method for Augmenting *Hibiscus sabdariffa* Bioactive Compounds and Antioxidant Activity

Athanasia Kourelatou [1], Theodoros Chatzimitakos [2], Vassilis Athanasiadis [2], Konstantina Kotsou [2], Ioannis Makrygiannis [2], Eleni Bozinou [2] and Stavros I. Lalas [2,*]

1 Department of Chemical Engineering, University of Western Macedonia, 50100 Kozani, Greece; athanasiak83@hotmail.com
2 Department of Food Science & Nutrition, University of Thessaly, Terma N. Temponera Str., 43100 Karditsa, Greece; tchatzimitakos@uth.gr (T.C.); vaathanasiadis@uth.gr (V.A.); kkotsou@agr.uth.gr (K.K.); ioanmakr1@uth.gr (I.M.); empozinou@uth.gr (E.B.)
* Correspondence: slalas@uth.gr; Tel.: +30-24410-64783

Abstract: The dried flowers of *Hibiscus sabdariffa* (HS), available worldwide, have various applications in both non-medicinal and medicinal fields. The growing global interest in the health benefits of HS is linked to its potential prevention or management of non-communicable diseases. The aim of this research was to find the optimal extraction method that ensures the maximum yield of multiple beneficial bioactive components, such as polyphenols, anthocyanins, vitamin C, β-carotene, antioxidant activity, free radical scavenging activity DPPH and ferric reducing antioxidant power (FRAP). To this end, stirring, pulsed electric field, and ultrasound-assisted extraction were evaluated, either alone or in combination. Under optimized extraction conditions, the obtained extract exhibited an elevated total polyphenol content (37.82 mg of gallic acid equivalents/g dry weight (dw)), total anthocyanin content (610.42 μg of cyanidin equivalents/g dw), total carotenoids content (921.84 μg of β-carotene equivalents/g dw), and ascorbic acid content (507.44 mg/100 g dw). Remarkably, the extracts exhibited strong antioxidant properties (487.51 μmol of ascorbic acid equivalents (AAE)/g dw and 243.42 μmol AAE/g dw as evidenced by FRAP and DPPH assays, respectively). This research advances the parameters that should be employed to produce the optimal and nutritionally enhanced HS flower extracts, that can be used in the commercial sector.

Keywords: hibiscus; extraction; ultrasonication; pulsed electric field; antioxidants; polyphenols; anthocyanins; response surface methodology; principal component analysis; partial least squares analysis

Citation: Kourelatou, A.; Chatzimitakos, T.; Athanasiadis, V.; Kotsou, K.; Makrygiannis, I.; Bozinou, E.; Lalas, S.I. Seeking Optimal Extraction Method for Augmenting *Hibiscus sabdariffa* Bioactive Compounds and Antioxidant Activity. *Processes* **2024**, *12*, 581. https://doi.org/10.3390/pr12030581

Academic Editors: Yanlin Zhang and Prashant K. Sarswat

Received: 21 February 2024
Revised: 12 March 2024
Accepted: 12 March 2024
Published: 14 March 2024

1. Introduction

The *Hibiscus sabdariffa* (HS), a member of the Malvaceae family, is commonly found in tropical and subtropical regions, including China, Thailand, Indonesia, Egypt, Sudan, Saudi Arabia, Taiwan, Vietnam, Nigeria, and Mexico [1]. This annual plant is well-known for its production of red flowers and is recognized by various names such as roselle, Jamaica, Red Sorrel, Indian Sorrel, wonjo, and karkade [2]. In addition to its ornamental appeal, the HS is valued for its diverse bioactive components, including phenolic acids, flavonoids, anthocyanins, and organic acids [3]. These compounds contribute to its various therapeutic applications, as evidenced by its use in traditional medicine, suggesting management of type 2 diabetes [4,5], as well as beneficial effects over nephropathy [6,7], iron-deficiency anemia [8], hypertension [9], and cardiovascular diseases [10]. Furthermore, the intense red color of HS and its extracts makes it a sought-after natural dye in the cosmetic and food industries [11]. Moreover, dehydrated HS flowers are extensively utilized in the formulation of an array of consumables, encompassing beverages, tea, jellies, jams, sauces, wines, syrups, ice cream, and chutneys [12]. This versatility underscores the relevance of HS in both botanical and commercial domains.

The extracts from the HS flowers have garnered significant attention from researchers due to their potent antioxidant activity [13]. To maximize the isolation of bioactive compounds and antioxidant potential of these extracts, various studies have been undertaken, exploring various extraction parameters. For instance, water-based extractions have been conducted employing substance-to-solvent ratios ranging from 1:10 to 1:100, temperatures spanning 25 to 100 °C, and extraction durations ranging from 5 to 600 min [14–16]. Additionally, extractions utilizing methanol at concentrations of 100% and 80% have been performed at both elevated (64 °C) and ambient (25 °C) temperatures, respectively [17,18]. Notably, alongside conventional extraction with stirring, environmentally sustainable extraction approaches have been investigated, including microwave-assisted extraction, employing 70% ethanol as solvent, with a substance-to-solvent ratio of 1:10, and an extraction duration of 2 min [19]. However, a comprehensive extraction method encompassing all critical factors (temperature, solvent, and extraction time) for the preparation of HS extracts with maximal content in bioactive compounds and antioxidant capacity, remains elusive.

A modern research area is the development of green extraction methods, which aim at improving the performance and environmental sustainability of extracting natural products. Green extraction methods are characterized by low energy consumption, high selectivity and efficiency, minimal use of harmful solvents, and safe and high-quality outcomes [20]. Advantages include minimized solvent consumption, absence of hazardous substances, shortened extraction durations, and low energy consumption [21–23]. Some of the most widely used green extraction techniques include pulsed electric field (PEF) extraction [24], ultrasound-assisted extraction (UAE) [25], pressurized liquid extraction (PLE) [26], microwave-assisted extraction (MAE) [27], supercritical fluid extraction (SFE) [28], and enzyme-assisted extraction (EAE) [29]. These techniques have been successfully applied to various aromatic and medicinal plants for various purposes. Examples include the use of MAE for extracting mint (*Mentha piperita* L.) [30], PLE for extracting two species of eucalyptus (*Eucalyptus marginata* and *Eucalyptus pauciflora*) [31], and the extraction of *Clitoria ternatea* through the use of UAE [32].

The aim of this study was to develop an optimized procedure for the preparation of extracts from HS flowers, rich in total polyphenol content (TPC). As such, pivotal extraction parameters, such as the composition of the solvent, as well as the extraction time and temperature were studied. Moreover, alongside the conventional extraction process with stirring (ST), the use of PEF or ultrasound (US) as pretreatment steps was examined, either alone or in combination. To aid the above study, a response surface methodology (RSM) was employed. The extract obtained under the optimum conditions is expected to contain a high amount of TPC and as such, high antioxidant activity.

2. Materials and Methods

2.1. Chemicals and Reagents

The details about all the chemicals and reagents that were used to accomplish this experimental survey are presented in the Supplementary Material.

2.2. Flower Collection and Preparation

The flowers of HS were gathered in late September of 2023, from a field in the Larissa region of Thessaly, central Greece (39°32′43.0″ N 22°25′46.0″ E, as per Google Earth coordinates). The flowers were manually harvested at their mature stage. Subsequently, the flowers were washed with tap water, rinsed with deionized water, and dried using a paper towel before being subjected to lyophilization in a Biobase BK-FD10P lyophilizer (Jinan, China) for 24 h at 7 Pa and temperature around −54 °C. The lyophilized flowers were finely ground to a fine powder and sieved using an Analysette 3 PRO (Fritsch GmbH, Oberstein, Germany). Particles with an average diameter of 106 μm were obtained and used for further experiments.

2.3. Extraction Procedure

One gram of powdered HS flowers was introduced into a 25 mL glass bottle, and 20 mL of an extraction solvent was added. Table 1 presents the composition of the extraction solvent, which includes varying concentrations of ethanol—C_{EtOH} (%, *v/v*)—ranging from 0 to 100% *v/v*. The convention extraction process was stirring (ST) at 500 rpm using a magnetic stirrer (Heidolph Instruments GmbH & Co. KG, Schwabach, Germany) at different temperatures (*T*) and durations (*t*), as outlined in Table 1. Some samples underwent PEF and US treatments, noting four different extraction techniques, as indicated in Table 1. The coded variable level 5 is concerned only C_{EtOH}, *t* and *T*. In all cases, the sample underwent hydration by immersing the dry powder in the different C_{EtOH} solvent for a *t* of 10 min. The extraction steps, for means of simplicity, are depicted in Figure 1.

Table 1. The actual and coded levels of the independent variables were used to optimize the process.

Independent Variables	Code Units	Coded Variable Level				
		1	2	3	4	5
Technique	X_1	ST	PEF + ST	US + ST	PEF + US + ST	–
C_{EtOH} (%, *v/v*)	X_2	0	25	50	75	100
t (min)	X_3	30	60	90	120	150
T (°C)	X_4	20	35	50	65	80

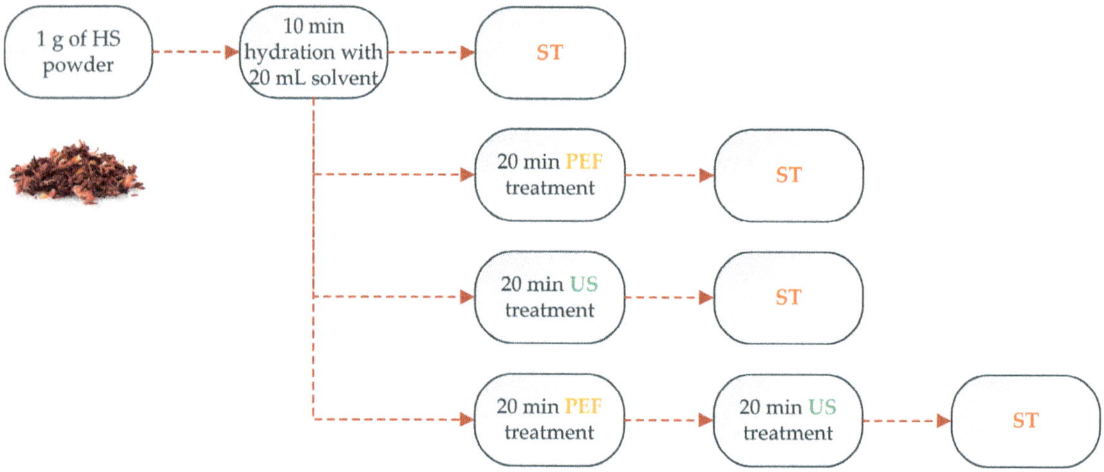

Figure 1. The extraction process of HS (*Hibiscus sabdariffa*) dry flowers using different techniques (ST: stirring, PEF: pulsed electric field, US: ultrasound).

To process the samples using PEF, two custom stainless-steel chambers (Val-Electronic, Athens, Greece), a mode/arbitrary waveform generator (UPG100, ELV Elektronik AG, Leer, Germany), a digital oscilloscope (Rigol DS1052E, Beaverton, OR, USA), and a high-voltage power generator (Leybold, LDDidactic GmbH, Huerth, Germany) were employed [33]. The electric field density was set at 1.0 kV/cm, with a pulse period of 1 ms (frequency: 1 kHz) and a pulse length of 10 µs. The US treatment was carried out by placing the Duran bottle in an Elmasonic P unit (Elma Schmidbauer GmbH, Singen, Germany), the temperature was maintained at 30 °C, and the frequency was set to 37 kHz and 180 W.

In all cases, after the extraction step was completed, the sample underwent centrifugation (Remi Elektrotechnik Ltd., Palghar, India) for 10 min at 3600× *g*. Subsequently, the supernatant was collected and preserved at −40 °C until further analysis.

2.4. Response Surface Methodology (RSM) Optimization of Extraction and Experiment Design

The Supplementary Material contains comprehensive details on how Response Surface Methodology (RSM) was utilized to enhance the sample's effectiveness in bioactive compounds.

2.5. HPLC-Based Analysis of the Pelargonin Chloride Content and the Other Polyphenolic Compounds

The detailed procedure is given in the Supplementary Material.

2.6. Analyses of Extracts

Detailed information on the analyses of the extracts, including TPC, TAC, antioxidant activity through the DPPH and the FRAP assays, TCC, and AAC, can be found in the Supplementary Material [34–38]. All results are expressed to the dry weight (dw) of the flower.

2.7. Statistical Analysis

The detailed presentation of the statistical analysis program used for this study is included in the Supplementary Material.

3. Results and Discussion

All the collected and studied flowers exhibited red calyces with five large petals interconnected by the corolla. The flowers had a length ranging from 5.10 to 5.15 cm, empirically and scientifically suitable for full bloom [39]. To prepare extracts from the HS flowers with a high concentration of bioactive compounds an RSM approach was applied. This method facilitated the systematic examination of various factors influencing the extraction process, including conventional extraction method (ST), green extraction methods (US and PEF), and their combinations. In addition, more extraction parameters such as extraction *t*, *T*, and C_{EtOH} (%, *v/v*) solvent mixtures were taken into consideration. In accordance with earlier investigations and preliminary experiments, the most suitable ratio of solid-to-C_{EtOH} (%, *v/v*) solvent ratio was determined to be 1:20 [34].

3.1. Extraction Optimization

Employing the RSM approach, the influence of each factor was evaluated and adjusted to attain optimal extraction efficiency. The recorded outcomes for each extract are outlined in Table 2. It can be seen that the design points 5, 15, and 20 are not particularly promising. On the contrary, design point 13 exhibited the most promising results. This signifies that the treatment with PEF and US constitutes the most optimal combination. Also, avoidance of excessively high *T* (<65 °C) coupled with a one-hour extraction *t* creates the ideal extraction conditions, yielding a highly enriched extract, as exemplified by design point 13.

Table 2. Experimental results for the four explored independent variables and the corresponding responses of the dependent variable.

Design Point	Independent Variables				Responses					
	X_1	X_2	X_3	X_4	TPC [1]	TAC [2]	FRAP [3]	DPPH [4]	TCC [5]	AAC [6]
1	3	1	3	4	25.30	341.08	295.60	164.82	710.02	381.32
2	3	2	1	3	29.62	604.39	477.68	243.42	763.65	480.75
3	2	3	4	3	21.29	543.96	226.20	119.58	921.84	441.42
4	2	4	5	4	12.27	347.47	149.64	106.69	708.51	327.26
5	3	5	4	2	3.90	135.53	82.72	66.59	256.92	267.87
6	4	1	4	5	32.33	285.56	354.95	105.94	578.57	448.03
7	4	2	3	1	34.43	465.42	358.60	108.00	690.92	401.23
8	1	3	3	2	29.42	578.07	274.41	81.21	847.53	484.04

Table 2. *Cont.*

Design Point	Independent Variables				Responses					
	X_1	X_2	X_3	X_4	TPC [1]	TAC [2]	FRAP [3]	DPPH [4]	TCC [5]	AAC [6]
9	1	4	4	1	23.38	398.22	301.91	82.18	689.48	386.66
10	1	5	1	4	12.75	162.97	216.52	32.14	231.75	303.81
11	1	1	2	3	32.26	468.95	386.91	119.12	545.52	461.45
12	1	2	5	5	35.97	430.21	487.51	123.12	594.45	487.20
13	4	3	2	4	37.82	539.70	467.69	135.80	768.40	507.19
14	3	4	2	5	24.51	479.91	472.39	109.09	450.20	507.44
15	2	5	3	5	6.60	80.51	100.24	25.01	215.26	239.33
16	2	1	1	1	14.78	444.93	155.15	102.84	428.31	351.04
17	2	2	2	2	23.50	610.42	269.15	137.85	769.89	480.28
18	3	3	5	1	22.39	599.57	474.49	125.81	750.33	481.11
19	4	4	1	2	25.11	361.03	265.08	76.82	427.64	393.68
20	4	5	5	3	8.66	145.46	111.72	1.44	176.06	298.24

[1] Total polyphenol content (TPC) in mg gallic acid equivalents (GAE)/g dw; [2] Total anthocyanin content (TAC) in μg cyanidin equivalents (CyE)/g dw; [3] Ferric reducing antioxidant power (FRAP) in μmol ascorbic acid equivalents (AAE)/g dw; [4] 2,2-Diphenyl-1-picrylhydrazyl (DPPH) in μmol AAE/g dw; [5] Total carotenoid content (TCC) in μg β-carotene equivalents (CtE)/g dw; [6] Ascorbic acid content (AAC) in mg/100 g dw.

Figure 2 depicts an HPLC-DAD chromatogram at 280 nm and 320 nm for an HS extract. As can be seen, various polyphenolic compounds such as 3-hydroxytyrosol, pelargonin chloride, rutin, and hesperidin were identified at 280 nm [40–43] as well as chlorogenic acid and luteolin at 320 nm [44].

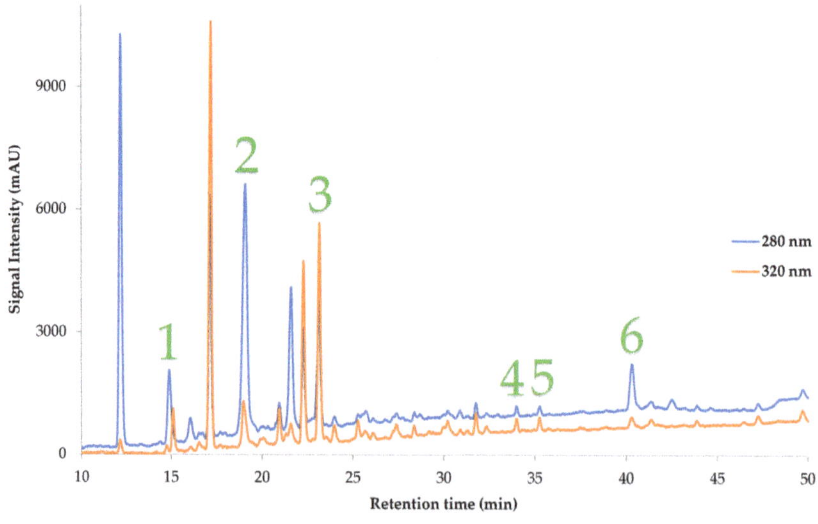

Figure 2. Representative HPLC chromatogram at 280 nm and 320 nm of an HS extract showcasing the identified polyphenolic compounds. 1: 3-Hydroxytyrosol; 2: Pelargonin chloride; 3: Chlorogenic acid; 4: Rutin; 5: Luteolin-7-glucoside; 6: Hesperidin.

Optimizing extraction parameters was important, as it resulted in achieving the highest extraction efficiency while minimizing the utilization of energy, *t*, and solvents. Enhancing these parameters is crucial for minimizing the environmental footprint of the entire extraction process [45]. C_{EtOH} selection significantly affects compound extraction [46]. For example, bioactive compounds characterized by medium polarity, such as the polyphenols abundantly present in the present samples, are not efficiently extracted with water owing to their lower polarity [47]. As a result, ethanol has been extensively utilized to enhance extraction efficiency [48]. This outcome was further corroborated in the present study, as regarding the solvent parameter, the presence of ethanol was deemed indispensable.

Table 3 presents the statistical parameters, second-order polynomial equations (models), and coefficients (with values exceeding 0.96) for each model. These findings indicate a strong alignment between the developed models and the data. Accompanying this, Figures S1–S6 illustrate plots comparing the actual response against the predicted response for each examined parameter, along with desirability functions. For a comprehensive understanding of the TPC, three-dimensional response plots are presented in Figure 3, while corresponding three-dimensional response plots for the remaining parameters can be found in Figures S7–S11.

According to the Pareto plots, the positive or negative influence of each extraction factor on the isolation efficiency of the respective bioactive element is presented in Figure 4. The prominently negative correlation observed in the C_{EtOH} (%, *v/v*) across all bioactivities (TPC, TAC, TCC, and AAC), as well as antioxidant capacity, regardless of the determination method, is intriguing. This implies that neither 100% water nor 100% ethanol as solvents assist in achieving maximal values of the aforementioned bioactivities, confirming results presented in Table 2, where the highest bioactivity values were achieved with ethanol-water mixtures. In contrast, the extraction method appears to positively influence both the isolation of TPC and TAC, suggesting a potential association between these methods. However, this result conflicts with the DPPH antioxidant method, which is negatively affected by the extraction method factor. Notably, a positive correlation is observed with the extraction *T* for the FRAP antioxidant method, whereas *T*, in combination with C_{EtOH} (%, *v/v*) and extraction *t*, exhibits a negative correlation for the values of DPPH and FRAP, respectively.

Table 3. Mathematical models developed through RSM were employed to optimize the extraction of HS. These models included only significant terms.

Responses	Second-Order Polynomial Equations (Models)	R^2 Predicted	R^2 Adjusted	*p*-Value	Equation
TPC [1]	$Y = 17.33 - 18.64X_1 + 15.67X_2 + 1.46X_3 + 6.2X_4 + 4.43X_1^2 - 3.31X_2^2 - 0.53X_3^2 + 1.0X_4^2 + 0.09X_1X_2 - 0.06X_1X_3 - 1.12X_1X_4 + 1.01X_2X_3 - 1.22X_2X_4 - 1.1X_3X_4$	0.9738	0.9004	0.0050	(2)
TAC [2]	$Y = 294.78 - 66.1X_1 + 466.69X_2 - 128.22X_3 + 14.9X_4 - 16.9X_1^2 - 93.75X_2^2 + 2.13X_3^2 + 8.85X_4^2 + 15.6X_1X_2 + 28.93X_1X_3 - 0.13X_1X_4 + 11.21X_2X_3 - 12.1X_2X_4 - 8.57X_3X_4$	0.9795	0.9221	0.0027	(3)
FRAP [3]	$Y = 71.14 - 120.69X_1 + 316.19X_2 - 151.96X_3 + 120.44X_4 + 26.76X_1^2 - 60.63X_2^2 + 13.76X_3^2 + 44.87X_4^2 + 8.92X_1X_2 + 15.58X_1X_3 - 31.57X_1X_4 + 32.22X_2X_3 - 40.81X_2X_4 - 41.59X_3X_4$	0.9652	0.8676	0.0097	(4)
DPPH [4]	$Y = -192.21 + 141.65X_1 + 107.58X_2 - 79.76X_3 + 107.46X_4 - 14.19X_1^2 - 18.06X_2^2 + 3.26X_3^2 + 3.36X_4^2 - 10.9X_1X_2 - 0.22X_1X_3 - 13.91X_1X_4 + 20.1X_2X_3 - 17.83X_2X_4 - 7.57X_3X_4$	0.9626	0.8580	0.0114	(5)
TCC [5]	$Y = -458.87 + 164.06X_1 + 344.69X_2 + 251.8X_3 + 215.7X_4 - 10.31X_1^2 - 67.16X_2^2 - 6.52X_3^2 - 55.52X_4^2 - 35.63X_1X_2 - 31.84X_1X_3 + 26.19X_1X_4 - 10.7X_2X_3 + 24.33X_2X_4 - 13.78X_3X_4$	0.9636	0.8618	0.0107	(6)
AAC [6]	$Y = 376.88 - 137.83X_1 + 313.01X_2 - 95.12X_3 + 4.76X_4 + 0.03X_1^2 - 54.98X_2^2 - 1.87X_3^2 + 18.3X_4^2 + 17.99X_1X_2 + 28.4X_1X_3 - 2.75X_1X_4 + 5.54X_2X_3 - 22.93X_2X_4 - 4.29X_3X_4$	0.9717	0.8925	0.0059	(7)

[1] TPC: Total polyphenol content; [2] TAC: Total anthocyanin content; [3] FRAP: Ferric reducing antioxidant power; [4] DPPH: 2,2-Diphenyl-1-picrylhydrazyl; [5] TCC: Total carotenoid content; [6] AAC: Ascorbic acid content.

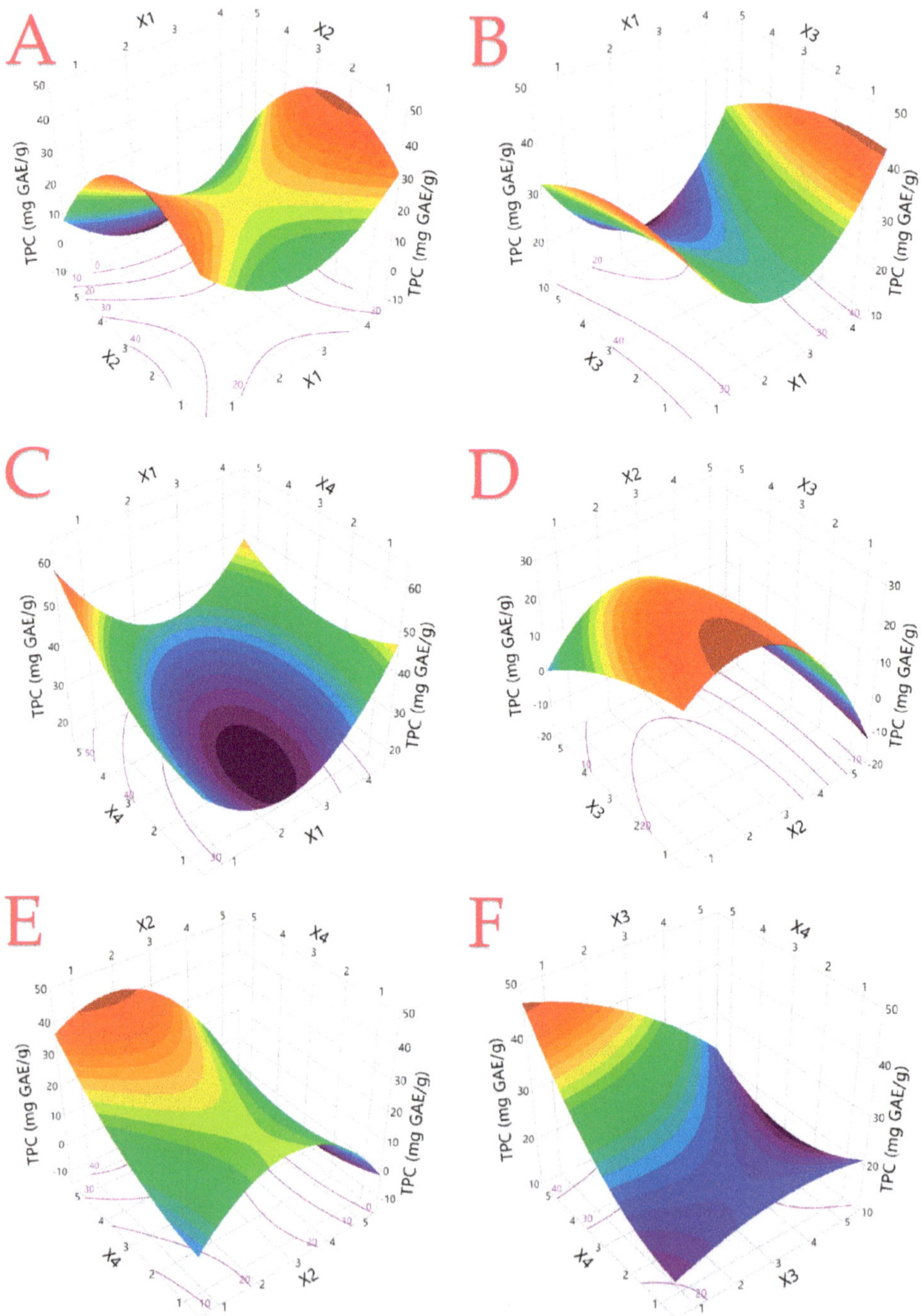

Figure 3. The optimized extraction of HS extracts is depicted in 3D graphs illustrating the influence of the process variables on the response (Total polyphenol content—TPC, mg GAE/g). Plot (**A**) shows the interaction between X_1 and X_2; plot (**B**) demonstrates the interaction between X_1 and X_3; plot (**C**) illustrates the interaction between X_1 and X_4; plot (**D**) showcases the interaction between X_2 and X_3; plot (**E**) displays the interaction between X_2 and X_4; plot (**F**) represents the interaction between X_3 and X_4.

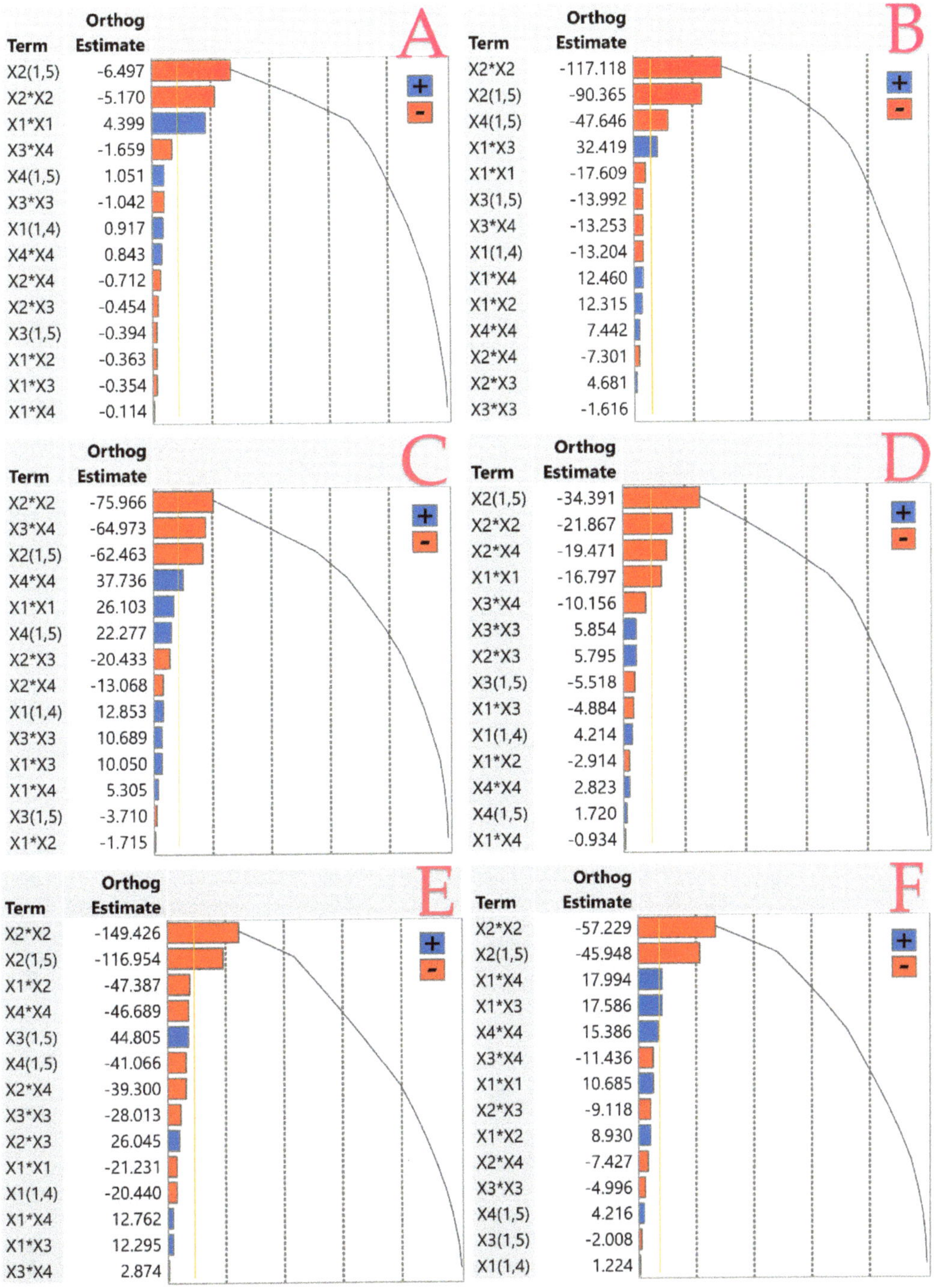

Figure 4. Pareto plots of transformed estimates for TPC (**A**), TAC (**B**), FRAP (**C**), DPPH (**D**), TCC (**E**), and AAC (**F**) assays. A reference gold line is drawn on the plot to indicate the level of significance ($p < 0.05$).

3.2. Analysis of the Extracts

3.2.1. TPC and TAC of the Extracts

The adopted extraction techniques and parameters notably affected the TPC and TAC in the extracts, as outlined in Table 2. The TPC ranged from 3.90 mg gallic acid equivalents (GAE)/g dw to 37.82 mg GAE/g dw, thus the quantity could increase by up to 869.83%. The highest extraction yields of phenolics were attained through the utilization of a 25% C_{EtOH} coupled with a short extraction *t*, as outlined in Table 4. Notably, the increase in *T*, along with the integration of both the green extraction processings prior to stirring, results in a further enhancement of the TPC content. Comparing the results with recent studies that employed conventional extraction methods, either with boiling water or ethanol solutions at 23 °C, notable differences were observed. Specifically, in the former case, a value of 14.24 mg GAE/g dw of the flower was recorded [49], while in the latter, using a 50% ethanolic solution, the value was 35.71 mg GAE/g dw. However, when employing a 75% ethanolic solution, the value decreased to 35.30 mg GAE/g dw [50]. By comparing all the results, it is once again evident that ethanol is indeed a suitable solvent for extracting bioactive compounds of HS, and complementary techniques preceding conventional extraction can increase the TPC to very significant levels.

Table 4. Maximum predicted responses and optimum extraction conditions for the dependent variables.

Responses	Optimal Conditions				
	Maximum Predicted Response	**Technique (X_1)**	***C* (%, *v*/*v*) (X_2)**	***t* (min) (X_3)**	***T* (°C) (X_4)**
TPC (mg GAE/g dw)	41.17 ± 7.73	PEF + US + ST (4)	25 (2)	60 (2)	65 (4)
TAC (μg CyE/g dw)	655.67 ± 113.56	US + ST (3)	50 (3)	30 (1)	65 (4)
FRAP (μmol AAE/g dw)	603.11 ± 121.21	US + ST (3)	50 (3)	30 (1)	65 (4)
DPPH (μmol AAE/g dw)	243.42 ± 46.76	US + ST (3)	25 (2)	30 (1)	50 (3)
TCC (μg CtE/g dw)	944.36 ± 194.3	ST (1)	50 (3)	120 (4)	50 (3)
AAC (mg/100 g dw)	561.11 ± 67.68	US + ST (3)	50 (3)	30 (1)	65 (4)

Concerning TAC, the quantities exhibited a range of values from 80.51 to 610.42 μg cyanidin equivalents (CyE)/g dw, contingent upon the employed extraction method. It was demonstrated that the optimum C_{EtOH} (%, *v*/*v*) was a 50% solution, with a minimal extraction *t*. The inclusion of US prior to the conventional stirring was also deemed imperative. In a previous investigation examining the enhancement of TAC levels in HS flowers through green extraction techniques such as maceration and ultrasonic turbolization, acidified water emerged as the preferred solvent. Maximum quantities for each method were documented as 96.30 and 87.30 μg CyE/g dw, respectively [51]. These recorded values were notably 533.87% and 599.22% lower compared to the levels achievable by following the recommended extraction parameters for obtaining an anthocyanin-rich extract from HS. Moreover, in a previous study where an ethanolic extract of HS flowers was studied, the amount of TAC was recorded at 359.3 mg CyE/100 g dw (i.e., 3593 μg CyE/g dw) [52]. Despite the impressive recorded amount, the content could be enhanced by 41.14% if the appropriate extraction conditions were applied.

As indicated in Table 4, the predicted values obtained by Partial Least Squares (PLS) for the optimal extraction stand at 41.17 mg GAE/g dw for TPC and 655.67 μg CyE/g dw for TAC. Nevertheless, considering the maximum values outlined in Table 5, the predicted responses, following the same extraction method and parameters for the preparation of a single optimal extract, contained 36.56 mg GAE/g dw for TPC and 681.96 μg CyE/g dw for TAC. These findings underscore the significance of meticulously choosing extraction techniques and C_{EtOH} (%, *v*/*v*), as they can substantially augment the efficiency and yield of bioactive compounds. This approach not only proves to be more energy-efficient but also enhances overall effectiveness in the extraction process.

Table 5. Highest desirability for all variables using the partial least squares (PLS) prediction profiler under the optimal extraction conditions (X_1:3, X_2:2, X_3:1, X_4:5).

Variables	PLS Model Values	Experimental Values
TPC (mg GAE/g dw)	36.56	40.49 ± 2.67
TAC (μg CyE/g dw)	681.96	642.18 ± 34.68
FRAP (μmol AAE/g dw)	690.18	697.1 ± 23.7
DPPH (μmol AAE/g dw)	227.12	216.14 ± 14.91
TCC (μg CtE/g dw)	781.14	676.86 ± 41.29
AAC (mg/100 g dw)	595.96	536.05 ± 15.55

3.2.2. Polyphenolic Compounds of the Optimum Extract

Pelargonin emerged as the principal polyphenolic compound within the optimal extract, as presented in Table 6, comprising approximately 87.80% of the total polyphenolic compounds. Pelargonin is classified as an anthocyanin, commonly present in red fruits and vegetables, functioning as a natural colorant [53]. Identified mainly in dates, pelargonin has been investigated for its anti-allergic properties [54]. Moreover, it is reported to exhibit oxygen radical absorbance capacity [55]. Pelargonin chloride has been previously identified in HS flowers from a variety in Mexico in minimal quantities. More specifically, it was found in amounts of 0.00217 mg/g dw [56], which is much lower than what can be obtained by applying ideal extraction procedures. According to previous studies, the leaves of HS are rich in chlorogenic acid [57], known for their pharmacological activities, encompassing anticancer, antidiabetic, antiviral, and antipyretic properties [58,59]. Besides the leaves of the HS, it is evident (Table 6) that the flowers also contain a considerable amount of chlorogenic acid. Additionally, it is apparent that hesperidin is present in the flowers, which has neuroprotective effects observed across various models of central nervous system disorders [60]. In a preceding investigation, the calyces of HS exhibited a hesperidin content of 0.14 mg/g dw, resulting in a 600% reduced content compared to the optimal extract of HS presented herein. This means that the enriched extract, besides excelling in polyphenolic compounds, may also possess all the beneficial properties of these compounds. Figure 5 shows the chemical structures of the polyphenolic compounds that were identified in the extracts of HS.

Table 6. Polyphenolic compounds extracted under optimal conditions (X_1:3, X_2:2, X_3:1, X_4:5).

Polyphenolic Compound	Optimal Extract (mg/g dw)
3-Hydroxytyrosol	1.33 ± 0.05
Pelargonin chloride	34.63 ± 1.94
Chlorogenic acid	1.27 ± 0.09
Rutin	0.68 ± 0.03
Luteolin-7-glucoside	0.54 ± 0.03
Hesperidin	0.98 ± 0.06
Total identified	39.44 ± 2.20

3.2.3. Antioxidant Properties of the Extracts

The antioxidant properties of the HS extracts acquired in this study were assessed through two distinct methodologies, namely, FRAP and DPPH. Optimal extraction methods for both approaches necessitate ultrasonic treatment prior to agitation, a relatively elevated T ranging from 50 to 65 °C, a C_{EtOH} of up to 50%, and a short t, as depicted in Table 2.

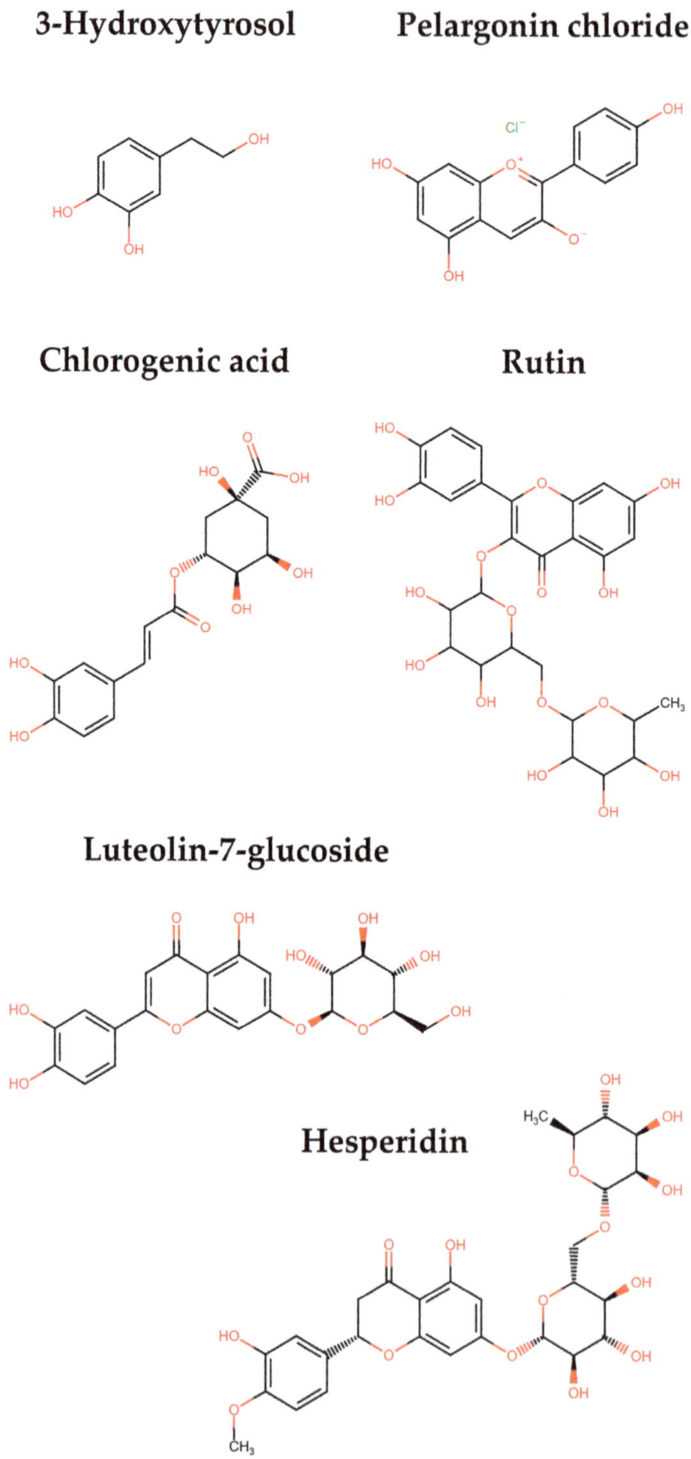

Figure 5. Chemical structures of all the polyphenolic compounds identified in HS extracts.

Regarding the antioxidant capacity via both methods, it can be seen that the samples exhibit significant variability in their capacity depending on the method employed. Specifically, the values range from 82.72 to 487.51 μmol ascorbic acid equivalents (AAE)/g dw according to Table 2, indicating that the antioxidant capacity can increase by up to 489.32%, in the FRAP method. Simultaneously, in the DPPH method the free radical scavenging capability increased by up to 16,800%. The experimental value of 216.14 ± 14.91 μmol AAE/g dw for the DPPH assay corresponds to an ascorbic acid concentration (AAC) of 10,826 ± 670 μmol/L. The IC_{50} value of the antioxidant ascorbic acid is 854.24 μmol/L, while the HS optimal extract has an IC_{50} value of 8545.32 μmol/L. The lower the IC_{50} value, the higher the antioxidant capacity. In a previous method, an attempt to optimize the antioxidant capacity of HS leaves using various solvents (methanol, 80% alcohol, hot and cold water) and drying methods (room, sun, oven, microwave, cross flow, infra-red, and freeze-dryer) was made. It was found that with 80% alcoholic solvent on freeze-dried leaves, the antioxidant capacity reached 202.93 μmol AAE/g dw for the FRAP method. Conversely, the highest total antioxidant activity was exhibited from the sun-dried leaf cold water extract at 394.39 μmol AAE/g dw with the room-dried leaf 80% alcohol extract following closely at 369.12 μmol AAE/g dw [61].

As per Table 4, the highest predicted values for each antioxidant capacity measurement method were 603.11 ± 121.21 μmol AAE/g dw for FRAP, whereas for DPPH, the corresponding value was 243.42 ± 46.76 μmol AAE/g dw, with desirability of 0.86 and 0.87, respectively, compared to the maximum value of 1. Comparing the results, it is evident that the HS extract can achieve and surpass the antioxidant capacity of the leaves using suitable methods, creating an extract with pronounced multifaceted antioxidant capability.

3.2.4. TCC and AAC of the Extracts

Carotenoids offer a plethora of health benefits and aesthetic therapeutic implications, including immune system strengthening, resilience against serious maladies such as cancer, and noteworthy anti-aging attributes [62,63]. Additionally, ascorbic acid, a natural antioxidant [64], not only serves as an antioxidant but also improves the absorption of non-heme iron [65]. A noteworthy outcome is that the extract of HS flowers is quite rich in TCC and AAC, with values ranging from 176.06 to 921.84 μg β-carotene equivalents (CtE)/g dw (a 423.59% increase) and 239.33 to 507.44 mg/100 g dw (indicating a 112.03% increase) after various processing conditions. This finding aligns with prior investigations where the abundant presence of both TCC and AAC in HS flowers has been emphasized [66,67]. In another study examining the nutrient content of HS flowers, the quantities recorded in TCC and AAC were 0.53 ± 0.27 mg/100 g dw (530 μg/100 g dw) and 141.09 mg/100 g dw, respectively. Therefore, once again, the importance of finding suitable extraction conditions for all plant extracts becomes evident [68]. Of equal significance is that, according to Table 4, achieving the maximum TCC quantity (944.36 μg CtE/g dw) requires simple stirring and a 50% ethanol concentration, while obtaining the maximum AA quantity (561.11 mg/100 g dw) necessitates pre-processing with US, the same C_{EtOH} (%, *v/v*), and minimal extraction *t*. The implementation of these extraction methodologies grants the formulation of an exceptionally enriched HS extract, characterized by heightened levels of AA and TC suitable for immediate and beneficial applications in the fields of medical science, cosmetic therapeutics, and nutritional sciences. The chemical structure of β-carotene and vitamin C are shown in Figure 6.

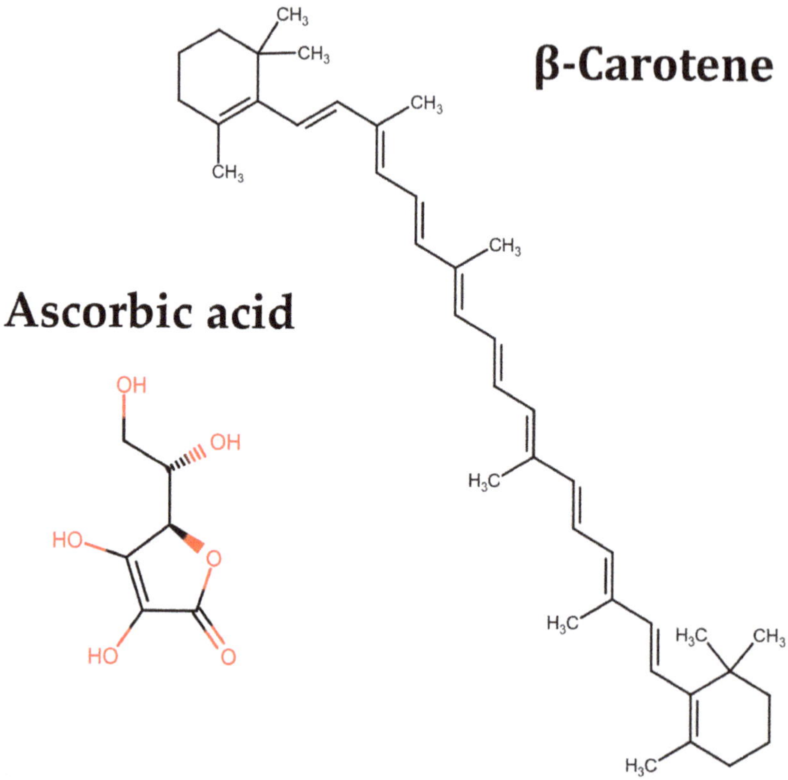

Figure 6. Chemical structure of β-carotene and ascorbic acid.

3.3. Principal Component Analysis (PCA) and Multivariate Correlation Analysis (MCA)

In Figures 7 and 8, the correlation of values among various bioactive compounds is depicted, revealing noteworthy outcomes. In Figure 7 it can be seen that component 1 explained 76.9% of the variability, presenting a positive correlation with all of the examined variables TPC, TAC, FRAP, DPPH, TCC, and AAC. Furthermore, it is discernible that among the extraction factors, the one that predominantly influences the increase in various bioactive compounds is X_2. It is extensively noted that ethanol plays a significant role in enhancing the isolation of bioactive compounds. Regarding Figure 8, it is worth noting that in numerous studies, the correlation among variables is often low or even negative. However, in the current research, not only a negative correlation is not observed, but the lowest correlation is 0.6, with a maximum value of 1, and this is observed in only one variable, that of antioxidant capacity through the scavenging of free radicals. Most bioactive substances exhibit a maximum correlation coefficient of 1. Specifically, the variables TPC, FRAP, and AAC recorded the highest correlation, demonstrating that both TP and AA exhibit strong antioxidant activity. Additionally, a robust correlation was observed between the TAC variable and both AAC and TCC. These results are highly encouraging, as it is rare to observe such a high degree of correlation in multivariate systems with multiple sensitive variables.

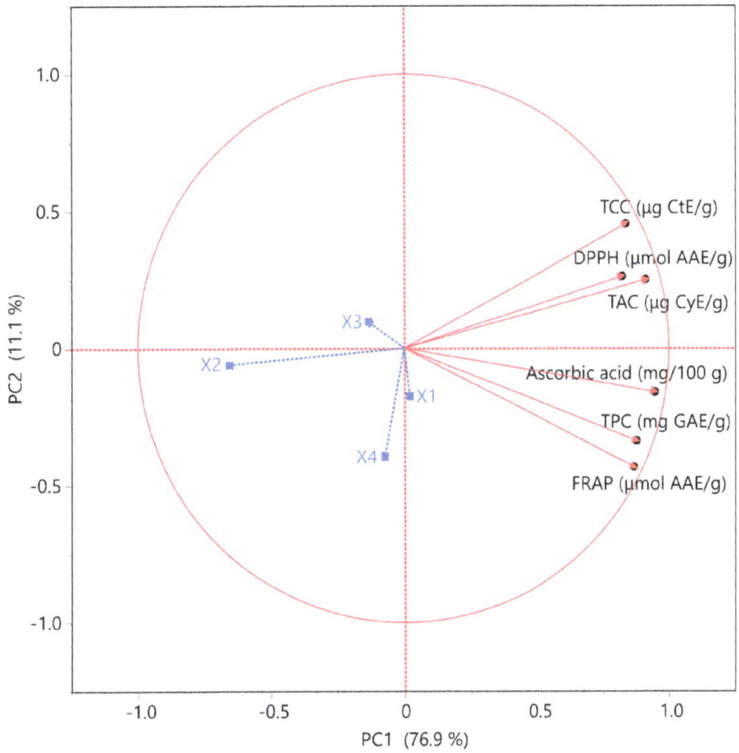

Figure 7. Principal component analysis (PCA) for the measured variables. Each *X* variable is presented with a blue color.

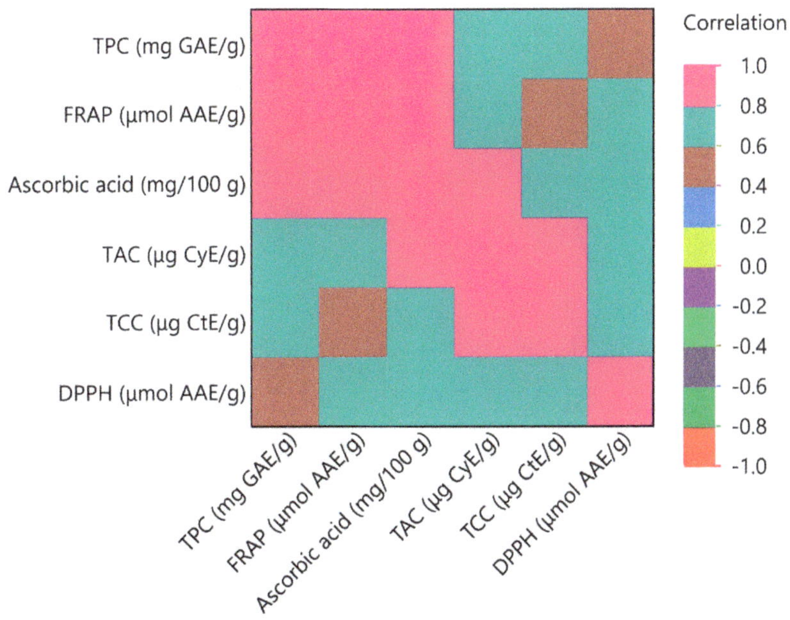

Figure 8. Multivariate correlation analysis of measured variables.

3.4. Partial Least Squares (PLS) Analysis

A PLS analysis, as depicted in Figure 9, was carried out to identify the key factors among the investigated extraction variables (X_1, X_2, X_3, and X_4). It is crucial to emphasize that PLS offers valuable insights for optimizing extraction conditions and contributes to comprehending how different factors interact to influence the yields of bioactive compounds. Figure 9 illustrates the quantities of all potentially extractable bioactive compounds by following the suggested extraction conditions based on the PLS analysis. The desirability of the results is notably high, registering a value of approximately 0.8930. The result, as indicated in Table 4, is logical and it is demonstrated that for almost all bioactive elements and antioxidant capacity, the maximum values are ensured through the application of the same parameters.

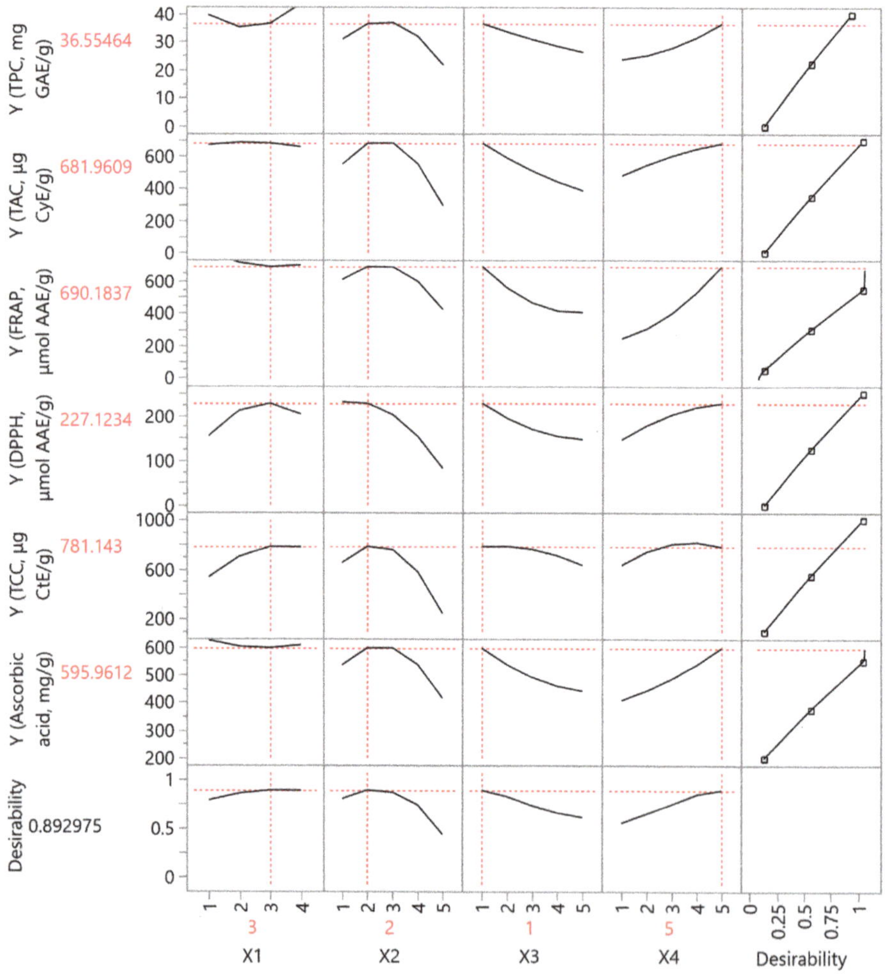

Figure 9. Partial least squares (PLS) prediction profiler of each variable and desirability function with extrapolation control for the optimization of HS extracts.

Upon comparing the values provided by the PLS model with those obtained through experimental analysis, a correlation of 0.9888 was observed, indicating no deviations, with a p-value of <0.0001.

4. Conclusions

The dried flowers of HS are highlighted as a versatile source with extensive applications in both the non-pharmaceutical and pharmaceutical sectors. The global interest in HS revolves around perceived health benefits, particularly in preventing or managing non-communicable diseases. This study, acknowledging the diverse bioactive compounds responsible for the myriad properties of HS, delves into three distinct extraction methods, including conventional ST and two environmentally friendly approaches, US and PEF, both individually and in combination. Through stochastic optimization, this research aims to produce the most nutrient-rich HS extract using more advanced and eco-friendly methods. Indeed, according to our results, the use of US before ST (at least for t 30 min) along with 25% C_{EtOH}, at 80 °C T was deemed the most suitable for the abundance of bioactive substances (36.56 mg GAE/g dw for TPC, 681.96 μg CyE/g dw for TAC, 781.14 μg CtE/g dw for TCC, 595.96 mg/100 g dw for AAC) and antioxidant capacity (690.18 μmol AAE/g dw and 227.12 μmol AAE/g dw for FRAP and DPPH method, respectively). Simultaneously, the application of the aforementioned extraction parameters offered a remarkable increase in TPC, TAC, TCC, and AAC reaching 837.46, 747.02, 343.67, and 149.02%, respectively. This study not only provides optimized extraction processes to maximize the overall nutritional content or individual nutrients of HS flower extracts but also underscores the diverse applications of this extract.

Supplementary Materials: The following supporting information can be downloaded at: https://www.mdpi.com/article/10.3390/pr12030581/s1, Supplementary Material File—Materials and Methods. Figures S1–S6 comprise plots that illustrate the comparison between the actual response and the predicted response for each parameter under examination, accompanied by the desirability functions. Figures S7–S11 present three-dimensional response plots for the remaining responses.

Author Contributions: Conceptualization, T.C., V.A. and S.I.L.; methodology, V.A. and T.C.; software, V.A.; validation, T.C., V.A. and K.K.; formal analysis, T.C. and V.A.; investigation, A.K., T.C., V.A., K.K., I.M. and E.B.; resources, S.I.L.; data curation, V.A.; writing—original draft preparation, A.K. and K.K.; writing—review and editing, A.K., T.C., V.A., K.K., I.M., E.B. and S.I.L.; visualization, V.A.; supervision, S.I.L.; and funding acquisition, S.I.L. All authors have read and agreed to the published version of the manuscript.

Funding: This research received no external funding.

Data Availability Statement: All related data and methods are presented in this paper. Additional inquiries should be addressed to the corresponding author.

Conflicts of Interest: The authors declare no conflicts of interest.

References

1. Bedi, P.S.; Bekele, M.; Gure, G. Phyto-Chemistry and Pharmacological Activities of *Hibiscus sabdariffa* Linn.—A Review. *Int. Res. J. Pure Appl. Chem.* **2020**, *21*, 41–54. [CrossRef]
2. Solangi, A.H.; Junejo, S.; Siddiqui, A.A.; Arain, M.Y.; Ansari, M.A.; Talpur, U.A.; AhadKolach, A. Roselle (*Hibiscus sabdariffa* L.) a Multipurpose Medicinal Plant and Its Uses: A Review. *Int. J. Biol. Res.* **2017**, *5*, 21–24.
3. Izquierdo-Vega, J.A.; Arteaga-Badillo, D.A.; Sánchez-Gutiérrez, M.; Morales-González, J.A.; Vargas-Mendoza, N.; Gómez-Aldapa, C.A.; Castro-Rosas, J.; Delgado-Olivares, L.; Madrigal-Bujaidar, E.; Madrigal-Santillán, E. Organic Acids from Roselle (*Hibiscus sabdariffa* L.)—A Brief Review of Its Pharmacological Effects. *Biomedicines* **2020**, *8*, 100. [CrossRef] [PubMed]
4. Sarbini, D.W.I.; Huriyati, E.M.Y.; Sadewa, H.; Wahyuningsih, M.S.H. The Effect of Rosella (*Hibiscus sabdariffa* Linn) on Insulin Resistance in Patients with Type 2 Diabetes Mellitus: A Randomized Clinical Trial. *Int. J. Pharm. Res.* **2019**, *11*, 547–557.
5. Abubakar, S.M.; Ukeyima, M.T.; Spencer, J.P.E.; Lovegrove, J.A. Acute Effects of Hibiscus sabdariffa Calyces on Postprandial Blood Pressure, Vascular Function, Blood Lipids, Biomarkers of Insulin Resistance and Inflammation in Humans. *Nutrients* **2019**, *11*, 341. [CrossRef]
6. Nwachukwu, D.C.; Aneke, E.I.; Nwachukwu, N.Z.; Azubike, N.; Obika, L.F. Does Consumption of an Aqueous Extract of *Hibiscus sabdariffa* Affect Renal Function in Subjects with Mild to Moderate Hypertension? *J. Physiol. Sci.* **2017**, *67*, 227–234. [CrossRef]
7. Sakhaei, R.; Nadjarzadeh, A.; Esmaeili, A.; Mohammadi, M.; Hemayati, R.; Reza, J.Z.; Mozaffari-Khosravi, H.; Ramezani-Jolfaie, N. Cardiovascular and Renal Effects of *Hibiscus sabdariffa* Linnaeus. in Patients with Diabetic Nephropathy: A Randomized, Double-Blind, Controlled Trial. *J. Nutr. Food Secur.* **2021**, *6*, 116–126. [CrossRef]

8. Peter, E.L.; Rumisha, S.F.; Mashoto, K.O.; Minzi, O.M.; Mfinanga, S. Efficacy of Standardized Extract of *Hibiscus sabdariffa* L. (Malvaceae) in Improving Iron Status of Adults in Malaria Endemic Area: A Randomized Controlled Trial. *J. Ethnopharmacol.* **2017**, *209*, 288–293. [CrossRef] [PubMed]
9. Nwachukwu, D.C.; Aneke, E.I.; Nwachukwu, N.Z.; Azubike, N.; Obika, L.F.O. Thirst Perception in Mild to Moderate Hypertensive Nigerians Treated with Aqueous Extract of *Hibiscus sabdariffa* L. *Afr. J. Pharm. Pharmacol.* **2016**, *10*, 403–410. [CrossRef]
10. Talal Al-Malki, A.; Abbas Ahmed Sindi, H.; Hussain Al-Qahtani, M. Regulation of CD36 Gene Expression by *Hibiscus sabdariffa* Tea Extracts to Affect the Atherosclerosis Biomarkers in Saudi Women. *J. Food Nutr. Res.* **2021**, *9*, 45–49. [CrossRef]
11. Islam, M. Food and Medicinal Values of Roselle (*Hibiscus sabdariffa* L. Linne Malvaceae) Plant Parts: A Review. *Open J. Nutr. Food Sci.* **2019**, *1*, 14–20.
12. Montalvo-González, E.; Villagrán, Z.; González-Torres, S.; Iñiguez-Muñoz, L.E.; Isiordia-Espinoza, M.A.; Ruvalcaba-Gómez, J.M.; Arteaga-Garibay, R.I.; Acosta, J.L.; González-Silva, N.; Anaya-Esparza, L.M. Physiological Effects and Human Health Benefits of *Hibiscus sabdariffa*: A Review of Clinical Trials. *Pharmaceuticals* **2022**, *15*, 464. [CrossRef]
13. Sindi, H.A.; Marshall, L.J.; Morgan, M.R.A. Comparative Chemical and Biochemical Analysis of Extracts of *Hibiscus sabdariffa*. *Food Chem.* **2014**, *164*, 23–29. [CrossRef]
14. Cisse, M.; Vaillant, F.; Acosta, O.; Claudie, D.M.; Dornier, M. Thermal Degradation Kinetics of Anthocyanins from Blood Orange, Blackberry, and Roselle Using the Arrhenius, Eyring, and Ball Models. *J. Agric. Food Chem.* **2009**, *57*, 6285–6291. [CrossRef]
15. Wong, S.P.; Leong, L.P.; William Koh, J.H. Antioxidant Activities of Aqueous Extracts of Selected Plants. *Food Chem.* **2006**, *99*, 775–783. [CrossRef]
16. Oboh, G.; Rocha, J.B.T. Antioxidant and Neuroprotective Properties of Sour Tea (*Hibiscus sabdariffa*, Calyx) and Green Tea (*Camellia sinensis*) on Some pro-Oxidant-Induced Lipid Peroxidation in Brain In Vitro. *Food Biophys.* **2008**, *3*, 382–389. [CrossRef]
17. Farombi, E.O.; Fakoya, A. Free Radical Scavenging and Antigenotoxic Activities of Natural Phenolic Compounds in Dried Flowers of *Hibiscus sabdariffa* L. *Mol. Nutr. Food Res.* **2005**, *49*, 1120–1128. [CrossRef] [PubMed]
18. Mohd-Esa, N.; Hern, F.S.; Ismail, A.; Yee, C.L. Antioxidant Activity in Different Parts of Roselle (*Hibiscus sabdariffa* L.) Extracts and Potential Exploitation of the Seeds. *Food Chem.* **2010**, *122*, 1055–1060. [CrossRef]
19. Amin, A.; Hamza, A.A.; Kambal, A.; Daoud, S. Herbal Extracts Counteract Cisplatin-Mediated Cell Death in Rat Testis. *Asian J. Androl.* **2008**, *10*, 291–297. [CrossRef] [PubMed]
20. Raj, B.; John, S.; Chandrakala, V.; Kumari, G.H. Green Extraction Techniques for Phytoconstituents from Natural Products. In *Medicinal Plants*; Kumar, S., Ed.; IntechOpen: Rijeka, Croatia, 2022.
21. Picot-Allain, C.; Mahomoodally, M.F.; Ak, G.; Zengin, G. Conventional versus Green Extraction Techniques—A Comparative Perspective. *Curr. Opin. Food Sci.* **2021**, *40*, 144–156. [CrossRef]
22. Romano, R.; De Luca, L.; Aiello, A.; Rossi, D.; Pizzolongo, F.; Masi, P. Bioactive Compounds Extracted by Liquid and Supercritical Carbon Dioxide from Citrus Peels. *Int. J. Food Sci. Technol.* **2022**, *57*, 3826–3837. [CrossRef]
23. Putra, N.R.; Yustisia, Y.; Heryanto, R.B.; Asmaliyah, A.; Miswarti, M.; Rizkiyah, D.N.; Yunus, M.A.C.; Irianto, I.; Qomariyah, L.; Rohman, G.A.N. Advancements and Challenges in Green Extraction Techniques for Indonesian Natural Products: A Review. *S. Afr. J. Chem. Eng.* **2023**, *46*, 88–98. [CrossRef]
24. Redondo, D.; Venturini, M.E.; Luengo, E.; Raso, J.; Arias, E. Pulsed Electric Fields as a Green Technology for the Extraction of Bioactive Compounds from Thinned Peach By-Products. *Innov. Food Sci. Emerg. Technol.* **2018**, *45*, 335–343. [CrossRef]
25. Somwongin, S.; Sirilun, S.; Chantawannakul, P.; Anuchapreeda, S.; Yawootti, A.; Chaiyana, W. Ultrasound-Assisted Green Extraction Methods: An Approach for Cosmeceutical Compounds Isolation from *Macadamia integrifolia* Pericarp. *Ultrason. Sonochem.* **2023**, *92*, 106266. [CrossRef]
26. Otero, P.; Quintana, S.E.; Reglero, G.; Fornari, T.; García-Risco, M.R. Pressurized Liquid Extraction (PLE) as an Innovative Green Technology for the Effective Enrichment of Galician Algae Extracts with High Quality Fatty Acids and Antimicrobial and Antioxidant Properties. *Mar. Drugs* **2018**, *16*, 156. [CrossRef]
27. Llompart, M.; Garcia-Jares, C.; Celeiro, M.; Dagnac, T. Microwave-Assisted Extraction. In *Reference Module in Chemistry, Molecular Sciences and Chemical Engineering*; Worsfold, P., Poole, C., Townshend, A., Miró, M., Eds.; Elsevier: Oxford, UK, 2018; pp. 67–77. ISBN 978-0-08-101984-9.
28. Singh, S.; Verma, D.K.; Thakur, M.; Tripathy, S.; Patel, A.R.; Shah, N.; Utama, G.L.; Srivastav, P.P.; Benavente-Valdés, J.R.; Chávez-González, M.L.; et al. Supercritical Fluid Extraction (SCFE) as Green Extraction Technology for High-Value Metabolites of Algae, Its Potential Trends in Food and Human Health. *Food Res. Int.* **2021**, *150*, 110746. [CrossRef]
29. Koina, I.M.; Sarigiannis, Y.; Hapeshi, E. Green Extraction Techniques for the Determination of Active Ingredients in Tea: Current State, Challenges, and Future Perspectives. *Separations* **2023**, *10*, 121. [CrossRef]
30. Kunhermanti, D.; Mahfud, M. Optimization of Peppermint (*Mentha Piperita* L.) Extraction Using Solvent-Free Microwave Green Technology. *Adv. Food Sci. Sustain. Agric. Agroindustrial Eng.* **2023**, 33–40.
31. Hasni, S.; Riguene, H.; Rigane, G.; Ghazghazi, H.; Salem, R.B.; Cifuentes, A.; Mendiola, J.A.; Ibáñez, E. Extraction of Neuroprotective Compounds from Eucalyptus Leaves Using Green Technologies. Ternary Mixture Design. In Proceedings of the 19th European Meeting on Supercritical Fluids, Budapest, Hungary, 21–24 May 2023.
32. Santos, L.G.; Martins, V.G. Optimization of the Green Extraction of Polyphenols from the Edible Flower Clitoria Ternatea by High-Power Ultrasound: A Comparative Study with Conventional Extraction Techniques. *J. Appl. Res. Med. Aromat. Plants* **2023**, *34*, 100458. [CrossRef]

33. Haida, Z.; Ab Ghani, S.; Juju Nakasha, J.; Hakiman, M. Determination of Experimental Domain Factors of Polyphenols, Phenolic Acids and Flavonoids of Lemon (*Citrus Limon*) Peel Using Two-Level Factorial Design. *Saudi J. Biol. Sci.* **2022**, *29*, 574–582. [CrossRef]
34. Chatzimitakos, T.; Athanasiadis, V.; Kotsou, K.; Bozinou, E.; Lalas, S.I. Response Surface Optimization for the Enhancement of the Extraction of Bioactive Compounds from *Citrus limon* Peel. *Antioxidants* **2023**, *12*, 1605. [CrossRef]
35. Bozinou, E.; Lakka, A.; Poulianiti, K.; Lalas, S.; Makris, D.P. Cyclodextrins as High-Performance Green Co-Solvents in the Aqueous Extraction of Polyphenols and Anthocyanin Pigments from Solid Onion Waste. *Eur. Food Res. Technol.* **2021**, *247*, 2831–2845. [CrossRef]
36. Shehata, E.; Grigorakis, S.; Loupassaki, S.; Makris, D.P. Extraction Optimisation Using Water/Glycerol for the Efficient Recovery of Polyphenolic Antioxidants from Two Artemisia Species. *Sep. Purif. Technol.* **2015**, *149*, 462–469. [CrossRef]
37. Ayour, J.; Alahyane, A.; Harrak, H.; Neffa, M.; Taourirte, M.; Benichou, M. Assessment of Nutritional, Technological, and Commercial Apricot Quality Criteria of the Moroccan Cultivar "Maoui" Compared to Introduced Spanish Cultivars "Canino" and "Delpatriarca" towards Suitable Valorization. *J. Food Qual.* **2021**, *2021*, 6679128. [CrossRef]
38. Dani, H.M.; Jagota, S.K. A New Calorimetric Technique for the Estimation of Vitamin C Using Folin Phenol Reagent. *Anal. Biochem.* **1982**, *127*, 178–182.
39. Ghosh, P.; Priyadarshini, S.; Munshi, M.; Beula, P.; Atthoti, L.; Kruthi, G.S.; Sravan, B.S.; Irshaan, S.; Rana, S.S.; Abdullah, S. Physicochemical Characterization and Mass Modelling of Roselle (*Hibiscus sabdariffa* L.) Calyx and Seeds. *Food Chem. Adv.* **2023**, *3*, 100400. [CrossRef]
40. Uddin, R.; Saha, M.R.; Subhan, N.; Hossain, H.; Jahan, I.A.; Akter, R.; Alam, A. HPLC-Analysis of Polyphenolic Compounds in Gardenia Jasminoides and Determination of Antioxidant Activity by Using Free Radical Scavenging Assays. *Adv. Pharm. Bull.* **2014**, *4*, 273–281. [CrossRef]
41. Kuntić, V.; Pejić, N.; Mićić, S. Direct Spectrophotometric Determination of Hesperidin in Pharmaceutical Preparations. *Acta Chim. Slov.* **2012**, *59*, 436–441.
42. Miralles, P.; Chisvert, A.; Salvador, A. Determination of Hydroxytyrosol and Tyrosol by Liquid Chromatography for the Quality Control of Cosmetic Products Based on Olive Extracts. *J. Pharm. Biomed. Anal.* **2015**, *102*, 157–161. [CrossRef]
43. Lee, J.; Rennaker, C.; Wrolstad, R.E. Correlation of Two Anthocyanin Quantification Methods: HPLC and Spectrophotometric Methods. *Food Chem.* **2008**, *110*, 782–786. [CrossRef]
44. Pauliuc, D.; Dranca, F.; Oroian, M. Antioxidant Activity, Total Phenolic Content, Individual Phenolics and Physicochemical Parameters Suitability for Romanian Honey Authentication. *Foods* **2020**, *9*, 306. [CrossRef]
45. Kotsou, K.; Stoikou, M.; Athanasiadis, V.; Chatzimitakos, T.; Mantiniotou, M.; Sfougaris, A.I.; Lalas, S.I. Enhancing Antioxidant Properties of *Prunus Spinosa* Fruit Extracts via Extraction Optimization. *Horticulturae* **2023**, *9*, 942. [CrossRef]
46. Pappas, V.M.; Lakka, A.; Palaiogiannis, D.; Athanasiadis, V.; Bozinou, E.; Ntourtoglou, G.; Makris, D.P.; Dourtoglou, V.G.; Lalas, S.I. Optimization of Pulsed Electric Field as Standalone "Green" Extraction Procedure for the Recovery of High Value-Added Compounds from Fresh Olive Leaves. *Antioxidants* **2021**, *10*, 1554. [CrossRef] [PubMed]
47. Athanasiadis, V.; Chatzimitakos, T.; Bozinou, E.; Kotsou, K.; Palaiogiannis, D.; Lalas, S.I. Maximizing the Extraction of Bioactive Compounds from *Diospyros kaki* Peel through the Use of a Pulsed Electric Field and Ultrasound Extraction. *Biomass* **2023**, *3*, 422–440. [CrossRef]
48. Wardy, W.; Saalia, F.K.; Steiner-Asiedu, M.; Budu, A.S.; Sefa-Dedeh, S. A Comparison of Some Physical, Chemical and Sensory Attributes of Three Pineapple (*Ananas comosus*) Varieties Grown in Ghana. *Afr. J. Food Sci.* **2009**, *3*, 22–25.
49. Amaya-Cruz, D.; Peréz-Ramírez, I.F.; Pérez-Jiménez, J.; Nava, G.M.; Reynoso-Camacho, R. Comparison of the Bioactive Potential of Roselle (*Hibiscus sabdariffa* L.) Calyx and Its by-Product: Phenolic Characterization by UPLC-QTOF MSE and Their Anti-Obesity Effect in Vivo. *Food Res. Int.* **2019**, *126*, 108589. [CrossRef]
50. Singh, M.; Thrimawithana, T.; Shukla, R.; Adhikari, B. Extraction and Characterization of Polyphenolic Compounds and Potassium Hydroxycitrate from *Hibiscus sabdariffa*. *Future Foods* **2021**, *4*, 100087. [CrossRef]
51. dos Santos Martins, D.R.; Lescano, C.H.; Justo, A.F.O.; Vicente, J.M.; Santos, S.H.S.; Aguilar, C.M.; Borges, A.; Pires de Oliveira, I.; Sanjinez-Argandoña, E.J. Effect of Different Extraction Methods on Anthocyanin Content in *Hibiscus sabdariffa* L. and Their Antiplatelet and Vasorelaxant Properties. *Plant Foods Hum. Nutr.* **2023**, *78*, 342–350. [CrossRef]
52. Amer, S.A.; Al-Khalaifah, H.S.; Gouda, A.; Osman, A.; Goda, N.I.A.; Mohammed, H.A.; Darwish, M.I.M.; Hassan, A.M.; Mohamed, S.K.A. Potential Effects of Anthocyanin-Rich Roselle (*Hibiscus sabdariffa* L.) Extract on the Growth, Intestinal Histomorphology, Blood Biochemical Parameters, and the Immune Status of Broiler Chickens. *Antioxidants* **2022**, *11*, 544. [CrossRef]
53. Li, S.; Li, W.; Wang, C.; Wu, R.; Yin, R.; Kuo, H.-C.; Wang, L.; Kong, A.-N. Pelargonidin Reduces the TPA Induced Transformation of Mouse Epidermal Cells -Potential Involvement of Nrf2 Promoter Demethylation. *Chem. Biol. Interact.* **2019**, *309*, 108701. [CrossRef]
54. Karasawa, K.; Otani, H. Anti-Allergic Properties of a Matured Fruit Extract of the Date Palm Tree (*Phoenix dactylifera* L.) in Mite-Sensitized Mice. *J. Nutr. Sci. Vitaminol.* **2012**, *58*, 272–277. [CrossRef] [PubMed]
55. Wang, H.; Cao, G.; Prior, R.L. Oxygen Radical Absorbing Capacity of Anthocyanins. *J. Agric. Food Chem.* **1997**, *45*, 304–309. [CrossRef]
56. Hapsari, B.W.; Manikharda; Setyaningsih, W. Methodologies in the Analysis of Phenolic Compounds in Roselle (*Hibiscus sabdariffa* L.): Composition, Biological Activity, and Beneficial Effects on Human Health. *Horticulturae* **2021**, *7*, 35. [CrossRef]

57. Wang, J.; Cao, X.; Jiang, H.; Qi, Y.; Chin, K.L.; Yue, Y. Antioxidant Activity of Leaf Extracts from Different *Hibiscus sabdariffa* Accessions and Simultaneous Determination Five Major Antioxidant Compounds by LC-Q-TOF-MS. *Molecules* **2014**, *19*, 21226–21238. [CrossRef] [PubMed]
58. Santana-Gálvez, J.; Cisneros-Zevallos, L.; Jacobo-Velázquez, D.A. Chlorogenic Acid: Recent Advances on Its Dual Role as a Food Additive and a Nutraceutical against Metabolic Syndrome. *Molecules* **2017**, *22*, 358. [CrossRef]
59. Haider, K.; Haider, M.R.; Neha, K.; Yar, M.S. Free Radical Scavengers: An Overview on Heterocyclic Advances and Medicinal Prospects. *Eur. J. Med. Chem.* **2020**, *204*, 112607. [CrossRef]
60. Li, X.; Huang, W.; Tan, R.; Xu, C.; Chen, X.; Li, S.; Liu, Y.; Qiu, H.; Cao, H.; Cheng, Q. The Benefits of Hesperidin in Central Nervous System Disorders, Based on the Neuroprotective Effect. *Biomed. Pharmacother.* **2023**, *159*, 114222. [CrossRef]
61. Kumar, S.S.; Manoj, P.; Shetty, N.P.; Giridhar, P. Effect of Different Drying Methods on Chlorophyll, Ascorbic Acid and Antioxidant Compounds Retention of Leaves of *Hibiscus sabdariffa* L. *J. Sci. Food Agric.* **2015**, *95*, 1812–1820. [CrossRef]
62. Cooperstone, J.L.; Schwartz, S.J. Recent Insights Into Health Benefits of Carotenoids. In *Handbook on Natural Pigments in Food and Beverages*; Carle, R., Schweiggert, R.M., Eds.; Woodhead Publishing Series in Food Science, Technology and Nutrition; Woodhead Publishing: Cambridge, UK, 2016; pp. 473–497. ISBN 978-0-08-100371-8.
63. Black, H.S.; Boehm, F.; Edge, R.; Truscott, T.G. The Benefits and Risks of Certain Dietary Carotenoids That Exhibit Both Anti- and Pro-Oxidative Mechanisms-A Comprehensive Review. *Antioxidants* **2020**, *9*, 264. [CrossRef]
64. Doseděl, M.; Jirkovský, E.; Macáková, K.; Krčmová, L.K.; Javorská, L.; Pourová, J.; Mercolini, L.; Remião, F.; Nováková, L.; Mladěnka, P. Vitamin c—Sources, Physiological Role, Kinetics, Deficiency, Use, Toxicity, and Determination. *Nutrients* **2021**, *13*, 615. [CrossRef]
65. Yin, X.; Chen, K.; Cheng, H.; Chen, X.; Feng, S.; Song, Y.; Liang, L. Chemical Stability of Ascorbic Acid Integrated into Commercial Products: A Review on Bioactivity and Delivery Technology. *Antioxidants* **2022**, *11*, 153. [CrossRef]
66. Maganha, E.G.; da Costa Halmenschlager, R.; Rosa, R.M.; Henriques, J.A.P.; de Paula Ramos, A.L.L.; Saffi, J. Pharmacological Evidences for the Extracts and Secondary Metabolites from Plants of the Genus Hibiscus. *Food Chem.* **2010**, *118*, 1–10. [CrossRef]
67. Enedina, R.; Luisa, M.; Iván, E.; Antonio, J.; Teresa, M. Free Radical Scavenging Properties and Their Relationship with Bioactive Compounds Content of Dehydrated Calyces of Roselle (*Hibiscus sabdariffa* L.). *Afr. J. Agric. Res.* **2015**, *10*, 1203–1210. [CrossRef]
68. Wong, P.-K.; Yusof, S.; Ghazali, H.M.; Che Man, Y.B. Physico-chemical Characteristics of Roselle (*Hibiscus sabdariffa* L.). *Nutr. Food Sci.* **2002**, *32*, 68–73. [CrossRef]

processes

Article

Comparing Quality and Functional Properties of Protein Isolates from Soybean Cakes: Effect of De-Oiling Technologies

Giulia Cestonaro [1,†], Rodrigo Gonzalez-Ortega [2,†], Antonella L. Grosso [2], Ksenia Morozova [2], Giovanna Ferrentino [2,*], Matteo Scampicchio [2] and Enrico Costanzo [1]

1 Cereal Docks Ss.p.A., Dipartimento Ricerca & Innovazione, via Innovazione 1, 36043 Camisano Vicentino, Italy; giuliacestonaro@cerealdocks.it (G.C.); enricocostanzo@cerealdocks.it (E.C.)
2 Faculty of Agriculture, Environmental and Food Sciences, Free University of Bozen-Bolzano, Piazza Università 1, 39100 Bolzano, Italy; rgortega@ubu.es (R.G.-O.); antonellaluciana.grosso@unibz.it (A.L.G.); ksenia.morozova@unibz.it (K.M.); matteo.scampicchio@unibz.it (M.S.)
* Correspondence: giovanna.ferrentino@unibz.it
† These authors contributed equally to this work.

Abstract: Driven by growing concerns about food supply and the environment, research on alternative protein sources has become increasingly important. In this context, de-oiled seed cakes, particularly soybean cakes, have emerged as a promising option. However, the conventional methods, such as organic solvent extraction, from which these cakes are obtained present several limitations. This study aims to evaluate the efficiency of supercritical fluid extraction (SFE) as an alternative method for de-oiling soybean seeds and obtaining related protein isolates. By using SFE for de-oiling, it was possible to achieve 19% more protein isolates from soybean cakes than the conventional de-oiling method using hexane. Moreover, protein isolates from the SFE de-oiled cake reported significantly improved ($p < 0.05$) emulsifying abilities and water absorption capacity. Gel electrophoresis and differential scanning calorimetry indicated the presence of a higher concentration of proteins in their native state in the SFE de-oiled flour. Finally, results from the sulfhydryl group content, surface hydrophobicity, and protein dispersibility index also supported these conclusions. The SFE process produced de-oiled soybean cakes with superior functional characteristics and lower environmental impact. Thus, this study provided important information for the food industry to develop more sustainable and healthier production methods.

Keywords: solvent extraction; green extraction; alternative protein sources; oilseed co-products; functional properties

Citation: Cestonaro, G.; Gonzalez-Ortega, R.; Grosso, A.L.; Morozova, K.; Ferrentino, G.; Scampicchio, M.; Costanzo, E. Comparing Quality and Functional Properties of Protein Isolates from Soybean Cakes: Effect of De-Oiling Technologies. *Processes* **2024**, *12*, 600. https://doi.org/10.3390/pr12030600

Academic Editors: Prashant K. Sarswat and Yanlin Zhang

Received: 24 February 2024
Revised: 14 March 2024
Accepted: 15 March 2024
Published: 17 March 2024

1. Introduction

The growing demand for environmental sustainability has led research and industry to a paradigm shift from animal to plant-based proteins. Given the need to adopt sustainable and environmentally friendly technologies, there is a growing interest in extraction technologies applied to vegetable seeds. Renowned as a rich source of plant-based protein, oil extraction from vegetable seeds leads to interesting co-products, commonly called "cakes." These defatted cakes are characterized by high protein content and significant nutritional value and represent an attractive and low-cost alternative to animal-based proteins [1–4].

In the production of de-oiled cakes, the extraction of oil from vegetable seeds plays a critical role. Conventional methods of oil extraction, such as mechanical pressing and solvent extraction with hexane, are prevalent but fraught with inefficiencies and environmental concerns. For instance, mechanical pressing is notably inefficient for soybean seeds, which are characterized by low oil content. Conversely, solvent extraction, despite its effectiveness in oil removal, often leads to protein denaturation, thereby affecting the functional quality of the resultant cakes [5–9]. This study investigates the potential of supercritical fluid extraction (SFE) as an alternative to conventional methods to de-oil soybean cake. SFE exploits the unique properties of CO_2 in its supercritical state, exhibiting both liquid and gaseous properties. This condition facilitates the penetration of CO_2 into the vegetable cake and the efficient extraction of oil. Using CO_2 as a solvent, SFE offers several advantages: its non-toxic nature ensures environmental compatibility, and the moderate temperatures involved in the process preserve the integrity and functionality of the obtained extracts [10–12]. Studies demonstrated the feasibility and effectiveness of using SFE for oil extraction from seeds [13–15]. The potential functional properties associated with oils extracted by SFE include high nutritional value, high antioxidant activity, and bioactive compounds. The process allows to preservation of the natural antioxidants present in the oils, such as phenolic compounds and tocopherols, which can help neutralize harmful free radicals in the body. Moreover, oils obtained through SFE contain bioactive compounds, such as sterols, phytosterols, and polyphenols, which may offer various health benefits, such as anti-inflammatory, antimicrobial, and anti-cancer properties [16]. Several studies also showed the potential of SFE for the recovery of oils from soybean seeds [17]. However, just a few investigations were carried out on the defatted residue obtained from SFE. Recently, a research study proved that the defatted soybean cake presented interesting physicochemical and functional properties, suggesting the potential of using the powder as an alternative ingredient to a costly soy protein isolate commonly used in the food industry [18].

To the best of the authors' knowledge, no studies have investigated and compared the physicochemical and functional properties of both the defatted soybean cake and the protein isolate obtained from the cake by de-oiling it using two different technologies, namely SFE and conventional solvent extraction.

The hypothesis of this study is that SFE, given its moderate processing conditions, is more efficient and environmentally sustainable than conventional extraction methods. Thus, this study aims to comprehensively compare the functional properties of proteins derived from de-oiled cakes obtained using conventional solvent extraction and SFE. The proximate composition of the de-oiled cakes obtained via both hexane extraction and SFE was determined, focusing on the extraction yield of protein isolates. Differential scanning calorimetry was applied to assess the thermal stability and denaturation behavior of both protein isolates. Moreover, the study delved into the functional properties of protein flours and isolates, including their foaming and emulsifying abilities and their water and oil-holding capacities. A subsequent correlation analysis explored the interrelationships between these functional properties.

Through this comprehensive approach, the study aimed to enhance the understanding of SFE as an efficient and sustainable method for producing high-quality de-oiled soybean cakes.

2. Materials and Methods

2.1. Materials

Soybean seeds (not GMO and harvested in Europe) were kindly provided by Cereal Docks S.p.A. (Camisano Vicentino, Italy). All chemicals used were of analytical grade and were purchased from Sigma-Aldrich (Milan, Italy). Distilled water (Milli-Q, Merck Millipore, Darmstadt, Germany) was used to prepare all solutions and dispersions.

2.2. Samples Preparation

Soybean seeds were ground with a hammer mill (Laboratory Mill 3100, Perten Instruments, with 800 μm mesh size), resulting in fine flour with the following particle size measured with an orbital shaker composed of four trays with mesh size ranging from 1000 to 100 micron (Retsch GmbH, Verder Scientific, Germany): 14.8 ± 1.5% between 250 and 500 μm, 31.0 ± 2.3% between 100 and 250 μm and 53.5 ± 3.2% lower than 100 μm. The proximate composition of soybean seeds was 9.2% moisture, 37.4% protein, 21.6% crude fiber, 19.7% lipids, and 4.6% ashes.

2.3. Soybean De-Oiling Methods

2.3.1. Solvent Extraction

De-oiling of soybean seeds was performed using an industrial plant at Cereal Docks S.p.A. (Camisano Vicentino, Vicenza, Italy). Initially, the seeds underwent a softening step. To this purpose, the seeds were passed through a horizontal softening pot at 70 °C for a period of up to 30 min. In a second step, softened seeds were subjected to flaking using an extruder to destroy the cell walls and increase the surface area, making the seeds more porous. The material was then extracted using hexane in a continuous countercurrent extractor at 68 °C. After extraction, the meal was desolventized using steam at 120 °C. The process was conducted in a toaster, with the steam flow running countercurrent to the meal flow to improve the efficiency of hexane desorption. Finally, the meal was cooled to ambient temperature and stored.

2.3.2. Supercritical Fluid Extraction

Ground soybean seeds were de-oiled using a high-pressure pilot plant (Superfluidi s.r.l., Padova, Italy) equipped with a 1 L capacity extractor vessel and two gravimetric separators. The high-pressure vessel included an 800-mL extraction cylinder sealed with porous stainless steel mesh filters on both ends, allowing carbon dioxide to pass through while retaining solids inside the cylinder. Approximately 120 g of soybean flour was loaded into the extraction cylinder for each batch. The solvent, liquid carbon dioxide (CO_2), was supplied from a storage vessel pre-cooled to 4.0 °C using a cooling heat exchanger. It was then pumped into the extraction chamber using a high-pressure diaphragm pump (Lewa LDC–M–9XXV1, Milan, Italy) at a flow rate of 2 L/h. The extraction conditions were set to 40 °C and 30 MPa for 4 h, as reported in previous studies [19].

2.4. Soybean Protein Isolates

Soybean protein isolates were prepared by dispersing 30 g of defatted soybean flour in 750 mL of distilled water. The mixture was then transferred into a 1 L jacketed glass reactor maintained at a constant temperature of 60 °C. After temperature stabilization, the mixture was stirred for 30 min at 500 rpm using an overhead stirrer (IKA Eurostar 20, IKA, Staufen, Germany).

Subsequently, an aliquot of this mixture was then centrifuged at 12,000× *g* for 15 min at 20 °C (Thermo Scientific SL 16 R Centrifuge, Waltham, MA, USA). The supernatant was then filtered through Whatman filter paper grade 1 and concentrated to obtain the molasse. The pellet collected on the filter paper was resuspended in distilled water at a 1:25 ratio, and the pH was readjusted to 9.0 using 0.1 M NaOH solution. A second extraction cycle was repeated under the same conditions, and the two molasses were combined. The second collected pellet was again resuspended in distilled water to yield a solubilized protein mixture. The pH of this mixture was then adjusted to pH = 4.5 by adding HCl solution (0.5 M). The solution was then cooled to 10 °C by recirculating cold water in a jacked vessel and allowed to stand for 10 min. The precipitated material was collected by centrifugation at 8000× *g* for 15 min, resuspended, and homogenized in distilled water using an UltraTurrax (T25, IKA). Finally, the resulting protein suspension was freeze-dried (Epsilon 2-6D LSCplus, Martin Christ, Osterode am Harz, Germany). The process was performed by freezing the samples at −40 °C for 4 h and subsequently setting a primary drying for 24 h at a temperature of 5 °C and a vacuum of 1.01 mbar, followed by a secondary drying phase for 24 h at a temperature of 15 °C and a vacuum of 0.05 mbar. The dried protein isolates were then pulverized using a pestle and mortar.

2.5. Proximate Analysis of De-Oiled Soybean Flours

The proximate composition of the defatted soybean flours, including moisture, ash, fat, crude fiber, and protein content, was determined using standard methods as outlined in [20]. For moisture content, 5 g of the sample was heated at 103 °C for 4 h in a ventilated oven. The sample was then allowed to cool to ambient temperature, stored for 45 min in a dryer, and subsequently weighed.

Ash content was determined by weighing 5 g of the sample in a crucible and heating it to 550 °C for at least 6 h in a muffle furnace. Then, the crucible was cooled down in a dryer and weighed.

For crude fiber analysis, 1 g of the sample was treated with boiling sulfuric acid solution to remove sugars and starches. The mixture was filtered and treated with a potassium hydroxide solution. The residue from this mixture was obtained by filtration using sintered glass. It was then washed with water and acetone, dried, weighed, and incinerated at 550 °C for 3 h. The weight loss from incineration indicated the crude cellulose content.

Crude protein content was determined using Kjeldahl's method [20]. The method indirectly quantified the amount of protein by the total nitrogen content. To obtain the crude protein content, the nitrogen content was multiplied by the factor 6.25 on a dry basis.

2.6. Protein Profiling via Sodium Dodecyl Sulfate-Polyacrylamide Gel Electrophoresis

Protein profiling of the defatted flours and their corresponding protein isolates was accomplished using sodium dodecyl sulfate-polyacrylamide gel electrophoresis (SDS-PAGE). To initiate this procedure, defatted flours were dehydrated for 4 h in 0.05 M phosphate buffer at pH 7.0, achieving a protein concentration of 10 mg/mL. Subsequently, the rehydrated flours were homogenized using an Ultraturrax device at 12,000 rpm for 2 min. The resulting slurry was designated as the total protein fraction. The remaining supernatant, after centrifugation at 12,000× *g* for 15 min at 20 °C, was designed as the soluble protein fraction and rehydrated in 0.05 M phosphate buffer at pH 7.0, resulting in an approximate protein concentration of 4 mg/mL. The total protein fraction and soluble protein isolates were combined in a 1:1 volume ratio. These combined fractions were mixed with Laemmli buffer consisting of 65.8 mM Tris-HCl (pH 6.8), 26.3% glycerol, 2.1% sodium dodecyl sulfate, and 0.01% bromophenol blue. In addition, 100 mM dithiothreitol (DTT) was included in the buffer for reducing conditions, whereas DTT was omitted for non-reducing conditions. Subsequently, the mixtures were heated at 95 °C for 5 min with agitation at 300 rpm using an Eppendorf thermomixer. The samples were then allowed to cool and then subjected to centrifugation at 2000× *g* for 5 min. Ten microliters of the resulting supernatants were loaded into pre-cast 4–20% Mini-PROTEAN® TGX gels (Bio-Rad Laboratories, Hercules, CA, USA), which were positioned within a Miniprotean® TetraCell electrophoresis cell. The cell was filled with running buffer (10× Tris/Glycine/SDS, Bio-Rad Laboratories) and connected to a PowerPac™ basic power supply (Bio-Rad Laboratories, Hercules, CA, USA). Molecular weight standards, specifically Precision Plus Protein™ Kaleidoscope™ prestained protein standards (Bio-Rad Laboratories), were employed, covering the range of 10–250 kDa. Electrophoresis was performed at 150 V for a duration of 1 h. Following electrophoresis, the gels were subjected to fixation using a solution consisting of 40% MeOH and 10% acetic acid for 15 min. Subsequently, they were stained with a staining solution comprising 0.1% Coomassie Brilliant Blue G-250 in 20% MeOH and 10% acetic acid for 30 min. After staining, the gels underwent destaining through multiple cycles employing the same fixing solution until a clear background was observed in the gels. Finally, the gels were scanned using a Gel Doc EQ system.

2.7. Differential Scanning Calorimetry

For DSC analysis, the defatted flour (150 mg/mL) and soy protein isolates (100 mg/mL) were dispersed in phosphate buffer (50 mM, pH 7.0). The dispersions were allowed to hydrate for 4 h and then homogenized using Ultraturrax operating at 12,000 rpm for 2 min. A differential scanning calorimeter (DSC 250, TA Instruments, New Castle, DE, USA) equipped with a refrigerated cooling system (Refrigerated Cooling System 90, TA Instruments) and operated with TRIOS 5.2 software (TA Instruments) was used. High-purity nitrogen gas served as the purging gas, flowing at a 50 mL/min rate throughout the analysis. The DSC instrument was calibrated using indium (melting temperature of 156.6 °C). For the analysis, approximately 10 mg of each sample was accurately weighed and placed into aluminum pans (Tzero cells), which were then hermetically sealed. The samples were initially equilibrated at 35 °C for 5 min. Subsequently, a heating ramp was applied, increasing the temperature from 30 to 125 °C at a rate of 10 °C/min. An empty pan was used as a reference. The resulting thermograms were analyzed using the TRIOS software (TA Instruments).

2.8. Surface Hydrophobicity

Soy protein isolates and defatted flour samples were prepared at 30 and 70 mg/mL concentrations to assess surface hydrophobicity. These samples were first dispersed in distilled water and then hydrated for 4 h. Following hydration, the flour samples were homogenized for 10 min at 8500 rpm using Ultraturrax. Centrifugation was then performed at 800× *g* for 10 min to separate the protein precipitate from the isolates and flour samples. The protein concentration in the isolates and flour supernatants was adjusted to 0.5–0.005 mg protein/mL. This was achieved through serial dilutions employing a 50 mM phosphate buffer (pH 7.0).

The surface hydrophobicity was determined using a fluorescence assay [21]. Briefly, 2 mL of each diluted sample was mixed with 10 µL of an 8 mM solution of 1-anilino-8-naphthalene-sulfonate (ANS), which was used as a fluorescence probe. This mixture was then incubated at 20 °C in a dark environment. Fluorescence intensity was quantified using an Infinite 200 PRO multimode plate reader (Tecan Trading AG, Männedorf, Switzerland). Excitation and emission wavelengths of 380 and 490 nm, respectively, were employed for the measurements. The H0 index was determined by calculating the slope of the fluorescence intensity versus protein concentration plot. Each sample was subjected to triplicate measurements for robustness and accuracy.

2.9. Free Sulfhydryl Groups

Using a spectrophotometric assay, free sulfhydryl groups were quantified using Ellman's reagent, 5,5-dithio-bis-2-nitrobenzoic acid (DTNB). Both the flour supernatant and protein isolates were prepared following the procedure described for the surface hydrophobicity measurement. However, the protein concentration was adjusted to 2 mg/mL using 50 mM Tris-HCl buffer with a pH of 7.0. A volume of 1 mL of the sample was then mixed with 20 µL of a 10 mM DTNB solution prepared in methanol. This mixture was then kept in the dark for 15 min. Subsequently, centrifugation at 10,000× *g* for 3 min was performed, and the absorbance was recorded at 412 nm using a spectrophotometer (Agilent Technologies, Milan, Italy). The concentration of sulfhydryl groups was calculated using a molar extinction coefficient of 13,600 $M^{-1}cm^{-1}$ for DTNB based on the following equation:

$$\mu mol\ SH\ groups/g = \frac{\left(\frac{ln\frac{I_z}{I_0}}{-\varepsilon x \lambda}\right) \times D \times V}{m} \tag{1}$$

where I_z/I_0 is the transmittance, ε is the molar extinction coefficient, λ is the path length, D is the dilution factor, V is the cell volume, and m is the protein mass in grams. Blank samples were measured and subtracted from the absorbance values to calculate net absorbance.

2.10. Pasting Properties of Defatted Flours and Isolates

To evaluate the pasting properties, a Discovery Hybrid Rheometer HR-2, equipped with a starch pasting cell and a Peltier Concentric Cylinder (TA-Instruments, Milan, Italy), featuring an impeller rotor (bob diameter: 32.40 mm; bob length: 12 mm) was employed. Temperature control was achieved using a recirculating chiller (LT ecocooler 150, Grant Instruments, Cambridge, UK). Defatted soy flours were initially dispersed in distilled water (15 wt.%), stirred for 1 h, and then hydrated for 16 h at 4 °C. The pH was then adjusted to 7.0, and the resulting slurries were transferred to a stainless-steel cup (36 mm diameter) for analysis. The gap between the rotor and cup was 5.5 mm. The experimental procedure consisted of several steps: an initial isothermal holding at 50 °C for 5 min, a subsequent heating ramp to 95 °C at a rate of 2 °C/min, and a final isothermal step for 5 min at 95 °C. A cooling phase was then executed, bringing the temperature down to 50 °C at a rate of 2 °C/min. An isothermal step was then performed for 5 min at 50 °C. The viscosity of the samples was measured throughout both the heating and cooling phases. A fixed shear rate of 15 rad/s was applied.

2.11. Protein Dispersibility Index

The protein dispersibility index (PDI) was determined, as reported by other authors [22]. In summary, the samples of soy protein isolates (30 mg/mL) and defatted flour (70 mg/mL) were dispersed in distilled water and allowed to hydrate for 4 h. The samples were homogenized for 10 min at 8500 rpm using Ultraturrax. After homogenization, both flour and isolate samples were centrifuged at 20 °C, 800× *g* for 10 min, and the resulting supernatants were collected.

The total nitrogen content was measured using the Kjeldahl method to assess the crude protein content. The protein dispersibility index (PDI) was calculated as the percentage of soluble protein relative to the initial protein content in the defatted flour or protein isolate according to the following equation:

$$\text{PDI}(\%) = \frac{\textit{Dispersible protein (supernatant)}}{\text{Total protein in the sample(defatted flour or isolate)}} \times 100 \quad (2)$$

2.12. Foaming Capacity and Stability

The measurement of foaming capacity and stability was performed using a method adapted from a research study [23]. Approximately 2.5 g of samples were suspended in 50 mL distilled water and homogenized using an Ultraturrax operating at 10,000 rpm for 3 min at room temperature. The samples were then transferred to a graduated cylinder, and the total volume was recorded immediately. Changes in foam volume within the graduated cylinder were recorded at 0, 15, and 60 min after homogenization. Foaming capacity (FC) and stability (FS) were calculated using the following equations:

$$\text{FC}(\%) = \frac{V_2 - V_1}{V_1} \times 100 \quad (3)$$

$$\text{FS}(\%) = \frac{V_3}{V_2} \times 100 \quad (4)$$

where V_1 is the volume of the suspension (mL) before stirring, V_2 is the volume of the suspension (mL) after stirring, and V_3 is the volume of the foam (mL) after it was allowed to stand.

2.13. Emulsifying Ability and Stability

The emulsifying ability and stability were determined by adapting a method previously described [23,24]. In brief, 5 g solution containing 0.5% (*w*/*w*) protein and 5 g of sunflower oil was homogenized at 7000 rpm for 30 s. Subsequently, 50 µL of the solution was extracted from the bottom of the samples at intervals of 0, 10, and 30 min. The extracted samples were then diluted with 7.5 mL of sodium phosphate buffer (10 mM, pH 7.0) containing 0.1% sodium dodecyl sulfonate (SDS) and mixed for 10 s. The absorbance of the mixed samples was measured by spectrophotometry at 500 nm in plastic cuvettes (1 cm path length) using 0.1% SDS solution as a blank reference. The emulsifying ability (EA) and stability (ES) were calculated at time intervals of 0, 10, and 30 min using the equations [25]:

$$\text{EA}(\text{m}^2 \times \text{g}^{-1}) = \frac{2 \times 2.303 \times \text{A}_0 \times D}{\text{C} \times \varphi \times 10{,}000} \quad (5)$$

$$ES(\%) = \frac{A_t}{A_0} \times 100 \quad (6)$$

where C is the protein concentration (g/mL) before emulsification, φ is the oil volume fraction (v/v) of the emulsion, D is the dilution factor, A_0 is the initial absorbance, and A_t the absorbance at time t.

2.14. Water and Oil Holding Capacity

To measure water (WHC) and oil holding capacity (OHC), 100 mg of soybean protein sample and 200 mg of defatted soybean flour were thoroughly mixed in 4 g of water (for water holding capacity) and 4 g of soybean oil (for oil holding capacity). The mixtures were homogenized at 7000 rpm for 30 s to ensure uniform dispersion. For WHC measurements, the homogenized samples were centrifuged at 18,000× g for 15 min at room temperature. The weight of the resulting wet pellet was recorded. The pellets were then dried for 24 h at 40 °C and weighed. *WHC* was calculated using the following equation:

$$WHC(g/g) = \frac{W_{wet\ pellet} - W_{dry\ pellet}}{W_{powder}} \quad (7)$$

where $W_{wet\ pellet}$ and $W_{dry\ pellet}$ are the weights of the wet and dry pellets, respectively, and W_{powder} is the weight of the initial sample.

For the *OHC* measurements, homogenized samples were centrifuged at 4800× g for 15 min at room temperature. The weight of the resulting oily pellet was recorded, and the OHC was calculated as follows:

$$OHC(g/g) = \frac{W_{oily\ pellet} - W_{dry\ pellet}}{W_{powder}} \quad (8)$$

where $W_{oily\ pellet}$ and $W_{dry\ pellet}$ are the weights of the oily pellet and dry pellets, respectively, and W_{powder} is the weight of the initial sample.

2.15. Statistical Analysis

The experimental measurements were conducted in a completely randomized design (CRD) with at least three replicates for each test. Data were expressed as mean ± standard deviation, calculated from the replicate measurements. Differences between mean values were assessed using one-way ANOVA followed by Tukey's multiple-range test. The significance of differences was defined at the $p < 0.05$ level.

3. Results and Discussion

3.1. Proximate Composition of De-Oiled Flours and Protein Isolates

Figure 1 shows the processing steps for producing soybean isolates using hexane and SFE de-oiled methods. Starting from soybean seeds, both methods produced de-oiled flours with comparable protein content on a dry basis ($p > 0.05$). Table 1 presents the proximate composition of soybean flours and isolates obtained by the two defatting methods. Results indicate that solvent extraction did not significantly affect the flour's protein content. These findings are consistent with existing literature, which reported negligible differences in protein content between these two defatting methods, suggesting that both extraction techniques did not compromise the protein content of soybean flour [8,9]. The other components quantified in the de-oiled flours and the protein isolates were also not significantly different.

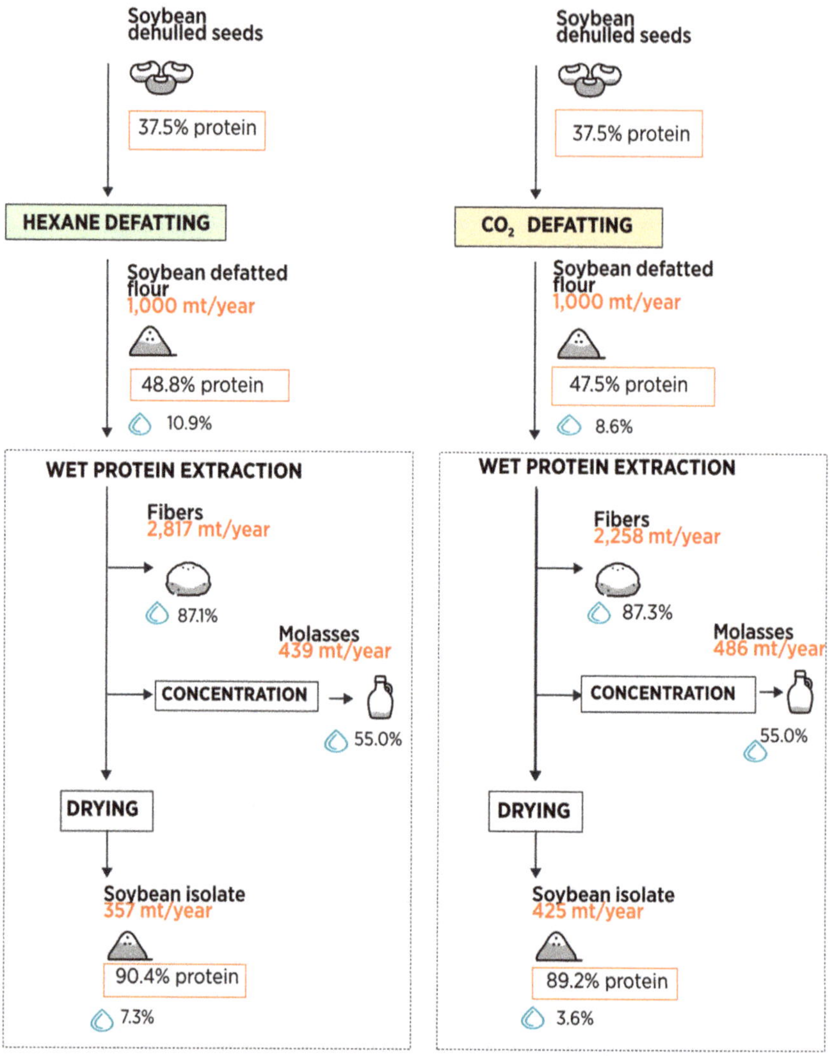

Figure 1. Mass balance of soybean protein isolates obtained using hexane and SFE de-oiled methods.

Table 1. Proximate composition of defatted soybean flours.

Composition (%)	Hexane De-Oiled	SFE De-Oiled	Hexane Isolate	Sfe Isolate
Moisture	10.8 ± 1.4 [a]	8.6 ± 1.2 [a]	7.3 ± 0.6 [a]	3.5 ± 1.2 [a]
Protein	48.8 ± 2.1 [a]	47.5 ± 1.2 [a]	83.7 ± 0.7 [a]	86.0 ± 0.4 [b]
Lipids	1.0 ± 0.2 [a]	0.5 ± 0.4 [a]	0.3 ±0.1 [a]	0.2 ± 0.01 [a]
Carbohydrate	29.1 ± 2.2 [a]	31.7 ± 1.2 [a]	3.8 ± 0.6 [a]	5.3 ± 0.1 [b]
Crude fiber	4.2 ± 1.0 [a]	5.8 ± 0.8 [a]	0.6 ± 0.01 [a]	0.6 ± 0.1 [a]
Ash	6.1 ± 0.2 [a]	5.9 ± 0.2 [a]	4.3 ± 0.6 [a]	4.4 ± 0.5 [a]

Results are expressed as mean values ± standard deviations. Different letters within columns indicate significant differences ($p < 0.05$).

Regardless of the de-oiled method, the mass balance analysis revealed three economically viable streams after protein extraction: fibers, molasses (a liquid stream obtained after the protein extraction), and protein isolates. In particular, when using the flour de-oiled by SFE, the yield of protein isolates was 19% higher than that obtained from the hexane de-oiled flour. The results were linked to the higher percentage (86.0 ± 0.4%) of proteins solubilized using the SFE de-oiled flour. The values were significantly higher ($p < 0.05$) compared to that of the hexane de-oiled flour (83.7 ± 0.7%). The higher extraction efficiency of the SFE method could compensate for and bypass the drawbacks related to its higher capital expenditures (CapEx) and operating expenses (OpEx) compared with traditional solvent extraction. This suggested a promising potential for high-quality protein production within the food industry while balancing economic and environmental considerations.

3.2. Protein Profile

Electrophoretic analysis, as shown in Figure 2a,b, compares the protein isolates and flours obtained using both hexane and SFE de-oiling methods. The protein isolates from both methods mainly consisted of glycinin and β-conglycinin bands. Further analysis of the defatted flours revealed protein bands with higher intensity in the SFE de-oiled flour, consistent with the increased percentage of solubilized proteins. Notably, during the wet protein extraction phase, samples from both de-oiled flours showed differences in the intensities of some bands, as shown in lines 4 and 5 of Figure 2a. The soluble fraction obtained from the SFE-defatted flour showed more intense bands, particularly at 35 kDa and 20 kDa, which were indicative of the presence of glycinin. In addition, more intense signals were observed at 67, 71, and 50 kDa, corresponding to β-conglycinin. These distinctions in band intensity and protein composition underscore the effectiveness of SFE in protein isolation. The enhanced solubility and extractability of proteins from SFE de-oiled flour likely contributed to these observed differences, reinforcing the potential advantages of SFE in protein extraction and isolation processes.

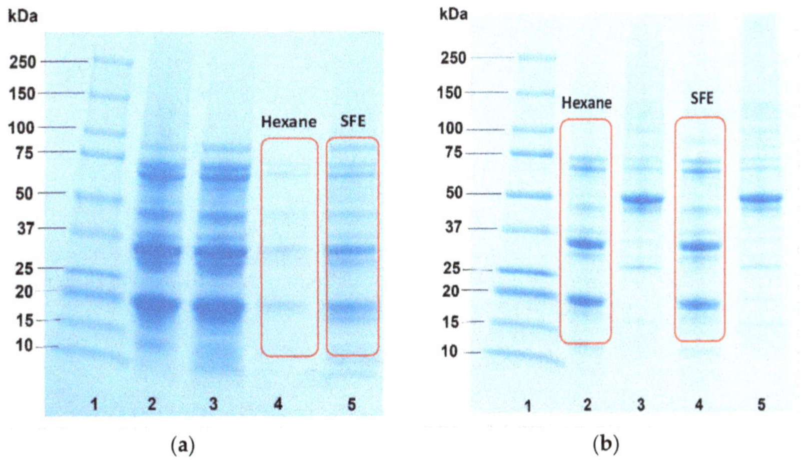

Figure 2. (**a**) SDS-PAGE of soy flours under reducing conditions. (1) Mw marker, (2) hexane de-oiled flour, (3) SFE de-oiled flour, (4) soluble fraction of hexane de-oiled flour, (5) soluble fraction of SFE de-oiled flour; (**b**) SDS-PAGE of soy protein isolates from hexane (2, 3) and SFE de-oiled flour (4, 5) under reducing (2, 3) and non-reducing conditions (3, 5).

3.3. Differential Scanning Calorimetry Analysis

Differential scanning calorimetry analysis was performed to assess the thermal stability and denaturation behavior of proteins in both de-oiled soybean flours and isolates. Figure 3 shows the thermogram obtained for the four samples, while Table 2 summarizes the key thermal parameters, including the onset and peak temperatures and the enthalpy change that quantifies protein denaturation. The results revealed significant differences for β-conglycinin enthalpy and the peak temperature measured for the flours de-oiled using SFE and hexane ($p < 0.05$).

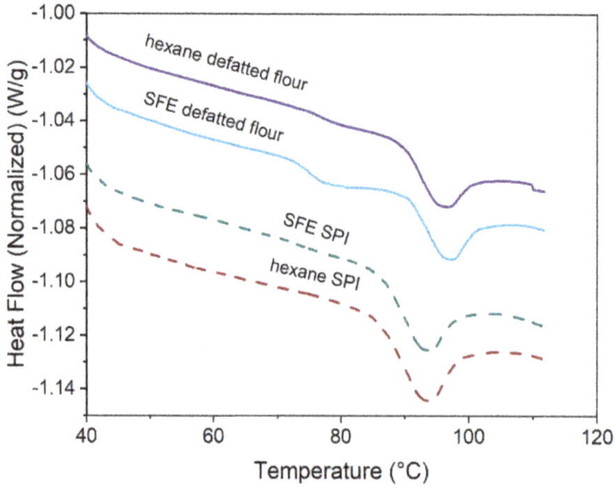

Figure 3. DSC thermograms of de-oiled flours and soybean protein isolates obtained from the same flours.

Table 2. Thermal properties of de-oiled soybean flours and protein isolates.

Sample	T_d Onset (°C)		T_d Peak (°C)		ΔH (J/g Protein)	
	β-Conglycinin (7S)	Glycinin (11S)	β-Conglycinin (7S)	Glycinin (11S)	β-Conglycinin (7S)	Glycinin (11S)
Hexane de-oiled cake	75.1 ±0.5 [a]	90.7 ±0.5 [a]	79.0 ±0.1 [b]	95.8 ±0.3 [b]	0.5 ±0.1 [a]	7.4± 1.3 [a]
SFE de-oiled cake	73.2 ±0.8 [a]	91.1 ±0.3 [a]	77.6 ±0.1 [a]	96.7 ±0.2 [b]	2.4 ±0.7 [b]	7.3 ±0.2 [a]
Hexane protein isolate	nd	86.7 ±0.3 [b]	nd	93.2 ±0.2 [a]	nd	9.9 ±2.0 [b]
SFE protein isolate	nd	89.1 ±2.4 [b]	nd	93.2 ±0.1 [a]	nd	8.8 ±2.0 [b]

nd = not detected. Results are expressed as mean values ± standard deviations. Different letters within rows indicate significant differences ($p < 0.05$).

Scanning the temperature from 40 to 110 °C, both SFE and hexane de-oiled flours exhibited two distinctive endothermic peaks at temperatures corresponding to the denaturation of β-conglycinin (78 °C) and glycinin (90–95 °C). However, a higher level of denaturation, as evidenced by the lower enthalpy values, was observed in the hexane de-oiled flour. In detail, the enthalpy of β-conglycinin denaturation of SFE de-oiled flour (2.4 ± 0.7 J/g protein) was significantly higher than that in hexane de-oiled flour (0.5 ±0.1 J/g protein). Similar significant differences were observed in glycinin denaturation, supporting the hypothesis that proteins in SFE de-oiled flours experienced less denaturation, thus preserving properties closer to their native states. This finding highlighted the SFE's ability to better retain the native state of proteins during the de-oiling.

Figure 3 also shows the thermograms of the protein isolates obtained from the two types of defatted flours. Only one thermal event was observed in these protein isolates, corresponding to glycinin denaturation. The absence of a second distinguishable thermal transition peak suggested the denaturation of a protein fraction during the wet protein extraction process or post-extraction treatment, where pH, ionic strength conditions, and temperature played major roles [26,27]. This observation also suggested that the drying conditions and temperature or pH during the wet extraction were important factors to consider for the production of protein isolates retaining their maximum native structure and functionality.

3.4. Sulfhydryl Group Content, Surface Hydrophobicity, and Protein Dispersibility Index

The protein dispersibility index (PDI) is a crucial indicator of the functional properties of food proteins, particularly reflecting their solubilization efficacy in aqueous solutions. Table 3 presents data indicating that SFE de-oiled soybean flour exhibited a PDI that was 3.5 times higher than that of hexane de-oiled flour. This significant difference suggested a higher concentration of native protein structures in SFE de-oiled flour, which had significant implications for protein functionality. As outlined in the existing literature, factors influencing PDI include flour composition, lipid removal method, thermal exposure, pH levels, ionic strength, and mixing speed [8,28]. Protein denaturation decreases PDI values. The high PDI values of SFE de-oiled soybean flour could enhance the water-holding capacity and broaden the applications of this defatted flour in diverse food formulations.

Table 3. Functional properties of de-oiled soy flours and protein isolates obtained from them.

	Units	Hexane De-Oiled Cake	SFE De-Oiled Cake	Hexane Protein Isolate	SFE Protein Isolate
Surface hydrophobicity	$\times 10^3$	134.6 ± 0.6 ^{c}	102.4 ± 0.7 ^{b}	84.7 ± 7.4 ^{b}	62.5 ± 5.7 ^{a}
Free sulfhydryl groups	µmol/g protein	9.3 ± 0.2 ^{d}	6.8 ± 0.2 ^{c}	5.8 ± 0.2 ^{b}	3.9 ± 0.1 ^{a}
Protein dispersibility index	%	19.1 ± 0.7 ^{a}	68.2 ± 0.6 ^{b}	91.5 ± 3.5 ^{c}	93.9 ± 0.9 ^{c}
Emulsifying ability	m^2/g protein	52.5 ± 3.5 ^{a}	107.5 ± 2.12 ^{d}	74.5 ± 0.7 ^{b}	90.1 ± 2.8 ^{c}
Emulsifying stability (at 15 min)	%	48.3 ± 2.8 ^{a}	87.4 ± 2.8 ^{c}	59.1 ± 1.4 ^{b}	64.2 ± 2.5 ^{b}
Emulsifying stability (at 60 min)	%	9.1 ± 1.5 ^{a}	71.2 ± 1.3 ^{c}	15.5 ± 2.1 ^{a}	51.1 ± 1.4 ^{b}
Foaming ability	%	110.0 ± 2.1 ^{a}	148.0 ± 2.8 ^{c}	136.0 ± 2.12 ^{b}	134.0 ± 0.7 ^{b}
Foaming stability (at 15 min)	%	92.9 ± 3.5 ^{b}	74.2 ± 2.8 ^{c}	68.6 ± 2.8 ^{a,b}	60.7 ± 0.7 ^{a}
Foaming stability (at 60 min)	%	62.6 ± 3.5 ^{a}	58.7 ± 2.8 ^{a}	52.5 ± 3.5 ^{a}	55.5 ± 0.7 ^{a}
Water Holding Capacity	g H_2O/g powder	1.75 ± 0.5 ^{b,c}	2.45 ± 0.2 ^{c}	0.55 ± 0.1 ^{a}	0.87 ± 0.1 ^{a,b}
Oil Holding Capacity	g oil/g powder	1.05 ± 0.2 ^{a}	1.55 ± 0.2 ^{a}	1.25 ± 0.1 ^{a}	1.38 ± 0.1 ^{a}

Results are expressed as mean values ± standard deviations. Different letters within columns indicate significant differences ($p < 0.05$).

Protein isolates derived from both de-oiled methods showed high PDI values, consistent with SDS-PAGE and DSC results. Similar correlations between PDI values, DSC, and SDS-PAGE results were obtained on peanut protein isolates [29,30]. This consistency underscored the minimal presence of denatured protein fractions, indicating the potential of these isolates to improve water holding capacity, extending their applications in various food products.

Table 3 also compares the surface hydrophobicity and concentration of free sulfhydryl groups in de-oiled flours and their corresponding protein isolates. Notably, the hexane de-oiled flour exhibited high values of surface hydrophobicity and sulfhydryl groups, indicating protein denaturation and molecular aggregation. An increase in surface hydrophobicity is typically correlated with a rise in free sulfhydryl group content, reflecting protein unfolding and the exposure of hydrophobic amino acid residues.

3.5. Functional Properties

3.5.1. Foaming and Emulsifying Capacity

Foaming capacity and stability, which are important factors for the suitability of de-oiled flours and protein isolates in food formulations, were significantly higher in SFE de-oiled flour (148 ± 2.8%, $p < 0.05$) than in the hexane one (Table 3). This superior foaming capacity was attributed to the higher concentration of native proteins in SFE de-oiled flour, which acted as effective surface-active agents enhancing foam stability through intermolecular forces [22]. Studies reported that defatted soybean cake possessed the highest foaming stability than other leguminous flours [31,32]. Another study compared different oilseed meals of almond, chestnut, hazelnut, pine nut, pistachio nut, and soybean. The authors found that the foaming capacity and stability of the defatted soybean cake was

40% higher compared to the other oilseed cakes [33]. This aspect is of great relevance to the food industry [34].

Importantly, protein isolates from both de-oiling methods exhibited similar foaming properties, likely because their comparable protein content levels aligned with previous studies [8,28,35].

The emulsifying capacity of the SFE de-oiled flour (107.5 ± 2.1 m^2/g) was significantly higher than that of the hexane de-oiled flour (52.5 ± 3.5 m^2/g of protein), as shown in Table 3. This suggested that SFE enhanced emulsifying abilities because of the preservation of protein structure and amino acid composition. These properties are crucial for the development of stable and high-quality emulsified food products [36]. In a previous study, it was reported that SFE-defatted wheat germ protein reported higher emulsifying properties, suggesting it is a good protein additive for use in emulsified foods [37]. Other authors also reported similar results, which showed that the higher protein content of the extracts prepared from SFE oats presented superior emulsifying properties [28].

3.5.2. Water and Oil Holding Capacity

SFE de-oiled flour demonstrated superior water-holding capacity compared with hexane de-oiled flour, as detailed in Table 3. This was attributed to the presence of a higher amount of native proteins in the SFE flour, which enhanced the water absorption capacity [8]. On the other side, the oil holding capacities did not show significant differences between the samples ($p > 0.05$). The improved water-holding capacity of SFE-defatted flour could enhance the texture and shelf life of various food products [38]. The higher water holding capacity was in line with previously published studies where a defatted soybean cake was obtained after oil extraction by SFE. The values were linked to the higher surface hydrophobicity of the protein content of the de-oiled cake [18,39].

3.5.3. Viscoamylographic Profiles

Viscoamylographic analysis revealed that the SFE de-oiled flour displayed a three-fold increase in peak viscosity at 95 °C and a two-fold increase in final viscosity at 55 °C compared to the hexane de-oiled flour, as shown in Figure 4. These findings correlated with the enhanced water-holding capacity of SFE de-oiled flour, which has significant implications for food products such as bakery, pasta, and sauces. The increased viscosity could improve product texture and consistency, potentially reducing the need for additional thickeners and leading to cost savings in food production [38].

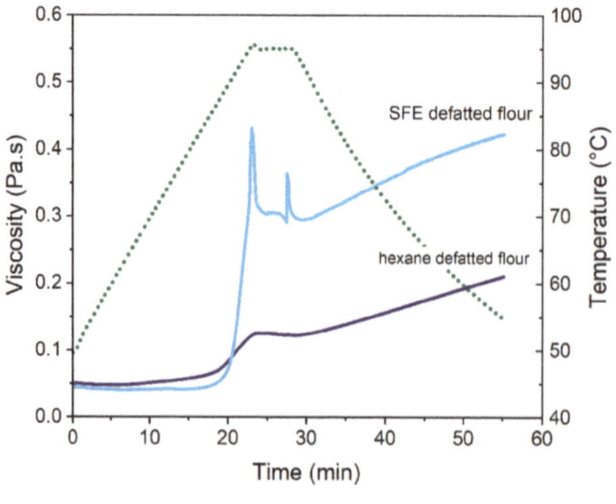

Figure 4. Pasting profile of defatted soybean flours.

3.6. Correlation Analysis

A correlation analysis was conducted to further understand the interrelationships among the measured functional properties. Figure 5 shows the correlation plot highlighting these relationships. SFE-defatted flour exhibited significantly higher emulsifying ability and stability ($p < 0.05$) with a positive correlation with the oil holding capacity ($R^2 = 0.98$). This study suggested that factors enhancing emulsification could also improve oil retention in these samples.

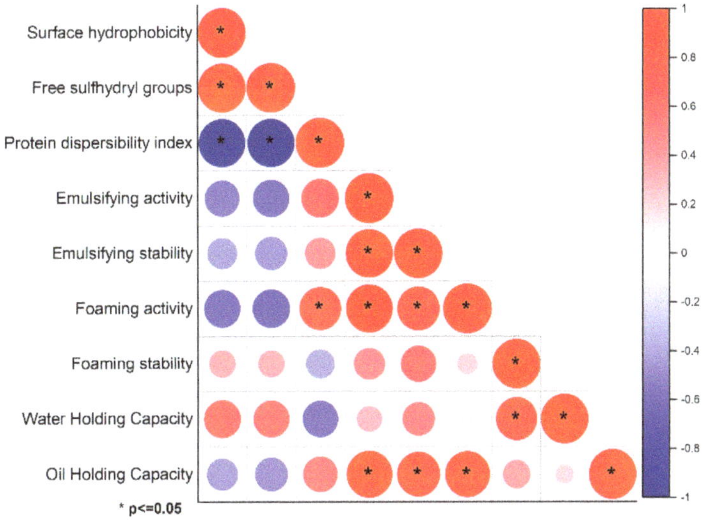

Figure 5. Correlation plot between the attributes measured for the samples. On the diagonal, the attributes of the y-axis are also reported. Significant or insignificant correlations could be assessed by matching the attributes vertically.

Interestingly, although it was not statistically significant, a correlation was also observed between the water holding capacity and protein dispersibility index. Conversely, an inverse significant correlation was observed between the protein dispersibility index, the surface hydrophobicity, and the free sulfhydryl group content. Samples with higher ease of dissolution in water exhibited lower surface hydrophobicity and reduced free sulfhydryl groups. These findings implied that samples with lower concentrations of native proteins reported increased surface hydrophobicity and free sulfhydryl groups, indicative of protein unfolding and exposure of hydrophobic residues.

4. Conclusions

This study demonstrated that SFE is an effective method for obtaining fully de-oiled soybean cakes, yielding a product with a proximate composition comparable to that of industrially extracted flour with hexane. In addition, a significant improvement in protein solubility and extraction yield was observed with soybean cake de-oiled using the SFE method. This improvement was mainly attributed to the reduced protein denaturation obtained at the end of the extraction process, which was supported by our findings.

Furthermore, while protein isolates from both SFE and hexane de-oiled methods shared similar protein profiles, several differences were evident in their functional properties. In particular, the flour de-oiled by SFE exhibited superior characteristics, including higher foaming and emulsifying capacity and stability and better water holding capacity.

Overall, the results of this study provided valuable insights into the preservation of protein quality during soybean cake processing, highlighting the advantages of SFE in obtaining de-oiled flours with unique functional attributes. Furthermore, the potential of

SFE-derived proteins in innovative food formulations is an exciting area for future studies. The findings supported the use of SFE as a green technology for obtaining protein-rich de-oiled plant matrices and encouraged further research to explore the potential of SFE in the field of alternative food proteins. This research aligned with the ongoing goal of applying technologies to sustainably exploit novel protein sources, achieving targeted functional properties. The study contributed to the broader analysis of sustainable practices in food production, answering the needs of the modern food industry to design environmentally safe processes.

Author Contributions: G.C. and R.G.-O.: Methodology, Formal analysis, Data curation, Investigation. A.L.G.: Formal analysis, Data curation, Investigation. K.M.: Formal analysis. G.F.: Writing—original draft, Resources, Funding acquisition, Project administration. M.S.: Writing—original draft, Project administration. E.C.: Resources, Funding acquisition, Project administration. All authors have read and agreed to the published version of the manuscript.

Funding: Funding for this research was received from Fondazione Cariverona. The authors would like to thank Fondazione Cariverona and Cereal Docks S.p.A. for financing the project. The work was also performed with the co-financing of the European Union FESR, PON Research and Innovation 2014–2020. The work was supported by the Project Award Number: Project code PE00000003, Concession Decree No. 1550 of 11 October 2022, adopted by the Italian Ministry of University and Research, CUP D93C22000890001, Project title "ON Foods—Research and innovation network on food and nutrition Sustainability, Safety and Security—Working ON Foods." Project funded under the National Recovery and Resilience Plan (NRRP), Mission 4 Component 2 Investment 1.3—Call for tender No. 341 of 15 March 2022 of Italian Ministry of University and Research funded by the European Union—NextGenerationEU.

Data Availability Statement: The data that support the findings of this study are available from the corresponding author, G.F., upon reasonable request.

Acknowledgments: The authors would like to thank Fondazione Cariverona and Cereal Docks S.p.A. for financing the project.

Conflicts of Interest: Authors Giulia Cestonaro and Enrico Costanzo are employed by the company Cereal Docks S.p.A. The remaining authors declare that the research was conducted in the absence of any commercial or financial relationships that could be construed as a potential conflict of interest.

References

1. Richardson, K.; Steffen, W.; Lucht, W.; Bendtsen, J.; Cornell, S.E.; Donges, J.F.; Drüke, M.; Fetzer, I.; Bala, G.; von Bloh, W.; et al. Earth beyond Six of Nine Planetary Boundaries. *Sci. Adv.* **2023**, *9*, eadh2458. [CrossRef]
2. Mota da Silva, A.M.; Souza Almeida, F.; Kawazoe Sato, A.C. Functional Characterization of Commercial Plant Proteins and Their Application on Stabilization of Emulsions. *J. Food Eng.* **2021**, *292*, 110277. [CrossRef]
3. Arrutia, F.; Binner, E.; Williams, P.; Waldron, K.W. Oilseeds beyond Oil: Press Cakes and Meals Supplying Global Protein Requirements. *Trends Food Sci. Technol.* **2020**, *100*, 88–102. [CrossRef]
4. Dhiman, A.; Thakur, K.; Parmar, V.; Sharma, S.; Sharma, R.; Kaur, G.; Singh, B.; Suhag, R. New Insights into Tailoring Physicochemical and Techno-Functional Properties of Plant Proteins Using Conventional and Emerging Technologies. *J. Food Meas. Charact.* **2023**, *17*, 3845–3873. [CrossRef]
5. Russin, T.A.; Boye, J.I.; Arcand, Y.; Rajamohamed, S.H. Alternative Techniques for Defatting Soy: A Practical Review. *Food Bioprocess Technol.* **2011**, *4*, 200–223. [CrossRef]
6. Ferrentino, G.; Morozova, K.; Mosibo, O.K.; Ramezani, M.; Scampicchio, M. Biorecovery of Antioxidants from Apple Pomace by Supercritical Fluid Extraction. *J. Clean. Prod.* **2018**, *186*, 253–261. [CrossRef]
7. Singh, P.; Kumar, R.; Sabapathy, S.N.; Bawa, A.S. Functional and Edible Uses of Soy Protein Products. *Compr. Rev. Food Sci. Food Saf.* **2008**, *7*, 14–28. [CrossRef]
8. Kang, S.W.; Rahman, M.S.; Kim, A.N.; Lee, K.Y.; Park, C.Y.; Kerr, W.L.; Choi, S.G. Comparative Study of the Quality Characteristics of Defatted Soy Flour Treated by Supercritical Carbon Dioxide and Organic Solvent. *J. Food Sci. Technol.* **2017**, *54*, 2485–2493. [CrossRef]
9. Politiek, R.G.A.; Dijkink, B.H.; van den Boogaard, L.M.; Keppler, J.K.; Schutyser, M.A.I.; Bruins, M.E. Comparing Electrostatic Separation of Soy and Lupin: Effect of de-Oiling by Solvent Extraction. *LWT* **2023**, *187*, 115290. [CrossRef]
10. Toledo, F.R.; del Valle, J.M.; Opazo, Á.P.; Núnez, G.A. Supercritical CO_2 Extraction of Pelletized Oilseeds: Representation Using a Linear Driving Force Model with a Nonlinear Sorption Isotherm. *J. Food Eng.* **2021**, *288*, 110241. [CrossRef]

11. Sweers, L.J.H.; Politiek, R.G.A.; Lakemond, C.M.M.; Bruins, M.E.; Boom, R.M.; Fogliano, V.; Mishyna, M.; Keppler, J.K.; Schutyser, M.A.I. Dry Fractionation for Protein Enrichment of Animal By-Products and Insects: A Review. *J. Food Eng.* **2022**, *313*, 110759. [CrossRef]
12. Yu, J.; Liu, Y.F.; Qiu, A.Y.; Wang, X.G. Preparation of Isoflavones Enriched Soy Protein Isolate from Defatted Soy Hypocotyls by Supercritical CO_2. *LWT* **2007**, *40*, 800–806. [CrossRef]
13. Bozan, B.; Temelli, F. Supercritical CO_2 Extraction of Flaxseed. *JAOCS J. Am. Oil Chem. Soc.* **2002**, *79*, 231–235. [CrossRef]
14. Sun, Q.; Shi, J.; Scanlon, M.; Xue, S.J.; Lu, J. Optimization of Supercritical-CO_2 Process for Extraction of Tocopherol-Rich Oil from Canola Seeds. *LWT* **2021**, *145*, 111435. [CrossRef]
15. Natolino, A.; Da Porto, C.; Scalet, M. Broken and Intact Cell Model for Supercritical Carbon Dioxide Extraction of Tea *Camellia sinensis* (L) Seed Oil. *J. Supercrit. Fluids* **2022**, *180*, 105422. [CrossRef]
16. Ben Mohamed, H.; Duba, K.S.; Fiori, L.; Abdelgawed, H.; Tlili, I.; Tounekti, T.; Zrig, A. Bioactive Compounds and Antioxidant Activities of Different Grape (*Vitis vinifera* L.) Seed Oils Extracted by Supercritical CO_2 and Organic Solvent. *LWT* **2016**, *74*, 557–562. [CrossRef]
17. Quitain, A.T.; Oro, K.; Katoh, S.; Moriyoshi, T. Recovery of Oil Components of Okara by Ethanol-Modified Supercritical Carbon Dioxide Extraction. *Bioresour. Technol.* **2006**, *97*, 1509–1514. [CrossRef]
18. Aussanasuwannakul, A.; Boonbumrung, S.; Pantoa, T. Valorization of Soybean Residue (Okara) by Supercritical Carbon Dioxide Extraction: Compositional, Physicochemical, and Functional Properties of Oil and Defatted Powder. *Foods* **2023**, *12*, 2698. [CrossRef]
19. Ferrentino, G.; Giampiccolo, S.; Morozova, K.; Haman, N.; Spilimbergo, S.; Scampicchio, M. Supercritical Fluid Extraction of Oils from Apple Seeds: Process Optimization, Chemical Characterization and Comparison with a Conventional Solvent Extraction. *Innov. Food Sci. Emerg. Technol.* **2020**, *64*, 102428. [CrossRef]
20. AOAC. *Official Methods of Analysis*, 1st ed.; Association of Official Analytical Chemists: Gaithersburg, MD, USA, 2003.
21. Ren, J.; Ma, C.; Li, M.; Dang, Y.; Yu, X.; Du, S. Physicochemical, Structural Structural and Functional Properties of Non-Waxy and Waxy Proso Millet Protein. *Foods* **2023**, *12*, 1116. [CrossRef]
22. Iwe, M.O.; Van Zuilichem, D.J.; Ngoddy, P.O.; Lammers, W. Amino Acid and Protein Dispersibility Index (PDI) of Mixtures of Extruded Soy and Sweet Potato Flours. *LWT* **2001**, *34*, 71–75. [CrossRef]
23. Zhang, Z.; He, S.; Liu, H.; Sun, X.; Ye, Y.; Cao, X.; Wu, Z.; Sun, H. Effect of PH Regulation on the Components and Functional Properties of Proteins Isolated from Cold-Pressed Rapeseed Meal through Alkaline Extraction and Acid Precipitation. *Food Chem.* **2020**, *327*, 126998. [CrossRef] [PubMed]
24. Pearce, K.N.; Kinsella, J.E. Emulsifying Properties of Proteins: Evaluation of a Turbidimetric Technique. *J. Agric. Food Chem.* **1978**, *26*, 716–723. [CrossRef]
25. Karaca, A.C.; Low, N.; Nickerson, M. Emulsifying Properties of Canola and Flaxseed Protein Isolates Produced by Isoelectric Precipitation and Salt Extraction. *Food Res. Int.* **2011**, *44*, 2991–2998. [CrossRef]
26. Shand, P.J.; Ya, H.; Pietrasik, Z.; Wanasundara, P.K.J.P.D. Physicochemical and Textural Properties of Heat-Induced Pea Protein Isolate Gels. *Food Chem.* **2007**, *102*, 1119–1130. [CrossRef]
27. Shrestha, S.; van 't Hag, L.; Haritos, V.; Dhital, S. Rheological and Textural Properties of Heat-Induced Gels from Pulse Protein Isolates: Lentil, Mungbean and Yellow Pea. *Food Hydrocoll.* **2023**, *143*, 108904. [CrossRef]
28. Konak, Ü.I.; Ercili-Cura, D.; Sibakov, J.; Sontag-Strohm, T.; Certel, M.; Loponen, J. CO_2-Defatted Oats: Solubility, Emulsification and Foaming Properties. *J. Cereal Sci.* **2014**, *60*, 37–41. [CrossRef]
29. Wang, X.; Zhao, Y.; Wang, T.; Li, Y.; Wang, C.; Zhao, X.; Liang, Y.; Wang, H. The Structural and Functional Properties Changing of Peanut Protein Isolate (PPI) via Different Treatments and Their Application in Pork Meatballs Production. *LWT* **2023**, *184*, 115046. [CrossRef]
30. Yang, K.; Xu, T.R.; Fu, Y.H.; Cai, M.; Xia, Q.; Guan, R.F.; Zou, X.G.; Sun, P.L. Effects of Ultrasonic Pre-Treatment on Physicochemical Properties of Proteins Extracted from Cold-Pressed Sesame Cake. *Food Res. Int.* **2021**, *139*, 109907. [CrossRef]
31. Ren, X.; Yang, Y.; Liu, Q.; Wang, Y.; Jin, Z.; Jiao, A. Effects of Enzymatic Extrusion on the Structure and Physicochemical Properties of Oat Flour and Its Application in Oat Milk Production. *Int. J. Food Sci. Technol.* **2023**, *58*, 4638–4651. [CrossRef]
32. Chau, C.F.; Cheung, P.C.K. Functional Properties of Flours Prepared from Three Chinese Indigenous Legume Seeds. *Food Chem.* **1998**, *61*, 429–433. [CrossRef]
33. Sharma, G.M.; Su, M.; Joshi, A.U.; Roux, K.H.; Sathe, S.K. Functional Properties of Select Edible Oilseed Proteins. *J. Agric. Food Chem.* **2010**, *58*, 5457–5464. [CrossRef] [PubMed]
34. Herneke, A.; Lendel, C.; Karkehabadi, S.; Lu, J.; Langton, M. Protein Nanofibrils from Fava Bean and Its Major Storage Proteins: Formation and Ability to Generate and Stabilise Foams. *Foods* **2023**, *12*, 521. [CrossRef] [PubMed]
35. Sun, M.; Xu, L.; Saldaña, M.D.A.; Temelli, F. Comparison of Canola Meals Obtained with Conventional Methods and Supercritical CO_2 with and without Ethanol. *JAOCS J. Am. Oil Chem. Soc.* **2008**, *85*, 667–675. [CrossRef]
36. Espinosa-Pardo, F.A.; Savoire, R.; Subra-Paternault, P.; Harscoat-Schiavo, C. Oil and Protein Recovery from Corn Germ: Extraction Yield, Composition and Protein Functionality. *Food Bioprod. Process.* **2020**, *120*, 131–142. [CrossRef]
37. Ge, Y.; Sun, A.; Ni, Y.; Cai, T. Some Nutritional and Functional Properties of Defatted Wheat Germ Protein. *J. Agric. Food Chem.* **2000**, *48*, 6215–6218. [CrossRef] [PubMed]

38. Capellini, M.C.; Chiavoloni, L.; Giacomini, V.; Rodrigues, C.E.C. Alcoholic Extraction of Sesame Seed Cake Oil: Influence of the Process Conditions on the Physicochemical Characteristics of the Oil and Defatted Meal Proteins. *J. Food Eng.* **2019**, *240*, 145–152. [CrossRef]
39. Puechkamut, Y.; Panyathitipong, W. Characteristics of Proteins from Fresh and Dried Residues of Soy Milk Production. *Kasetsart J.—Nat. Sci.* **2012**, *46*, 804–811.

processes

Article

Antioxidant Activity of Carob Tree (*Ceratonia siliqua* L.) Leaf Extracts Obtained by Advanced Extraction Techniques

Ena Cegledi [1], Erika Dobroslavić [1,*], Zoran Zorić [2], Maja Repajić [1] and Ivona Elez Garofulić [1,*]

[1] Faculty of Food Technology and Biotechnology, University of Zagreb, Pierottijeva 6, 10000 Zagreb, Croatia; ecegledi@pbf.hr (E.C.); maja.repajic@pbf.unizg.hr (M.R.)
[2] Department of Ecology, Agronomy and Aquaculture, University of Zadar, Trg Kneza Višeslava 9, 23000 Zadar, Croatia; zzoric@unizd.hr
* Correspondence: edobroslavic@pbf.hr (E.D.); ivona.elez@pbf.unizg.hr (I.E.G.)

Abstract: Carob (*Ceratonia siliqua* L.) is a widely spread Mediterranean evergreen tree whose plant parts are rich in bioactive compounds with potential for application in functional food production. Carob leaves are the least explored part of the plant, and the main compounds of interest are polyphenols. Advanced extraction techniques, such as pressurized liquid (PLE), microwave-assisted (MAE) and ultrasound-assisted (UAE) extraction have not been sufficiently explored for their potential in extracting these compounds. Therefore, the aim of this paper was to optimize the parameters (temperature and time) of PLE, MAE and UAE of carob leaf polyphenols, characterize the individual compounds by ultra-high performance liquid chromatography tandem mass spectrometry (UPLC-MS^2) and determine the antioxidant activity of the extracts. The optimal temperature and time were 160 °C/5 min for PLE, resulting in a total phenolic content of 68.21 mg of gallic acid equivalents (GAE) g^{-1}. For MAE and UAE, the optimal conditions were 70 °C/10 min, under which total phenolic contents of 78.80 and 55.98 mg GAE g^{-1} were achieved, respectively. A total of 26 compounds, represented mainly by myricetin, quercetin-3-rhamnoside and gallic acid, were identified in extracts obtained by all three extraction techniques. All extracts showed high antioxidant activity (0.46–1.05 and 0.50–0.58 mmol TE g^{-1} for FRAP and DPPH, respectively), which correlated with the polyphenolic content and was affected by the concentration of ascorbic acid (0.03–0.52 mg mL^{-1}), confirming that carob leaf extracts obtained by advanced extraction possess high potential for application in functional food products.

Keywords: carob leaves; green extraction; pressurized liquid extraction; microwave-assisted extraction; ultrasound-assisted extraction; polyphenols; antioxidant activity

Citation: Cegledi, E.; Dobroslavić, E.; Zorić, Z.; Repajić, M.; Elez Garofulić, I. Antioxidant Activity of Carob Tree (*Ceratonia siliqua* L.) Leaf Extracts Obtained by Advanced Extraction Techniques. *Processes* **2024**, *12*, 658. https://doi.org/10.3390/pr12040658

Academic Editors: Yanlin Zhang, Prashant K. Sarswat and Abraham Kabutey

Received: 9 February 2024
Revised: 21 March 2024
Accepted: 25 March 2024
Published: 26 March 2024

1. Introduction

In light of growing consumer demand for functional products enhanced with natural elements, there has been a surge in interest to explore novel sources of natural ingredients. One of the plant species that proved to be an excellent basis for obtaining functional products, due to its rich chemical composition, is carob.

Carob (*Ceratonia siliqua* L.) is a perennial, evergreen tree from the Fabaceae family, native to the Mediterranean region, but now widely distributed across the world. It often grows in dry and rocky areas, preferring warm climates. Carob is resistant to drought and soil salinity and can tolerate high temperatures, enabling it to thrive in challenging environments, expanding agricultural possibilities and contributing to food security in areas prone to arid conditions [1,2]. Its significance is appreciated not only for its resilience and economic value, but also for the nutritional value of all parts of the plant (pods, pulp, seeds, leaves), which are used in the food, pharmaceutical and cosmetic industries due to their biological activities [3]. Carob pods consist of 90% pulp and 10% seeds. They contain large amounts of carbohydrates, fibers, proteins, minerals, fats and polyphenols and have been used in human nutrition and as animal feed since ancient times [4,5]. The pulp is used to make molasses and syrup, while carob seed powder is commonly utilized as a cocoa substitute in various food applications, as it has a similar sweet flavor but contains less fat and caffeine [6]. In addition, the carob leaves, which are often overshadowed by prominent pods, are a rich source of valuable nutrients, such as carbohydrates, protein and fiber, and contain even more polyphenols, compared to pods [4,7]. Polyphenols have an antioxidant effect and can contribute to various health benefits [8]. As a result of the high content of bioactive molecules, carob is not only a source of powerful antioxidants, but also has antimicrobial, antiproliferative and anti-inflammatory effects, being used as a diuretic, an anti-diarrheal agent and for treating obesity [3,6,9].

In order to isolate highly valuable bioactive compounds such as polyphenols from the matrix, an extraction process is required. It is therefore important to apply and optimize a suitable extraction technique that enables a high yield and stability of the target components during extraction and storage [10,11]. Various techniques can be applied to obtain the extracts, and advanced extraction techniques, such as pressurized liquid extraction (PLE), microwave-assisted extraction (MAE) and ultrasound-assisted extraction (UAE) are increasingly being applied. These techniques, compared to conventional extraction, enable rapid and environmentally friendly extraction by improving solvent penetration into leaf sample pores and enhancing mass transport. This results in increased leachability of target components, leading to higher yields, while consuming a lower amount of energy and solvent, which is why they are considered as "green techniques" [12,13]. Optimizing multiple process parameters is crucial for the efficient extraction and isolation of bioactive molecules such as polyphenols having in mind optimal yield and superior quality of the extracts. The most important parameters include the selection of the suitable solvent, temperature and extraction time. Following the extraction process, the subsequent stage involves the identification and quantification of targeted compounds. Ultra-performance liquid chromatography, coupled with tandem mass spectrometry (UPLC-MS^2), has proven to be an excellent technique for the chemical characterization of polyphenols, as it integrates chromatographic and spectral techniques [14]. Overall, the combination of advanced extraction techniques with UPLC-MS^2 enables reliable and precise analysis of polyphenols in a wide range of samples.

The influence of advanced extraction techniques on polyphenols from carob leaves has not yet been sufficiently investigated and, to our knowledge, MAE and PLE have never been applied to carob leaves. Consequently, the aim of this work was to examine the influence of different extraction conditions of PLE, MAE and UAE on the polyphenol content in carob leaf extracts and to determine the optimal extraction conditions for their isolation. An additional aim was to characterize individual polyphenols from carob leaf extracts obtained by advanced techniques under optimal conditions using UPLC-MS^2 and to examine their antioxidant activity by DPPH and FRAP assays.

2. Materials and Methods

2.1. Chemicals and Reagents

Purified water was prepared utilizing the advanced Milli-Q system manufactured by Millipore Corp. (Bedford, NY, USA), ensuring the highest quality for experimental procedures. High-performance liquid chromatography (HPLC) grade acetonitrile (CH_3CN) and acetic acid (CH_3COOH) of exceptional purity were procured from J.T. Baker Chemicals in Deventer, The Netherlands. The Folin-Ciocalteu reagent was acquired from Merck (Darmstadt, Germany). Ethanol (C_2H_5OH, 96%) of analytical grade was sourced from Kemika d.d. (Zagreb, Croatia). Iron (III) chloride hexahydrate ($FeCl_3{\cdot}6H_2O$), ascorbic acid, sodium acetate trihydrate ($CH_3COONa{\cdot}3H_2O$, 99%) and sodium carbonate (Na_2CO_3, 99.5%) were obtained from Kemika d.d. (Zagreb, Croatia). Antioxidants reference standards, such as Trolox ($C_{14}H_{18}O_4$) and 2,4,6-tri(2-pyridyl)-s-triazine (TPTZ), vital for assessing radical scavenging activity, were procured from Acros Organics (Geel, Belgium). High purity methanol (CH_3OH) was sourced from Honeywell (Riedel-de-Haën, Bucharest, Romania).

2,2-Diphenyl-1-picrylhydrazyl (DPPH) and 2,6-dichlorophenolindophenol (DCPI, 97%) were acquired from Sigma-Aldrich (St. Louis, MO, USA). Hydrochloric acid (HCl, 37%) was sourced from Carlo Erba Reagents (Val-de-Reuil, France). Commercial HPLC standards, including rosmarinic acid ($C_{18}H_{16}O_8$), gallic acid ($C_7H_6O_5$), syringic acid ($C_9H_{10}O_5$), *p*-coumaric acid ($C_9H_8O_3$), caffeic acid ($C_9H_8O_4$), chlorogenic acid ($C_{16}H_{18}O_9$), protocatechuic acid ($C_7H_6O_4$), ferulic acid ($C_{10}H_{10}O_4$), myricetin ($C_{15}H_{10}O_8$) and quercetin-3-glucoside ($C_{21}H_{20}O_{12}$), were selected from Sigma–Aldrich (St. Louis, MO, USA) to serve as precise standards for analytical calibration. Additionally, supplemental compounds, such as procyanidin B2, catechin, epicatechin gallate, epigallocatechin gallate, rutin ($C_{27}H_{30}O_{16}$), kaempferol-3-glucoside and apigenin were obtained from Extrasynthese (Genay, France). Quercetin-3-rutinoside ($C_{27}H_{30}O_{16}$) was sourced from Thermo Fisher Scientific (Waltham, MA, USA). Standard stock solutions were produced in methanol. For the specific compound apigenin, dissolution in ethanol with dimethyl sulfoxide (0.5%) was carried out. Working standard solutions (10–200 mg L^{-1}) were then prepared by diluting the prepared stock solutions to achieve five determined concentrations, ensuring precise calibration for accurate analytical measurements.

2.2. Plant Material

Wild carob leaves were collected in Grgurići, Croatia (42.784227 N, 17.876950 E), in August 2023 and identified by the Laboratory for Biology and Microbial Genetics at the Faculty of Food Technology and Biotechnology, University of Zagreb. After harvest, the material was air-dried in shade at ambient temperature until dry matter content >95%, as determined by drying to constant mass at 103 ± 2 °C, was achieved. Before the extraction procedures, the leaves were ground into a coarse powder (particle size 0.45–1 mm) using the GT 1108 electric grinder (Tefal, Rumily, France).

2.3. Pressurized Liquid Extraction (PLE)

Polyphenols were extracted from carob leaf sample using PLE with a Dionex™ ASE™ 350 extractor from Thermo Fisher Scientific Inc. (Sunnyvale, CA, USA), employing 50% aqueous ethanol as the extraction solvent. Ground carob leaves (1 g) were combined with 2 g of Dionex™ ASE™ diatomaceous earth (Thermo Fisher Scientific Inc, Sunnyvale, CA, USA; lot no. 171) and placed into 34 mL stainless steel cells equipped with two cellulose filters. To optimize polyphenols isolation, variations in temperature (ranging from 60 to 160 °C) and static time (5 to 15 min) were applied (refer to Table 1), while maintaining other extraction parameters constant: one extraction cycle, pressure of 10.34 MPa, 30 s of nitrogen purge, and 30% volume flush. After the extraction, the obtained extract was filtered using Whatman No. 40 filter paper into a 50 mL volumetric flask and supplemented with the extraction solvent. The resulting extracts were stored in plastic Falcon tubes at −18 °C. All extractions were performed in duplicate (n = 2).

2.4. Microwave-Assisted Extraction (MAE)

The polyphenolic extracts from carob leaves were obtained by MAE in an Ethos Easy reactor (Milestone, Italy). The extraction was carried out with a constant power of 500 W, stirring at 50% and a post-extraction ventilation time of 1 min. The varied parameters were temperature (30–80 °C) and extraction time (5–15 min) (Table 1). The time required to achieve the extraction temperature was varied as follows: 1 min for temperatures of 30 and 40 °C, 2 min for temperatures of 50 and 60 °C and 3 min for temperatures of 70 and 80 °C. For every extraction run, 1 g of ground sample was blended with 40 mL of 50% aqueous ethanol in an extraction vessel equipped with a magnetic stirrer, which was placed in the microwave extractor. Subsequently, the process after extraction was the same as described in Section 2.3.

Table 1. Experimental design and values of total phenolic content in carob leaf extracts obtained by PLE, MAE and UAE.

Extraction Technique	Temperature (°C)	Extraction Time (min)	Total Phenols (mg GAE g^{-1} Leaf)
PLE	60	5	40.81 ± 0.63
		10	34.96 ± 0.70
		15	33.72 ± 0.29
	80	5	44.17 ± 0.22
		10	44.91 ± 0.31
		15	43.72 ± 0.50
	100	5	42.90 ± 0.49
		10	37.09 ± 0.52
		15	42.94 ± 0.98
	120	5	46.86 ± 0.42
		10	53.60 ± 0.18
		15	42.34 ± 0.83
	140	5	62.73 ± 0.28
		10	53.47 ± 0.13
		15	67.73 ± 0.46
	160	5	68.21 ± 0.31
		10	68.94 ± 0.54
		15	80.73 ± 0.20
MAE	30	5	60.78 ± 0.90
		10	56.81 ± 0.61
		15	62.84 ± 0.30
	40	5	51.25 ± 0.59
		10	69.20 ± 0.60
		15	61.60 ± 0.68
	50	5	61.74 ± 0.34
		10	63.18 ± 0.73
		15	71.00 ± 0.43
	60	5	63.27 ± 0.68
		10	71.58 ± 0.53
		15	73.93 ± 0.43
	70	5	65.50 ± 0.41
		10	78.80 ± 1.29
		15	90.67 ± 0.93
	80	5	75.88 ± 0.76
		10	83.02 ± 0.72
		15	82.41 ± 1.43
UAE	30	5	31.64 ± 0.22
		10	51.44 ± 0.85
		15	42.21 ± 0.57
	40	5	45.23 ± 0.29
		10	32.46 ± 0.62
		15	46.52 ± 0.59
	50	5	30.23 ± 0.35
		10	45.60 ± 0.09
		15	42.04 ± 0.24
	60	5	25.87 ± 0.23
		10	44.39 ± 0.79
		15	53.27 ± 0.95
	70	5	43.75 ± 1.41
		10	55.98 ± 0.53
		15	55.82 ± 0.48
	80	5	40.01 ± 0.13
		10	43.82 ± 0.51
		15	51.21 ± 1.23

PLE = pressurized liquid extraction. MAE = microwave-assisted extraction. UAE = ultrasound-assisted extraction. GAE = gallic acid equivalents. Values are expressed as mean ± SD.

2.5. Ultrasound-Assisted Extraction (UAE)

For the UAE of carob leaves polyphenols, an ultrasound bath (Elmasonic P, Elma Schmidbauer GmbH, Singen, Germany) equipped with a temperature control unit was used at the standard device frequency of 37 kHz. The temperature (30–80 °C) and the sonication time (5–15 min) were varied as shown in Table 1. Briefly, a glass beaker containing 1 g of ground sample was combined with 40 mL of 50% aqueous ethanol pre-heated at the selected temperatures and placed in the ultrasonic bath during the varied extraction times. The process after extraction was the same as described in Section 2.3.

2.6. Total Phenolic Content (TPC)

Total phenolic content (TPC) was determined by a slightly modified method previously described [15]. A volume of 100 µL of appropriately diluted extract (50% aqueous ethanol for blank) was mixed with 200 µL of Folin-Ciocalteu reagent and 2000 µL of distilled water in a test tube. After 3 min, 1000 µL of 20% Na_2CO_3, w/v aqueous solution was added to the mixture, which was then tempered in a water bath at the temperature of 50 °C for 25 min. Absorbance was read at the wavelength of 765 nm using the VWR UV 1600-PC spectrophotometer (VWR International GmbH, Darmstadt, Germany). Working standard solutions of gallic acid with a concentration range of 50–500 mg L^{-1} were used to produce a calibration curve from which the TPC was calculated and expressed as mg gallic acid equivalents (GAE) g^{-1} of dry carob leaf. Each measurement was conducted in duplicate.

2.7. Antioxidant Activity

The determination of antioxidant activity in the extracts was carried out using two methodologies, as detailed by Dobroslavić et al. [16].

2.7.1. FRAP-Ferric Reducing Antioxidant Power Assay

To prepare the FRAP reagent, 20 mM aqueous iron (III) chloride solution, 10 mM TPTZ reagent in 40 mM hydrochloric acid and 300 mM acetate buffer (pH 3.8) were mixed in a ratio of 1:1:10. Additionally, 240 µL distilled water, 80 µL sample and 2080 µL FRAP reagent were added into the glass tubes, mixed well (with a vortex mixer) and incubated at 37 °C for 5 min. Afterwards, the absorbance was read at 593 nm on the spectrophotometer. The calibration curve was prepared using working standard solutions of Trolox in the concentration range of 25–1000 µmol L^{-1}. The results were expressed as mmol equivalents of Trolox (TE) per g of dry carob leaf. Each measurement was conducted in duplicate ($n = 4$).

2.7.2. DPPH Radical Scavenging Assay

Briefly, 750 µL of extract and 1500 µL of 0.2 mM DPPH solution in methanol were added into the test tube. The test tubes with the contents were kept for 20 min in the dark at room temperature, after which the absorbance was measured at 517 nm using the VWR UV 1600-PC spectrophotometer (VWR International GmbH, Darmstadt, Germany), with methanol as a blank. The calibration curve was prepared using working standard solutions of Trolox in the concentration range of 10–150 µmol L^{-1}. The results were expressed as mmol TE per g of dry carob leaf. Each measurement was conducted in duplicate.

2.8. Determination of Ascorbic Acid Content

The content of ascorbic acid was determined according to standard simple redox titration of ascorbic acid by 2,6-dichlorophenolindophenol (DCPIP) [17], which served as an oxidant and acid-base indicator, aiding in pinpointing the end point of titration, marked by the appearance of a pink hue in the solution, caused by the surplus of DCPIP reagent. Briefly, DCPIP reagent was standardized by titration of 5 mL of ascorbic acid solution (0.2 mg mL^{-1}), which was mixed with 150 mL distilled water and 5 mL of 10% (v/v) aqueous acetic acid, thus obtaining f_{AA}, which reflects the concentration of ascorbic acid responsive to 1 mL of DCPIP. After standardization, the same procedure was carried out with blank (distilled water) and the extracts (in quadruplet), thus obtaining the mean volumes of DCPI spent for titration (V_B and V_E) and f_{DCPIP}, calculated as follows (Equation (1)):

$$f_{DCPIP} = \frac{5\ \text{mL}}{V_B} \tag{1}$$

where f_{DCPIP} is the DCPIP factor, and V_B is the volume of DCPIP spent for titration of blank.

The concentration of ascorbic acid in the extract was then calculated according to Equation (2):

$$\frac{(V_E - V_B) \times f_{DCPIP} \times f_{AA}}{V} \tag{2}$$

where V_E is the volume of DCPIP spent for titration of the extract, V_B is the volume of DCPIP spent for titration of blank, f_{DCPIP} is the DCPIP factor calculated from V_B, f_{AA} is the concentration of ascorbic acid responsive to 1 mL of DCPIP and V is the volume of extract used in titration.

2.9. Polyphenolic Characterization of the Extracts by UPLC-MS^2

The carob leaf extracts, obtained by PLE, MAE and UAE at optimal conditions, were filtered into glass vials through 0.45 μm syringe filters and analyzed by UPLC-MS^2 on the Agilent series 1290 RRLC instrument (Agilent, Santa Clara, CA, USA) paired with an Agilent 6430 Triple Quad LC/MS mass spectrometer (Agilent, Santa Clara, CA, USA) and Agilent MassHunter Workstation Software (ver. B.04.01, Agilent, Santa Clara, CA, USA) for data processing and instrument control. The analytes were ionized by ESI ion source in positive and negative modes. Desolvation and collision were performed using nitrogen at a flow rate of 11 L h^{-1}, temperature of 300 °C, nebulizer pressure of 2.76 bar and capillary voltage of 4/−3.5 kV. Separation was carried out using a C18 column (Agilent Zorbax Eclipse Plus; 100 × 2.1 mm; particle size 1.8 μm, Agilent, Santa Clara, CA, USA) at the temperature 35 °C and an injection volume of 2.5 μL. The solvent composition, gradient conditions, collision energy for individual compounds and the analytical quality parameters, including detection and quantification limits, were described previously [18]. Compounds were identified based on the standard calibration or, for the compounds lacking reference standards, through comparison of mass fragmentation patterns with literature reports, as described in detail previously [18]. Concentrations of identified polyphenols were expressed as mg g^{-1} of dry carob leaf. All analyses were performed in duplicate.

2.10. Experimental Design and Statistical Analysis

Statistica version 10.0 software (Statsoft Inc., Tulsa, OK, USA) was employed for comprehensive data analysis. To optimize each extraction technique, a full factorial design, as outlined in Table 1, was conducted. This design involved 18 experimental trials, each performed in duplicate, to evaluate how extraction parameters influenced the TPC of the resulting extracts. The independent variables (X), which included temperature and extraction time, were varied across six and three levels, respectively, while TPC was designated as the dependent variable (Y). To ensure the robustness of the analysis, assessments for normality and homogeneity of data and residuals were carried out utilizing Shapiro–Wilk's and Levene's tests, respectively. For datasets meeting normality and homogeneity criteria, a one-way analysis of variance (ANOVA) was employed, followed by Tukey's Honest Significant Difference (HSD) multiple comparison test. Conversely, for datasets not meeting these criteria, the Kruskal–Wallis one-way ANOVA, coupled with multiple comparisons of mean ranks, was utilized. Moreover, the values obtained from UPLC-MS^2 analysis and antioxidant activity assays were compared using a one-way ANOVA. All statistical analyses considered significance at $p \leq 0.05$.

3. Results and Discussion

This paper studied the influence of temperature and time during the PLE, MAE and UAE of polyphenols from carob leaves. The raw data were statistically analyzed, and the results are shown in Table 2. The extracts, which were chosen as optimal based on the results of the statistical analysis, were further analyzed by UPLC-MS^2 in order to determine the individual polyphenolic content, while the antioxidant activity in the selected extracts was determined by the DPPH and FRAP methods. As shown in Table 1, the TPC of carob leaf extracts ranged from 25.87 to 90.67 mg GAE g^{-1}, which is in the range reported in the literature for carob leaves. The few existing studies reported TPC values of 6.28 [19], 21.41 [20], 45.26 [7] and up to 680 mg GAE g^{-1}, achieved through successive extraction and fractionation with different solvents [21]. As for differences between the techniques, the obtained ranges overlapped (PLE 33.72–80.73 mg GAE g^{-1}, MAE 51.25–90.67 mg GAE g^{-1} and UAE 25.87–55.98 mg GAE g^{-1}), with UAE yielding the lowest value and MAE yielding the highest value. Since similar values were achieved by all three techniques, the differences were likely a result of the applied extraction parameters, the effects of which are discussed further in text.

Table 2. Influence of temperature and time on the phenolic content of carob leaf extracts obtained by PLE, MAE and UAE.

	Source of Variation	N	Total Phenols (mg GAE g^{-1})
PLE	Temperature (°C)		$p < 0.01$ *
	60	12	36.50 ± 0.94 ^{a}
	80	12	44.27 ± 0.18 ^{b}
	100	12	40.98 ± 0.85 ^{a,b}
	120	12	47.60 ± 1.40 ^{b}
	140	12	61.31 ± 1.78 ^{c}
	160	12	72.63 ± 1.73 ^{d}
	Static time (min)		$p = 0.49$
	5	24	50.95 ± 2.20 ^{a}
	10	24	48.83 ± 2.40 ^{a}
	15	24	51.87 ± 3.46 ^{a}
MAE	Temperature (°C)		$p < 0.01$ *
	30	12	60.14 ± 0.77 ^{a}
	40	12	60.68 ± 2.22 ^{a}
	50	12	65.31 ± 1.23 ^{a,b}
	60	12	69.59 ± 1.39 ^{b}
	70	12	78.32 ± 3.11 ^{c}
	80	12	80.43 ± 1.01 ^{c}
	Time (min)		$p < 0.01$ *
	5	24	63.07 ± 1.52 ^{a}
	10	24	70.43 ± 1.85 ^{b}
	15	24	73.74 ± 2.15 ^{b}
UAE	Temperature (°C)		$p < 0.01$ *
	30	12	41.76 ± 2.44 ^{a}
	40	12	41.40 ± 1.92 ^{a}
	50	12	39.29 ± 1.98 ^{a}
	60	12	41.18 ± 3.44 ^{a}
	70	12	51.85 ± 1.74 ^{b}
	80	12	45.01 ± 1.42 ^{a}
	Time (min)		$p < 0.01$ *
	5	24	36.12 ± 1.52 ^{a}
	10	24	45.61 ± 1.52 ^{b}
	15	24	48.51 ± 1.11 ^{b}

PLE = pressurized liquid extraction. MAE = microwave-assisted extraction. UAE = ultrasound-assisted extraction. N = number of trials. GAE = gallic acid equivalents. Values are expressed as mean ± SE. Values within group marked with different letters are statistically different at $p \leq 0.05$. * Statistically significant at $p < 0.05$.

3.1. Influence of Temperature on the Total Phenolic Content of Carob Leaf Extracts

Temperature plays a vital role in the extraction process, as it directly impacts the rate of mass transfer and the solubility of targeted compounds. Typically, higher temperatures are anticipated to augment the diffusion of solvents into the plant matrix, consequently increasing the solubility of the targeted compounds and leading to their greater desorption [22]. However, high temperatures pose a risk of degradation thermosensitive compounds, which is why optimization is crucial. In the present research, temperatures applied for MAE and UAE were in the range of 30–80 °C, since the boiling point of 50% of aqueous ethanol is around 80 °C [23]. In the case of PLE, elevated temperatures were employed, due to the capability of high pressure to enable the utilization of solvents at temperatures surpassing their boiling points [24].

The results of PLE showed that higher applied temperatures had a statistically significant effect on the TPC of carob leaves, with the highest temperature of 160 °C emerging as the optimal one. In this study, the yield obtained at the highest applied temperature was 50% higher than that at the lowest temperature during PLE. According to the literature, PLE at temperatures above 150 °C leads to a significant increase in TPC yield due to the release of polyphenols from the plant material and the degradation of lignin, while at higher temperatures (above 180 °C), thermal degradation may occur [22,25]. Other studies on the leaves of *Urtica dioica* L. [26], *Elaeagnus rhamnoides* L. [27] and *Laurus nobilis* L. [16] also concluded that increasing the temperature in PLE increased the yield of TPC.

As for MAE, an increase of the TPC was observed at temperatures above 50 °C, with the highest value obtained at 70 °C. Further increase of temperature to 80 °C did not yield a higher TPC. Therefore, 70 °C emerged as optimal for the MAE of carob leaf polyphenols. Dobrinčić et al. [28] also observed a rise in the TPC of *Olea europaea* L. leaf extracts with the increase in temperature from 45 °C to 80 °C, which is consistent with the results of this study. The results were also partially in line with the results achieved during MAE *Salvia officinalis* L. using 30% ethanol, where a steady increase of TPC was observed with the increase in temperature from 30 to 80 °C [28,29], with the difference of the optimal temperature in the mentioned study being 80 °C. The differences could be a result of varying chemical composition of plant materials, namely the presence of other constituents, such as proteins or carbohydrates, which influence the yield of extraction and therefore require different extraction conditions. In the case of UAE, results were similar to MAE, with no differences observed in the range of 30–60 °C, while the highest TPC was achieved at 70 °C. As opposed to MAE, further increasing the temperature to 80 °C resulted in a lower TPC, indicating thermal degradation. In addition, it was shown that during UAE, higher temperatures increase the solvent's vapor pressure, resulting in a lower cavitation force and a consequent lower yield of the targeted compounds [30].

3.2. Influence of Time on the Total Phenolic Content of Carob Leaf Extracts

Time is another important parameter during extraction of bioactive compounds. Prolonged exposure of the plant material to solvent is expected to enhance the extraction yield by promoting the diffusion of the compounds from the plant matrix [16]. However, prolonged exposure to the solvent might cause oxidation, or in case of higher temperatures, thermal degradation of the targeted compounds [31], therefore finding optimal time for each plant material is essential for a successful extraction process.

Regarding PLE, time had no statistically significant effect on the extraction of TPC from carob leaves, so the optimal extraction time was 5 min, offering both time and energy savings. Dobroslavić et al. [16] also concluded that time had no significant effect in optimizing the PLE conditions for the isolation of phenols from *L. nobilis* L. leaves. Similarly, other studies [32–34] have confirmed that a static time of 5 min during PLE proved to be effective for the isolation of polyphenols from various plant materials. Briefly, a short extraction time at a high temperature was optimal for efficient isolation and no degradation of the target components occurred.

In the case of both MAE and UAE, the statistical analysis showed an increase of TPC between 5 and 10 min, while further prolongation to 15 min had no significant influence indicating that, following the Fick's law of diffusion, equilibrium was achieved in that time [35]. Therefore, 10 min can be considered the optimal time for both techniques. This is consistent with the literature data on other plant materials. For example, 10 min was the optimal extraction time during the MAE of *Sideritis scardica*, *Sideritis raeseri* and *Origanum vulgare* L. [36], as well as *L. nobilis* L. polyphenols [11]. The same was observed during the UAE of polyphenols from *Mesembryanthemum edule* L. and *L. nobilis* L. [11,37].

Based on the results of the statistical analysis, the optimal temperature and time were defined for the three advanced techniques as follows: 160 °C and 5 min for PLE, 70 °C and 10 min for both MAE and UAE.

3.3. Individual Polyphenolic Content of the Carob Leaf Extracts

Once the optimal temperature and extraction time were established for each of the three extraction techniques used, the resulting optimal extracts underwent UPLC-MS^2 analysis (Table 3). A total of 26 compounds, comprising flavonols, phenolic acids, flavones, flavan-3-ols and proanthocyanidins, were identified in the extracts obtained through all three techniques. Their identification was based on comparisons with authentic standards or by matching fragmentation patterns described in prior literature [10]. Chromatograms are shown in Figure 1.

Flavonols were the predominant group of carob leaf polyphenols, represented mostly by myricetin and quercetin-3-rhamnoside, which are known to usually be the most abundant flavonoids in carob [6,38,39]. Phenolic acids were the second most significant group of polyphenols, with gallic acid making up approximately 90% of them, which is consistent with the literature, since gallic acid was shown to be the main representative of phenolic acids in carob leaves [3,6]. As for the influence of the extraction techniques, UAE yielded the lowest concentration of both flavonols and phenolic acids even though the optimal temperature and time were the same as in MAE. It is possible that the degradation of the compounds occurred due to application of a frequency higher than 20 kHz (37 kHz), which can increase the generation of hydroxyl radicals that promote the decomposition of polyphenols [40]. The content of flavonols was the same in the extracts obtained by PLE and MAE, while PLE was more successful in yielding phenolic acids, possibly due to the application of high temperature, above 150 °C, which was shown to result in the release of bound phenolic acids due to the breakage of lignin bonds [22]. The content of other groups was significantly lower than that of flavonols and phenolic acids. Flavan-3-ols were mainly represented by epigallocatechin gallate, which was found to be a significant flavan-3-ol in carob leaves [6], while the flavones whose content is expected to be low in carob leaves were represented by apigenin and luteolin [3]. Unlike the proanthocyanidin content in carob pods, which can reach up to 20% of dry weight, the content of proanthocyanidins in carob leaves was reported to be only 0.7% of dry weight [41]. Therefore, the low content of the detected procyanidin trimer type B is consistent with literature data. A slightly higher concentration of proanthocyanidins was observed in the extract obtained by UAE, likely due to the cavitation-induced mechanochemical breakage of polymer units [42]. The TPC determined by UPLC-MS^2 was lower than the one determined by spectrophotometric analysis and showed different trends between techniques where MAE was not dominant over PLE, likely due to the presence of other extracted non-phenolic compounds, such as ascorbic acid, carbohydrates, minerals, other organic acids, or pigments such as chlorophyll, which can all be detected by spectrophotometer [43,44] and cause an apparently higher result.

3.4. Antioxidant Activity of the Carob Leaf Extracts

The antioxidant activity of carob leaf extracts obtained at the optimal extraction parameters was evaluated using two in vitro model systems: FRAP and DPPH assays. The results, presented in Figure 2a,b, demonstrated that the antioxidant activity of extracts obtained by PLE, MAE, and UAE ranged between 0.46–1.02 mmol TE g^{-1} for the FRAP assay and 0.50–0.58 mmol TE g^{-1} for the DPPH assay. Custodio et al. [45] investigated the antioxidant activity of the extracts obtained by the decoction of various parts of the carob tree and found that the leaves showed the highest activity according to the FRAP and DPPH test, which can be attributed to the high polyphenolic content. Other studies [46,47] that compared the antioxidant activity of plant extracts using the above-mentioned methods also concluded that the highest antioxidant activity was shown by the extracts with the highest level of polyphenols. In order to investigate the effect of polyphenolic content in the optimal extracts obtained in this study and antioxidant activity, Pearson's correlation factors were calculated (Table 4).

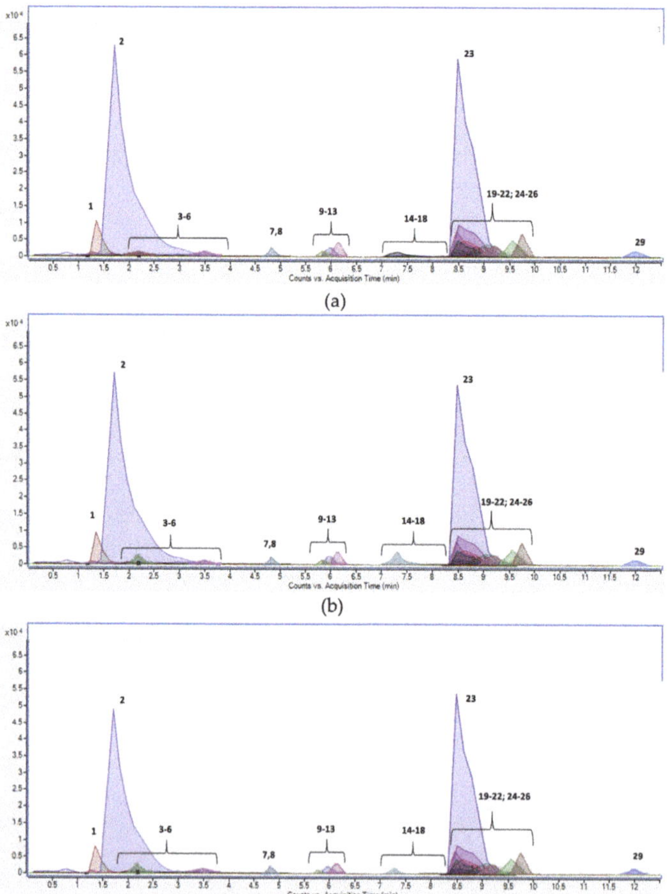

Figure 1. UPLC MS^2 chromatogram of optimal carob leaf extracts for (**a**) PLE, (**b**) MAE and (**c**) UAE in MRM acquisition mode. (1) Myricetin, (2) gallic acid, (3) 3,4-dihidrobenzoic acid hexoside, (4) apigenin-6-C(O-deoxyhexosyl)-hexoside, (5) syringic acid, (6) protocatechuic acid, (7) rosmarinic acid, (8) p-hydroxybenzoic acid, (9) chlorogenic acid, (10) caffeic acid, (11) catechin, (12) epicatechin, (13) epigallocatechin gallate, (14) procyanidin trimer, (15) luteolin-6-C glucoside, (16) p-coumaric acid, (17) rutin, (18) ferulic acid, (19) quercetin-3-glucoside, (20) apigenin, (21) kaempferol-3-rutinoside, (22) kaempferol-3-O-hexoside, (23) quercetin-3-pentoside, (24) isorhamnetin-3-hexoside, (25) quercetin-3-rhamnoside, (26) kaempferol-3-O-pentoside, (27) kaempferol-3-O-deoxyhexoside, (28) luteolin, (29) epigallocatechin gallate.

Table 3. UPLC-MS^2 analysis and ascorbic acid content of the carob leaf extracts obtained by PLE, MAE and UAE.

Compound Number	Retention Time	Tentative Identification	Concentration (mg g^{-1} Carob Leaf)		
			PLE	MAE	UAE
		Flavonols			
1	1.37	Myricetin *	9.57 ± 0.15 [b]	9.55 ± 0.19 [b]	8.66 ± 0.18 [a]
15	7.692	Rutin *	0.10 ± 0.00 [c]	0.06 ± 0.00 [a]	0.09 ± 0.00 [b]
17	7.969	Quercetin glucoside	0.76 ± 0.02 [a]	0.74 ± 0.02 [a]	0.90 ± 0.02 [b]
19	8.48	Kaempferol rutinoside	0.00 ± 0.00 [b]	0.00 ± 0.00 [a]	0.01 ± 0.00 [c]
20	8.51	Kaempferol-*O*-hexoside	0.18 ± 0.00 [c]	0.17 ± 0.00 [b]	0.16 ± 0.00 [a]
21	8.52	Quercetin pentoside	0.49 ± 0.01 [b]	0.50 ± 0.01 [b]	0.44 ± 0.01 [a]
22	8.877	Isorhamnetin hexoside	0.29 ± 0.01 [a]	0.32 ± 0.01 [b]	0.32 ± 0.01 [b]
23	8.897	Quercetin rhamnoside	5.96 ± 0.12 [b]	6.17 ± 0.13 [b]	5.73 ± 0.12 [a]
24	9.178	Kaempferol-*O*-pentoside	0.03 ± 0.00 [a]	0.06 ± 0.00 [b]	0.03 ± 0.00 [a]
		Total flavonols	17.39 ± 0.35 [b]	17.58 ± 0.36 [b]	16.33 ± 0.33 [a]
		Phenolic acids			
2	1.679	Gallic acid *	9.09 ± 0.19 [b]	8.78 ± 0.18 [b]	6.61 ± 0.13 [a]
3	2.313	3.4-Dihidrobenzoic acid hexoside	0.01 ± 0.00 [b]	0.01 ± 0.00 [c]	0.00 ± 0.00 [a]
4	3.488	Syringic acid *	0.04 ± 0.00 [a]	0.04 ± 0.00 [a]	0.06 ± 0.00 [b]
5	3.508	Protocatehuic acid *	0.08 ± 0.00 [a]	0.08 ± 0.00 [a]	0.07 ± 0.00 [a]
6	4.259	Rosmarinic acid *	0.01 ± 0.00 [a]	0.02 ± 0.00 [c]	0.01 ± 0.00 [b]
7	4.813	*p*-hydroxybenzoic acid	0.44 ± 0.01 [a]	0.41 ± 0.01 [a]	0.42 ± 0.01 [a]
8	5.043	Chlorogenic acid *	0.04 ± 0.00 [c]	0.01 ± 0.00 [a]	0.03 ± 0.00 [b]
9	5.711	Caffeic acid *	0.76 ± 0.02 [c]	0.03 ± 0.00 [b]	0.02 ± 0.00 [a]
14	7.28	*p*-coumaric acid *	0.28 ± 0.01 [b]	0.40 ± 0.01 [c]	0.23 ± 0.00 [a]
16	7.787	Ferulic acid *	0.01 ± 0.00 [a]	0.03 ± 0.00 [c]	0.02 ± 0.00 [b]
		Total phenolic acids	10.75 ± 0.22 [c]	9.8114 ± 0.2003 [b]	7.4583 ± 0.1522 [a]
		Flavones			
18	8.29	Apigenin *	0.05 ± 0.00 [a]	0.05 ± 0.00 [a]	0.05 ± 0.00 [a]
25	9.849	Luteolin *	0.08 ± 0.00 [a]	0.08 ± 0.00 [b]	0.09 ± 0.00 [b]
		Total flavones	0.13 ± 0.00 [a]	0.13 ± 0.00 [a]	0.14 ± 0.00 [b]
		Flavan-3-ols			
10	5.93	Catechin *	0.03 ± 0.00 [a]	0.05 ± 0.00 [c]	0.04 ± 0.00 [b]
11	5.937	Epicatechin	0.04 ± 0.00 [a]	0.07 ± 0.00 [b]	0.04 ± 0.00 [a]
12	6.02	Epigallocatechin gallate *	1.39 ± 0.03 [b]	1.25 ± 0.03 [b]	0.96 ± 0.02 [a]
26	12.159	Epicatechin gallate *	0.46 ± 0.01 [b]	0.37 ± 0.01 [a]	0.36 ± 0.01 [a]
		Total flavan-3-ols	1.93 ± 0.04 [c]	1.73 ± 0.04 [b]	1.39 ± 0.03 [a]
		Proanthocyanidins			
13	6.249	Procyandinin trimer type B	0.01 ± 0.00 [b]	0.01 ± 0.00 [a]	0.11 ± 0.00 [c]
		Total polyphenols	30.20 ± 0.62 [b]	29.25 ± 0.60 [b]	25.43 ± 0.52 [a]
		Ascorbic acid (mg/mL)	0.03 ± 0.00 [a]	0.44 ± 0.02 [b]	0.52 ± 0.02 [c]

PLE = pressurized liquid extraction. MAE = microwave-assisted extraction. UAE = ultrasound-assisted extraction. Values are expressed as mean ± SD. Values within row marked with different letters are statistically different at $p \leq 0.05$. * Identification confirmed by comparison with authentic standard.

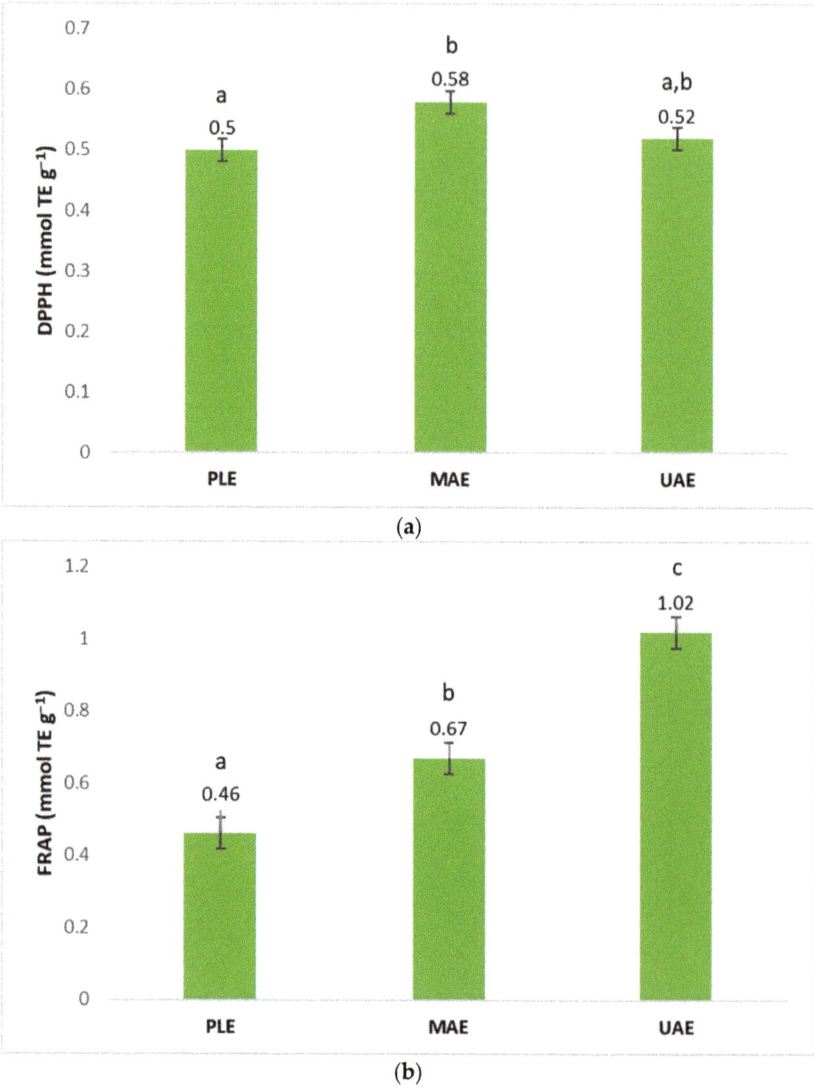

Figure 2. Antioxidant activity of carob leaves extract obtained by (**a**) DPPH assay and (**b**) FRAP assay at optimal PLE, MAE and UAE conditions. Values with different letter are statistically different at $p \leq 0.05$. PLE = pressurized liquid extraction. MAE = microwave assisted extraction. UAE = ultrasound assisted extraction. TE = trolox equivalents.

Considering the DPPH assay, the results were similar between the extracts obtained by all three techniques; however, MAE yielded a slightly higher DPPH value compared to the PLE extract (Figure 2a). As shown in Table 4, the correlation between antioxidant activity and the content of polyphenolic groups or their main representatives varied significantly. In the case of antioxidant activity determined by DPPH, only the content of myricetin, as the main representative of flavonols in the extracts, showed a statistically significant and highly positive correlation, indicating that myricetin has the most significant effect on the DPPH radical and significantly contributes to the antioxidant activity of carob leaf extracts. This was expected, since Gerhäuser [48] stated that flavonols such as myricetin are powerful scavengers of DPPH radicals, due to the presence of hydroxyl groups at certain positions. In this study, all three optimal extracts contain high concentrations of myricetin, as shown by UPLC-MS^2 quantification, with the UAE extract having a slightly lower content of the mentioned compound (Table 3). Nevertheless, since the values were very similar, the consistent activity can be explained by the fact that all three extraction procedures were effective in the extraction of antioxidant polyphenols present in carob leaves. Other studies [49–51], which measured antioxidant activity in extracts obtained from different parts of carob, obtained lower DPPH values, which could be due to the use of different materials, solvents or extraction techniques.

Table 4. Correlation between antioxidant activity and the content of polyphenols in carob leaf extracts.

Group of Compounds	Pearson for DPPH	Pearson for FRAP
Flavonols	0.41	−0.87 *
Phenolic acids	0.00	−1.00 *
Flavones	0.54	0.89 *
Flavan-3-ols	−0.08	−1.00 *
Proanthocyanidins	−0.30	0.92 *
Gallic acid	0.17	−0.97 *
Myricetin	0.70 *	−0.64 *
Quercetin-3-rhamnoside	0.26	−0.93 *
Total phenols UPLC-MS^2	0.09	−0.98 *

* Statistically significant at $p < 0.05$.

In the case of FRAP, significant differences were observed between the three applied techniques, and, in this case, the extract obtained by UAE showed the highest antioxidant activity, while the PLE extract showed the lowest antioxidant activity (Figure 2b). All of the polyphenolic groups and individual representatives showed significant correlations with the antioxidant activity. However, only flavones and proanthocyanidins, whose content in the extracts is relatively low, showed a high positive correlation, indicating that their presence was followed by higher antioxidant activity. All of the other groups and individual representatives, including myricetin, quercetin, rhamnoside and gallic acid showed a high negative correlation, indicating that higher content was associated with lower antioxidant activity, as determined by the FRAP method. Since it is known that polyphenols effectively reduce the Fe^{3+} in the FRAP reagent [52], the negative correlation could potentially be explained by the fact that other constituents of carob leaf that were present in the extract had a stronger reaction with the Fe^{3+}. For example, ascorbic acid, which was shown to be an important factor in the antioxidant and antitumor activity of carob [53] and is well known for its ferric ion reducing properties [54], could have been the main contributor to the antioxidant activity determined by FRAP, therefore overshadowing the effects of polyphenols. Considering this, the differences in results obtained by different extraction techniques could also be explained by the content of ascorbic acid (Table 3), which was consistent with the results of the FRAP assay. The lowest content of ascorbic acid was determined in the extract obtained by PLE, followed by MAE and UAE, which had the highest content. The significantly lower content of ascorbic acid in the extract obtained by PLE could be a result of thermal degradation at the temperature of 160 °C applied during PLE, resulting in the lowest antioxidant activity in this extract [55]. The highest antioxidant activity of the extract obtained by UAE is supported by the content of flavones and proanthocyanidins, which was slightly higher in this extract, but it could also potentially be a result of the higher content of ascorbic acid and its oxidation protecting effects on polyphenols [56]. This is because oxidation often occurs during the application of ultrasound, leading to the generation of hydroxyl radicals and resulting in less efficient antioxidant activity [57]. Since the mechanisms of the FRAP and DPPH methods differ, it is likely that the effect of ascorbic acid and other constituents was less significant in the DPPH method compared to the FRAP method.

4. Conclusions

This study optimized the temperature and time of PLE, MAE and UAE as advanced techniques for the efficient isolation of polyphenols from carob leaf, marking the first application of PLE and MAE for this purpose. The determined optimal temperature and time for MAE and UAE were 70 °C/10 min, and 160 °C/5 min for PLE. The phenolic profile of carob leaf extracts, acquired through all three techniques, encompassed 26 compounds, categorized into flavonols, phenolic acids, flavones, flavan-3-ols and proanthocyanidins groups. The predominant polyphenols were myricetin, gallic acid and quercetin-3-rhamnoside. The optimized PLE and MAE procedures showed slightly higher amounts of total and individual polyphenols compared to UAE. All examined extracts showed high antioxidant activity, as assessed by the FRAP and DPPH assays, indicating that carob leaf extracts obtained by advanced extraction techniques represent a rich source of polyphenols, which is supported by the content of ascorbic acid, making them promising ingredients for incorporation into functional foods. In order to achieve this, future research should focus on environmental growth factors and harvesting season during the cultivation of carob, all of which can contribute to obtaining higher yields of polyphenols. In addition, future research on micro- and nano-encapsulation, aimed at achieving maximum retention, quality and bioaccessibility of encapsulated carob leaf polyphenols, represents the next step toward the efficient utilization of this valuable plant source.

Author Contributions: Conceptualization, E.C. and E.D.; methodology, E.C., E.D. and Z.Z.; formal analysis, E.C., E.D. and Z.Z.; investigation, E.C. and E.D.; data curation, E.C. and E.D., writing—original draft preparation, E.C. and E.D.; writing—review and editing, M.R. and I.E.G.; supervision, I.E.G. All authors have read and agreed to the published version of the manuscript.

Funding: This work was supported by the project "Bioactive molecules of medical plant as natural antioxidants, microbicides and preservatives" (KK.01.1.1.04.0093), co-financed by the Croatian Government and the European Union through the European Regional Development Fund—Operational Programme Competitiveness and Cohesion (KK.01.1.1.04).

Data Availability Statement: The dataset is available on request from the authors.

Conflicts of Interest: The authors declare no conflicts of interest. The funders had no role in the design of the study; in the collection, analyses or interpretation of data; in the writing of the manuscript; or in the decision to publish the results.

References

1. El Hajaji, H.; Lachkar, N.; Alaoui, K.; Cherrah, Y.; Farah, A.; Ennabili, A.; El Bali, B.; Lachkar, M. Antioxidant Properties and Total Phenolic Content of Three Varieties of Carob Tree Leaves from Morocco. *Rec. Nat. Prod.* **2010**, *4*, 193–204.
2. Amine, E.; Taibi, M.; Ouassou, H.; Ouahhoud, S.; Ou-Yahia, D.; Loukili, E.H.; Aherkou, M.; Mansouri, F.; Bencheikh, N.; Laarej, S.; et al. Exploring the Multi-Faceted Potential of Carob Leaves from Morocco: A Comprehensive Analysis of Polyphenols Profile, Antimicrobial Activity, Cytotoxicity against Breast Cancer Cell Lines, and Genotoxicity. *Pharmaceuticals* **2023**, *16*, 840. [CrossRef]
3. Dahmani, W.; Elaouni, N.; Abousalim, A.; Akissi, Z.L.E.; Legssyer, A.; Ziyyat, A.; Sahpaz, S. Exploring Carob (*Ceratonia siliqua* L.): A Comprehensive Assessment of Its Characteristics, Ethnomedicinal Uses, Phytochemical Aspects, and Pharmacological Activities. *Plants* **2023**, *12*, 3303. [CrossRef]
4. Nechchadi, H.; Benhssaine, K.; Boulbaroud, S.; Berrougui, H.; Ramchoun, M. Factors of Variation and the Techniques for Improving Extraction and Bioaccessibility of Carob Polyphenol: A Review. *J. Food Meas. Charact.* **2023**, *17*, 4775–4799. [CrossRef]
5. Rtibi, K.; Selmi, S.; Grami, D.; Amri, M.; Eto, B.; El-benna, J.; Sebai, H.; Marzouki, L. Chemical Constituents and Pharmacological Actions of Carob Pods and Leaves (*Ceratonia siliqua* L.) on the Gastrointestinal Tract: A Review. *Biomed. Pharmacother.* **2017**, *93*, 522–528. [CrossRef]
6. Stavrou, I.J.; Christou, A.; Kapnissi-Christodoulou, C.P. Polyphenols in Carobs: A Review on Their Composition, Antioxidant Capacity and Cytotoxic Effects, and Health Impact. *Food Chem.* **2018**, *269*, 355–374. [CrossRef]
7. Brahim, E.B. Total Polyphenols and Gallic Acid Contents in Domesticated Carob (*Ceratonia siliqua* L.) Pods and Leaves. *Int. J. Pure Appl. Biosci.* **2017**, *5*, 22–30. [CrossRef]

8. Singh, S.V.; Manhas, A.; Singh, S.P.; Mishra, S.; Tiwari, N.; Kumar, P.; Shanker, K.; Srivastava, K.; Sashidhara, K.V.; Pal, A. A Phenolic Glycoside from Flacourtia Indica Induces Heme Mediated Oxidative Stress in Plasmodium Falciparum and Attenuates Malaria Pathogenesis in Mice. *Phytomedicine* **2017**, *30*, 1–9. [CrossRef]
9. Aissani, N.; Coroneo, V.; Fattouch, S.; Caboni, P. Inhibitory Effect of Carob (*Ceratonia siliqua*) Leaves Methanolic Extract on *Listeria monocytogenes*. *J. Agric. Food Chem.* **2012**, *60*, 9954–9958. [CrossRef]
10. Savic, I.M.; Savic Gajic, I.M. Optimization of Ultrasound-Assisted Extraction of Polyphenols from Wheatgrass (*Triticum aestivum* L.). *J. Food Sci. Technol.* **2020**, *57*, 2809–2818. [CrossRef]
11. Dobroslavić, E.; Elez Garofulić, I.; Zorić, Z.; Pedisić, S.; Dragović-Uzelac, V. Polyphenolic Characterization and Antioxidant Capacity of *Laurus nobilis* L. Leaf Extracts Obtained by Green and Conventional Extraction Techniques. *Processes* **2021**, *9*, 1840. [CrossRef]
12. Ameer, K.; Shahbaz, H.M.; Kwon, J. Green Extraction Methods for Polyphenols from Plant Matrices and Their Byproducts: A Review. *Compr. Rev. Food Sci. Food Saf.* **2017**, *16*, 295–315. [CrossRef] [PubMed]
13. Arya, P.; Kumar, P. Comparison of Ultrasound and Microwave Assisted Extraction of Diosgenin from Trigonella Foenum Graceum Seed. *Ultrason. Sonochem.* **2021**, *74*, 105572. [CrossRef] [PubMed]
14. López-Fernández, O.; Domínguez, R.; Pateiro, M.; Munekata, P.E.S.; Rocchetti, G.; Lorenzo, J.M. Determination of Polyphenols Using Liquid Chromatography–Tandem Mass Spectrometry Technique (LC–MS/MS): A Review. *Antioxidants* **2020**, *9*, 479. [CrossRef] [PubMed]
15. Shortle, E.; O'Grady, M.N.; Gilroy, D.; Furey, A.; Quinn, N.; Kerry, J.P. Influence of Extraction Technique on the Anti-Oxidative Potential of Hawthorn (*Crataegus monogyna*) Extracts in Bovine Muscle Homogenates. *Meat Sci.* **2014**, *98*, 828–834. [CrossRef] [PubMed]
16. Dobroslavić, E.; Elez Garofulić, I.; Šeparović, J.; Zorić, Z.; Pedisić, S.; Dragović-Uzelac, V. Pressurized Liquid Extraction as a Novel Technique for the Isolation of *Laurus nobilis* L. Leaf Polyphenols. *Molecules* **2022**, *27*, 5099. [CrossRef] [PubMed]
17. Voća, S.; Dobričević, N.; Skendrović Babojelić, M.; Družić, J.; Duralija, B.; Levačić, J. Differences in Fruit Quality of Strawberry Cv. Elsanta Depending on Cultivation System and Harvest Time. *Agric. Conspec. Sci.* **2007**, *72*, 285–288.
18. Elez Garofulić, I.; Zorić, Z.; Pedisić, S.; Brnčić, M.; Dragović-Uzelac, V. UPLC-MS^2 Profiling of Blackthorn Flower Polyphenols Isolated by Ultrasound-Assisted Extraction. *J. Food Sci.* **2018**, *83*, 2782–2789. [CrossRef] [PubMed]
19. Corsi, L.; Avallone, R.; Cosenza, F.; Farina, F.; Baraldi, C.; Baraldi, M. Antiproliferative Effects of *Ceratonia siliqua* L. on Mouse Hepatocellular Carcinoma Cell Line. *Fitoterapia* **2002**, *73*, 674–684. [CrossRef]
20. Meziani, S.; Oomah, B.D.; Zaidi, F.; Simon-Levert, A.; Bertrand, C.; Zaidi-Yahiaoui, R. Antibacterial Activity of Carob (*Ceratonia siliqua* L.) Extracts against Phytopathogenic Bacteria Pectobacterium Atrosepticum. *Microb. Pathog.* **2015**, *78*, 95–102. [CrossRef]
21. Ben Hsouna, A.; Trigui, M.; Mezghani Jarraya, R.; Damak, M.; Jaoua, S. Identification of Phenolic Compounds by High Performance Liquid Chromatography/Mass Spectrometry (HPLC/MS) and in Vitro Evaluation of the Antioxidant and Antimicrobial Activities of *Ceratonia siliqua* Leaves Extracts. *J. Med. Plant Res.* **2015**, *9*, 479–485. [CrossRef]
22. Antony, A.; Farid, M. Effect of Temperatures on Polyphenols during Extraction. *Appl. Sci.* **2022**, *12*, 2107. [CrossRef]
23. Kumar, S.; Singh, N.; Prasad, R. Anhydrous Ethanol: A Renewable Source of Energy. *Renew Sust. Energy Rev.* **2010**, *14*, 1830–1844. [CrossRef]
24. Mustafa, A.; Turner, C. Pressurized Liquid Extraction as a Green Approach in Food and Herbal Plants Extraction: A Review. *Anal. Chim. Acta* **2011**, *703*, 8–18. [CrossRef]
25. Luthria, D.L. Optimization of Extraction of Phenolic Acids from a Vegetable Waste Product Using a Pressurized Liquid Extractor. *J. Funct. Foods* **2012**, *4*, 842–850. [CrossRef]
26. Repajić, M.; Cegledi, E.; Kruk, V.; Pedisić, S.; Çinar, F.; Kovačević, D.B.; Žutić, I.; Dragović-Uzelac, V. Accelerated Solvent Extraction as a Green Tool for the Recovery of Polyphenols and Pigments Fromwild Nettle Leaves. *Processes* **2020**, *8*, 803. [CrossRef]
27. Čulina, P.; Repajić, M.; Elez Garofulić, I.; Dragović-Uzelac, V.; Pedisić, S. Evaluation of Polyphenolic Profile and Antioxidant Activity of Sea Buckthorn (*Elaeagnus Rhamnoides* (L.) A. Nelson) Leaf and Berry Extracts Obtained via Optimized Microwave-Assisted and Accelerated Solvent Extraction. *Processes* **2024**, *12*, 126. [CrossRef]
28. Dobrinčić, A.; Repajić, M.; Garofulić, I.E.; Tuđen, L.; Dragović-Uzelac, V.; Levaj, B. Comparison of Different Extraction Methods for the Recovery of Olive Leaves Polyphenols. *Processes* **2020**, *8*, 1008. [CrossRef]
29. Putnik, P.; Kovačević, D.B.; Penić, M.; Fegeš, M.; Dragović-Uzelac, V. Microwave-Assisted Extraction (MAE) of Dalmatian Sage Leaves for the Optimal Yield of Polyphenols: HPLC-DAD Identification and Quantification. *Food Anal. Methods* **2016**, *9*, 2385–2394. [CrossRef]
30. Santos, F.F.P.; Rodrigues, S.; Fernandes, F.A.N. Optimization of the Production of Biodiesel from Soybean Oil by Ultrasound Assisted Methanolysis. *Fuel Process Technol.* **2009**, *90*, 312–316. [CrossRef]
31. Naczk, M.; Shahidi, F. Extraction and Analysis of Phenolics in Food. *J. Chromatogr. A* **2004**, *1054*, 95–111. [CrossRef] [PubMed]
32. Martín-García, B.; Pimentel-Moral, S.; Gómez-Caravaca, A.M.; Arráez-Román, D.; Segura-Carretero, A. Box-Behnken Experimental Design for a Green Extraction Method of Phenolic Compounds from Olive Leaves. *Ind. Crops Prod.* **2020**, *154*, 112741. [CrossRef]

33. Lama-Muñoz, A.; del Mar Contreras, M.; Espínola, F.; Moya, M.; de Torres, A.; Romero, I.; Castro, E. Extraction of Oleuropein and Luteolin-7-O-Glucoside from Olive Leaves: Optimization of Technique and Operating Conditions. *Food Chem.* **2019**, *293*, 161–168. [CrossRef] [PubMed]
34. Hossain, M.B.; Barry-Ryan, C.; Martin-Diana, A.B.; Brunton, N.P. Optimisation of Accelerated Solvent Extraction of Antioxidant Compounds from Rosemary (*Rosmarinus Officinalis* L.), Marjoram (*Origanum Majorana* L.) and Oregano (*Origanum Vulgare* L.) Using Response Surface Methodology. *Food Chem.* **2011**, *126*, 339–346. [CrossRef]
35. Silva, E.; Rogez, H.; Larondelle, Y. Optimization of Extraction of Phenolics from Inga Edulis Leaves Using Response Surface Methodology. *Sep. Purif. Technol.* **2007**, *55*, 381–387. [CrossRef]
36. Sarakatsianos, I.; Adamopoulos, K.; Samanidou, V.; Goula, A.; Ninou, E. Optimization of Microwave-Assisted Extraction of Phenolic Compounds from Medicinal and Aromatic Plants: *Sideritis raeseri*, *Sideritis scardica* and *Origanum vulgare*. *Curr. Anal. Chem.* **2020**, *16*, 106–111. [CrossRef]
37. Falleh, H.; Ksouri, R.; Lucchessi, M.-E.; Abdelly, C.; Magné, C. Ultrasound-Assisted Extraction: Effect of Extraction Time and Solvent Power on the Levels of Polyphenols and Antioxidant Activity of *Mesembryanthemum edule* L. Aizoaceae Shoots. *Trop. J. Pharm. Res.* **2012**, *11*, 243–249. [CrossRef]
38. Goulas, V.; Stylos, E.; Chatziathanasiadou, M.; Mavromoustakos, T.; Tzakos, A. Functional Components of Carob Fruit: Linking the Chemical and Biological Space. *Int. J. Mol. Sci.* **2016**, *17*, 1875. [CrossRef] [PubMed]
39. Vaya, J.; Mahmood, S. Flavonoid Content in Leaf Extracts of the Fig (*Ficus carica* L.), Carob (*Ceratonia siliqua* L.) and Pistachio (*Pistacia lentiscus* L.). *BioFactors* **2006**, *28*, 169–175. [CrossRef]
40. Dzah, C.S.; Duan, Y.; Zhang, H.; Wen, C.; Zhang, J.; Chen, G.; Ma, H. The Effects of Ultrasound Assisted Extraction on Yield, Antioxidant, Anticancer and Antimicrobial Activity of Polyphenol Extracts: A Review. *Food Biosci.* **2020**, *35*, 100547. [CrossRef]
41. Makris, D.P.; Kefalas, P. Carob Pods (*Ceratonia siliqua* L.) as a Source of Polyphenolic Antioxidants. *Food Technol. Biotechnol.* **2004**, *42*, 105–108.
42. Lv, J.-M.; Gouda, M.; Zhu, Y.-Y.; Ye, X.-Q.; Chen, J.-C. Ultrasound-Assisted Extraction Optimization of Proanthocyanidins from Kiwi (*Actinidia chinensis*) Leaves and Evaluation of Its Antioxidant Activity. *Antioxidants* **2021**, *10*, 1317. [CrossRef] [PubMed]
43. El-Hamidi, M.; Zaher, F.A. Comparison Between Some Common Clays as Adsorbents of Carotenoids, Chlorophyll and Phenolic Compounds from Vegetable Oils. *Am. J. Food Technol.* **2016**, *11*, 92–99. [CrossRef]
44. Prior, R.L.; Wu, X.; Schaich, K. Standardized Methods for the Determination of Antioxidant Capacity and Phenolics in Foods and Dietary Supplements. *J. Agric. Food Chem.* **2005**, *53*, 4290–4302. [CrossRef]
45. Custódio, L.; Patarra, J.; Alberício, F.; Neng, N.R.; Nogueira, J.M.F.; Romano, A. In Vitro Antioxidant and Inhibitory Activity of Water Decoctions of Carob Tree (*Ceratonia siliqua* L.) on Cholinesterases, α-Amylase and α-Glucosidase. *Nat. Prod. Res.* **2015**, *29*, 2155–2159. [CrossRef]
46. Milić, A.; Daničić, T.; Tepić Horecki, A.; Šumić, Z.; Teslić, N.; Bursać Kovačević, D.; Putnik, P.; Pavlić, B. Sustainable Extractions for Maximizing Content of Antioxidant Phytochemicals from Black and Red Currants. *Foods* **2022**, *11*, 325. [CrossRef]
47. Irakli, M.; Skendi, A.; Bouloumpasi, E.; Christaki, S.; Biliaderis, C.G.; Chatzopoulou, P. Sustainable Recovery of Phenolic Compounds from Distilled Rosemary By-Product Using Green Extraction Methods: Optimization, Comparison, and Antioxidant Activity. *Molecules* **2023**, *28*, 6669. [CrossRef] [PubMed]
48. Gerhäuser, C. Phenolic Beer Compounds to Prevent Cancer. In *Beer in Health and Disease Prevention*; Elsevier: Amsterdam, The Netherlands, 2008; pp. 669–684; ISBN 9780123738912.
49. Khelouf, I.; Jabri Karoui, I.; Abderrabba, M. Chemical Composition, in Vitro Antioxidant and Antimicrobial Activities of Carob Pulp (*Ceratonia siliqua* L.) from Tunisia. *Chemical* **2023**, *77*, 6125–6134. [CrossRef]
50. Carbas, B.; Salinas, M.V.; Serrano, C.; Passarinho, J.A.; Puppo, M.C.; Ricardo, C.P.; Brites, C. Chemical Composition and Antioxidant Activity of Commercial Flours from *Ceratonia siliqua* and *Prosopis* spp. *J. Food Meas. Charact.* **2019**, *13*, 305–311. [CrossRef]
51. Martić, N.; Zahorec, J.; Stilinović, N.; Andrejić-Višnjić, B.; Pavlić, B.; Kladar, N.; Šoronja-Simović, D.; Šereš, Z.; Vujčić, M.; Horvat, O.; et al. Hepatoprotective Effect of Carob Pulp Flour (*Ceratonia siliqua* L.) Extract Obtained by Optimized Microwave-Assisted Extraction. *Pharmaceutics* **2022**, *14*, 657. [CrossRef]
52. Danilewicz, J.C. Folin-Ciocalteu, FRAP, and DPPH• Assays for Measuring Polyphenol Concentration in White Wine. *Am. J. Enol. Vitic.* **2015**, *66*, 463–471. [CrossRef]
53. Amessis-Ouchemoukh, N.; Ouchemoukh, S.; Meziant, N.; Idiri, Y.; Hernanz, D.; Stinco, C.M.; Rodríguez-Pulido, F.J.; Heredia, F.J.; Madani, K.; Luis, J. Bioactive Metabolites Involved in the Antioxidant, Anticancer and Anticalpain Activities of *Ficus carica* L., *Ceratonia siliqua* L. and *Quercus ilex* L. Extracts. *Ind. Crops Prod.* **2017**, *95*, 6–17. [CrossRef]
54. Kontoghiorghes, G.J.; Kolnagou, A.; Kontoghiorghe, C.N.; Mourouzidis, L.; Timoshnikov, V.A.; Polyakov, N.E. Trying to Solve the Puzzle of the Interaction of Ascorbic Acid and Iron: Redox, Chelation and Therapeutic Implications. *Medicines* **2020**, *7*, 45. [CrossRef] [PubMed]
55. Kadakal, Ç.; Duman, T.; Ekinci, R. Thermal Degradation Kinetics of Ascorbic Acid, Thiamine and Riboflavin in Rosehip (*Rosa canina* L.) Nectar. *Food Sci. Technol.* **2017**, *38*, 667–673. [CrossRef]

56. Kaack, K.; Austed, T. Interaction of Vitamin C and Flavonoids in Elderberry (*Sambucus nigra* L.) during Juice Processing. *Plant Foods Hum. Nutr.* **1998**, *52*, 187–198. [CrossRef]
57. Handa, S.S.; Khanuja, S.P.; Longo, G.; Rakesh, D.D. *Extraction Technologies for Medicinal and Aromatic Plants*; Earth, Environmental and Marine Sciences and Technologies: Trieste, Italy, 2008.

processes

Article

Efficiency of Supercritical CO_2 and Ultrasound-Assisted Extraction Techniques for Isolation of Bioactive Molecules from Sea Buckthorn (*Elaeagnus rhamnoides* (L.) A. Nelson) Berry Oils and Cakes

Patricija Čulina [1], Sandra Balbino [2], Stela Jokić [3], Verica Dragović-Uzelac [2] and Sandra Pedisić [1,*]

[1] Laboratory for Drying Technologies and Monitoring of Biologically Active Compounds, Centre for Food Technology and Biotechnology, Faculty of Food Technology and Biotechnology, University of Zagreb, P. Kasandrića 3, 23000 Zadar, Croatia; patricijaculina@gmail.com

[2] Department of Food Engineering, Faculty of Food Technology and Biotechnology, University of Zagreb, Pierottijeva 6, 10000 Zagreb, Croatia; sandra.balbino@pbf.unizg.hr (S.B.); vdragov@pbf.hr (V.D.-U.)

[3] Department of Process Engineering, Faculty of Food Technology Osijek, Josip Juraj Strossmayer University of Osijek, Franje Kuhača 18, 31000 Osijek, Croatia; stela.jokic@ptfos.hr

* Correspondence: sandra.pedisic@pbf.unizg.hr; Tel.: +385-23-331-077

Abstract: This study compared the efficiency of advanced supercritical CO_2 extraction (SC-CO_2) and conventional Soxhlet extraction (SE) in the isolation of lipophilic bioactive molecules (BAMs) from the oils of different sea buckthorn (SB) berries (Leikora and Ascola). A SB cake, a valuable by-product of oil extraction, was analyzed for phenolic compounds and antioxidant activity (AA). For this purpose, ultrasound-assisted extraction (UAE) with 70% (v/v) ethanol was optimized, and the following optimal UAE conditions were determined: ultrasonic power of 60%, temperature of 50 °C, and extraction time of 20 min. The individual BAMs in the oils and cake extracts were analyzed chromatographically, while the AAs were determined using the ORAC method. The oils extracted with SC-CO_2 had a higher content of lipophilic BAMs and a higher AA value than the oils extracted with SE. Palmitic and palmitoleic acids, β-sitosterol, and α-tocopherol dominated the Leikora oil, while a higher content of oleic, linoleic, and α-linolenic acids and carotenoids was found in the Ascola oil. The highest phenolic content was found in the Ascola cake, with the flavonols isorhamnetin-3-hexoside, isorhamnetin-3-rutinoside, and quercetin-3-glucoside being the most abundant. Oil and cake extracts of the Ascola variety also had a higher AA.

Keywords: sea buckthorn lipophilic and hydrophilic bioactive molecules; sea buckthorn berry oil and cake; supercritical CO_2 extraction; Soxhlet extraction; ultrasound-assisted extraction; antioxidant activity

Citation: Čulina, P.; Balbino, S.; Jokić, S.; Dragović-Uzelac, V.; Pedisić, S. Efficiency of Supercritical CO_2 and Ultrasound-Assisted Extraction Techniques for Isolation of Bioactive Molecules from Sea Buckthorn (*Elaeagnus rhamnoides* (L.) A. Nelson) Berry Oils and Cakes. *Processes* **2024**, *12*, 698. https://doi.org/10.3390/pr12040698

Academic Editors: Yanlin Zhang and Prashant K. Sarswat

Received: 5 March 2024
Revised: 27 March 2024
Accepted: 28 March 2024
Published: 29 March 2024

1. Introduction

Sea buckthorn (SB), lat. *Elaeagnus rhamnoides* (L.) A. Nelson, is a shrubby plant with yellow or orange colored berries, naturally distributed in Asia and Europe [1]. Various parts of the plant, especially the berries, are used in traditional medicine for their health-promoting effects, such as antioxidant, anti-inflammatory, antitumor, immunomodulatory, antimicrobial, and other properties [2–5]. SB berries have a high nutritional and biological value, which is due to the presence of both lipophilic and hydrophilic antioxidants such as carotenoids, tocopherols, fatty acids, sterols, phenolic compounds, and ascorbic acid [4,6–8]. It is known that factors such as variety, time of harvest, climate, growing conditions, ripeness at harvest, storage conditions, and processing technologies have a significant influence on the content and composition of BAMs [9,10]. The oil is the most valuable part of SB berries due to the presence of omega-3 (linolenic acid), omega-6 (linoleic acid) and omega-7 fatty acids (vaccenic acid, palmitooleic acid), α-tocopherol, sterols such as β-sitosterol, campesterol, and stigmasterol, and carotenoids such as zeaxanthin, zeaxanthin dipalmitate and its esters, β-carotene, lycopene, and β-cryptoxanthin palmitate [7,11–13]. The oil quality is significantly influenced by pre-harvest factors such as the variety and the growing region, as well as by processing factors [14]. For the isolation of targeted lipophilic BAMs, i.e., plant oils, advanced extraction techniques such as supercritical CO_2 extraction (SC-CO_2) as a promising alternative to conventional extraction and mechanical pressing of oils have received considerable attention [15]. SC-CO_2 is an effective solvent for dissolving non-polar or slightly polar compounds, although its dissolving power for hydrophilic substrates can be increased by adding a polar co-solvent such as ethanol [16]. The main advantage of SC-CO_2 is the lower viscosity, better diffusion, and surface tension of CO_2, which allow better penetration into the material from which the target compounds are extracted. In addition, as an environmentally friendly and safe solvent (GRAS), CO_2 can be easily removed from the oil as it is inert and does not interact with the extract, so it serves as a carrier fluid, allowing the extract to be free of contaminants [15]. Most importantly, SC-CO_2 protects the antioxidants during extraction as it is carried out in the absence of oxygen, preventing the formation of oxidation products, which is a very important aspect for the preservation of BAMs such as tocopherols, carotenoids, unsaturated fatty acids, and sterols [17–19]. The extraction of oils generates large quantities of by-products that are currently used inefficiently or even disposed of as waste, but which represent a cheap source of valuable nutrients such as phenolic acids and flavonoids [20,21]. Ultrasound-assisted extraction (UAE) is one of the most commonly used extraction techniques for polyphenols, which allows higher reproducibility in a shorter time and the use of less solvent at lower processing temperatures compared with conventional extraction [22]. The efficiency of UAE depends on its parameters, mainly temperature, extraction time, and ultrasonic power, and it is of great importance to optimize it with regard to the plant material used and the desired target compounds. In addition, studies have shown that the biological properties of extracts obtained under optimal UAE conditions are preserved, which increases their application in various fields [23]. Goldsmith et al. [24] and Nunes et al. [25] reported that UAE is a very effective method for phenolic extraction from olive cake obtained after oil extraction, as the yield of phenolic compounds is high in a short time. Nowadays, there is a great need for the characterization of BAMs isolated from plant material and the evaluation of their antioxidant activity (AA). The most effective techniques for the characterization of polyphenols and pigments are liquid chromatographic techniques (HPLC/UPLC) coupled with diode-array detection (DAD), fluorescence detector (FLD), and/or mass spectrometry (MS), while gas chromatography (GC) coupled with MS or FID is used for sterol and fatty acid identification [26,27]. Due to the presence of different BAMs in SB, a suitable method for determining AA is the oxygen reducing antioxidant capacity assay (ORAC), which can measure both lipophilic and hydrophilic antioxidants.

To our knowledge, data on the efficiency of advanced extraction techniques to isolate lipophilic and hydrophilic BAMs with AA from SB berries is limited. In this study, SB berry oil was produced by conventional Soxhlet extraction (SE) and advanced extraction techniques (SC-CO_2) from different SB berry varieties (Leikora and Ascola). Therefore, the aim of this study was to compare the efficiency of SC-CO_2 and SE in the isolation of different lipophilic BAMs in oils. In addition, a SB cake obtained after oil extraction represents a valuable source of BAMs such as phenolic compounds. For this purpose, UAE was optimized with regard to total phenolic content (TPC) and AA. After determining the optimal UAE conditions, the polyphenolic composition of the extracts obtained from the cake under optimal UAE conditions was evaluated by UPLC/ESI-MS^2. The AA of the oils and the cake extracts were also evaluated using the ORAC method.

2. Materials and Methods

2.1. Plant Material

Different SB berry varieties, Leikora and Ascola, grown in Croatia, near Zagreb (Latitude, 45°58′53.05″ N; Longitude, 15°58′1.90″ E) in 2020, were used for this study. The botanical identification of the plants was carried out by the experts of the Faculty of Agriculture, University of Zagreb, Croatia.

2.2. Sample Preparation

Samples were freeze-dried (CoolSafe 55-9 PRO, Labogene, Allerød, Denmark) to a dry matter content of 90%, ground in a commercial grinder (GT11, Tefal, Rumilly, France), sieved through a 60-mesh sieve, placed in closed plastic containers, and stored at 4 °C until use.

2.3. Chemicals and Reagents

All solvents used for the extraction and analysis were HPLC grade and were obtained from different manufacturers; ethanol, formic acid and acetonitrile were obtained from BDH (Prolabo, Lutterworth, UK), hexane from J. T. Baker (Phillipsburg, NJ, USA), distilled water from Millipore Corp. (Bedford, NY, USA), anhydrous sodium carbonate (≥99.5%) and sodium phosphate (96%) were obtained from Kemika (Zagreb, Croatia), Folin-Ciocalteu reagent from Merck (Darmstadt, Germany), fluorescein sodium salt from Honeywell Riedel-de-Haën (Bucharest, Romania), Trolox (6-hydroxy-2,5,7,8-tetramethylchroman-2-carboxylic acid) from Acros Organics (Thermo Fisher Scientific, Geel, Belgium), 2,20-Azobis (2-amidinopropane) hydrochloride and methilated β-cyclodextrine were obtained from Sigma-Aldrich (Steinheim, Germany) and CO_2 (99.97%) (*w*/*w*) was obtained from the Messer (Osijek, Croatia).

2.4. Supercritical CO_2 Extraction

Supercritical CO_2 extraction of oil from SB berries was performed in the SC-CO_2 system at an extraction pressure of 276 Bar, a temperature of 35 °C, and a mass flow rate of 2 kg/h according to the optimal conditions obtained in our previous research by Pavlović et al. [28] with minor modifications. The extraction process took 90 min until all the amount of oil was extracted (each 15 min the amount of obtained extracts were collected in glass tubes weighed beforehand using a balance with a precision of ±0.0001 g).

2.5. Soxhlet Extraction

Soxhlet extraction of oil from SB berries was performed in Soxtherm (Gerhardt, Germany) using *n*-hexane as extraction solvent, according to the study by Harkat et al. [29], with minor modifications. The Soxhlet procedure consisted of placing 10 g of freeze-dried and ground SB berries in a thimble into the Soxhlet extraction system heated with a hot-plate, in which 50 mL of *n*-hexane was refluxed over the sample for 90 min at increasing temperatures (up to 180 °C). After extraction, the solvent was removed at 60 °C using a rotary evaporator. The obtained oil was filtered and stored in a colored bottle at −18 °C in a nitrogen gas atmosphere until analysis (no longer than 10 days).

2.6. Ultrasound-Assisted Extraction of by-Products of Oil Extraction

A SB cake of the Leikora variety obtained after SC-CO_2 extraction was used to determine the optimal extraction conditions for phenolic compounds. A mass of SB cake of Leikora variety (1 ± 0.001 g) obtained after SC-CO_2 extraction was mixed with 50 mL of aqueous ethanol solution (70%, *v*/*v*), and extraction was performed in an ultrasonic bath (Elmasonic P, Elma-Hans Schmidbauer GmbH & Co., Ltd., Singen, Germany) with 120 W ultrasonic power. A full factorial design comprising 27 experimental trials was chosen to evaluate the effect of three independent variables—ultrasonic power, temperature, and extraction time. According to the experimental design, ultrasonic power (30, 60, and 90%), temperature (35, 50, and 65 °C), and extraction time (10, 20, and 30 min) were varied in order to find the best extraction conditions with the maximum yield of total phenolic compounds. After the extraction, the obtained mixture was centrifuged at 4500× *g* rpm for 15 min, and the supernatants were filtered (Whatman No. 4) into 50 mL volumetric flasks and made up with extraction solvent. Extraction was performed in duplicate (n = 2), and the extracts were stored at −18 °C in a nitrogen gas atmosphere until analysis (no longer than 10 days).

The extraction of the phenolic compounds in all SB cakes obtained after oil extraction was carried out under the determined optimal conditions (60% ultrasonic power, 50 °C, 20 min). Prior to UPLC/ESI-MS^2 analysis the extracts were filtered through a 0.45 μm membrane filter (Macherey-Nagel GmbH, Düren, Germany).

2.7. Characterization of Fatty Acids and Sterols in SB Oils

Gas chromatography according to the ISO 12966-2:2017 method [30] and ISO 12228-1:2014 [31] was used for the characterization of fatty acids and sterols.

Fatty acid methyl esters were separated on a TRACE TR-FAME capillary column (30 m × 0.22 mm × 0.25 μm) using a stationary phase of 70% cyanopropyl polysilphenylene–siloxane (Thermo Scientific, Waltham, MA, USA), and the peaks of fatty acid methyl esters were identified by comparing their retention times with the retention times of the 37 Component FAME Mix and petroselinic acid standard (Supelco, Sigma-Aldrich, St. Louis, MO, USA). Results were expressed as weight percentages of total fatty acids (%).

Sterol content and composition analysis were performed using the Agilent Technologies 6890N Network GC System gas chromatograph (Santa Clara, CA, USA) equipped with a sample injection system, a flame ionization detector, and an Agilent Technologies 5973 inert mass selective detector. Peak identification was performed by comparing retention times and mass spectra with authentic standards. All sterols were quantified using an internal standard method with α-cholestanol (1 mg/mL). Results were expressed in mg per 100 g of oil (mg/100 g).

2.8. Characterization of Carotenoids in Oils by HPLC Analysis

High-performance liquid chromatography (HPLC) was used for the characterization of carotenoids using an Agilent 1260 Infinity quaternary LC system (Agilent Technologies, Santa Clara, CA, USA) coupled with a diode array detector (DAD). The reverse-phase separations were performed on a Develosil RP-Aqueous (C30) reversed-phase column (250 mm × 4.6 mm i.d., 5 μm particle size) (Phenomenex, Torrance, CA, USA). For the determination of carotenoids, 1 mL of oil was diluted in 5 mL of *n*-hexane. The solvent composition and gradient program for the analysis were used as described by Casto Puyana et al. [32], and wavelength detection was at 450 nm. Identification of the carotenoids and their derivatives was based on the DAD absorption spectra and retention times of the authentic standards of lutein, zeaxanthin, and β-carotene and literature data [33], while quantification was based on their calibration curves. All results were expressed in mg per 100 g of oil (mg/100 g).

2.9. Determination of α-Tocopherol

The tocopherol content was determined according to ISO method 9936 [34]. Samples were prepared by dissolving 0.1 g of oil in 10 mL of *n*-hexane and then analyzed using the Agilent 1260 Infinity quaternary LC system (Agilent Technologies, Santa Clara, CA, USA) equipped with a fluorescence detector. Separation was performed on a LiChroCART Silica 60 column (250 mm 94.6 mm, 5 L; Merck, Darmstadt, Germany), and a standard calibration curve for α-tocopherol was used for quantification. Results were expressed in mg per 100 g of oil (mg/100 g).

2.10. Determination of Total Phenolic Content

The UV-VIS spectrophotometer (UV-VIS UviLine 9400; Secomam, Ales, France) was used for the determination of total phenol content (TPC) according to the spectrophotometric Folin-Ciocalteu method, as previously described by Čulina et al. [35]. All measurements were performed in duplicate, and the results were expressed as mg gallic acid equivalents (GAE) per 100 g of dry matter (mg/100 g dm).

2.11. Characterization of Phenolic Compounds by UPLC/ESI-MS^2 Analysis

The UPLC/ESI-MS^2 analysis using an Agilent 1290 RRLC instrument (Agilent, Santa Clara, CA, USA) coupled with an Agilent triple quadrupole mass spectrometer (6430 QqQ) with an ESI ion source was used for the characterization of phenolic compounds according to the method previously described by Čulina et al. [32]. The reverse-phase separations were performed on a C18 column of Agilent Zorbax Eclipse Plus (100 × 2.1 mm, 1.8 μm particle size) at 35 °C. Identification of phenolic compounds was performed by comparing the mass spectra and fragmentation patterns of the authentic standards and using previously reported data [35,36]. All measurements were done in duplicate, and the obtained results were expressed in mg per 100 g of dried matter (mg/100 g dm).

2.12. Determination of Antioxidant Activity by the ORAC Method

The ORAC assay was conducted on a 96-well microplate using a fluorescence plate reader (Clariostar BMG LABTECH, Offenburg, Germany) according to the method previously described by Elez Garofulić et al. [37] for hydrophilic ORAC (H-ORAC) and according to the method previously described by Naguib et al. [38] for lipophilic ORAC (L-ORAC). The blank for H-ORAC was sodium phosphate buffer (75 mM, pH 7.4), and for L-ORAC was 7% RMCD solvent (*w*/*v*) made in a 50% acetone-water mixture (*v*/*v*). Trolox (25 μL) was used as the standard, and results were expressed in μmol of Trolox equivalent per g of dried matter (μmol TE/g dm, $n = 2$).

2.13. Statistical Analysis

For statistical analysis, Statistica ver. 10.0 software (Statsoft Inc., Tulsa, OK, USA) was applied. One-way ANOVA and Tukey's HSD test were performed to compare the mass concentrations of individual fatty acids, sterols, and carotenoids in the SBO of Leikora and Ascola obtained by SC-CO_2 and SE and to evaluate the differences between the SB varieties and the efficiency of SC-CO_2 and SE in terms of total fatty acid content (total saturated, unsaturated, monounsaturated, and polyunsaturated fatty acids), total sterol content, total carotenoid content, α-tocopherol content, and antioxidant activity. All tests were carried out at a significance level of $p \leq 0.05$.

For the optimization of the UAE, a three-level full factorial design comprising 27 experimental trials was chosen to evaluate the effect of three independent variables—temperature, ultrasound power, and extraction time—during the UAE on the total phenolic content of the cake. The Shapiro-Wilk and Levene tests were performed to determine the homoscedasticity and normality of the data. A one-way ANOVA and Tukey's HSD test were performed to compare the mass concentrations of individual and total phenolic compounds between the Leikora and Ascola SB cakes. All tests were carried out with a significance level of $p < 0.05$.

3. Results and Discussion

3.1. Fatty Acid Composition of SB Berry Oil

GC-MS analysis was used to compare the fatty acid profile of sea buckthorn oil (SBO) of the Leikora and Ascola varieties obtained by SC-CO_2 and SE. The results of the individual compounds identified are shown in Table 1, while the total content of saturated fatty acids (SFAs) and unsaturated fatty acids (UFAs), consisting of monounsaturated fatty acids (MUFAs) and polyunsaturated fatty acids (PUFAs), is shown in Table 2. A total of 14 compounds were identified, including 7 SFAs and 7 UFAs (5 MUFAs and 2 PUFAs) (Table 1).

According to the results presented in Table 2, the predominant fatty acids in the studied berry oils were UFAs (54.75–54.78% in the SBO of the Leikora variety and 72.22–72.25% in the SBO of the Ascola variety), which were mainly composed of MUFAs (42.93–42.96% in the SBO of the Leikora variety and 55.78–55.80% in the SBO of the Ascola of variety) and PUFAs (11.82% in the SBO of the Leikora variety and 16.44% in the SBO of the Ascola variety). The total content of SFAs was 45.22–45.25% in the SBO of the Leikora variety and 27.75–27.78% in the SBO of the Ascola variety.

Table 1. Fatty acid composition (weight percentages of total fatty acids) of SBO of Leikora and Ascola varieties obtained by different extraction methods.

Fatty Acids	Leikora		Ascola	
	$SC\text{-}CO_2$	SE	$SC\text{-}CO_2$	SE
SATURATED				
Myristic	0.26 ± 0.01 [b]	0.25 ± 0.01 [b]	0.15 ± 0.01 [a]	0.15 ± 0.01 [a]
Pentadecanoic acid	0.06 ± 0.01 [a]	0.05 ± 0.01 [a]	0.07 ± 0.01 [a]	0.08 ± 0.01 [a]
Palmitic	42.86 ± 1.01 [b]	42.85 ± 1.14 [b]	25.44 ± 1.05 [a]	25.43 ± 0.05 [a]
Heptadecanoic acid	1.04 ± 0.01 [b]	1.03 ± 0.01 [b]	0.11 ± 0.01 [a]	0.12 ± 0.01 [a]
Stearic	0.77 ± 0.01 [a]	0.77 ± 0.01 [a]	1.59 ± 0.02 [b]	1.59 ± 0.02 [b]
Arachidic acid	0.18 ± 0.01 [a]	0.18 ± 0.01 [a]	0.28 ± 0.02 [b]	0.29 ± 0.01 [b]
Docosanoic acid	0.09 ± 0.01 [a]	0.09 ± 0.01 [a]	0.12 ± 0.01 [a]	0.12 ± 0.01 [a]
UNSATURATED				
MONOUNSATURATED				
Palmitoleic acid	37.47 ± 1.11 [b]	37.46 ± 2.01 [b]	31.85 ± 1.15 [a]	31.83 ± 2.08 [a]
Elaidic acid	0.05 ± 0.01 [a,b]	0.07 ± 0.01 [b]	0.01 ± 0.01 [a]	0.01 ± 0.00 [a]
Gondoic acid	0.03 ± 0.01 [a]	0.03 ± 0.01 [a]	0.13 ± 0.01 [b]	0.14 ± 0.00 [b]
Oleic acid	5.35 ± 0.02 [a]	5.36 ± 0.01 [a]	16.38 ± 1.04 [b]	16.37 ± 1.24 [b]
Vaccenic acid	0.03 ± 0.00 [a]	0.04 ± 0.00 [a]	7.43 ± 1.03 [b]	7.43 ± 1.07 [b]
POLYUNSATURATED				
Linoleic acid	11.16 ± 0.05 [a]	11.16 ± 0.05 [a]	11.20 ± 0.45 [a]	11.19 ± 0.63 [a]
α-Linolenic acid	0.66 ± 0.01 [a]	0.66 ± 0.01 [a]	5.24 ± 0.14 [b]	5.25 ± 0.01 [b]

Results are expressed as mean ± standard deviation. Values with different letters within a row are significantly different at $p \leq 0.05$. $SC\text{-}CO_2$= supercritical CO_2 extraction; SE= Soxhlet extraction.

According to the results presented in Table 2, SBO had a higher content of UFAs than SFAs. A similar trend was observed in studies by Dulf [39], Barkhuu et al. [40], Teleszko et al. [41], Xu et al. [42], and Xu et al. [43]. The fatty acid profile of the Ascola variety is consistent with the study by Vaitkevičiene et al. [44], in which the proportion of total fatty acids was as follows: SFAs (25.01–25.39%), UFAs (74.43–74.70%), MUFAs (46.34–51.27%), and PUFAs (9.37–11.06%). Similar results were obtained by Dulf et al. [39] and Barkhuu et al. [40]. On the other hand, Telesko et al. [41] reported a higher proportion of SFAs (35.05–41.05%), which is consistent with the proportion of SFAs in the Leikora variety.

As shown in Table 1, 14 fatty acids were identified in the SBO samples: myristic, pentadecanoic, palmitic, heptadecanoic, stearic, arachidic, docosanoic, palmitoleic, elaidic, oleic, linolelaidic, linoleic, α-linolenic, and gondoic acid. The most abundant fatty acid in SBO of the Leikora variety was palmitic (42.85–42.86%), followed by palmitoleic (37.46–37.47%), linoleic (11.16%), and oleic (5.36–5.36%) fatty acids. On the other hand, the most abundant fatty acid in SBO of the Ascola variety was palmitoleic (31.83–31.85%), followed by palmitic (25.43–25.44%), oleic (16.37–16.38%), linoleic (11.19–11.20%), vaccenic (7.43%), and α-linolenic acid (5.24–5.25%). The fatty acids determined are consistent with previously reported data [43–45], although in varying proportions, reflecting differences between varieties, geographical origin, fruit development stages, etc. Xu et al. [43] found that the most abundant fatty acids in SBO obtained with $SC\text{-}CO_2$ and SE were palmitoleic (38.8–39.3%), palmitic (28.8–29.4%), and oleic acid (16.4–16.5%), which is consistent with our results. In the study by Yang and Kallio (2001) [45], the following fatty acids were identified in SB berries of different origins: linoleic acid (10.2–33.9), palmitic acid (14.1–29.2%), palmitoleic acid (8.9–31.0%), oleic acid (12.6–24.8%), α-linolenic acid (6.5–18.1%), vaccenic acid (5.1–8.5%), and stearic acid (1–2%). According to Vaitkevičienė et al. [44], the most common fatty acids in SB berries were oleic acid (25.5%), palmitic acid (22.72%), palmitoleic acid (18.5%), linoleic acid (12.7%), and α-linolenic acid (10%). In our study, the SBOs of the Leikora variety had a higher content of palmitic and palmitoleic acid and a lower content of oleic acid than the reported data [44,45]. In addition, significantly lower levels of α-linolenic acid and oleic acid were found.

Table 2. Total fatty acid composition (%) of Leikora and Ascola SBO obtained by different extraction methods.

Source of Variation:		SUM SFAs: SC-CO_2	SUM SFAs: SE		SUM UFAs: SC-CO_2	SUM UFAs: SE		SUM MUFAs: SC-CO_2	SUM MUFAs: SE		SUM PUFAs: SC-CO_2	SUM PUFAs: SE
		$p < 0.001$	$p < 0.001$		$p < 0.001$	$p < 0.001$		$p < 0.05$	$p < 0.05$		$p < 0.05$	$p < 0.05$
Leikora	$p = 0.90$	45.25 ± 0.02 b,A	45.22 ± 0.21 b,A	$p = 0.92$	54.75 ± 0.03 a,A	54.78 ± 0.26 a,A	$p = 0.95$	42.93 ± 0.37 a,A	42.96 ± 0.24 a,A	$p = 1.00$	11.82 ± 0.17 a,A	11.82 ± 0.30 a,A
Ascola	$p = 0.93$	27.75 ± 0.20 a,A	27.78 ± 0.26 a,A	$p = 0.85$	72.25 ± 0.10 b,A	72.22 ± 0.09 b,A	$p = 0.99$	55.80 ± 0.82 b,A	55.78 ± 0.66 b,A	$p = 1.00$	16.44 ± 0.27 b,A	16.44 ± 0.47 b,A

Results are expressed as mean ± standard error. Different lowercase letters indicate statistically significant differences ($p < 0.05$) between SBO samples within each extraction technique (column), while the same uppercase letters indicate no statistically significant differences ($p > 0.05$) between extraction techniques for individual SBO samples within each fatty acid (row). SC-CO_2 = supercritical CO_2 extraction; SE = Soxhlet extraction.

To evaluate the differences between the SB varieties and the efficiency of SC-CO_2 and SE in terms of individual and total fatty acid content, the mean values were compared (Tables 1 and 2). A statistically significant difference was found in the fatty acid distribution of Leikora and Ascola SBO ($p < 0.05$). SBO of the Ascola variety had a statistically significantly higher total content of MUFAs and PUFAs than Leikora SBO (Table 2). The MUFAs were mainly represented by palmitoleic acid and oleic acid. The content of palmitoleic acid was higher in the SBO of the Leicora variety than in the SBO of the Ascola variety, while higher levels of oleic acid were found in the SBO of the Ascola variety. Other MUFAs, such as elaidic acid and gondoic acid, were detected in traces (<0.15%). Regarding PUFAs, no difference was found in linoleic acid between the two varieties, while a higher proportion of linoleic and α-linolenic acid was found in the Ascola variety, i.e., traces of linoleic and α-linolenic acid were found in Leikora SBO (0.03–0.04 and 0.66%). The total content of SFAs was significantly higher in Leikora SBO, and the most represented fatty acid was palmitic acid. Other SFAs found in trace levels (<1.6%), such as pentadecanoic acid, stearic, arachidic, and docosanoic acid, were found in higher proportions in Ascola SBO, while myristic acid and heptadecanoic acid were more abundant in Leikora SBO. The differences between the varieties can be attributed to morphological characteristics (seed, pulp, and peel content), as the pulp is characterized by palmitic and palmitoleic acids [46], while the seeds contain linoleic, linolenic, and oleic acids [45]. According to Mate et al. [47], the berry and seed weight of SB berries are between 0.25 and 0.64 g and 0.019 and 0.036 g, respectively, depending on the variety. The Leikora variety is characterized by larger berries, i.e., a higher content of pulp and skin in relation to the seeds, which could be one of the reasons for the different fatty acid composition, i.e., a higher proportion of palmitic acids. The studies by Kuhkhei et al. [48], Yang and Kallio [45], Vaitkevičiene et al. [44], and Teleszko et al. [41] also confirmed the differences in fatty acid content between different varieties. On the other hand, the extraction method used had no influence ($p < 0.05$) on the proportion of identified fatty acids (Table 2). Despite the fact that a significantly higher temperature was used in the SE than in the SC-CO_2 extraction, the fatty acid content did not change significantly. The possible reason could be the presence of a high proportion of palmitic acid in both varieties of SB berries, which is known to be less susceptible to oxidation than UFAs. A similar trend was observed in the study by Pavlović et al. [28], Xu et al. [42], and Xu et al. [43] when comparing the influence of SC-CO_2 and SE on the fatty acid profile of SBO.

3.2. Sterol Composition of SB Berry Oil

GC-MS analysis was used to compare the sterol profiles of the SBO of Leikora and Ascola varieties obtained by SC-CO_2 and SE. The mass concentration of individual identified compounds and the total sterol content are shown in Tables 3 and 4.

A total of 16 compounds were identified, including desmosterol, kampesterol, stigmasterol, 5-α-colestanol, sitosterol, stigmast-8-en-3β-ol, 24(Z)-stigmasta-5,24(241)-dien-3β-ol, obtusifoliol, Δ5-avenasterol, stigmas-ta-7,24(28)-dien-3-ol, anosta-7,9(11)-dien-3-ol, uvaol, eritrodiol, citrostadienol, Δ7-avenasterol, and ursolic aldehyde (Table 3). The statistical analysis of the individual sterols revealed a statistically significant difference ($p \leq 0.05$) between the oils of the different varieties obtained by different extraction methods (Table 3). The most abundant sterol in SBO was β-sitosterol which ranged from 473.85 to 589.96 mg/100 g in Leikora SBO and from 352.56 to 433.82 mg/100 g in Ascola SBO. In various studies, the content of β-sitosterol has been reported in the range of 320–1377 mg/100 g, depending on the varieties, the plant part of the plant, the geographical origin, the fruit development stage, the extraction method used, etc. [13,49–51]. According to Yang et al. [13], β-sitosterol as the main sterol compound accounts for 57–83% of total sterols and 576.9 mg/100 g of oil, which is consistent with our results. Similar levels of β-sitosterol in the pulp oil (504–557 mg/100 g dm) were found in the study by Zheng et al. [49], while a significantly higher level of β-sitosterol (990–1031 mg/100 g dm) was found in the seed oil. SBO contains a higher content of β-sitosterol compared to sunflower and virgin olive oil [1,49]. Overall ranges of campesterol (22.31–32.3 mg/100 g), uvaol (19.01–22.02 mg/100 g), and obtusifoliol (12.30–13.31 mg/100 g) were determined in the Leikora SBO, and 24(Z)-stigmasta-5,24(241)-dien-3β-ol (7.81–8.83 mg/100 g), uvaol (19.01–22.02 mg/100 g), obtusifoliol (12.30–13.31 mg/100 g), Δ5-avenasterol (4.10–5.09 mg/100 g) and campesterol (22.31–32.3 mg/100 g) were more abundant in the Ascola SBO. For the other sterols identified, the order of decreasing concentrations was as follows: 24(Z)-stigmasta-5,24(241)-dien-3β-ol > stigmasta-7,24(28)-dien-3-ol > cirtostadienol > stigmast-8-en-3β-ol > Δ5-avenasterol > eritrodiol > Δ5-avenasterol > stigmasterol > ursolic aldehyde > lanosta-7,9(11)-dien-3-ol > 5-α-colestanol > demosterol in the variety Lekora and stigmasta-7,24(28)-dien-3-ol > stigmast-8-en-3β-ol > eritrodiol > citrostadienol > Δ5-avenasterol > ursolic aldehyde > demosterol > lanosta-7,9(11)-dien-3-ol > 5-α-colestanol in the variety Ascola. Sterols determined in our study are consistent with previously reported data [12,41,52–54]. The most abundant sterols found in the seeds were β-sitosterol, Δ5-avenasterol, and obtusifoliol (approx. 15–17% of total sterols), which, together with stigmasta-8-en-3β-ol, constitute approximately 5–6% and 8–10% of the total sterols found in soft tissues and whole fruits, respectively [1,49].

Table 3. Sterol content (mg/100 g) of Leikora and Ascola SBO obtained by SC-CO_2 and SE.

Compound	Leikora		Ascola	
	SC-CO_2	SE	SC-CO_2	SE
Desmosterol	0.53 ± 0.01 ^{a}	0.31 ± 0.01 ^{a}	2.68 ± 0.01 ^{b}	2.20 ± 0.01 ^{b}
Campesterol	32.30 ± 0.15 ^{d}	22.31 ± 0.14 ^{c}	16.91 ± 0.12 ^{a,b}	11.89 ± 0.11 ^{a}
Stigmasterol	3.08 ± 0.01 ^{d}	1.11 ± 0.05 ^{a,b}	1.43 ± 0.01 ^{c}	0.45 ± 0.01 ^{a}
5-α-kolestanol	0.76 ± 0.01 ^{b}	0.22 ± 0.00 ^{a}	1.36 ± 0.01 ^{c}	0.85 ± 0.01 ^{b}
β-Sitosterol	589.96 ± 2.45 ^{c}	473.85 ± 2.14 ^{b}	433.82 ± 2.11 ^{b}	352.56 ± 2.41 ^{a}
Stigmast-8-en-3β-ol	6.51 ± 0.08 ^{c}	3.45 ± 0.02 ^{a}	7.28 ± 0.07 ^{c}	5.26 ± 0.02 ^{b}
24(Z)-stigmasta-5,24(241)-dien-3β-ol	8.83 ± 0.05 ^{a}	7.81 ± 0.05 a	30.26 ± 0.45 ^{b}	23.23 ± 0.52 ^{b}
Obtusifoliol	13.31 ± 0.12 ^{a}	12.30 ± 0.05 ^{a}	24.27 ± 0.25 ^{c}	18.51 ± 0.12 ^{b}
Δ5-avenasterol	5.09 ± 0.01 ^{a}	4.10 ± 0.01 ^{a}	17.67 ± 0.23 ^{b}	15.21 ± 0.11 ^{b}
Stigmasta-7,24(28)-dien-3-ol	7.50 ± 0.02 ^{b}	3.10 ± 0.01 ^{a}	9.85 ± 0.05 ^{c}	6.86 ± 0.05 ^{b}
Lanosta-7,9(11)-dien-3-ol	1.92 ± 0.01 ^{c}	0.58 ± 0.00 ^{a}	1.43 ± 0.00 ^{a,b}	1.12 ± 0.01 ^{a,b}
Uvaol	22.02 ± 0.05 ^{a}	19.01 ± 0.04 ^{a}	26.23 ± 0.11 ^{b}	20.25 ± 0.06 ^{a}
Eritrodiol	5.12 ± 0.03 ^{b}	3.02 ± 0.01 ^{a}	6.84 ± 0.01 ^{c}	5.85 ± 0.00 ^{b}
Citrostadienol	7.05 ± 0.02 ^{b}	6.01 ± 0.01 ^{a,b}	5.91 ± 0.01 ^{a,b}	4.78 ± 0.01 ^{a}
Δ7-avenasterol	3.35 ± 0.01 ^{a,b}	3.14 ± 0.01 ^{a}	4.36 ± 0.01 ^{c}	3.95 ± 0.00 ^{b,c}
Ursolic aldehyde	1.99 ± 0.04 ^{a,b}	1.01 ± 0.01 ^{a}	3.04 ± 0.01 ^{b}	2.52 ± 0.00 ^{b}

Results are expressed as mean ± standard deviation. Values with different letters within a row are significantly different at $p \leq 0.05$. SC-CO_2 = supercritical CO_2 extraction; SE = Soxhlet extraction.

Table 4. Total sterol content (mg/100 g) of Leikora and Ascola SBO obtained by SC-CO_2 and SE.

Source of Variation:		Total Sterol Content (mg/100 g) SC-CO_2	SE
		$p < 0.05$	$p < 0.05$
Leikora	$p < 0.05$	709.32 ± 4.20 ^{b,B}	561.33 ± 3.10 ^{b,A}
Ascola	$p < 0.05$	593.34 ± 4.18 ^{a,B}	475.49 ± 4.20 ^{a,A}

Results are expressed as mean ± standard error. Different lowercase letters indicate statistically significant differences ($p < 0.05$) between SBO samples within each extraction technique (column), while different uppercase letters indicate statistically significant differences ($p < 0.05$) between extraction techniques for individual SBO samples within each fatty acid (row). SC-CO_2 = supercritical CO_2 extraction; SE = Soxhlet extraction.

As shown in Table 4, the total content of sterols in the SBOs ranged from 561.33 to 709.32 mg/100 g in Leikora and from 475.49 to 593.34 mg/100 g in Ascola SBO. Our results are consistent with previously reported data [49,51,55]. In the study by Yang et al. [55], the total sterol content in the seed oil, soft parts, and whole fruit was 12–23 g/kg, 10–29 g/kg, and 13–33 g/kg, respectively. Zheng et al. [49] reported a total sterol content in the range of

778–1847 mg/100 g, depending on the parts of the berry (pulp/seed) and the extraction method used. A similar range of total sterols (440.3–1688.7 mg/100 g) in oil obtained from different parts of the berries and by different extraction methods was reported in the study by Arimboor et al. [51].

To evaluate the differences between the SB varieties and the efficiency of SC-CO_2 and SE in terms of total sterol content, the mean values were compared (Table 4). A statistically significant difference was found in the total sterol content of Leikora and Ascola SBO ($p < 0.05$) within each extraction method. SBO of the Leikora variety had a statistically significantly higher total sterol content than Ascola SBO. This is due to a significantly higher content of β-sitosterol, while the other identified sterols were mainly present in higher concentrations in the Ascola variety. Teleszko et al. [41] also reported a statistically significant difference in the sterol content in the berries of eight Russian SB cultivars. The extraction method used also had a statistically significant effect ($p < 0.05$) on the total sterol content within each SB variety. A similar trend was observed in the study by Sajfrtova et al. [56] when the content of β-sitosterol was compared in extracts obtained with SC-CO_2 and SE. SC-CO_2 resulted in up to five-fold higher concentrations of β-sitosterol in the extract (0.50% *w/w*) than the conventional SE with hexane (0.10% *w/w*). The lower selectivity of hexane and the high extraction temperature, in combination with the influence of atmospheric oxygen, probably led to a partial degradation of β-sitosterol. Li et al. [12] reported that SC-CO_2 extraction was better than extraction with hexane and cold pressing in terms of total and individual sterol content in SBO. Arimboor et al. [51] also reported a higher content of sterols in extracts obtained by SC-CO_2. Yang et al. [57] pointed out that the supercritical fluid (CO_2) has a high penetrating and dissolving capacity and thus offers advantages in the extraction of lipids and other bioactive substances from various types of raw materials. As it is an environmentally friendly process, supercritical solvent extraction is usually carried out at low temperatures and under vacuum, which protects the BAMs from thermal and oxidative degradation [58].

3.3. Carotenoid Content of Leikora and Ascola SBO Obtained by SC-CO_2 and SE

The carotenoid profiles of the SBO of Leikora and Ascola varieties obtained by SC-CO_2 and SE were analyzed by HPLC-DAD and compared. The mass concentration of the individual compounds identified and the total carotenoid content are shown in Tables 5 and 6.

Table 5. Carotenoid content of Leikora and Ascola SBO obtained by SC-CO_2 and SE.

Carotenoids:	Mass Concentration (mg/100 g dm)			
	Leikora		Ascola	
	SC-CO_2	SE	SC-CO_2	SE
Zeaxanthin	8.50 ± 0.15 ᵃ	5.42 ± 0.16 ᵃ	8.77 ± 0.01 ᵃ	8.45 ± 0.05 ᵃ
Lutein	3.38 ± 0.05 ᵃ	4.33 ± 0.01 ᵇ	nd	nd
β-cryptoxanthin	5.21 ± 0.06 ᵇ	3.76 ± 0.01 ᵃ	6.85 ± 0.02 ᶜ	5.75 ± 0.01 ᵇ
γ-carotene	0.07 ± 0.00 ᵃ	0.05 ± 0.00 ᵃ	0.18 ± 0.00 ᵇ	0.09 ± 0.00 ᵃ
cis γ-carotene	0.44 ± 0.01 ᵇ	0.19 ± 0.00 ᵃ	1.18 ± 0.01 ᵈ	0.99 ± 0.01 ᶜ
β-carotene	2.09 ± 0.01 ᵇ	1.73 ± 0.00 ᵃ	2.09 ± 0.01 ᵇ	1.95 ± 0.15 ᵃ,ᵇ
β-cryptoxanthin palmitate	0.73 ± 0.02 ᵇ	0.78 ± 0.01 ᵇ	0.14 ± 0.00 ᵃ	nd
Zeaxanthin-myristate	26.66 ± 0.15 ᵃ	24.76 ± 0.25 ᵃ	28.65 ± 0.75 ᵃ	27.52 ± 0.25 ᵃ
Lutein-palmitate	5.85 ± 0.01 ᵇ	3.07 ± 0.01 ᵃ	3.30 ± 0.00 ᵃ	2.95 ± 0.00 ᵃ
Zeaxanthin-pamitate	10.86 ± 0.14 ᵃ	8.18 ± 0.15 ᵃ	31.01 ± 0.20 ᵇ	30.82 ± 0.45 ᵇ
Lutein di-myristate	nd	nd	10.39 ± 0.15 ᵇ	9.12 ± 0.12 ᵃ
Zeaxanthin-palmitate-myristate	19.97 ± 0.25 ᵇ	12.87 ± 0.16 ᵃ	36.22 ± 1.05 ᶜ	35.98 ± 1.21 ᶜ
Lutein di-palmitate	6.44 ± 0.05 ᵃ	5.40 ± 0.05 ᵃ	24.17 ± 0.95 ᵇ	23.25 ± 0.85 ᵇ
Zeaxanthin-di-palmitate	56.73 ± 1.12 ᵇ	43.38 ± 1.25 ᵃ	58.08 ± 1.75 ᵇ	57.05 ± 1.44 ᵇ
Lutein-palmitate-stearate	35.05 ± 1.15 ᶜ	26.94 ± 0.45 ᵇ	4.37 ± 0.00 ᵃ	3.75 ± 0.00 ᵃ

Results are expressed as mean ± standard deviation. Values with different letters within a row are significantly different at $p \leq 0.05$. SC-CO_2 = supercritical CO_2 extraction; SE = Soxhlet extraction; nd-not detected.

Table 6. Total carotenoids of Leikora and Ascola SBO obtained by SC-CO_2 and SE.

Source of Variation:		Total Carotenoid Content (mg/100 g) SC-CO_2	SE
		$p < 0.05$	$p < 0.001$
Leikora	$p < 0.05$	181.98 ± 1.73 ^{a,B}	140.86 ± 1.50 ^{a,A}
Ascola	$p = 0.09$	215.4 ± 2.18 ^{b,A}	207.67 ± 1.11 ^{b,A}

Results are expressed as mean ± standard error. Different lowercase letters indicate statistically significant differences ($p < 0.05$) between SBO samples within each extraction technique (column), while different uppercase letters indicate statistically significant differences ($p < 0.05$) between extraction techniques for individual SBO samples within total carotenoid content (row). SC-CO_2 = supercritical CO_2 extraction; SE = Soxhlet extraction.

The total carotenoid content of SBO was relatively high, ranging from 140.88–187.99 mg/100 g for the Leikora variety and 217.93–226.84 mg/100 g for the Ascola variety, depending on the extraction method (Table 6). The high concentration of pigments in SBO was also confirmed in previous studies [59,60]. As shown in Table 5, a total of 16 compounds were identified, including free form carotenoids (zeaxanthin, lutein, β-cryptoxanthin, γ-carotene, cis-γ-carotene, and β-carotene) and the esterified forms of zeaxanthin, lutein, and β-cryptoxanthin. Esterified forms (with one or two fatty acids) of zeaxanthin and lutein corresponded to 71% of the total identified carotenoids [33]. The presence of individual carotenoids differs between the SB varieties (Table 5). In SBO of Leikora variety, zeaxanthin-di-palmitate (43.38–56.73 mg/100 g) was the most abundant, followed by lutein-palmitate-stearate (26.94–35.05 mg/100 g) and zeaxanthin myristate (24.76–26.66 mg/100 g), while zeaxanthin-di-palmitate (57.05–58.08 mg/100 g) was the most abundant in SBO of Ascola variety, followed by zeaxanthin-palmitate-myristate (35.98–36.22 mg/100 g), zeaxanthin-palmitate (30.82–31.01 mg/100 g), zeaxanthin-myristate (27.52–28.65 mg/100 g) and lutein-dipalmitate. Tudor et al. [59] also reported that the most common compounds in SBO are esterified forms of zeaxanthin, with zeaxanthin-di-palmitate being the most abundant. Pop et al. [33] also found zeaxanthin di-palmitate in the highest amount in SB berries. In our study, β-Cryptoxanthin palmitate was the only β-cryptoxanthin ester detected in significantly lower concentrations than zeaxanthin and lutein esters (0.14–1.12 mg/100 g). The zeaxanthin content was between 5.42 and 8.50 mg/100 g for the Leikora variety and between 8.45 and 8.77 mg/100 g for the Ascola variety. Lutein was only detected in the extracts of the Leikora variety (3.38–4.33 mg/100 g). The lower content of carotenoids in free form could be due to the fact that xanthophylls such as zeaxanthin are converted to esterified forms during ripening, which are more stable forms of carotenoid storage in fruits [61]. In contrast, β-cryptoxanthin was present at a higher concentration than its ester, ranging from 3.76 to 5.21 mg/100 g in SBO of the Leikora variety and from 5.75 to 6.85 in SBO of the Ascola variety. Among the carotenes, β-, γ-, and cis-γ carotene were detected in SBOs in a range of 0.99–4.15 mg/100 g, 0.05–0.21 mg/100 g and 0.07–1.52 mg/100 g, respectively, depending on the variety and extraction method.

To evaluate the differences between the SB varieties and the efficiency of SC-CO_2 and SE in terms of total carotenoid content, the mean values were compared (Table 6). A statistically significant difference was found in the total carotenoid content of Leikora and Ascola SBO ($p < 0.05$) within each extraction method. SBO of the Ascola variety had a statistically significantly higher total carotenoid content than SBO of the Leikora variety. Various studies have found considerable differences in carotenoid composition and content depending on variety, geographical location, growing conditions, harvest maturity, storage conditions, and analytical methods [33,41,62–64]. As SB is a wind-pollinated species, a high degree of genetic variation is also observed [64]. The extraction method also had a statistically significant effect ($p < 0.05$) on the total carotenoid content of the SBO of the Leikora variety, and a higher content of carotenoids was found when SC-CO_2 was used. The high temperatures and long extraction time used in SE lead to thermal degradation and cis to trans isomerization of carotenoids, while the higher diffusivity and lower viscosity of supercritical fluids, shorter extraction time, and lower temperatures used in SC-CO_2 prevent the degradation of BAMs such as the carotenoids [58]. Arimboor et al. [51] also reported that SC-CO_2 contributed to the higher carotenoid content in SBO than solvent extraction with hexane.

3.4. *α-Tocopherol Content in SB Berry Oil*

The α-tocopherol content of SBO of the Leikora and Ascola varieties obtained with SC-CO_2 and SE was determined by HPLC-FID and compared in Table 7.

Table 7. Mass concentration of α-tocopherol (mg/100 g) determined in Leikora and Ascola SBO obtained by different extraction methods.

Source of Variation:		α-Tocopherol (mg/100 g dm) SC-CO₂	SE
		$p = 0.33$	$p = 0.14$
Leikora	$p < 0.05$	106.86 ± 2.61 ^{a,B}	84.23 ± 1.00 ^{a,A}
Ascola	$p < 0.05$	102.45 ± 2.33 ^{a,B}	81.06 ± 0.85 ^{a,A}

Results are expressed as mean ± standard error. The same lowercase letters indicate no statistically significant differences ($p > 0.05$) between SBO samples within each extraction technique (column), while different uppercase letters indicate statistically significant differences ($p < 0.05$) between extraction techniques for individual SBO samples within each fatty acid (row). SC-CO_2 = supercritical CO_2 extraction; SE = Soxhlet extraction.

As shown in Table 7, the α-tocopherol content in SBO of the Leikora variety ranged from 84.23 to 106.86 mg/100 g and in SBO of the Ascola variety from 81.06 to 102.45 mg/100 g. Our results are in agreement with the studies of Zadernowski et al. [65] and Bal et al. [66]. On the other hand, Otgonbayar et al. [67] reported a lower content of α-tocopherol in seeds (34.57 mg/100 g oil) and pulp oil (59.02 mg/100 g oil). Similar results were obtained in the study by Andersson et al. [10]. Tocopherol content depends on the extraction and analysis methods used, the part of the berry, genetic variation, climate, growing conditions, annual variation, ripeness at harvest, storage conditions, and harvesting method [9,10].

To evaluate the differences between the SB varieties and the efficiency of SC-CO_2 and SE in terms of α-tocopherol content, the mean values were compared (Table 7). A statistically significant difference ($p < 0.05$) in α-tocopherol content was found between the two extraction methods used. In addition, more tocopherols were found in oil extracted with SC-CO_2. Xu et al. [43] and Arimboor et al. [51] also reported a higher α-tocopherol content in oils obtained using SC-CO_2. Several studies have confirmed that the oil obtained by SC-CO_2 from different matrices has a higher tocopherol content than the oil obtained by conventional extraction with hexane [68–71]. Although it is stable at higher temperatures, the oxidation rate of α-tocopherol increases at high temperatures and in the presence of oxygen [72,73]. This explains the higher content of α-tocopherol in SC-CO_2, which is carried out under vacuum and at significantly lower temperatures compared to SE.

3.5. *Antioxidant Activity of SB Berry Oil*

The lypophilic ORAC method was used to evaluate the AA of the oils obtained by SC-CO_2 and SE (Table 8).

Table 8. Antioxidant activity of SBOs obtained by SC-CO_2 and SE.

Source of Variation:		L-ORAC (μmolTE/100 g dm) SC-CO₂	SE
		$p = 0.20$	$p = 0.55$
Leikora	$p < 0.05$	1537.10 ± 4.19 ^{a,B}	1356.58 ± 19.62 ^{a,A}
Ascola	$p < 0.05$	1558.50 ± 10.34 ^{a,B}	1372.63 ± 10.47 ^{a,A}

Results are expressed as mean ± standard error. The same lowercase letters indicate no statistically significant differences ($p < 0.05$) between SBO samples within each extraction technique (column), while different uppercase letters indicate statistically significant differences ($p < 0.05$) between extraction techniques for individual SBO samples within the ORAC value (row). SC-CO_2 = supercritical CO_2 extraction; SE = Soxhlet extraction.

According to the results presented in Table 8, a higher AA (1558.50 μmolTE/100 g) was observed in the SBO of the Leikora variety obtained with SC-CO_2 than in the SBO of the Ascola variety (1537.10 μmolTE/100 g), although no significant difference ($p = 0.20$) was observed. On the other hand, the lower AA was determined in the SBO of Ascola (1356.58 μmolTE/100 g) and Leikora (1372.63 μmolTE/100 g) obtained by SE. The high AA of the oils could be related to the high content of antioxidants such as α-tocopherol and carotenoids in both varieties. Andrei et al. [74] have shown that carotenoids from SB fruits have a good antioxidant effect and protect vegetable oils from peroxidation processes induced in the presence of AAPH. In addition, vitamin E is one of the most important antioxidants in SB berries [10]. The statistical analysis (Table 8) showed that the oils obtained by SC-CO_2 had a significantly ($p < 0.05$) higher AA, which correlates with the higher content of BAMS in the oils obtained by SC-CO_2.

3.6. Polyphenolic Compounds in SB Cake Obtained after Oil Extraction

SB cake of the Leikora variety was used to determine the optimal UAE conditions, i.e., temperature (35, 50, and 65 °C), ultrasound power (30, 60, and 90%), and extraction time (10, 20 and 30 min).

The results of TPC and AA in SB cake extracts produced under different UAE conditions are shown in Table 9. To determine the optimal UAE conditions, an ANOVA statistical analysis of the effects of ultrasound power, temperature, and extraction time was performed (Table 10). The extraction of phenolic compounds from the cakes of Leikora and Ascola SB obtained after SC-CO_2 and SE was carried out under the determined optimal extraction conditions, and the extracts were analyzed by the UPLC/ESI-MS^2 method (Table 11). The AA of SB cake extracts obtained at optimal UAE conditions is shown in Figure 1.

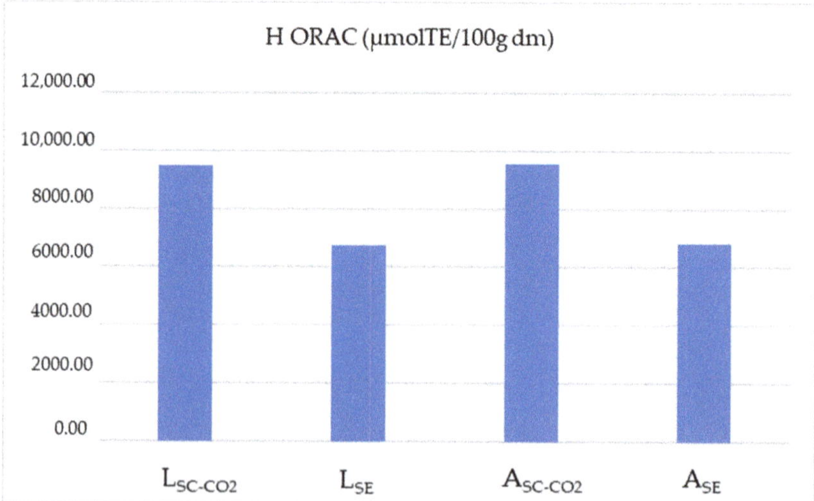

Figure 1. AA of SB cake extracts obtained under optimal UAE. L_{SC-CO2} = cake of Leikora variety obtained after supercritical CO_2 extraction, L_{SE} = cake of Leikora variety obtained after Soxhlet extraction, A_{SC-CO2} = cake of Ascola variety obtained after supercritical CO_2 extraction, A_{SE} = cake of Ascola variety obtained after Soxhlet extraction.

Table 9. Total phenolic content of SB cake extracts obtained using different UAE conditions.

Source of Variation		Ultrasound Power (%)					
		30	60	90	30	60	90
Temperature (°C)	**Time (min)**	**TPC mg/100 g dm**			**AA μmol TE/100 g dm**		
35	10	1123.25 ± 25.02	1254.77 ± 22.15	1242.25 ± 24.81	6871.34 ± 121.45	7524.12 ± 65.74	7192.18 ± 132.14
	20	1256.36 ± 31.45	1596.75 ± 18.45	1591.90 ± 35.62	7322.90 ± 147.23	8194.90 ± 111.41	8280.29 ± 101.81
	30	1499.83 ± 33.15	1366.05 ± 21.75	1437.08 ± 85.41	8759.13 ± 150.74	8113.61 ± 121.71	8189.16 ± 108.45
50	10	1572.21 ± 45.18	1851.72 ± 75.41	1749.01 ± 58.41	8983.91 ± 101.41	8756.20 ± 125.41	8751.92 ± 102.71
	20	1795.40 ± 52.41	1998.98 ± 33.45	1792.11 ± 62.35	8700.93 ± 121.74	9504.85 ± 125.36	8795.53 ± 125.85
	30	1735.74 ± 65.85	1965.86 ± 25.41	1846.55 ± 55.24	8440.58 ± 122.52	9448.77 ± 145.87	8931.60 ± 123.52
65	10	1303.05 ± 25.42	1381.00 ± 25.74	1281.03 ± 55.24	8887.88 ± 95.41	8631.72 ± 155.71	8427.25 ± 185.41
	20	1753.91 ± 75.41	1853.88 ± 65.41	1829.31 ± 52.36	8994.28 ± 127.34	9332.67 ± 191.74	9485.58 ± 155.24
	30	1808.65 ± 48.26	1793.14 ± 75.61	1761.33 ± 75.14	8931.72 ± 154.25	8912.14 ± 185.74	9104.90 ± 151.74

Results are expressed as mean ± standard deviation. TPC = total phenolic content, AA = antioxidant activity.

Table 10. Influence of UEA parameters on TPC of SB cake extracts.

Source of Variation	TPC mg/100 g dm	AA μmol TE/100 g dm
Ultrasound power (%)	$p = 0.29$	$p = 0.39$
30	1538.71 ± 59.40 ^{a}	8432.52 ± 179.75 ^{a}
60	1673.57 ± 64.51 ^{a}	8735.44 ± 161.66 ^{a}
90	1614.51 ± 54.90 ^{a}	8706,49 ± 167.81 ^{a}
Temperature (°C)	$p \leq 0.001$	$p \leq 0.001$
35	1374.25 ± 38.75 ^{a}	7827.51 ± 142.20 ^{a}
50	1811.95 ± 30.04 ^{c}	8946.03 ± 90.92 ^{b}
65	1640.59 ± 55.80 ^{b}	9100.90 ± 67.87 ^{b}
Time (min)	$p \leq 0.001$	$p = 0.11$
10	1417.59 ± 57.62 ^{a}	8336.28 ± 205.74 ^{a}
20	1718.73 ± 49.37 ^{b}	8756.88 ± 170.40 ^{a}
30	1690.47 ± 47.21 ^{b}	8781.29 ± 105.64 ^{a}

Results are expressed as mean value ± standard error. Means with different letters within the column are significantly different at $p \leq 0.05$. TPC = total phenolic content; AA = antioxidant activity.

Table 11. Mass spectrometric data and identification of phenolic compounds in SB cake extracts obtained using optimized UAE conditions.

Phenolic Compounds	Precursor Ion (*m*/*z*)	Fragment Ions (*m*/*z*)	Ionization Mode	Mass Concentration (mg/100 g dm)			
				$L_{SC\text{-}CO_2}$	L_{SE}	$A_{SC\text{-}CO_2}$	A_{SE}
			FLAVONOLS				
Isorhamnetin	317	201	positive	0.45 ± 0.01 ^{a}	0.93 ± 0.02 ^{b}	1.05 ± 0.01 ^{b}	0.98 ± 0.01 ^{b}
Isorhamnetin-3-sinapoyglucose-glucoside-7-rhamnoside	993	463, 317	positive	1.71 ± 0.05 ^{b}	0.62 ± 0.01 ^{a}	1.80 ± 0.02 ^{b}	0.53 ± 0.00 ^{a}
Ishorhamnetin-3-sophoroside-7-rhamnoside	787	463, 317	positive	11.67 ± 0.12 ^{c}	6.79 ± 0.01 ^{b}	8.17 ± 0.02 ^{b}	3.20 ± 0.01 ^{a}
Isorhamnetin-3-rutinoside-7-glucoside	787	479, 317	positive	1.98 ± 0.05 ^{b,c}	1.46 ± 0.01 ^{b}	2.25 ± 0.01 ^{c}	0.33 ± 0.00 ^{a}
Isorhamnetin-3-hexoside	479	317	positive	65.37 ± 2.14 ^{c}	35.90 ± 1.45 ^{a}	79.44 ± 2.74 ^{d}	56.54 ± 1.85 ^{b}
Isorhamnetin-3-rhamnoside	463	317	positive	2.20 ± 0.00 ^{a}	4.01 ± 0.01 ^{b}	4.67 ± 0.01 ^{b}	2.68 ± 0.00 ^{a}
Isorhamnetin-3.7-dihexoside	641	479, 317	positive	0.55 ± 0.00 ^{a,b}	1.25 ± 0.01 ^{b}	0.51 ± 0.00 ^{a}	2.91 ± 0.01 ^{c}
Isorhamnetin-3-rutinoside	625	479, 317	positive	16.01 ± 0.54 ^{b}	9.64 ± 0.65 ^{a}	30.88 ± 0.17 ^{c}	15.87 ± 0.28 ^{b}
Kaempferol	287	145	positive	7.56 ± 0.01 ^{b}	3.53 ± 0.00 ^{a}	8.53 ± 0.01 ^{b}	2.97 ± 0.00 ^{a}
Kaempferol-3-O-sophorose-7-O-rhamnoside	757	287	positive	1.76 ± 0.01 ^{a}	1.71 ± 0.01 ^{a}	1.36 ± 0.01 ^{a}	3.98 ± 0.01 ^{b}
Kaemferol-3-O-glucoside-7-O-rhamnoside	595	433, 287	positive	15.71 ± 0.54 ^{b}	12.38 ± 0.25 ^{a}	18.72 ± 0.23 ^{c}	13.89 ± 0.08 ^{a,b}
Kaempferol-3-rutinoside	595	287	positive	1.97 ± 0.01 ^{a}	0.42 ± 0.02 ^{a}	2.03 ± 0.02 ^{a}	1.22 ± 0.02 ^{a}
Kaempferol-rhamnoside	433	287	positive	3.32 ± 0.01 ^{b}	3.12 ± 0.01 ^{b}	9.39 ± 0.05 ^{c}	1.27 ± 0.05 ^{a}
Quercetin-3-sophoroside-7-rhamnoside	773	611, 303	positive	2.32 ± 0.00 ^{b}	0.20 ± 0.00 ^{a}	2.11 ± 0.01 ^{b}	0.33 ± 0.00 ^{a}
Quercetin-3-rhamnosylglucoside-7-rhamnoside	757	303	positive	1.97 ± 0.02 ^{b}	1.65 ± 0.01 ^{b}	0.75 ± 0.02 ^{a}	0.55 ± 0.01 ^{a}
Quercetin-3-rutinoside (rutin)	611	303	positive	47.14 ± 2.45 ^{b}	32.49 ± 1.24 ^{a}	41.74 ± 0.98 ^{b}	33.27 ± 0.98 ^{a}
Quercetin-3-glucoside	465	303	positive	20.93 ± 0.04 ^{b}	13.46 ± 0.02 ^{a}	48.16 ± 0.51 ^{c}	16.43 ± 0.09 ^{a}
Quercetin-3-rhamnoside (quercitrin)	449	303	positive	1.70 ± 0.01 ^{b}	1.60 ± 0.01 ^{b}	0.95 ± 0.00 ^{a}	2.95 ± 0.01 ^{c}
Quercetin-3-pentoside	435	303	positive	0.81 ± 0.00 ^{b}	0.50 ± 0.00 ^{a}	0.54 ± 0.00 ^{a}	0.38 ± 0.00 ^{a}
SUM:				205.13 ± 6.01 ^{c}	131.66 ± 3.52 ^{a}	263.05 ± 4.81 ^{d}	160.28 ± 4.92 ^{b}

Table 11. *Cont.*

Phenolic Compounds	Precursor Ion (*m/z*)	Fragment Ions (*m/z*)	Ionization Mode	Mass Concentration (mg/100 g dm) $L_{SC\text{-}CO2}$	L_{SE}	$A_{SC\text{-}CO2}$	A_{SE}
			FLAVAN-3-OLS				
Catechin	291	139	positive	2.15 ± 0.00 [b]	0.64 ± 0.00 [a]	3.16 ± 0.00 [c]	2.10 ± 0.00 [b]
Epicatechin	291	165	positive	0.99 ± 0.00 [a]	0.51 ± 0.00 [a]	3.82 ± 0.00 [c]	2.50 ± 0.00 [b]
SUM:				3.14 ± 0.01 [b]	1.15 ± 0.00 [a]	6.98 ± 0.01 [c]	4.6 ± 0.01 [b]
			PHENOLIC ACIDS				
Caffeic acid	179	135	negative	23.14 ± 1.41 [a]	21.01 ± 0.95 [a]	23.03 ± 1.01 [a]	22.05 ± 1.11 [a]
Chorogenic acid	353	191	negative	2.33 ± 0.01 [a]	14.35 ± 0.85 [b]	3.46 ± 0.01 [a]	17.63 ± 1.14 [c]
Gallic acid	169	125	negative	23.92 ± 1.05 [a]	21.82 ± 0.98 [a]	21.30 ± 1.75 [a]	20.71 ± 1.14 [a]
p-hydroxybenzoic acid	137	93	negative	31.49 ± 2.14 [b]	18.75 ± 0.85 [a]	38.13 ± 1.74 [c]	29.72 ± 1.12 [b]
p-coumaric acid	163	119	negative	14.67 ± 0.85 [b]	11.51 ± 0.08 [b]	13.13 ± 0.75 [b]	7.55 ± 0.05 [a]
Protocatechuic acid	153	109	negative	38.08 ± 1.74 [c]	9.95 ± 0.83 [a]	29.53 ± 1.41 [b]	9.55 ± 0.81 [a]
Vanillic acid	169	125	positive	46.80 ± 1.28 [a]	42.63 ± 1.95 [a]	60.46 ± 2.53 [b]	48.32 ± 1.75 [a]
SUM:				180.43 ± 8.48 [b]	140.02 ± 6.49 [a]	189.04 ± 8.94 [b]	155.53 ± 6.21 [a,b]

Results are expressed as mean ± standard deviation. Values with different letters within a row are significantly different at $p \leq 0.05$. $L_{SC\text{-}CO2}$ = cake of Leikora variety obtained after supercritical CO_2 extraction, L_{SE} = cake of Leikora variety obtained after Soxhlet extraction, $A_{SC\text{-}CO2}$ = cake of Ascola variety obtained after supercritical CO_2 extraction, A_{SE} = cake of Ascola variety obtained after Soxhlet extraction.

The TPC of the SB cake extracts ranged from 1123.25 to 1998.98 mg/100 g dm (Table 9). The lower TPC content in SB cake defatted with SC-CO_2 (4.71–7.87 mg/g dm) was reported in the study by Dienaite et al. [20]. On the other hand, our results of TPC were similar with studies on cakes obtained after oil extraction from other raw materials such as sunflower (751.00–1851.00 mg/100 g dm), hemp (911.30–1542.03 mg/100 g), and flax (873.97–1257.35), confirming that the cake obtained after oil extraction is a valuable source of BAMs such as phenolic compounds [21,75]. In our previous study [36], a significantly lower TPC was found in freeze-dried SB berries compared to the TPC of the cake obtained after oil extraction. However, the content and type of phenols depend on the extraction methods used, their chemical nature, the particle size, the presence of interfering compounds, and the storage conditions [76]. SB berries are rich in fats and pigments, which can impair the extraction of phenolic compounds. Therefore, the removal of fats and pigments after oil extraction probably led to an increase in the yield of TPC in cake extracts. In addition, the cavitation effect of UAE promotes the damage of cell walls, which facilitates the release of phenolic compounds into the extraction solvent as done by Rodriguez et al. [77].

The complete recovery of antioxidants and other BAMs is very important for the development of effective processes for the utilization of by-products [20]. Therefore, the ORAC method was used to determine the effect of ultrasound on the AA of the extracted phenolic compounds. The AA of the SB cake extracts ranged from 6871.34 to 9504.85 μmol TE/100 g dm (Table 9). The ORAC values were significantly higher than the ORAC values determined in our previous study Čulina et al. [35] on freeze-dried SB berries, which is consistent with the higher TPC content in SB cake extracts. Also, lower ORAC values in SB cake obtained after SC-CO_2 (15.84–35.26 μmolTE/g dm) were reported in the study by Dienaite et al. [20].

3.6.1. Optimization of the UAE of Polyphenols from SB Cake

To achieve a high extraction yield, the variables that influence UAE, such as ultrasonic power, extraction time, and temperature, must be optimized. According to the results of the statistical analysis in Table 10, temperature and extraction time had a statistically significant effect ($p < 0.001$) on the TPC, while temperature had a statistically significant effect on the AA of the SB cake extracts.

Ultrasonic power is considered one of the critical parameters that need to be optimized, as the use of high values usually improves the extraction yield due to the generation of strong shear forces. In addition, a higher ultrasonic power shortens the extraction time [78]. As shown in Table 10, increasing the ultrasonic power from 30 to 60% increased the TPC and AA of SB cake extracts. The ultrasonic waves cause cavitation and thermal and mechanical phenomena leading to cell wall disruption, particle size reduction, and a higher reaction rate due to mass transfer through the cell wall, which improves solvent penetration and accelerates the extraction of BAMs. A similar trend of increase in TPC and AA, with an increase in ultrasound power from 20 to 60%, was observed in the study by Sharayei et al. [79] on pomegranate peels. On the other hand, increasing the ultrasonic power from 60 to 90% led to a decrease in TPC and AA in SB cake extracts. At higher ultrasound power, cavitation phenomena can initiate new reaction mechanisms, i.e., bubble destruction, degassing, and free radical formation, and cause changes in temperature, pressure, and physical interaction between solid and liquid interfaces that favor the degradation of BAMs [80]. Borrás-Enríquez et al. [81] and Pedisić et al. [82] found that the optimal ultrasonic power for obtaining extracts with high TPC and AA in mango residues and monofloral honeys is 60%, which is consistent with our study. In addition, temperature can affect the integrity of BAMs, as most of them are thermolabile. Considering that high ultrasonic power combined with long extraction times can lead to sample degradation, temperature control is essential for the proper design of the cooling reactor and the optimization of the UAE. Increasing the temperature from 35 to 50 °C led to an increase in the TPC and AA in the SB cake extracts (Table 10). The application of high temperatures leads to a softening of the plant tissue and a weakening of the interactions between phenolic compounds and proteins, which ultimately leads to a better migration of phenolic compounds into the extraction solvent [83]. Although a higher temperature leads to a higher yield, above a certain temperature, the TPC decreases due to a lower effect of cavitation, i.e., a large number of bubbles implode with less intensity, causing less damage to the plant material [84]. Also, excessively high temperatures can impair the propagation of ultrasound in the medium [85]. Thus, increasing the temperature to 65 °C led to a decrease in the TPC (Table 10). The study by Setyaningsih et al. [86] reported that many phenolic compounds were stable at an extraction temperature of 70 °C. In contrast, some phenolic compounds started to degrade at 50 °C. To achieve a yield of more than 90%, it is recommended to perform ultrasonic extraction at temperatures between 10 °C, and 50 °C. Increasing the temperature from 50 °C to 65 °C resulted in a higher AA, but no significant difference was found between the values obtained. The study by Shehata et al. [87] has shown that 50 °C is the optimal temperature to achieve the highest TPC and AA, which is consistent with our study. UAE enables a good extraction yield with relatively short extraction times (maximum 60 min), as longer times can lead to undesirable changes in the extracted compounds. The optimized extraction time is usually between 20 and 60 min, which minimizes energy consumption and reduces the exposure of the compounds to the process [88]. TPC and AA in SB cake extracts increased with prolongation of the extraction time from 10 to 20 min. However, excessively long extraction times can lead to a decrease in yield due to BAM degradation [84]. Thus, increasing the extraction time to 30 min resulted in a decrease in TPC and an increase in AA; however, no statistically significant difference was observed between the values obtained (Table 10). Our results are in accordance with the studies by Teh and Birch [21], Dobrinčić et al. [89], Galviz-Quezada et al. [90], and Borrás-Enríquez et al. [81], in which an extraction time of 20 min proved to be the most suitable to obtain extracts with high TPC and AA. The results of the statistical analysis show that the optimal UAE conditions for obtaining SB cake extracts with high TPC and AA were: an ultrasonic amplitude of 60%, a temperature of 50 °C, and a sonication time of 20 min.

3.6.2. Polyphenolic Characterization of Different SB Cake Extracts

UPLC/ESI-MS^2 was used to compare the polyphenolic profile of SB cake extracts obtained under optimal UAE conditions. A total of 29 compounds were identified, including 19 flavonols, 2 flavan-3-ols, and 7 phenolic acids (Table 11). Spectral mass data for identified compounds have already been reported in our previous study [36].

According to the results presented in Table 11, the most abundant phenolic compounds in both varieties of SB cake were flavonols such as isorhamnetin, kaempferol and its glycosides, and quercetin glycosides. In general, most flavonols in SB are either 3-glycosides or 3,7-diglycosides [91], with large differences in the sugar units within each aglycone group. The most common sugar residues are glucose and galactose, although rhamnose, xylose, and arabinose have also been found [92]. In the SB cake extracts, 19 flavonols were identified, as shown in Table 11. Isorhamnetin-3-hexoside was determined to have the highest concentration (35.90–79.44 mg/100 g dm), followed by quercetin-3-glucoside (13.46–48.16 mg/100 g dm), quercetin-3-rutinoside (32.49–47.14 mg/100 g dm), isorhamnetin-3-rutinoside (9.66–30.88 mg/100 g dm), kaemferol-3-O-glucoside-7-O-rhamnoside (12.38–18.72 mg/100 g dm), and isorhamnetin-3-sophoroside-7-rhamnoside (3.20–11.67 mg/100 g dm). Similar findings were reported by Rösch et al. [91] and Yang et al. [92], who found that isorhamnetin and quercetin glycosides, such as isorhamnetin-3-O-rutinoside, isorhamnetin-3-O-glucoside-7-O-rhamnoside, quercetin-3-O-glucoside, and quercetin-3-O-rutinoside, were the most prevalent flavonol glycosides in SB. Kallio et al. [93] reported that the most abundant phenolic compounds determined in the berries were isorhamnetin-3-sophoroside-7-rhamnoside (32%), followed by quercetin-3-glucoside (15%), quercetin-3-rutinoside (9%), isorhamnetin-3-rutinoside (6%), and isorhamnetin-3-glucoside (4%). Other flavonol glycosides such as isorhamnetin-3-sinapoyglucose-glucoside-7-rhamnoside, isorhamnetin-3-rutinoside-7-glucoside, isorhamnetin-3-rhamnoside, isorhamnetin-3,7-dihexoside, kaempferol-3-O-sophoroside-7-O-rhamnoside, kaempferol-3-rutinoside, kaempferol-rhamnoside, quercetin-3-sophoroside-7-rhamnoside, quercetin-3-rhamnosylglucoside-7-rhamnoside, quercetin-3-rhamnoside (quercitrin), and quercetin-3-pentoside were determined in a significantly lower concentration (less than 10 mg/100 g dm). These compounds were also detected in the study by Rosch et al. [91] and Pop et al. [94]. In the analyzed extracts, the aglycones isorhamnetin and kaempferol were determined in a range of 0.45–0.98 mg/100 g dm and 2.97–8.53 mg/100 g dm. According to Fatima et al. [95] and Guo et al. [96], the most important aglycones in berries are kaempferol (32–72 mg/kg fresh weight (fw)), quercetin (67–175 mg/kg fw), myricetin (36–172 mg/kg fw), and isorhamnetin (45–106 mg/kg fw). The higher content of kaempferol and isorhamnetin compared to our results is probably due to the acid hydrolysis used in the extraction [95]. Among the flavan-3-ols, catechin and epicatechin were detected in the SB cake extracts in a range of 0.64–2.15 mg/100 g dm and 0.51–2.82 mg/100 g dm, respectively (Table 10). The presence of catechin in SB berries was also confirmed by Bittova et al. [97] in a range of 4.1–22.2 mg/kg dm, which is in accordance with our study. Dienaite et al. [20] reported a similar concentration of epicatechin in defatted SB berry cake, while the concentration of catechin was significantly higher than in our study. In the study of Guo et al. [96], the content of catechin and epicatechin in four SB subspecies varied from 0.82 to 4.51 and from 7.60 to 8.99 mg/100 g dm, respectively. In the analyzed SB cake extracts, seven phenolic acids were identified, as shown in Table 11. Vanillic acid was determined to have the highest concentration (42.63–60.46 mg/100 g dm), followed by *p*-hydroxybenzoic (18.75–38.13 mg/100 g dm), protocatechuic (9.55–38.08 mg/100 g dm), gallic (20.71–23.92 mg/100 g dm), caffeic (21.01–23.14 mg/100 g dm), chlorogenic (2.33–17.63 mg/100 g dm), and *p*-coumaric acid (7.55–14.67 mg/100 g dm). The presence of the mentioned phenolic acids in SB berries was confirmed by literature data [35,98], but their contribution differs from our results. In the study of Zadernowski et al. [98], the most represented phenolic acid was salicylic acid, followed by gallic, cinnamic, protocatechuic, *p*-coumaric, and other minor phenolic acids. In contrast, the study by Arimboor et al. (2008) [99] showed that gallic acid was identified as the predominant phenolic acid in both free and bound forms, accounting for 66.0% of the total phenolic acids in SB berry pulp. In addition, considerable amounts of protocatechuic acid (136 mg/kg), ferulic acid (69 mg/kg), salicylic acid (54 mg/kg), *p*-hydroxybenzoic acid (40 mg/kg), and *p*-coumaric acid (37 mg/kg) were determined in the SB berry pulp.

In the analyzed SB cake extracts, the total content of phenolic groups is shown in Table 11. The total content of flavonols ranged from 131.66 to 205.13 mg/100 g dm for the Leikora variety and 160.28 to 263.05 mg/10 g dm for the Ascola variety; the total content of flavan-3-ols ranged from 1.15 to 3.14 mg/100 g dm for the Leikora variety and from 4.60 to 6.98 mg/100 g dm for the Ascola variety; and the total content of phenolic acids ranged from

140.02 to 180.43 mg/100 g dm for the Leikora variety and from 155.53 to 189.04 mg/100 g dm for the Ascola variety. The results presented in Table 11 show that the Ascola variety has a statistically significant ($p \leq 0.05$) higher content of individual and total flavonols and flavan-3-ols than the Leikora variety. A large variability in phenolic composition between different species and varieties has been confirmed by various studies [100–103]. According to earlier studies, SB berries have a high polyphenol content. According to Ercisli et al. [104], the TPC value of SB berries was between 213.1 and 553.8 mg GAE/100 g. In contrast, Korekar et al. [64] reported an 11-fold difference (from 964 to 10704 mg GAE/100 g) in TPC content. The significant difference in the TPC of the cultivated varieties under the same climatic conditions can be clearly explained by the different genetic material [103]. Even the same varieties grown in different areas can have different levels of phenolic compounds. According to Rop et al. [105], Leikora berries collected near Brno in the Czech Republic had more than three-fold higher TPC content than 'Leikora' berries grown under Hungarian growing and climatic conditions. In addition, the cake remaining after SE had a statistically significant ($p \leq 0.05$) lower content of individual and total flavonols and flavan-3-ols. Since the phenols are heat-sensitive and easily oxidizable compounds [106], the SE of the oil had an influence on the degradation of the phenolic compounds in the residual cake due to the high extraction temperature and the long extraction time in the presence of oxygen.

3.7. *Antioxidant Activity of SB Cake*

The AA of SB cake extracts obtained under optimal UAE conditions was determined using hydrophilic (H) ORAC (Figure 1).

The AA ranged from 6825.15 to 9550.54 μmolTE/100 g dm in Ascola SB cake and from 6750.15 to 9504.85 μmolTE/100 g dm in Leikora SB cake. The cakes produced after SC-CO_2 had a higher AA than cakes produced with SE, which is consistent with the higher mass concentration of phenolic compounds in SC-CO_2 cakes. SE can lead to the degradation of thermolabile bioactive compounds during a long extraction time when heated, thus reducing AA [107]. According to Guo et al. [96], the total phenols and flavonoid aglycones in SB extracts had antioxidant and antiproliferative activities. Previous studies had also shown that the fruits of SB varieties have strong AA, but there were significant differences between varieties [63,64].

4. Conclusions

The method of oil extraction and the variety influenced the content of BAMs and AA in the SB oils and cakes. SC-CO_2 proved to be the better method for obtaining oils and cakes with a higher content of BAMs and AA than SE. The oil of the Leikora variety had a higher content of palmitic and palmitoleic acids, β-sitosterol, and α-tocopherol, while the Ascola variety had a higher content of oleic, linoleic, and α-linolenic acids and carotenoids. In addition, the Ascola cake had the highest content of phenols, with flavonols being the most abundant and isorhamnetin-3-hexoside, isorhamnetin-3-rutinoside, and quercetin-3-glucoside dominating. SB oils and cake extracts showed high AA, but the Ascola variety had higher ORAC values. A combination of SC-CO_2 for oil extraction and UAE for other bioactive compound extraction is a good choice for maximal valorization of the SB berries, which have great potential for the food, nutritional, and cosmetic industries.

Author Contributions: Conceptualization, P.Č. and S.P.; methodology, P.Č. and S.P.; formal analysis, P.Č., S.J. and S.B.; investigation, P.Č. and S.P.; data curation, V.D.-U. and S.B.; writing—original draft preparation, P.Č. and S.P.; writing—review and editing, S.P., S.B. and V.D.-U. All authors have read and agreed to the published version of the manuscript.

Funding: This work was supported by the project "Bioactive molecules of medical plant as natural antioxidants, microbicides and preservatives" (KK.01.1.1.04.0093), co-financed by the Croatian Government and the European Union through the European Regional Development Fund-Operational Programme Competitiveness and Cohesion (KK.01.1.1.04.).

Data Availability Statement: Data are contained within the article.

Conflicts of Interest: The authors declare no conflicts of interest.

References

1. Li, T.S.C.; Beveridge, T.H.J. *Sea Buckthorn: Hippophae rhamnoides L.: Production and Utilization*; NRC Research Press: Ottawa, ON, Canada, 2003.
2. Guo, R.X.; Guo, X.B.; Li, T.; Fu, X.; Liu, R.H. Comparative assessment of phytochemical profiles, antioxidant and antiproliferative activities of Sea buckthorn (*Hippophae rhamnoides* L.) berries. *Food Chem.* **2017**, *221*, 997–1003. [CrossRef] [PubMed]
3. Sharma, U.K.; Sharma, K.; Sharma, N.; Sharma, A.; Singh, H.P.; Sinha, A.K. Microwave-assisted efficient extraction of different parts of for the comparative evaluation of antioxidant activity and quantification of its phenolic constituents by reverse-phase high-performance liquid chromatography (RP-HPLC). *J. Agric. Food Chem.* **2008**, *56*, 374–379. [CrossRef] [PubMed]
4. Jasniewska, A.; Diowksz, A. Wide Spectrum of Active Compounds in Sea Buckthorn (*Hippophae rhamnoides* L.) for Disease Prevention and Food Production. *Antioxidants* **2021**, *10*, 1279. [CrossRef] [PubMed]
5. Chandra, S.; Dwivedi, P.; Borkar, P.; Shinde, L. Pharmacological and nutritional importance of sea buckthorn (*Hippophae*). *Pharma Innov.* **2018**, *7*, 258.
6. Zielinska, A.; Nowak, I. Abundance of active ingredients in sea-buckthorn oil. *Lipids Health Dis.* **2017**, *16*, 95. [CrossRef] [PubMed]
7. Ciesarová, Z.; Murkovic, M.; Cejpek, K.; Kreps, F.; Tobolková, B.; Koplík, R.; Belajová, E.; Kukurová, K.; Daško, L'.; Panovská, Z.; et al. Why is sea buckthorn (*Hippophae rhamnoides* L.) so exceptional? A review. *Food Res. Int.* **2020**, *133*, 109170. [CrossRef] [PubMed]
8. Liu, Y.; Fan, G.; Zhang, J.; Zhang, Y.; Li, J.J.; Xiong, C.; Zhang, Q.; Li, X.D.; Lai, X.R. Metabolic discrimination of sea buckthorn from different species by H NMR based metabolomics. *Sci. Rep.* **2017**, *7*, 1585. [CrossRef] [PubMed]
9. Kallio, H.; Yang, B.R.; Peippo, P. Effects of different origins and harvesting time on vitamin C, tocopherols, and tocotrienols in sea buckthorn (*Hippophae rhamnoides*) berries. *J. Agric. Food Chem.* **2002**, *50*, 6136–6142. [CrossRef] [PubMed]
10. Andersson, S.C.; Rumpunen, K.; Johansson, E.; Olsson, M.E. Tocopherols and tocotrienols in sea buckthorn (*Hippophae rhamnoides* L.) berries during ripening. *J. Agric. Food. Chem.* **2008**, *56*, 6701–6706. [CrossRef] [PubMed]
11. Yang, B.R.; Kallio, H. Composition and physiological effects of sea buckthorn (*Hippophae rhamnoides* L.) lipids. *Trends Food Sci. Technol.* **2002**, *13*, 160–167. [CrossRef]
12. Li, T.S.C.; Beveridge, T.H.J.; Drover, J.C.G. Phytosterol content of sea buckthorn (*Hippophae rhamnoides* L.) seed oil: Extraction and identification. *Food Chem.* **2007**, *101*, 1633–1639. [CrossRef]
13. Baoru, Y.; Kallio, H.P.; Koponen, J.; Tahvonen, R.L.; Pfannhauser, W.; Fenwick, G.R.; Khokhar, S. Free and esterified sterols in seed oil and pulp/peel oil of sea buckthorn (*Hippophaë rhamnoides* L.). In Proceedings of the EUROFOODCHEM XI Meeting, Norwich, UK, 26–28 September 2001.
14. Kaseke, T.; Opara, U.L.; Fawole, O.A. Fatty acid composition, bioactive phytochemicals, antioxidant properties and oxidative stability of edible fruit seed oil: Effect of preharvest and processing factors. *Heliyon* **2020**, *6*, e04962. [CrossRef]
15. Herrero, M.; Cifuentes, A.; Ibañez, E. Sub- and supercritical fluid extraction of functional ingredients from different natural sources: Plants, food-by-products, algae and microalgae: A review. *Food Chem.* **2006**, *98*, 136–148. [CrossRef]
16. Bubalo, M.C.; Vidovic, S.; Redovnikovic, I.R.; Jokic, S. Green solvents for green technologies. *J. Chem. Technol. Biot.* **2015**, *90*, 1631–1639. [CrossRef]
17. Rawson, A.; Tiwari, B.K.; Brunton, N.; Brennan, C.; Cullen, P.J.; O'Donnell, C.P. Application of Supercritical Carbon Dioxide to Fruit and Vegetables: Extraction, Processing, and Preservation. *Food Rev. Int.* **2012**, *28*, 253–276. [CrossRef]
18. Uquiche, E.; Romero, V.; Ortíz, J.; del Valle, J.M. Extraction of Oil and Minor Lipids from Cold-Press Rapeseed Cake with Supercritical Co. *Braz. J. Chem. Eng.* **2012**, *29*, 585–597. [CrossRef]
19. Miekus, N.; Iqbal, A.; Marszalek, K.; Puchalski, C.; Swiergiel, A. Green Chemistry Extractions of Carotenoids from *Hippophae rhamnoides* L.-Supercritical Carbon Dioxide and Enzyme-Assisted Methods. *Molecules* **2019**, *24*, 4339. [CrossRef] [PubMed]
20. Dienaitė, L.; Pukalskas, A.; Pukalskienė, M.; Pereira, C.V.; Matias, A.A.; Venskutonis, P.R. Phytochemical Composition, Antioxidant and Antiproliferative Activities of Defatted Sea Buckthorn (*Hippophae rhamnoides* L.) Berry Pomace Fractions Consecutively Recovered by Pressurized Ethanol and Water. *Antioxidants* **2020**, *9*, 274. [CrossRef]
21. Teh, S.S.; Birch, E.J. Effect of ultrasonic treatment on the polyphenol content and antioxidant capacity of extract from defatted hemp, flax and canola seed cakes. *Ultrason. Sonochem* **2014**, *21*, 346–353. [CrossRef] [PubMed]
22. Shen, L.P.; Pang, S.X.; Zhong, M.M.; Sun, Y.F.; Qayum, A.; Liu, Y.X.; Rashid, A.; Xu, B.G.; Liang, Q.F.; Ma, H.L.; et al. A comprehensive review of ultrasonic assisted extraction (UAE) for bioactive components: Principles, advantages, equipment, and combined technologies. *Ultrason. Sonochem* **2023**, *101*, 106646. [CrossRef]
23. Dzah, C.S.; Duan, Y.Q.; Zhang, H.H.; Wen, C.T.; Zhang, J.X.; Chen, G.Y.; Ma, H.L. The effects of ultrasound assisted extraction on yield, antioxidant, anticancer and antimicrobial activity of polyphenol extracts: A review. *Food Biosci.* **2020**, *35*, 100547. [CrossRef]
24. Goldsmith, C.D.; Vuong, Q.V.; Stathopoulos, C.E.; Roach, P.D.; Scarlett, C.J. Ultrasound increases the aqueous extraction of phenolic compounds with high antioxidant activity from olive pomace. *LWT-Food Sci. Technol.* **2018**, *89*, 284–290. [CrossRef]

25. Nunes, M.A.; Costa, A.S.G.; Bessada, S.; Santos, J.; Puga, H.; Alves, R.C.; Freitas, V.; Oliveira, M.B.P.P. Olive pomace as a valuable source of bioactive compounds: A study regarding its lipid- and water-soluble components. *Sci. Total Environ.* **2018**, *644*, 229–236. [CrossRef]
26. Mikulic-Petkovsek, M.; Slatnar, A.; Stampar, F.; Veberic, R. HPLC-MS identification and quantification of flavonol glycosides in 28 wild and cultivated berry species. *Food Chem.* **2012**, *135*, 2138–2146. [CrossRef]
27. Petrarca, M.H.; Vicente, E.; Tfouni, S.A.V. Single-run gas chromatography-mass spectrometry method for the analysis of phthalates, polycyclic aromatic hydrocarbons, and pesticide residues in infant formula based on dispersive microextraction techniques. *Microchem. J.* **2024**, *197*, 109824. [CrossRef]
28. Pavlović, N.; Valek Lendić, K.; Miškulin, M.; Moslavac, T.; Jokić, S. Supercritical CO_2 Extraction of Sea Buckthorn. *Food Health Dis.* **2016**, *5*, 55–61.
29. Harkat, H.; Bousba, R.; Benincasa, C.; Atrouz, K.; Gueltekin-oezgueven, M.; Altuntas, U.; Demircan, E.; Zahran, H.A.; Oezcelik, B. Assessment of Biochemical Composition and Antioxidant Properties of Algerian Date Palm (*Hippophae rhamnoides* L.) Seed Oil. *Plants* **2022**, *11*, 381. [CrossRef]
30. *ISO 12966-22017*; Animal and Vegetable Fats and Oils—Gas Chromatography of Fatty Acid Methyl Esters—Part 2: Preparation of Methyl Esters of Fatty Acids [WWW Document]. ISO: Geneva, Switzerland, 2017. Available online: https://www.iso.org/standard/72142.html (accessed on 1 October 2022).
31. Determination of Individual and Total Sterols Contents—Gas Chromatographic Method—Part 1: Animal and Vegetable Fats and Oils [WWW Document]. ISO: Geneva, Switzerland, 2014. Available online: https://www.iso.org/standard/60248.html (accessed on 20 October 2022).
32. Castro-Puyana, M.; Pérez-Sánchez, A.; Valdés, A.; Ibrahim, O.H.M.; Suarez-Alvarez, S.; Ferragut, J.A.; Micol, V.; Cifuentes, A.; Ibáñez, E.; García-Cañas, V. Pressurized liquid extraction of Neochloris oleoabundans for the recovery of bioactive carotenoids with anti-proliferative activity against human colon cancer cells. *Food Res. Int.* **2017**, *99*, 1048–1055. [CrossRef]
33. Pop, R.M.; Weesepoel, Y.; Socaciu, C.; Pintea, A.; Vincken, J.-P.; Gruppen, H. Carotenoid composition of berries and leaves from six Romanian sea buckthorn (*Hippophae rhamnoides* L.) varieties. *Food Chem.* **2014**, *147*, 1–9. [CrossRef] [PubMed]
34. *ISO 9936-1:2016*; Animal and Vegetable Fats and Oils—Determination of Tocopherol and Tocotrienol Contents by Highperformance Liquid Chromatography [WWW Document]. ISO: Geneva, Switzerland, 2016. Available online: https://www.iso.org/standard/69595.html (accessed on 20 October 2022).
35. Culina, P.; Repajic, M.; Garofulic, I.E.; Dragovic-Uzelac, V.; Pedisic, S. Evaluation of Polyphenolic Profile and Antioxidant Activity of Sea Buckthorn (*Eleagnus rhamnoides* (L.) A Nelson) Extraction. *Processes* **2024**, *12*, 126. [CrossRef]
36. Culina, P.; Cvitkovic, D.; Pfeifer, D.; Zoric, Z.; Repajic, M.; Garofulic, I.E.; Balbino, S.; Pedisic, S. Phenolic Profile and Antioxidant Capacity of Selected Medicinal and Aromatic Plants: Diversity upon Plant Species and Extraction Technique. *Processes* **2021**, *9*, 2207. [CrossRef]
37. Garofulic, I.E.; Kruk, V.; Martic, A.; Martic, I.; Zoric, Z.; Pedisic, S.; Dragovic, S.; Dragovic-Uzelac, V. Evaluation of Polyphenolic Profile and Antioxidant Activity of L. Leaves and Fruit Extract Obtained by Optimized Microwave-Assisted Extraction. *Foods* **2020**, *9*, 1556. [CrossRef]
38. Naguib, Y.M.A.; Hari, S.P.; Passwater, R.; Huang, D.J. Antioxidant activities of natural vitamin E formulations. *J. Nutr. Sci. Vitaminol.* **2003**, *49*, 217–220. [CrossRef]
39. Dulf, F.V. Fatty acids in berry lipids of six sea buckthorn (*Hippophae rhamnoides* L., subspecies carpatica) cultivars grown in Romania. *Chem. Cent. J.* **2012**, *6*, 106. [CrossRef]
40. Barkhuu, B.; Lodonjav, M.; Ganzorig, O.; Tumurtogoo, N. *The Physicochemical Composition of Sea Buckthorn (Hippophae rhamnoides L.) Oil and Its Treatment Characteristics*; Atlantis Press: Amsterdam, The Netherlands, 2021.
41. Teleszko, M.; Wojdylo, A.; Rudzinska, M.; Oszmianski, J.; Golis, T. Analysis of Lipophilic and Hydrophilic Bioactive Compounds Content in Sea Buckthorn (*Hippophae rhamnoides* L.) Berries. *J. Agric. Food Chem.* **2015**, *63*, 4120–4129. [CrossRef]
42. Xu, X.; Gao, Y.X.; Liu, G.M.; Wang, Q.; Zhao, H. Optimization of supercritical carbon dioxide extraction of sea buckthorn (*Hippophae rhamnoides* L.) oil using response surface methodology. *LWT-Food Sci. Technol.* **2008**, *41*, 1223–1231. [CrossRef]
43. Xu, X.; Gao, Y.X.; Liu, G.M.; Zheng, Y.Y.; Zhao, J. Supercritical CO extraction of whole berry oil from sea buckthorn fruit. *Food Sci. Biotechnol.* **2008**, *17*, 470–474.
44. Vaitkeviciene, N.; Jariene, E.; Danilcenko, H.; Kulaitiene, J.; Mazeika, R.; Hallmann, E.; Blinstrubiene, A. Comparison of Mineral and Fatty Acid Composition of Wild and Cultivated Sea Buckthorn Berries from Lithuania. *J. Elem.* **2019**, *24*, 1101–1113. [CrossRef]
45. Yang, B.R.; Kallio, H.P. Fatty acid composition of lipids in sea buckthorn (*Hippophae rhamnoides* L.) berries of different origins. *J. Agric. Food Chem.* **2001**, *49*, 1939–1947. [CrossRef]
46. Cakir, A. Essential oil and fatty acid composition of the fruits of *Hippophae rhamnoides* L. (Sea Buckthorn) and *Myrtus communis* L. from Turkey. *Biochem. Syst. Ecol.* **2004**, *32*, 809–816. [CrossRef]
47. Máté, M.; Selimaj, G.; Simon, G.; Szalóki-Dorkó, L.; Gitta, F. Assessment of Berries of Some Sea Buckthorn Genotypes by Physicochemical Properties and Fatty Acid Content of the Seed. *Plants* **2022**, *11*, 3412. [CrossRef]
48. Kuhkheil, A.; Badi, H.N.; Mehrafarin, A.; Abdossi, V. Chemical constituents of sea buckthorn (*Hippophae rhamnoides* L.) fruit in populations of central Alborz Mountains in Iran. *Res. J. Pharmacogn.* **2017**, *4*, 1–12.

49. Zheng, L.; Shi, L.-K.; Zhao, C.-W.; Jin, Q.-Z.; Wang, X.-G. Fatty acid, phytochemical, oxidative stability and in vitro antioxidant property of sea buckthorn (*Hippophaë rhamnoides* L.) oils extracted by supercritical and subcritical technologies. *LWT* **2017**, *86*, 507–513. [CrossRef]
50. St George, S.; Cenkowski, S. Influence of Harvest Time on the Quality of Oil-Based Compounds in Sea Buckthorn (*Hippophae rhamnoides* L. ssp. sinensis) Seed and Fruit. *J. Agric. Food Chem.* **2007**, *55*, 8054–8061. [CrossRef]
51. Arimboor, R.; Venugopalan, V.; Sarinkumar, K.; Arumughan, C.; Sawhney, R.C. Integrated processing of fresh Indian sea buckthorn (*Hippophae rhamnoides*) berries and chemical evaluation of products. *J. Sci. Food Agric.* **2006**, *86*, 2345–2353. [CrossRef]
52. Madawala, S.R.P.; Brunius, C.; Adholeya, A.; Tripathi, S.B.; Hanhineva, K.; Hajazimi, E.; Shi, L.; Dimberg, L.; Landberg, R. Impact of location on composition of selected phytochemicals in wild sea buckthorn (*Hippophae rhamnoides*). *J. Food Compos. Anal.* **2018**, *72*, 115–121. [CrossRef]
53. Shi, L.; Zheng, L.; Zhao, C.; Jin, Q.; Wang, X. Chemical composition and antioxidant capacity of extracts from the whole berry, pulp and seed of *Hippophaë rhamnoides* ssp. yunnanensis. *Nat. Prod. Res.* **2019**, *33*, 3596–3600. [CrossRef]
54. Seglina, D.; Krasnova, I.; Grygier, A.; Radziejewska-Kubzdela, E.; Rudzińska, M.; Górnaś, P. Unique bioactive molecule composition of sea buckthorn (*Hippophae rhamnoides* L.) oils obtained from the peel, pulp, and seeds via physical "solvent-free" approaches. *J. Am. Oil Chem. Soc.* **2021**, *98*, 1009–1020. [CrossRef]
55. Yang, B.; Karlsson, R.M.; Oksman, P.H.; Kallio, H.P. Phytosterols in sea buckthorn (*Hippophaë rhamnoides* L.) berries: Identification and effects of different origins and harvesting times. *J. Agric. Food Chem.* **2001**, *49*, 5620–5629. [CrossRef]
56. Sajfrtová, M.; Licková, I.; Wimmerová, M.; Sovová, H.; Wimmer, Z. β-Sitosterol: Supercritical carbon dioxide extraction from sea buckthorn (*Hippophae rhamnoides* L.) seeds. *Int. J. Mol. Sci.* **2010**, *11*, 1842–1850. [CrossRef]
57. Yang, B.; Ahotupa, M.; Määttä, P.; Kallio, H. Composition and antioxidative activities of supercritical CO_2-extracted oils from seeds and soft parts of northern berries. *Food Res. Int.* **2011**, *44*, 2009–2017. [CrossRef]
58. Sovová, H.; Galushko, A.A.; Stateva, R.P.; Rochová, K.; Sajfrtová, M.; Bártlová, M. Supercritical fluid extraction of minor components of vegetable oils: β-Sitosterol. *J. Food Eng.* **2010**, *101*, 201–209. [CrossRef]
59. Tudor, C.; Bohn, T.; Iddir, M.; Dulf, F.V.; Focsan, M.; Rugina, D.O.; Pintea, A. Sea Buckthorn Oil as a Valuable Source of Bioaccessible Xanthophylls. *Nutrients* **2020**, *12*, 76. [CrossRef]
60. Mihalcea, L.; Turturica, M.; Ghinea, I.O.; Barbu, V.; Ionita, E.; Cotârlet, M.; Stanciuc, N. Encapsulation of carotenoids from sea buckthorn extracted by CO supercritical fluids method within whey proteins isolates matrices. *Innov. Food Sci. Emerg.* **2017**, *42*, 120–129. [CrossRef]
61. Rodriguez-Amaya, D.B. Quantitative analysis, assessment of bioavailability and antioxidant activity of food carotenoids-A review. *J. Food Compos. Anal.* **2010**, *23*, 726–740. [CrossRef]
62. Ranjith, A.; Kumar, K.S.; Venugopalan, V.V.; Arumughan, C.; Sawhney, R.C.; Singh, V. Fatty acids, tocols, and carotenoids in pulp oil of three sea buckthorn species grown in the Indian Himalayas. *J. Am. Oil Chem. Soc.* **2006**, *83*, 359–364. [CrossRef]
63. Gao, X.Q.; Ohlander, M.; Jeppsson, N.; Björk, L.; Trajkovski, V. Changes in antioxidant effects and their relationship to phytonutrients in fruits of sea buckthorn (*Hippophae rhamnoides* L.) during maturation. *J. Agric. Food Chem.* **2000**, *48*, 1485–1490. [CrossRef]
64. Korekar, G.; Dolkar, P.; Singh, H.; Srivastava, R.B.; Stobdan, T. Variability and the genotypic effect on antioxidant activity, total phenolics, carotenoids and ascorbic acid content in seventeen natural population of Seabuckthorn (*Hippophae rhamnoides* L.) from trans-Himalaya. *LWT-Food Sci. Technol.* **2014**, *55*, 157–162. [CrossRef]
65. Zadernowski, R.; Naczk, M.; Amarowicz, R. Tocopherols in sea buckthorn (*Hippophaë rhamnoides* L.) berry oil. *J. Oil Fat. Ind.* **2003**, *80*, 55–58. [CrossRef]
66. Bal, L.M.; Meda, V.; Naik, S.N.; Satya, S. Sea buckthorn berries: A potential source of valuable nutrients for nutraceuticals and cosmoceuticals. *Food Res. Int.* **2011**, *44*, 1718–1727. [CrossRef]
67. Otgonbayar, C.; Matthaus, B.; Odonmajig, P. Fatty acid, Tocopherol and Sterol Composition in Sea buckthorn (*Hippophae rhamnoides* L.) of Mongolia. *Mong. J. Chem.* **2014**, *12*, 126–130. [CrossRef]
68. Eisenmenger, M.; Dunford, N.T. Bioactive Components of Commercial and Supercritical Carbon Dioxide Processed Wheat Germ Oil. *J. Am. Oil Chem. Soc.* **2008**, *85*, 55–61. [CrossRef]
69. Gracia, I.; Rodríguez, J.F.; de Lucas, A.; Fernandez-Ronco, M.P.; García, M.T. Optimization of supercritical CO_2 process for the concentration of tocopherol, carotenoids and chlorophylls from residual olive husk. *J. Supercrit. Fluids* **2011**, *59*, 72–77. [CrossRef]
70. Cuco, R.P.; Cardozo-Filho, L.; Silva, C.D. Simultaneous extraction of seed oil and active compounds from peel of pumpkin (*Cucurbita maxima*) using pressurized carbon dioxide as solvent. *J. Supercrit. Fluids* **2019**, *143*, 8–15. [CrossRef]
71. Sun, Q.; Shi, J.; Scanlon, M.; Xue, S.J.; Lu, J. Optimization of supercritical-CO_2 process for extraction of tocopherol-rich oil from canola seeds. *LWT* **2021**, *145*, 111435. [CrossRef]
72. Verleyen, T.; Verhe, R.; Huyghebaert, A.; Dewettinck, K.; De Greyt, W. Identification of α-Tocopherol Oxidation Products in Triolein at Elevated Temperatures. *J. Agric. Food Chem.* **2001**, *49*, 1508–1511. [CrossRef]
73. Sabliov, C.; Fronczek, C.; Astete, C.; Khachaturyan, M.; Khachatryan, L.; Leonardi, C. Effects of Temperature and UV Light on Degradation of α-Tocopherol in Free and Dissolved Form. *JAOCS J. Am. Oil Chem. Soc.* **2009**, *86*, 895–902. [CrossRef]
74. Andrei, S.; Bunea, A.; Pop, E.A.; Dulf, F.; Pintea, A. The Protective Effect of *Hippophae rhamnoides* Carotenoid Extract Against Lipid Peroxidation in Crude Vegetable Oils. *Bull. Univ. Agric. Sci. Vet. Med. Cluj-Napoca* **2014**, *71*, 101–106. [CrossRef]

75. Zardo, I.; de Espíndola Sobczyk, A.; Marczak, L.D.F.; Sarkis, J. Optimization of Ultrasound Assisted Extraction of Phenolic Compounds from Sunflower Seed Cake Using Response Surface Methodology. *Waste Biomass Valorization* **2019**, *10*, 33–44. [CrossRef]
76. Alara, O.R.; Abdurahman, N.H.; Ukaegbu, C.I. Extraction of phenolic compounds: A review. *Curr. Res. Food Sci.* **2021**, *4*, 200–214. [CrossRef]
77. Rodríguez Barragan, Ó.; Bona, S.; Rodríguez-Turienzo, L.; Staebler, A. Ultrasound-Assisted Extraction of Polyphenols from Olive Pomace: Scale up from Laboratory to Pilot Scenario. *Processes* **2022**, *10*, 2481. [CrossRef]
78. Carreira-Casais, A.; Rodríguez, M.; González Pereira, A.; Rivo, F.N.; Soria López, A.; García-Pérez, P.; Otero, P.; Cao, H.; Xiao, J.; Simal-Gandara, J.; et al. Critical Variables Influencing the Ultrasound-Assisted Extraction of Bioactive Compounds—A Review. *Chem. Proc.* **2021**, *5*, 50. [CrossRef]
79. Sharayei, P.; Azarpazhooh, E.; Zomorodi, S.; Ramaswamy, H.S. Ultrasound assisted extraction of bioactive compounds from pomegranate peel. *LWT-Food Sci. Technol.* **2019**, *101*, 342–350. [CrossRef]
80. Gomes, W.F.; Tiwari, B.K.; Rodriguez, O.; de Brito, E.S.; Fernandes, F.A.N.; Rodrigues, S. Effect of ultrasound followed by high pressure processing on prebiotic cranberry juice. *Food Chem.* **2017**, *218*, 261–268. [CrossRef]
81. Borrás-Enríquez, A.; Reyes-Ventura, E.; Villanueva, S.; Moreno-Vilet, L. Effect of Ultrasound-Assisted Extraction Parameters on Total Polyphenols and Its Antioxidant Activity from Mango Residues (*Mangifera indica* L. var. *Manililla*). *Separations* **2021**, *8*, 94. [CrossRef]
82. Pedisic, S.; Culina, P.; Pavlesic, T.; Vahcic, N.; Garofulic, I.E.; Zoric, Z.; Dragovic-Uzelac, V.; Repajic, M. Efficiency of Microwave and Ultrasound-Assisted Extraction as a Green Tool for Polyphenolic Isolation from Monofloral Honeys. *Processes* **2023**, *11*, 3141. [CrossRef]
83. Mokrani, A.; Madani, K. Effect of solvent, time and temperature on the extraction of phenolic compounds and antioxidant capacity of peach fruit. *Sep. Purif. Technol.* **2016**, *162*, 68–76. [CrossRef]
84. Kumar, K.; Srivastav, S.; Sharanagat, V.S. Ultrasound assisted extraction (UAE) of bioactive compounds from fruit and vegetable processing by-products: A review. *Ultrason. Sonochem.* **2021**, *70*, 105325. [CrossRef] [PubMed]
85. Martini, S. Sonocrystallization of Fats: Considerations for Industrial Applications. *J. Am. Oil Chem. Soc.* **2021**, *98*, 5.
86. Setyaningsih, W.; Saputro, I.E.; Palma, M.; Barroso, C.G. Stability of 40 Phenolic Compounds during Ultrasound-Assisted Extractions (UAE). *AIP Conf. Proc.* **2016**, *1755*, 080009. [CrossRef]
87. Shehata, M.G.; Aziz, N.M.A.; Youssef, M.M.; El-Sohaimy, S.A. Optimization conditions of ultrasound-assisted extraction of phenolic compounds from orange peels using response surface methodology. *J. Food Process Preserv.* **2021**, *45*, e15870. [CrossRef]
88. Aguilar-Hernández, G.; Vivar-Vera, M.D.; García-Magaña, M.D.; González-Silva, N.; Pérez-Larios, A.; Montalvo-González, E. Ultrasound-Assisted Extraction of Total Acetogenins from the Soursop Fruit by Response Surface Methodology. *Molecules* **2020**, *25*, 1139. [CrossRef] [PubMed]
89. Dobrincic, A.; Repajic, M.; Garofulic, I.E.; Tuden, L.; Dragovic-Uzelac, V.; Levaj, B. Comparison of Different Extraction Methods for the Recovery of Olive Leaves Polyphenols. *Processes* **2020**, *8*, 1008. [CrossRef]
90. Galviz-Quezada, A.; Ochoa, A.; Zabala, M.; Ochoa, S.; Osorio-Tobón, J. Obtaining phenolic compounds from iraca waste (*Carludovica palmata* Ruiz & Pav) through ultrasound-assisted extraction. *Biomass Convers. Biorefinery* **2021**, *13*, 4965–4976. [CrossRef]
91. Rösch, D.; Krumbein, A.; Mügge, C.; Kroh, L.W. Structural investigations of flavonol glycosides from sea buckthorn (*Hippophaë rhamnoides*) pomace by NMR spectroscopy and HPLC-ESI-MS(n). *J. Agric. Food Chem.* **2004**, *52*, 4039–4046. [CrossRef] [PubMed]
92. Yang, B.; Halttunen, T.; Raimo, O.; Price, K.; Kallio, H. Flavonol glycosides in wild and cultivated berries of three major subspecies of and changes during harvesting period. *Food Chem.* **2009**, *115*, 657–664. [CrossRef]
93. Kallio, H.; Yang, B.; Halttunen, T. Flavonol glycosides of berries of three major sea buckthorn subspecies, *Hippophaë rhamnoides* ssp. *Rhamnoides*, ssp. *Sinensis* and ssp. Mongolica. In Proceedings of the 2nd International Seabuckthorn Association Conference, ISA2005, Beijing, China, 26–29 August 2005; pp. 29–35.
94. Pop, R.M.; Socaciu, C.; Pintea, A.; Buzoianu, A.D.; Sanders, M.G.; Gruppen, H.; Vincken, J.P. UHPLC/PDA-ESI/MS Analysis of the Main Berry and Leaf Flavonol Glycosides from Different Carpathian *Hippophae rhamnoides* L. Varieties. *Phytochem. Anal.* **2013**, *24*, 484–492. [CrossRef]
95. Fatima, T.; Kesari, V.; Watt, I.; Wishart, D.; Todd, J.F.; Schroeder, W.R.; Paliyath, G.; Krishna, P. Metabolite profiling and expression analysis of flavonoid, vitamin C and tocopherol biosynthesis genes in the antioxidant-rich sea buckthorn (*Hippophae rhamnoides* L.). *Phytochemistry* **2015**, *118*, 181–191. [CrossRef]
96. Guo, X.; Guo, R.; Chang, X.; Brennan, C.; Li, T.; Fu, X.; Liu, R. Phenolic compounds, antioxidant activity, antiproliferative activity and bioaccessibility of Sea buckthorn (*Hippophaë rhamnoides* L.) berries as affected by in vitro digestion. *Food Funct.* **2017**, *8*, 4229–4240. [CrossRef]
97. Bittová, M.; Krejzová, E.; Roblova, V.; Kubáň, P.; Kubáň, V. Monitoring of HPLC profiles of selected polyphenolic compounds in sea buckthorn (*Hippophae rhamnoides* L.) plant parts during annual growth cycle and estimation of their antioxidant potential. *Cent. Eur. J. Chem.* **2014**, *12*, 1152–1161. [CrossRef]
98. Zadernowski, R.; Naczk, M.; Czaplicki, S.; Rubinskiene, M.; Szlakiewicz, M. Composition of phenolic acids in sea buckthorn (*Hippophae rhamnoides* L.) berries. *J. Am. Oil Chem. Soc.* **2005**, *82*, 175–179. [CrossRef]

99. Arimboor, R.; Kumar, K.S.; Arumughan, C. Simultaneous estimation of phenolic acids in sea buckthorn (*Hippophaë rhamnoides*) using RP-HPLC with DAD. *J. Pharm. Biomed. Anal.* **2008**, *47*, 31–38. [CrossRef] [PubMed]
100. Wang, Z.; Zhao, F.; Wei, P.; Chai, X.; Hou, G.; Meng, Q. Phytochemistry, health benefits, and food applications of sea buckthorn (*Hippophae rhamnoides* L.): A comprehensive review. *Front. Nutr.* **2022**, *9*, 1036295. [CrossRef]
101. Urcan, A.; Criste, A.; Dezmirean, D.; Mărgăoan, R.; Caeiro, A.; Campos, M. Similarity of Data from Bee Bread with the Same Taxa Collected in India and Romania. *Molecules* **2018**, *23*, 2491. [CrossRef] [PubMed]
102. Tańska, M.; Roszkowska, B.; Czaplicki, S.; Borowska, E.J.; Bojarska, J.; Dąbrowska, A. Effect of Fruit Pomace Addition on Shortbread Cookies to Improve Their Physical and Nutritional Values. *Plant Foods Hum. Nutr.* **2016**, *71*, 307–313. [CrossRef] [PubMed]
103. Ficzek, G.; Mátravölgyi, G.; Furulyas, D.; Рэнцэндава, Ч.; Jócsák, I.; Papp, D.; Simon, G.; Végvári, G.; Stéger-Máté, M. Analysis of bioactive compounds of three sea buckthorn cultivars (*Hippophaë rhamnoides* L. 'Askola', 'Leikora', and 'Orangeveja') with HPLC and spectrophotometric methods. *Eur. J. Hortic. Sci.* **2018**, *84*, 31–38. [CrossRef]
104. Orhan, E.; Çakir, Ö.; Şengül, M. The genotypic effects on the chemical composition and antioxidant activity of sea buckthorn (*Hippophae rhamnoides* L.) berries grown in Turkey. *Sci. Hortic.* **2007**, *115*, 27–33. [CrossRef]
105. Rop, O.; Ercisli, S.; Mlcek, J.; Jurikova, T.; Hoza, I. Antioxidant and radical scavenging activities in fruits of 6 sea buckthorn (*Hippophae rhamnoides* L.) cultivars. *Turk. J. Agric.* **2014**, *38*, 224–232. [CrossRef]
106. Robards, K. Strategies for the determination of bioactive phenols in plants, fruit and vegetables. *J. Chromatogr. A* **2003**, *1000*, 657–691. [CrossRef]
107. Shi, L.; Zhao, W.; Yang, Z.; Subbiah, V.; Suleria, H.A.R. Extraction and characterization of phenolic compounds and their potential antioxidant activities. *Env. Sci. Pollut. Res. Int.* **2022**, *29*, 81112–81129. [CrossRef] [PubMed]

processes

Article

Preliminary Study on the Chemical and Biological Properties of Propolis Extract from Stingless Bees from the Northern Region of Brazil

Lindalva Maria de Meneses Costa Ferreira [1], Poliana Dimsan Queiroz de Souza [2], Rayanne Rocha Pereira [3], Edilene Oliveira da Silva [2], Wagner Luiz Ramos Barbosa [1], José Otávio Carréra Silva-Júnior [1], Attilio Converti [4,*] and Roseane Maria Ribeiro-Costa [1]

1 Institute of Health Sciences, Federal University of Pará, Belem 66075-110, Brazil; lindalva.costa.ferreira@ics.ufpa.br (L.M.d.M.C.F.); barbosa@ufpa.br (W.L.R.B.); carrera@ufpa.br (J.O.C.S.-J.); rmrc@ufpa.br (R.M.R.-C.)

2 Institute of Biological Sciences, Federal University of Pará, Belem 66075-110, Brazil; polianadimsan@gmail.com (P.D.Q.d.S.); edilene@ufpa.br (E.O.d.S.)

3 Institute of Public Health, Federal University of Western Pará (UFOPA), Santarem 68040-255, Brazil; rayanne.pereira@ufopa.edu.br

4 Department of Civil, Chemical and Environmental Engineering, Pole of Chemical Engineering, University of Genoa, Via Opera Pia 15, 16145 Genoa, Italy

* Correspondence: converti@unige.it

Abstract: Natural products are historically regarded as the main sources of bioactive compounds to fight the most diverse diseases; among them, propolis deserves to be highlighted due to several biological activities, such as antioxidant, anti-inflammatory, healing, antibacterial and antileishmanial activity. In this study, the antioxidant and antipromastigote activities and the cytotoxicity of propolis of the *Scaptotrigona* aff. *postica* stingless bee were evaluated. Propolis extract was analyzed by infrared spectroscopy, thermogravimetry and spectrophotometry for total polyphenols and flavonoid contents. Antioxidant activity was assessed spectrophotometrically by the DPPH, ABTS and FRAP methods, while antipromastigote activity and cytotoxicity were assessed by the MTT assay. The infrared spectra showed the characteristic bands of both phenols and alcohols. The thermogravimetry study revealed two events and thermal stability around 370 °C, while differential scanning calorimetry showed two endothermic events and an exothermic event. Total polyphenol and flavonoid contents were 21.29 ± 0.003 mg/g and 8.17 ± 0.00 mg/g, respectively, and the antioxidant activities identified by the DPPH, ABTS and FRAP methods were 910.09 ± 0.0264 μM Trolox (IC_{50} 18.9 ± 0.01 μg/mL), 859.97 ± 0.0854 μM Trolox (IC_{50} 692 ± 0.01 μg/mL) and 1613.67 ± 0.2610 μM Trolox (IC_{50} 80 ± 0.1 μg/mL), respectively. The satisfactory antioxidant activity of propolis can be related to the high content of phenolic compounds, which adds value to this product and can contribute to the development of meliponiculture in the Brazilian state of Pará. The propolis extract caused a toxic effect on promastigote forms of *Leishmania amazonensis* at all concentrations tested, with an IC_{50} value of 1.50 μg/mL and a statistically significant difference compared to the negative control ($p < 0.001$). These results show that the propolis extract from *S. postica* bees may be a promising alternative for treatment against promastigote forms of *L. amazonensis*.

Keywords: propolis; total phenols; cytotoxicity; *Scaptotrigona*; antioxidant activity; antipromastigote activity

Citation: Ferreira, L.M.d.M.C.; Souza, P.D.Q.d.; Pereira, R.R.; da Silva, E.O.; Barbosa, W.L.R.; Silva-Júnior, J.O.C.; Converti, A.; Ribeiro-Costa, R.M. Preliminary Study on the Chemical and Biological Properties of Propolis Extract from Stingless Bees from the Northern Region of Brazil. *Processes* **2024**, *12*, 700. https://doi.org/10.3390/pr12040700

Academic Editors: Yanlin Zhang, Prashant K. Sarswat and Agnieszka Zgoła-Grześkowiak

Received: 2 February 2024
Revised: 21 March 2024
Accepted: 27 March 2024
Published: 29 March 2024

1. Introduction

Stingless bees (meliponines), which are part of the Meliponinae subfamily (Apidae family), have different characteristics from stinging bees, such as colony size, indication biology, brood selection, production of queen bees, method of bee storage and bee recruitment instrument [1]. Meliponine females do not have a stinger or have it in an atrophied form, as they have created different ways of protection, such as a strong bite [2,3]. It is reported that around 600 varieties have been identified and are widespread throughout tropical and subtropical areas [3], including South and Central America, southern North America, Africa, Southeast Asia and northern Oceania [1]. Around 200 of them, belonging to 29 genera, are distributed throughout Brazil, with 89 endemic species, which represents around 20% of neotropical stingless bees [2,3].

The genera *Plebeia*, *Trigona*, *Melipona*, *Scaptotrigona* and *Trigonisca* contain a wide variety of popular species [3]. The genus *Scaptotrigona*, composed of 24 species commonly known as canudo bees, mandaguari, tubi and tubiba, is widespread throughout the Neotropical region and includes species that build their nests in pre-determined locations [4]. The species *Scaptotrigona affinis postica* Latreille (*S. postica*), popular as canudo and tucanaíra, is geographically located in Bolivia, Brazil and Peru [4].

S. postica is a stingless bee commonly found in the Amazon, which has broad potential for making honey and is used to pollinate native crop species [5], due to the visit to a large number of botanical species, tolerance to management, and ability to reproduce on a large scale [6–8]. Stingless bees produce geopropolis by adding plant resin to wax and soil [9,10]; however, *S. postica* does not use soil to produce propolis [11].

Stingless bee propolis is a substance made up of oils and resins collected by bees in different parts of plants, from exudates from cuts or buds and flower buds, which are mixed with wax they produce in their oral apparatus [12,13]. This substance is used in nest structures (inlet tubes, casing), in gap seals, and in small portions to protect against intruders [7]. Propolis is an apicultural product largely applied in traditional medicine since ancient times to carry out asepsis and to heal wounds [5,14,15]. More recently, thanks to its biological activities related to its complex chemical composition [16], it has been considered one of the most promising products in the chemoprevention and treatment of various diseases, especially cancer [17].

Some studies have demonstrated that the presence of compounds such as flavonoids, aromatic acids and their esters, terpenoids and carotenoids, among others [18–20], is related to biological properties such as the antioxidant, antimicrobial and antileishmanial activities found in propolis [21–24]. In this sense, American Cutaneous Leishmaniasis (ACL) is a polymorphic, spectral disease of the skin and mucous membranes, which is divided into several clinical forms, such as the cutaneous, mucosal cutaneous and diffuse cutaneous ones [24]. Its therapy still faces challenges due to low patient adherence to the first-choice medications used for its treatment, which are responsible for numerous adverse reactions [24]. As a way to minimize this problem, natural products can be used, including propolis.

Propolis is traditionally used in wound healing and as an antiseptic [17]. It is important to highlight that, despite the biological properties of propolis, contact with it can cause anaphylaxis and local irritations, due to the allergic effect that occurs not because of the intrinsic nature of propolis but rather because of the vulnerability of the individual [14]. A study carried out on the hydroethanolic extract of *S. postica* propolis indicated that it may be important in controlling tumor growth in models of experimental solid tumors [17]. The effects of propolis from *Scaptotrigona* sp. were studied in human brain tumor (glioblastoma) cells for cell proliferation, clonogenic capacity and cell death, and the extract was observed to inhibit proliferation in glioblastoma and fibroblast cell lines [25].

Studies have shown the potential of propolis to combat electron transfer mediating molecules in the skin, which makes its use safe in the treatment of burns. This natural product acts speeding up the tissue repair of burned skin and leads to the remodeling of damaged tissue due to the ability of flavonoids to reduce lipid peroxidation and prevent cell death [26,27]. Polyphenols act as antioxidants through a complex mechanism, preventing the formation of reactive oxygen species (ROS), chelating metal ions involved in both the production and elimination of ROS, affecting the reactions that induce lipid peroxidation and collaborating synergistically with other antiradical compounds [28].

Research has revealed that stingless bee propolis has great potential in preventing harmful changes in cellular metabolism due to its antiradical action and inflammatory processes, in addition to having immunomodulatory potential [18,29]. Stingless bee propolis has an antioxidant effect and acts by directly scavenging free radicals and inhibiting lipid peroxidation [30–32]. Lipid peroxidation, due to the attack of ROS on cell lipids, promotes the generation of carcinogenic compounds in the body and causes insertions, removals and substitutions of nucleotides [33].

Therefore, studies on quality assurance of extracts from natural sources by physicochemical characterization techniques become necessary to ensure product effectiveness, quality and safety. There are several methods used for this purpose, such as thermogravimetric analysis (TG/DTG), differential scanning calorimetry (DSC), Fourier-transform infrared (FT-IR) spectroscopy, and ultraviolet–visible (UV-Vis) spectrophotometry.

In search of new bioproducts focused on the cosmetic and pharmaceutical sectors, studies are required on extracts from bee products such as propolis, which has a complex chemical composition and several biological properties. Most research to date has focused on propolis produced by *Apis mellifera* and *Melipona scutellaris*, while only a few studies were carried out on that produced by the stingless bee *S. postica*. So, research efforts to assess its physicochemical profile by TG/DTG and DSC as well as its antioxidant potential and biological activities are expected to expand the possibility of obtaining new bioproducts to be exploited in the most diverse pharmaceutical areas.

From this perspective, in this study, the propolis extract of *S. postica* was characterized physicochemically and evaluated for its antioxidant potential and in vitro antipromastigote activity, seeking its possible use as a leishmanicidal agent.

2. Materials and Methods

2.1. Propolis Samples

The *S. postica* propolis was provided by Embrapa Eastern Amazon. The material was collected in Igarapé Açu, a city of the Brazilian Pará state, in an irrigated area of açaí (*Euterpe oleracea*) monoculture (1°01′46′′ S 47°35′03′′ W). The samples (two samples), taken in a commercial apiary, were divided into small portions, packed in plastic bags, and kept frozen (−18 °C) for 24 h before the extract preparation.

2.2. Reagents and Standards

L-ascorbic acid, gallic acid, quercetin, 2,2-diphenyl-1-picrylhydrazyl (DPPH), 2,2′-azinobis (3-ethylbenzothiazoline-6)-sulfonic acid (ABTS), 6-hydroxy-2,5,7,8-tetramethylchroman-2-carboxylic acid (Trolox), 3-(4,5-dimethylthiazol-2yl)-2,5-diphenyl tetrazoline bromide (MTT) and 2,4,6-Tris(2-pyridyl)-s-triazine (TPTZ) were obtained from Sigma-Aldrich (St. Louis, MO, USA), while potassium persulfate, Folin–Ciocalteu reagent, sodium carbonate, aluminum chloride, potassium bromide, sodium acetate trihydrate and ferric chloride hexahydrate were acquired from LabSynth (São Paulo, Brazil).

2.3. Extract Preparation and Physicochemical Characterization

2.3.1. Propolis Extract Preparation

Extraction was carried out using the maceration method using 100 g of crushed propolis and 1 L of 70% ethanolic solution in a closed amber bottle placed in the dark at room temperature (25 °C) for 7 days. The obtained tincture was vacuum-filtered and placed in a rotary evaporator (Buchi R-210, Geneva, Switzerland) at 40 °C. The vacuum was controlled for a period of 3 h until ethanol was completely removed, thus obtaining the concentrated extract, which was kept in an amber bottle under refrigeration. Then, the extract was dried by lyophilization for later analyses [34].

2.3.2. Fourier-Transform Infrared Spectroscopy

The Fourier-Transform infrared (FTIR) spectroscopy spectra of the extract were obtained in a spectrometerIR Prestige-21 (Shimadzu®, Kyoto, Japan). For this purpose, 1 mg of the lyophilized extract was mixed in a smooth agate mortar with 0.99 mg of dry and sprayed potassium bromide (KBr). The mixture was then pressed in special molds, under pressure of around 70,000–100,000 kPa, until the formation of a transparent tablet. The spectra were analyzed in the wavenumber range from 4000 to 400 cm^{-1}, with 32 scans and a resolution of 4 cm^{-1} [35,36].

2.3.3. Thermogravimetric Behavior

Thermogravimetric curves (TGs) of extract samples (3 mg) were recorded in a TGA-50 thermal analyzer (Shimadzu®) using an aluminum crucible under a nitrogen atmosphere (flowrate of 50 mL/min), temperature of 600 °C and heating rate of 10 °C/min. The TA-50W software, version 2.21 (Shimadzu®) was used to investigate the results of thermogravimetric analysis [37].

2.3.4. Thermal Behavior by Differential Scanning Calorimetry

Propolis extract differential scanning calorimetry (DSC) curve was recorded using DSC-60 plus equipment (Shimadzu®). Samples (3 mg) were deposited in an aluminum crucible and analyzed under a nitrogen atmosphere (flowrate of 50 mL/min) at a heating rate of 10 °C/min and temperature of 350 °C [37].

2.3.5. Total Polyphenols

The content of total polyphenols was quantified in a UV 1800 spectrophotometer (Shimadzu®) using a standard curve of gallic acid at different concentrations. The extract was prepared by weighing 20 mg of the lyophilized in a beaker and solubilizing it in 10 mL of distilled water contained in a volumetric balloon, in order to obtain the final concentration of 2 mg/mL. The extract was then placed in an ultrasonic bath (Cleaner Kondentech® CD-4820, São Paulo, Brazil) for 1 min to complete homogenization. Quantification was carried out on a mixture of 100 µL of extract, 500 µL of Folin–Ciocalteu reagent, 6 mL of distilled water and 2 mL of 20% (*w*/*v*) sodium carbonate solution. After 2 h of reaction, readings were obtained in the spectrophotometer at 760 nm. Gallic acid standard curve was obtained under the same conditions using 100 µL of solutions at different concentrations (5 to 75 mg/mL) [38]. The concentration of total polyphenols was calculated based on the straight-line equation $y = 0.0106\,x + 0.094$ ($R^2 = 0.9927$) and expressed in mg/mL.

2.3.6. Total Flavonoids

Total flavonoid determination was performed in the same spectrophotometer as above at a wavelength of 425 nm, using a standard curve of quercetin at different concentrations. Flavonoids were quantified on a mixture of 800 µL of the extract and 1 mL of 2.5% (*w*/*v*) aluminum chloride solution. After 30 min of reaction, readings were made in the spectrophotometer at 425 nm. Quercetin standard curve was obtained under the same conditions using 800 µL of solutions at different concentrations (5 to 30 mg/mL) [38]. The concentration of total flavonoids was calculated based on the straight-line equation $y = 0.0258\,x + 0.0838$ ($R^2 = 0.9905$) and expressed in mg/mL.

2.4. Antioxidant Activity

2.4.1. ABTS Radical Cation Scavenging Assay

Antioxidant activity by capture of $ABTS^{\bullet+}$ was performed in the same spectrophotometer. After 16 h from the preparation of the radical cation solution from the 7 mM ABTS stock solution and the 140 mM potassium persulfate solution, an aliquot of 30 µL of the diluted extract and 3000 µL of the $ABTS^{\bullet+}$ solution was vortexed in a dark environment for 6 min, and readings were then taken at a wavelength of 734 nm. Antioxidant capacity was measured using a Trolox standard curve (100–2000 µM) described by the straight-line equation $y = -0.0003\,x + 0.7131$ ($R^2 = 0.9988$). The percentage of $ABTS^{\bullet+}$ inhibition was calculated according to Equation (1) [39], using the initial absorbance of the radical cation solution as a control. The half maximal inhibitory concentration (IC_{50}) was calculated after stabilization of the absorbance read every 6 min [39].

$$\% \text{ inhibition} = \frac{(\text{Absorbanceofthecontrol} - \text{Absorbanceofthesample})}{\text{Absorbanceofthecontrol}} \times 100 \quad (1)$$

2.4.2. DPPH Radical Scavenging Assay

Antioxidant activity by capture of the DPPH• radical was performed in the same spectrophotometer. After preparing the DPPH• solution, a 150 µL aliquot of the diluted extract was mixed with 5850 µL of the DPPH• solution, and the mixture was vortexed. After 30 min of reaction, readings were then taken at a wavelength of 515 nm. Antioxidant capacity was measured using a Trolox standard curve (50–1000 µM) described by the straight line equation $y = -0.0005\,x + 0.6516$ ($R^2 = 0.9984$). The inhibition percentage was calculated according to Equation (1) [40], using the initial absorbance of the radical solution as a control. IC_{50} was calculated after stabilization of the absorbance read every 30 min.

2.4.3. FRAP Assay

Antioxidant activity by the FRAP assay was determined in the same spectrophotometer at a wavelength of 595 nm. After preparing the FRAP reagent, aliquots of 90 µL of the diluted extract were mixed with 270 µL of distilled water and 2700 µL of the FRAP reagent, vortexed for 1 min and kept in a water bath at 37 °C for 30 min. After the necessary time, readings were carried out at a wavelength of 595 nm, using the FRAP reagent to calibrate the spectrophotometer. Antioxidant capacity was measured using a standard Trolox curve (160–1600 µM) described by the straight-line equation $y = 0.0009\,x - 0.1088$ ($R^2 = 0.9948$), using the initial absorbance of the FRAP reagent solution as a control. The IC_{50} value was also calculated similarly to the previous two sections [41].

2.5. *In Vitro Cytotoxicity Assay*

2.5.1. Cell Culture

The kidney epithelial fibroblast cell line from the African green monkey (Vero) was kindly provided by the Structural Biology and Parasitology Laboratory of the Federal University of Pará (UFPA). Initially, the cells were placed in a humidified atmosphere with CO_2 (5%) in an oven at a temperature of 37 °C in Dulbecco's Modified Eagle Medium (DMEM) supplemented with 10% Fetal Bovine Serum (FBS). After replacing the media until the formation of a cell monolayer, cells were washed with trypsinized phosphate-buffered saline (PBS), centrifuged at 1200 rpm for 10 min and counted in a Neubauer chamber. Then, in order to stabilize them, cells were divided into 96-well plates at a concentration of 1×10^4 cells/mL of DMEM and incubated in an atmosphere of CO_2 (5%) and air (95%) for 24 h [42].

2.5.2. Analysis of Cell Viability

Cell viability analysis was performed using the MTT method. To this end, Vero cells at a concentration of 1×10^4 cells/mL were cultivated in microplates and treated with propolis extract at different concentrations (2.5, 5, 10, 20, 50, 100, 250 and 500 µg/mL), in an incubator at 37 °C under 5% CO_2 atmosphere for 24 h. After this period, the supernatant was removed, PBS was used to wash the microplate, MTT diluted in PBS (0.5 mg/mL) was added, and cells were incubated for 3 h in an oven at 37 °C under 5% CO_2 atmosphere. The supernatant was then removed, and the microplate washed again with PBS. The solution was placed on another microplate, and readings were performed at a wavelength of 570 nm on a microplate reader (model 450, Bio-Rad Laboratories, Richmond, CA, USA). A suspension of dead cells in 15% formaldehyde was used as a control [42].

2.6. *In Vitro Antipromastigote Activity Assay*

2.6.1. Obtaining and Cultivation of the Parasite

The promastigotes of *Leishmania amazonensis* (strain MHOM/BR/26361), kindly provided in Novy–Nicolle–McNeal (NNN) medium by the Evandro Chagas Institute Leishmaniasis Program, Belém/Pará, were cultivated at 24 °C in Roswell Park Memorial Institute (RPMI) 1640 medium supplemented with FBS (10%) within a Biological Oxygen Demand incubator. To perform the experiments, the strain was maintained by six weekly transfers, and promastigotes in growth stage (4th day) were used. After centrifugation of promastigote forms for 10 min at 2500 rpm, the cell precipitate was suspended in 1 mL of RPMI medium. After counting in a Neubauer chamber, 1×10^6 parasites/mL were collected and placed in the wells of 24-well culture plates for 72 h incubation at different extract concentrations [43].

2.6.2. Viability According to the MTT Method

The MTT technique was also used to evaluate the mitochondrial viability of promastigote forms of *L. amazonensis* treated with the extract. For this purpose, parasite promastigotes in the exponential phase were incubated at 25 °C for 24 h together with the extract at different concentrations of (2.5, 5, 10, 20, 50, 100, 250 and 500 µg/mL). Aliquots (200 µL) of the protozoan suspensions were then transferred to the microplate, treated with aliquots (20 µL) of MTT (2 mg/mL) and then incubated for 4 h. After incubation, 20 µL of DMSO per well was added to solubilize the crystals of formazan, and the plate was stirred for 30 min. The color intensity of the final solution was quantified using the microplate reader mentioned above under the same conditions. Cultures of *L. amazonensis* promastigotes not treated with extract were used as the experimental control of cell viability, while a 0.5 µg/mL amphotericin B solution was used as the experimental positive control [43]. After experiments, the data were analyzed to determine IC_{50}.

2.7. *Statistical Analysis*

The analysis of variance (ANOVA) and Tukey's post hoc test were used to verify the level of significance ($p < 0.05$) of the data obtained in the assessment of cell viability and antipromastigote activity using the Graphpad 5.0 software package.

3. Results and Discussion

3.1. *Physicochemical Characterization of Propolis*

3.1.1. FTIR Spectrum

The FTIR spectra of propolis extract are shown in Figure 1. The propolis extract showed an absorption band at 3296 cm^{-1} referring to the symmetrical axial deformation vibration of hydrogen bonding in primary amines, an absorption signal at 2927 cm^{-1} due to the C-H axial deformation oscillation in alkanes, the symmetrical angular strain in the N-H plane in amines close to 1600 cm^{-1}, a band at 1416 cm^{-1} ascribable to symmetrical angular deformation in the plane of the terminal methylene group characteristic of the vinyl group absorption, the axial deformation of C-H in alkanes at 1372 cm^{-1}, C-N axial strain vibrations at 1293 cm^{-1} ascribable to secondary aromatic amines, the absorption signal at 1022 cm^{-1} due to C-O-C symmetric axial deformation in ethers (alkyl–aryl–ethers), the out-of-plane angular deformation of C-H in vinyl at 889 cm^{-1} and the C-H angular deformation at 771 cm^{-1} [44].

In short, the spectra of propolis extract highlighted a set of absorption bands referring to groups of important classes such as phenols, esters, aldehydes, ketones and carboxylic acids present in flavonoids, isoflavones and waxes, among others [35,44].

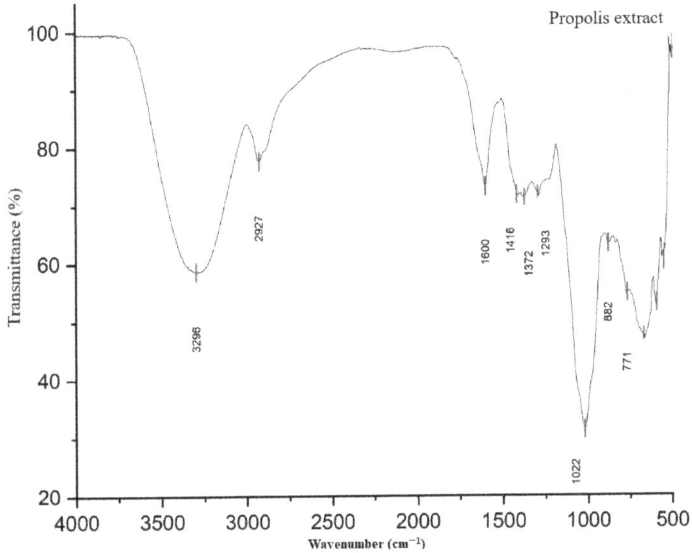

Figure 1. FTIR spectra of propolis extract in the wavenumber range from 4000 to 600 cm^{-1}, with 32 scans and 4 cm^{-1} resolution.

3.1.2. Thermal Behavior by Thermogravimetry

The TG/DTG curves up to 600 °C of propolis extract are shown in Figure 2. A single mass loss of 30.77% was observed in the temperature range of 143–500 °C due to successive reactions of degradation of organic compounds [45–47] including sugars [48].

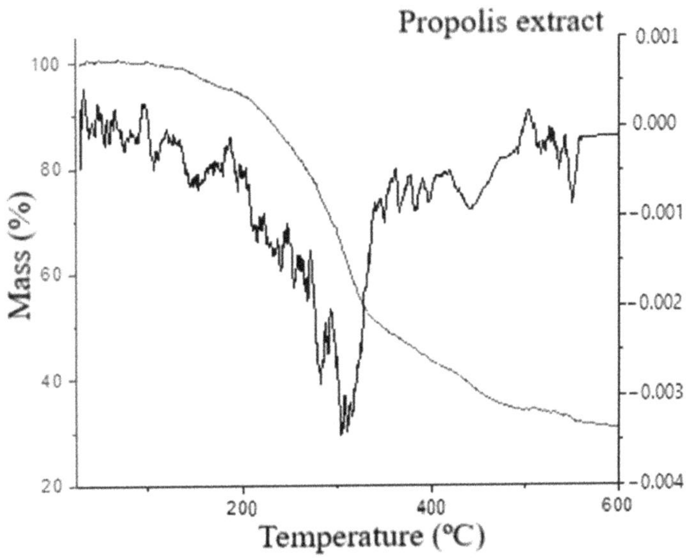

Figure 2. TG/DTG curves of propolis extract. Conditions: nitrogen atmosphere, flowrate of 50 mL/min, heating rate of 10 °C/min, temperature of 600 °C.

The study of the thermal stability of the natural product used as raw material is necessary to guarantee the safety and effectiveness of the herbal medicine obtained.

Thermogravimetry is not a selective technique, so this event may have resulted from the simultaneous decomposition of various substances present in the propolis extract. In the DTG curve, it is possible to identify the existence of various simultaneous and overlapping mass loss events, without a clear definition of their beginning and end, whose complexity is typical of plant extracts. However, it can be assumed that low-molecular-weight substances can evaporate at low temperatures, while more complex substances can evaporate or even decompose at higher temperatures; for example, aromatic phenolic compounds may decompose slowly due to their rigid chemical structures [49,50]. Although the TG/DTG curve of the propolis extract did not give clear insights, the events observed in this work occurred in a temperature range close to the one reported for red propolis (115–620 °C) [49–51].

3.1.3. Thermal Behavior Shown by Differential Scanning Calorimetry

The DSC curves of propolis extract are shown in Figure 3. The extract showed two endothermic events: one in the temperature range 31.67–116.91 °C (ΔH = 249.19 J/g) and the other in the range 160.53–170.59 °C (ΔH = 3.58 J/g), both due to thermal decomposition of constituents [35], and an exothermic one between 293.13 and 297.29 °C (ΔH = 1.51 J/g) likely due to material degradation [37]. The peak temperature of the extract was between 295.7 and 300.0 °C.

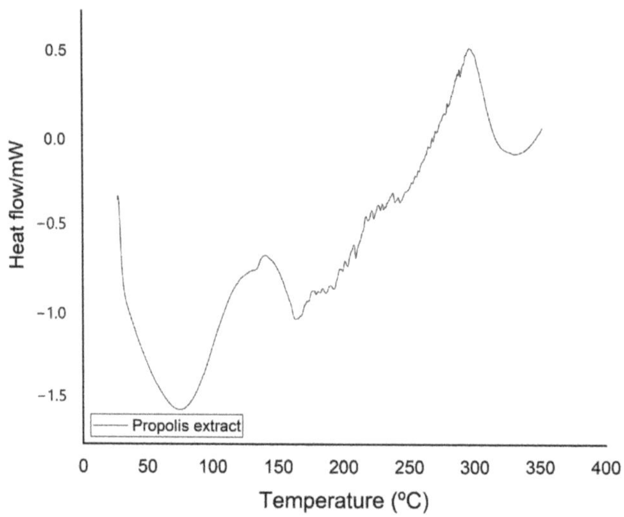

Figure 3. DSC curve of propolis extract. Conditions: nitrogen atmosphere, flowrate of 50 mL/min, heating rate of 10 °C/min, temperature of 350 °C.

DSC study showed extract stability around 300 °C, consistent with the results of thermogravimetry.

3.2. *Total Polyphenols and Total Flavonoids*

Total phenolic compounds in propolis can be quantified using a simple and quick tool such as UV-Vis spectroscopy [52,53]. Phenolics are compounds found in a wide range of plant species, which are considered the main group with potential to combat free radicals [54] and to exert a large number of therapeutic properties [54,55]. Research has revealed that flavonoids are bioactive components abundant in the propolis of meliponines living in temperate and tropical areas [20,29]. The contents of polyphenols and flavonoids found in propolis extract are within the limits established in current legislation for these compounds in propolis and its derivatives (Table 1).

Table 1. Contents of phenolic compounds found in *S. postica* propolis extract in this study and results reported in the literature for propolis from other species.

Species	Total Polyphenols (mg GAE/g)	Total Flavonoids (mg QE/g)
Scaptotrigona postica	21.29 ± 0.003	8.17 ± 0.01
Tetrigona apicalis [56]	28.57	——
Heterotrigona itama [56]	34.17	——
Geniotrigona thoracica [56]	55.16	——
Melipona quadrifasciata [57]	84.4–189.5	0.74–0.79
Tetragonisca angustula [57]	34.9–42.0	0.82–0.93
Melipona quadrifasciata quadrifasciata [58]	3.87	0.14
Tetragonisca angustula [58]	1.26	0.15

Results of tests performed in triplicate are reported as mean and standard deviation. GAE = gallic acid equivalent, QE = quercetin equivalent.

It is hard to compare these values with those reported in the literature not only due to the paucity of studies on *S. postica* propolis but also because the chemical constituents of propolis vary in concentration depending on the botanical origin of the resin; the phytogeography; the region climate [56–61]; the geographic location, that is, the visits of bees to different vegetation; and even the type of soil collected for propolis production [62,63]. Another important factor is the range of bee action. For instance, *S. postica* bumblebees [64] have a flight radius of just over 600 m, being the only report for male meliponines [65], while Araújo et al. [66] estimated a maximum flight distance for medium-sized individuals of the same species to be in the 1159–1710 m range.

Nonetheless, there are only a few reports on the chemical characterization of *S. postica* propolis. Isidorov et al. [67] identified, using GC-MS in an Argentinian propolis extract of this species, diterpenes, abietic and dehydroabietic phenolic acids and pentacyclic triterpenes. Some studies have revealed the substances contained in propolis from several other species by different methods (Table 2).

Table 2. Chemical components reported in the literature for propolis from different meliponines.

Species	Chemical Components
Scaptotrigona mexicana [68]	Phenolic lipids, anacardic and phenolic acids, terpenes, lignans
Melipona orbignyi [69]	Benzoic acids, phenolic acid, dihydrocinnamic acids and Cinnamic acids
Tetragonula sapiens [70]	Mangiferonic and ambolic acids, cycloartenol
Chilean propolis [71]	Galangin, quercetin, apigenin, caffeic acid phenethyl ester
Lithuanian propolis [72]	*p*-Cumaric, ferulic, caffeic and vanillic acids, apigenin, galangin, vanillin
Tetrigona melanoleuca [73]	3-O-acetylursolic acid, dipterocarpol, ocotillones I and II, oleanolic aldehydes
Tetragonula laeviceps [73]	α-Mangostin, 8-desoxigartanin, gartanin, garcinone B, methylpinoresinol

In this study, it was only possible to determine the phenolic compounds of the *S. postica* propolis extract by UV–Vis spectrophotometry. However, a more detailed study of its chemical composition is in progress, and the complete data will be presented in a new publication.

In general, the content of phenolic compounds found in the propolis extract investigated in this study is consistent with both the FTIR and thermogravimetry profiles discussed in the previous two sections and may be relevant in view of the antioxidant and antipromastigote activities demonstrated in this study.

3.3. Antioxidant Activity

The values of antioxidant activities of the propolis extract determined by the DPPH, ABTS and FRAP methods are listed in Table 3 and compared with those reported in the literature for both ascorbic acid taken as a standard and other propolis samples. The antioxidant activity was expressed either as concentration (µM Trolox equivalent/g) or as percentage of radical inhibition (%) by the DPPH and ABTS methods, while only the concentration (µM Trolox equivalent/g) was used for the FRAP method.

Table 3. Antioxidant activities of *S. postica* propolis extract and ascorbic acid (standard) assessed in this study, as well as of different propolis extracts reported in the literature.

	DPPH•		ABTS•+		FRAP
	µM TE/g	Inhibition %	µM TE/g	Inhibition %	µM TE/g
Scaptotrigona postica (Brazil, this study)	910.09 ± 0.0264	63.95	859.97 ± 0.0854	34.43	1613.67 ± 0.2610
Ascorbic acid	1132.68 ± 0.005	85.54	2380.04 ± 0.0000	91.31	2408.57 ± 0.058
Canadian propolis [73]	———	64.0–93.7	———		———
Tetragonisca angustula (Brazil) [74]	———	19.6–30.5	———		———
Scaptotrigona mexicana [68]	———	1.8–17.7	———		———
Scaptotrigona depilis (Brazil) [32]	———	14.9	———	73.4	———
Melipona quadrifasciata anthidioides (Brazil) [32]	———	97.5	———	99.3	———
Melipona orbignyi (Brazil) [75]	———	96	———		———
Tetragonisca fiebrigi (Brazil) [31]	———		———	86.5	———

Results of tests performed in triplicate are expressed as mean and standard deviation. TE = Trolox equivalent.

It can be seen that the results referring to the extract and the ascorbic acid obtained using the DPPH and FRAP assays are not very different, which confirms the antioxidant potential of the extract. On the other hand, the wide variation in the antioxidant activity of propolis from different origins can be attributed, similarly to what was observed for the contents of bioactive compounds, to several factors related to seasonality [76].

The food industry uses synthetic antioxidants for food preservation, but their use raises concerns about safe doses and toxicity; therefore, the interest of researchers in the search for natural substitutes has increased [74]. Secondary metabolites, including flavonoids, are the compounds mainly responsible for the antiradical power of plant extracts and bee products [74]. Antioxidants are compounds that, in small quantity compared to the oxidizable substrate, delay the start or prevent the onset or propagation of the oxidation reaction chain, in addition to inhibiting lipid peroxidation as well as the oxidation of other macromolecules and DNA [52]. Therefore, they prevent the gap between the formation and the deactivation of ROS, which triggers several changes in the body [28].

In this study, the IC_{50} of propolis extract and ascorbic acid standard were also evaluated using the three assays selected to determine the antioxidant activity (Table 4). As expected, the IC_{50} value of the extract was higher than that of the standard. It is noteworthy that, usually, the lower the IC_{50}, the greater the antioxidant potential [60]. The IC_{50} value identified by the FRAP assay is worth highlighting, being close to that of the selected antioxidant standard. This suggests that the antioxidant substances present in the extract have a greater capacity to reduce iron. The IC_{50} values of the extract by the DPPH and ABTS assays were more than twice that observed for ascorbic acid. Based on the IC_{50} values, the highest antioxidant activity was detected using the FRAP method, which corroborates the result of the evaluation of concentration (µM Trolox equivalent/g). Therefore, the IC_{50}

values together with those of the concentration confirm the antioxidant potential of the propolis extract assessed by the three different methods.

Table 4. Values of IC_{50} of the *S. postica* propolis extract and ascorbic acid (standard) assessed in this study, as well as of different propolis extracts reported in the literature.

	IC_{50} (μg/mL)		
	DPPH•	**ABTS•+**	**FRAP**
Scaptotrigona postica (Brazil, this study)	18.9 ± 0.01	692 ± 0.01	80 ± 0.1
Ascorbic acid	6.80 ± 0.02	310 ± 0.02	60 ± 0.2
Mandaçaia [77]	11.05 ± 0.6	———	———
Melipona quadrifasciata [57]	117.5	———	
Heterotrigona itama [78]	630.31 ± 0.76	321.58 ± 3.67	———
Tunisian propolis extract [79]	20.1–43	244–616	375–780
Melipona fasciculata (Viana, Brazil) [9]	76.16 ± 1.05	13.28 ± 0.11	———
Melipona fasciculata (Pinheiros, Brazil) [9]	265.91 ± 0.29	58.94 ± 0.09	———

Resuming, the propolis extract showed antioxidant potential and IC_{50} values that confirm its antiradical power with the three methods evaluated, as the likely result of its high phenolic compound content. They are within the range reported in the literature for other propolis extracts and, according to Al Naggar et al. [73], may be exploited to stop the emergence of diseases related to oxidative stress. In addition, these promising results suggest its use in the wound-healing process due to the benefits offered by antioxidants in fighting free radicals.

3.4. Cell Viability Test

The preliminary test to assess the toxicity of any material intended to be used for the manufacture of biomedical devices is in vitro cytotoxicity. Once its non-toxicity is proven, the study of its biocompatibility follows [50].

Propolis extract cytotoxicity was evaluated in fibroblast cells from monkey (Vero) by the MTT method. This cell line was chosen because it proved successful in previous cytotoxicity studies [80,81]. Figure 4 shows the results in terms of cell viability obtained using eight different concentrations of the propolis extract.

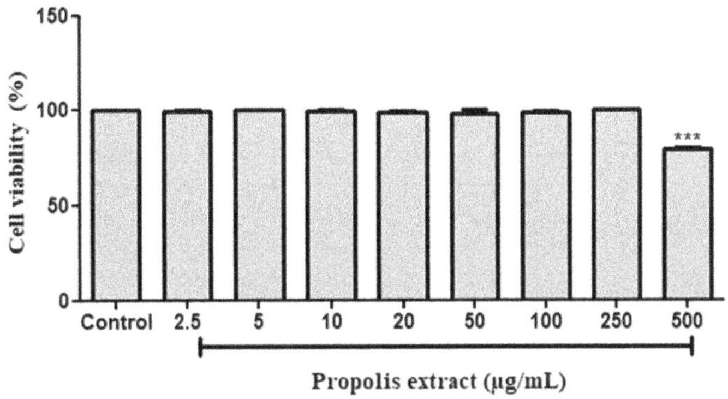

Figure 4. Cytotoxicity of propolis extract against fibroblast cells treated for 24 h in an oven at a temperature of 37 °C (Control = control without treatment). The results were compared using the analysis of variance followed by Tukey's post hoc test, *** $p < 0.01$.

The extract, which proved to be moderately cytotoxic at the maximum tested concentration (500 µg/mL), did not lead to any decrease in cell viability up to the concentration of 250 µg/mL, where it may be safely used without causing cell damage. These results allowed us to estimate a concentration that caused cytotoxicity by 50% (CC50) of 321.4 µg/mL. The groups treated with the extract showed no statistically significant difference compared to the control group after 24 h of the cell viability test, which indicates that the propolis extract did not induce cell death in fibroblasts at concentrations lower than the maximum tested one.

3.5. *In Vitro Antipromastigote Assay*

Leishmaniasis is a disease with high global impact responsible for a public health problem whose treatment is still limited, mainly due to drug toxicity [82]. In the search for alternative treatments, propolis has shown promising leishmanicidal activity against different species of *Leishmania* [83]. In particular, in vitro experiments showed propolis' ability to reduce the diameter of injuries, cause morphological changes in promastigotes or amastigotes or even improve immune response to *Leishmania* spp.-activating macrophages [22].

Based on these studies, the antileishmanicidal effect of propolis extract was investigated in in vitro tests carried out at different concentrations (2.5–500 µg/mL) against promastigotes of *L. amazonensis*, whose results are illustrated in Figure 5. Propolis extract caused a toxic effect to promastigotes across the entire concentration range tested and had a IC_{50} value of only 1.50 µg/mL. The groups treated with propolis extract showed statistically significant differences compared to both the negative control ($p < 0.001$) and the group treated with amphotericin B ($p < 0.001$, IC_{50} = 0.50 µg/mL).

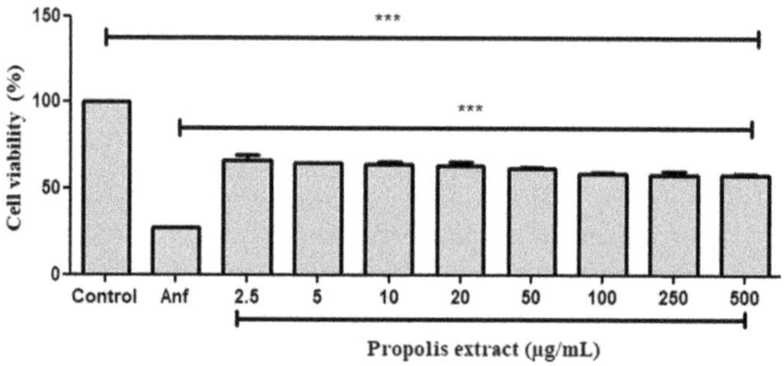

Figure 5. Cell viability of promastigotes of *L. amazonensis* treated with propolis extract at different concentrations for 24 h in an oven at temperature of 25 °C. Control = control without treatment; Anf = treatment with amphotericin B. The results were compared using the analysis of variance followed by Tukey's post hoc test, *** $p < 0.001$.

Over 75 years, the therapeutic window for treating leishmaniasis has been quite limited due to resistance and toxicity, which has led to the search for new drug options [84]. The results of the present study demonstrated that a natural product such as propolis, used since ancient times to treat other illnesses, non-toxic to humans and easy to acquire, is capable of exerting a toxic effect on promastigote forms of *L. amazonensis*. This may have a positive impact on the search for new drugs for the treatment of leishmaniasis. In particular, propolis extract can be used as an alternative ingredient not only to obtain new drugs for oral use, but also to prepare topical formulations, such as gels and hydrogels, which act directly at the site of lesions caused by leishmaniasis. This extract may also be used in association with drugs already used to treat leishmaniasis as a way of enhancing their antileishmanial effect.

Cavalcante et al. [83] observed comparable activity against *L. amazonensis* amastigotes (IC_{50} = 1.3 ± 0.1 µg/mL) of the hexane fraction of propolis collected at the Bahian semiarid region (Brazil). Ayres et al. [85] observed that different propolis ethanolic extracts were able to reduce the parasitic load of *L. amazonensis* promastigotes and amastigotes in parasite-infected macrophages. The extract of Adana propolis was able to inhibit the growth of *Leishmania tropica* [86], and Machado et al. [24] reported the leishmanicidal potential of extracts of propolis collected in Brazil and Bulgaria against four different species belonging to the *Leishmania* genus, namely *L. amazonensis*, *L. braziliensis*, *L. chagasi* and *L. major*. Finally, the green propolis exhibited antileishmanial potential against *L. braziliensis* [54].

The ways in which propolis exerts its potential to cause death in amastigote and promastigote forms of *Leishmania* species requires elucidation. However, it has been suggested that it may be the result of the activation of macrophages, cells that participate in various activities, including the removal of pathogens and cell debris, generation of metabolites and inflammation mediators, among others [22].

Despite the excellent results of parasite growth inhibition against promastigote forms of *L. amazonesis*, it is necessary to study the extract's effect against amastigote forms to confirm its leishimanicidal potential, and in vivo studies with amastigotes and promastigote forms are needed to prove its potential in the healing of skin lesions caused by *L. amazonesis*.

4. Conclusions

Quality control parameters of propolis from *Scaptotrigona postica* from the northeast of the Brazilian Pará state showed values consistent with those obtained in previous studies for propolis produced in other geographic areas, within the standards established by the Brazilian legislation and suitable for obtaining pharmacognostic specifications. The extract showed total polyphenol content of 21.29 $\pm$ 0.003 mg/g and total flavonoid content of 8.17 $\pm$ 0.00 mg/g, thermal stability around 370 °C and antioxidant activities identified by the DPPH, ABTS and FRAP assays of 910.09 $\pm$ 0.0264 µM Trolox (IC_{50} 18.9 $\pm$ 0.01 µg/mL), 859.97 $\pm$ 0.0854 µM (IC_{50} 692 $\pm$ 0.01 µg/mL) and 1613.67 $\pm$ 0.2610 µM Trolox (IC_{50} 80 $\pm$ 0.1 µg/mL), respectively. Finally, cell viability and antipromastigote activity tests allowed us to determine CC_{50} and IC_{50} values of 321.4 µg/mL and 1.50 µg/mL, respectively. These results provide the basis for research aimed at identifying and quantifying the major substances that are part of the chemical composition of *S. postica* propolis, as well as verifying other biological activities for its possible incorporation in innovative formulations for the pharmaceutical and cosmetic industries. The propolis extract, with its excellent antipromastigote activity, proved to have promising potential to be used in the development of new herbal drugs to treat leishmaniasis, a neglected disease that requires a constant search for new drugs due to the particularities of existing treatments. In addition, they can add value to propolis and contribute to the development of meliponiculture in the state of Pará.

Author Contributions: Conceptualization, L.M.d.M.C.F.; methodology, P.D.Q.d.S., L.M.d.M.C.F. and E.O.d.S.; formal analysis, L.M.d.M.C.F.; investigation, L.M.d.M.C.F.; data curation, L.M.d.M.C.F., R.R.P. and P.D.Q.d.S.; writing—original draft preparation, L.M.d.M.C.F.; writing—review and editing, L.M.d.M.C.F., W.L.R.B. and A.C.; supervision, R.M.R.-C.; resources, A.C. and J.O.C.S.-J.; project administration, R.M.R.-C. All authors have read and agreed to the published version of the manuscript.

Funding: This research was funded by the Brazilian Coordination for the Improvement of Higher Education Personnel (CAPES) through postgraduate scholarship funding at PhD level (L.M.M.C.F). Financial code: 001.

Data Availability Statement: Data are contained within the article.

Acknowledgments: The authors are grateful to the Conselho Nacional de Desenvolvimento Científico e Tecnológico (CNPq), Embrapa Amazônia Oriental in the figure of Daniel Santiago Pereira and Mozaniel Santana de Oliveira.

Conflicts of Interest: The authors declare no conflicts of interest.

References

1. Hrncir, M.; Jarau, S.; Barth, F.G. Stingless bees (meliponini): Senses and behavior. *J. Comp. Physiol. A* **2016**, *202*, 597–601. [CrossRef]
2. De Menezes Pedro, S.R. The stingless bee fauna in Brazil (Hymenoptera: Apidae). *Sociobiology* **2014**, *61*, 348–354. [CrossRef]
3. Conceição, V.D.; dos Santos, A.M.; da Conceição, C.A. Polinizadores que visitam a espécie arbórea *Myracrodrun urundeuva* (Anacardiaceae) na borda oeste do pantanal, assentamento taquaral em Corumbá-MS. *Realização* **2019**, *6*, 128–140. [CrossRef]
4. Loiola, N.D.S.; Ethur, E.M.; Weber, A.C. Atividade antimicrobiana in vitro de extrato etanólico de própolis da abelha *Scaptotrigona aff. postica* (Latreille, 1807). *Rev. Ibero-Am. Ciências Ambient.* **2020**, *11*, 612–620. [CrossRef]
5. Filho, J.S.P.; Bicalho, L.; da Silva, D.A. Uso de própolis associada a outros componentes no tratamento de feridas oncológicas após excisão. *Acta Apic. Bras.* **2012**, *3*, 15–25.
6. Menezes, C.; Vollet-Neto, A.; Fonseca, V.L.I. An advance in the in vitro rearing of stingless bee queens. *Apidologie* **2013**, *44*, 491-500. [CrossRef]
7. Oliveira, F.F.; Richers, B.T.T.; da Silva, J.R.S.; Farias, R.C.; Matos, T.A.L. *Guia Ilustrado das Abelhas "Sem-Ferrão" das Reservas Amaná e Mamirauá, Amazonas, Brasil*, 1st ed.; Eliete Amador Alves Silva; IDSM: Salvador, Brazil, 2013; pp. 13–17; ISBN 978-85-88758-27-8.
8. Vollet-Neto, A.; Blochtein, B.; Viana, B.; dos Santos, C.F.; Menezes, C.; Silva, P.N.; Jaffé, R.; Amoedo, S. *Desafios e Recomendações para o Manejo e Transporte de Polinizadores Desafios e Recomendações para o Manejo e o Transporte de Polinizadores*, 1st ed.; Vollet Neto, A., Menezes, C., Eds.; Associação Brasileira de Estudos das Abelhas: São Paulo, Brazil, 2018; pp. 39–46; ISBN 978-85-69982-03-6.
9. Barboza, J.R.; Pereira, F.A.N.; Fernandes, R.A.; Vasconcelos, C.C.; Cartágenes, M.d.S.d.S.; Lopes, A.J.O.; de Melo, A.C.; Guimarães, I.D.S.; da Rocha, C.Q.; Ribeiro, M.N.d.S. Cytotoxicity and pro-apoptotic, antioxidant and anti-inflammatory activities of geopropolis produced by the stingless bee *Melipona fasciculata* Smith. *Biology* **2020**, *9*, 292. [CrossRef]
10. Barboza, J.R.; Pereira, F.A.N.; Leite, J.A.C.; Coutinho, D.F.; Ribeiro, M.N.d.S. Abordagem química e toxicidade em modelo zebrafish de geoprópolis de *Melipona fasciculata* Smith. *Braz. J. Heal. Rev.* **2019**, *2*, 5582–5594. [CrossRef]
11. Barbiéri, C.; Francoy, T.M. Modelo teórico para análise interdisciplinar de atividades humanas: A meliponicultura como atividade promotora da sustentabilidade. *Ambient. Soc.* **2020**, *23*, e00202. [CrossRef]
12. Araujo, M.; Bufalo, M.; Conti, B.; Jr, A.; Trusheva, B.; Bankova, V.; Sforcin, J. The Chemical Composition and Pharmacological Activities of Geopropolis Produced by *Melipona fasciculata* Smith in Northeast Brazil. *J. Mol. Pathophysiol.* **2015**, *4*, 12. [CrossRef]
13. Bankova, V.; Popova, M.; Trusheva, B. Propolis volatile compounds: Chemical diversity and biological activity: A review. *Chem. Cent. J.* **2014**, *8*, 28. [CrossRef] [PubMed]
14. Pereira, D.S.; Iberê, C.; Freitas, A.; Freitas, M.O.; Berg, J.; Agra, R. Histórico e principais usos da própolis apícola. *Acsa* **2015**, *11*, 1–21.
15. Bankova, V.; Popova, M.; Trusheva, B. New emerging fields of application of propolis. *Maced. J. Chem. Chem. Eng.* **2016**, *35*, 1–11. [CrossRef]
16. Sanches, M.A.; Pereira, A.M.S.; Serrão, J.E. Acciones farmacológicas de extractos de propóleos de abejas sin aguijón (Meliponini). *J. Apic. Res.* **2017**, *56*, 50–57. [CrossRef]
17. Araújo, M.J.A.M.; Dutra, R.P.; Costa, G.C.; Reis, A.S.; Assunção, A.K.M.; Libério, S.A.; Maciel, M.C.G.; Silva, L.A.; Guerra, R.N.M.; Ribeiro, M.N.S.; et al. Efeito do tratamento com própolis de *Scaptotrigona aff. postica* sobre o desenvolvimento do tumor de ehrlich em camundongos. *Rev. Bras. Farm.* **2010**, *20*, 580–587. [CrossRef]
18. Lavinas, F.C.; Macedo, E.H.B.C.; Sá, G.B.L.; Amaral, A.C.F.; Silva, J.R.A.; Azevedo, M.M.B.; Vieira, B.A.; Domingos, T.F.S.; Vermelho, A.B.; Carneiro, C.S.; et al. Brazilian stingless bee propolis and geopropolis: Promising sources of biologically active compounds. *Rev. Bras. Farm.* **2019**, *29*, 389–399. [CrossRef]
19. Turco, J.F.; do Nascimento, C.L.; de Lima, V.A.; Torres, Y.R. Could antioxidant capacity and flavonoid content of ethanolic extracts of geopropolis from Brazilian native bees be estimated from digital photos and NIR spectra? *Microchem. J.* **2020**, *157*, 105031. [CrossRef]
20. Miquez Souza, C.; Cavalcante da Silva, S.M.P.; do Nascimento, A.S.; Silva, F.d.L.; Bagaldo, A.R.; de Carvalho, C.A.L. Chemical and microbiological characterization of "Samburá" from two stingless bee species. *J. Apic. Res.* **2022**, *61*, 70–78. [CrossRef]
21. Kasote, D.M.; Pawar, M.V.; Gundu, S.S.; Bhatia, R.; Nandre, V.S.; Jagtap, S.D.; Mahajan, S.G.; Kulkarni, M.V. Chemical profiling, antioxidant, and antimicrobial activities of Indian stingless bees propolis samples. *J. Apic. Res.* **2019**, *58*, 617–625. [CrossRef]
22. Oda, J.M.M.; Fujita, T.C.; de Faveri Pitz, A.; Amarante, M.K.; Felipe, I.; Saridakis, H.O.; Sforcin, J.M.; Watanabe, M.A.E.; Costa, I.C. Ação do extrato de própolis na leishmaniose. *Semin. Ciências Biológicas E Da Saúde* **2011**, *32*, 111–121. [CrossRef]
23. Segueni, N.; Akkal, S.; Benlabed, K.; Nieto, G. Potential use of propolis in phytocosmetic as phytotherapeutic constituent. *Molecules* **2022**, *27*, 5833. [CrossRef] [PubMed]
24. Machado, G.M.D.C.; Leon, L.L.; Lisboa, S.; Castro, D. Activity of Brazilian and Bulgarian propolis against different species of *Leishmania*. *Mem. Inst. Oswaldo Cruz* **2007**, *102*, 73–77. [CrossRef] [PubMed]
25. Borges, K.S.; Brassesco, M.S.; Scrideli, C.A.; Soares, A.E.E.; Tone, L.G. Antiproliferative effects of tubi-bee propolis in glioblastoma cell lines. *Genet. Mol. Biol.* **2011**, *34*, 310–314. [CrossRef]
26. Olczyk, P.; Komosinska-Vassev, K.; Winsz-Szczotka, K.; Stojko, J.; Klimek, K.; Kozma, E.M. Propolis induces chondroitin/dermatan sulphate and hyaluronic acid accumulation in the skin of burned wound. *Evid.-Based Complement. Altern. Med.* **2013**, *2013*, e290675. [CrossRef] [PubMed]

27. Olczyk, P.; Wisowski, G.; Komosinska-Vassev, K.; Stojko, J.; Klimek, K.; Olczyk, M.; Kozma, E.M. Propolis modifies collagen types I and III accumulation in the matrix of burnt tissue. *Evid.-Based Complement. Altern. Med.* **2013**, *2013*, e423809. [CrossRef] [PubMed]
28. Martinotti, S.; Ranzato, E. Propolis: A new frontier for wound healing? *Burn. Trauma* **2015**, *3*, 9. [CrossRef] [PubMed]
29. Bankova, V. Recent trends and important developments in propolis research. *Evid.-Based Complement. Altern. Med.* **2005**, *2*, 29–32. [CrossRef] [PubMed]
30. Hochheim, S.; Guedes, A.; Faccin-Galhardi, L.; Rechenchoski, D.Z.; Nozawa, C.; Linhares, R.E.; da Silva, H.H.; Rau, M.; Siebert, D.A.; Micke, G.; et al. Determination of phenolic profile by HPLC–ESI-MS/MS, antioxidant activity, in vitro cytotoxicity and anti-herpetic activity of propolis from the Brazilian native bee *Melipona quadrifasciata*. *Rev. Bras. Farm.* **2019**, *29*, 339–350. [CrossRef]
31. Campos, J.F.; dos Santos, U.P.; da Rocha, P.d.S.; Damião, M.J.; Balestieri, J.B.P.; Cardoso, C.A.L.; Paredes-Gamero, E.J.; Estevinho, L.M.; de Picoli Souza, K.; dos Santos, E.L. Antimicrobial, antioxidant, anti-inflammatory, and cytotoxic activities of propolis from the stingless bee *Tetragonisca fiebrigi* (Jataí). *Evid.-Based Complement. Altern. Med.* **2015**, *2015*, e296186. [CrossRef]
32. Bonamigo, T.; Campos, J.F.; Alfredo, T.M.; Balestieri, J.B.P.; Cardoso, C.A.L.; Paredes-Gamero, E.J.; de Picoli Souza, K.; dos Santos, E.L. Antioxidant, cytotoxic, and toxic activities of propolis from two native bees in Brazil: *Scaptotrigona depilis* and *Melipona quadrifasciata anthidioides*. *Oxid. Med. Cell. Longev.* **2017**, *2017*, 1038153. [CrossRef]
33. Gentile, F.; Arcaro, A.; Pizzimenti, S.; Daga, M.; Cetrangolo, G.P.; Dianzani, C.; Lepore, A.; Graf, M.; Ames, P.R.J.; Barrera, G. DNA damage by lipid peroxidation products: Implications in cancer, inflammation and autoimmunity. *Aims Genet.* **2017**, *4*, 103–137. [CrossRef]
34. Ministério da Saúde; Agência Nacional de Vigilância Sanitária—Anvisa. Farmacopeia Brasileira 2019, v. 1, 6ª ed. Available online: http://portal.anvisa.gov.br (accessed on 20 January 2023).
35. De Cássia Almeida Sampaio, R.; Da Costa, R.S.; De Souza, C.R.F.; Duarte Júnior, A.P.; Ribeiro-Costa, R.M.; Da Costa, C.E.F.; De Oliveira, W.P.; Converti, A.; Silva Júnior, J.O.C. Thermal characterization of *Arrabidaea chica* (Humb. & Bonpl.) B. Verl. dry extracts obtained by spray dryer. *J. Therm. Anal. Calorim.* **2016**, *123*, 2469–2475. [CrossRef]
36. Silverstein, R.M.; Webster, F.X.; Kiemle, D.J. *Spectrometric Identification of Organic Compounds*, 7th ed.; David, L.B., Ed.; LTC: Rio de Janeiro, Brazil, 2007; pp. 71–107; ISBN 978-85-216-3637-3.
37. Silva da Costa, R.; Pinheiro, W.B.d.S.; Arruda, M.S.P.; Costa, C.E.F.; Converti, A.; Ribeiro Costa, R.M.; Silva Júnior, J.O.C. Thermoanalytical and phytochemical study of the cupuassu (*Theobroma grandiflorum* Schum.) Seed by-Product in Different Processing Stages. *J. Therm. Anal. Calorim.* **2020**, *147*, 275–284. [CrossRef]
38. Funari, C.S.; Ferro, V.O. Análise de Própolis. *Food Sci. Technol.* **2006**, *26*, 171–178. [CrossRef]
39. Re, R.; Pellegrini, N.; Proteggenete, A.; Pannala, A.; Yang, M.; Rice-Evans, C. Antioxidant activity applying an improved ABTS radical cation decolorization assay. *Free Radic. Biol. Med.* **1999**, *26*, 1231–1237. [CrossRef] [PubMed]
40. Brand-Williams, W.; Cuvelier, M.; Berset, C. Use of a free radical method to evaluate antioxidant activity. *LWT-Food Sci. Technol.* **1995**, *28*, 25–30. [CrossRef]
41. Benzie, I.F.; Strain, J. The Ferric Reducing Ability of Plasma (FRAP) as a measure of "Antioxidant Power": The FRAP Assay. *Anal. Biochem.* **1996**, *239*, 70–76. [CrossRef] [PubMed]
42. Fotakis, G.; Timbrell, J.A. In vitro cytotoxicity assays: Comparison of LDH, neutral red, MTT and protein assay in hepatoma cell lines following exposure to cadmium chloride. *Toxicol. Lett.* **2006**, *160*, 171–177. [CrossRef] [PubMed]
43. Moraes, L.S.; Donza, M.R.H.; Rodrigues, A.P.D.; Silva, B.J.M.; Brasil, D.S.B.; Zoghbi, M.D.G.B.; Andrade, E.H.A.; Guilhon, G.M.S.P.; Silva, E.O.; Schmidt, T.J. Leishmanicidal Activity of (+)-Phyllanthidine and the Phytochemical profile of *Margaritaria nobilis* (Phyllanthaceae). *Molecules* **2015**, *20*, 22157–22169. [CrossRef]
44. Forato, L.A.; Filho, R.B.; Osiro, D.; Bicudo, T.D.C.; Colnago, L.A. A Espectroscopia na região do infravermelho e algumas aplicações. *Embrapa Instrumentação* **2010**, *1*. Available online: https://www.infoteca.cnptia.embrapa.br/bitstream/doc/884592/1/DOC512010.pdf (accessed on 15 December 2023).
45. Li, F.; Awale, S.; Zhang, H.; Tezuka, Y.; Esumi, H.; Kadota, S. Chemical constituents of propolis from myanmar and their preferential cytotoxicity against a human pancreatic cancer cell line. *J. Nat. Prod.* **2009**, *72*, 1283–1287. [CrossRef] [PubMed]
46. Zhang, L.; Zhao, X.; Ji, J.; Dai, Y. TG-DTG as an effective method for the characterization of rutin extracted from the buds of *Sophora japonica* L. *J. Therm. Anal. Calorim.* **2009**, *95*, 917–922. [CrossRef]
47. da Costa, R.S.; Santos, O.V.D.; Lannes, S.C.d.S.; Casazza, A.A.; Aliakbarian, B.; Perego, P.; Ribeiro-Costa, R.M.; Converti, A.; Silva Júnior, J.O.C. Bioactive compounds and value-added applications of cupuassu (*Theobroma grandiflorum* Schum.) agroindustrial by-product. *J. Food Sci. Technol.* **2020**, *40*, 401–407. [CrossRef]
48. Gabbay Alves, T.V.; Silva da Costa, R.; Aguiar Gomes, A.T.; Ferreira da Costa, C.E.; Perego, P.; Carréra Silva Júnior, J.O.; Converti, A.; Ribeiro Costa, R.M. Quality control of Amazonian cocoa (*Theobroma cacao* L.) by-products and microencapsulated extract by thermal analysis. *J. Therm. Anal. Calorim.* **2018**, *134*, 993–1000. [CrossRef]
49. Pardauil, J.J.R.; de Molfetta, F.A.; Braga, M.; de Souza, L.K.C.; Filho, G.N.R.; Zamian, J.R.; da Costa, C.E.F. Characterization, thermal properties and phase transitions of Amazonian vegetable oils. *J. Therm. Anal. Calorim.* **2017**, *127*, 1221–1229. [CrossRef]
50. do Nascimento, T.G.; do Nascimento, N.M.; Ribeiro, A.S.; de Almeida, C.P.; dos Santos, J.I.Z.; Basílio-Júnior, I.D.; Calheiros-Silva, F.G.; Lira, G.M.; Escodro, P.B.; de Moraes Porto, I.C.C.; et al. Preparation and characterization of chitosanates loaded with Brazilian red propolis extract. *J. Therm. Anal. Calorim.* **2022**, *147*, 7837–7848. [CrossRef]

51. Pereira, E.; Ferreira, M.C.; Sampaio, K.A.; Grimaldi, R.; Meirelles, A.J.d.A.; Maximo, G.J. Physical properties of Amazonian fats and oils and their blends. *Food Chem.* **2019**, *278*, 208–215. [CrossRef] [PubMed]
52. Al-Hatamleh, M.A.I.; Boer, J.C.; Wilson, K.L.; Plebanski, M.; Mohamud, R.; Mustafa, M.Z. Antioxidant-based medicinal properties of stingless bee products: Recent progress and future directions. *Biomolecules* **2020**, *10*, 923. [CrossRef] [PubMed]
53. Maldonado, L.; Marcinkevicius, K.; Borelli, R.; Gennari, G.; Salomón, V.; Isla, M.I.; Vera, N.; Borelli, V. Differentiation of Argentine propolis from different species of bees and geographical origins by UV spectroscopy and chemometric analysis. *J. Saudi Soc. Agric. Sci.* **2020**, *19*, 185–191. [CrossRef]
54. Deus, V.L.; Santos, A.P.C.; Walker, J.F.; Neta, L.G.S.; Souza, L.S. Compostos fenólicos em hortaliças cultivadas nos sistemas convencional e orgânico: Uma revisão. *Braz. J. Heal. Pharm.* **2019**, *1*, 70–84. [CrossRef]
55. Melo, A.R.; Alves, L.A.; Morais, N.R.L.; Silva, F.F.M.; Bertini, L.M.; Oliveira Neto, F.B. Prospecção fitoquímica e avaliação do potencial antioxidante de *Cnidoscolus phyllacanthus* (Müll. Arg.) Pax & k.Hoffm. oriundo de Apodi—RN. *Rev. Bras. Plantas. Med.* **2016**, *18*, 180–185. [CrossRef]
56. Asem, N.; Abdul Gapar, N.A.; Abd Hapit, N.H.; Omar, E.A. Correlation between Total phenolic and flavonoid contents with antioxidant activity of malaysian stingless bee propolis extract. *J. Apic. Res.* **2020**, *59*, 437–442. [CrossRef]
57. dos Santos, L.; Hochheim, S.; Boeder, A.M.; Kroger, A.; Tomazzoli, M.M.; Dal Pai Neto, R.; Maraschin, M.; Guedes, A.; de Cordova, C.M.M. Caracterización química, antioxidante, actividad citotóxica y antibacteriana de extractos de propóleos y compuestos aislados de las abejas sin aguijón brasileñas *Melipona quadrifasciata* y *Tetragonisca angustula*. *J. Apic. Res.* **2017**, *56*, 543–558. [CrossRef]
58. Torres, A.R.; Sandjo, L.P.; Friedemann, M.T.; Tomazzoli, M.M.; Maraschin, M.; Mello, C.F.; Santos, A.R.S. Chemical characterization, antioxidant and antimicrobial activity of propolis obtained from *Melipona quadrifasciata quadrifasciata* and *Tetragonisca angustula* stingless bees. *Braz. J. Med. Biol. Res.* **2018**, *51*, e7118. [CrossRef]
59. Lopes, A.J.O.; Vasconcelos, C.C.; Garcia, J.B.S.; Pinheiro, M.S.D.; Pereira, F.A.N.; Camelo, D.d.S.; Morais, S.V.d.; Freitas, J.R.B.; Rocha, C.Q.d.; Ribeiro, M.N.d.S.; et al. Anti-Inflammatory and Antioxidant Activity of Pollen Extract Collected by *Scaptotrigona affinis postica*: In silico, in vitro, and in vivo Studies. *Antioxidants* **2020**, *9*, 103. [CrossRef]
60. Mohtar, L.G.; Messina, G.A.; Bertolino, F.A.; Pereira, S.V.; Raba, J.; Nazareno, M.A. Comparative study of different methodologies for the determination the antioxidant activity of Venezuelan propolis. *Microchem. J.* **2020**, *158*, 105244. [CrossRef]
61. Ferreira, B.L.; Gonzaga, L.V.; Vitali, L.; Micke, G.A.; Baggio, D.; de Oliveira Costa, A.C.; Fett, R. Dataset about southern-Brazilian geopropolis: Physical and chemical perspectives. *Data Br.* **2020**, *29*, 105109. [CrossRef] [PubMed]
62. De Oliveira, A.C.; Valentim, I.B.; Goulart, M.O.F.; Silva, C.A.; Bechara, E.J.H.; Trevisan, M.T.S. Fontes vegetais naturals de antioxidantes. *Quim. Nova* **2009**, *32*, 689–702. [CrossRef]
63. Carneiro, M.J.; López, B.G.C.; Lancellotti, M.; Franchi, G.C.; Nowill, A.E.; Sawaya, A.C.H.F. Evaluación de la composición química y la actividad biológica de los extractos de propóleos de *Tetragonisca angustula* y *Schinus terebinthifolius* Raddi (Anacardiaceae). *J. Apic. Res.* **2016**, *55*, 315–323. [CrossRef]
64. Guimarães, S.; Souza, X.D.; Fabian, A.; Teixeira, R.; Mariana, A.; Melo, C.D. As abelhas sem ferrão (Apidae; Meliponina) residentes no campus Federação/Ondina da Universidade Federal da Bahia, Salvador, Bahia, Brasil. *Candombá* **2005**, *1*, 57–69.
65. Carvalho-zilse, G.A.; Kerr, W.E. Natural substitutions of queens and flight distance ofmales in tiuba (*Melipona compressipes fasciculata* Smith,1854) and uruçu (*Melipona scutellaris Latreille*, 1811) (*Apidae, Meliponini*). *Acta Amaz.* **1854**, *34*, 649–652. [CrossRef]
66. Araújo, M.J.A.M.; Bosco, S.d.M.G.; Sforcin, J.M. *Pythium insidiosum*: Inhibitory effects of propolis and geopropolis on hyphal growth. *Braz. J. Microbiol.* **2016**, *47*, 863–869. [CrossRef] [PubMed]
67. Isidorov, V.A.; Maslowiecka, J.; Szoka, L.; Pellizzer, N.; Miranda, D.; Olchowik-Grabarek, E.; Zambrzycka, M.; Swiecicka, I. Chemical composition and biological activity of Argentinian propolis of four species of stingless bees. *Molecules* **2022**, *27*, 7686. [CrossRef] [PubMed]
68. Gerginova, D.; Popova, M.; Chimshirova, R.; Trusheva, B.; Shanahan, M.; Guzmán, M.; Solorzano-Gordillo, E.; López-Roblero, E.; Spivak, M.; Simova, S.; et al. The chemical composition of *Scaptotrigona mexicana* honey and propolis collected in two locations: Similarities and differences. *Foods* **2023**, *12*, 3317. [CrossRef] [PubMed]
69. Pujirahayu, N.; Suzuki, T.; Katayama, T. Cycloartane-type triterpenes and botanical origin of propolis of stingless Indonesian bee *Tetragonula sapiens*. *Plants* **2019**, *8*, 57. [CrossRef] [PubMed]
70. Alvear, M.; Santos, E.; Cabezas, F.; Pérez-SanMartín, A.; Lespinasse, M.; Veloz, J. Geographic area of collection determines the chemical composition and antimicrobial potential of three extracts of Chilean propolis. *Plants* **2021**, *10*, 1543. [CrossRef] [PubMed]
71. Stanciauskaite, M.; Marksa, M.; Liaudanskas, M.; Ivanauskas, L.; Ivaskiene, M.; Ramanauskiene, K. Extracts of *Poplar buds* (*Populus balsamifera* L., *Populus nigra* L.) and Lithuanian propolis: Comparison of their composition and biological activities. *Plants* **2021**, *10*, 828. [CrossRef] [PubMed]
72. Sanpa, S.; Popova, M.; Bankova, V.; Tunkasiri, T.; Eitssayeam, S.; Chantawannakul, P. Antibacterial compounds from propolis of *Tetragonula laeviceps* and *Tetrigona melanoleuca* (Hymenoptera: Apidae) from Thailand. *PLoS ONE* **2015**, *10*, e0126886. [CrossRef] [PubMed]
73. Al Naggar, Y.; Sun, J.; Robertson, A.; Giesy, J.P.; Wiseman, S. Caracterización química y propiedades antioxidantes del propóleos Canadiense. *J. Apic. Res.* **2016**, *55*, 305–314. [CrossRef]
74. Freitas, C.P.; Bilibio, D. Atividade antioxidante da própolis de abelhas Jataí antioxidant activity of propolis of jataí bees. *Braz. J. Anim. Environ. Res.* **2021**, *4*, 989–996. [CrossRef]

75. Campos, J.F.; dos Santos, U.P.; Macorini, L.F.B.; de Melo, A.M.M.F.; Balestieri, J.B.P.; Paredes-Gamero, E.J.; Cardoso, C.A.L.; de Picoli Souza, K.; dos Santos, E.L. Antimicrobial, antioxidant and cytotoxic activities of propolis from *Melipona orbignyi* (Hymenoptera, Apidae). *Food Chem. Toxicol.* **2014**, *65*, 374–380. [CrossRef]
76. Zhang, C.; Shen, X.; Chen, J.; Jiang, X.; Hu, F. Identification of free radical scavengers from Brazilian green propolis using off-line HPLC-DPPH Assay and LC-MS. *J. Food Sci.* **2017**, *82*, 1602–1607. [CrossRef]
77. Pazin, W.M.; Mônaco, L.d.M.; Egea Soares, A.E.; Miguel, F.G.; Berretta, A.A.; Ito, A.S. Actividad antioxidante de tres tipos de propóleos de abeja sin aguijón y propóleos verdes. *J. Apic. Res.* **2017**, *56*, 40–49. [CrossRef]
78. Zhang, W.; Cai, Y.; Chen, X.; Ji, T.; Sun, L. Optimized extraction based on the terpenoids of *Heterotrigona itama* propolis and their antioxidative and anti-inflammatory activities. *J. Food Biochem.* **2020**, *44*, e13296. [CrossRef] [PubMed]
79. Béji-Srairi, R.; Younes, I.; Snoussi, M.; Yahyaoui, K.; Borchard, G.; Ksouri, R.; Frachet, V.; Wided, M.K. Ethanolic extract of Tunisian propolis: Chemical composition, antioxidant, antimicrobial and antiproliferative properties. *J. Apic. Res.* **2020**, *59*, 917–927. [CrossRef]
80. de Sousa, J.P.A.; de Sousa, J.M.S.; Rodrigues, R.R.L.; de Lima Nunes, T.A.; Machado, Y.A.A.; de Araujo, A.C.; da Silva, I.G.M.; Barros-Cordeiro, K.B.; Báo, S.N.; de Moraes Alves, M.M.; et al. Antileishmanial activity of 2-amino-thiophene derivative SB-200. *Int. Immunopharmacol.* **2023**, *123*, 110750. [CrossRef] [PubMed]
81. Pessotti, J.H.; Zaverucha Do Valle, T.; Corte-Real, S.; Gonçalves Da Costa, S.C. Interaction of *Leishmania* (L.) *chagasi* with the vero cell line. *Parasite* **2004**, *11*, 99–102. [CrossRef] [PubMed]
82. Dutra, R.P.; Bezerra, J.L.; da Silva, M.C.P.; Batista, M.C.A.; Patrício, F.J.B.; Nascimento, F.R.F.; Ribeiro, M.N.S.; Guerra, R.N.M. Antileishmanial activity and chemical composition from Brazilian geopropolis produced by stingless bee *Melipona fasciculata*. *Rev. Bras. Farm.* **2019**, *29*, 287–293. [CrossRef]
83. Cavalcante, G.M.; Camara, C.A.; Silva, E.M.S.D.; Santos, M.S.; Leite, A.B.; Queiroz, A.C.; Evelyn Da Silva, A.; Araújo, M.V.; Alexandre-Moreira, M.S.; Silva, T.M.S. Leismanicidal activity of propolis collected in the semiarid region of Brazil. *Front. Pharmacol.* **2021**, *12*, 702032. [CrossRef]
84. Branquinha, M.H.; Araújo, P.S.S.; Oliveira, S.S.C.; Sangenito, L.S.; Gonçalves, D.S.; Seabra, S.H.; d'Avila-Levy, C.M.; Santos, A.L.S. Antileishmanial efficacy of the calpain inhibitor MDL28170 in combination with amphotericin B. *Trop. Med. Infect. Dis.* **2022**, *7*, 29. [CrossRef]
85. Ayres, D.C.; Marcucci, M.C.; Giorgio, S. Effects of Brazilian propolis on *Leishmania amazonensis*. *Mem. Inst. Oswaldo Cruz* **2007**, *102*, 215–220. [CrossRef]
86. Duran, G.; Duran, N.; Culha, G.; Ozcan, B.; Oztas, H.; Ozer, B. In vitro antileishmanial activity of Adana propolis samples on *Leishmania tropica*: A preliminary study. *Parasitol. Res.* **2008**, *102*, 1217–1225. [CrossRef]

processes

MDPI

Article

Studies on Reactive Extraction of Itaconic Acid from Fermentation Broths

Alexandra Cristina Blaga [1], Lenuta Kloetzer [1,*], Dan Cascaval [1], Anca-Irina Galaction [2] and Alexandra Tucaliuc [1,*]

[1] "Cristofor Simionescu" Faculty of Chemical Engineering and Environmental Protection, "Gheorghe Asachi" Technical University of Iasi, D. Mangeron 73, 700050 Iasi, Romania; alexandra-cristina.blaga@academic.tuiasi.ro (A.C.B.); dan.cascaval@academic.tuiasi.ro (D.C.)

[2] Faculty of Medical Bioengineering, "Grigore T. Popa" University of Medicine and Pharmacy, M. Kogalniceanu 9–13, 700454 Iasi, Romania; anca.galaction@umfiasi.ro

* Correspondence: lenuta.kloetzer@academic.tuiasi.ro (L.K.); alexandra.tucaliuc@academic.tuiasi.ro (A.T.)

Abstract: Itaconic acid is a high-value organic acid that serves as a platform molecule in different industries. This research focuses on the separation of itaconic acid using reactive extraction as a sustainable and efficient method for acid recovery from fermentation broth. Itaconic acid was produced through fungal fermentation processes involving *Aspergillus terreus* ATCC® 32588™, obtaining a concentration of 47 g/L in the final broths. For the reactive extraction system, the organic phase included tri-n-octylamine as an extractant dissolved in dichloromethane or n-heptane and 1-octanol as a phase modifier. The effect of the main influencing factors (pH of the aqueous phase, extractant concentrations in the organic phase, and the addition of 1-octanol) on extraction efficiency was investigated. The highest extraction degree (97%) was achieved using an organic phase with tri-n-octylamine dissolved in dichloromethane and with 20% 1-octanol. Conversely, in the reactive extraction system with tri-n-octylamine dissolved in n-heptane and 20% 1-octanol, the extraction efficiency reached 67%. This finding suggests a promising separation system that is less toxic for microorganisms. The study results highlight the feasibility of employing reactive extraction systems for the direct separation of itaconic acid.

Keywords: itaconic acid; *Aspergillus terreus*; fermentation broths; tri-n-octylamine; reactive extraction

Citation: Blaga, A.C.; Kloetzer, L.; Cascaval, D.; Galaction, A.-I.; Tucaliuc, A. Studies on Reactive Extraction of Itaconic Acid from Fermentation Broths. *Processes* **2024**, *12*, 725. https://doi.org/10.3390/pr12040725

Academic Editors: Prashant K. Sarswat and Yanlin Zhang

Received: 3 March 2024
Revised: 31 March 2024
Accepted: 1 April 2024
Published: 3 April 2024

1. Introduction

In the context of sustainable development, many important chemicals are suitable to be replaced by natural alternatives through biobased processes. Itaconic acid (IA) is a dicarboxylic acid (Figure 1) with various applications across different industries. These include chemical applications (as a building block for products like resins, biodegradable and sustainable polymers, and coatings), cosmetic uses (as a thickening agent, emulsion stabilizer, and moisturizing agent), textile applications (as a dye-binding agent and as a modifier for enhancing the dye ability of fibers), and even in the food industry (as an acidulant and flavor enhancer) [1].

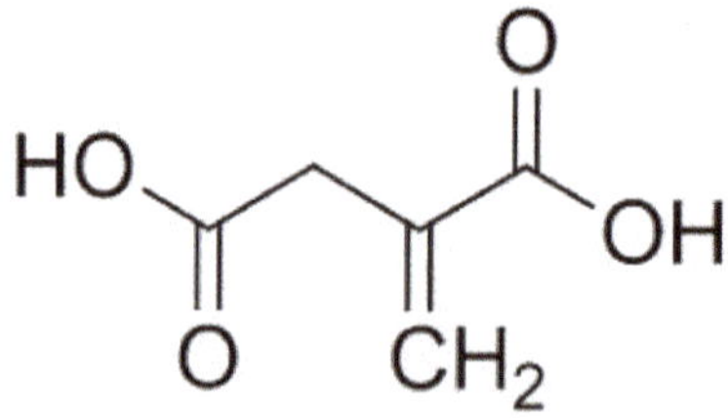

Figure 1. Itaconic acid chemical structure.

Itaconic acid has the advantage that it can be obtained by fermentation processes. Biotechnological production of itaconic acid involves various microorganisms, including filamentous fungi such as *Aspergillus itaconicus*, *Aspergillus terreus*, *Pseudozyma antarctica*, and *Ustilago maydis* [2–4], as well as yeasts like *Candida* sp., *Rhodotorula* sp. [5], and even bacteria such as engineered *E. coli* [1]. The fermentation process for itaconic acid production provides a sustainable, cost-effective, and environmentally friendly approach that aligns with the growing demand for biobased chemicals and renewable manufacturing processes. Furthermore, it generates fewer greenhouse gas emissions and requires less energy input, reducing the carbon footprint associated with itaconic acid production. Additionally, it can use renewable feedstocks such as sugars derived from biomass, agricultural waste, or even lignocellulosic materials as carbon sources [6].

Notably, commercial production of IA often relies on the fermentation of *A. terreus* using glucose as a substrate, resulting in a maximum concentration of 129 g/L in the fermentation broth [2].

Following fermentation, designing a cost-effective downstream process becomes a key step in biobased itaconic acid production. Several separation processes, including ion exchange, adsorption, evaporative crystallization, precipitation, electrodialysis, and extraction) have been used for IA recovery. Among these, crystallization is the most commonly employed recovery method, yielding approximately 75%. However, it requires further downstream steps to improve the purity of the final product. The choice of separation method for itaconic acid recovery from fermentation broth—whether through physical or chemical-based separation and whether it is coupled with fermentation—comes with its own set of advantages and disadvantages. As the demand for itaconic acid continues to grow in various industries, having a reliable and efficient separation method becomes crucial for meeting market requirements while promoting environmentally friendly products. Considering important criteria such as yield, equipment costs, product purity, energy consumption, and downstream complexity, the reactive extraction method emerges as a favorable choice for IA separation. Also, this method can be directly coupled with fermentation if the chosen solvents are non-toxic to microorganisms. By expanding the availability of premium-grade IA, manufacturers can enhance their competitiveness in the market. This approach not only meets customer demand but also positions them as reliable suppliers, ultimately driving business growth.

As a promising method for IA separation, reactive extraction, derived from liquid-liquid separation methods, is a process that involves the transfer of a solute from one liquid phase to another immiscible phase, facilitated by an extractant. This method is especially useful for separating compounds with similar physical properties or when conventional separation techniques like distillation or crystallization are not effective or economically viable [7]. Reactive extraction is commonly employed for efficiently separating organic acids, leading to improved results [8–12]. Various types of extractants have been evaluated for IA separation [12–18]. These include organophosphates and aliphatic amines dissolved in different solvents [11–16]. Among them, tributyl phosphate (TBP) [13,16], trioctylmethylammonium chloride (Aliquat 336) [15,16], trioctylphosphineoxide [10], dioctylamine, N-methyldioctylamine [8], and tri-n-octylamine (TOA) [8,13,17,18] were used and the efficiency of reactive extraction from the aqueous phase followed the order of quaternary ammonium > tertiary > secondary > primary amines [11]. Different solvents, both active and inactive, were analyzed for extraction, including alcohols, esters, and alkanes. Aliquat 336 was dissolved in ethylacetate, hexane, toluene, and kerosene. The system containing ethyl acetate provided the highest extraction degree of 72.66% [15]. Considering the solvent phase toxicity, Aliquat 336 and TBP were compared by dissolving them in a biocompatible solvent, sunflower oil, and Aliquat 336 exhibited sustainable enhancement in itaconic acid extraction in the organic phase, surpassing TBP. The distribution coefficient was 0.95 for Aliquat 336 and 0.4 for TBP [16]. In a study by Pal et al. (2016) tributyl phosphate dissolved in alcohols (1-butanol, 1-octanol, 1-dodecanol) resulted in the highest extraction degree (82.47%) when 50% TBP was dissolved in 1-butanol [19]. Tri-n-octylamine, an aliphatic amine with a long carbon chain, has been identified as an effective extractant for itaconic acid separation. Its advantages include low solubility in water and thermal stability, allowing for easy recycling through distillation [13]. The stoichiometry of reactive extraction for IA using TOA dissolved in various solvents, such as alkane, esters, ketone, and alcohols, was established by Uslu and Datta (2015). Among these systems, TOA combined with 1-octanol demonstrated the most favorable performance in terms of distribution coefficient (KD = 18.20) and extraction degree (% E = 94.80%) [18].

Considering that the applications of IA are constantly evolving as research and development efforts continue to explore its potential uses in various industries, and the fact that the downstream processing for IA manufacturing constitutes approximately 40% of overall production expenses, like for other important organic acids, ensuring the creation and implementation of effective recovery methods remains necessary [20,21]. The commercial production of IA continues to draw interest due to its high high-end application potential. However, the critical barrier to overcome is finding an efficient and environmentally friendly separation process. Taking into account the trends in direct separations of the carboxylic acids, the research objectives of this study focused on two distinct steps: (i) itaconic acid production by controlling the fungal fermentation of *Aspergillus terreus* (ATCC® 32588™), and (ii) the investigation of the efficiency of different reactive extraction systems for IA recovery from filtered fermentation broths; two solvents, n-heptane and dichloromethane, were employed for diluting the aminic extractant, TOA. The research aim of this study was to establish a suitable solvent system and investigate the impact of various factors on efficiency during the reactive extraction of IA, including aqueous phase pH (from the clear fermentation broth), TOA concentrations in the organic phase, and the addition of 1-octanol as a phase modifier.

2. Materials and Methods

2.1. Microorganism and Culture Media

For fungal fermentation processes, *Aspergillus terreus* (ATCC® 32588™, ATCC, Manassas, VA, USA) was cultivated on malt agar medium at 35 °C. To create spore suspension, the spores were washed twice with 10 mL of sterilized water each time. Conidiospores were then quantified using a hemocytometer (specifically, a CE Neubauer-improved Marienfeld hemocytometer, Fisher Scientific, Göteborg, Sweden) and an Optika microscope (manufactured by, OPTIKA S.r.l., Ponteranica (BG), Italy). The resulting spore suspensions, containing approximately 10^8 or 10^9 spores/mL, served as the inoculum for the fermentation process. The inoculum size was considered 10% of the fermentation culture medium volume. The inoculum medium used in the spore-germination experiment had the following composition (in grams per liter): glucose 45, KH_2PO_4 1, $MgSO_4$ 1, $ZnSO_4$ 0.05, NH_4SO_4 5, and $FeSO_4$ 0.04. The inoculum medium compounds were provided from Carl ROTH, Karlsruhe, Germany.

2.2. Fermentation Processes

The fermentation experiments were conducted in a 2 L laboratory stirred bioreactor (manufactured by Fermac, Electrolab, Mumbai, India), equipped with a computer system for controlling and recording key parameters such as temperature, pH, and dissolved oxygen. The working volume for all fermentation experiments was 1 L. *Aspergillus terreus* cultivation was carried out in a batch mode at 35 °C, a fermentation time of 190 h, an agitation rate of 200 rpm using a turbine-type impeller, and an aeration rate of 1 L/minute. The acid production medium contained (in grams per liter) glucose 180, KH_2PO_4 0.2, NH_4NO_3 2.5, $FeSO_4 \cdot 7H_2O$ 0.04, $MgSO_4 \cdot 7H_2O$ 1.9, and $ZnSO_4 \cdot 7H_2O$ 0.15, and the pH value was adjusted to 3.5 [22]. The culture medium compounds were provided from Carl ROTH, Karlsruhe, Germany. The pH value was set at the beginning of fermentation and was not controlled during the fermentation processes. The bioreactor and the culture medium were sterilized at 120 °C for 21 min in a vertical autoclave (type Raypa AES-110, Raypa, Terrassa (Barcelona), Spain). During the fermentation processes, samples were collected at 24-h interval. Upon completion of the fermentation step, the broths underwent the procedure described in reference [22]. To determine the cell dry weight (d.w.) gravimetrically, 2 mL of the fermentation broth was centrifuged at 3000 rpm for 10 min at room temperature. The experimental conditions were similar to those reported in the literature [23,24]. After fermentation, the biomass was separated through filtration, and the supernatant was used for subsequent IA reactive extraction.

2.3. Reactive Extraction Procedure

For the reactive extraction experiments, we utilized a laboratory extraction column equipped with vibratory mixing. Detailed specifications of this equipment were previously documented [25]. The extraction column consists of a glass column with specific geometric characteristics: an internal diameter of 46 mm and a height of 250 mm. These dimensions ensure a high interfacial area and facilitate rapid attainment of the equilibrium state. For temperature control during experiments, a thermostatic jacket was employed, circulating a thermal agent (a mixture of water and ethylene glycol, SERVA Electrophoresis GmbH, Heidelberg, Germany) at 25 °C. To maintain the desired interfacial area, a perforated disk made of stainless steel, with a diameter of 45 mm and a 20% free section, was precisely positioned at the initial contact interface between the aqueous and organic phases. The vibration applied had a frequency of 50 s^{-1} and an amplitude of 5 mm. The extraction process was conducted at 25 °C, with an extraction time of 1 min. Following extraction, the resulting emulsion was separated using a centrifugal separator (a Biobase Laboratory Centrifuge high-speed TG16-WS, Biobase Biodustry Co., Ltd., Jinan, China) operating at 5000 rpm. The aqueous phase used for extraction originated from the fermentation broth, containing 50 g/L (0.384 M) IA. The IA separation by reactive extraction was performed using TOA (Merck KGaA Billerica, MA, USA) as an extractant, dissolved in two organic solvents with distinct dielectric constants: dichloromethane (DCM) (ε = 9.08) (Fluka, Paris, France, GC 99.9%) or n-heptane (ε = 1.90) (Merck, GC 99%) [9]. The concentration of extractant in the organic phase (either pure solvent or solvent mixed with 1-octanol) ranged between 0 and 240 g/L (0.678 M). Additionally, a phase modifier, 1-octanol with purity exceeding 98% (from Merck, ε = 10.3) was mixed into the aforementioned solvents. The volumetric ratio between the aqueous and organic phases was 1:1 (20 mL of each phase).

The pH range of the aqueous phase with itaconic acid was between 1 and 6 and was measured using a digital pH meter (type C836, Consort bvba, Turnhout, Belgium). To obtain the desired pH value of the aqueous phase, solutions of 3% H_2SO_4 or 3% NaOH (Merck KGaA, Darmstadt, Germany) were used. During the experiments, the pH of the aqueous phase was monitored, and no variations were observed.

The extraction results were analyzed based on the IA extraction degrees, calculated as the ratio of the itaconic acid concentration in the extracted phase ($[(H_2I)_{(org)}]$) to the initial acid concentration in the aqueous phase ($[(H_2I)_{(aq)i}]$) while assuming no change in volume at equilibrium (Equation (1)):

$$\eta = \frac{\left[(H_2I)_{(org)}\right]}{\left[(H_2I)_{(aq)_i}\right]} \times 100 \quad (1)$$

The equilibrium for the IA extraction process with TOA occurs when the rate of formation equals the rate of dissolution. It can be described as a series of reactions between m molecules of acid (H_2I), and n molecules of extractant (E), resulting in the formation of various acid:extractant complexes (m:n) (Equation (2)):

$$m\,H_2I + n\overline{E} \leftrightarrow \overline{(H_2I)_m(E)_n} \quad (2)$$

The studies on the stoichiometry (m, n) of the reactive extraction of IA with TOA suggested that different types of acid:amine complexes like $[H_2I.E]$ (1:1), and $[(H_2I)_2.E_2]$ (2:1) are formed [11].

The IA concentration in the aqueous phase was determined by the HPLC method. For the acid concentration in the solvent phase, the mass balance calculation has been done. The HPLC equipment used was the UltiMate 3000 Dionex system, equipped with an Acclaim™ OA column (4 mm diameter, 150 mm length, 5 μm porous particle size, Dionex—Thermo Fisher Scientific, Sunnyvale, CA, USA) and a UV detector set at 210 nm [12]. The mobile phase consisted of a 100 mM Na_2SO_4 solution with a pH of 2.65 (adjusted using methanesulfonic acid). The flow rate of the mobile phase was 0.6 mL/min. The analysis has been carried out at 30 °C. Each experiment was repeated three times under identical conditions, and the average value of the considered parameters was used. The maximum experimental error observed was 6.1%. The methods and equipment used in this study were chosen based on the accuracy of results and the advantages offered, as mentioned in previously published papers [9,12,25].

3. Results and Discussion

3.1. Fermentation Process

Aspergillus terreus (ATCC® 32588™) stands out as a highly promising strain for itaconic production through aerobic cultivation. This strain leads to high concentrations of itaconic acid in fermentation broths. The production of itaconic acid was significantly influenced by several factors:

(i) Fungi morphology and growth form: In this study, the dominant growth form of *A. terreus* was filamentous, characterized by short hyphae. This morphology was advantageous because it avoids high viscosities that could otherwise influence the oxygen transfer rate from the gas phase to the cells;

(ii) Bioreactor design and operational parameters: The successful production of itaconic acid hinges on a well-designed bioreactor and carefully controlled parameters during the fermentation process.

From the experimental data depicted in Figure 2, it can be observed that itaconic acid production started at 45 h of *A. terreus* cultivation. As the glucose concentration decreased, itaconic acid accumulated in the broth gradually. However, beyond 120 h from the start of fermentation, the itaconic acid concentration stagnates and reaches a maximum value of 47 g/L.

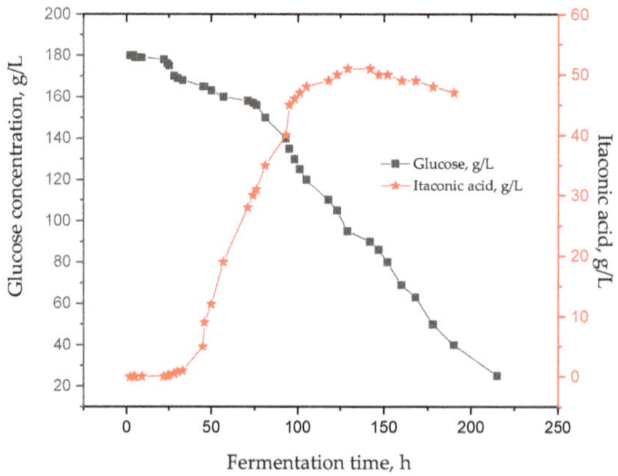

Figure 2. Glucose consumption and itaconic acid accumulation during *A. terreus* fermentation.

Regarding glucose consumption and itaconic acid accumulation, Figure 3 indicated the following presumptions: (i) *A. terreus* might be sensitive to acidic conditions, leading to reduced glucose utilization [26], (ii) the low pH affects the enzymatic reactions involved in glucose metabolism, and, additionally, the accumulation of itaconic acid could further impact the microorganism's metabolism due to the product inhibition mechanism [27,28]. According to the experimental results, the glucose consumption gradually decreased during the first 80 h of fermentation, while there was an accumulation of itaconic acid in the fermentation broth. At the end of fermentation, the residual glucose concentration was 19 g/L, which indicated that the metabolic activity of the microorganism *A. terreus* was influenced by the presence of IA [26–28]. The initial pH value was set at 3.5, and by the end of fermentation, the pH value dropped below 2. This initial pH value was adopted according to the literature to achieve IA production due to an improved performance of *A. terreus* [29–31]. Such a significant decrease in pH can have several implications for microbial metabolism and enzyme activity [28].

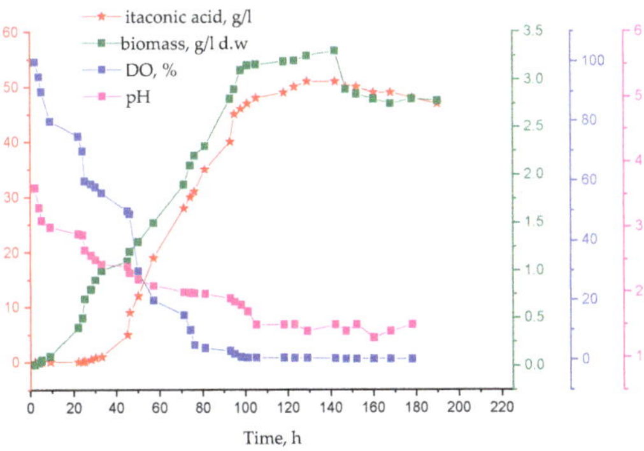

Figure 3. Evolution of parameters (itaconic acid, cells dry weight, dissolved oxygen, pH) during aerobic cultivation of *A. terreus*.

A key parameter with a strong impact on *A. terreus* growth and itaconic acid production was represented by dissolved oxygen (DO). Dissolved oxygen directly influences itaconic acid biosynthesis. In the first stage of fermentation, DO decreased rapidly to a minimum value due to fungus growth and biomass accumulation in the broth. However, in the late stage of fermentation, the mycelial growth slowed down, and itaconic acid production increased. Thus, the significant decrease in DO can be correlated with itaconic acid accumulation in the broth (Figure 3).

Additionally, this effect is related to the possibility that molecular oxygen may serve as a co-substrate for IA synthesis, representing a key element for the growth and metabolism of aerobic fungi. The final biomass concentration of 3.3 g/L d.w was attained after 142 h of fermentation. Moreover, the DO level of 15% was related to the maximum biomass concentration after 80 h of fermentation. This result also highlights the limitation of IA production due to oxygen supply during fermentation. A critical DO value was observed to be 20%. Therefore, during the IA production phase in *A. terreus* fermentation, the concentration of DO significantly affected the production of IA. Interestingly, the variation in the dissolved oxygen level between 20% and 60% did not significantly impact IA accumulation in the broth. However, a slight increase in IA production was observed at a low level of DO [32]. Also, the literature indicated that a dissolved oxygen concentration of 2% saturation was sufficient for maximal biomass formation. Increasing it to 30% saturation had no effect on IA production [33].

Following the maximum accumulation of itaconic acid in the broth (47 g/L) and with the pH value remaining constant, the fermentation broths were subject to separation. Consequently, the filtered fermentation broths were subsequently utilized for IA separation by reactive extraction.

3.2. Reactive Extraction without Phase Modifier

The initial step was to investigate the effectiveness of reactive extraction without the addition of a phase modifier in the organic phase. The filtered fermentation broths were then brought into contact with the organic phase containing varying concentrations of extractant.

The efficiency of reactive extraction for carboxylic acid is highly influenced by the pH value of the aqueous solutions. Two traditional mechanisms—ion-pair formation and hydrogen bonding—have been mentioned in the literature as responsible for the formation of interfacial hydrophobic complexes [34].

Itaconic acid has two dissociation steps with pKa values of 3.84 and 5.55 at 25 °C [35], which significantly impacts the extraction mechanisms at various pH values of the aqueous phase. The undissociated acid molecule and the aminic extractant form a hydrogen bond only at aqueous solution pH values lower than pKa (below pH 3, with the undissociated form of itaconic acid, H_2I). On the other hand, ion-pair formation can occur independent of the pH value. This is due to proton transfer at pH values below the acid's pKa value or ionic interactions between charged species above the pKa value [34]. In systems without the addition of a phase modifier (as shown in Figure 4), an increase in the aqueous phase's pH value led to a lower value of the itaconic acid's extraction degree. This result is attributed to increased itaconic acid dissociation and weaker interactions between the extractant and dissociated IA. Moreover, organic solvents like n-heptane and dichloromethane are hydrophobic, and dissociated IA is less soluble in these, an aspect that contributes to a lower extraction degree and weaker interaction.

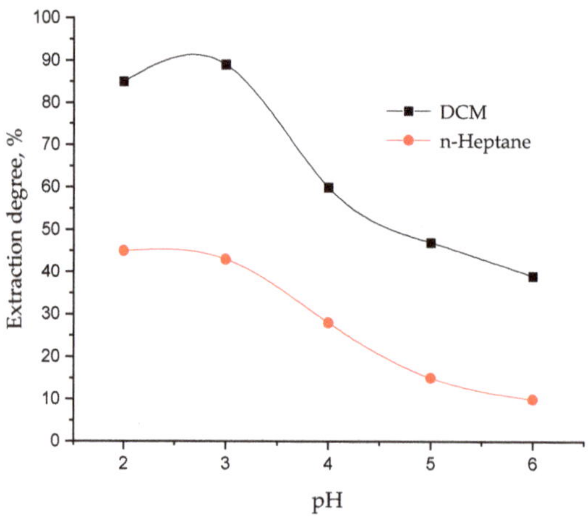

Figure 4. Influence of aqueous phase's pH value on itaconic acid extraction degree (extractant concentration, TOA = 120 g/L).

The pH directly impacted the dissociation behavior of itaconic acid. Thus, for pH values before 4, itaconic acid undergoes partial dissociation into two forms (HI^- and I^{2-}) and the interfacial reaction of the partially ionized acid with TOA influenced the extraction efficiency. For pH values above 4, itaconic acid continues to dissociate but its ability to form intermolecular hydrogen bonds decreases. This effect was more pronounced for solvents with a lower dielectric constant, such as n-heptane. The extraction degree of itaconic acid reached low values and was almost impractical for the alkaline pH domain. In this reactive extraction system with TOA and no addition of 1-octanol, the superior dielectric constant of dichloromethane offers the highest value of itaconic acid's extraction degree (88%), the lowest value being reached for n-heptane (43%). For pH values higher than 5, the extraction degree decreased for the system with dichloromethane and became very low for extraction with n-heptane due to the complete dissociation of itaconic acid at higher pH values. Therefore, the optimum pH value for the extraction was selected as 3, when only 11% of itaconic acid was dissociated to a single functional group and 89% was in the undissociated form, as calculated by the measured pKa values [36]. This pH value corresponds to the initial stage of itaconic acid formation during fermentation, making it appropriate as a direct extraction method [37].

The increase of amine extractant concentration in the organic phase had a positive impact on the extraction degree due to the increased interfacial amount of one of the reactants. In Figure 5, it becomes evident that the impact of TOA concentration is significant only when it varies below a specific threshold (120 g/L). Beyond this threshold, its effect becomes negligible.

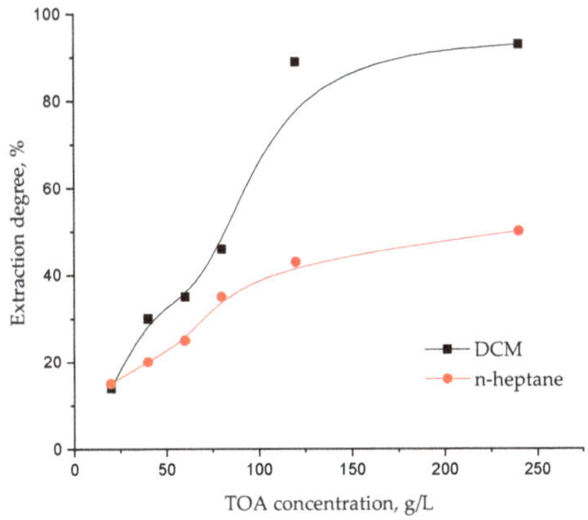

Figure 5. Influence of amine extractant concentration on itaconic acid extraction degree (aqueous phase pH = 3).

The magnitude of this effect was influenced by the polarity of the organic phase. Thus, the amine concentration threshold is 120 g/L for the system with dichloromethane, while for n-heptane, it is 80 g/L. This is the result of the modification of the number of TOA molecules involved in the formation of the complex with itaconic acid.

By using non-polar solvents such as n-heptane, for the separation of carboxylic acid, a stable emulsion and dimer might be formed in the organic phase [11]. To counteract this effect and ensure an increase in the solubility of the formed itaconic acid–TOA complex, a phase modifier, 1-octanol, was added to the organic phase.

3.3. Reactive Extraction with Phase Modifier

The addition of 1-octanol in the organic phase does not change the shapes of the dependence between reactive extraction efficiency and the aqueous phase's pH values. However, the addition of the phase modifier increased the extraction degree, an effect that was more pronounced for n-heptane, the solvent with lower polarity (Figure 6).

At pH 2, the elevated extraction degrees for both solvents primarily result from the increased solubility of the interfacial compound. This effect is further amplified by the additional solubilization of ionized molecules facilitated by 1-octanol. Notably, when 1-octanol was added to n-heptane, it significantly intensified the formation of an amine:acid complex, surpassing the effect observed in dichloromethane. Also, the positive effect of adding 1-octanol on the reactive extraction system with n-heptane was accentuated by using different volume fractions of phase modifier and TOA concentrations (Figure 7).

The 1-octanol addition improved the extraction efficiency regardless of the extractant concentration. This effect was more significant in the n-heptane system due to the supplementary solubilization of itaconic acid dissociated molecules by increasing this solvent polarity. The substantial enhancement in the reactive extraction degree of itaconic acid, achieved by adding 1-octanol in n-heptane, stems from the reduction in the formation of intermolecular hydrogen bonds between carboxylic groups of the acid. Consequently, this facilitates the reaction of carboxylic groups with TOA, leading to the formation of compounds with greater solubility in the organic phase [9]. The most important increase (67%) was recorded for 20% 1-octanol dissolved in n-heptane at a 120 g/L TOA concentration.

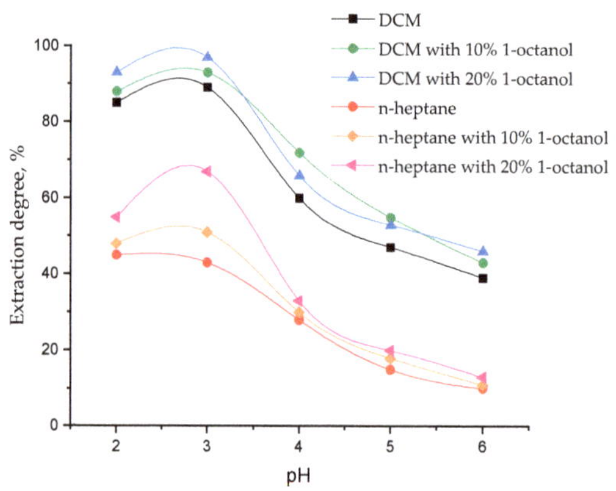

Figure 6. Influence of pH values of the aqueous phase on itaconic acid extraction degree in the presence of a phase modifier (TOA = 120 g/L).

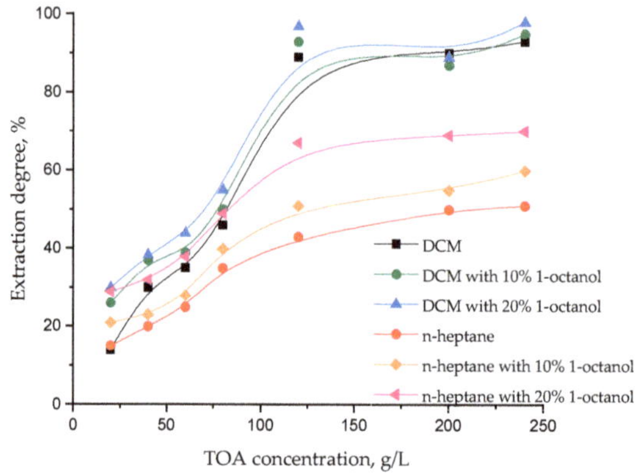

Figure 7. The influence of TOA concentration on the extraction degree of itaconic acid in the presence of phase modifier (aqueous phase's pH = 3).

To underline the beneficial effect of 1-octanol addition, the variation of amplification factor with pH and extractant concentration was graphically represented in Figures 8 and 9 (the amplification factor represents the ratio between the extraction degrees of itaconic acid with and without 1-octanol in the organic phase, respectively, obtained under identical experimental conditions).

Therefore, according to Figures 8 and 9, the increase of 1-octanol concentration determines the increase of the amplification factor. The amplification factor values were higher in the presence of phase modifier, especially for n-heptane, indicating an improved reactive extraction degree for itaconic acid especially in systems with a volumetric 1-octanol fraction of 0.2. At higher 1-octanol concentrations, the increased effect of the amplification factor can be attributed to ester formation at the interfacial reaction between itaconic acid and alcohol. Thus, the reaction rate was positively influenced by the high-phase modifier concentration in the extraction system with low polarity, which considerably improved the solubilization capacity of n-heptane. In the case of using DCM, the presence of 1-octanol in high concentration also induced the increase of the amplification factor, but the obtained values were lower than in the organic phase with high polarity.

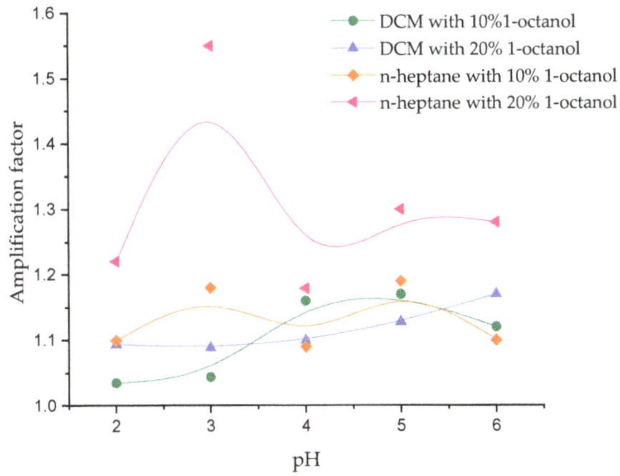

Figure 8. The variation of amplification factor with pH value of aqueous phase (TOA concentration = 120 g/L).

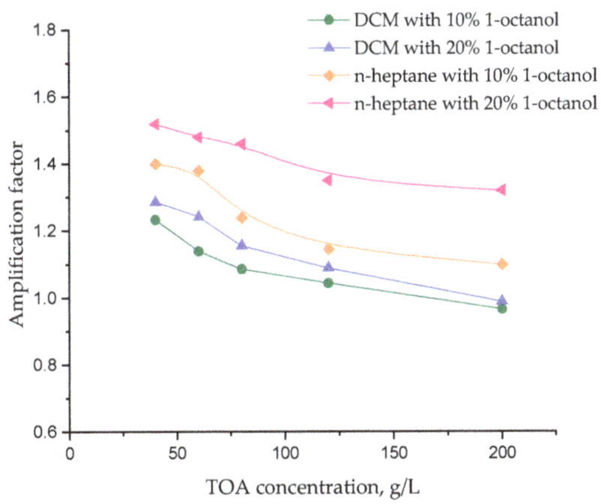

Figure 9. The variation of amplification factor with extractant concentration (pH of aqueous phase = 3).

The extraction of partially or totally dissociated forms of IA is closely related to the pH of the aqueous phase and the polarity of the organic phase. In the system without 1-octanol, the extraction degrees decrease with increasing pH. However, the addition of a phase modifier allows for high extraction degrees at pH values above 4. The maximum value of the amplification factor was reached for pH 5 of clarified broth. Contrarily, the minimum value of the amplification factor at pH = 3 confirmed the enhancement of the positive influence of a phase modifier on IA reactive extraction for solvents with lower polarity.

The graphical representation in Figure 9 underscores the positive impact of a phase modifier—specifically, 1-octanol—on IA reactive extraction. When compared to extraction systems without 1-octanol, the addition of this modifier significantly enhanced the extraction process. At a volumetric fraction of 1-octanol equal to 0.2, this effect was most pronounced. For dichloromethane, the extraction degree of itaconic acid increased by approximately 1.28 times, while for n-heptane, the enhancement was even more substantial, reaching 1.55 times.

Remarkably, regardless of solvent type and alcohol concentration, the highest amplification factor occurs in extraction systems containing 40 g/L TOA. This highlights the critical role of the phase modifier in improving the extraction process, particularly concerning solvent phase polarity. For the analyzed extraction system, the n-heptane solvent phase combined with 40 g/L yields the maximum amplification factor.

4. Conclusions

Considering the current trends in developing efficient separation processes for biobased products, the current study analyzed the possibility of obtaining a biocompatible separation system of itaconic acid from fermentation media. According to the analyzed extraction systems, the experimental conditions led to remarkable results. The highest extraction degree, reaching an impressive 97%, was achieved using TOA dissolved in dichloromethane with 20% 1-octanol. This combination proved exceptionally effective for extracting IA.

An alternative system involving TOA with n-heptane and 20% 1-octanol yielded a lower extraction degree (67%). However, this system has distinct advantages. Particularly, n-heptane exhibits low toxicity to microorganisms, making it suitable for direct IA extraction from fermentation broth.

The positive effect of the phase modifier (1-octanol) was further emphasized by the amplification factor variation with extractant concentration. The maximum amplification factor occurred in the system with 20% 1-octanol and n-heptane at 40 g/L TOA concentration in the solvent phase. Interestingly, the system with DCM required a higher TOA concentration (80 g/L) to achieve similar results.

This study highlights the possibility of IA production through fungal fermentation of *A. terreus* and underscores the importance of optimizing both fermentation and downstream extraction processes. The results of this study offer a viable pathway for in situ reactive extraction using a non-polar solvent and a phase modifier system. Upcoming research will explore the feasibility of combining fermentation and extraction steps for further process improvement. Additionally, scalability, cost-effectiveness, and environmental impact should be investigated to advance IA production and separation processes.

Author Contributions: Conceptualization, A.C.B. and A.T.; data curation, L.K.; formal analysis, A.C.B. and A.T.; investigation, A.T.; methodology, L.K. and A.T.; supervision, A.-I.G.; validation, A.C.B., L.K. and A.T.; visualization, A.C.B. and D.C.; Writing—original draft, A.T.; Writing—review and editing, A.C.B., L.K. and D.C. All authors have read and agreed to the published version of the manuscript.

Funding: This research received no external funding.

Data Availability Statement: Data are contained within the article.

Conflicts of Interest: The authors declare no conflicts of interest.

References

1. Diankristanti, P.A.; Ng, I.S. Microbial itaconic acid bioproduction towards sustainable development: Insights, challenges and prospects. *Bioresour. Technol.* **2023**, *384*, 129280. [CrossRef] [PubMed]
2. Hevekerl, A.; Kuenz, A.; Vorlop, K.D. Influence of the pH on the itaconic acid production with *Aspergillus terreus*. *Appl. Microbiol. Biotechnol.* **2014**, *98*, 10005–10012. [CrossRef]
3. Levinson, W.E.; Kurtzman, C.P.; Kuo, T.M. Production of itaconic acid by Pseudozyma antarctica NRRL Y-7808 under nitrogen-limited growth conditions. *Enzym. Microb. Technol.* **2006**, *39*, 824–827. [CrossRef]
4. Tabuchi, T. Manufacture of Itaconic Acid with Ustilago. Japan Patent JP-H-03-357-85A, 15 September 1991.
5. Kawamura, D.; Furuhashi, M.; Saito, O.; Matsui, H. Production of Itaconic Acid by Fermentation. Japan Patent JP-A-56 137 861, 28 October 1981.
6. Saha, B.C.; Kennedy, G.J.; Bowman, M.J.; Qureshi, N.; Nichols, N.N. Itaconic acid production by *Aspergillus terreus* from glucose up to pilot scale and from corn stover and wheat straw hydrolysates using new manganese tolerant medium. *Biocatal. Agric. Biotechnol.* **2022**, *43*, 102418. [CrossRef]
7. Yıldız, E.; Lalikoglu, M.; Aşçı, Y.S.; Tarım, B.S. Investigation of reactive extraction of monocarboxylic acids with menthol-based hydrophobic deep eutectic solvent by response surface methodology. *Sep. Sci. Technol.* **2023**, *58*, 1450–1459. [CrossRef]

8. Kaur, G.; Elst, K. Development of reactive extraction systems for itaconic acid: A step towards in situ product recovery for itaconic acid fermentation. *RSC Adv.* **2014**, *4*, 45029–45039. [CrossRef]
9. Poştaru, M.; Cârlescu, A.; Galaction, A.I.; Caşcaval, D. Direct separation of propionic acid from Propionibacterium acidipropionici broths by reactive extraction 1. Interfacial mechanism and influencing factors. *Environ. Eng. Manag. J.* **2012**, *11*, 709–7016.
10. Hano, T.; Matsumoto, M.; Ohtake, T.; Sasaki, K.; Hori, F.; Kawano, Y. Extraction equilibria of organic acids with tri-n-octylphosphineoxide. *J. Chem. Eng. Jpn.* **1990**, *6*, 734–738. [CrossRef]
11. Bressler, E.; Braun, S. Separation mechanisms of citric and itaconic acids by water-immiscible amines. *J. Chem. Technol. Biot.* **1999**, *74*, 891–896. [CrossRef]
12. Kloetzer, L.; Poştaru, M.; Galaction, A.I.; Blaga, A.C.; Caşcaval, D. Comparative study on rosmarinic acid separation by reactive extraction with Amberlite LA-2 and D2EHPA. 1. Interfacial reaction mechanism and influencing factors. *Ind. Eng. Chem. Res.* **2013**, *52*, 13785–13794. [CrossRef]
13. Matsumoto, M.; Otono, T.; Kondo, K. Synergistic extraction of organic acids with tri-n-octylamine and tri-n-butylphosphate. *Sep. Purif. Technol.* **2001**, *24*, 337–342. [CrossRef]
14. Aşçi, Y.S.; Inci, I. A novel approach for itaconic acid extraction: Mixture of trioctylamine and tridodecylamine in different diluents. *J. Ind. Eng. Chem.* **2012**, *18*, 1705–1709. [CrossRef]
15. Wasewar, K.L.; Shende, D.; Keshav, A. Reactive extraction of itaconic acid using quaternary amine Aliquat 336 in ethyl acetate, toluene, hexane, and kerosene. *Ind. Eng. Chem. Res.* **2011**, *50*, 1003–1011. [CrossRef]
16. Wasewar, K.L.; Shende, D.; Keshav, A. Reactive extraction of itaconic acid using tri-n-butyl phosphate and aliquat 336 in sunflower oil as a non-toxic diluent. *J. Chem. Technol. Biotechnol.* **2011**, *86*, 319–323. [CrossRef]
17. Datta, D.; Kumar, S.; Uslu, H. Status of the reactive extraction as a method of separation. *J. Chem.* **2015**, *2015*, 853789. [CrossRef]
18. Uslu, H.; Datta, D. Experimental and theoretical investigations on the reactive extraction of itaconic (2-methylidenebutanedioic) acid using trioctylamine (N,N-dioctyloctan-1-amine). *J. Chem. Eng. Data* **2015**, *60*, 1426–1433. [CrossRef]
19. Pal, S.; Shende, D.Z.; Anjum, S.; Rathore, A.K. Reactive extraction of itaconic acid using tri-n-butyl phosphate in 1-butanol, 1-octanol and 1-dodecanol. *Int. J. Innov. Sci. Eng. Technol.* **2016**, *5*, 1425–1433.
20. Magalhães, A.I.; de Carvalho, J.C.; Medina, J.D.C.; Soccol, C.R. Downstream process development in biotechnological itaconic acid manufacturing. *Appl. Microbiol. Biotechnol.* **2017**, *101*, 1–12. [CrossRef] [PubMed]
21. Rózsenberszki, T.; Komáromy, P.; Kőrösi, E.; Bakonyi, P.; Nemestóthy, N.; Bélafi-Bakó, K. Investigation of itaconic acid separation by operating a commercialized electrodialysis unit with bipolar membranes. *Processes* **2020**, *8*, 1031. [CrossRef]
22. Blaga, A.C.; Ciobanu, C.; Cascaval, D.; Galaction, A.I. Enhancement of ergosterol production by Saccharomyces cerevisiae in batch and fed-batch fermentation processes using n-dodecane as oxygen-vector. *Biochem. Eng. J.* **2018**, *131*, 70–76. [CrossRef]
23. Elnaghy, M.A.; Megalla, S.E. Itaconic-acid production by a local strain of *Aspergillus terreus*. *Eur. J. Appl. Microbiol.* **1975**, *1*, 159–172. [CrossRef]
24. Kautola, H.; Rymowicz, W.; Linko, Y.Y.; Linko, P. Itaconic acid production by immobilized *Aspergillus terreus* with varied metal additions. *Appl. Microbiol. Biot.* **1991**, *35*, 154–158. [CrossRef]
25. Kloetzer, L.; Tucaliuc, A.; Galaction, A.-I.; Caşcaval, D. Fractionation of dicarboxylic acids produced by Rhizopus oryzae using reactive extraction. *Sci. Rep.* **2022**, *12*, 2020. [CrossRef] [PubMed]
26. Cordes, T.; Michelucci, A.; Hiller, K. Itaconic Acid: The surprising role of an industrial compound as a mammalian antimicrobial metabolite. *Annu. Rev. Nutr.* **2015**, *35*, 451–473. [CrossRef] [PubMed]
27. Yahiro, K.; Takahama, T.; Park, Y.S.; Okabe, M. Breeding of *Aspergillus terreus* mutant TN-484 for itaconic acid production with high yield. *J. Ferment. Bioeng.* **1995**, *79*, 506–508. [CrossRef]
28. Gao, Q.; Liu, J.; Liu, L. Relationship between morphology and itaconic acid production by *Aspergillus terreus*. *J. Microbiol. Biotechnol.* **2014**, *24*, 168–176. [CrossRef] [PubMed]
29. Komáromy, P.; Bakonyi, P.; Kucska, A.; Tóth, G.; Gubicza, L.; Bélafi-Bakó, K.; Nemestóthy, N. Optimized pH and its control strategy lead to enhanced itaconic acid fermentation by *Aspergillus terreus* on glucose substrate. *Fermentation* **2019**, *5*, 31. [CrossRef]
30. Kuenz, A.; Gallenmüller, Y.; Willke, T.; Vorlop, K.D. Microbial production of itaconic acid: Developing a stable platform for high product concentrations. *Appl. Microbiol. Biotechnol.* **2012**, *9*, 1209–1216. [CrossRef] [PubMed]
31. Rychtera, M.; Wase, D.J. The growth of *Aspergillus terreus* and the production of itaconic acid in batch and continuous cultures. The influence of pH. *J. Chem. Technol. Biotechnol.* **1981**, *31*, 509–521. [CrossRef]
32. Park, Y.S.; Ohta, N.; Okabe, M. Effect of dissolved oxygen concentration and impeller tip speed on itaconic acid production by *Aspergillus terreus*. *Biotechnol. Lett.* **1993**, *15*, 583–586. [CrossRef]
33. Molnár, Á.P.; Németh, Z.; Kolláth, I.S.; Fekete, E.; Flipphi, M.; Ág, N.; Soós, Á.; Kovács, B.; Sándor, E.; Kubicek, C.P.; et al. High oxygen tension increases itaconic acid accumulation, glucose consumption, and the expression and activity of alternative oxidase in *Aspergillus terreus*. *Appl. Microbiol. Biotechnol.* **2018**, *102*, 8799–8808. [CrossRef] [PubMed]
34. Nolte, L.; Brandenbusch, C. Monitoring and investigating reactive extraction of dicarboxylic acids using online FTIR—Part II: Reaction equilibria, reaction kinetics and competition within the complex formation between itaconic acid and several amine extractants. *J. Mol. Liq.* **2022**, *366*, 120223. [CrossRef]
35. Kyuchoukov, G.; Morales, A.F.; Albet, J.; Malmary, G.; Molinier, J. On the possibility of predicting the extraction of dicarboxylic acids with tributylphosphate dissolved in a diluent. *J. Chem. Eng. Data* **2008**, *53*, 639–647. [CrossRef]

36. Kreyenschulte, D.; Heyman, B.; Eggert, A.; Maßmann, T.; Kalvelage, C.; Kossack, R.; Regestein, L.; Jupke, A.; Büchs, J. In situ reactive extraction of itaconic acid during fermentation of *Aspergillus terreus*. *Biochem. Eng. J.* **2018**, *135*, 133–141. [CrossRef]
37. Krull, S.; Hevekerl, A.; Kuenz, A.; Prüße, U. Process development of itaconic acid production by a natural wild type strain of *Aspergillus terreus* to reach industrially relevant final titers. *Appl. Microbiol. Biotechnol.* **2017**, *101*, 4063–4072. [CrossRef]

processes

MDPI

Article

Comparative Analysis of Enzyme-, Ultrasound-, Mechanical-, and Chemical-Assisted Extraction of Biflavonoids from Ginkgo Leaves

Anita Šalić [1], Lina Šepić [1], Iva Turkalj [1], Bruno Zelić [1,2] and Dunja Šamec [2,*]

1 Faculty of Chemical Engineering and Technology, University of Zagreb, Marulićev trg 19, HR-10000 Zagreb, Croatia; asalic@fkit.unizg.hr (A.Š.); lsepic@fkit.unizg.hr (L.Š.); iturkalj@fkit.unizg.hr (I.T.); bzelic@fkit.unizg.hr (B.Z.)

2 University North, Trg dr. Žarka. Dolinara 1, HR-48000 Koprivnica, Croatia

* Correspondence: dsamec@unin.hr

Abstract: The biflavonoid extraction from ginkgo (*Ginkgo biloba* L.) leaves using solvent-based extraction with 70% ethanol, alone and in combination with enzyme-assisted, ultrasound-assisted, mechanical-assisted, and chemically assisted methods was investigated and the influence of extraction duration was explored. The total content of polyphenols, flavonoids, and phenolic acids in the extracts was determined spectrophotometrically, while individual biflavonoids were identified and quantified using HPLC-DAD. Amentoflavone, bilobetin, ginkgetin, isoginkgetin, and sciadopitysin were identified in all our extracts. Among these, sciadopitysin emerged as the most prevalent biflavonoid with an amount above 1 mg g^{-1} dw, followed by isoginkgetin. Comparative analysis of the extraction methods revealed that, except for chemically assisted extraction, similar levels of compounds were obtained after 45 min of extraction. However, enzymatic (EAE) and mechanical-assisted extraction (MAE) exhibited significantly higher individual (EAE: 19–41% higher; MAE: 22–67% higher) and total biflavonoid (EAE: 29% higher; MAE 50% higher) levels after just 5 min, suggesting their potential to abbreviate extraction duration and facilitate the efficient retrieval of target compounds. However, as extraction efficiency varies between individual biflavonoids, our findings also underscore the importance of considering specific compounds and extraction kinetics in the optimization of ginkgo leaf extraction processes.

Keywords: amentoflavone; assisted extraction; biflavonoids; bilobetin; ginkgo; ginkgetin; isoginkgetin; sciadopitysin

Citation: Šalić, A.; Šepić, L.; Turkalj, I.; Zelić, B.; Šamec, D. Comparative Analysis of Enzyme-, Ultrasound-, Mechanical-, and Chemical-Assisted Extraction of Biflavonoids from Ginkgo Leaves. *Processes* **2024**, *12*, 982. https://doi.org/10.3390/pr12050982

Academic Editors: Yanlin Zhang and Prashant K. Sarswat

Received: 19 April 2024
Revised: 10 May 2024
Accepted: 10 May 2024
Published: 12 May 2024

1. Introduction

The ginkgo (*Ginkgo biloba* L.), often referred to as the "maidenhair tree", is a living fossil that has withstood the test of time for over 270 million years [1]. Revered for its unique fan-shaped leaves and remarkable adaptability, this ancient tree has not only survived but has also become a fascinating topic in modern research and wellness practice. Ginkgo extract, derived from the leaves of the ancient ginkgo tree, has a rich history steeped in traditional medicine and cultural significance [2]. This herbal remedy has endured through the centuries, evolving from its roots in traditional Eastern medicine to a globally recognized dietary supplement with a variety of potential health benefits [1,3]. The standardized extract preparation of ginkgo leaves contains 24% flavonoids, 6% terpene trilactones (TTLs), and less than 5 ppm ginkgolic acid [4]. To date, 110 different flavonoid structures have been identified in a ginkgo, including flavonol and its glycosides, flavone and its glycosides, flavanone and its glycosides, isoflavone and its glycosides, flavan-3-ols, biflavonoids and biginkgosides [4]. Biflavonoids are dimeric structures of flavonoids that have recently gained attention as antimicrobial, especially antiviral agents, and as molecules with the potential for the prevention and treatment of cancer and neurological diseases [3,5]. The first characterized biflavonoid, ginkgetin, was isolated from ginkgo leaves and so far 13 different biflavonoids have been identified in a ginkgo, with amentoflavone, bilobetin, ginkgetin, isoginkgetin, and sciadopitysin being the most abundant [3,6] (Figure 1).

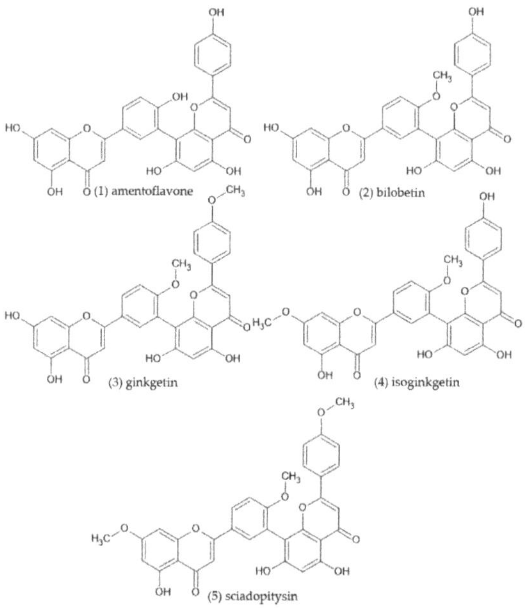

Figure 1. Chemical structure of five biflavonoids from ginkgo leaves (1) amentoflavone, (2) bilobetin, (3) ginkgetin, (4) isoginkgetin, and (5) sciadopitysin.

Flavonoids, a diverse group of polyphenolic compounds consisting of 6000 different structures, are known for their strong antioxidant properties and numerous health benefits [7]. However, the biological activity of flavonoids can vary considerably due to their structure [7,8]. For example, biflavonoids have significantly lower antioxidant activity than monomeric subunits, but higher neuroprotective potential [9]. The extraction of flavonoids from plant sources requires careful optimization to increase both extraction efficiency and yield without overtaxing resources [10]. Optimizing extraction processes for flavonoids from plant sources is paramount for ensuring maximum yield, minimizing production costs, and facilitating reliable results in research while maintaining product quality [11,12]. By carefully controlling extraction parameters, such as temperature and solvent choice, the degradation of bioactive compounds is minimized, thus preserving their health benefits [12]. Furthermore, optimized processes contribute to reducing environmental impact by minimizing solvent usage and waste generation [13]. Selective extraction techniques enable the isolation of specific compounds tailored to desired applications, thereby unlocking the full potential of plant-derived compounds across various fields [12]. The extraction technique used for flavonoids extraction may be divided into two groups: Conventional techniques (maceration, reflux, and Soxhlet extraction) and modern techniques (extraction by ultrasound, microwaves, pressurized liquid, supercritical fluid, enzyme assistance, matrix solid-phase dispersion, etc.) [12]. Ultrasound-assisted extraction (UAE) uses high-frequency sound waves to break down the cell walls and accelerate the release of bioactive compounds from the plant material. In this method, the plant material is immersed in a solvent and simultaneously sonicated with ultrasonic waves, creating cavitation that breaks down the cell walls and facilitates the extraction of the compounds. UAE is known for its ability to speed up processes, shorten extraction time, and increase yields. It is often used to extract polyphenols, flavonoids, and essential oils [14–16]. In enzyme-assisted extraction (EAE), specific enzymes are used to break down the cell walls and release the target compounds. It is selective, environmentally friendly, improves efficiency, and is often used to extract polyphenols, polysaccharides, and proteins [17,18]. Mechanically assisted extraction (MAE) uses mechanical force such as crushing or pressing to break up the plant tissue and improve the release of the substances. It is simple, cost-effective, and suitable for large-scale processes and is generally used for the extraction of oils and lipophilic compounds [19,20]. In chemically assisted extraction (CAE), solvents or reagents are used to dissolve and extract bioactive compounds. This versatile method can be tailored to specific compounds and is efficient for the extraction of polyphenols, alkaloids, and terpenoids from plant material on an industrial scale. Each of the mentioned technique offer various advantages, which are selected based on the target compounds, the properties of the plant material, and the desired extraction efficiency.

Over the years, numerous research endeavors have been dedicated to refining the extraction methods for flavonoids from ginkgo leaves, with a predominant focus on total flavonoids or monomeric flavonoids [21–27]. Information regarding the targeted extraction of biflavonoids from ginkgo leaves remains scarce. As per our current knowledge, only one study [28] has addressed this issue. Lei et al. [28] optimized the extraction process for four biflavonoids—bilobetin, ginkgetin, isoginkgetin, and sciadopitysin—from ginkgo using ultrasonic-assisted ionic liquid extraction (UAILE). However, it is noteworthy that the authors did not specify the specific part of the ginkgo plant from which the biflavonoids were extracted in their paper. Considering the significant pharmacological potential associated with biflavonoids and the accessibility of ginkgo leaves as a cost-effective source of these compounds, our study aims to assess various extraction techniques, including solvent-based extraction, enzyme-assisted extraction, ultrasound-assisted extraction, mechanical-assisted extraction, and chemically assisted extraction, for isolating biflavonoids from ginkgo leaves. We meticulously examined parameters such as solvent composition and extraction duration to evaluate their impact on the content of total phenolics, total flavonoids, total phenolic acids determined spectrophotometrically, as well as the five most abundant flavonoids from ginkgo leaves—amentoflavone, bilobetin, ginkgetin, isoginkgetin, and sciadopitysin—along with the total biflavonoid content using HPLC-DAD analysis.

2. Materials and Methods

2.1. Plant Material and Extraction

Ginkgo leaves were collected from a ginkgo tree in an urban area in Croatia, air-dried, and stored in a drying cabinet until analysis.

Solid–liquid extraction (SLE): Extracts were prepared by mixing 60 mg of dried and ground ginkgo leaves with 2 mL of 70% ethanol. The samples were briefly mixed and placed in an ultrasonic bath for 10 min, followed by 45 min in the rotator. This is a commonly used method in our laboratory for the preparation of various plant extracts. At the end, the samples were centrifuged and the supernatant was used for further analysis.

Enzyme-assisted extraction (EAE): 60 mg amount of dried and ground leaves were mixed with 0.6 mL of 0.1 M acetate buffer pH 4 with 6% (v/v) enzyme (*Viscozyme L* (Novozymes, Bagsvaerd, Denmark) a blend of beta-glucanases, pectinases, hemicellulases and xylanases with declared activity ≥ FBGU/g. Each FBG is the amount of enzyme required to hydrolyze β-glucans under the standard condition (30 °C, pH 5.0). The content was briefly homogenized with a vortex and the enzymatic degradation of the lignocellulose was carried out for 4 and 24 h at 50 °C and 200 rpm. After enzymatic treatment, 1.4 mL of 100% ethanol was added to the sample to obtain 70% ethanol in the sample. The procedure that followed was the same as for solid–liquid extraction.

Ultrasound-assisted extraction (UAE): 60 mg amounts of dried and ground leaves were mixed with 0.6 mL of distilled water. The content was briefly homogenized and treated with ultrasound (Sonoplus HD 3100, Berlin, Germany, horn type SH 70 G, 20 kHz, Bandelin GmbH, 62% amplitude) for 10 min with constant cooling. After the ultrasound treatment,

1.4 mL of 100% ethanol was added to the sample to obtain 70% ethanol in the sample. The procedure that followed was the same as for solid–liquid extraction.

Mechanically assisted extraction (MAE): 60 mg of dried and ground leaves were mixed with 0.6 mL of distilled water followed by adding 120 mg of 4 mm glass beads to the sample. The content was shaken at maximum speed of 600 rpm for 20 min. After the mechanical treatment, 1.4 mL of 100% ethanol was added to the sample to obtain 70% ethanol in the sample. The procedure that followed was the same as for solid–liquid extraction.

Chemically assisted extraction (CAE): 60 mg of dried and ground leaves were mixed with 0.5 mL of 0.1% TritonX solution and 10% NaClO solution. The samples were stirred for 48 h at 25 °C at 200 rpm. After chemical treatment, 1.4 mL of 100% ethanol was added to the sample to obtain 70% ethanol in the sample. The procedure that followed was the same as for solid–liquid extraction.

Schematically, extraction process is shown in Figure 2.

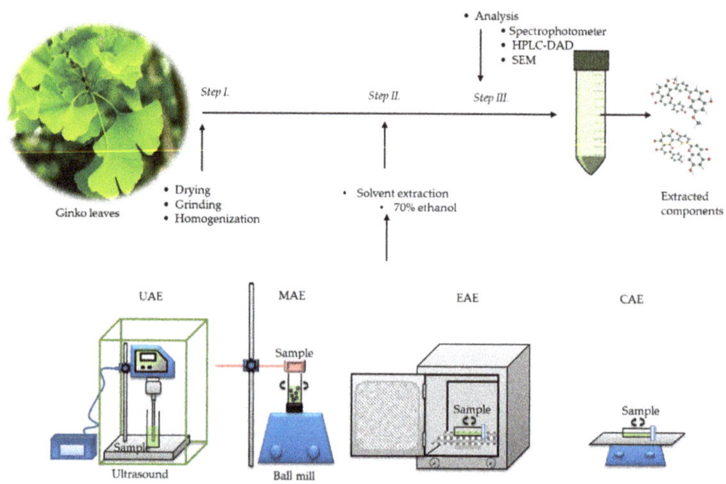

Figure 2. Schematic representation of extraction process.

2.2. Determination of Total Polyphenols, Flavonoids and Phenolic Acids Content

Determination of total phenolic compounds was carried out through a colorimetric reaction involving Folin–Ciocalteu reagent [29]. Absorbance readings at 765 nm were taken after a 2 h incubation period. Gallic acid was employed for the calibration curve, and the outcomes are presented as gallic acid equivalents per dry weight (μg GAE mg^{-1} dw).

For the determination of total flavonoids, the methodology outlined by Zhinshen et al. [30], adapted to a reduced scale, was employed. The absorbance of the resulting mixture was measured at 510 nm. Catechin was utilized for the calibration curve, and the findings are expressed as catechin equivalents per dry weight (μg CE mg^{-1} dw).

Total phenolic acids were determined using Arnow's reagent [31]. Absorbance readings were taken at 505 nm. Caffeic acid served as the standard for constructing the calibration curve, and the results were expressed as caffeic acid equivalents per dry weight (μg CAE mg^{-1} dw).

2.3. Determination of Individual Biflavonoids Using HPLC-DAD

The analysis of individual biflavonoids, amentoflavone, bilobetin, ginkgetin, isoginkgetin, and sciadopitysin, was performed utilizing an Agilent 1260 Infinity II High-Performance Liquid Chromatography (HPLC) system (Agilent, Santa Clara, CA, USA) equipped with a quaternary pump, autosampler, column compartment, and a diode array detector (DAD). The acquisition and subsequent data processing were carried out employing the Agilent OpenLAB CDS software (version 2.6, Agilent, Santa Clara, CA, USA). Chromatographic separation was achieved using a Zorbax 300Extend-C18 column (Agilent, Santa Clara, CA, USA). Analysis parameters were similar to those published elsewhere [6]. Identification and quantification were accomplished employing commercially available biflavonoid standards, as previously detailed in our report where we also reported curve equation, R^2, LOD, and LOQ [6]. Results were expressed per dry weight (dw).

2.4. Scanning Electron Microscopy (SEM)

Scanning electron microscopy (SEM) was used to examine the morphology of the samples using a Tescan Vega 3 SEM (Tescan Brno S.R.O., Brno, Czech Republic) apparatus. The samples were coated for 60 s in an argon plasma with a gold/platinum conductive coating prior to scanning. A 10 kV voltage and a secondary electron (SE) detector were utilized in the study.

2.5. Statistical Analysis

All extractions and measurements were conducted in at least in triplicate, and the results are expressed as mean ± standard deviation (SD). All statistical analyses were performed using XLSTAT Excel data add-on, version 2024.1. One-way ANOVA and subsequent multiple mean comparisons with Tukey's HSD test were executed, and distinctions between measurements were considered significant at $p < 0.05$.

3. Results

3.1. Influence of Pretreatment on Total Polyphenols, Flavonoids, Phenolic Acids and Individual and Total Biflavonoids

In the initial phase of this research, the impact of various pretreatment processes (EAE, MAE, CAE, and UAE) on the levels of total polyphenols, phenolic acids, and flavonoids was analyzed. The findings from this investigation are shown in Figure 3.

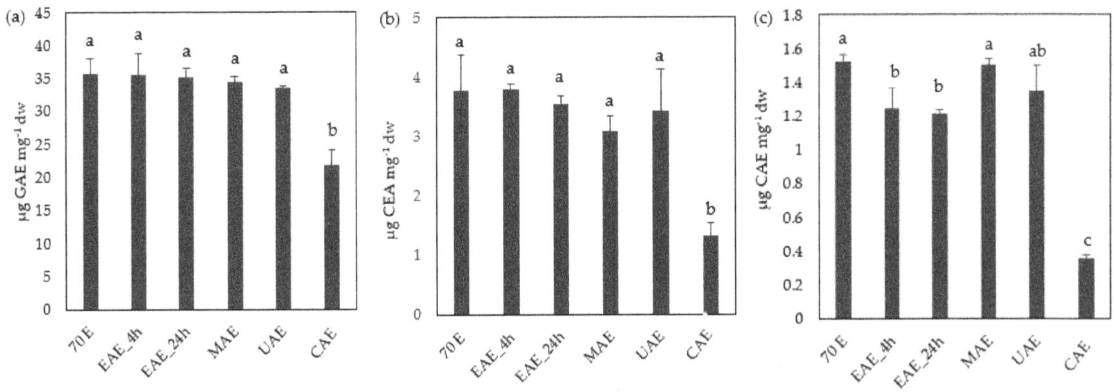

Figure 3. Influence of different pretreatment methods, enzyme-assisted for 4 h (EAE_4h), enzyme-assisted for 24 h (EAE_24h), mechanical-assisted (MAE), chemically assisted (CAE), and ultrasound-assisted (UAE), and time of extraction on the content of total polyphenols (**a**), total flavonoids (**b**) and total phenolic acids (**c**) in comparison with commonly used solid–liquid extraction using 70% ethanol (70 E). Values marked with different letters are significantly different at $p < 0.05$.

Overall, the pretreatment methods UAE, MAE, and EAE combined with 70% extraction exhibited comparable outcomes in terms of total polyphenols and total flavonoids when compared to extraction with 70% ethanol alone. However, regarding total phenolic acids, extraction with 70% ethanol alone and MAE yielded the highest content, while CAE and EAE proved to be less effective. Moreover, a comparison of the total polyphenols, total flavonoids, and total phenolic acids obtained after 4 h and 24 h enzymatic treatments of the sample showcased similar extraction efficiencies. Conversely, CAE extraction resulted in significantly lower yields for total polyphenols, flavonoids, and total phenolic acids compared to other methods.

A similar trend across both individual (Table 1) and total biflavonoids (Figure 3) was observed, where UAE, MAE, and EAE yielded comparable results, while CAE resulted in lower amounts. The only exception was amentoflavone, which exhibited the highest concentration, significantly surpassing extraction solely in 70% ethanol, MAE, and CAE in an extract after EAE extraction over 24 h. Notably, in the analyzed sample, amentoflavone appeared in the lowest concentration among the biflavonoids, whereas sciadopitysin followed by isoginkgetin was the most abundant (Table 1).

Table 1. Influence of different pretreatment methods, enzyme-assisted for 4 h (EAE_4h), enzyme-assisted for 24 h (EAE_24h), mechanical-assisted (MAE), chemically assisted (CAE), and ultrasound-assisted (UAE), and time of extraction on the content of individual biflavonoids in comparison with commonly used solid–liquid extraction using 70% ethanol (70 E). Values marked with different letters are significantly different at $p < 0.05$.

	70 E	EAE_4 h	EAE_24 h	MAE	UAE	CAE
Amentoflavone $\mu g\ g^{-1}$ dw	64.36 ± 4.17 [b]	66.81 ± 2.77 [ab]	69.95 ± 2.77 [a]	64.08 ± 0.38 [b]	65.95 ± 1.84 [ab]	44.72 ± 7.03 [c]
Bilobetin $\mu g\ g^{-1}$ dw	163.89 ± 14.50 [a]	166.01 ± 3.14 [a]	172.22 ± 1.58 [a]	167.21 ± 0.97 [a]	177.36 ± 12.12 [a]	108.52 ± 23.73 [b]
Ginkgetin $\mu g\ g^{-1}$ dw	607.68 ± 50.29 [a]	627.45 ± 10.33 [a]	646.03 ± 7.37 [a]	622.57 ± 3.25 [a]	634.54 ± 9.23 [a]	466.39 ± 55.55 [a]
Isoginkgetin $\mu g\ g^{-1}$ dw	945.83 ± 90.12 [a]	974.23 ± 18.60 [a]	1007.06 ± 13.21 [a]	969.24 ± 4.69 [a]	994.31 ± 15.45 [a]	630.82 ± 123.53 [b]
Sciadopitysin $\mu g\ g^{-1}$ dw	1387.56 ± 105.51 [a]	1430.17 ± 21.34 [a]	1461.27 ± 105.51 [a]	1419.03 ± 6.87 [a]	1450.39 ± 18.57 [a]	1054.89 ± 99.93 [b]

The individual biflavonoid contents mirror those of total biflavonoids (Figure 4), which ranged between 3169 and 3322 $\mu g\ mg^{-1}$ dw in extracts prepared by all tested methods except CEA, which yielded significantly lower amounts of total biflavonoids (2305.33 ± 308.44 $\mu g\ g^{-1}$ dw). Consequently, based on the obtained results, the CAE + 70% ethanol and EAE + 70% ethanol (24 h) methods were excluded from further experiments.

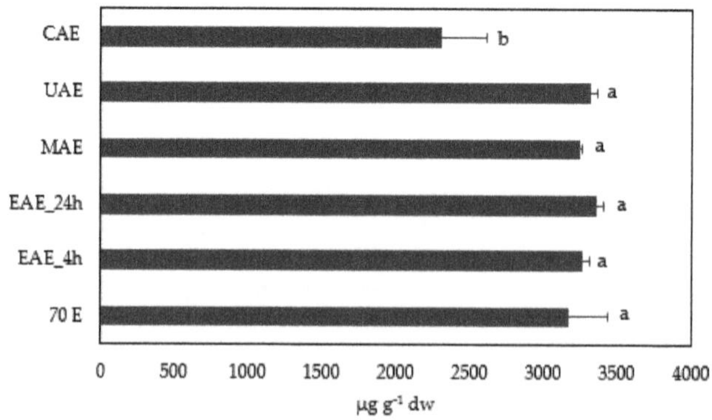

Figure 4. Influence of different pretreatment methods, enzyme-assisted for 4 h (EAE_4h), enzyme-assisted for 24 h (EAE_24h), mechanical-assisted (MAE), chemically assisted (CAE), and ultrasound-assisted (UAE), and time of extraction on the content of total biflavonoids in comparison with commonly used solid–liquid extraction using 70% ethanol (70 E). Values marked with different letters are significantly different at $p < 0.05$.

3.2. Influence of Extraction Time on Total Polyphenols, Flavonoids, Phenolic Acids and Individual and Total Biflavonoids

In the second phase of the research, the impact of varying durations of the extraction process with 70% ethanol on the levels of total polyphenols, phenolic acids, flavonoids, individual biflavonoids (amentoflavone, bilobetin, ginkgetin, isoginkgetin, and sciadopitysin), and total bioflavonoids was explored. The results for total polyphenols, total flavonoids, and total phenolic acids can be observed in Figure 5.

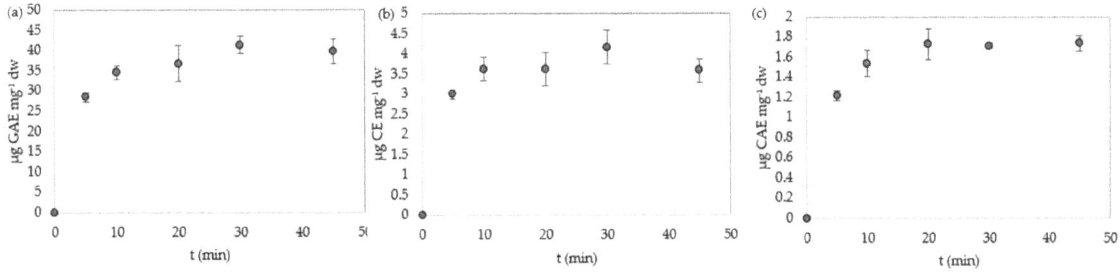

Figure 5. Dynamic change in total polyphenols (**a**), total flavonoids (**b**), and total phenolic acids (**c**) in extract of ginkgo leaves prepared with 70% ethanol.

As illustrated in Figure 5, within the first 10 min, the majority of total polyphenols, flavonoids, and phenolic acids are extracted in 70% ethanol, reaching their peak content within 20 min under the tested material–solvent ratio. Additionally, the extraction dynamics of biflavonoids, with results for individual (a–e) and total biflavonoids (f) were examined and presented in Figure 6.

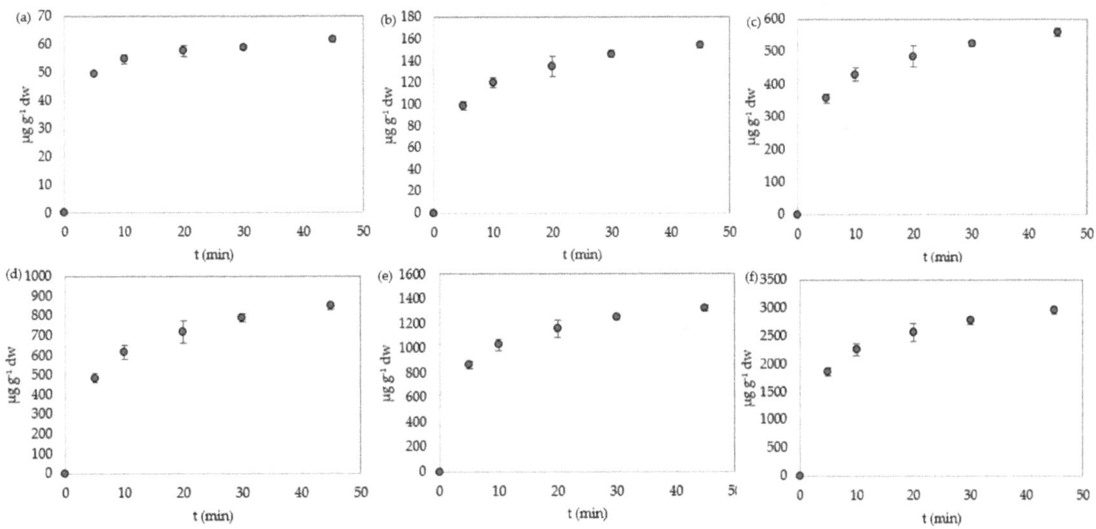

Figure 6. Dynamic change in amentoflavone (**a**), bilobetin (**b**), ginkgetin (**c**), isoginkgetin (**d**), sciadopitisyn (**e**), and total biflavonoids (**f**) in an extract of ginkgo leaves prepared with 70% ethanol.

The obtained results revealed a positive correlation between the duration of the extraction process and the efficiency of extracting both individual and total biflavonoids content. This was evidenced by the consistent increase in the content of all tested biflavonoids within the extract. While intensive extraction occurred predominantly within the first 10–20 min, performed experiments demonstrated a time-dependent increase in biflavonoids throughout the 45 min duration, suggesting that prolonged extraction may enhance biflavonoids content.

Furthermore, the extraction efficiency of total polyphenols, total flavonoids, and total phenolic acids, along with individual and total biflavonoids, using assisted extraction methods such as UAE, MAE, and EAE was explored. Morphological SEM analyses of ginkgo leaf powder before and after extraction were also conducted (Figure 7). The results unveiled distinct morphological alterations on the surface of the ginkgo leaf powder induced by various treatments. Across all treated samples, microcracks and cellular-level changes were observed, resulting in a visibly different surface structure of the sample.

A comparative analysis of assisted extraction efficiency, employing UAE, MAE, and EAE, against extraction with 70% ethanol alone at both 5 min and 45 min durations was conducted (Figure 8, Table 2). Consistent with earlier observations, after 45 min of extraction using all three tested assisted methods, results comparable to those obtained using 70% ethanol alone were observed. Additionally, at the 5 min mark, comparable results for total polyphenols, flavonoids, and phenolic acids were observed.

Table 2. Comparison of the content of individual biflavonoids in extracts for different durations (5 min and 45 min) of the solid–liquid extraction using 70% ethanol (70 E), ultrasound-assisted (UAE), mechanical-assisted (MAE), and enzyme-assisted (EAE) extraction processes. Values marked with different letters are significantly different at $p < 0.05$.

		Amentoflavone $\mu g\ g^{-1}$ dw	Bilobetin $\mu g\ g^{-1}$ dw	Ginkgetin $\mu g\ g^{-1}$ dw	Isoginkgetin $\mu g\ g^{-1}$ dw	Sciadopitysin $\mu g\ g^{-1}$ dw
70 E	5 min	49.69 ± 0.58^{d}	99.21 ± 3.70^{e}	357.37 ± 13.49^{e}	486.72 ± 24.81^{d}	866.35 ± 26.68^{d}
	45 min	64.36 ± 4.17^{ab}	163.90 ± 14.5^{ab}	607.68 ± 50.30^{ab}	945.83 ± 90.12^{a}	1387.56 ± 105.51^{a}
UAE	5 min	56.82 ± 1.59^{c}	124.89 ± 5.55^{de}	439.37 ± 21.24^{de}	632.18 ± 37.92^{cd}	999.26 ± 72.14^{cd}
	45 min	65.95 ± 1.84^{a}	177.36 ± 12.12^{a}	634.54 ± 9.23^{a}	994.31 ± 15.45^{a}	1450.34 ± 18.57^{a}
MAE	5 min	59.07 ± 0.70^{bc}	133.89 ± 1.09^{cd}	467.50 ± 7.23^{cd}	684.37 ± 11.44^{bc}	1061.32 ± 19.88^{c}
	45 min	64.08 ± 0.38^{ab}	167.21 ± 0.98^{ab}	622.57 ± 3.25^{a}	969.24 ± 4.69^{a}	1419.03 ± 6.87^{a}
EAE	5 min	60.39 ± 1.94^{bc}	150.83 ± 11.24^{bc}	538.48 ± 40.63^{bc}	812.89 ± 74.21^{b}	1223.56 ± 30.96^{b}
	45 min	66.81 ± 2.77^{a}	166.01 ± 3.14^{ab}	627.45 ± 10.33^{a}	975.23 ± 18.60^{a}	1430.17 ± 21.30^{a}

However, assisted extraction (UAE, MAE, and EAE) exhibited higher amounts of amentoflavone, bilobetin, ginkgetin, isoginkgetin, sciadopitysin, and total biflavonoids compared to extraction with 70% ethanol alone at the 5 min mark (Table 2, Figure 9).

At 5 min, both EAE and MAE resulted in significantly higher amounts of total biflavonoids compared to extraction with 70% ethanol alone (Figure 9). This suggests that the tested assisted extraction methods may effectively reduce the extraction time for biflavonoids.

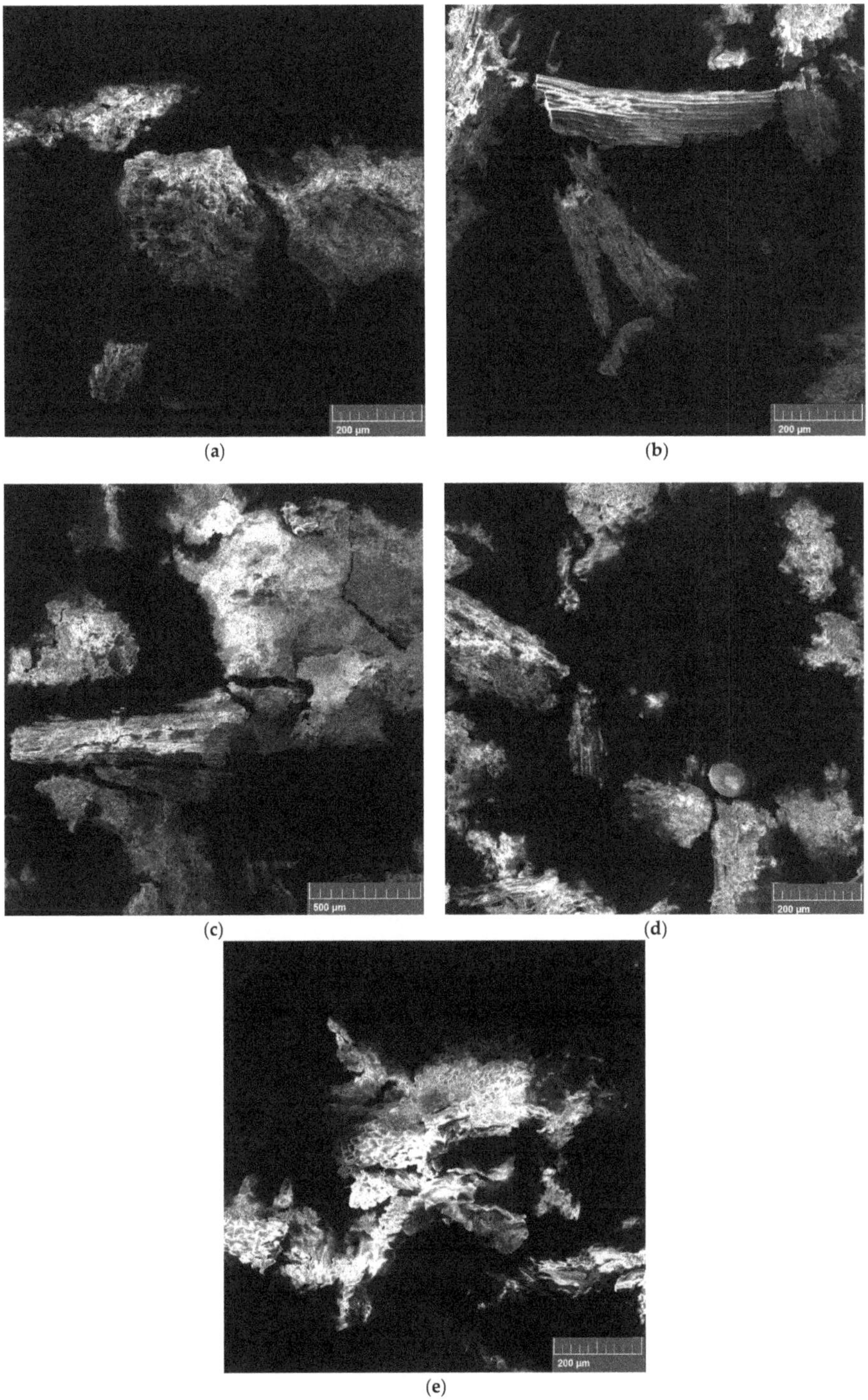

Figure 7. Morphological analysis of ginkgo sample (**a**) before extraction and after extraction with (**b**) 70% ethanol, (**c**) ultrasound-assisted (UAE) + 70% ethanol, (**d**) mechanical-assisted (MEA) + 70% ethanol and (**e**) enzyme-assisted (EAE) + 70% ethanol.

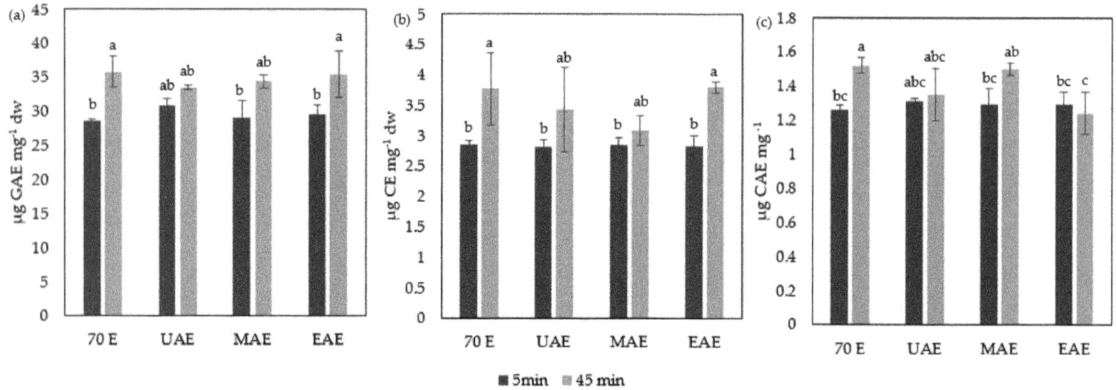

Figure 8. Comparison of the content of total polyphenols (**a**), flavonoids (**b**), and phenolic acids (**c**) in solid–liquid extraction using 70% ethanol (70 E), ultrasound-assisted (UAE), mechanical-assisted (MAE), and enzyme-assisted (EAE) extracts for different durations (5 min and 45 min) of the extraction process. Values marked with different letters are significantly different at $p < 0.05$.

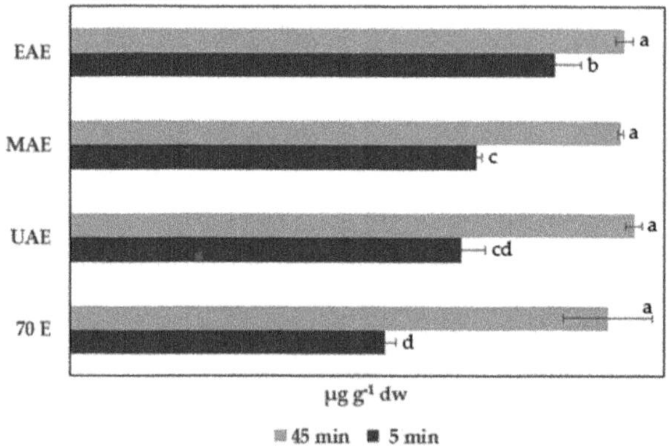

Figure 9. Comparison of the content of total biflavonoids in solid–liquid extraction using 70% ethanol (70 E), ultrasound-assisted (UAE), mechanical-assisted (MAE), and enzyme-assisted (EAE) extracts for different durations (5 min and 45 min) of the extraction process. Values marked with different letters are significantly different at $p < 0.05$.

4. Discussion

In recent years, traditional extraction methods have undergone a significant transformation, giving way to modern techniques meticulously designed to reduce reliance on organic solvents and streamline extraction processes [32]. These innovations are underpinned by a dual commitment to enhancing efficiency and promoting environmental sustainability. In this study, SLE, EAE, UAE, MEA, and CAE were compared. Different treatments induced varied morphological changes on the surface of ginkgo leaf powder (Figure 7), aligning with the existing literature findings [33]. The analysis of SEM photos (Figure 7) of ginkgo leaf powder before and after the application of various extraction methods revealed significant changes in morphological structure. While the structure of the powder before treatment was smooth with adjacent holes of the cell wall arranged densely, the surface of ginkgo leaf powder was damaged severely after the treatments. After the mechanical and ultrasound treatment many cracks appeared on the samples due to cavitation and mechanical friction. When it comes to samples treated with enzymes, lots of new holes appeared and adjacent holes were loosely arranged like in the results reported by Zhang et al. [34]. In the sample treated with 70% ethanol, no significant morphological change was noticed.

Among all evaluated assisted methods, except CAE, comparable amounts of total polyphenols and total flavonoids were obtained, significantly higher than those from CAE (Figure 3). Furthermore, extracts prepared with CAE exhibited significantly lower total phenolic acids content compared to those from SLE with 70% ethanol and UAE and MAE, indicating the superior efficiency of the latter methods. Thus far, it has been reported that mechanochemical-assisted extraction (MCAE), employing 21% solid reagent ($NaHCO_3$) followed by milling, enhances the content of both total flavonoids and monomeric ones in ginkgo extracts [26]. Conversely, in contrast to our findings, CAE yielded significantly lower amounts and was not investigated further regarding extraction time. For EAE in our study, a blend of beta-glucanases, pectinases, hemicellulases, and xylanases was employed, which did not yield higher amounts of total polyphenols, flavonoids, and phenolic acids than in an extraction using just 70% ethanol for 45 min. This contrasts with the findings by Chen et al. [24], who reported increased total flavonoid content in ginkgo leaves pretreated with cellulose and pectinase compared to the control without enzymes. Han et al. [23] also highlighted the significant role of enzymes like polygalacturonase and β-glucosidase in flavonoid extraction. UAE is often recommended in the literature for extracting total flavonoids from ginkgo leaves [35,36], and presented results show comparable efficacy after 45 min as with extraction using 70% ethanol alone. However, as evident from Figure 8, UAE yielded statistically similar results after 5 min as after 45 min, suggesting accelerated extraction of total polyphenols, flavonoids, and total phenolic acids. The literature data showed the beneficial effects of UAE also for extracting other compounds from ginkgo leaves such as polysaccharides [37,38] and ginkgolides [39]. Also, UAE has shown promising results for extraction compounds such as terpene lactones from ginkgo seeds [40]. Similar to UAE, in our experiments MAE demonstrated comparable results for total polyphenols, flavonoids, and phenolic acids after 5 min and 45 min, indicating accelerated extraction with this method as well. In the case of extraction with 70% ethanol alone, the majority of total polyphenols, flavonoids, and phenolic acids were extracted within the first 20 min, after which the amount plateaued (Figure 5). The literature varies regarding extraction times for these compounds from ginkgo without assisted extraction, with Cui et al. [41] performing extractions for 4 h by shaking ginkgo leaves material on a laboratory rotary shaker with different solvents, reporting the highest amounts of total polyphenols and flavonoids in ethanol and methanol extracts. Kobus et al. [42] used maceration for 16 h hours for total flavonoid extraction, but results obtained in our study suggest that such prolonged extraction may not be necessary.

While measurements of total polyphenols, flavonoids, and phenolic acid content are commonly used for optimizing flavonoid extraction from ginkgo leaves [22,26,35,36,41,42], these methods are known to lack specificity [43]. Considering the diverse chemical structures of flavonoids, it is imperative to evaluate the extraction efficiency of each individual compound. In this study, we focused on a less studied group of flavonoids—biflavonoids, specifically the five most abundant biflavonoids in ginkgo leaves [6]. In all the extracts analyzed, amentoflavone, bilobetin, ginkgetin, isoginkgetin, and sciadopitysin were identified. Sciadopitysin was the most abundant, followed by isoginkgetin, which is consistent with the literature data [6,21,28,44]. Similar to total polyphenols and flavonoids, UAE, MAE, and EAE yielded comparable results, while CEA resulted in lower amounts of individual (Table 1) and total biflavonoids (Figure 4) after 45 min of extraction. The exception was amentoflavone, which showed significantly higher amounts in extracts after EAE for 24 h compared to extraction in 70% ethanol alone, MAE, and CEA. This suggests that the localization of amentoflavone within plant tissue/cells benefits from applied enzyme treatments, enhancing its extraction. According to the literature data and MALDI imaging, amentoflavone is localized in the outer part of leaves/tissues [44,45]. Time significantly influences the extraction of all biflavonoids when using only 70% ethanol without pretreatments; a constant increase in individual as well as total biflavonoid content over a 45 min period, with the highest increase occurring in the first 10 min was observed in this study (Figure 6). With assisted extractions (UAE, MAE, and EAE), longer extraction durations also resulted in higher amounts of all individual and total biflavonoids (Table 2, Figure 9). However, similar to the results for total polyphenols, flavonoids, and phenolic acids, higher amounts of all biflavonoids in extracts prepared with pretreatments compared to those without at the 5 min mark were observed. Significantly higher amounts were observed, especially for MAE and EAE. According to the results of this study, these techniques reduce extraction time, enabling the extraction of the highest amounts of biflavonoids in a shorter period using these methods. This aligns with the existing literature, where these techniques are acknowledged as green extraction methods due to their ability to shorten extraction duration and thereby decrease processing costs [20,46]. According to our results, UAE demonstrated higher total and individual biflavonoids content after 5 min of extraction compared to extraction with 70% ethanol alone. However, this increase was statistically significant only for amentoflavone. Lei et al. [28] reported that ultrasonic-assisted ionic liquid extraction may be utilized for extracting four biflavonoids from ginkgo: bilobetin, isoginkgetin, ginkgetin, and sciadopitysin. Similar to our findings, they noted sciadopitysin as the most abundant biflavonoids, followed by isoginkgetin and ginkgetin, with bilobetin being the least abundant. Our findings concerning individual biflavonoids underscore slight variability in extraction dynamics and kinetics due to their diverse chemical structures (Table 2) which is consistent with the finding of Lei et al. [28]. Consequently, for extraction optimization, it is preferable to employ methods capable of identifying individual compounds rather than relying on nonspecific techniques that measure the total content of compounds within a group, as is commonly seen in the literature.

5. Conclusions

In recent years, advancements in extraction techniques have aimed to enhance efficiency and environmental sustainability by minimizing reliance on organic solvents. In this study, various assisted extraction methods, including SLE, EAE, UAE, MAE, and CAE were compared focusing on ginkgo leaf powder. Distinct morphological changes induced by different treatments, consistent with previous research, were observed. Except for CAE extraction, comparable levels of total polyphenols and flavonoids were obtained, surpassing those from CAE. UAE showed comparable efficiency to SLE after 45 min, while EAE and MAE demonstrated higher efficiency after 5 min of extraction. Time significantly influenced extraction efficiency, particularly for biflavonoids, with the highest increase observed within the first 10 min. As per our results, assisted extraction methods such as EAE and MAE may reduce extraction time, facilitating efficient extraction of target compounds. This marks the initial report on biflavonoids, and we advocate for further studies delving into the kinetics of each assisted extraction method and its impact on specific biflavonoids. This study also underscores the importance of considering specific compounds and extraction dynamics when optimizing extraction processes for ginkgo leaves.

Author Contributions: Conceptualization, A.Š. and D.Š.; methodology, A.Š.; formal analysis, L.Š. and I.T.; investigation, L.Š. and I.T.; writing—original draft preparation, D.Š.; writing—review and editing, D.Š., A.Š. and B.Z.; visualization, D.Š. and A.Š.; supervision, A.Š. and B.Z.; project administration, D.Š.; funding acquisition, D.Š. All authors have read and agreed to the published version of the manuscript.

Funding: This research was funded by the Croatian Science Foundation project "Biflavonoids role in plants: *Ginkgo biloba* L. as a model system" under Project No. UIP-2019-04-1018.

Data Availability Statement: The raw data supporting the conclusions of this article will be made available by the authors on request.

Conflicts of Interest: The authors declare no conflicts of interest.

References

1. Akaberi, M.; Baharara, H.; Amiri, M.S.; Moghadam, A.T.; Sahebkar, A.; Emami, S.A. Ginkgo Biloba: An Updated Review on Pharmacological, Ethnobotanical, and Phytochemical Studies. *Pharmacol. Res.-Mod. Chin. Med.* **2023**, *9*, 100331. [CrossRef]
2. Lin, H.; Li, W.; Lin, C.; Wu, H.; Zhao, Y. International Biological Flora: Ginkgo Biloba. *J. Ecol.* **2022**, *110*, 951–982. [CrossRef]
3. Šamec, D.; Karalija, E.; Dahija, S.; Hassan, S.T.S. Biflavonoids: Important Contributions to the Health Benefits of Ginkgo (*Ginkgo biloba* L.). *Plants* **2022**, *11*, 1381. [CrossRef] [PubMed]
4. Liu, L.; Wang, Y.; Zhang, J.; Wang, S. Advances in the Chemical Constituents and Chemical Analysis of Ginkgo Biloba Leaf, Extract, and Phytopharmaceuticals. *J. Pharm. Biomed. Anal.* **2021**, *193*, 113704. [CrossRef] [PubMed]
5. Tatlı Çankaya, İ.İ.; Devkota, H.P.; Zengin, G.; Šamec, D. Neuroprotective Potential of Biflavone Ginkgetin: A Review. *Life* **2023**, *13*, 562. [CrossRef] [PubMed]
6. Kovač Tomas, M.; Jurčević, I.; Šamec, D. Tissue-Specific Profiling of Biflavonoids in Ginkgo (*Ginkgo biloba* L.). *Plants* **2022**, *12*, 147. [CrossRef]
7. Shen, N.; Wang, T.; Gan, Q.; Liu, S.; Wang, L.; Jin, B. Plant Flavonoids: Classification, Distribution, Biosynthesis, and Antioxidant Activity. *Food Chem.* **2022**, *383*, 132531. [CrossRef]
8. Kumar, S.; Pandey, A.K. Chemistry and Biological Activities of Flavonoids: An Overview. *Sci. World J.* **2013**, *2013*, 162750. [CrossRef] [PubMed]
9. Jurčević Šangut, I.; Šarkanj, B.; Karalija, E.; Šamec, D. A Comparative Analysis of Radical Scavenging, Antifungal and Enzyme Inhibition Activity of 3′-8″-Biflavones and Their Monomeric Subunits. *Antioxidants* **2023**, *12*, 1854. [CrossRef]
10. Jurinjak Tušek, A.; Šamec, D.; Šalić, A. Modern Techniques for Flavonoid Extraction—To Optimize or Not to Optimize? *Appl. Sci.* **2022**, *12*, 11865. [CrossRef]
11. Chávez-González, M.L.; Sepúlveda, L.; Verma, D.K.; Luna-García, H.A.; Rodríguez-Durán, L.V.; Ilina, A.; Aguilar, C.N. Conventional and Emerging Extraction Processes of Flavonoids. *Processes* **2020**, *8*, 434. [CrossRef]
12. Tzanova, M.; Atanasov, V.; Yaneva, Z.; Ivanova, D.; Dinev, T. Selectivity of Current Extraction Techniques for Flavonoids from Plant Materials. *Processes* **2020**, *8*, 1222. [CrossRef]
13. Kumar, M.; Dahuja, A.; Tiwari, S.; Punia, S.; Tak, Y.; Amarowicz, R.; Bhoite, A.G.; Singh, S.; Joshi, S.; Panesar, P.S.; et al. Recent Trends in Extraction of Plant Bioactives Using Green Technologies: A Review. *Food Chem.* **2021**, *353*, 129431. [CrossRef] [PubMed]
14. Yi Ling, Y.; Sook Fun, P.; Yeop, A.; Yusoff, M.M.; Gimbun, J. Assessment of Maceration, Ultrasonic and Microwave Assisted Extraction for Total Phenolic Content, Total Flavonoid Content and Kaempferol Yield from Cassia Alata via Microstructures Analysis. *Mater. Today Proc.* **2019**, *19*, 1273–1279. [CrossRef]
15. Liao, J.; Guo, Z.; Yu, G. Process Intensification and Kinetic Studies of Ultrasound-Assisted Extraction of Flavonoids from Peanut Shells. *Ultrason. Sonochem.* **2021**, *76*, 105661. [CrossRef] [PubMed]
16. Zimare, S.B.; Mankar, G.D.; Barmukh, R.B. Optimization of Ultrasound-Assisted Extraction of Total Phenolics and Flavonoids from the Leaves of Lobelia Nicotianifolia and Their Radical Scavenging Potential. *Curr. Res. Green Sustain. Chem.* **2021**, *4*, 100109. [CrossRef]
17. Saeed, R.; Ahmed, D.; Mushtaq, M. Ultrasound-Aided Enzyme-Assisted Efficient Extraction of Bioactive Compounds from Gymnema Sylvestre and Optimization as per Response Surface Methodology. *Sustain. Chem. Pharm.* **2022**, *29*, 100818. [CrossRef]
18. Krakowska, A.; Rafińska, K.; Walczak, J.; Buszewski, B. Enzyme-Assisted Optimized Supercritical Fluid Extraction to Improve Medicago Sativa Polyphenolics Isolation. *Ind. Crops Prod.* **2018**, *124*, 931–940. [CrossRef]
19. Wu, K.; Ju, T.; Deng, Y.; Xi, J. Mechanochemical Assisted Extraction: A Novel, Efficient, Eco-Friendly Technology. *Trends Food Sci. Technol.* **2017**, *66*, 166–175. [CrossRef]
20. Fan, L.; Fan, W.; Mei, Y.; Liu, L.; Li, L.; Wang, Z.; Yang, L. Mechanochemical Assisted Extraction as a Green Approach in Preparation of Bioactive Components Extraction from Natural Products—A Review. *Trends Food Sci. Technol.* **2022**, *129*, 98–110. [CrossRef]
21. Sati, P.; Dhyani, P.; Bhatt, I.D.; Pandey, A. Ginkgo Biloba Flavonoid Glycosides in Antimicrobial Perspective with Reference to Extraction Method. *J. Tradit. Complement. Med.* **2019**, *9*, 15–23. [CrossRef] [PubMed]
22. Yang, M. Efficient Extraction of Bioactive Flavonoids from Ginkgo Biloba Leaves Using Deep Eutectic Solvent/Water Mixture as Green Media. *Chem. Biochem. Eng. Q.* **2018**, *32*, 315–324. [CrossRef]
23. Han, S.; Chio, C.; Ma, T.; Mokale Kognou, A.L.; Shrestha, S.; Chen, F.; Qin, W. Extracting Flavonoid from Ginkgo Biloba Using Lignocellulolytic Bacteria *Paenarthrobacter* sp. and Optimized via Response Surface Methodology. *Biofuels Bioprod. Biorefining* **2021**, *15*, 867–878. [CrossRef]
24. Chen, S.; Xing, X.-H.; Huang, J.-J.; Xu, M.-S. Enzyme-Assisted Extraction of Flavonoids from Ginkgo Biloba Leaves: Improvement Effect of Flavonol Transglycosylation Catalyzed by Penicillium Decumbens Cellulase. *Enzyme Microb. Technol.* **2011**, *48*, 100–105. [CrossRef] [PubMed]
25. Wang, J.; Cao, F.; Su, E.; Wu, C.; Zhao, L.; Ying, R. Improving Flavonoid Extraction from Ginkgo Biloba Leaves by Prefermentation Processing. *J. Agric. Food Chem.* **2013**, *61*, 5783–5791. [CrossRef] [PubMed]
26. Zhu, X.-Y.; Mang, Y.-L.; Xie, J.; Wang, P.; Su, W.-K. Response Surface Optimization of Mechanochemical-Assisted Extraction of Flavonoids and Terpene Trilactones from Ginkgo Leaves. *Ind. Crops Prod.* **2011**, *34*, 1041–1052. [CrossRef]

27. Amoussa, A.M.O.; Zhang, L.; Lagnika, C.; Riaz, A.; Zhang, L.; Liu, X.; Beta, T. Effects of Preheating and Drying Methods on Pyridoxine, Phenolic Compounds, Ginkgolic Acids, and Antioxidant Capacity of *Ginkgo Biloba* Nuts. *J. Food Sci.* **2021**, *86*, 4197–4208. [CrossRef] [PubMed]
28. Lei, J.; Jiang, Y.; Luo, X.; Zheng, Y.; Zhu, L.; Sun, C.; Linghu, L.; Qin, C.; Gang, W. Ultrasonic-Assisted Ionic Liquid Extraction of Four Biflavonoids from *Ginkgo biloba* L. *ChemistrySelect* **2021**, *6*, 3297–3307. [CrossRef]
29. Singleton, V.L.; Rossi, J.A. Colorimetry of Total Phenolics with Phosphomolybdic-Phosphotungstic Acid Reagents. *Am. J. Enol. Vitic.* **1965**, *16*, 144–158. [CrossRef]
30. Zhishen, J.; Mengcheng, T.; Jianming, W. The Determination of Flavonoid Contents in Mulberry and Their Scavenging Effects on Superoxide Radicals. *Food Chem.* **1999**, *64*, 555–559. [CrossRef]
31. *European Pharmacopoeia*; Council of Europe: Strasbourg, France, 2004; pp. 2377–2378.
32. López-Cruz, R.; Sandoval-Contreras, T.; Iñiguez-Moreno, M. Plant Pigments: Classification, Extraction, and Challenge of Their Application in the Food Industry. *Food Bioprocess Technol.* **2023**, *16*, 2725–2741. [CrossRef]
33. Chemat, F.; Rombaut, N.; Sicaire, A.-G.; Meullemiestre, A.; Fabiano-Tixier, A.-S.; Abert-Vian, M. Ultrasound Assisted Extraction of Food and Natural Products. Mechanisms, Techniques, Combinations, Protocols and Applications. A Review. *Ultrason. Sonochem.* **2017**, *34*, 540–560. [CrossRef] [PubMed]
34. Zhang, C.-W.; Wang, C.-Z.; Tao, R. Analysis on the Physicochemical Properties of Ginkgo Biloba Leaves after Enzymolysis Based Ultrasound Extraction and Soxhlet Extraction. *Molecules* **2016**, *21*, 97. [CrossRef] [PubMed]
35. Gao, H.; Liu, B.; Nan, H.; Chen, C.; Li, L. Optimization of Ultrasonic-Assisted Extraction of Flavonoids with Ethanol from Ginkgo Leaves by Response Surface Methodology. In Proceedings of the 2009 IEEE International Symposium on IT in Medicine & Education, Albuquerque, NM, USA, 3–4 August 2009; pp. 1154–1157.
36. Zhou, G.; Ma, J.; Tang, Y.; Wang, X.; Zhang, J.; Yao, X.; Jiang, W.; Duan, J.-A. Optimization of Ultrasound-Assisted Extraction Followed by Macroporous Resin Purification for Maximal Recovery of Functional Components and Removal of Toxic Components from Ginkgo Biloba Leaves. *Biomed. Res. Int.* **2018**, *2018*, 4598067. [CrossRef] [PubMed]
37. Li, J.; Chen, Z.; Shi, H.; Yu, J.; Huang, G.; Huang, H. Ultrasound-Assisted Extraction and Properties of Polysaccharide from Ginkgo Biloba Leaves. *Ultrason. Sonochem.* **2023**, *93*, 106295. [CrossRef] [PubMed]
38. Zhang, L.; Guo, S.; Wang, M.; He, L. PEG-Based Ultrasound-Assisted Enzymatic Extraction of Polysaccharides from Ginkgo Biloba Leaves. *Int. J. Biol. Macromol.* **2015**, *80*, 644–650. [CrossRef] [PubMed]
39. Zhou, G.; Ma, J.; Tang, Y.; Wang, X.; Zhang, J.; Duan, J.-A. Multi-Response Optimization of Ultrasonic Assisted Enzymatic Extraction Followed by Macroporous Resin Purification for Maximal Recovery of Flavonoids and Ginkgolides from Waste Ginkgo Biloba Fallen Leaves. *Molecules* **2018**, *23*, 1029. [CrossRef] [PubMed]
40. Zhou, G.; Yao, X.; Tang, Y.; Qian, D.; Su, S.; Zhang, L.; Jin, C.; Qin, Y.; Duan, J. An Optimized Ultrasound-Assisted Extraction and Simultaneous Quantification of 26 Characteristic Components with Four Structure Types in Functional Foods from Ginkgo Seeds. *Food Chem.* **2014**, *158*, 177–185. [CrossRef] [PubMed]
41. Cui, N.; Zhang, L.; Quan, M.; Xu, J. Profile of the Main Bioactive Compounds and in Vitro Biological Activity of Different Solvent Extracts from Ginkgo Biloba Exocarp. *RSC Adv.* **2020**, *10*, 45105–45111. [CrossRef]
42. Kobus, J.; Flaczyk, E.; Siger, A.; Nogala-Kałucka, M.; Korczak, J.; Pegg, R.B. Phenolic Compounds and Antioxidant Activity of Extracts of Ginkgo Leaves. *Eur. J. Lipid Sci. Technol.* **2009**, *111*, 1150–1160. [CrossRef]
43. Ignat, I.; Volf, I.; Popa, V.I. A Critical Review of Methods for Characterisation of Polyphenolic Compounds in Fruits and Vegetables. *Food Chem.* **2011**, *126*, 1821–1835. [CrossRef] [PubMed]
44. Beck, S.; Stengel, J. Mass Spectrometric Imaging of Flavonoid Glycosides and Biflavonoids in Ginkgo Biloba L. *Phytochemistry* **2016**, *130*, 201–206. [CrossRef] [PubMed]
45. Šamec, D.; Pierz, V.; Srividya, N.; Wüst, M.; Lange, B.M. Assessing Chemical Diversity in Psilotum Nudum (l.) Beauv., a Pantropical Whisk Fern That Has Lost Many of Its Fern-like Characters. *Front. Plant Sci.* **2019**, *10*, 459690. [CrossRef] [PubMed]
46. Nadar, S.S.; Rao, P.; Rathod, V.K. Enzyme Assisted Extraction of Biomolecules as an Approach to Novel Extraction Technology: A Review. *Food Res. Int.* **2018**, *108*, 309–330. [CrossRef] [PubMed]

processes

Article

NADES-Based Extracts of Selected Medicinal Herbs as Promising Formulations for Cosmetic Usage

Djurdja Ivkovic [1], Ilija Cvijetic [2], Aleksandra Radoicic [1], Jelena Stojkovic-Filipovic [3], Jelena Trifkovic [2], Maja Krstic Ristivojevic [4,*] and Petar Ristivojevic [2,*]

1 Innovation Centre of Faculty of Chemistry Ltd., University of Belgrade-Faculty of Chemistry, Studentski Trg 12-16, 11158 Belgrade, Serbia; djurdja@chem.bg.ac.rs (D.I.); aradoicic@chem.bg.ac.rs (A.R.)
2 Department of Analytical Chemistry, University of Belgrade-Faculty of Chemistry, Studentski Trg 12-16, 11158 Belgrade, Serbia; ilija@chem.bg.ac.rs (I.C.); jvelicko@chem.bg.ac.rs (J.T.)
3 Clinic of Dermatovenereology, Clinical Center of Serbia, Department of Dermatovenereology, University of Belgrade-Faculty of Medicine, Dr Subotića Starijeg 8, 11000 Belgrade, Serbia; sfjelena@gmail.com
4 Department of Biochemistry, University of Belgrade-Faculty of Chemistry, Studentski Trg 12-16, 11158 Belgrade, Serbia
* Correspondence: krstic_maja@chem.bg.ac.rs (M.K.R.); ristivojevic@chem.bg.ac.rs (P.R.); Tel.: +381-11-3336676 (M.K.R.); +381-11-333-67-66 (P.R.)

Abstract: As a functional extraction medium, natural deep eutectic solvents (NADESs) can dissolve various plant metabolites. Those solvents provide untapped potential for creating novel green extracts with distinctive phytochemical signatures and unique biological activities. This is particularly relevant given the rising need for eco-friendly and sustainable skin care products. The main aim of this work was to optimize the most efficient natural deep eutectic solvents (NADESs) for extracting bioactives from 18 medicinal herbs applicable to the cosmetic industry. Selection of the most potent herbs involved assessing their conventional extracts for tyrosinase inhibition, antioxidant activity, and keratinocyte cytotoxicity. Moreover, we analyzed the phenolic profile using ultra-high-pressure liquid chromatography/mass spectrometry (UHPLC/MS) and spectrophotometric assays such as total phenolic (TPC) and flavonoid content (TFC). Using the COSMO-RS method, we modeled the solubility of 12 phenolics in 64 virtual NADESs and selected the 7 most promising ones for further experimental validation. NADESs, including betaine-urea, betaine-proline, and betaine-lysine, were computationally chosen and demonstrated the highest levels of TPC and antioxidative capacity, as confirmed by in vitro assays. The proposed combination of NADES herbal extracts represents a promising natural constituent for the cosmetic industry.

Keywords: medicinal herbs from Serbia; tyrosinase inhibition; radical scavenging activity; keratinocytes; UHPLC-MS; COSMO-RS; natural deep eutectic solvents; skin anti-aging

Citation: Ivkovic, D.; Cvijetic, I.; Radoicic, A.; Stojkovic-Filipovic, J.; Trifkovic, J.; Krstic Ristivojevic, M.; Ristivojevic, P. NADES-Based Extracts of Selected Medicinal Herbs as Promising Formulations for Cosmetic Usage. *Processes* **2024**, *12*, 992. https://doi.org/10.3390/pr12050992

Academic Editors: Yanlin Zhang and Prashant K. Sarswat

Received: 25 April 2024
Revised: 7 May 2024
Accepted: 10 May 2024
Published: 13 May 2024

1. Introduction

The growing global consciousness regarding the use of natural, renewable, and sustainable ingredients in the cosmetics industry has reached remarkable levels. Conventional organic solvents, once commonplace, pose substantial challenges due to their toxicity, flammability, and adverse environmental impact. In response, the industry is shifting toward more sustainable and eco-friendly alternatives. Green cosmetics products not only prioritize natural raw materials and ingredients but also emphasize the sustainability of the entire production process. By doing so, they contribute to a reduced carbon footprint, aligning with the urgent need for responsible practices in the beauty sector [1,2].

Natural deep eutectic solvents (NADESs) are a new generation of green solvents that meet all principles of green chemistry [3]. NADESs are two or multi-component mixtures designed from different plant secondary metabolites acting as hydrogen bond donors (HBDs) and acceptors (HBAs). The mixture's melting point decreases by forming intermolecular interactions, resulting in a liquid homogeneous phase of different viscosities—the eutectic mixture. By tailoring the components of NADESs, such as organic acids, amino acids, and sugars, as well as the molar ratios of NADES components, solvents of different physical and chemical properties can be obtained, thus leading to high selectivity in extraction [4–6]. The main advantages of NADESs are their low cost, non-toxicity, biodegradability, maximum atom economy, and ease of preparation. NADESs function as solubilizing agents, serving as carriers of active compounds and improving the skin permeability of polar compounds [7]. In addition, it has been observed that NADESs can often enhance the biological activity of the extracts compared to conventional solvents [8]. These facts suggest that NADESs should be used more in the cosmetic industry and could eventually replace traditional organic solvents completely.

The toxicity and cytotoxicity of deep eutectic solvents (DESs) have undergone extensive investigation through numerous in vitro and in vivo assays [9,10]. It is noteworthy that the toxicity profile is influenced by the DES's inherent characteristics, the molar ratio of its components, and the dosage administered [9]. Notably, certain components of DESs, such as choline chloride, are prohibited in European cosmetics due to their toxic properties [9,11]. Moreover, organic acids like tartaric, oxalic, and malonic acids are not preferred due to their propensity to significantly increase the cytotoxicity in specific DES formulations [9]. Given that sugars present in NADES are entirely natural and safe, and amino acids are commonly used in cosmetics as safe ingredients, the selection of NADES components holds promise for safe cosmetic use [12]. Additionally, betaine has emerged as a suitable and entirely natural substitute for choline chloride, being utilized as a food additive and found to be non-toxic at all studied concentrations [13].

Computational models play a crucial role in understanding and enhancing natural product extraction processes, providing valuable insights into the fundamental mechanisms involved [14,15]. The application of computational models to predict the performance of NADES extraction ensures efficiency, economic viability, and environmental sustainability. One notable model for this purpose is COSMO-RS (Conductor-like Screening MOdel for Real Solvents), which utilizes quantum mechanics and statistical thermodynamics to simulate molecular behavior in a solution. COSMO-RS predicts various thermodynamic and transport properties, including solubility, partitioning, and viscosity [16].

Skin aging follows various transformations in skin physiology, and these changes can significantly influence self-perception and self-confidence. Enzymes within the extracellular matrix, including collagenase, elastase, and tyrosinase, become more active over time. These enzymes are involved in the degradation of crucial skin molecules responsible for maintaining elasticity, hydration, and structural integrity. The effects of their increased activity include thinner skin, sagging, wrinkles, dryness, reduced elasticity, and hyperpigmentation. The overactivity of tyrosinase enzyme contributes to accelerated skin pigmentation during aging. While tyrosinase normally produces a protective skin pigment known as melanin, excessive activation leads to issues like melasma, hyperpigmentation age spots, and freckles, which some may find aesthetically undesirable [17–21].

Free radicals are reactive molecules generated in organisms during normal physiological processes but also due to a variety of external factors, including UV radiation, exposure to xenobiotics, pollution, smoking, etc. Further, these molecules play a significant role in various biological processes but can also lead to oxidative stress and damage when their levels become excessive [22]. Oxidative stress occurs when free radicals interact with biomacromolecules such as DNA, proteins, and lipids in cells, leading to their damage. This process contributes to premature aging [23]. Furthermore, free radicals directly influence the activation of the mentioned enzymes, further inducing a loss of skin elasticity and the appearance of wrinkles [24].

Herbs consist of diverse specialized compounds, such as phenolic compounds and terpenoids, renowned for their abilities to combat melanin production, act as antioxidants, and provide valuable means in the strategy against aging [25–28]. They are useful in controlling many diseases, including the prevention and treatment of skin disorders, particularly skin aging processes. Given the significance of adopting green chemistry principles in the cosmetic sector, we recognized the need for developing an effective eco-friendly extraction method enriched with phytochemicals with potential anti-aging effects on the skin.

In this regard, 18 herbal extracts were chosen and tested using five in vitro spectrophotometric assays: antioxidative (AO) capacity (DPPH and ABTS), total flavonoid (TFC), total phenolic content (TPC), and tyrosinase inhibition. Green tea (GT) was used as a reference sample in these experiments. Characterization of the extracts included the quantification of 26 phenolic compounds present in the plants using UHPLC-MS/MS. Given the critical importance of dosage in cosmetic product formulation, we characterized plant extracts by assessing their cytotoxicity on human keratinocytes (HaCaT cells).

Therefore, we aimed to select the six most potent herbs and design the NADESs for their mixture using the COSMO-RS approach. The AO capacity and TPC were conducted to compare the designed NADES solvents with a conventional and three NADESs already employed in cosmetics. The most efficient NADESs were identified as valuable ingredients for cosmetics products.

2. Materials and Methods

2.1. Chemicals and Reagents

All chemicals and reagents used in this study were provided by commercial suppliers. Ethanol, methanol, dimethylsulfoxide, glycerol, and dichloromethane were purchased from ZORKA Pharma (Šabac, Serbia). Free radical 2,2-diphenyl-1-picrylhydrazyl radical (DPPH) was purchased from Fluka (Buchs, Switzerland). Phosphate buffer (disodium phosphate and monosodium phosphate), 2,2′-azino-bis(3-ethylbenzothiazoline-6-sulfonic acid) diammonium salt (ABTS), tyrosinase enzyme from mushrooms, 6-hydroxy-2,5,7,8-tetramethyl chroman-2-carboxylic (Trolox, T), protocatechuic acid (PCA), quercetin 3-*O*-glucoside (Q-3-*O*-G), 3,4-dihydroxy-L-phenylalanine (L-DOPA), caffeic acid (CAFA), aesculin, syringic acid (SYA), chlorogenic acid (CHA), rutin (RU), *p*-hydroxybenzoic acid (*p*-hydroxyBA), aesculetin, isorhamnetin (ISORH), isoorientin (ISOOR), kaempferol (K), vitexin (VIT), rosmarinic acid (RA), p-coumaric acid (*p*-COUM), kaempferol-3-*O*-glucoside (K-3-*O*-G), taxifolin (TAX), naringenin, ellagic acid (ELLA), quercitrin, aromadendrin (AD), luteolin, eriodictyol (ED), hispidulin (HP), quercetin, naringin gallic acid (GA), betaine, urea, L-proline, L-lysine, and kojic acid were bought from Sigma-Aldrich (Darmstadt, Germany). D-glycose, D-fructose, xylitol, and D-sorbitol were purchased from Thermo Scientific (Waltham, MA, USA). Ascorbic acid was provided by Betahem (Belgrade, Serbia). Folin–Ciocalteu (FC) reagent, potassium persulfate, and sodium carbonate were obtained from Merck (Darmstadt, Germany). Sucrose was bought from a local grocery store (Maxi Supermarket, Belgrade, Serbia).

2.2. Herbal Material

2.2.1. Conventional Extractions

Medicinal herbs were selected and classified into six different families (*Lamiaceae*: wild thyme (1), basil (2), peppermint (3), oregano (4), rosemary (5), winter savoury (6), mountain germander (17), and wall germander (19); *Rosaceae*: midland hawthorn (7), agrimony (9), red raspberry (13), blackberry (14), and dog rose (16); *Plantaginaceae*: broadleaf plantain (10); *Hypericaceae*: St John's wort (11); *Moraceae*: white mulberry (12); and *Asteraceae*: chamomile (15) and yarrow (18)). The medicinal herbs investigated in the current study were obtained by the Institute of Medicinal Plant Research "Dr. Josif Pancic", Belgrade, Serbia, while mulberry was cultivated in Ogladjenovac (Valjevo), Serbia (Table S1). The green tea plant, a well-known anti-aging plant (*Theaceae*, 8), obtained by "Dr. Josif Pancic", was used as the reference sample.

The raw herbal starting materials were milled into powder for 5 min using a home miller (Gorenje, Belgrade, Serbia), then 1 g of herbs material was mixed with 10 mL of extraction solvent, methanol:dichloromethane (1:1, *v*/*v*), and ultrasound-assisted extraction was applied as an eco-friendly method to extract plant metabolites. The extracts were collected and centrifuged, and the solvents were evaporated to dry rest using a distilling rotary (IKA-Werke, Breisgau, Germany). The plant-dried residues were dissolved in methanol and stored in the fridge until further use.

2.2.2. NADESs Preparation and Green Extractions of Plant Mixture

The components forming the eutectic mixture were initially mixed and heated using a mechanical stirrer at 80 °C until the formation of a transparent phase was observed. Subsequently, a specific water content was added, and stirring continued for an additional 30 min. After forming the extraction solvent, 10 mL of the NADES was transferred to a flask, and precisely 1 g of the plant mixture was added. The selection of plants for the mixture was based on the results of spectrophotometric assays and cytotoxicity assessments. Phenolics from the plant material were then extracted for 30 min at a temperature of 50 °C. Following extraction, the plant material was separated by centrifugation at 12,326× *g* for 10 min, while the NADES extract was purified using solid-phase extraction (SPE) for characterization. Methanol and water were used as the reference conventional solvents.

2.2.3. Solid-Phase Extraction of Plant Mixture NADES Extracts

In brief, the sorbent (silica RP-C18 gel, Agilent Technologies–Bond Elut C18) underwent conditioning with water (5 mL) followed by methanol (5 mL), and the resulting supernatant (4 mL) was loaded. For the removal of non-adsorbed compounds and NADES components, water (10 mL) was used. The elution of the phenolics was carried out using methanol (6 mL), and the resultant methanolic extracts were stored in opaque glass vials at −20 °C until analysis.

2.3. LC-MS

Quantification of the phenolic compounds in herbal extracts was performed using the Dionex Ultimate 3000 UHPLC system (Thermo Fisher Scientific, Waltham, MA, USA) configured with a diode array detector (DAD) and a triple quadrupole mass spectrometer (TSQ Quantum Access Max, Thermo Fisher Scientific). A Hypersil gold C18 column (50 × 2.1 mm, 1.9 µm particle size) was used for the separation of the phenolic compounds. The mobile phase consisted of acetic acid (0.2%) in water (*v*/*v*) (A) and acetonitrile (B). The injection volume was 10 µL, while the flow rate of the mobile phase was 0.4 mL/min.

A triple quadrupole mass spectrometer (qqqMS) equipped with a heated electrospray ionization (HESI) source was set as follows: sheet gas (N_2) pressure of 50 AU, spray voltage of 4000 V, vaporizer temperature of 450 °C, auxiliary gas pressure of 20 AU, and ion sweep gas pressure of 0 AU. The skimmer offset was 0 V, and the capillary temperature was set to 320 °C. For the phenolic compounds, a negative ion mode was applied, while the collision energy was 30 eV.

Working standard solutions were prepared by dissolving the stock solution of a pure compound to obtain a concentration of 100 mg/mL, and the other calibration levels were obtained by diluting the solutions with methanol. The dilution ranged from 0.025 to 1.000 mg/L. Calibration curves were constructed by plotting the peak areas of the standards against their concentrations ($R^2 > 0.9998$), and quantification of the phenolic compounds was performed using the external standard quantification procedure [14].

2.4. Spectrophotometric Assays

For the 96-well microplate spectrophotometric assays, herbal extracts were standardized to a concentration of 100 µg DW per gram of plant material (100 µg DW/g) to determine the AO capacity (DPPH and ABTS). A tyrosinase inhibition assessment was performed on the herbal extracts at a concentration of 100 µg/mL of extract. To obtain the most potent enriched extract (NADES), the efficiency of green extractions of the herbal mixture was determined by measuring the AO capacity (DPPH) and the quantification of the TP content.

2.4.1. Total Phenolic Content (TPC)

The Folin–Ciocalteu method was used for the determination of the total phenolic content (TPC) of the herbal (NADES) extracts [29]. A 0.5 mL aliquot of diluted NADES extract was mixed with 0.5 mL distilled water and 2.5 mL of 10% FC reagent (*v*/*v*). The mixture was incubated for 5 min at room temperature, followed by the addition of 2 mL of 7.5% Na_2CO_3 (*w*/*v*) to the reaction mixture. After shaking and 2 h of incubation at room temperature, the absorbance was measured at 765 nm using a GBC UV–Visible Cintra 6 spectrophotometer. The standard curve for the total phenols was constructed using gallic acid standard solutions (20–120 mg/L). The results were expressed as milligrams of gallic acid equivalents per gram of dried weight (mg GAE/g).

2.4.2. Total Flavonoid Content (TFC)

The total flavonoid content of the analyzed samples was determined using the Pekal and Pyrzynska method with some modifications [30]. Then, 0.3 mL of a diluted herbal extract was mixed with 3.4 mL of 30% methanol (*v*/*v*). Further, 0.15 mL of $NaNO_2$ (0.5 mol/L) and 0.15 mL of $AlCl_3$ (0.15 mol/L) were added to the reaction solution. After 5 min of incubation, a NaOH solution (1 mL of 1 mol/L) was added, and each tube was thoroughly shaken. The absorbance was measured at 506 nm (GBC UV–Visible Cintra 6 spectrophotometer). A calibration curve based on standard rutin solutions was used to estimate the TFC, expressed in milligrams of rutin equivalents per gram of dried weight (mg RUE/g).

2.4.3. DPPH Radical Scavenging Assay

The DPPH• scavenging capacity of the plant NADES extracts was determined by the standardized in vitro protocol [31]. In a 1.5 mL vial, 280 µL of ethanolic DPPH• solution (0.24 mg/mL, initial absorbance of 1.09) was combined with 10 µL of diluted herbal (NADES) extract and 10 µL of absolute ethanol. Five replicates were prepared for each sample. Following a 24 h incubation period, 250 µL from each replicate was transferred into microtiter plates. Subsequently, the decrease in absorbance at 517 nm was measured using a BioTek 800 TS spectrophotometer (Agilent Technologies, Inc., Headquarters, Santa Clara, CA, USA). Standard solutions of Trolox were utilized for the calibration curve construction. The results were expressed as a percentage of DPPH radical inhibition (%) and micromoles of Trolox equivalents per gram of dried weight (mmol(mM)TE/g).

2.4.4. ABTS•+ Radical Scavenging Assay

The reduction power of 2,2′-azino-bis(3-ethylbenzothiazoline-6-sulfonic acid) (ABTS•) radical cation was applied as another method for assessing the radical scavenging capacity. The ABTS•+ solution was prepared by mixing 7 mmol/L of ABTS solution in water and 245 mmol/L of potassium persulfate in proportions that yielded a final concentration of 2.45 mmol/L of potassium persulfate and stored in the dark at room temperature for 16 h [32]. After dilution, the reaction solution was adjusted to an absorbance of 0.7028 at 734 nm, then 10 µL of diluted herbal extracts were mixed with 190 µL of ABTS•+ solution, followed by an immediate absorbance measurement at 734 nm using a BioTek 800 TS spectrophotometer. To construct a calibration curve, standard solutions of Trolox (ranging from 10 to 500 µmol/L) were employed. The results were reported as both a percentage of ABTS radical inhibition and micromoles of Trolox equivalents per gram of dried weight (mmol(mM)TE/g).

2.4.5. Mushroom Tyrosinase Inhibition Assay

The inhibitory activity of the tyrosinase enzyme was determined using a 96-well microtiter plate spectrophotometric protocol, with slight modifications [33]. L-DOPA served as the substrate for the assay. A 140 µL phosphate buffer solution (pH 6.8; 20 mmol/L) was mixed with 20 µL of herbal extract (diluted in phosphate buffer) and 20 µL of enzyme solution in buffer (480 IU/mL). Further, the mixture was incubated for 10 min at room temperature. Additionally, the reaction mixture was mixed with 20 µL of 5 mmol/L

L-DOPA, which resulted in the formation of a brown color in the wells. We measured the decreasing absorbance at 475 nm after 20 min using a BioTek 800 TS spectrophotometer (Agilent Technologies, Inc., Headquarters, USA). The obtained results were expressed as a percentage of the enzyme inhibition (%):

$$I(\%) = 100 \cdot [(A - B) - (C - D)]/[A - B],$$

where A: negative control of the sample, B: blank, C: positive control of the sample, and D: negative control of an enzyme. Kojic acid was used as a positive control for this experiment.

2.5. Cytotoxicity Assay

2.5.1. Cell Culture

Human keratinocyte (HaCaT; AddexBio, No. T0020001) cells were cultured as monolayers in Dulbecco's Modified Eagle's Medium (DMEM). The DMEM was supplemented with 1% non-essential amino acid (NEAA), 100 IU/mL penicillin, 0.1 mg/mL streptomycin, 2 mM glutamine, 1 mM sodium pyruvate, and 10% fetal bovine serum (FBS). The cells were incubated at 37 °C in a humidified atmosphere containing 5% CO_2.

2.5.2. Determination of Target Cell Survival

The HaCaT cells were plated at a density of 10,000 cells per well in 96-well microtiter plates. After allowing the cells to adhere for 24 h, polyphenolic extracts were added into the wells at a final concentration of 100 μg/mL. Control wells received only a nutrient medium. All experiments were conducted in triplicate. Cell viability was assessed using the MTT assay after 24 h of treatment, following a slightly modified version of Mosmann's procedure [34]. In brief, 20 μL of MTT solution (5 mg/mL in phosphate-buffered saline, PBS) was added to each well and incubated for 2 h at 37 °C in a humidified atmosphere containing 5% CO_2. Subsequently, the medium was carefully aspirated, and 200 μL of dimethyl sulfoxide (DMSO) was added to dissolve the formazan complexes. Absorbance was measured at 530 nm for the test wavelength and 670 nm for the reference wavelength using a BioTek 800 TS spectrophotometer (Agilent Technologies, Inc., Headquarters, USA). The data were presented as a percentage of viability, with untreated cells considered as 100%.

2.6. COSMO-RS Modeling

The solubility of 12 compounds, namely quercetin, luteolin, hispidulin, rutin, dihydrokaempferol, eriodyctiol, naringenin, caffeic acid, 4-hydroxybenzoic acid, syringic acid, rosmarinic acid, and chlorogenic acid, was predicted in 64 virtual NADESs. These NADESs comprised 1:1 mixtures of betaine, carnitine, creatine, and ornithine (serving only as hydrogen bond acceptors—HBAs) and glycerol, dextrose, sorbitol, xylitol, sucrose, urea, ascorbic acid, L-proline, lysine, tyrosine, and fructose (acting as both hydrogen bond donors—HBDs and HBAs). Additionally, for comparative analysis, the solubility of these polyphenolics in water and methanol was predicted.

The σ-profiles for 4-hydroxybenzoic acid, betaine, carnitine, ornithine, creatine, tyrosine, and proline were retrieved from the ADF sigma profile database. For all other compounds' screening charges, their densities were computed by first optimizing their geometry at the semiempirical PM7 level [35] with the COSMO water model. Subsequently, single-point calculations were performed employing a small core TZP basis set, the Becke-Perdew (GGA:BP86) functional, and the relativistic scalar ZORA method for gas phase calculations [14]. These charges were then converted into activity coefficients (γ) using the COSMO-RS method implemented in the ADF2022.101 program [36] (https://www.scm.com/product/cosmo-rs/ (accessed on 14 March 2023)). The solubility parameter (β), directly proportional to the mutual solvent–solute affinity, was calculated as $\beta = 1/\gamma$ [37].

3. Results and Discussion

3.1. Total Phenolics and Flavonoid Contents (TPCs and TFCs)

The phenolic acids, flavonoids, and glycosides contents in the herbal extracts were evaluated by measuring the total phenolic (TPC) and total flavonoid content (TFC) (Table 1). The TPC varied between 0.21 ± 0.19 and 7.65 ± 1.23 mg GAE/g, with the GT extract showing the highest TPC at 12.37 ± 1.38 mg GAE/g, serving as the reference. Red raspberry extract had the lowest TPC, while rosemary extract exhibited the highest TPC. Notably high TPCs, surpassing half of GT's TPC, was observed in oregano (6.67 ± 0.77 mg GAE/g), agrimony (6.75 ± 0.50 mg GAE/g), and wall germander (7.59 ± 0.92 mg GAE/g) extracts.

Table 1. The bioactive content of the 18 herbal extracts and GT: TPC expressed as mg of gallic acid equivalents per g of plant material and TFC expressed as mg of rutin equivalents per g of plant material.

Number Designation	Common Name	TPC mg GAE/g	TFC mg RUE/g
1	Wild thyme	2.95 ± 0.41	13.89 ± 0.42
2	Basil	5.76 ± 0.45	2.64 ± 0.24
3	Peppermint	2.07 ± 0.41	4.72 ± 0.39
4	Oregano	6.67 ± 0.77	19.30 ± 0.80
5	Rosemary	7.65 ± 1.23	13.53 ± 1.23
6	Winter savoury	0.56 ± 0.25	1.99 ± 0.22
7	Midland hawthorn	4.48 ± 1.54	9.31 ± 0.85
8	Green tea	12.37 ± 1.38	78.09 ± 4.34
9	Agrimony	6.75 ± 0.50	21.33 ± 1.07
10	Broadleaf plantain	2.88 ± 0.26	17.97 ± 0.82
11	St John's wort	3.11 ± 0.41	8.82 ± 0.42
12	White mulberry	0.88 ± 0.11	1.50 ± 0.12
13	Red raspberry	0.21 ± 0.19	1.24 ± 0.41
14	Blackberry	1.03 ± 0.06	1.32 ± 0.14
15	Chamomile	1.85 ± 0.47	11.58 ± 0.43
16	Dog rose	1.53 ± 0.84	5.36 ± 0.77
17	Mountain germander	0.82 ± 0.35	4.71 ± 0.31
18	Yarrow	0.93 ± 0.02	2.57 ± 0.26
19	Wall germander	7.59 ± 0.92	28.61 ± 1.00

In terms of TFC, the range in the herbal extracts varied from 1.24 ± 0.41 to 28.61 ± 1.00 mg RUE/g, with GT extract displaying the highest TFC of 78.09 ± 4.34 mg RUE/g. Similar to TPC, red raspberry extract had the lowest TFC, while wall germander extract had the highest. Notably high TFCs, around more than a quarter of the reference extract's values, were found in broadleaf plantain (17.97 ± 0.82 mg RUE/g), oregano (19.30 ± 0.80 mg RUE/g), and agrimony (21.33 ± 1.07 mg RUE/g) extracts. Oregano, agrimony, and wall germander extracts demonstrated notable levels of both TP and TF content.

3.2. Antioxidative (AO) Capacity Assays

The herbal extracts' AO capacity was determined using DPPH and ABTS spectrophotometric assays measuring the radical scavenging activity (Table 2). The results from the DPPH and ABTS assays indicated that all extracts possessed an AO capacity, albeit with varying degrees of radical inhibition potential. Both assays showed a generally similar trend in activity, supported by a high correlation coefficient ($R^2 = 0.8305$). Among the herbal extracts, midland hawthorn extract exhibited the lowest radical inhibition (0.27 ± 0.03%, DPPH; 12 ± 1%, ABTS). Conversely, wall germander (72 ± 3%, DPPH; 60 ± 2%, ABTS), winter savoury (47 ± 2%, DPPH; 45 ± 3%, ABTS), mountain germander (45 ± 2%, DPPH; 28 ± 2%, ABTS), peppermint (66 ± 2%, DPPH; 40 ± 2%, ABTS), agrimony (36 ± 2%, DPPH; 26 ± 1%, ABTS), and yarrow (36 ± 5, DPPH; 26 ± 2, ABTS) were identified as the most potent radical scavengers. Notably, wall germander, winter savoury, and peppermint demonstrated a higher AO capacity compared to GT (42 ± 5%, DPPH; 41 ± 3%, ABTS), while agrimony exhibited significant inhibition with a value comparable to GT.

Table 2. The AO capacity of the 18 herbal extracts and GT (100 μg of DW/g of plant material): expressed as a percent of DPPH and ABTS radical inhibition and μM of Trolox equivalents per g of plant material.

Number Designation	Common Name	DPPH RSA (%)	ABTS RSA (%)	DPPH mM TE/g	ABTS mM TE/g
1	Wild thyme	25 ± 2	26 ± 3	0.18 ± 0.01	0.18 ± 0.02
2	Basil	10 ± 4	13 ± 1	0.06 ± 0.01	0.11 ± 0.01
3	Peppermint	66 ± 2	40 ± 2	0.46 ± 0.01	0.27 ± 0.01
4	Oregano	6.8 ± 3	18 ± 3	0.06 ± 0.01	0.14 ± 0.01
5	Rosemary	4.8 ± 1.9	13 ± 2	0.04 ± 0.01	0.14 ± 0.06
6	Winter savoury	47 ± 2	45 ± 3	0.33 ± 0.01	0.32 ± 0.01
7	Midland hawthorn	0.27 ± 0.03	12 ± 1	0.01 ± 0.01	0.10 ± 0.01
8	Green tea	42 ± 5	41 ± 3	0.29 ± 0.03	0.29 ± 0.02
9	Agrimony	36 ± 2	26 ± 1	0.24 ± 0.01	0.19 ± 0.01
10	Broadleaf plantain	10 ± 1	11 ± 1	0.07 ± 0.0	0.09 ± 0.01
11	St John's wort	24 ± 3	24 ± 2	0.17 ± 0.02	0.18 ± 0.01
12	White mulberry	12 ± 1	19 ± 2	0.08 ± 0.01	0.15 ± 0.01
13	Red raspberry	18 ± 3	24 ± 2	0.13 ± 0.02	0.18 ± 0.01
14	Blackberry	24 ± 4	24 ± 1	0.15 ± 0.01	0.18 ± 0.01
15	Chamomile	25 ± 7	21 ± 1	0.19 ± 0.02	0.16 ± 0.01
16	Dog rose	14 ± 4	13 ± 1	0.08 ± 0.01	0.11 ± 0.01
17	Mountain germander	45 ± 2	28 ± 2	0.32 ± 0.01	0.20 ± 0.01
18	Yarrow	36 ± 5	26 ± 2	0.24 ± 0.01	0.19 ± 0.01
19	Wall germander	72 ± 3	60 ± 2	0.51 ± 0.02	0.42 ± 0.01

It is widely acknowledged that the reference sample of green tea possesses a strong AO capacity due to compounds known as catechins [38]. However, agrimony demonstrated potent AO activity comparable to the reference sample [39]. The high AO capability observed in wall germander and agrimony can be attributed to their heightened levels of TP and TF content, probably due to the presence of active phenols and flavonoids within these extracts (as outlined in Table 1). The highest quantities of quercetin (50 ± 4 mg/kg) and chlorogenic acid (43 ± 3 mg/kg) were measured in wall germander, consistent with values reported in the literature (Table 3) [40]. These compounds have been recognized as strong inhibitors of radicals, potentially accounting for the elevated antioxidant capacity of the wall germander extract [41,42]. The AO effect of winter savoury is influenced by other classes of compounds besides phenolics. Furthermore, free radicals have been implicated in skin hyperpigmentation resulting from oxidative stress induced by ultraviolet and visible light exposure. Therefore, the potential importance of herbal extracts lies in their capacity to scavenge these radicals, which could be pivotal in hindering such skin concerns.

Table 3. List of quantified phenols determined in the herbal extracts and GT (mg/kg ± SD). Measurements were performed in triplicate. NF—not found.

	Aesculin	*p*-HydroxyBA	Aesculetin	ISOOR	Q-3-*O*-G	TAX	PCA	SYA	CHA	CAFA	RU	VIT	*p*-COUM
1	3.62 ± 0.18	84 ± 5	30.7 ± 1.1	7 ± 1	24 ± 1	32 ± 2	17.2 ± 0.9	18.53 ± 0.24	23 ± 3	38 ± 2	NF	2.4 ± 0.2	3.03 ± 0.12
2	5.50 ± 0.28	70 ± 3	41 ± 2	0.49 ± 0.03	28 ± 2	NF	7.91 ± 0.37	9.45 ± 0.18	8.9 ± 0.3	53 ± 1	61 ± 3	NF	7 ± 1
3	0.67 ± 0.03	42.4 ± 2.4	12.99 ± 1.29	NF	1.85 ± 0.03	0.97 ± 0.03	2.99 ± 0.2	2.99 ± 0.06	12.9 ± 0.8	17.2 ± 0.7	33 ± 2	NF	3.0 ± 0.5
4	5.80 ± 0.28	64 ± 4	36 ± 2	6.63 ± 0.39	3.91 ± 0.18	100 ± 5	12.1 ± 1.0	13.9 ± 0.3	25.7 ± 1.9	48 ± 3	41 ± 6	9.7 ± 0.4	7 ± 1
5	1.05 ± 0.06	20.1 ± 1.0	13.03 ± 0.51	NF	0.39 ± 0.03	NF	3.21 ± 0.34	2.90 ± 0.14	0.49 ± 0.02	11.7 ± 0.8	0.89 ± 0.04	NF	NF
6	0.56 ± 0.03	10.7 ± 0.7	2.28 ± 0.02	NF	2.61 ± 0.09	8.8 ± 0.5	1.99 ± 0.11	1.76 ± 0.03	0.58 ± 0.05	3.17 ± 0.07	0.82 ± 0.04	NF	1.34 ± 0.04
7	3.40 ± 0.17	10.1 ± 0.4	0.98 ± 0.05	0.80 ± 0.04	NF	201 ± 9	1.34 ± 0.07	5.18 ± 0.10	6.3 ± 0.1	101 ± 5	0.99 ± 0.03	99 ± 5	NF
8	10.80 ± 0.54	22.5 ± 0.6	1.00 ± 0.05	0.80 ± 0.04	NF	29 ± 2	0.6 ± 0.1	6.77 ± 0.17	NF	5.26 ± 0.47	3.8 ± 0.2	137 ± 7	12.7 ± 0.5
9	14 ± 1	2.47 ± 0.14	5 ± 1	3.37 ± 0.20	8.68 ± 0.34	167 ± 9	0.29 ± 0.02	12 ± 1	12.9 ± 0.52	4.66 ± 0.48	0.01 ± 0.01	150 ± 8	11.5 ± 0.4
10	8.15 ± 0.29	4.61 ± 0.23	7.0 ± 0.5	6 ± 1	1.51 ± 0.06	0.78 ± 0.03	0.71 ± 0.11	4.64 ± 0.11	24 ± 2	23 ± 3	NF	0.83 ± 0.03	NF
11	36 ± 4	2.12 ± 0.10	2.88 ± 0.05	1.91 ± 0.11	0.69 ± 0.03	179 ± 10	0.38 ± 0.07	39 ± 1	8.8 ± 1.1	14 ± 1	1.31 ± 0.06	552 ± 9	NF
12	0.39 ± 0.02	3.80 ± 0.16	0.57 ± 0.01	0.67 ± 0.02	NF	52 ± 3	0.28 ± 0.03	0.28 ± 0.02	15.4 ± 2.0	53 ± 4	NF	81 ± 5	0.02 ± 0.01
13	18 ± 1	5.32 ± 0.27	35 ± 3	23 ± 2	11.51 ± 0.58	3.39 ± 0.17	0.19 ± 0.01	18.6 ± 0.4	13.68 ± 1.14	19 ± 1	NF	NF	NF
14	2.97 ± 0.08	2.84 ± 0.08	3.33 ± 0.09	2.26 ± 0.24	1.29 ± 0.06	3.28 ± 0.17	0.13 ± 0.02	3.21 ± 0.11	4.66 ± 0.21	3.00 ± 0.02	0.01 ± 0.01	9 ± 1	NF
15	7.98 ± 0.19	26.4 ± 0.9	15 ± 2	11 ± 1	1.29 ± 0.07	19.70 ± 0.99	0.39 ± 0.01	8.50 ± 0.32	14.8 ± 0.4	77 ± 4	NF	4.40 ± 0.15	NF
16	36 ± 2	4.71 ± 0.10	0.58 ± 0.05	0.60 ± 0.05	8.92 ± 0.55	22 ± 2	0.69 ± 0.03	36 ± 1	10 ± 1	2.38 ± 0.06	NF	4.40 ± 0.23	0.01 ± 0.01
17	11 ± 1	1.02 ± 0.05	1.37 ± 0.06	1.12 ± 0.06	NF	2.38 ± 0.08	0.29 ± 0.02	0.98 ± 0.02	15.9 ± 0.7	3.80 ± 0.18	NF	4.08 ± 0.13	NF
18	1.52 ± 0.06	1.22 ± 0.08	6.97 ± 0.48	5.02 ± 0.31	1.25 ± 0.06	1.89 ± 0.09	0.24 ± 0.02	1.22 ± 0.06	16.8 ± 0.5	33 ± 4	0.02 ± 0.01	8.15 ± 0.39	0.02 ± 0.01
19	28 ± 2	16.5 ± 0.4	10.5 ± 0.9	7.99 ± 0.43	3.82 ± 0.07	6.01 ± 0.29	0.48 ± 0.03	9.6 ± 0.5	43 ± 3	26.9 ± 1.8	0.02 ± 0.01	7 ± 1	NF

	K	HP	ISORH	RA	AD	ED	LUT	Q	Naringenin	ELLA	Naringin	K-3-*O*-G	Quercitrin
1	12 ± 1	1.45 ± 0.08	0.9 ± 0.1	470 ± 9	12 ± 1	56 ± 3	49 ± 4	30 ± 3	32.8 ± 2.8	40 ± 2	42 ± 3	14.92 ± 1.15	154 ± 6
2	3.1 ± 0.4	3 ± 1	1.6 ± 0.2	446 ± 6	0.02 ± 0.01	4.76 ± 0.05	12.7 ± 1.2	0.7 ± 0.1	4.09 ± 1.04	1.49 ± 0.07	NF	12.2 ± 0.89	2.88 ± 0.08
3	2.44 ± 0.18	0.94 ± 0.03	NF	345 ± 7	0.02 ± 0.01	8.93 ± 0.76	10.6 ± 0.6	1.31 ± 0.04	12.16 ± 1.04	NF	310 ± 9	2.65 ± 0.08	0.09 ± 0.02
4	22 ± 3	0.58 ± 0.03	NF	542 ± 13	142.5 ± 6.65	254 ± 9	183 ± 6	14 ± 1	269 ± 5	69 ± 5	9.91 ± 0.50	2.36 ± 0.08	0.53 ± 0.04
5	6.06 ± 0.38	25 ± 3	NF	134 ± 11	NF	6.51 ± 0.96	9 ± 1	0.8 ± 0.1	1.49 ± 0.11	NF	NF	1.70 ± 0.06	0.29 ± 0.03
6	1.68 ± 0.17	0.12 ± 0.02	0.11 ± 0.01	95.1 ± 434	22.8 ± 1.9	15.96 ± 2.19	12 ± 1	0.75 ± 0.05	43.8 ± 2.4	6.6 ± 0.4	0.41 ± 0.01	1.17 ± 0.05	0.30 ± 0.01
7	1.99 ± 0.02	8 ± 1	NF	3.41 ± 0.12	3.99 ± 0.38	NF	7 ± 1	2.63 ± 0.20	10.45 ± 0.95	4.48 ± 0.14	NF	NF	42 ± 4
8	0.86 ± 0.03	8 ± 2	0.05 ± 0.01	NF	2.98 ± 0.04	1.08 ± 0.76	0.5 ± 0.1	NF	33.4 ± 3.1	5.7 ± 0.4	3.33 ± 0.17	2.02 ± 0.08	26 ± 1
9	1.31 ± 0.03	5.70 ± 0.95	0.03 ± 0.01	640 ± 9	3.88 ± 0.07	NF	1.0 ± 0.2	0.50 ± 0.02	23 ± 2	6. ± 1	29 ± 2	NF	48 ± 4
10	0.32 ± 0.03	NF	2 ± 1	1.65 ± 0.08	15.81 ± 0.85	NF	1.05 ± 0.08	24 ± 2	0.44 ± 0.02	NF	NF	NF	1.90 ± 0.10
11	1.03 ± 0.03	7.12 ± 0.38	0.13 ± 0.02	150 ± 4	5.32 ± 0.66	0.81 ± 0.04	0.99 ± 0.12	0.91 ± 0.02	430 ± 9	4.11 ± 0.18	1.06 ± 0.07	0.53 ± 0.03	12 ± 2
12	0.04 ± 0.01	3.0 ± 0.3	NF	0.32 ± 0.01	1.08 ± 0.04	0.39 ± 0.02	0.3 ± 0.1	0.68 ± 0.04	1.93 ± 0.03	6.0 ± 0.5	2.93 ± 0.09	0.18 ± 0.02	60 ± 4
13	0.2 ± 0.1	2.91 ± 0.17	0.04 ± 0.01	2.0 ± 0.5	1.31 ± 0.02	NF	0.47 ± 0.09	NF	5.41 ± 0.54	2.85 ± 0.09	55 ± 3	NF	7 ± 1
14	1.00 ± 0.03	2.17 ± 0.19	0.04 ± 0.01	0.30 ± 0.02	5.93 ± 0.36	NF	0.60 ± 0.05	0.24 ± 0.04	2.56 ± 0.22	1.01 ± 0.04	49 ± 3	NF	10 ± 1
15	2.0 ± 0.1	8.3 ± 0.67	1.55 ± 0.02	NF	2.0 ± 0.5	0.76 ± 0.02	5 ± 1	7 ± 1	89 ± 3	NF	NF	0.3 ± 0.1	14 ± 1
16	1.02 ± 0.03	6 ± 1	0.10 ± 0.01	78 ± 5	3.59 ± 0.12	0.71 ± 0.01	2.5 ± 0.5	NF	32 ± 3	8.0 ± 0.5	179 ± 10	NF	18.5 ± 0.5
17	4.0 ± 0.5	NF	0.30 ± 0.02	9 ± 1	2.27 ± 0.02	NF	1.0 ± 0.1	23 ± 4	1.08 ± 0.06	0.61 ± 0.03	NF	0.11 ± 0.08	11.6 ± 0.9
18	0.15 ± 0.02	2.63 ± 0.14	1.69 ± 0.08	0.20 ± 0.03	0.69 ± 0.04	NF	0.29 ± 0.03	29 ± 3	6.89 ± 0.40	NF	1.49 ± 0.07	0.65 ± 0.07	10.45 ± 0.95
19	30 ± 3	6.12 ± 0.23	0.47 ± 0.02	NF	6.71 ± 0.08	NF	1.59 ± 0.04	50 ± 4	1.24 ± 0.03	0.76 ± 0.03	0.85 ± 0.03	NF	5.35 ± 0.31

3.3. Cytotoxicity and Tyrosinase Inhibition Assays

HaCaT cells serve as a valuable in vitro model for evaluating cytotoxicity, especially in toxicology, pharmaceuticals, and cosmetics, since there is a good correlation with the results obtained using in vivo experimental models [43]. The cytotoxicity results (Figure 1A) revealed different effects on the HaCaT cell viability of the examined extracts in the applied concentration of 100 μg/mL. Peppermint, midland hawthorn, and yarrow induced a notable but not statistically significant reduction in cell viability (viability: 72.5 ± 10.9%, 80.0 ± 4.7%, and 76.4 ± 7.9%, respectively) compared to the untreated cells. Importantly, rosemary, St John's wort, and white mulberry displayed significant cytotoxicity (viability: 31.4 ± 8.5%, 51.4 ± 13.4%, and 67.0 ± 1%, respectively). Interestingly, agrimony, broadleaf plantain, dog rose, and wall germander showed significant proliferative effects on HaCaT cells (viability: 130.8 ± 13.9%, 132.4 ± 14.5%, 148.7 ± 4.7%, and 160.0 ± 11.1%, respectively). The plant extracts with no notable effects on HaCaT cell viability were the wild thyme, basil, oregano, winter savoury, red raspberry, blackberry, chamomile, mountain germander, and GT reference extracts.

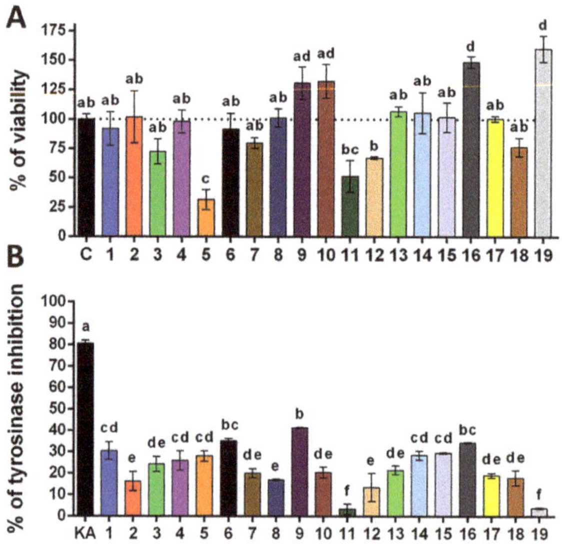

Figure 1. (**A**) The viability percentage of HaCaT cells following exposure to phenolic extracts at a concentration of 100 μg/mL (1–19) and a control (C). (**B**) Tyrosinase inhibition of 18 investigated herbs (1–19), GT (8), and a standard of kojic acid (KA). Different letters above the columns denote significant differences at $p < 0.05$ according to Tukey's multiple comparisons test.

Milutinovic and colleagues studied wall germander extract's impact on colon carcinoma and HaCaT cells, finding no cytotoxic effects on HaCaT cells and even a slight increase in proliferation, which aligns with our findings [44].

The tyrosinase inhibition of the examined extracts at a 100 μg/mL concentration and KA (81 ± 6%) as a known tyrosinase inhibitor was tested (Figure 1B). Among the extracts, agrimony displayed the highest tyrosinase inhibition (42 ± 1%), followed by winter savoury (35 ± 1%), dog rose (34 ± 1%), and wild thyme (31 ± 6%). All other tested extracts exhibited approximately the same potential for tyrosinase inhibition, whereas St John's wort and wall germander extracts demonstrated negligible potential for tyrosinase inhibition (3 ± 2% and 4 ± 2%, respectively).

This study aimed to identify new natural resources that could be utilized for cosmetic purposes in skin lightening. It is known that flavonoids with a hydroxyl group at C-3′ of ring B, a hydroxyl group at C-3, and a carbonyl group at C-4 of ring C enhance tyrosinase inhibitory effects [45,46]. Our study quantified taxifolin, vitexin, and quercitrin as the primary metabolites in agrimony. These phenolics meet one or more conditions from the statement above for potent tyrosinase inhibition, likely causing high inhibition in this plant extract. Indeed, the data indicate a high inhibition of these compounds against tyrosinase [47–50]. Additionally, in the aforementioned sample, a high amount of caffeic acid, a confirmed tyrosinase inhibitor, was quantified [51]. To the best of our knowledge, there is limited available data regarding the tyrosinase inhibition potential of agrimony. The existing information suggests that it exhibits low potency in tyrosinase inhibition, while, in our study, this sample, even at low concentration, acted as the most potent inhibitor among these plants (42 ± 1%) [52]. Wild thyme also represents a relatively underexplored plant, with only one result in the literature indicating moderate inhibition. Our research confirms this finding (31 ± 6%) [52].

According to the available literature, there has been only one study that explored the impact of winter savory on tyrosinase inhibitory activity [53]. In our research, this herb demonstrated moderate tyrosinase inhibition, showing the second-best inhibitory activity (35 ± 1%) and exhibiting characteristics of a potent in vitro anti-pigmentation agent. Limited information exists on the tyrosinase inhibitory activity of midland hawthorn. Our results demonstrated moderate inhibitory activity at low concentrations (21 ± 3%). Other samples, such as peppermint, broadleaf plantain, and chamomile, exhibited almost identical inhibition as found in the literature [53]. Surprisingly, red raspberry (22 ± 3%) and dog rose (35 ± 1%) showed moderate levels of tyrosinase inhibition in our study, contrary to descriptions in the literature where they are reported to have weak or no inhibitory properties [53].

A correlation was observed between high tyrosinase inhibition activity and elevated concentrations of specific phenolics in the most effective extracts. Wild thyme and dog rose contained high levels of naringenin and naringin (Table 3). Wild thyme showed a noteworthy amount of eriodictyol, ellagic acid, and luteolin. Previous studies have demonstrated that luteolin effectively inhibits tyrosinase-catalyzed oxidation in both cell-free extracts and living cells [54]. Additionally, ellagic acid has been described in the literature as an effective inhibitor of this enzyme both in vivo and in vitro [55]. Notably, Fan and colleagues investigated flavonoids from food sources and found that all the mentioned flavonoids exhibited satisfactory binding energies to tyrosinase. Among them, luteolin demonstrated the highest binding affinity, which aligns with our findings [46]. In the agrimony extract, the highest level of rosmarinic acid (640 ± 9 mg/kg) was found. The content of the same significant phenols in this plant has been confirmed in the literature [56]. Rosmarinic acid, found in high concentrations in the samples of wild thyme, agrimony, and dog rose, may be associated with their significant tyrosinase inhibition, considering that this compound is known to act as a potent tyrosinase inhibitor [57]. Similarly, quercitrin, found in the highest concentration in wild thyme and agrimony, has demonstrated potent anti-melanogenic properties [50].

3.4. UHPLC-DAD-MS/MS

Using UHPLC-MS analysis, 26 phenolic compounds were quantified in our extracts (Table 3). The major phenolic acids found in the herbal extracts were syringic acid, *p*-hydroxybenzoic acid, protocatechuic acid, caffeic acid, chlorogenic acid, and rosmarinic acid. The main flavonoid glycosides, isoquercitrin, quercitrin, and kaempferol-3-O-glucoside, were found in the herbal extracts.

Among the extracts, agrimony and dog rose exhibited significant inhibitory effects on tyrosinase and proliferative effects on the HaCaT cells. Agrimony extract contained the highest levels of rosmarinic acid (640 ± 9 mg/kg) and notable amounts of *p*-coumaric acid (11.5 ± 0.4 mg/kg), vitexin (150 ± 8 mg/kg), taxifolin (167 ± 9 mg/kg), and quercitrin (48 ± 4 mg/kg). Dog rose extract showed the highest aesculin content (36 ± 2 mg/kg) and substantial levels of syringic acid (36 ± 1 mg/kg), rosmarinic acid (78 ± 5 mg/kg), and naringin (179 ± 10 mg/kg). Wild thyme and winter savoury also displayed considerable potency in tyrosinase inhibition. Wild thyme was rich in rosmarinic acid (470 ± 9 mg/kg), protocatechuic acid (17.2 ± 0.9 mg/kg), *p*-hydroxybenzoic acid (84 ± 5 mg/kg), kaempferol-3-O-glucoside (14.92 ± 1.15 mg/kg), and quercitrin (154 ± 6 mg/kg). Winter savoury, although not notable for any specific phenolic compound, exhibited significant tyrosinase inhibition. Broadleaf plantain extract, containing notable amounts of chlorogenic acid (24 ± 2 mg/kg) and isorhamnetin (6 ± 1 mg/kg), demonstrated significant proliferative effects on HaCaT cells. However, wall germander extract showed the highest potency for HaCaT cell proliferation despite poor tyrosinase inhibition and contained substantial levels of quercetin (50 ± 4 mg/kg) and kaempferol (30 ± 3 mg/kg). These results suggest that the observed effects arise from the harmonized complex combination of phenolic compounds obtained through specific established procedures and solvents utilized during plant processing and extraction rather than the efficacy of any individual compound present in significant quantities.

3.5. Selection of the Most Potent Plants for Cosmetics Formulations

As a result of previous experiments, it was concluded that herbal extracts of wild thyme, agrimony, and wall germander proved to be effective radical scavengers. Additionally, herbal extracts of wild thyme, winter savoury, agrimony, and dog rose demonstrated exceptional depigmentation properties, while agrimony, broadleaf plantain, dog rose, and wall germander showed effectiveness as skin cell protectors. Furthermore, a significant proliferative effect on keratinocyte cells was a criterion for selecting herbs for incorporation into the herbal mixture, given that keratinocytes are important for skin re-epithelialization and recovery. Bearing this in mind, plants exhibiting the highest degree of proliferation were chosen for the herbal blend: agrimony, broadleaf plantain, dog rose, and wall germander. Individual extracts exhibiting multiple biological activities and proliferative or at least no cytotoxic effects on HaCaT cells were selected for the plant mixture. Finally, the plant mixture contained wild thyme, winter savoury, agrimony, broadleaf plantain, dog rose, and wall germander, all mixed in equal proportions.

3.6. NADESs Designs for Phenolic Extraction from the Herbal Mixture

Addressing the complexities of diverse mixtures of natural phenolic compounds derived from plant origins is highly in demand, particularly for the initial pre-screening and predicting their extractability to obtain mixtures with desired biological activity using natural deep eutectic solvents (NADESs) instead of conventional solvents. For this purpose, the *in silico* COSMO-RS method was employed.

3.6.1. COSMO-RS Screening of NADESs

The screening charge densities of hispidulin and syringic acid (Figure 2a), representing flavones and carboxylic acids found in the studied plant materials, illustrate the distribution of the negative electrostatic potential (ESP) around carbonyl and hydroxyl oxygens, with a positive ESP near hydroxyl group hydrogens. These screening charge densities were then converted into σ-profiles. Subsequently, the activity coefficients and their reciprocal values (solubility parameter β) were calculated for each of the 12 phenolics across 64 theoretical NADESs. Detailed data are available in the Electronic Supplementary Information, Table S2. Figure 2b,c reveal that carnitine-based and betaine-based NADESs exhibit a higher affinity for hispidulin and syringic acid compared to other mixtures. Additionally, these high-affinity NADESs show higher β values for all phenolics when compared to conventional solvents like water and methanol. This observation can be attributed to the complementarity of hydrogen-bonding regions for solvents and solutes, along with the significant non-polar area in σ-profiles (Figure 2d). These findings suggest the significant role of hydrogen bonding and van der Waals interactions in solubilizing these compounds in NADESs. After analyzing the results of the COSMO-RS screening, we further focused on investigating only betaine-based NADESs due to their usage in cosmetics [58].

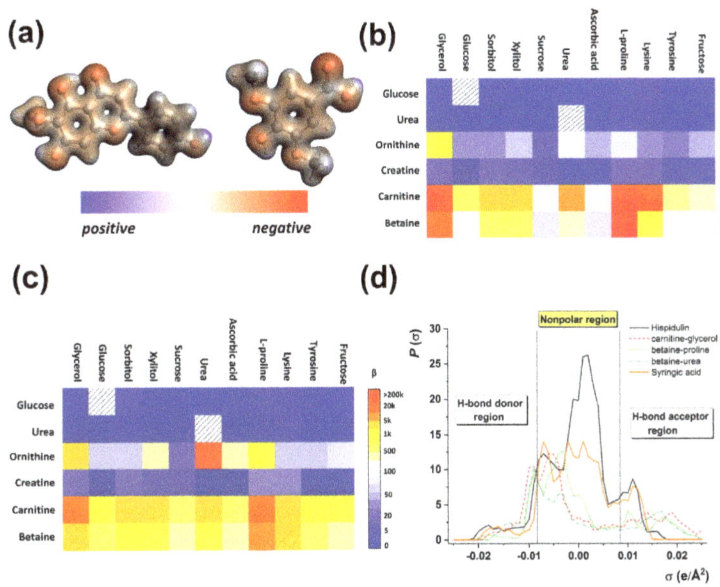

Figure 2. (**a**) Screening charge densities for hispidulin (**left**) and syringic acid (**right**); Heatmaps representing the solubility of (**b**) hispidulin and (**c**) syringic acid in 1:1 mixtures of corresponding NADES constituents. (**d**) Sigma profiles of two phenolics and three NADESs with the highest affinities for these compounds. For additional data, including the solubility parameters for 10 other studied compounds, refer to Table S2 in the Electronic Supplementary Information (ESI).

3.6.2. NADES Selection, Preparation, and the Extraction of Phenolic Compounds

Considering that ionic liquids are often incompatible for use in cosmetics, we have opted for NADESs as a green alternative to conventional organic solvents for cosmetic applications [59]. Bearing in mind the results of the COSMO-RS screening and previously reported NADESs for cosmetic applications, we synthesized ten betaine-based NADESs (Table 4).

Table 4. The abbreviations and components of the investigated NADESs.

	Green Solvents	
Abbreviation	**NADES Composition**	**Molar Ratio (Water, *w*/*w*)**
N1	Betaine/Glycerol	1/1 (30%)
N2	Betaine/Xylitol	1/1 (30%)
N3	Betaine/Sucrose	1/1 (30%)
N4	Betaine/Urea	1/1 (30%)
N5	Betaine/Ascorbic acid	1/1 (30%)
N6	Betaine/L-proline	1/1 (30%)
N7	Betaine/Lysine	1/1 (30%)
GF1	Betaine/Glucose	1/1 (23%)
GF2	Betaine/Sorbitol	1/1 (23%)
GF3	Betaine/Fructose	1/1 (23%)
	Conventional solvents	
M	Methanol	
W	Water	

We prepared seven proposed (N1–N7) and three reference NADESs (GF1–GF3) [11] constructed from plant metabolites such as glycerol; ascorbic acid; urea; sugars (glucose, fructose, sucrose, xylitol, and sorbitol); and amino acids (lysine and proline) that acted like hydrogen bond donors (HBDs). Also, raw materials compatible with the cosmetic industry, such as betaine (a greener substitute for choline chloride), were used as bio-based and cost-effective materials [11].

The main drawback of NADESs is their high viscosity. The viscosity can be reduced by adding water to the NADESs or by increasing the temperature. To avoid disrupting the hydrogen bonds responsible for NADES formation or extraction, it is permissible to add a maximum of 50% (*w*/*w*) water [56]. The addition of a specific amount of water plays a crucial role in certain extractions [60]. Therefore, we added 23–30% water for easier handling of the NADESs.

Through the experimental validation of COSMO-RS, the proposed NADESs proved to be moderate to excellent green extraction media for biologically active compounds from the herbal material. The NADES extracts demonstrated a superior ability to extract phenolics as compared to both conventional solvents and reference NADESs, making them a more efficient and eco-friendlier alternative for cosmetics solvents (see Table 5).

Table 5. DPPH and TPC values of the herbal mixture NADES extracts (N1–N7), reference NADES extracts used in cosmetics (GF1–GF3), and conventional extracts (M and W). The results are presented as the mean ± standard deviation.

Extracts Designation	TPC mg GAE/g	DPPH mM TE/g
N1	16.0 ± 0.2	5.28 ± 0.46
N2	16.3 ± 0.5	5.09 ± 0.26
N3	10.1 ± 0.2	3.80 ± 0.34
N4	33 ± 1	15.06 ± 0.39
N5	11.8 ± 0.6	0.25 ± 0.15
N6	36.6 ± 0.6	18.14 ± 0.26
N7	82 ± 4	19.58 ± 0.09
GF1	0.58 ± 0.03	0.25 ± 0.01
GF2	9.5 ± 0.6	2.67 ± 0.01
GF3	0.62 ± 0.03	0.28 ± 0.02
M	1.85 ± 0.08	0.47 ± 0.05
W	1.87 ± 0.01	0.45 ± 0.03

It was concluded that the nature of the HBD played a significant role in the extraction efficiency. NADESs containing a component with basic nitrogen as the HBD component in their structure extracted the highest amounts of phenolics (N4, N6, and N7), with N7, featuring lysine as the second component, showing the most promising performance. NADES N7 extracted more than twice the amount of phenols compared to N4 and N6. Additionally, N7 extracted between 8.6 times and 141 times more phenols compared to the reference NADESs, while extracting 44 times more phenols compared to the water and methanol extracts, representing a new promising alternative to organic solvents used in cosmetics. These findings are consistent with the predicted activity coefficients obtained after COSMO-RS screening. Additionally, mixtures GF1 (glucose), GF3 (fructose), N3 (sucrose), and N5 (ascorbic acid) proved to be the poorest extractors of phenolic compounds, which aligns with the COSMO-RS predictions. An exception was GF2 (sorbitol), which extracted slightly more phenolics, as also predicted by COSMO-RS.

Betaine, as a component of NADES, would be an effective addition to cosmetic products, because it acts as a confirmed hydrating agent due to the formation of strong hydrogen bonds with water, making it an ideal addition for skin hydration. Additionally, betaine has been confirmed to reduce skin irritation, further enhancing its appeal for cosmetic use [61].

3.7. Quality Assessment of 18 Investigated NADESs from the Plant Mixture

TP Content and AO Capacity of the Investigated NADESs

The TP content and AO capacity for seven designed NADESs (N1–N7) were compared with three extracts obtained using patented cosmetical NADESs (GF1–GF3) and conventional methanolic (M) and water extracts (W) (Table 5).

The good correlation (R^2 = 0.8139) between the results of the TPC and DPPH tests indicates that the AO capacity is strongly influenced by the phenolic compounds present in the samples. In the examined NADES extracts, phenolic compounds were detected in a range of 10.1 ± 0.2 mg GAE/g to 82 ± 4 mg GAE/g. NADES N7 quantified the highest amounts of phenols, while N3 quantified the least. It was observed that NADES N1 (16.0 ± 0.2 mg GAE/g) and N2 (16.3 ± 0.5 mg GAE/g), followed by N3, N5 (11.8 ± 0.6 mg GAE/g), and GF2 (9.5 ± 0.6 mg GAE/g), as well as N4 (33 ± 1 mg GAE/g) and N6 (36.6 ± 0.6 mg GAE/g), quantified similar amounts of phenolics from the plant mixture. The reference NADES extracts proved to be inferior extractors of phenols, ranging between 0.58 ± 0.03 mg GAE/g and 9.5 ± 0.6 mg GAE/g, with GF1 quantifying the lowest amount of phenols and GF2 quantifying the highest. Methanol (1.85 ± 0.08 mg GAE/g) and water (1.87 ± 0.01 mg GAE/g) extracted more phenolics compared to the GF1 and GF3 mixtures, albeit significantly less than any of the other investigated NADES extracts.

A similar trend was observed in the DPPH test, where NADES N7 (19.58 ± 0.09 mM TE/g) emerged as the most potent radical inhibitor, while N5 (0.25 ± 0.15 mM TE/g) exhibited the lowest radical inhibition. N7 demonstrated a significantly higher AO capacity, displaying between 7 to 78 times greater radical inhibitory ability compared to the reference NADESs, 43 times higher AO capacity than water, and 42 times higher capacity than methanol. NADES N4 (15.06 ± 0.39 mM TE/g) and N6 (18.14 ± 0.26 mM TE/g) also showed a high AO capacity consistent with their quantified TP. The AO capacity of GF1–GF3 ranged from 0.25 ± 0.01 to 2.67 ± 0.01 mM TE/g, with GF2 exhibiting the highest and GF1 the lowest AO capacity. All proposed NADES mixtures demonstrated superior radical inhibition compared to those currently used in cosmetics, except for the proposed N5 (ascorbic acid, HBD). The conventional solvents exhibited almost identical AO capacities (0.47 ± 0.05 mM TE/g, M; 0.45 ± 0.03 mM TE/g, W), which were higher than GF1, GF3, and N5 but significantly lower than all the other proposed NADESs.

Proposed NADESs N4, N6, and N7 contained the highest TP contents, aligning with their highest values of AO capacity.

4. Conclusions

This study highlights the potential of natural deep eutectic solvents (NADESs) as a valuable extraction medium for specialized plant metabolites. By dissolving bioactives from 18 medicinal herbs, NADESs offer a green and sustainable approach to creating novel cosmetic extracts. Experimental validation through the green extraction of an herb mixture confirmed NADESs N4 (betaine-urea), N6 (betaine-proline), and N7 (betaine-lysine) as the most effective solvents, surpassing the reference NADESs (GF1–GF3) and conventional extracts in the AO assays and TP content. These promising natural constituents hold significant promise for the cosmetics industry, addressing the growing demand for eco-friendly and effective skincare products. Future research should explore their practical applications and potential benefits in formulations aimed at combating skin aging and promoting overall skin health. Moreover, carnitine-based and ornithine-based NADESs identified through COSMO-RS models represent promising solvents for experimental validation. In summary, NADESs offer a green pathway to harnessing the power of medicinal herbs. Their potential impact on skincare formulations could revolutionize the industry, providing consumers with sustainable bioactive cosmetic products.

Supplementary Materials: The following supporting information can be downloaded at https://www.mdpi.com/article/10.3390/pr12050992/s1: Table S1: List of plants with data on their common and Latin names, plant family names, and plant parts used for extraction. Table S2: Activity coefficients and solubility parameters for 12 phenolic compounds in the designed NADESs.

Author Contributions: Conceptualization, M.K.R. and P.R.; Methodology, D.I., I.C., A.R., J.S.-F., M.K.R. and P.R.; Software, D.I., I.C., M.K.R. and P.R.; Validation, J.T., M.K.R. and P.R.; Formal analysis, I.C.; Investigation, D.I., A.R. and M.K.R.; Resources, J.T., M.K.R. and P.R.; Data curation, D.I., I.C., J.S.-F. and P.R.; Writing—original draft preparation, D.I. and I.C.; Writing—review and editing, I.C., J.S.-F., J.T., M.K.R. and P.R.; Visualization, D.I.; Supervision, M.K.R. and P.R.; Project Administration, P.R.; Funding acquisition, J.T., M.K.R. and P.R. All authors have read and agreed to the published version of the manuscript.

Funding: This work was supported by The Science Fund of the Republic of Serbia, Serbian Science and Diaspora Collaboration Program, No. 6389927, and the Ministry of Science, Technological Development and Innovation of Republic of Serbia, contract numbers: 451-03-66/2024-03/200168 and 451-03-66/2024-03/200288.

Data Availability Statement: The data are contained within the article and the Supplementary Materials.

Conflicts of Interest: The authors declare no conflicts of interest.

References

1. Carré, P. About Solvents Used in the Preparation of Oils for Cosmetic Products Complying with the Cosmos Standard. *OCL* **2021**, *28*, 16. [CrossRef]
2. Martins, A.M.; Marto, J.M. A Sustainable Life Cycle for Cosmetics: From Design and Development to Post-Use Phase. *Sustain. Chem. Pharm.* **2023**, *35*, 101178. [CrossRef]
3. Espino, M.; de los Ángeles Fernández, M.; Gomez, F.J.V.; Silva, M.F. Natural Designer Solvents for Greening Analytical Chemistry. *TrAC Trends Anal. Chem.* **2016**, *76*, 126–136. [CrossRef]
4. Obluchinskaya, E.D.; Pozharitskaya, O.N.; Zakharova, L.V.; Daurtseva, A.V.; Flisyuk, E.V.; Shikov, A.N. Efficacy of Natural Deep Eutectic Solvents for Extraction of Hydrophilic and Lipophilic Compounds from Fucus Vesiculosus. *Molecules* **2021**, *26*, 4198. [CrossRef] [PubMed]
5. Benvenutti, L.; Sanchez-Camargo, A.d.P.; Zielinski, A.A.F.; Ferreira, S.R.S. NADES as Potential Solvents for Anthocyanin and Pectin Extraction from Myrciaria Cauliflora Fruit by-Product: In Silico and Experimental Approaches for Solvent Selection. *J. Mol. Liq.* **2020**, *315*, 113761. [CrossRef]
6. Mero, A.; Koutsoumpos, S.; Giannios, P.; Stavrakas, I.; Moutzouris, K.; Mezzetta, A.; Guazzelli, L. Comparison of Physicochemical and Thermal Properties of Choline Chloride and Betaine-Based Deep Eutectic Solvents: The Influence of Hydrogen Bond Acceptor and Hydrogen Bond Donor Nature and Their Molar Ratios. *J. Mol. Liq.* **2023**, *377*, 121563. [CrossRef]
7. Punzo, A.; Porru, E.; Silla, A.; Simoni, P.; Galletti, P.; Roda, A.; Tagliavini, E.; Samorì, C.; Caliceti, C. Grape Pomace for Topical Application: Green NaDES Sustainable Extraction, Skin Permeation Studies, Antioxidant and Anti-Inflammatory Activities Characterization in 3D Human Keratinocytes. *Biomolecules* **2021**, *11*, 1181. [CrossRef] [PubMed]
8. Radošević, K.; Ćurko, N.; Gaurina Srček, V.; Cvjetko Bubalo, M.; Tomašević, M.; Kovačević Ganić, K.; Radojčić Redovniković, I. Natural Deep Eutectic Solvents as Beneficial Extractants for Enhancement of Plant Extracts Bioactivity. *LWT* **2016**, *73*, 45–51. [CrossRef]
9. Lomba, L.; Ribate, M.P.; Sangüesa, E.; Concha, J.; Garralaga, M.P.; Errazquin, D.; García, C.B.; Giner, B. Deep Eutectic Solvents: Are They Safe? *Appl. Sci.* **2021**, *11*, 10061. [CrossRef]
10. Benlebna, M.; Ruesgas-Ramón, M.; Bonafos, B.; Fouret, G.; Casas, F.; Coudray, C.; Durand, E.; Cruz Figueroa-Espinoza, M.; Feillet-Coudray, C. Toxicity of Natural Deep Eutectic Solvent Betaine:Glycerol in Rats. *J. Agric. Food Chem.* **2018**, *66*, 6205–6212. [CrossRef]
11. Benoit, C.; Virginie, C.; Boris, V. Chapter Twelve—The Use of NADES to Support Innovation in the Cosmetic Industry. In *Advances in Botanical Research*; Verpoorte, R., Witkamp, G.-J., Choi, Y.H., Eds.; Eutectic Solvents and Stress in Plants; Academic Press: Cambridge, MA, USA, 2021; Volume 97, pp. 309–332. [CrossRef]
12. Burnett, C.; Heldreth, B.; Bergfeld, W.F.; Belsito, D.V.; Hill, R.A.; Klaassen, C.D.; Liebler, D.C.; Marks, J.G.; Shank, R.C.; Slaga, T.J.; et al. Safety Assessment of Animal- and Plant-Derived Amino Acids as Used in Cosmetics. *Int. J. Toxicol.* **2014**, *33*, 5S–12S. [CrossRef]
13. Hayes, K.C.; Pronczuk, A.; Cook, M.W.; Robbins, M.C. Betaine in Sub-Acute and Sub-Chronic Rat Studies. *Food Chem. Toxicol.* **2003**, *41*, 1685–1700. [CrossRef]
14. Lazović, M.; Cvijetić, I.; Jankov, M.; Milojković-Opsenica, D.; Trifković, J.; Ristivojević, P. COSMO-RS in Prescreening of Natural Eutectic Solvents for Phenolic Extraction from Teucrium Chamaedrys. *J. Mol. Liq.* **2023**, *387*, 122649. [CrossRef]

15. Jankov, M.; Ristivojević, P.; Cvijetić, I.; Milojković-Opsenica, D. Assessing Radical Scavenging Capacity of *Sempervivum tectorum* L. Leaf Extracts: An Integrated High-Performance Thin-Layer Chromatography/in Silico/Chemometrics Approach. *J. Chromatogr. A* **2023**, *1703*, 464082. [CrossRef]
16. Klamt, A.; Eckert, F.; Arlt, W. COSMO-RS: An Alternative to Simulation for Calculating Thermodynamic Properties of Liquid Mixtures. *Annu. Rev. Chem. Biomol. Eng.* **2010**, *1*, 101–122. [CrossRef]
17. Bonté, F.; Girard, D.; Archambault, J.-C.; Desmoulière, A. Skin Changes During Ageing. In *Biochemistry and Cell Biology of Ageing: Part II Clinical Science*; Harris, J.R., Korolchuk, V.I., Eds.; Springer: Singapore, 2019; pp. 249–280. ISBN 9789811336812.
18. Jiratchayamaethasakul, C.; Ding, Y.; Hwang, O.; Im, S.-T.; Jang, Y.; Myung, S.-W.; Lee, J.M.; Kim, H.-S.; Ko, S.-C.; Lee, S.-H. In Vitro Screening of Elastase, Collagenase, Hyaluronidase, and Tyrosinase Inhibitory and Antioxidant Activities of 22 Halophyte Plant Extracts for Novel Cosmeceuticals. *Fish. Aquat. Sci.* **2020**, *23*, 6. [CrossRef]
19. Villarama, C.D.; Maibach, H.I. Glutathione as a Depigmenting Agent: An Overview. *Int. J. Cosmet. Sci.* **2005**, *27*, 147–153. [CrossRef]
20. Kim, H.H.; Kim, J.K.; Kim, J.; Jung, S.-H.; Lee, K. Characterization of Caffeoylquinic Acids from Lepisorus Thunbergianus and Their Melanogenesis Inhibitory Activity. *ACS Omega* **2020**, *5*, 30946–30955. [CrossRef]
21. Jimbow, K.; Minamitsuji, Y. Topical Therapies for Melasma and Disorders of Hyperpigmentation. *Dermatol. Ther.* **2001**, *14*, 35–45. [CrossRef]
22. Nakai, K.; Tsuruta, D. What Are Reactive Oxygen Species, Free Radicals, and Oxidative Stress in Skin Diseases? *Int. J. Mol. Sci.* **2021**, *22*, 10799. [CrossRef]
23. Emerit, I. Free Radicals and Aging of the Skin. In *Free Radicals and Aging*; Emerit, I., Chance, B., Eds.; Birkhäuser: Basel, Switzerland, 1992; pp. 328–341. ISBN 978-3-0348-7460-1.
24. Wittenauer, J.; Mäckle, S.; Sußmann, D.; Schweiggert-Weisz, U.; Carle, R. Inhibitory Effects of Polyphenols from Grape Pomace Extract on Collagenase and Elastase Activity. *Fitoterapia* **2015**, *101*, 179–187. [CrossRef]
25. Tu, P.T.B.; Tawata, S. Anti-Oxidant, Anti-Aging, and Anti-Melanogenic Properties of the Essential Oils from Two Varieties of Alpinia Zerumbet. *Molecules* **2015**, *20*, 16723–16740. [CrossRef]
26. Lee, S.-G.; Karadeniz, F.; Seo, Y.; Kong, C.-S. Anti-Melanogenic Effects of Flavonoid Glycosides from Limonium Tetragonum (Thunb.) Bullock via Inhibition of Tyrosinase and Tyrosinase-Related Proteins. *Molecules* **2017**, *22*, 1480. [CrossRef] [PubMed]
27. Sarkhail, P.; Salimi, M.; Sarkheil, P.; Kandelous, H.M. Anti-Melanogenic Activity and Cytotoxicity of Pistacia Vera Hull on Human Melanoma SKMEL-3 Cells. *Acta Med. Iran.* **2017**, *55*, 422–428. [CrossRef]
28. Zemour, K.; Labdelli, A.; Adda, A.; Dellal, A.; Talou, T.; Merah, O. Phenol Content and Antioxidant and Antiaging Activity of Safflower Seed Oil (*Carthamus tinctorius* L.). *Cosmetics* **2019**, *6*, 55. [CrossRef]
29. Tawaha, K.; Alali, F.Q.; Gharaibeh, M.; Mohammad, M.; El-Elimat, T. Antioxidant Activity and Total Phenolic Content of Selected Jordanian Plant Species. *Food Chem.* **2007**, *104*, 1372–1378. [CrossRef]
30. Pękal, A.; Pyrzynska, K. Evaluation of Aluminium Complexation Reaction for Flavonoid Content Assay. *Food Anal. Methods* **2014**, *7*, 1776–1782. [CrossRef]
31. Sachett, A.; Gallas-Lopes, M.; Conterato, G.M.M.; Herrmann, A.; Piato, A. Antioxidant Activity by DPPH Assay: In Vitro Protocol. 2021. Available online: https://www.protocols.io/view/antioxidant-activity-by-dpph-assay-in-vitro-protoc-q26g783n9lwz/v1 (accessed on 25 February 2024).
32. Re, R.; Pellegrini, N.; Proteggente, A.; Pannala, A.; Yang, M.; Rice-Evans, C. Antioxidant Activity Applying an Improved ABTS Radical Cation Decolorization Assay. *Free Radic. Biol. Med.* **1999**, *26*, 1231–1237. [CrossRef] [PubMed]
33. Chan, E.W.C.; Lim, Y.Y.; Wong, L.F.; Lianto, F.S.; Wong, S.K.; Lim, K.K.; Joe, C.E.; Lim, T.Y. Antioxidant and Tyrosinase Inhibition Properties of Leaves and Rhizomes of Ginger Species. *Food Chem.* **2008**, *109*, 477–483. [CrossRef]
34. Mosmann, T. Rapid Colorimetric Assay for Cellular Growth and Survival: Application to Proliferation and Cytotoxicity Assays. *J. Immunol. Methods* **1983**, *65*, 55–63. [CrossRef]
35. Stewart, J.J.P. Optimization of Parameters for Semiempirical Methods VI: More Modifications to the NDDO Approximations and Re-Optimization of Parameters. *J. Mol. Model.* **2013**, *19*, 1–32. [CrossRef] [PubMed]
36. Klamt, A.; Jonas, V.; Bürger, T.; Lohrenz, J.C.W. Refinement and Parametrization of COSMO-RS. *J. Phys. Chem. A* **1998**, *102*, 5074–5085. [CrossRef]
37. Qin, Z.; Cheng, H.; Song, Z.; Ji, L.; Chen, L.; Qi, Z. Selection of Deep Eutectic Solvents for Extractive Deterpenation of Lemon Essential Oil. *J. Mol. Liq.* **2022**, *350*, 118524. [CrossRef]
38. Rusak, G.; Komes, D.; Likić, S.; Horžić, D.; Kovač, M. Phenolic Content and Antioxidative Capacity of Green and White Tea Extracts Depending on Extraction Conditions and the Solvent Used. *Food Chem.* **2008**, *110*, 852–858. [CrossRef]
39. Venskutonis, P.R.; Škėmaitė, M.; Ragažinskienė, O. Radical Scavenging Capacity of Agrimonia Eupatoria and Agrimonia Procera. *Fitoterapia* **2007**, *78*, 166–168. [CrossRef] [PubMed]
40. Vlase, L.; Benedec, D.; Hanganu, D.; Damian, G.; Csillag, I.; Sevastre, B.; Mot, A.C.; Silaghi-Dumitrescu, R.; Tilea, I. Evaluation of Antioxidant and Antimicrobial Activities and Phenolic Profile for Hyssopus Officinalis, Ocimum Basilicum and Teucrium Chamaedrys. *Molecules* **2014**, *19*, 5490–5507. [CrossRef]
41. Naveed, M.; Hejazi, V.; Abbas, M.; Kamboh, A.A.; Khan, G.J.; Shumzaid, M.; Ahmad, F.; Babazadeh, D.; FangFang, X.; Modarresi-Ghazani, F.; et al. Chlorogenic Acid (CGA): A Pharmacological Review and Call for Further Research. *Biomed. Pharmacother.* **2018**, *97*, 67–74. [CrossRef]

42. Boots, A.W.; Haenen, G.R.M.M.; Bast, A. Health Effects of Quercetin: From Antioxidant to Nutraceutical. *Eur. J. Pharmacol.* **2008**, *585*, 325–337. [CrossRef]
43. Wilhelm, K.; Böttjer, B.; Siegers, C. Quantitative Assessment of Primary Skin Irritants in Vitro in a Cytotoxicity Model: Comparison with in Vivo Human Irritation Tests. *Br. J. Dermatol.* **2001**, *145*, 709–715. [CrossRef]
44. Milutinović, M.G.; Maksimović, V.M.; Cvetković, D.M.; Nikodijević, D.D.; Stanković, M.S.; Pešić, M.; Marković, S.D. Potential of *Teucrium chamaedrys* L. to Modulate Apoptosis and Biotransformation in Colorectal Carcinoma Cells. *J. Ethnopharmacol.* **2019**, *240*, 111951. [CrossRef]
45. Şöhretoğlu, D.; Sari, S.; Barut, B.; Özel, A. Tyrosinase Inhibition by Some Flavonoids: Inhibitory Activity, Mechanism by in Vitro and in Silico Studies. *Bioorg. Chem.* **2018**, *81*, 168–174. [CrossRef] [PubMed]
46. Fan, M.; Ding, H.; Zhang, G.; Hu, X.; Gong, D. Relationships of Dietary Flavonoid Structure with Its Tyrosinase Inhibitory Activity and Affinity. *LWT* **2019**, *107*, 25–34. [CrossRef]
47. Phu, D.H.; Tho, L.H.; Truong, D.V.N.; Hai, N.X.; Mai, N.T.T.; Nhan, N.T. Flavonoids from the Stem Barks of Swintonia Griffithii Kurz (Anacadiaceae) and Their Tyrosinase Inhibitory Activities. *Vietnam J. Chem.* **2019**, *57*, 581–584. [CrossRef]
48. JEONG, C.H.; SHIM, K.H. Tyrosinase Inhibitor Isolated from the Leaves of Zanthoxylum Piperitum. *Biosci. Biotechnol. Biochem.* **2004**, *68*, 1984–1987. [CrossRef]
49. Yao, Y.; Cheng, X.; Wang, L.; Wang, S.; Ren, G. Mushroom Tyrosinase Inhibitors from Mung Bean (*Vigna radiatae* L.) Extracts. *Int. J. Food Sci. Nutr.* **2012**, *63*, 358–361. [CrossRef]
50. HONG, C.-O.; LEE, H.A.; RHEE, C.H.; CHOUNG, S.-Y.; LEE, K.-W. Separation of the Antioxidant Compound Quercitrin from Lindera Obtusiloba Blume and Its Antimelanogenic Effect on B16F10 Melanoma Cells. *Biosci. Biotechnol. Biochem.* **2013**, *77*, 58–64. [CrossRef]
51. Garcia-Jimenez, A.; Teruel-Puche, J.A.; Garcia-Ruiz, P.A.; Saura-Sanmartin, A.; Berna, J.; Rodríguez-López, J.N.; Garcia-Canovas, F. Action of Tyrosinase on Caffeic Acid and Its N-Nonyl Ester. Catalysis and Suicide Inactivation. *Int. J. Biol. Macromol.* **2018**, *107*, 2650–2659. [CrossRef]
52. Chiocchio, I.; Mandrone, M.; Sanna, C.; Maxia, A.; Tacchini, M.; Poli, F. Screening of a Hundred Plant Extracts as Tyrosinase and Elastase Inhibitors, Two Enzymatic Targets of Cosmetic Interest. *Ind. Crops Prod.* **2018**, *122*, 498–505. [CrossRef]
53. Lianza, M.; Mandrone, M.; Chiocchio, I.; Tomasi, P.; Marincich, L.; Poli, F. Screening of Ninety Herbal Products of Commercial Interest as Potential Ingredients for Phytocosmetics. *J. Enzyme Inhib. Med. Chem.* **2020**, *35*, 1287–1291. [CrossRef]
54. An, S.M.; Kim, H.J.; Kim, J.-E.; Boo, Y.C. Flavonoids, Taxifolin and Luteolin Attenuate Cellular Melanogenesis despite Increasing Tyrosinase Protein Levels. *Phytother. Res.* **2008**, *22*, 1200–1207. [CrossRef]
55. Shimogaki, H.; Tanaka, Y.; Tamai, H.; Masuda, M. In Vitro and in Vivo Evaluation of Ellagic Acid on Melanogenesis Inhibition. *Int. J. Cosmet. Sci.* **2000**, *22*, 291–303. [CrossRef] [PubMed]
56. Lazović, M.; Cvijetić, I.; Jankov, M.; Milojković-Opsenica, D.; Trifković, J.; Ristivojević, P. Efficiency of Natural Deep Eutectic Solvents to Extract Phenolic Compounds from Agrimonia Eupatoria: Experimental Study and In Silico Modelling. *Plants* **2022**, *11*, 2346. [CrossRef] [PubMed]
57. Kang, H.-S.; Kim, H.-R.; Byun, D.-S.; Park, H.-J.; Choi, J.-S. Rosmarinic Acid as a Tyrosinase Inhibitors from Salvia Miltiorrhiza. *Nat. Prod. Sci.* **2004**, *10*, 80–84.
58. Burnett, C.L.; Bergfeld, W.F.; Belsito, D.V.; Hill, R.A.; Klaassen, C.D.; Liebler, D.C.; Marks, J.G.; Shank, R.C.; Slaga, T.J.; Snyder, P.W.; et al. Safety Assessment of Alkyl Betaines as Used in Cosmetics. *Int. J. Toxicol.* **2018**, *37*, 28S–46S. [CrossRef]
59. Zhao, D.; Liao, Y.; Zhang, Z. Toxicity of Ionic Liquids. *CLEAN—Soil Air Water* **2007**, *35*, 42–48. [CrossRef]
60. Kumar, A.K.; Shah, E.; Patel, A.; Sharma, S.; Dixit, G. Physico-Chemical Characterization and Evaluation of Neat and Aqueous Mixtures of Choline Chloride + lactic Acid for Lignocellulosic Biomass Fractionation, Enzymatic Hydrolysis and Fermentation. *J. Mol. Liq.* **2018**, *271*, 540–549. [CrossRef]
61. Nicander, I.; Åberg, P.; Ollmar, S. The Use of Different Concentrations of Betaine as a Reducing Irritation Agent in Soaps Monitored Visually and Non-Invasively. *Skin Res. Technol.* **2003**, *9*, 43–49. [CrossRef]

processes

Article

The Effects of Drying and Grinding on the Extraction Efficiency of Polyphenols from Grape Skin: Process Optimization

Lea Peternel, Tea Sokač Cvetnić, Jasenka Gajdoš Kljusurić, Tamara Jurina, Maja Benković, Ivana Radojčić Redovniković, Ana Jurinjak Tušek * and Davor Valinger

Faculty of Food Technology and Biotechnology, University of Zagreb, Pierottijeva 6, 10 000 Zagreb, Croatia; tamara.jurina@pbf.uniz.hr (T.J.); irredovnikovic@pbf.unizg.hr (I.R.R.)
* Correspondence: ana.tusek.jurinjak@pbf.unizg.hr

Abstract: Maximizing the yield of bioactive molecules extracted from plant materials requires the investigation of extraction process variables; therefore, in this research, a traditional aqueous solid–liquid extraction method was employed on two distinct grape pomace skin samples. The grape skin pomace represents a potentially valuable source of biologically active compounds, particularly polyphenols. Experiment 1 utilized ground grape pomace skin, whereas experiment 2 utilized grape pomace skin that had been both dried and ground beforehand. Employing a Box–Benkhen experimental design and response surface modeling in the Statistica 14.0 software package, this study evaluated the impact of temperature, extraction time, solid-to-liquid ratio (S/L), and mixing speed on extraction efficiency. The extracted compounds were assessed for both physical properties (conductivity, total dissolved solids, and pH) and chemical properties (total polyphenol content and antioxidant activity using 2,2-Diphenyl-1-picrylhydrazyl (DPPH) and ferric reducing antioxidant power (FRAP) assays). The optimization matrix design identified the specific conditions required to achieve the optimal physical and chemical properties of grape skin extract as follows: (i) for experiment 1, extraction time (t) = 15 min, temperature (T) = 80 °C, solid-to-liquid ratio (S/L) = 10 g/L, and mixing speed (rpm) = 500 1/min and (ii) for experiment 2, extraction time (t) = 15 min, temperature (T) = 80 °C, solid-to-liquid ratio (S/L) = 10 g/L, and mixing speed (rpm) = 375 1/min. Under optimal process conditions, 26.1284 $mg_{GAE}/g_{d.m.}$ and 25.1024 $mg_{GAE}/g_{d.m.}$, respectively, were obtained. These findings demonstrate the effectiveness of the optimization process in identifying precise extraction conditions that yield the optimal chemical properties of grape skin extracts.

Keywords: drying; grape skin; extraction condition optimization; grinding; polyphenol aqueous extraction; response surface modeling

Citation: Peternel, L.; Sokač Cvetnić, T.; Gajdoš Kljusurić, J.; Jurina, T.; Benković, M.; Radojčić Redovniković, I.; Jurinjak Tušek, A.; Valinger, D. The Effects of Drying and Grinding on the Extraction Efficiency of Polyphenols from Grape Skin: Process Optimization. *Processes* **2024**, *12*, 1100. https://doi.org/10.3390/pr12061100

Academic Editor: Agnieszka Zgoła-Grześkowiak

Received: 30 April 2024
Revised: 24 May 2024
Accepted: 25 May 2024
Published: 27 May 2024

1. Introduction

Grape pomace, the main solid organic waste generated during wine production, is estimated to produce approximately 1 kg of pomace for every 6 L of wine produced. Globally, annual pomace production is estimated at around 9 million tons [1], with Croatia alone producing about 40,600 tons [2]. The composition of grape pomace mainly consists of grape skin, seeds, and stems, with their proportions influenced by factors such as grape variety, maturity, climate, and processing methods [3]. Red grape pomace is separated after partial fermentation, while white grape pomace is separated before fermentation begins. The chemical composition of pomace is influenced by various factors, including the vinification or fermentation method, grape variety, harvest year, climatic conditions, geographical origin, cultivation method, storage method, and isolation technique, making comparisons between studies challenging [4].

Pomace moisture content typically ranges from 50% to 72% and varies based on grape variety and ripeness [4]. Lignin constitutes 16.8% to 24.2% of pomace components, while protein accounts for less than 4% [5]. Pectin compounds account for 37% to 54% of the total polysaccharides in grape cell walls, while cellulose makes up the remaining 27% to 37% of polymeric components. Various products such as ethanol, tartrate, citric acid, oil, hydrocolloid, and dietary fiber can be extracted from grape pomace. Additionally, grape pomace is rich in polyphenols (like resveratrol, anthocyanins, flavones, and tannins) that remain in the grape pomace during grape processing, and it can serve as an unconventional source of pectin [4]. Specified properties make the grape pomace an interesting raw material for the food industry. Several papers on the valorization of grape pomace have been published in the last few years. For example, Megías-Pérez et al. [6] optimized pectin extraction from grape pomace, and Filippi et al. [7] used grape pomace as the substrate for succinic acid production using the bacterial strain *Actinobacillus succinogenes*, while Elejade et al. [8] analyzed the potential of polyphenol supplementation for exercise-induced oxidation stress. Also, the effect of the addition of grape pomace to food products like bread [9–11], cookies [12], muffins [9,13], and/or biscuits [14,15] has been extensively analyzed. Grape pomace has also been used for the fortification of meat and fish products [16,17], as well as the fortification of dairy products [18,19].

Phenolic compounds are abundant and diverse natural products found in plants, exhibiting a wide range of physiological properties, including anti-allergic, anti-inflammatory, antimicrobial, antioxidant, cardioprotective, and vasodilating effects [20]. Many phenolic compounds have been identified in grape pomace, with anthocyanins, hydroxybenzoic and hydroxycinnamic acids, flavan-3-ols, flavonols, and stilbenes being the most prevalent [21,22].

Classical solid–liquid extraction serves as the basis for many analytical procedures in sample preparation [23]. Maximizing the yield of bioactive molecules extracted from plant materials requires the investigation of extraction process variables. Solvent extraction is favored for extracting bioactive molecules as it allows the recovery of thermally sensitive materials at low temperatures [24]. Extraction efficiency is influenced by variables such as temperature, solvent type, extraction time, solid-to-liquid ratio, and the sample's chemical composition and physical characteristics [25]. The solvent choice is critical, with commonly used solvents including water, methanol, ethanol, acetone, ethyl acetate, and aqueous alcohol solutions. The solvent choice affects both the amount and rate of polyphenol extraction [26]. Organic solvents used in plant extraction are typically those with boiling temperatures below 80 °C and exhibit low reactivity, low viscosity, cost-effectiveness, suitability for reuse, and heat, oxygen, and light stability, safety, and availability [27]. Considering these criteria, water emerges as a suitable solvent for solid–liquid extraction processes, being a low-cost, non-dangerous polar solvent capable of effectively extracting a broad range of phenolic molecules with notable antioxidant properties originating from a variety of plant sources [28].

Statistical and mathematical modeling techniques have been successfully utilized to analyze variables significant for extracting bioactive molecules from plant materials and predict optimal extraction conditions [29]. Traditional one-variable-at-a-time methods are time-consuming, expensive, and fail to consider the interaction effects of components, thus being less reliable. The Response Surface Methodology (RSM) addresses these limitations by simultaneously evaluating multiple factors and how they interact with one or more response variables, lowering the number of experimental procedures needed to extract bioactive substances [30–33]. There are several examples of the optimization of polyphenol extraction from grape pomace. For example, Moutinho et al. [34] applied the RSM to optimize polyphenol extraction from two grape varieties, taking into consideration the ethanol concentration, extraction temperature, and extraction pH. Ćurko et al. [35] optimized the effect of methanol concentration, temperature, and extraction time on the microwave-assisted extraction of different groups of polyphenolic compounds from grape skin pomace, while Putnik et al. [36] optimized the different acidities and extraction time effects on flavonoid recovery from grape skin pomace (*Vitis vinifera* L. cv. Merlot).

Based on the previously mentioned research gap, the primary goal of this study was to identify the ideal conditions for the conventional aqueous solid–liquid extraction of bioactive compounds from dried and ground grape skin using the Box–Behnken experimental design and RSM modeling. The novelty of this study lies in its systematic application of the Box–Behnken design and RSM modeling to optimize the extraction process, potentially transforming grape skin, a typically discarded byproduct, into a valuable source of bioactive compounds. This approach not only contributes to waste reduction but also adds economic value to the food industry by utilizing a resource that is often considered unprofitable.

2. Materials and Methods

2.1. Materials

2.1.1. Grape Skin Samples

This study analyzed the pomace skin of white Graševina grapes (*Vitis vinifera* cv. Graševina) harvested in Kutjevo, Croatia in 2021. To ensure consistency, the pomace was stored at −18 °C before experimentation. The seeds were separated from the skin before analysis. In order to establish a consistent sample, all batches were blended together to reduce the effects of variability within the pomace. The dry matter content of the grape skin was determined based on the AOAC method [37]. The physical and chemical characteristics of the grape skin used in the extraction experiments were as follows: moisture content of 65.07%, pH of 4.60, and C/N ratio of 32.16.

2.1.2. Chemicals and Reagents

The following materials were acquired from Sigma-Aldrich Chemie (St. Louis, MO, USA): 2,4,6-tris(2-pyridyl)-striazine (TPTZ), gallic acid (≥98% purity), iron (II) sulphate heptahydrate, 1,1-diphenyl-2-picrylhydrazyl (DPPH), and 6-hydroxy-2,5,7,8-tetramethylchromane-2 carboxylic acid (Trolox). Gram-Mol d.o.o. (Zagreb, Croatia) supplied the sodium carbonate, 30% hydrochloric acid, and hexahydrate iron (III) chloride. J.T. Baker (Deventer, The Netherlands) provided the sodium acetate trihydrate. The Folin–Ciocalteu reagent was sourced from Kemika d.d., located in Zagreb, Croatia. Acetic acid was supplied by T.T.T. d.o.o. in Sveta Nedjelja, Croatia. Methanol was obtained from Carlo Erba Reagents S.A.S. (Peypin, France).

2.2. Methods

2.2.1. Grape Skin Pretreatments

Grape skin, both fresh and dried at T = 50 °C for 24 h, was ground using an IKA Tube Mill (IKA Werk GmbH & Co. KG, Staufen, Germany) at a blade rotation speed of rpm = 15,000 1/min. The grinding time was adjusted according to the size of the particles to be obtained (t = 10–40 s). Both fresh grape skin and grape skin that had been previously dried were ground to a particle size ranging from 100 to 300 µm. The experimental procedure is presented in Figure 1.

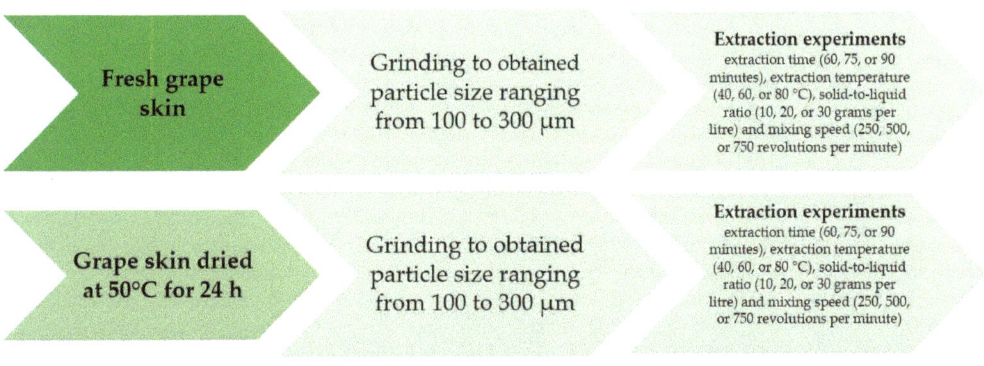

Figure 1. Experimental procedure.

2.2.2. Extraction Experiments

The experiments were performed according to the Box–Behnken design previously described by Sokač Cvetnić et al. [38]. This experimental design examined the impact of four variables on the extracts' polyphenol production: extraction time (60, 75, or 90 min), extraction temperature (40, 60, or 80 °C), solid-to-liquid ratio (10, 20, or 30 g per liter) and mixing speed (250, 500, or 750 revolutions per minute). The experimental conditions were selected based on the available data in the literature [39]. The extraction experiments were performed in a laboratory oil bath (IKA-Werk GmbH & Co. KG, Staufen, Germany) with precise control of the temperature and mixing speed. Following the specified extraction period, the mixture was filtered through a cellulose paper filter (pore size of 5–13 μm, LLG Labware, Meckenheim, Germany) in order to isolate the solid material from the liquid extract that contained polyphenols. Then, the physical and chemical characteristics of the extracts were examined.

2.2.3. Physical Properties of the Grape Skin Extracts

The pH values of the aqueous extracts were determined using a pH meter (Model 914, Metrohm, Switzerland). A conductometer (SevenCompact, Mettler Toledo, Switzerland) was used to test the extracts' conductivity and total dissolved solids. The AOAC method was used to evaluate the extraction yield [37]. The physical properties were measured three times, with mean values ± standard deviations reported for each measurement.

2.2.4. Total Phenolic Content and Antioxidant Activity of the Grape Skin Extracts

A spectrophotometric approach based on the colorimetric reaction of phenol with the Folin–Ciocalteu reagent was used to quantify the total amount of polyphenols [40]. The findings are given in milligrams of gallic acid equivalent (GAE) for each gram of the sample's dry matter. The DPPH and FRAP techniques were used to quantify antioxidant activity. By using the process outlined by Brand-Williams et al. [41], the 2,2-diphenyl-2-picrylhydrazyl (DPPH) radical was reduced in a methanol solution as part of the DPPH method. The results are shown as millimoles of Trolox equivalent for each gram of sample dry matter. According to the method of Benzie and Strain [42], the FRAP technique comprises reducing the colorless complex of iron (III) tri-pyridyltriazine (Fe^{3+}-TPTZ) to the ferrous form (Fe^{2+}), resulting in a vivid blue color. The FRAP results are expressed in millimoles of $FeSO_4 \cdot 7H_2O$ per gram of dry matter in the sample. The extracts' chemical properties were evaluated in triplicate, and the findings are shown as mean values ± standard deviations.

2.2.5. Statistical Analysis and Data Modeling

Three parallel experiments were used to assess all the physical and chemical characteristics of the extracted materials. The Statistica 14.0 software package (TIBCO® Statistica, Palo Alto, CA, USA) was used to perform basic statistical analysis, including mean and standard deviation. The Spearman correlation matrix function in Statistica 14.0 was used to look at correlations between the extraction conditions and the samples' physical and chemical characteristics.

Utilizing a Box–Behnken design integrated into the Statistica 14.0 software program, the effects of four independent variables (extraction time (X_1), extraction temperature (X_2), solid-to-liquid ratio (X_3), and mixing speed (X_4)) were evaluated. The aqueous grape skin extracts were subjected to the simultaneous optimization of three chemical parameters (total phenolic content, TPC, DPPH antioxidant activity, and FRAP antioxidant activity) and two physical properties (conductivity and total dissolved solids— TDS), expressed as Y. Thirty experiments were conducted at random in accordance with the experimental design, and the effect of each variable was studied at three levels (−1, 0, and 1). Equations of second-order polynomials were used to fit the experimental data. Utilizing the Statistica 14.0 software (TIBCO® Statistica, Palo Alto, CA, USA), response surface modeling was carried out. The suggested response surface models served as the basis for estimating the ideal extraction conditions.

3. Results

3.1. Aqueous Extracts from Grape Skin: Their Physical and Chemical Properties

In this study, two classical solid–liquid extraction processes were conducted. In the first experiment, biologically active components were extracted from ground grape pomace skin, while in the second experiment, extraction was performed using previously dried and ground grape pomace skin. Thirty trials were carried out in a variety of settings to guarantee the best extraction conditions. The produced extracts' physical characteristics were analyzed, and the findings are shown in Table 1. Upon analyzing the results, it was observed that the lowest pH value in the first experimental set was 3.94 for sample 6 (t = 27.5 min, T = 60 °C, S/L = 30 g/L, and rpm = 250 L/min), while the highest pH value was 4.15 for sample 20 (t = 40 min, T = 60 °C, S/L = 10 mg/L, and rpm = 500). Nonetheless, based on the other pH values identified it may be concluded that the extraction conditions had very little impact on the extracts' pH. The results indicate that the prior drying of grape pomace skin does not notably affect the pH value as the range of pH values falls between 3.93 and 4.14. This range is in line with the pH range (pH 3.4–5.8) for this kind of residue, which is frequently mentioned in the literature [23,43].

According to the obtained results for experimental set 1, the lowest values for electrical conductivity and total dissolved solids were measured in sample 7 (t = 27.5 min, T = 60 °C, S/L = 10 g/L, and rpm = 750 L/min), and the highest values in sample 17 (t = 27.5 min, T = 80 °C, S/L = 30 g/L, and rpm = 500 L/min), indicating that increasing the temperature and extracting a larger amount of sample increases both the total dissolved substance and the electrical conductivity. For experimental set 2, the highest values for TDS and S were also shown by sample 17 (t = 27.5 min, T = 80 °C, S/L = 30 g/L, and rpm = 500 L/min), and the lowest values by sample 20 (t = 40 min, T = 60 °C, S/L = 10 g/L, and rpm = 500 L/min), again indicating that at higher temperatures and larger amounts of sample, there is an increase in total dissolved matter and electrical conductivity. According to the work of Pinel et al. [44], increasing the temperature promotes extraction by improving the solubility of the substance. By comparing experiments 1 and 2, it can be concluded that experiment 2, in which a sample of dried and ground skin was used, shows higher TDS and electrical conductivity values because such a sample of grape pomace is exposed to a larger contact surface with the solvent than only ground grape pomace skin, which is evident from the obtained results.

At different temperatures, mixing speeds, extraction times, and solid-to-liquid ratios, the chemical properties of the extracts were examined: the total amount of polyphenols and the antioxidant activity of grape pomace extracts were measured according to the DPPH and FRAP methods. For experiment 1, the highest amount of polyphenols was observed in sample 15 (t = 27.5 min, T = 80 °C, S/L = 10 g/L, and rpm = 500 L/min), at 15.33 $mg_{GAE}/g_{d.m.}$, while the lowest value was in sample 16 (t = 27.5 min, T = 40 °C, S/L = 30 g/L, and rpm = 500 L/min) at 3.55 $mg_{GAE}/g_{d.m.}$, almost five times lower than in sample 15. A similar trend was noticed for extracts after experiment 2; the highest value of total polyphenols was measured in sample 24 (t = 27.5 min, T = 80 °C, S/L = 20 g/L, and rpm = 250 L/min) at 17.13 $mg_{GAE}/g_{d.m.}$, while the lowest value was in sample 25 (t = 27.5 min, T = 40 °C, S/L = 20 g/L, and rpm = 750 L/min) at 5.04 $mg_{GAE}/g_{d.m.}$. Moreover, the nature and polarity of the solvent, the methods used for isolation, the purity of the active ingredients, and the instrument employed to measure the antioxidant activity all affect the antioxidant activity of plant extracts [45]. When determining the antioxidant activity of extracts after experiment 1, the highest values for DPPH and FRAP were measured precisely in sample 15: DPPH = 0.0447 $mmol_{Trolox}/g_{d.m.}$ and FRAP = 0.0779 $mmol_{Trolox}/g_{d.m.}$, while the lowest values for DPPH were seen in sample 1 (DPPH = 0.0002 $mmol_{Trolox}/g_{d.m.}$) and for FRAP in sample 16 (FRAP = 0.0135 $mmol_{FeSO4\cdot 7H2O}/g_{d.m.}$). Small deviations are visible with DPPH; however, if we look at samples 1 and 16, we see that the temperature parameters and the solid-to-liquid ratio are the same in both cases, confirming that the extraction efficiency increases with increasing temperature, while other parameters are not significant. In the case of extracts prepared according to experiment 2, the highest values of DPPH and FRAP were measured in sample 24 (DPPH = 0.0611 $mmol_{Trolox}/g_{d.m.}$ and FRAP = 0.1256 $mmol_{FeSO4·7H2O}/g_{d.m.}$), and the lowest values were recorded for sample 1 (t = 15 min, T = 40 °C, S/L = 20 g/L, and rpm = 500 L/min) for DPPH and sample 25 (t = 27.5 min, T = 40 °C, S/L = 20 g/L, and rpm = 750 L/min) for FRAP.

Table 1. Physical and chemical properties of the aqueous extracts from grape skin after experiment 1 (green shading) and experiment 2. X_1—extraction time, X_2—extraction temperature, X_3—solid-to-liquid ratio, X_4—mixing speed, S—conductivity, TDS—total dissolved solids, TPC—total polyphenol content, DPPH—antioxidant activity measured by the DPPH method, and FRAP—antioxidant activity measured by the FRAP method.

Exp.	X_1	X_2	X_3	X_4	pH	S (µS/cm)	TDS (mg/L)	TPC ($mg_{GAE}/g_{d.m.}$)	DPPH ($mmol_{Trolox}/g_{d.m.}$)	FRAP ($mmol_{FeSO4·7H2O}/g_{d.m.}$)
1.	15	40	20	500	4.16 ± 0.06	188.20 ± 1.90	380.50 ± 0.71	4.82 ± 0.0001	0.0002 ± 0.0017	0.0168 ± 0.0018
					4.04 ± 0.01	370.50 ± 7.78	704.00 ± 8.49	7.27 ± 0.1754	0.0158 ± 0.0006	0.0349 ± 0.0022
2.	40	40	20	500	4.02 ± 0.03	167.70 ± 0.14	319.50 ± 17.68	5.23 ± 0.0002	0.0119 ± 0.0007	0.0186 ± 0.0081
					4.00 ± 0.01	431.50 ± 0.71	868.00 ± 1.41	5.84 ± 0.0877	0.0124 ± 0.0001	0.0276 ± 0.0025
3.	15	80	20	500	4.03 ± 0.03	156.8 ± 0.14	311.00 ± 1.41	9.11 ± 0.0005	0.0401 ± 0.0015	0.0412 ± 0.0075
					3.97 ± 0.01	368.50 ± 0.71	756.00 ± 8.49	14.71 ± 0.2631	0.0548 ± 0.0006	0.0985 ± 0.0021
4.	40	80	20	500	3.95 ± 0.02	257.00 ± 0.00	488.50 ± 21.92	9.90 ± 0.0007	0.0447 ± 0.0005	0.0399 ± 0.0199
					3.98 ± 0.01	397.00 ± 0.00	811.00 ± 2.83	15.70 ± 0.5262	0.0497 ± 0.0014	0.1032 ± 0.0022
5.	27.5	60	10	250	4.12 ± 0.01	102.70 ± 0.63	214.50 ± 4.95	7.87 ± 0.0004	0.0210 ± 0.0007	0.0244 ± 0.0241
					4.13 ± 0.03	229.00 ± 0.00	456.50 ± 0.71	10.57 ± 0.7600	0.0326 ± 0.0016	0.0442 ± 0.0193
6.	27.5	60	30	250	3.94 ± 0.00	211.50 ± 0.70	418.50 ± 4.95	3.96 ± 0.0001	0.0125 ± 0.0002	0.0186 ± 0.0059
					3.91 ± 0.01	416.00 ± 15.56	861.50 ± 6.36	7.41 ± 0.3508	0.0236 ± 0.0004	0.0470 ± 0.0029
7.	27.5	60	10	750	4.13 ± 0.04	97.75 ± 0.49	198.70 ± 0.71	8.43 ± 0.0011	0.0182 ± 0.0026	0.0295 ± 0.0018
					4.02 ± 0.04	238.5 ± 0.71	488.00 ± 9.89	13.17 ± 0.0000	0.0395 ± 0.0016	0.0739 ± 0.0082
8.	27.5	60	30	750	3.99 ± 0.01	216.00 ± 2.83	442.00 ± 1.41	5.67 ± 0.0006	0.0114 ± 0.0011	0.0236 ± 0.0004
					3.98 ± 0.02	380.5 ± 6.36	7850 ± 2.82	7.28 ± 0.2631	0.0212 ± 0.0001	0.0398 ± 0.0041
9.	27.5	60	20	500	4.01 ± 0.01	148.15 ± 1.06	286.50 ± 4.95	5.61 ± 0.0004	0.0162 ± 0.0015	0.0265 ± 0.0018
					3.93 ± 0.04	337.50 ± 0.71	664.50 ± 6.36	8.51 ± 0.0877	0.0229 ± 0.0025	0.0421 ± 0.0030
10.	15	60	20	250	3.97 ± 0.01	164.45 ± 0.35	330.00 ± 0.00	5.29 ± 0.0005	0.0146 ± 0.0005	0.0239 ± 0.0038
					3.99 ± 0.01	350.50 ± 10.61	724.00 ± 2.83	10.49 ± 0.0000	0.0330 ± 0.0025	0.0629 ± 0.0023
11.	40	60	20	250	4.04 ± 0.01	152.20 ± 0.28	305.50 ± 0.71	5.38 ± 0.0006	0.0041 ± 0.0002	0.0262 ± 0.0001
					3.98 ± 0.04	350.00 ± 0.00	659.50 ± 23.33	11.17 ± 0.0000	0.0363 ± 0.0003	0.0624 ± 0.0077
12.	15	60	20	750	4.07 ± 0.04	155.00 ± 0.85	293.50 ± 4.95	4.87 ± 0.0009	0.0030 ± 0.0012	0.0231 ± 0.0002
					4.02 ± 0.04	373.50 ± 0.71	757.00 ± 11.31	7.95 ± 0.0000	0.0178 ± 0.0008	0.0360 ± 0.0015
13.	40	60	20	750	4.07 ± 0.04	165.75 ± 0.77	304.00 ± 8.49	5.75 ± 0.0012	0.0055 ± 0.0008	0.0293 ± 0.0011
					4.03 ± 0.03	289.50 ± 7.78	615.00 ± 14.14	9.62 ± 0.0000	0.0269 ± 0.0005	0.0415 ± 0.0157
14.	27.5	40	10	500	4.14 ± 0.01	118.90 ± 0.85	216.20 ± 39.32	6.84 ± 0.0017	0.0062 ± 0.0013	0.0265 ± 0.0101
					4.07 ± 0.01	266.00 ± 1.41	538.50 ± 4.95	8.09 ± 0.0000	0.0185 ± 0.0013	0.0015 ± 0.0355
15.	27.5	80	10	500	4.00 ± 0.04	194.55 ± 0.91	392.50 ± 0.71	15.33 ± 0.0018	0.0447 ± 0.0007	0.0779 ± 0.0237
					4.11 ± 0.02	231.00 ± 0.00	465.50 ± 4.95	16.27 ± 0.0000	0.0556 ± 0.0000	0.0647 ± 0.0392
16.	27.5	40	30	500	4.02 ± 0.00	232.50 ± 7.77	485.00 ± 0.00	3.55 ± 0.0004	0.0019 ± 0.0006	0.0135 ± 0.0103
					3.97 ± 0.01	429.00 ± 1.41	896.50 ± 2.12	5.30 ± 0.0000	0.0124 ± 0.0007	0.0167 ± 0.0105
17.	27.5	80	30	500	4.00 ± 0.01	295.50 ± 0.70	585.50 ± 2.12	8.68 ± 0.0015	0.0407 ± 0.0183	0.0419 ± 0.0053
					3.96 ± 0.02	484.00 ± 1.41	971.50 ± 0.71	14.23 ± 0.0000	0.0431 ± 0.0003	0.0744 ± 0.0083
18.	27.5	60	20	500	4.02 ± 0.03	140.30 ± 0.42	282.50 ± 0.71	4.54 ± 0.0006	0.0098 ± 0.0000	0.0232 ± 0.0215
					4.08 ± 0.02	273.50 ± 6.37	572.50 ± 4.95	8.38 ± 0.0000	0.0227 ± 0.0003	0.0308 ± 0.0281
19.	15	60	10	500	4.11 ± 0.01	106.75 ± 0.91	202.50 ± 3.54	7.50 ± 0.0018	0.0069 ± 0.0013	0.0319 ± 0.0055
					4.14 ± 0.01	200.00 ± 0.00	365.50 ± 6.36	11.56 ± 0.0000	0.0298 ± 0.0039	0.0453 ± 0.0029
20.	40	60	10	500	4.15 ± 0.07	122.95 ± 2.47	204.00 ± 5.66	9.92 ± 0.0015	0.0152 ± 0.0017	0.0459 ± 0.0068
					4.11 ± 0.02	195.80 ± 0.35	391.50 ± 0.71	8.71 ± 0.0000	0.0289 ± 0.0016	0.0589 ± 0.0310
21.	15	60	30	500	4.05 ± 0.00	173.50 ± 1.41	351.50 ± 0.71	5.05 ± 0.0005	0.0064 ± 0.0006	0.0173 ± 0.0021
					4.03 ± 0.01	376.50 ± 0.71	741.00 ± 11.31	6.79 ± 0.4092	0.0201 ± 0.0008	0.0392 ± 0.0017
22.	40	60	30	500	3.99 ± 0.01	208.00 ± 0.00	414.00 ± 1.41	5.57 ± 0.0007	0.0133 ± 0.0043	0.0245 ± 0.0074
					4.05 ± 0.01	446.00 ± 7.07	504.00 ± 2.83	7.49 ± 0.0000	0.0198 ± 0.0004	0.0401 ± 0.0073
23.	27.5	40	20	250	4.00 ± 0.02	202.00 ± 2.82	410.00 ± 0.00	4.63 ± 0.0001	0.0037 ± 0.0003	0.0174 ± 0.0006
					3.99 ± 0.01	405.00 ± 5.66	784.50 ± 4.95	5.84 ± 0.3508	0.0172 ± 0.0010	0.0294 ± 0.0050
24.	27.5	80	20	250	4.06 ± 0.01	191.85 ± 0.92	391.00 ± 5.66	8.27 ± 0.0018	0.0229 ± 0.0023	0.0420 ± 0.0056
					4.00 ± 0.01	386.00 ± 1.41	765.00 ± 5.65	17.13 ± 0.0877	0.0611 ± 0.0110	0.1256 ± 0.0028

Table 1. *Cont.*

Exp.	X_1	X_2	X_3	X_4	pH	S (μS/cm)	TDS (mg/L)	TPC ($mg_{GAE}/g_{d.m.}$)	DPPH ($mmol_{Trolox}/g_{d.m.}$)	FRAP ($mmol_{FeSO4·7H2O}/g_{d.m.}$)
25.	27.5	40	20	750	4.12 ± 0.03	147.75 ± 4.31	295.50 ± 9.19	5.52 ± 0.0002	0.0067 ± 0.0056	0.0167 ± 0.0003
					4.02 ± 0.00	374.50 ± 0.71	755.00 ± 1.41	5.04 ± 0.5262	0.0140 ± 0.0001	0.0266 ± 0.0068
26.	27.5	80	20	750	4.08 ± 0.02	250.50 ± 0.71	509.50 ± 4.95	11.07 ± 0.0012	0.0427 ± 0.0025	0.0534 ± 0.0076
					4.05 ± 0.01	389.00 ± 1.41	766.00 ± 7.07	14.15 ± 0.5262	0.0399 ± 0.0003	0.0828 ± 0.0029
27.	27.5	60	20	500	4.04 ± 0.03	158.75 ± 0.49	319.50 ± 0.71	6.92 ± 0.0002	0.0124 ± 0.0008	0.0291 ± 0.0026
					4.06 ± 0.00	398.50 ± 2.12	812.50 ± 7.78	8.32 ± 0.0000	0.0233 ± 0.0005	0.0421 ± 0.0000
28.	27.5	60	20	500	3.98 ± 0.02	153.85 ± 0.92	311.00 ± 0.00	5.71 ± 0.0004	0.0138 ± 0.0005	0.0310 ± 0.0003
					4.09 ± 0.01	355.50 ± 4.95	734.00 ± 5.66	7.45 ± 0.0877	0.0180 ± 0.0009	0.0387 ± 0.0052
29.	27.5	60	20	500	3.98 ± 0.02	151.15 ± 0.92	261.00 ± 9.89	5.85 ± 0.0005	0.0151 ± 0.0051	0.0300 ± 0.0006
					3.90 ± 0.01	394.50 ± 0.71	783.00 ± 2.83	8.51 ± 0.6139	0.0274 ± 0.0012	0.0496 ± 0.0020
30.	27.5	60	20	500	4.00 ± 0.01	144.75 ± 0.78	292.50 ± 0.71	6.17 ± 0.0009	0.0002 ± 0.0017	0.0245 ± 0.0004
					3.98 ± 0.01	374.50 ± 0.71	741.50 ± 12.02	9.62 ± 0.0000	0.0248 ± 0.0009	0.0546 ± 0.0029

According to the obtained results, we can conclude that an increase in temperature favored the extraction, while the mixing speed had a negative effect on the extraction efficiency. This negative effect might be due to the possible retention of the sample on the walls of the glass, or at a rotation speed of 750 L/min, the sample might have been ejected from the solution, resulting in a reduced amount of sample being extracted. Multiple researchers suggest that raising the extraction temperature improves the substance's solubility, which in turn promotes the rate of extraction.

Also, it is important to point out that temperature significantly influenced the obtained results while the mixing speed and the extraction time remained constant. Benchaachoua et al. [46] conducted a study on the influence of different solvents on the extraction of phenolic compounds from thistle root extract. They discovered that the water–ethanol (50:50, v/v) solvent extract had the highest concentration of phenolic chemicals, measuring 17.22 mgGAE/$g_{d.m.}$. The water extract had a slightly lower concentration of phenolic compounds (16.38 mgGAE/$g_{d.m.}$), but the difference was not statistically significant ($p < 0.05$). Comparable outcomes were found in the above study, indicating that the effectiveness of the solvent's extraction process is dependent on the type of phenolic chemicals present in the plant. Although the water–ethanol solvent is appropriate for acquiring biologically active compounds with a variety of polarizations due to the relatively polar environment created by the addition of organic solvents to water, the application of water as an extraction solvent generates a highly polar surrounding that is acceptable for extracting strongly polarizing biologically active substances [47]. Valinger et al. [24], who conducted the extraction of polyphenols from olive leaves, obtained the highest concentration of polyphenols at 80 °C and the lowest value at 40 °C while the other parameters of mixing speed and extraction time remained the same. They also found the highest antioxidant activity using the DPPH and FRAP methods in the same samples with the highest polyphenol concentration. These outcomes agree with this study's conclusions. By comparing the chemical properties of extracts prepared in experiment 1 and experiment 2, it can be seen that slightly higher values were obtained in experiment 2, where a dried and ground sample was used, which contributed to a larger contact surface with the solvent itself, compared to the case of only ground samples in experiment 1, which resulted in better extraction. The obtained polyphenol concentration from the extraction process aligns with the findings in the existing literature. For instance, Libran et al. [48] reported an extraction efficiency of 5 $mg_{GAE}/g_{d.m.}$ from white grape skin. Similarly, Yammine et al. [49] documented an extraction efficiency of 3.07 $mg_{GAE}/g_{d.m.}$, and Gerardi et al. [50] reported a range of 1.2 to 3.07 $mg_{GAE}/g_{d.m.}$ for polyphenol extraction efficiency. It is worth noting that ethanol was employed as the extracting agent in all the aforementioned experiments. Comparing our results to those presented by Sokač Cvetnić et al. [38], it can be noticed that grape skin grinding and drying have a significant effect on extraction efficiency. The abovementioned authors reported a value of 8.38 $mg_{GAE}/g_{d.m.}$ under optimal process conditions.

Optimizing extraction time and temperature is crucial for lowering energy costs and achieving high compound recovery percentages. There is broad agreement among authors that raising the working temperature improves the solubility of chemicals, which in turn improves extraction efficiency. High temperatures do, nonetheless, have a beneficial influence on extraction yields; however, these variables are unable to be elevated continuously since temperatures above 80 °C may degrade polyphenols. [51]. Some studies have indicated that a temperature of 60 °C enhances phenol extraction from grapes [52]. Additionally, according to Pinelo et al. [44], contact time is not as significant when water is used as a solvent.

3.2. Relationships between Extraction Conditions and Physical and Chemical Properties of the Aqueous Extracts from Grape Skin

To elucidate the relationship between the physical and chemical properties of the prepared grape skin extracts, a correlation matrix was constructed. Tables 2 and 3 present the correlation coefficients between temperature, solid-to-liquid ratio, extraction time, mixing speed, pH, total dissolved solids (TDS), electrical conductivity, total polyphenol concentration, and antioxidant activity determined by both the DPPH and FRAP methods. Significant correlations are highlighted in bold.

For extracts from experiment 1, total polyphenols exhibited positive correlations with extraction time (r = 0.110), temperature (r = 0.681), and mixing speed (r = 0.127), while a negative correlation was observed with the solid-to-liquid ratio (r = −0.502). This trend was similarly observed for antioxidant activities determined by the DPPH and FRAP methods. Comparable findings were reported by Pinelo et al. [44], who discovered that temperature and solid-to-liquid ratio were crucial factors in maximizing extraction effectiveness. Notably, pH values showed a positive correlation solely with total polyphenols (r = 0.103), while TDS exhibited positive correlations with all variables, except pH. Additionally, extraction time and temperature displayed positive correlations with all chemical and physical characteristics, except pH, while the solid-to-liquid ratio exhibited negative correlations with all parameters, except TDS and electrical conductivity.

Table 2. Spearman correlation matrix of physical and chemical characteristics of grape pomace skin extracts prepared by the classical extraction procedure for experiment 1. Significant correlations are shown in bold.

	t	T	S/L	rpm	pH	TDS	S	TPC	DPPH	FRAP
t	**1**									
T	0.000	**1**								
S/L	0.000	0.000	**1**							
rpm	0.000	0.000	0.000	**1**						
pH	**−0.138**	**−0.166**	**−0.390**	**0.163**	**1**					
TDS	**0.147**	**0.331**	**0.678**	0.009	**−0.384**	**1**				
S	**0.091**	**0.313**	**0.694**	−0.014	**−0.367**	**0.989**	**1**			
TPC	**0.110**	**0.681**	**−0.502**	**0.127**	**0.103**	**0.116**	**0.094**	**1**		
DPPH	**0.122**	**0.754**	**−0.391**	**0.094**	−0.005	**0.196**	**0.166**	**0.946**	**1**	
FRAP	**0.095**	**0.824**	**−0.105**	0.035	**−0.199**	**0.439**	**0.417**	**0.820**	**0.799**	**1**

In the case of extracts from experiment 2, extraction time positively correlated with antioxidant activity measured by both the DPPH (r = 0.034) and FRAP (r = 0.010) methods, as well as with TDS (r = 0.049) and pH values (r = 0.056), while negatively correlating with electrical conductivity (r = −0.069) and total polyphenols (r = −0.004). Positive correlations were observed between extraction temperature and chemical characteristics (total polyphenols, DPPH, and FRAP), with negative correlations evident with physical characteristics. Mixing speed (rpm) displayed negative correlations with all variables, except pH value. Furthermore, chemical characteristics (total polyphenols and antioxidant activity measured by both the FRAP and DPPH methods) exhibited positive correlations with each other and with temperature.

Table 3. Spearman correlation matrix of physical and chemical characteristics of grape pomace skin extracts prepared by the classical extraction procedure for experiment 2. Significant correlations are shown in bold.

	t	T	S/L	rpm	pH	TDS	S	TPC	DPPH	FRAP
t	**1**									
T	0.000	**1**								
S/L	0.000	0.000	**1**							
rpm	0.000	0.000	0.000	**1**						
pH	**0.056**	−0.012	**−0.641**	**0.100**	**1**					
TDS	**0.049**	**−0.093**	**0.832**	**−0.065**	**−0.709**	**1**				
S	**−0.069**	**−0.056**	**0.713**	−0.029	**−0.750**	**0.882**	**1**			
TPC	−0.004	**0.861**	**−0.312**	**−0.085**	**0.051**	**−0.259**	**−0.188**	**1**		
DPPH	0.034	**0.839**	**−0.063**	**−0.145**	**−0.138**	0.007	**0.051**	**0.873**	**1**	
FRAP	0.010	**0.858**	**−0.259**	**−0.178**	0.002	**−0.223**	**−0.149**	**0.970**	**0.905**	**1**

3.3. RSM Modeling and Optimization of the Grape Extraction Procedure

This investigation examined the effects of four process variables—temperature, time, mixing speed, and solid-to-liquid ratio—on extracts' chemical and physical characteristics, including total polyphenol content and antioxidant activity as measured by FRAP and DPPH assays, as well as their conductivity, total dissolved solids, and pH. To fit the model to the data, second-order polynomial equations were used in the course of this study.

The coefficients of determination (R^2) of the developed models ranged from 0.2709 to 0.9609 (see Table 4). The best agreement between the Response Surface Methodology (RSM) model and experimental data was observed for conductivity values in experimental set 1, while the lowest agreement was found for the pH value in experimental set 2. The R^2 values for total dissolved solids and electrical conductivity ranged from 0.9555 to 0.9609, indicating a very good fit of the model to the experimental data. For total polyphenols, an R^2 value of 0.8976 was obtained, while for antioxidant activity, the RSM-based models showed a high agreement with the experimental data, with R^2 = 0.9469 for DPPH and 0.8937 for FRAP.

Table 4. RSM models for the description of physical and chemical characteristics of grape pomace skin extracts prepared by the classical extraction procedure after experiment 1 (green shading) and experiment 2. Statistically significant coefficients are shown in bold.

Model Variable	RSM Equation	R^2	F-Value	*p*-Value
pH	$Y = \mathbf{4.091} - \mathbf{0.069}\cdot X_3$	0.2709	5.4624	0.0274
	$Y = \mathbf{4.027} - \mathbf{0.065}\cdot X_3$	0.4417	4.5495	0.0075
TDS	$Y = 153.597 + 10.740\cdot X_1 + \mathbf{30.133}\cdot X_2 + \mathbf{52.070}\cdot X_3 + 1.58\cdot X_4 - 4.293\cdot X_1^2 - \mathbf{46.784}\cdot X_2^2 - 3.901\cdot X_3^2 - 4.464\cdot X_4^2 + \mathbf{36.195}\cdot X_1X_2 + 5.49\cdot X_1X_3 + 9.071\cdot X_1X_4 - 3.163\cdot X_2X_3 + \mathbf{28.225}\cdot X_2X_4 + 2.35\cdot X_3X_4$	0.9555	19.9534	<0.0001
	$Y = 349.625 + 9.269\cdot X_1 - 12.541\cdot X_2 + \mathbf{95.401}\cdot X_3 - 10.907\cdot X_4 - 0.919\cdot X_1^2 + \mathbf{41.497}\cdot X_2^2 - \mathbf{29.575}\cdot X_3^2 - 5.442\cdot X_4^2 - 9.75\cdot X_1X_2 + 22.125\cdot X_1X_3 - 32.703\cdot X_1X_4 - 5\cdot X_2X_3 + 8.375\cdot X_2X_4 - 11.25\cdot X_3X_4$	0.8241	8.5362	0.0002

Table 4. *Cont.*

Model Variable	RSM Equation	R^2	F-Value	*p*-Value
S	$Y = 297.751 + 8.730 \cdot X_1 + \mathbf{59.533} \cdot X_2 + \mathbf{111.415} \cdot X_3 - 0.806 \cdot X_4 - 20.482 \cdot X_1^2 - \mathbf{98.472} \cdot X_2^2 - 11.961 \cdot X_3^2 - 15.837 \cdot X_4^2 + \mathbf{71.55} \cdot X_1 X_2 + 18.3 \cdot X_1X_3 + 13.776 \cdot X_1X_4 - 18.95 \cdot X_2X_3 + \mathbf{58.25} \cdot X_2X_4 + 9.825 \cdot X_3X_4$	0.9609	22.8023	<0.0001
	$Y = 725.076 - 49.924 \cdot X_1 - 18.908 \cdot X_2 + \mathbf{161.733} \cdot X_3 - 9.918 \cdot X_4 - 99.578 \cdot X_1^2 + \mathbf{88.446} \cdot X_2^2 - \mathbf{119.1908} \cdot X_3^2 + 1.485 \cdot X_4^2 - 32.7 \cdot X_1X_2 - 78.9 \cdot X_1 \cdot X_3 - 32.606 \cdot X_1X_4 - 0.5 \cdot X_2X_3 + 7.625 \cdot X_2X_4 - 27 \cdot X_3X_4$	0.9469	16.5601	<0.0001
TPC	$Y = 5.682 + 0.589 \cdot X_1 + \mathbf{2.663} \cdot X_2 - \mathbf{2.171} \cdot X_3 + 0.5407 \cdot X_4 + 0.145 \cdot X_1^2 + \mathbf{1.764} \cdot X_2^2 + 1.382 \cdot X_3^2 - 0.302 \cdot X_4^2 + 0.112 \cdot X_1X_2 - 0.568 \cdot X_1X_3 + 0.155 \cdot X_1X_4 - 0.839 \cdot X_2X_3 + 0.4778 \cdot X_2X_3 + 0.287 \cdot X_3X_4$	0.8976	8.1422	0.0003
	$Y = 8.640 + 0.071 \cdot X_1 + \mathbf{4.688} \cdot X_2 - 1.396 \cdot X_3 - 0.398 \cdot X_4 + 0.268 \cdot X_1^2 + \mathbf{1.845} \cdot X_2^2 + 0.136 \cdot X_3^2 + 0.622 \cdot X_4^2 + 0.725 \cdot X_1 X_2 + 1.066 \cdot X_1X_3 + 0.284 \cdot X_1X_4 + 0.186 \cdot X_2X_3 - 0.542 \cdot X_2X_4 - 0.682 \cdot X_3X_4$	0.9043	8.7757	0.0002
DPPH	$Y = 0.027 + 0.002 \cdot X_1 + \mathbf{0.015} \cdot X_2 - 0.008 \cdot X_3 + 0.002 \cdot X_4 - 0.002 \cdot X_1^2 + \mathbf{0.007} \cdot X_2^2 + 0.003 \cdot X_3^2 - 0.002 \cdot X_4^2 - 0.001 \cdot X_1X_2 - 0.002 \cdot X_1X_3 + 0.002 \cdot X_1X_4 - 0.006 \cdot X_2X_3 + 0.003 \cdot X_2X_4$	0.9469	16.5684	<0.0001
	$Y = 0.042 + 0.007 \cdot X_1 + \mathbf{0.034} \cdot X_2 - \mathbf{0.005} \cdot X_3 - \mathbf{0.005} \cdot X_4 + 0.012 \cdot X_1^2 + \mathbf{0.010} \cdot X_2^2 - 0.003 \cdot X_3^2 + 0.008 \cdot X_4^2 + 0.003 \cdot X_1X_2 - 0.004 \cdot X_1X_3 - 0.002 \cdot X_1X_4 - 0.001 \cdot X_2 X_3 - 0.009 \cdot X_2X_4 - 0.009 \cdot X_3X_4$	0.9455	16.1230	<0.0001
FRAP	$Y = 0.014 + 0.002 \cdot X_1 + \mathbf{0.016} \cdot X_2 - \mathbf{0.003} \cdot X_3 + 0.001 \cdot X_4 - 0.004 \cdot X_1^2 + \mathbf{0.012} \cdot X_2^2 + \mathbf{0.002} \cdot X_3^2 - 0.003 \cdot X_4^2 - 0.002 \cdot X_1X_2 - 0.0004 \cdot X_1X_3 + 0.002 \cdot X_1 X_4 + 0.0001 \cdot X_2X_3 + 0.004 \cdot X_2X_4 + 0.0004 \cdot X_3X_4$	0.8937	7.8000	0.0003
	$Y = 0.023 + 0.001 \cdot X_1 + \mathbf{0.018} \cdot X_2 - \mathbf{0.0058} \cdot X_3 - 0.003 \cdot X_4 + 0.002 \cdot X_1^2 + 0.007 \cdot X_2^2 + \mathbf{0.001} \cdot X_3^2 + 0.003 \cdot X_4^2 - 0.001 \cdot X_1X_2 + 0.0002 \cdot X_1X_3 + 0.001 \cdot X_1X_4 - \mathbf{0.002} \cdot X_2X_3 - 0.004 \cdot X_2X_4 - 0.002 \cdot X_3X_4$	0.9584	21.3800	<0.0001

Based on the F-values and *p*-values ($p < 0.05$), all models were found to be significant. Le Man et al. [53] stated that if a model shows a satisfactory fit between the experimental and predicted values and the coefficient of determination is greater than 0.75, it can be deemed relevant. Furthermore, as proposed by Teng et al. [54], smaller *p*-values and greater coefficients of regression for any given model component indicate a more significant impact on the variable under examination. The acquired values show that, within the examined range of variables, the suggested models are reliable.

Based on the developed Response Surface Methodology (RSM) models, extraction temperature and solid-to-liquid ratio are shown to be significant factors affecting the total polyphenol content and antioxidant activity measured by both the DPPH and FRAP methods. The beneficial impact of extraction temperature on the yield of polyphenol extraction has been highlighted by numerous authors [55–58]. In traditional extraction methods, the highest polyphenol yield is achieved within the temperature range of 60–80 °C, whereas the yield tends to decrease at temperatures exceeding 80 °C [59]. Furthermore, the solid-to-liquid ratio has been identified as a significant factor influencing polyphenol yield [38]. The relationships between these variables are graphically presented in Figure 2.

Based on the desirability profiles that were obtained from the RSM-projected values, the extraction conditions were optimized for the chemical characteristics of the extracts. The desirability scale was used, which ranges from 0 (very unpleasant) to 1. According to the optimization matrix design, the following circumstances must be met for the grape skin extract to have its ideal physical and chemical properties: (i) for experiment 1, extraction time (t) = 15 min, temperature (T) = 80 °C, solid-to-liquid ratio (S/L) = 10 g/L, and mixing speed (rpm) = 500 1/min and (ii) for experiment 2, extraction time (t) = 15 min, temperature (T) = 80 °C, solid-to-liquid ratio (S/L) = 10 g/L, and mixing speed (rpm)

= 375 1/min. The models were evaluated in a randomized study using predetermined optimal process settings, showing a satisfactory match between the model's estimated data and the experimental results, as shown in Table 5.

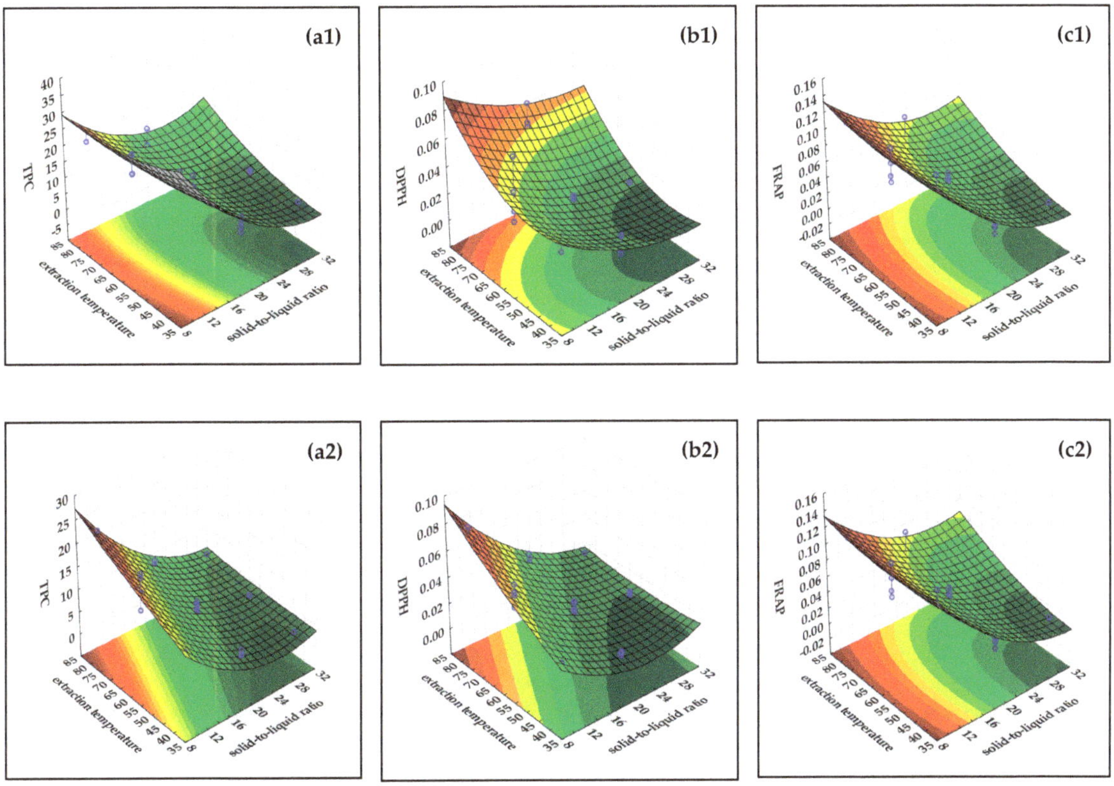

Figure 2. RSM models for the description of the chemical characteristics of grape pomace skin extracts prepared by the classical extraction procedure after experiment 1 and experiment 2. (**a**) Total polyphenol content, (**b**) antioxidant activity measured by the DPPH method, and (**c**) antioxidant activity measured by the FRAP method.

Table 5. Optimal extraction conditions and differences between the experimental data and model-predicted data under optimal conditions.

Model Output		Optimal Extraction Conditions	RMSE-Predicted Value of Output Variable	Experimental Value of Output Variable
Set 1	TPC	t = 15 min	26.6182	26.1284 ± 0.1287
	DPPH	T = 80 °C	0.0685	0.0601 ± 0.0011
	FRAP	S/L = 10 g/L rpm = 500 1/min	0.1210	0.1157 ± 0.0224
Set 2	TPC	t = 15 min	25.2930	25.1024 ± 0.0585
	DPPH	T = 80 °C	0.0831	0.0795 ± 0.0135
	FRAP	S/L = 10 g/L rpm = 375 1/min	0.0427	0.0440 ± 0.0022

4. Conclusions

The measured pH values of the extracts in both experiments did not differ much regarding the extraction conditions, and the obtained values corresponded to the pH range of grape pomace that is often mentioned in the literature (3.4–5.8). By increasing the temperature and extracting a larger amount of sample, TDS and electrical conductivity increased in both experiments, and the correlation between them was confirmed because a significant part of the dissolved substances are ions that conduct electricity. Higher values were obtained in experiment 2 because such a sample was exposed to a larger contact surface with the solvent, and the removal of water from the sample improved the extraction efficiency. For both the first and second experiments, the most suitable physical and chemical characteristics of the grape skin extract were achieved under specified parameters determined by the optimization matrix architecture. For experiment 1, the optimal conditions were determined as follows: extraction time (t) = 15 min, temperature (T) = 80 °C, solid-to-liquid ratio (S/L) = 10 g/L, and mixing speed (rpm) = 500 1/min. Similarly, for experiment 2, the optimal conditions were as follows: extraction time (t) = 15 min, temperature (T) = 80 °C, solid-to-liquid ratio (S/L) = 10 g/L, and mixing speed (rpm) = 375 1/min. These findings demonstrate the effectiveness of the optimization process in identifying precise extraction conditions that yield optimal chemical properties in grape skin extracts. The presented results contribute to the valorization of grape skin as grape skin represents a valuable source of polyphenols with diverse biological activities. They are of interest in both the food and pharmaceutical industries and in scientific research exploring their health-promoting properties.

Author Contributions: Conceptualization, T.S.C. and A.J.T.; methodology, M.B.; formal analysis, T.S.C. and L.P.; writing—original draft preparation, T.S.C., L.P. and T.J.; writing—review and editing, A.J.T. and M.B.; visualization, D.V.; supervision, J.G.K.; project administration, I.R.R.; funding acquisition, I.R.R. All authors have read and agreed to the published version of the manuscript.

Funding: This work was supported by the European Union through the European Regional Development Fund—Competitiveness and Cohesion 2014–2020 (KK.01.1.1.07.0007).

Data Availability Statement: Data are contained within the article.

Conflicts of Interest: The authors declare no conflicts of interest.

References

1. Spigno, G. State of the Art in Grape Processing By-Products. In *Handbook of Grape Processing By-Products*; Academic Press: Cambridge, MA, USA, 2017; pp. 1–27. [CrossRef]
2. Skračić, Z.; Ljubenkov, I.; Mimicac, N.; Generalić Mekinić, I. Valorizacija nusproizvoda proizvodnje vina. *Kem. Ind.* **2023**, *72*, 247–255. [CrossRef]
3. Almanza-Oliveros, A.; Bautista-Hernández, I.; Castro-López, C.; Aguilar-Zárate, P.; Meza-Carranco, Z.; Rojas, R.; Michel, M.R.; Martínez-Ávila, G.C.G. Grape pomace—Advances in its bioactivity, health benefits, and food applications. *Foods* **2024**, *13*, 580. [CrossRef] [PubMed]
4. Spinei, M.; Oroian, M. The potential of grape pomace varieties as a dietary source of pectic substances. *Foods* **2021**, *10*, 867. [CrossRef] [PubMed]
5. Mangione, R.; Simões, R.; Pereira, H.; Catarino, S.; Ricardo-da-Silva, J.; Miranda, I.; Ferreira-Dias, S. Potential use of grape stems and pomaces from two red grapevine cultivars as source of oligosaccharides. *Processes* **2022**, *10*, 1896. [CrossRef]
6. Megías-Pérez, R.; Ferreira-Lazarte, A.; Villamiel, M. Valorization of grape pomace as a renewable source of techno-functional and antioxidant pectins. *Antioxidants* **2023**, *12*, 957. [CrossRef] [PubMed]
7. Filippi, K.; Georgaka, N.; Alexandi, M.; Papapostolus, H.; Koutinas, A. Valorisation of grape stalks and pomace for the production of bio-based succinic acid by *Actinobacillus succinogenes*. *Ind. Crops Prod.* **2021**, *168*, 113578. [CrossRef]
8. Elejalde, E.; Villarán, M.C.; Alonso, R.M. Grape polyphenols supplementation for exercise-induced oxidative stress. *J. Int. Soc. Sports Nutr.* **2021**, *18*, 3. [CrossRef] [PubMed]
9. Ortega-Heras, M.; Gómez, I.; de Pablos-Alcalde, S.; González-Sanjosé, M.L. Application of the just-about-right scales in the development of new healthy whole-wheat muffins by the addition of a product obtained from white and red grape pomace. *Foods* **2019**, *8*, 419. [CrossRef] [PubMed]
10. Walker, R.; Tseng, A.; Cavender, G.; Ross, A.; Zhao, Y. Physicochemical, nutritional, and sensory qualities of wine grape pomace fortified baked goods. *J. Food Sci.* **2014**, *79*, S1811–S1822. [CrossRef]
11. Mildner-Szkudlarz, S.; Zawirska-Wojtasiak, R.; Szwengiel, A.; Pacyński, M. Use of grape by-product as a source of dietary fibre and phenolic compounds in sourdough mixed rye bread. *Int. J. Food Sci. Technol.* **2011**, *46*, 1485–1493. [CrossRef]
12. Theagarajan, R.; Malur Narayanaswamy, L.; Dutta, S.; Moses, J.A.; Chinnaswamy, A. Valorisation of grape pomace (cv. *Muscat*) for development of functional cookies. *Int. J. Food Sci. Technol.* **2019**, *54*, 1299–1305. [CrossRef]

13. Bender, A.B.; Speroni, C.S.; Salvador, P.R.; Loureiro, B.B.; Lovatto, N.M.; Goulart, F.R.; Lovatto, M.T.; Miranda, M.Z.; Silva, L.P.; Penna, N.G. Grape pomace skins and the effects of its inclusion in the technological properties of muffins. *J. Culin. Sci. Technol.* **2017**, *15*, 143–157. [CrossRef]
14. Mildner-Szkudlarz, S.; Bajerska, J.; Zawirska-Wojtasiak, R.; Górecka, D. White grape pomace as a source of dietary fibre and polyphenols and its effect on physical and nutraceutical characteristics of wheat biscuits. *J. Sci. Food Agric.* **2013**, *93*, 389–395. [CrossRef]
15. Aksoylu, Z.; Çağindi, Ö.; Köse, E. Effects of blueberry, grape seed powder and poppy seed incorporation on physicochemical and sensory properties of biscuit. *J. Food Qual.* **2015**, *38*, 164–174. [CrossRef]
16. Lee, H.J.; Lee, J.J.; Jung, M.O.; Choi, J.S.; Jung, J.T.; Choi, Y.I.; Lee, J.K. Meat Quality and Storage Characteristics of Pork Loin Marinated in Grape Pomace. *Korean J. Food Sci. Anim. Resour.* **2017**, *37*, 726. [CrossRef] [PubMed]
17. Özvural, E.B.; Vural, H. Grape seed flour is a viable ingredient to improve the nutritional profile and reduce lipid oxidation of frankfurters. *Meat Sci.* **2011**, *88*, 179–183. [CrossRef] [PubMed]
18. Marchiani, R.; Bertolino, M.; Belviso, S.; Giordano, M.; Ghirardello, D.; Torri, L.; Piochi, M.; Zeppa, G. Yogurt enrichment with grape pomace: Effect of grape cultivar on physicochemical, microbiological and sensory properties. *J. Food Qual.* **2016**, *39*, 77–89. [CrossRef]
19. Demirkol, M.; Tarakci, Z. Effect of grape (*Vitis labrusca* L.) pomace dried by different methods on physicochemical, microbiological and bioactive properties of yoghurt. *LWT* **2018**, *97*, 770–777. [CrossRef]
20. Zhang, Z.; Li, X.; Sang, S.; McClements, D.J.; Chen, L.; Long, J.; Jiao, A.; Jin, Z.; Qiu, C. Polyphenols as plant-based nutraceuticals: Health effects, encapsulation, nano-delivery, and application. *Foods* **2022**, *11*, 2189. [CrossRef]
21. Soto, M.L.; Falqué, E.; Domínguez, H. Relevance of natural phenolics from grape and derivative products in the formulation of cosmetics. *Cosmetics* **2015**, *2*, 259–276. [CrossRef]
22. Martinović, J.; Lukinac, J.; Jukić, M.; Ambrus, R.; Planinić, M.; Šelo, G.; Klarić, A.-M.; Perković, G.; Bucić-Kojić, A. In vitro bioaccessibility assessment of phenolic compounds from encapsulated grape pomace extract by ionic gelation. *Molecules* **2023**, *28*, 5285. [CrossRef] [PubMed]
23. Naviglio, D.; Scarano, P.; Ciaravolo, M.; Gallo, M. Rapid solid-liquid dynamic extraction (rslde): A powerful and greener alternative to the latest solid-liquid extraction techniques. *Foods* **2019**, *8*, 245. [CrossRef] [PubMed]
24. Valinger, D.; Kušen, M.; Benković, M.; Jurina, T.; Panić, M.; Radojčić Redovniković, I.; Kljusurić, J.G.; Tušek, A.J. Enhancement of the green extraction of bioactive molecules from *Olea europaea* leaves. *Separations* **2022**, *9*, 33. [CrossRef]
25. Che Sulaiman, I.S.; Basri, M.; Fard Masoumi, H.R.; Chee, W.J.; Ashari, S.E.; Ismail, M. Effects of temperature, time, and solvent ratio on the extraction of phenolic compounds and the anti-radical activity of *Clinacanthus nutans* Lindau leaves by response surface methodology. *Chem. Cent. J.* **2017**, *11*, 54. [CrossRef] [PubMed]
26. Xu, B.J.; Chang, S.K. A comparative study on phenolic profiles and antioxidant activities of legumes as affected by extraction solvents. *J. Food Sci.* **2007**, *72*, 159–166. [CrossRef] [PubMed]
27. Albu, S. Potential for the use of ultrasound in the extraction of antioxidants from Rosmarinus officinalis for the food and pharmaceutical industry. *Ultrason. Sonochem.* **2004**, *11*, 261–265. [CrossRef] [PubMed]
28. Goldsmith, C.D.; Vuong, Q.V.; Stathopoulos, C.E.; Roach, P.D.; Scarlett, C.J. Optimization of aqueous extraction of phenolic compounds from olive leaves. *Antioxidants* **2014**, *3*, 700–712. [CrossRef] [PubMed]
29. Tsegay, Z.T.; Agriopoulou, S.; Chaari, M.; Smaoui, S.; Varzakas, T. Statistical Tools to Optimize the Recovery of Bioactive Compounds from Marine Byproducts. *Mar. Drugs* **2024**, *22*, 182. [CrossRef]
30. Kumar, M.; Dahuja, A.; Sachdev, A.; Kaur, C.; Varghese, E.; Saha, S.; Sairam, K.V.S.S. Valorisation of black carrot pomace: Microwave assisted extraction of bioactive phytoceuticals and antioxidant activity using Box–Behnken design. *J. Food Sci. Technol.* **2019**, *56*, 995–1007. [CrossRef]
31. Kumar, M.; Dahuja, A.; Sachdev, A.; Kaur, C.; Varghese, E.; Saha, S.; Sairam, K.V.S.S. Evaluation of enzyme and microwave-assisted conditions on extraction of anthocyanins and total phenolics from black soybean (*Glycine* max L.) seed coat. *Int. J. Biol. Macromol.* **2019**, *135*, 1070–1081. [CrossRef]
32. Bakar, F.I.A.; Bakara, M.F.A.B.; Abdullah, N.; Endrini, S.; Fatmawati, S. Optimization of extraction conditions of phytochemical compounds and anti-gout activity of *Euphorbia hirta* L. (Ara Tanah) using response surface methodology and Liquid Chromatography-Mass Spectrometry (LC-MS) analysis. *Evid.-Based Complement. Altern. Med.* **2020**, *2020*, 4501261. [CrossRef] [PubMed]
33. Salih, A.M.; Al-Qurainy, F.; Nadeem, M.; Tarroum, M.; Khan, S.; Shaikhaldein, H.O.; Al-Hashimi, A.; Alfagham, A.; Alkahtani, J. Optimization method for phenolic compounds extraction from medicinal plant (*Juniperus procera*) and phytochemicals screening. *Molecules* **2021**, *26*, 7454. [CrossRef] [PubMed]
34. Moutinho, J.; Gouvinhas, I.; Domínguez-Perles, R.; Barros, A. Optimization of the Extraction Methodology of Grape Pomace Polyphenols for Food Applications. *Molecules* **2023**, *28*, 3885. [CrossRef] [PubMed]
35. Ćurko, N.; Kelšin, K.; Dragović-Uzelac, V.; Valinger, D.; Tomašević, M.; Kovačević Ganić, K. Microwave-assisted extraction of different groups of phenolic compounds from grape skin pomaces: Modeling and optimization. *Pol. J. Food Nutr. Sci.* **2019**, *69*, 235–246. [CrossRef]

36. Puntnik, P.; Bursać-KOvačević, D.; Radojčin, M.; Dragović-Uzelac, V. Influence of Acidity and Extraction Time on the Recovery of Flavonoids from Grape Skin Pomace Optimized by Response Surface Methodology. *Chem. Biochem. Eng. Q.* **2016**, *30*, 455–464. [CrossRef]
37. Association of Official Analytical Chemists (AOAC). *Official Methods of Analysis*, 16th ed.; AOAC: Washington, DC, USA; Arlington, VA, USA, 1998.
38. Sokač Cvetnić, T.; Krog, K.; Benković, M.; Jurina, T.; Valinger, D.; Gajdoš Kljusurić, J.; Radojčić Redovniković, I.; Jurinjak Tušek, A. Solid–liquid extraction of bioactive molecules from white grape skin: Optimization and near-infrared spectroscopy. *Separations* **2023**, *10*, 452. [CrossRef]
39. Jurinjak Tušek, A.; Benković, M.; Valinger, D.; Jurina, T.; Belščak-Cvitanović, A.; Gajdoš-Kljusurić, J. Optimizing bioactive compounds extraction from different medicinal plants and prediction through nonlinear and linear models. *Ind. Crops Prod.* **2016**, *126*, 449–458. [CrossRef]
40. Pinelo, M.; Rubilar, M.; Jerez, M.; Sineiro, J.; Núñez, M.J. Effect of solvent, temperature, and solvent-to-solid ratio on the total phenolic content and antiradical activity of extracts from different components of grape pomace. *J. Agric. Food Chem.* **2005**, *53*, 2111–2117. [CrossRef]
41. Brand-Williams, W.; Cuvelier, M.E.; Berset, C. Use of a free radical method to evaluate antioxidant activity. *LWT-Food Sci. Technol.* **1995**, *28*, 25–30. [CrossRef]
42. Benzie, I.F.F.; Strain, J.J. The Ferric Reducing Ability of Plasma (FRAP) as a Measure of "Antioxidant Power": The FRAP Assay. *Anal. Biochem.* **1996**, *239*, 70–76. [CrossRef]
43. Bordiga, M. *Valorization of Winemaking By-Products*; CRC Press, Taylor & Francis Group: Boca Raton, FL, USA, 2016.
44. El Tannir, H.; Houhou, D.; Debs, E.; Koubaa, M.; Jammoul, A.; Azakir, B.; Khalil, M.I.; El Darra, N.; Louka, N. Optimization of Aqueous Extraction of Polyphenols from *Cuminum cyminum* Seeds Using Response Surface Methodology and Assessment of Biological Activity. *BioTech* **2024**, *13*, 7. [CrossRef]
45. Zhang, Q.-W.; Lin, L.-G.; Ye, W.-C. Techniques for extraction and isolation of natural products: A comprehensive review. *Chin. Med.* **2018**, *13*, 20. [CrossRef] [PubMed]
46. Benchaachoua, A.; Bessam, H.M.; Saidi, I. Effects of different extraction methods and solvents on the phenolic composition and antioxidant activity of Silybum marianum leaves extracts. *Int. J. Med. Sci. Clin. Invent.* **2018**, *5*, 3641–3647. [CrossRef]
47. Ghasemzadeh, A.; Jaafar, H.; Juraimi, A.; Tayebi-Meigooni, A. Comparative evaluation of different extraction techniques and solvents for the assay of phytochemicals and antioxidant activity of Hashemi rice bran. *Molecules* **2015**, *20*, 10822–10838. [CrossRef] [PubMed]
48. Librán, C.; Mayor, L.; Garcia-Castello, E.M.; Vidal-Brotons, D. Polyphenol extraction from grape wastes: Solvent and pH Effect. *Agric. Sci.* **2013**, *4*, 56–62. [CrossRef]
49. Yammine, S.; Delsart, C.; Vitrac, X.; Peuchot, M.M.; Ghidossi, R. Characterisation of Polyphenols and antioxidant potential of red and white pomace by-product extracts using subcritical water extraction. *OENO One* **2020**, *54*, 263–278. [CrossRef]
50. Guerrero, M.S.; Torres, J.S.; Nuñez, M.J. Extraction of polyphenols from white distilled grape pomace: Optimization and modelling. *Bioresour. Technol.* **2008**, *99*, 1311–1318. [CrossRef]
51. Antony, A.; Farid, M. Effect of Temperatures on Polyphenols during Extraction. *Appl. Sci.* **2022**, *12*, 2107. [CrossRef]
52. Fontana, A.R.; Antoniolli, A.; Bottini, R. Grape pomace as a sustainable source of bioactive compounds: Extraction, characterization, and biotechnological applications of phenolics. *J. Agric. Food Chem.* **2013**, *61*, 8987–9003. [CrossRef]
53. Le Man, H.; Behera, S.K.; Park, H.S. Optimization of operational parameters for ethanol production from Korean food waste leachate. *Int. J. Environ. Sci. Technol.* **2010**, *7*, 157–164. [CrossRef]
54. Teng, D.; Fang, Y.; Song, X.; Gao, Y. Optimization of enzymatic hydrolysis parameters for antioxidant capacity of peptide from goat placenta. *Food Bioprod. Process.* **2011**, *89*, 202–208. [CrossRef]
55. Giovanna, B.; Giulio, F.; Cristina, M.; Carini, M.; Paolo, M.; Aldini, G. Effect of extraction solvent and temperature on polyphenol. *Molecules* **2021**, *26*, 5454. [CrossRef] [PubMed]
56. Gironi, F.; Piemonte, V. Temperature and solvent effects on polyphenol extraction process from chestnut tree wood. *Chem. Eng. Res. Des.* **2011**, *89*, 857–862. [CrossRef]
57. Begum, S.N.; Hossain, M.; Adnan, M.; Rahaman, C.H.; Reza, A. Optimization and Characterization of Phenolic Extraction Conditions and Antioxidant Activity Evaluation of *Adenanthera pavonina* L. Bark. *Plants* **2023**, *12*, 3902. [CrossRef] [PubMed]
58. Dimitrijević, A.; Marić, S.; Jocić, A.; Tekić, D.; Mušović, J.; Amaral, J.S. Green extraction strategy using bio-based aqueous biphasic systems for polyphenol valorization from grape by-product. *Foods* **2024**, *13*, 954. [CrossRef]
59. Ferriera-Sntos, P.; Nobre, C.; Rodrigues, R.M.; Genisheva, Z.; Botelho, C.; Teixeira, J.A. Extraction of phenolic compounds from grape pomace using ohmic heating: Chemical composition, bioactivity and bioaccessibility. *Food Chem.* **2024**, *436*, 137780. [CrossRef]

processes

Article

Comparison of *Tetraselmis suecica* Cell Disruption Techniques: Kinetic Study and Extraction of Hydrosoluble Compounds

Hussein Rida [1,2,*], Jérôme Peydecastaing [1], Hosni Takache [2,3], Ali Ismail [2] and Pierre-Yves Pontalier [1]

1 Laboratoire de Chimie Agro-Industrielle (LCA), Université de Toulouse, INRAE, Toulouse INP, CEDEX 4, 31030 Toulouse, France; jerome.peydecastaing@toulouse-inp.fr (J.P.); pierreyves.pontalier@ensiacet.fr (P.-Y.P.)
2 Platforme de Recherche et d'Analyses en Sciences de l'Environnement (PRASE), Université Libanaise, Beirut P.O. Box 6573/14, Lebanon; hosni.takache@ul.edu.lb (H.T.); aismail@ul.edu.lb (A.I.)
3 Bio-Information Research Laboratory (BIRL), The Higher Institute of Biotechnologies of Paris (Sup'biotech), 94800 Villejuif, France
* Correspondence: hussein.rida@toulouse-inp.fr; Tel.: +33-534-323-506

Abstract: The optimization of cell disruption is a critical step in microalgal biorefineries. We used the same batch of *Tetraselmis suecica* culture to compare two mechanical cell disruption techniques, focusing on the extraction yield of water-soluble molecules. The conditions for high-pressure homogenization (HPH) studied were two passes at a moderate pressure of 300 bars. For ultrasound (US) treatment, we used an amplitude of 20% (equivalent to 100 W) for 25 min. These conditions were chosen on the basis of a preliminary screen of extraction conditions. HPH extracted proteins and pigments more efficiently than US, whereas US was superior for uronic acid extraction. Interestingly, the two methods had similar extraction yields for carbohydrates under the studied conditions. We also analyzed the kinetics of molecule release by considering the centrifugation time lag for HPH and applying a first-order kinetic model for US. HPH outperformed US in terms of the immediate extraction and release of molecules.

Keywords: microalgae; high-pressure homogenization; ultrasonication; kinetics; extraction

Citation: Rida, H.; Peydecastaing, J.; Takache, H.; Ismail, A.; Pontalier, P.-Y. Comparison of *Tetraselmis suecica* Cell Disruption Techniques: Kinetic Study and Extraction of Hydrosoluble Compounds. *Processes* **2024**, *12*, 1139. https://doi.org/10.3390/pr12061139

Academic Editors: Prashant K. Sarswat and Yanlin Zhang

Received: 26 April 2024
Revised: 27 May 2024
Accepted: 29 May 2024
Published: 31 May 2024

1. Introduction

Microalgae are a prospective source of valuable nutrients, including pigments, proteins, carbohydrates, and lipid molecules [1]. They use carbon dioxide (CO_2) at rates 10–50 times higher than those of terrestrial plants, produce biomass faster than land plants, and have the capacity to remove nitrogen, phosphorus, and heavy metals from water [2]. Despite ongoing research efforts to overcome issues potentially hindering the release of microalgal molecules, efficient downstream processing remains a major challenge for commercial-scale implementation [3]. Cell wall disruption is a crucial step in the efficient recovery of intracellular molecules and the fractionation of cell wall components [4]. Several cell disruption methods have been applied to microalgae, with various degrees of success. In cell disruption, it is essential to differentiate between "permeabilized cells" and "lysed cells." The walls and membranes of the cells are partially ruptured in permeabilized cells, whereas the lysed cells are completely disintegrated [5]. Cell disruption methods can be classified as mechanical (bead milling, high-pressure homogenization, ultrasound, etc.) or non-mechanical (chemical and enzymatic) methods on the basis of the mechanisms used for cell disintegration [6].

The efficiency of cell disruption depends on the cell wall structure. Some microalgae can be easily disrupted by mild or more energy-efficient disruption techniques, facilitating the direct access of the solvent to intracellular compounds and thereby increasing the extraction efficiency [7]. The use of such methods ensures high product recovery, low operating costs, and a high degree of purity of the extract [8]. High levels of polysaccharides in the cell wall and the presence of a non-hydrolysable biopolymer composed of long ω-hydroxy fatty-acid chains render the cell wall rigid [9].

Tetraselmis suecica is a fast-growing marine green microalga that accumulates proteins, carbohydrates, lipids, and other molecules [10]. It can withstand a diverse range of salt concentrations [11] and is widely used in aquaculture as a feed for bivalves, shrimp larvae, rotifers, and brine shrimp [12]. *Tetraselmis suecica* is an ovoid unicellular flagellated green alga that is about 9–13 μm long and 7–8 μm wide. Its composition varies with the growing conditions, which influence the accumulation of the target components [13]. *Tetraselmis suecica* fractionation has been widely studied because this alga has a high protein content. However, these studies have highlighted difficulties in optimizing its valorization, particularly in terms of the selection of an efficient extraction technology.

High-pressure homogenization (HPH) is a mechanical cell disruption technique in which very high pressures (from about 200 to 1200 bars) are used to generate turbulence, liquid shear stress, and friction, leading to the release of the contents of the microalga. The efficiency of HPH depends on several parameters, including the operating pressure and the number of homogenization cycles [14,15]. This continuous system based on the use of algal suspensions or slurries is easy to scale up. During this process, the biomass is forced through a narrow nozzle outlet by a high-pressure pump [16]. The cell suspension passes radially through a valve, strikes a ring, and then passes through another valve or is discharged. The mode of action of HPH remains unclear, but it is thought that the combination of a high shear force, cavitation, and a sudden decrease in pressure between the nozzle and the external environment leads to cell disruption [8]. Magpusao et al. [17] performed studies to optimize pressure levels for the HPH treatments of *Tetraselmis* sp. They found that a pressure of 300 bars caused substantial disintegration, with complete disintegration occurring at 600 bars. However, few studies have reported the effect of HPH on the disruption of *Tetraselmis suecica*. Spiden et al. [18] reported that lower pressures were sufficient to cause high levels of *Tetraselmis suecica* cell disruption by HPH, with a degradation of metabolites occurring at pressures above 90 bars, according to measurements of UV absorbance. Safi et al. [13] showed that the rate of *Tetraselmis suecica* cell rupture increased with the homogenization pressure: cells were resistant at pressures up to 400 bars, and the effect of HPH began to occur at 600 bars.

In addition, previous research on microalgae has demonstrated enhanced cell disruption with varying numbers of passes. For instance, Zhang et al. [19] found that 10 passes resulted in the highest extraction yield of proteins and carbohydrates from *Parachlorella kessleri*, showcasing the importance of exploring different pass numbers. Delran et al. [20] studied the effect of the pressure applied and the number of passes on the destruction of *Tetraselmis suecica* cells and reported that a single pass at a pressure of 400 bars was sufficient for protein extraction. None of these studies investigated the effect of a pressure of 300 bars on *Tetraselmis suecica* cell disruption, and it is therefore possible that the threshold for disruption lies closer to 300 bars than to 400 bars. At such a low pressure of 300 bars, the effect of the number of passes was not investigated. Importantly, none of these studies performed diffusion kinetic modeling on the extracted molecules.

Unlike HPH, ultrasonication (US) is a physical process in which sound waves with frequencies above 20 kHz are used to induce acoustic cavitation in a liquid medium [5]. These waves generate a liquid shear that moves from the metallic tip to the concentrated cellular suspension [21], causing the formation of microbubbles, which increase in size and then collapse, creating cavitation conditions. Violent bubble collapse increases the temperature to above 4000 °C and the pressure by hundreds of atmospheres [6]. Various parameters, including temperature, ultrasound frequency, cell concentration, acoustic intensity, number of cycles, wave amplitude, viscosity, mode used, and treatment time, should be controlled during this process [4,22]. Most studies have focused on lipid extraction from *Tetraselmis suecica* by US [23–25], but a few have considered the effects of US parameters on cell rupture and the release of water-soluble molecules. Delran et al. [26] studied the effects of US frequency, power, and treatment time on the destruction of *Tetraselmis suecica* cells and reported that low-frequency US (20 kHz) treatment allowed the extraction of 90% of proteins and 70% of carbohydrates.

The kinetics of *Tetraselmis suecica* protein release by US were investigated by Delran et al. [26], who studied the effect of the treatment time on the protein extraction yield without using a kinetic model. Suarez Garcia et al. [27] investigated the kinetics of protein release from *Tetraselmis suecica* by bead milling and used a first-order kinetic model. Halim et al. [23] proposed a kinetic model for *Tetraselmis suecica* cell lysis by HPH and US, in which the concentration of lysed cells was determined by hemocytometry. No kinetic model has yet been established for the extraction of water-soluble proteins from *Tetraselmis suecica* by US, and there have been no kinetic modeling studies of biomolecule release from *Tetraselmis suecica* during US treatment.

Major studies of the disruption of *Tetraselmis suecica* cells have been performed independently for different techniques. In this study, we compared HPH and US on the same batch of cells. It is not appropriate to compare cell disruption between different batches of *Tetraselmis suecica* because the morphology and physiological state of the biomass received may differ between batches. Mear et al. [28] identified various differed morphological properties and physiological states in *Tetraselmis chui* cultures obtained from the same supplier and grown in the same medium, highlighting disparities in physiological states between different batches. Cell wall rigidity and the biochemical composition of the biomass in the two batches they studied were not comparable, and the observed differences had direct impacts on cell disruption and protein recovery.

The primary objective of this study was to determine the most effective cell disruption technique and conditions for extracting hydrosoluble compounds from *Tetraselmis suecica* after centrifugation. Specifically, our aim was to compare the performance of two mechanical techniques, HPH and US, across various conditions. This comparison will be based on two critical factors:

1. Yield of extracted molecules—Quantifying the extracted molecules in the supernatant after cell disruption.
2. Kinetic release rate—Determining the time required for the extraction process, reflecting the efficiency of extracting a defined biomass quantity.

The selection of the optimal cell disruption technique based on these factors was performed using two different batches of *Tetraselmis suecica*. We then compared the most promising conditions of each technique within the same batch of cells. This study will facilitate the comparison of techniques and the optimization of processes, thereby increasing efficiency across diverse scientific and industrial applications.

2. Materials and Methods

2.1. Microalgae

Tetraselmis suecica was purchased from Inalve (Nice, France). It was produced as a biofilm (patented rotating algal growth system, WO2021180713A1) and had a dry matter content ranging from 15% to 20%, depending on the batch. Fresh biomass was received two days after harvesting as a concentrated paste. It was stored at 4 °C in the dark, and diluted to a concentration of 5% dry weight (DW). A portion of the biomass was freeze-dried for further characterization of its total biomolecule content to prevent rapid degradation. The rest was diluted 1:3 for fractionation. Cell disruption and centrifugation were performed within 48 h of biomass reception.

2.2. Reagents

The Bradford assay kit (including BSA standards) was purchased from Thermo Fisher Scientific, Illkirch-Graffenstaden, France. All the chemicals used were of analytical grade and were purchased from Sigma Aldrich, Saint-Quentin-Fallavier, France.

2.3. Treatments

Before cell disruption experiments, the fresh paste was diluted 1:3 with distilled water by gentle shaking at 150 rpm for 30 min on an orbital shaker. After each disruption, the supernatant was recovered by centrifugation at 10,000× *g* for 10 min at 20 °C in a 6–16 K centrifuge (Sigma Laboratory Centrifuges, Osterode am Harz, Germany) and subjected to biochemical analysis.

2.3.1. Control

The biomass dilution, corresponding to a non-mechanical water extraction, was considered a blank for comparison with the other treatments in this study [20,29].

2.3.2. Ultrasonication (US)

Ultrasonic extraction was performed with a Sonics Vibra-Cell ultrasonic processor probe–500 W (Thermo Fisher Scientific, Illkirch-Graffenstaden, France), operating at a frequency of 22 kHz. Sonication was performed on 250 mL of biomass at two amplitudes, 20% (100 W) and 70% (350 W), in a plastic beaker, with a 13 mm-diameter probe placed 5 cm below the surface. The chosen tip depth in the volume was based on the recommendation of the manufacturer. The treatment was performed in cool conditions (10 °C) and in pulsed mode, with cycles of 5 s of US and 20 s of resting time to prevent overheating of the sample. The suspension was irradiated for various times (5, 10, 15, and 25 min), and several samples were collected at different time points.

2.3.3. High-Pressure Homogenization (HPH)

This treatment was performed with an APV-1000 Lab homogenizer (Søborg, Denmark), with a pressure range of 0 to 1000 bars and a flow rate of 22 L/h. All experiments were performed at room temperature (approximately 20 °C) without a cooling system. The temperature at the HPH outlet never exceeded 40 °C in any of the conditions used. A volume of 250 mL was used for experiments relating exclusively to this technique, whereas a volume of 300 mL was used for the experiments comparing HPH with US. The impact of HPH was assessed under three different sets of conditions, as outlined in Table 1.

Table 1. HPH operating conditions.

Condition	Pressure in Bars	Number of Passes
1	300	2
2	300	5
3	600	2

2.4. Extraction Kinetics

A kinetic study was also conducted to investigate the release of metabolites during US and HPH.

2.4.1. Ultrasonication (US)

A kinetic model was used to describe the diffusion of proteins and carbohydrates during *Tetraselmis suecica* cell lysis by US.

In this model, first-order release kinetics for proteins or carbohydrates were used to calculate the kinetic constant k as follows:

$$-r_{\text{molecules}} = -\frac{dC_{\text{molecules}}}{dt} = k.C_{\text{molecules}} \tag{1}$$

$$C_{\text{molecules}} = C_{0,\text{molecules}} - C_{\text{released molecules}} \tag{2}$$

where

- $R_{\text{molecules}}$—Diffusion rate in mg/mL.s;

- C—Concentration of the molecule remaining in the cell in mg/mL;
- $C_{0,\text{ molecules}}$—Total concentration of the molecule in the initial biomass in mg/mL;
- $C_{\text{released molecules}}$—Concentration of the molecule released into the supernatant in mg/mL;
- k—Molecule extraction rate constant in s^{-1};
- t—Treatment time for US in minutes.

The primary objective of kinetic modeling is to establish a quantitative relationship between the concentration of molecules released into the supernatant and the duration of US (treatment time). This relationship is derived from the solution of Equation (1) as follows:

$$\int_{C_{0,\text{molecules}}}^{C_{\text{molecules}}} \frac{dC_{\text{molecules}}}{C_{\text{molecules}}} = -k\int_{0}^{t} dt \tag{3}$$

to give

$$-\ln\frac{C_{\text{molecules}}}{C_{0,\text{molecules}}} = kt \tag{4}$$

Replacing Equation (2) in Equation (4) yields:

$$-\ln\frac{\left(C_{0,\text{molecules}} - C_{\text{released molecules}}\right)}{C_{0,\text{molecules}}} = kt \tag{5}$$

We define two dimension-less variables to simplify the above equations.

The fraction of molecules released, X, is calculated as follows:

$$X = \frac{C_{\text{released molecules}}}{C_{0,\text{molecules}}} \tag{6}$$

The time fraction, ζ, is calculated as follows:

$$\zeta = \frac{t}{t_{\text{max}}} \tag{7}$$

where t_{max} is the maximum treatment duration, which was 25 min of US.

Equation (5) can then be simplified as follows:

$$-\ln(1 - X) = kt \tag{8}$$

2.4.2. High-Pressure Homogenization (HPH)

The biomolecule extraction kinetics of *Tetraselmis suecica* were also determined on the basis of the centrifugation time lag, defined as the time between the completion of HPH and the beginning of centrifugation. This study was performed to ascertain whether the released molecules would continue to diffuse out of the residual solid matrix after cell rupture. We studied six extraction times after HPH extraction: 5 min, 20 min, 35 min, 1 h, 2 h, and 3 h. The extraction kinetics were studied for the three conditions shown in Table 1. This is the first study, to the best of our knowledge, to have considered the effect of the time lag to centrifugation. The kinetics of biomolecule diffusion into the aqueous supernatant were monitored for proteins, neutral carbohydrates, and pigments, including chlorophyll *a*, chlorophyll *b*, fucoxanthin, lutein, and *β*-carotene.

2.5. Biochemical Analysis

A biochemical analysis was performed to evaluate the biomass at different stages of treatment. We analyzed fresh biomass (dry weight and ash content), freeze-dried initial biomass (total proteins, total carbohydrates, and uronic acids exclusively for comparisons of HPH and US), diluted biomass (dry weight), freeze-dried disrupted biomass (total pigments exclusively for comparisons of HPH and US), and the supernatant (dry weight, ash content, total proteins, soluble proteins, carbohydrates, uronic acids, and pigments).

Dry weight was determined gravimetrically by heating the sample in an oven at 103 °C until its weight stabilized. The ash content was determined by calcination at 550 °C for 12 h.

For the total protein biomass analysis and extraction yield calculation, proteins were analyzed by the Kjeldahl method with a Kjeltec 8400 automatic analyzer (Nanterre, France) using a standard conversion factor (nitrogen to protein) of 5.95 [30]. For the analysis of soluble proteins and the comparison of disruption methods, the Bradford assay was used, with bovine serum albumin (BSA) as the standard [31].

The carbohydrate content was determined according to the protocol of the National Renewable Energy Laboratory (NREL) [32]. In brief, 25 mg of freeze-dried biomass was hydrolyzed with 250 µL of 72% sulfuric acid for 1 h at 30 °C in a water bath. We then added 7 mL of ultrapure water, and the biomass was hydrolyzed for 1 h in an autoclave at 121 °C. After cooling, the acidic hydrolysate was neutralized with calcium carbonate, filtered through a 0.22-micron filter, and analyzed. The monosaccharide distribution was determined with an HPIC DIONEX ICS 3000 DC-EG (Illkirch-Graffenstaden, France) with a Carbo-Pac PA-1 column, a post-column with 300 mM NaOH as the elution buffer, and an AS3000 automatic sampler. The device was equipped with a pulsed electrochemical detector.

Uronic acids were analyzed by the Blumenkrantz titration method [33]. The biomass was hydrolyzed for 5 min at 100 °C in a solution of sodium tetraborate (a reactant) in sulfuric acid. Once cooled, a basic metahydroxybiphenyl solution was added, and the absorption was recorded at 520 nm with a BMG-Labtech Spectrostar-Nano spectrophotometer (BMG LABTECH SARL, Champigny s/Marne, France). The concentration was determined by comparison with a calibration curve obtained with galacturonic acid as the standard. The uronic acid content was reported on a dry weight basis as follows:

$$\%\text{Uronic acid} = \frac{\text{Concentration} \times V_{\text{reactant}+H_2SO_4}}{m_{\text{sample}} \times \%\text{DW}} \times 100 \quad (9)$$

Total pigments (fucoxanthin, lutein, chlorophyll *a*, chlorophyll *b*, and *β*-carotene) were extracted from the freeze-dried biomass according to a slightly modified version of the NREL protocol [34]. The fresh biomass was disrupted by HPH, with two passes at a pressure of 300 bars, and immediately freeze-dried. Purified water (100 µL) was added to 15 mg of the dried lysate, which was then vortexed for 15 s and stored on ice for 30 min to limit biological activity during rehydration. We then added 0.5 mL of methanol and vortexed the mixture for 15 s before adding 0.5 mL of chloroform and vortexing for another 15 s. The samples were centrifuged at 5000 RCF, and the supernatant was recovered in another vial. Repeated extractions with methanol/chloroform were performed until the supernatant became transparent. The recovered supernatants were then dried under a gentle stream of clean nitrogen gas and redissolved in 1 mL of methanol/acetone (8:2 *v*/*v*). The dissolved samples were then passed through a 0.22 µm-pore polytetrafluoroethylene (PTFE) syringe filter and transferred to HPLC vials for analysis.

For the freeze-dried extract (supernatant), 3 mL of methanol was added to 20 mg of the freeze-dried sample in an inactinic vial. The mixture was shaken in an ultrasound bath for 5 min, passed through a 0.22 µm-pore PTFE syringe filter, and transferred to HPLC vials. For the liquid extract, pigments were extracted from the liquid supernatant (for comparison with the dried supernatant) as described by Safi et al. [13].

Pigment analysis was performed by HPLC with a Dionex UltiMate™ 3000 quaternary pump (Thermo Fisher Scientific, Illkirch-Graffenstaden, France) and a C18 column (150 mm × 4.6 mm inner diameter, 5 µm particle size, Agilent, Les Ulis, France). The pigments were eluted in acetone/methanol/water (55:25:20) at a flow rate of 1 mL/min.

2.6. Statistical Analysis

Experiments performed separately for the two techniques for kinetic studies and for the optimization of conditions were performed only once and analyzed at least twice due to the limited amount of biomass available and the number of experiments to be performed.

For a final comparison of HPH and US, experiments were conducted in triplicate, and each sample was analyzed at least twice. All analysis and characterization data are presented as the means ± standard deviations of three experiments. Statistical analysis was performed by one-way ANOVA in Microsoft Excel 2019. The results were reproducible to within ±5%.

3. Results and Discussion

3.1. Ultrasonication (US)

This batch of *Tetraselmis suecica* had a DW concentration of 20% on reception, and was diluted to 6.6% DW for further processing. The initial biomass had a total protein content of 20.14 ± 0.01% DW, as determined by the Kjeldahl method.

The protein concentration of the extract was greater at higher amplitude (70%), reflecting more intense cell lysis. The protein extraction yield increased with longer treatment times at low amplitude (20%), from 22% at 5 min to 51% at 25 min (Figure 1). These findings are consistent with those of Delran et al. [26], who observed similar trends with the rupture of *Tetraselmis suecica* cells (10% DW) at a power of 100 W. Progressive changes in the internal structure of microalgae over time facilitate the solubilization and release of proteins. At an amplitude of 70% (equivalent to 350 W), the protein concentration reached a peak of 2.2 mg/mL after 10 min, corresponding to the release of 68% of the protein present. However, the protein concentration decreased beyond this point, possibly because the harsh conditions led to protein denaturation [35]. Better results were obtained for an amplitude of 70% than for an amplitude of 20% at all time points studied, with a faster extraction of larger amounts of protein. These findings indicate that cell lysis is more potent at 70%, facilitating rapid protein release, albeit with a risk of protein denaturation. The increase in protein extraction was attributed to the ultrasound-induced perforation of microalgal cell walls. These pores enhance cell wall permeability, facilitating solvent penetration and protein release [36].

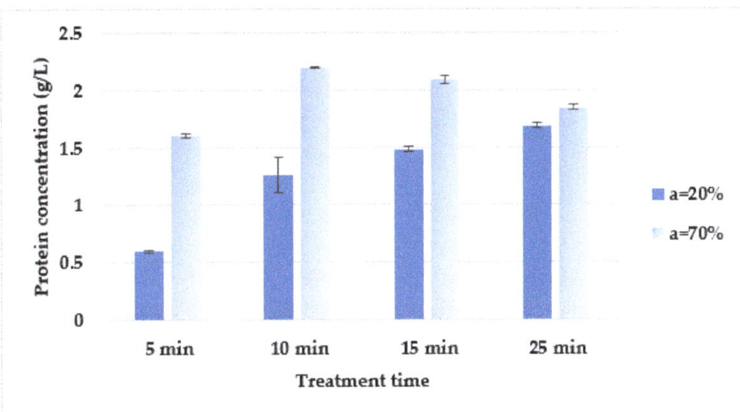

Figure 1. Variation in the protein concentration, as analyzed with the Bradford assay, under ultrasound (US) conditions. "a" is the amplitude.

The variation in the fraction of water-soluble proteins released over time followed a polynomial trend for both amplitudes (Figure 2). This trend fit the 20% amplitude data exceptionally well (R^2 = 0.9971), whereas the fit was slightly less accurate for an amplitude of 70% (R^2 = 0.9371). The decrease in the protein concentration should correspond to protein denaturation and may disturb the kinetic model for the 70% amplitude. Even if proteins are extracted at this time, the measured concentrations do not correspond to the actual level of protein release. Therefore, the kinetic parameters were also calculated for the initial points before the degradation, and the model was referred to as the "a = 70% modified".

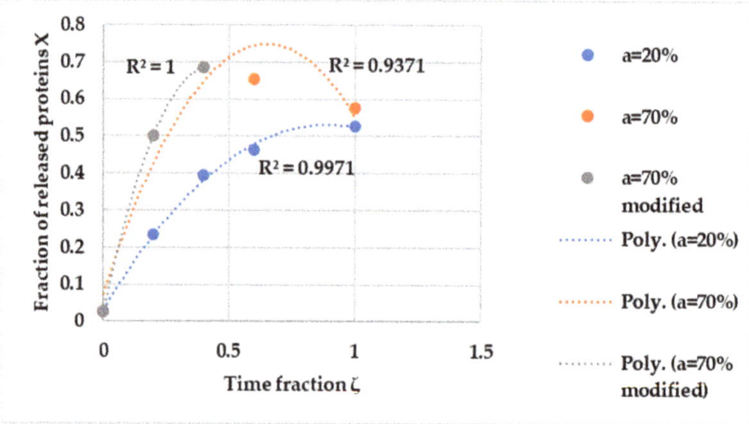

Figure 2. Fraction of proteins released over time. The dashed line represents a polynomial fit to the experimental data. "a" is the amplitude. "a = 70% modified" corresponds to the fraction released after only 10 min of US treatment.

The protein release fraction was time-dependent and this relationship was characterized by two different models based on the amplitude or power used:

$$X = -0.6345\zeta^2 + 1.1281\zeta + 0.0305 \quad \text{for a} = 20\% \tag{10}$$

$$X = -1.5902\zeta^2 + 3.0933\zeta + 0.026 \quad \text{for a} = 70\% \tag{11}$$

Upon limiting our model to $\zeta = 0.4$ for the 70% amplitude, corresponding to the observed increase in the protein concentration without denaturation, we obtained a well-fitted polynomial trend with $R^2 = 1$ (Equation (12)).

$$X = -3.605\zeta^2 + 3.0933\zeta + 0.0261 \quad \text{for a} = 70\% \tag{12}$$

For the determination of the diffusion kinetics of the proteins, we defined the release rate constant, k, which was obtained by the graphical solution of Equation (8). The difference in protein release rates was explained by the kinetic model. Proteins diffused from the cells four times faster at an amplitude of 70% than at an amplitude of 20%, with protein release constants of 0.0005 s^{-1} and 0.0019 s^{-1}, respectively. However, the fit of the kinetic model was less well suited for the 20% amplitude ($R^2 = 0.9091$). When the last concentration point was eliminated from the model, the fitted linear trend became more accurate, and the model was better validated ($R^2 = 0.9804$). In this work, we used the value of k calculated from all the data points without model adjustment, as this value represents a more realistic model for the kinetics of protein release. It is clear from Figure 2 that the variation in slope between the two amplitudes reflects the speed of protein extraction, which is greater at 70% than at 20%. This observation indicates that the access of proteins to water is greater at higher amplitudes, facilitating their release. The calculated release rate constant provides a qualitative explanation for the mechanism of protein diffusion. Suarez Garcia et al. [37] explained that aqueous proteins are rapidly released from the cytosol and internal organelles during cell lysis. Functional proteins are released later as the lysis conditions become more severe, due to their less accessible locations, with proteins involved in light capture located in the chloroplasts, for example. The concentration of the extracts at steady state indicated that increasing the severity of treatment led to an increase in the amount of protein that could be extracted. Nevertheless, structural proteins in the cell wall may undergo denaturation during cell disintegration because they are released by osmotic shock during the dilution of the paste.

The total carbohydrate content of this batch of *Tetraselmis suecica* was 9.1 ± 0.01% DW for samples analyzed in duplicate. This value is within the range of carbohydrate levels (8–57% DW) reported in previous studies on *Tetraselmis suecica* [13]. The main sugars were galactose, glucose, and mannitol, with traces of fucose. Ribose, arabinose, and mannose were also detected, but at concentrations below the limit of quantification, and xylose was entirely absent. The extraction of carbohydrates for the two amplitudes tested (20% and 70%) showed that 10 min of extraction was sufficient for the release of 35% of carbohydrates into the supernatant for the 20% amplitude and 22% for the 70% amplitude (Figure 3). The treatment at 70% is less efficient than at 20%, and a longer treatment time at 70% amplitude shows a decrease in the carbohydrate concentration due to the modification of the extract structure, leading to lower separation efficiency. The lower yield obtained at an amplitude of 20% after 15 min could also be due to these phenomena. Further investigations are required to confirm these hypotheses and rule out experimental error. We deduce from our findings that sonication for 10 min has an effect almost equivalent to that of 25 min of sonication at the 20% amplitude. At an amplitude of 70%, the carbohydrate concentration decreased after periods of exposure to US exceeding 10 min. We hypothesize that long periods of exposure to high-amplitude US may lead to carbohydrate degradation. These results are consistent with those of Di Caprio Fabrizio et al. [38], who demonstrated that carbohydrate extraction rates for *T. obliquus* reach a maximum at the start of US exposure, subsequently stabilizing during an equilibrium phase. Furthermore, in accordance with the findings of Delran et al. [26], 5 min of sonication was sufficient to extract a maximum of 60% carbohydrates from *Tetraselmis suecica* with a power of 200 Watts. A comparison of the two amplitudes, 20% and 70%, revealed that an amplitude of 20% (equivalent to 100 W) resulted in higher carbohydrate release rates than an amplitude of 70% (equivalent to 350 W). We can infer from these findings that an amplitude of 20% is optimal for carbohydrate extraction from *Tetraselmis suecica* (Figure 3).

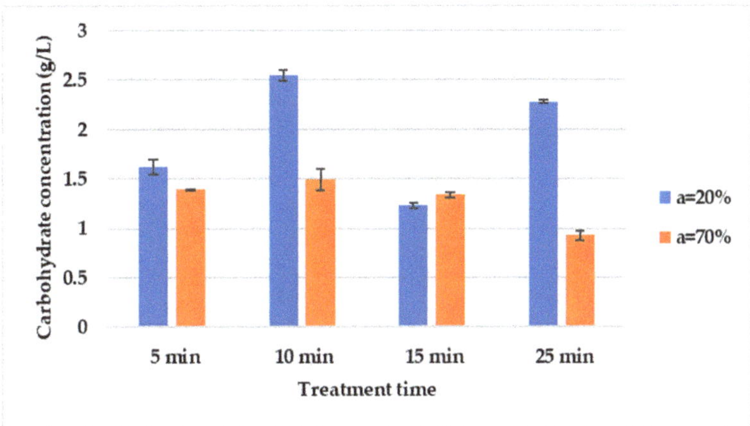

Figure 3. Variation in the carbohydrate concentration over time under US conditions. "a" is the amplitude applied during US treatment.

The fraction of carbohydrates released $X_{carbohydrates}$ at amplitudes of 20% and 70% over a period of 10 min followed a polynomial trend ($R^2 = 1$ for both amplitudes) and was based on three times: 0 (which is obtained from the control treatment), 5 and 10 min. This fraction was linked to the time fraction ζ by the following equations:

$$X_{carbohydrates} = 0.6474\zeta^2 + 0.3718\zeta + 0.1776 \quad \text{for a} = 20\% \tag{13}$$

$$X_{carbohydrates} = -0.4469\zeta^2 + 0.3558\zeta + 0.1776 \quad \text{for a} = 70\% \tag{14}$$

Based on first-order kinetics, the carbohydrate release rate constant was determined by graphically solving Equation (8), as for proteins. Over the 10 min period considered, carbohydrates diffused at a rate of 0.0006 s^{-1}, which was notably slower than the rate of protein diffusion observed at an amplitude of 20% over this period (0.0008 s^{-1}). This finding suggests that diffusion rates are significantly higher for proteins than for carbohydrates. Suarez Garcia et al. [37] explained that carbohydrates diffuse more slowly due to the solubilization of saccharides from the cell wall and starch granules. Interestingly, at an amplitude of 70%, carbohydrate release was much slower than at 20%, at a rate of 0.0002 s^{-1}. An amplitude of 20% therefore appeared to be better in terms of both the carbohydrate yield and carbohydrate release rate.

We analyzed the following pigments in this batch of *Tetraselmis suecica*: fucoxanthin, lutein, chlorophyll *b*, chlorophyll *a*, and β-carotene. We also compared the pigment concentrations of the aqueous extracts obtained under various US conditions. The variations in the pigment concentration with US treatment time for amplitudes of 20% and 70% are shown in Figures 4 and 5, respectively. The concentration of pigments gradually increased with treatment time at both amplitudes tested. During US treatment, the cells were progressively lysed over time until complete fragmentation occurred, allowing the release of pigments. The pigment concentration reached similar levels for the two amplitudes after 25 min. This observation suggests that the microalgal wall is fragile, with even low-amplitude treatments being sufficient for pigment extraction. Natarajan et al. [24] showed that the US power had a minimal effect on the efficiency of cell lysis in *Tetraselmis suecica*. The pigment content in the supernatant can be explained by the nature of these hydrophobic biomolecules and their localization in chloroplasts, necessitating major cellular disintegration for their release. It is thought that pores form under these conditions, providing the solvent (water) with access to these biomolecules. However, these lipophilic pigments have a limited solubility in water and their levels were not influenced by the use of a high power. We did not perform a kinetic study of the pigments because the total pigment content of the initial biomass was not analyzed for this batch.

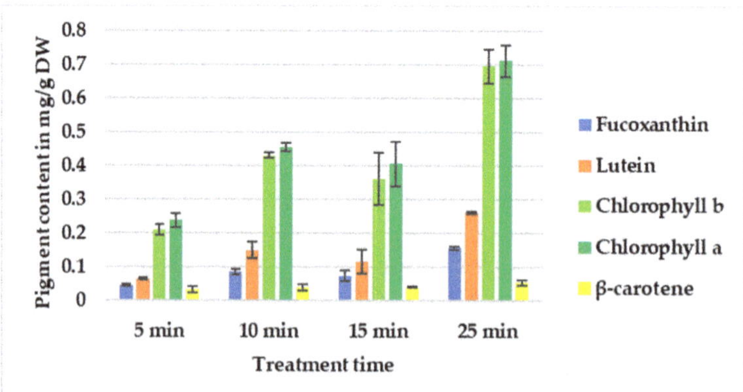

Figure 4. Variation in the pigment content of supernatants over time with an ultrasound amplitude of 20%.

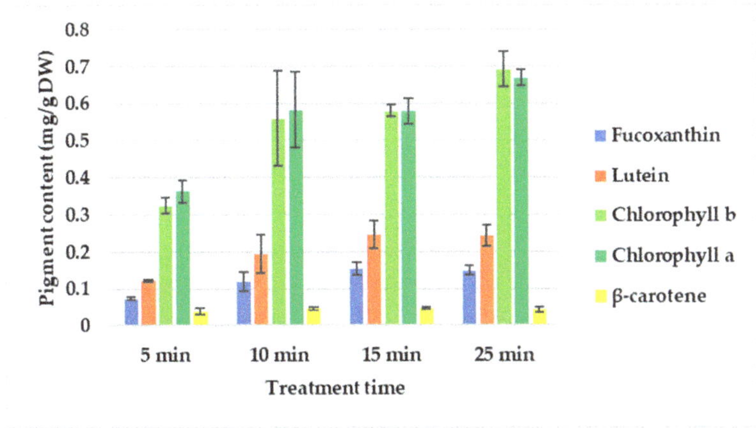

Figure 5. Variation in the pigment content of supernatants over time with an ultrasound amplitude of 70%.

3.2. High-Pressure Homogenization (HPH)

This batch of *Tetraselmis suecica* had an initial DW content of 15% and was diluted to 5% DW for further processing. The maximum temperatures reached 31 °C, 37 °C, and 43 °C for conditions 1, 2, and 3, respectively, without the need for cooling water. These temperatures are optimal for protein solubilization, facilitating protein extraction, and are sufficiently low to prevent carotenoid oxidation, protein denaturation, and chlorophyll degradation. We analyzed the kinetics of biomolecule diffusion in the aqueous supernatant for proteins, neutral carbohydrates, and pigments (chlorophyll *a*, chlorophyll *b*, fucoxanthin, lutein, and β-carotene). With the DW content of all the recovered supernatants maintained at 3.2 ± 0.3%, the extraction rates of these biomolecules can be expressed in terms of concentration (mg/L), concentration per g of dry matter (mg/g of DW), or purity.

Total protein accounted for 30.6 ± 0.07% of the DW of the initial biomass, as determined by the Kjeldahl method (equivalent to 5.15% DW of nitrogen; NTP = 5.95). The high levels of protein present necessitated appropriate fractionation to achieve the best possible yields and purities. We monitored the kinetics of protein diffusion at various centrifugation times, ranging from 5 min to 3 h, by analyzing both extracts centrifuged immediately at t_1 and extracts centrifuged after 3 h, at t_6. Such analyses can shed light on the mechanisms of molecule release and the effects of centrifugation, either facilitating molecule release from the residue or favoring molecule retention within the matrix. We used the Kjeldahl method to assess the variation in the total protein content with the centrifugation lag time for the three HPH extraction conditions.

All the extracts had constant protein contents, maintained at 33.1 ± 1.1% DW (Figure S1 in the Supplementary Materials). The protein concentration therefore remained constant at 7.9 ± 1.7 g/L, regardless of the timing of centrifugation after the end of lysis. Centrifugation therefore had no noticeable effect on the diffusion of proteins from the cytoplasm. The variations in the protein concentration observed at the time points studied can be attributed to differences in the dry weight of the prepared samples, as each sample was prepared separately. However, when the results for all samples were normalized for the same dry weight, the differences were negligible. Once the cells were lysed, all proteins were released rapidly. Similarly, the protein extraction yield remained constant across all three conditions at 64.4 ± 1.5% (Figure 6). This value is the mean extraction yield for the 18 samples, corresponding to six analyses for each of the three sets of conditions. Increasing the number of passes from two to five or increasing the pressure from 300 bars to 600 bars had no discernible impact on protein extraction. Proteins were released immediately, even under milder lysis conditions (P = 300 bars and two passes).

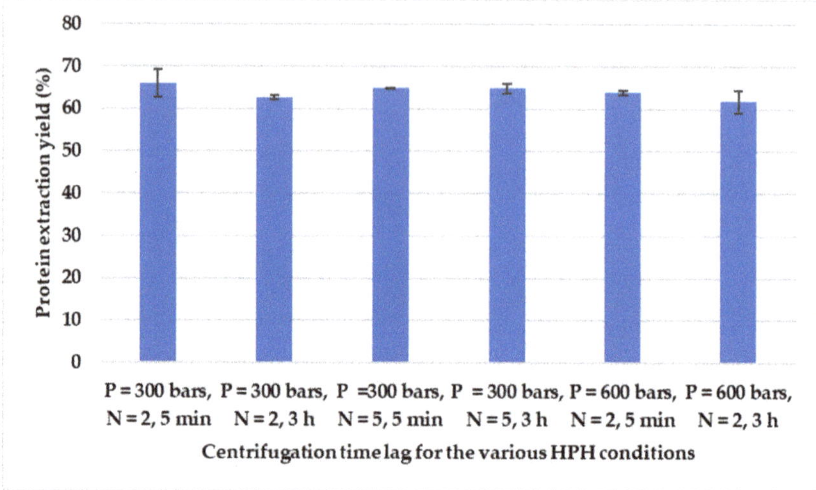

Figure 6. Variation in the protein extraction yield with the centrifugation time lag for the various HPH conditions analyzed by the Kjeldahl method (P is the pressure and N is the number of cycles or passes).

These findings suggest that significant cell lysis occurred under these conditions, releasing 64.4% of the proteins. Similar results were obtained by Rida et al. [39], who extracted 61.9 ± 3.9% of proteins from *Tetraselmis suecica* using 300 bars of pressure and two passes. Moreover, *Tetraselmis suecica* appears to have a weak cell wall. These results are consistent with those obtained by Safi et al. [40], who demonstrated that two passes were sufficient to extract 66% of the proteins from *Chlorella vulgaris,* with the protein yield remaining consistent regardless of the number of passes.

The concentration of water-soluble proteins remained relatively constant across all supernatants at 2 ± 0.3 mg/mL. It was not possible to compare the two analytical methods at this point due to differences in their underlying principles. In the Kjeldahl method, the total Kjeldahl organic nitrogen content was quantified, including molecules not necessarily associated with proteins. In contrast, the Bradford method primarily detects residues of three specific amino acids (arginine, lysine, and histidine).

Total carbohydrate accounted for 11 ± 0.1% of the DW of the initial biomass. A constant carbohydrate extraction yield of 30 ± 2% was obtained across all centrifugation time lags and extraction conditions (Figure 7), suggesting that similar amounts of carbohydrate were released by centrifugation immediately after cell lysis and centrifugation after a time lag of 3 h. Once the cell is disrupted, all the soluble carbohydrates are rapidly released, and centrifugation has no detectable effect on sugar diffusion. Furthermore, the carbohydrate content of the aqueous extract remained constant at 6.6 ± 0.6% DW for all three sets of lysis conditions.

Increasing the number of passes from two to five and the pressure from 300 to 600 bars did not affect carbohydrate extraction. Thus, the conditions used here were not harsh enough to extract carbohydrates efficiently from *Tetraselmis suecica*. These polymers may act as structural molecules of limited solubility in water or form insoluble complexes with other molecules. These complexes tend to settle in the pellet during separation by centrifugation. Alternatively, these water-soluble complexes may be entrapped within lipophilic emulsions, preventing their passage into the supernatant [41]. These findings are consistent with those of Delran et al. [20], who found that with a single pass at 400 bars, the carbohydrate extraction yield reached 60%, remaining constant thereafter, even if the conditions became more severe.

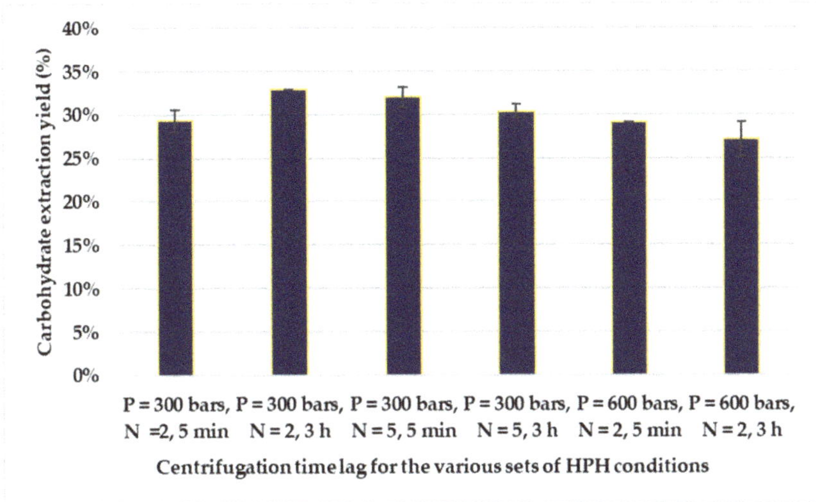

Figure 7. Variation in the carbohydrate extraction yield with the centrifugation time lag for the various sets of HPH conditions.

The pigment analysis was conducted on the liquid supernatant, which was stored at −20 °C for one year before analysis. Figures 8–10 show the differences in the pigment concentration in aqueous extracts between conditions 1, 2, and 3, respectively, of cell lysis by HPH. Pigment concentrations clearly remained relatively constant over the six centrifugation lag times tested, with only slight variations. These variations can be attributed to the degradation of pigments in water during the storage of the aqueous samples. Centrifugation at three hours after cell lysis did not result in significantly greater pigment diffusion than centrifugation at five minutes after lysis. Such behavior would typically be expected for lyophilic pigments, which tend to move toward the pellet during centrifugation. The relatively low pigment concentration in the supernatant can be attributed to the probable formation of complexes between certain pigments and water-soluble proteins during the centrifugation phase.

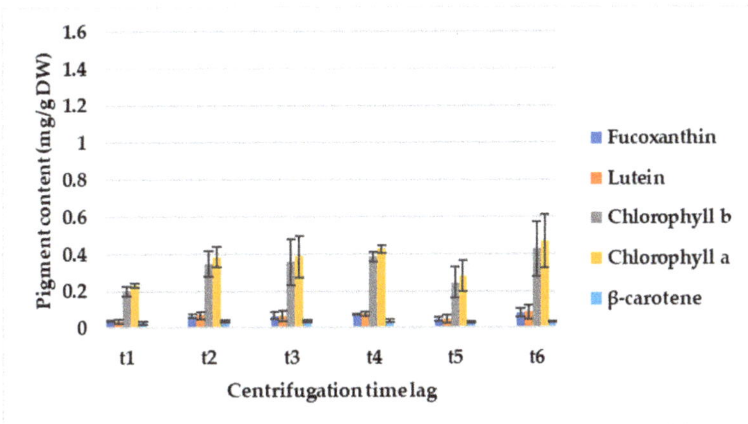

Figure 8. Pigment content variation as a function of the time to centrifugation (centrifugation time lag) for HPH condition 1 (P = 300 bars, N = 2 passes).

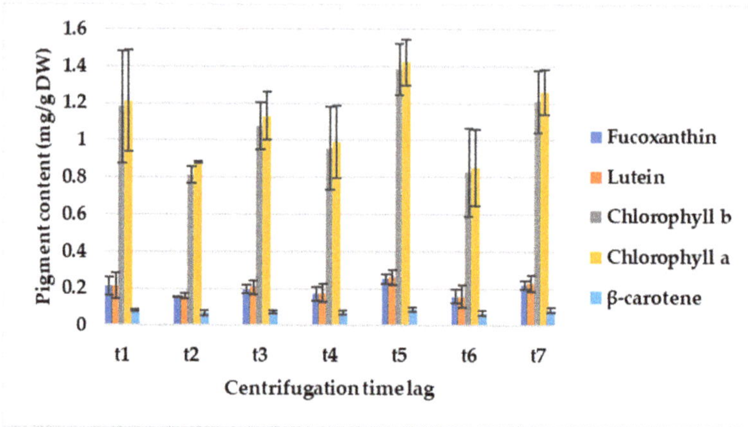

Figure 9. Pigment content variation with the centrifugation time lag for HPH condition 2 (P = 300 bars, N = 5 passes).

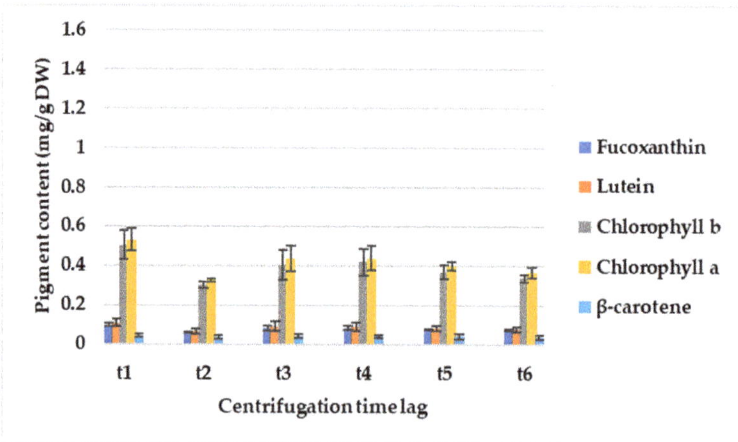

Figure 10. Pigment content variation with the centrifugation time lag for HPH condition 3 (P = 600 bars, N = 2 passes).

Increasing the number of passes from two to five resulted in higher concentrations of fucoxanthin and lutein, increasing from 0.06 to 0.2 mg/g DW for each pigment, of chlorophyll *b*, increasing from 0.3 to 1 mg/g DW, of chlorophyll *a*, increasing from 0.4 to 1.1 mg/g DW, and of *β*-carotene, increasing from 0.04 to 0.08 mg/g DW. However, increasing the pressure from 300 bars to 600 bars resulted in only a slight increase in pigment concentrations, except for chlorophyll *a* and *β*-carotene. The number of passes appeared to have a greater impact than homogenization pressure. This increase in pigment concentrations suggests that the chloroplasts were disrupted, allowing water unhindered access to the interthylakoid space, where certain pigments are located [13]. The extraction of pigments can be explained as a function of their cellular location. Chlorophyll *a* is typically found in photosystem I (PSI), which is situated on the surface of the thylakoid in contact with the stroma, making it more accessible to the solvent than chlorophyll *b*, which is typically located in photosystem II (PSII). This accounts for the concentration of chlorophyll *a* being higher than that of chlorophyll *b*.

3.3. Comparison of HPH and US

The preceding section addressed the comparison of various HPH conditions and their effects on biomolecule extraction. The optimal conditions for HPH were determined to be two passes at a pressure of 300 bars, with no improvement in extraction yield observed with further increases in the pass number and pressure. We then showed that an amplitude of 20% and a treatment time of 25 min were the ideal conditions for US, striking a balance between protein and carbohydrate yields while minimizing energy consumption. We then performed tests on a single batch of cells under the optimal conditions defined above to compare HPH and US in a meaningful manner and to identify the most efficient extraction method with the highest yield for use in the subsequent biorefining of *Tetraselmis suecica*.

3.3.1. Dry Matter

The batch of *Tetraselmis suecica* used had an initial DW content of 15% and was diluted to 5.65 ± 0.05% DW for further processing. We first compared the dry matter extraction yield between the two disruption techniques (HPH and US). HPH extracted 52.1 ± 1.7% of dry matter, whereas US extracted 46.4 ± 1.5% of dry matter. Both methods were more effective than the control, which extracted only 24.1 ± 0.3% of the dry matter (Figure 11). Dry matter extraction is crucial for microalgal valorization and for assessments of the efficiency of cell lysis. However, a comprehensive comparison between the two techniques also requires the characterization of the extract and a determination of the content of the biomolecules of interest within it. Many studies have focused on the extraction yield without considering the metabolite content. Suarez Garcia et al. [37] emphasized the importance of using both the yield and content to guide subsequent purification steps and to enhance the technical classification of these products.

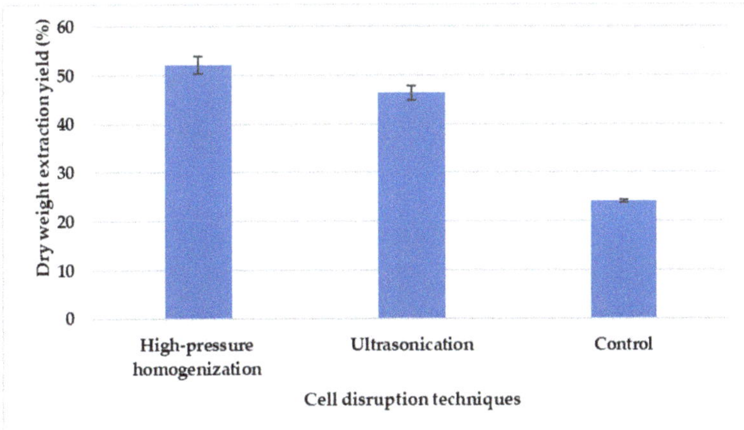

Figure 11. Comparison of dry matter extraction yields.

The amount of dry matter recovered in the supernatant corresponds to the biomolecules extracted, as these molecules tend to migrate to the aqueous phase during centrifugation. We can therefore conclude that HPH is a more effective cell lysis technique than US, resulting in higher extraction rates. These results contrast with those of Halim et al. [23], who found that US and HPH were similarly effective for the lysis of *Tetraselmis suecica* cells but that US gave better results for lipid extraction. They suggested that the extracted lipids might be degraded under the HPH conditions used. In contrast, Zhang et al. [19] compared HPH and US for the lysis of *P. kessleri* and concluded that US damaged the cells less effectively, whereas HPH led to the generation of large amounts of cell debris.

3.3.2. Minerals

The biomass received had a high mineral content, at 29.6 ± 0.6% DW. The primary goal of microalgal fractionation is to extract valuable molecules with important industrial applications, such as pigments, proteins, and carbohydrates. The control extract (obtained by lysis in water) had the highest mineral content, at 71.2 ± 1.02% DW (Figure 12), due to the hydrophilic nature and high solubility of these molecules in water. The US extract had a higher mineral content (43.8 ± 1.8% DW) than the HPH extract (35.5 ± 0.8% DW). These findings suggest that HPH is more effective for the extraction of intracellular biomolecules than US, leading to the release of fewer minerals during subsequent aqueous washing. This is an advantage of HPH over US, as HPH yields an extract with fewer minerals and higher concentrations of other compounds of interest.

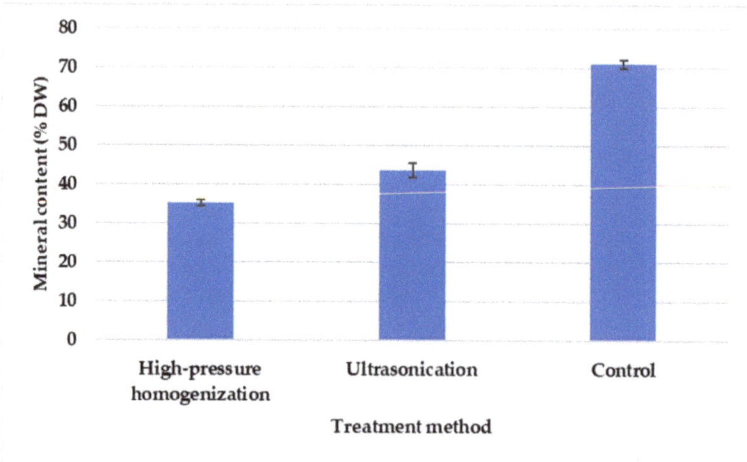

Figure 12. Comparison of the mineral content between disruption techniques.

3.3.3. Proteins

The total protein content of the initial biomass was 23.1 ± 0.3% DW, as analyzed by the Kjeldahl method. The protein concentration was determined by the Bradford method with BSA as the standard for the comparison of biomass treatments (Figure 13) and with the Kjeldahl method for the yield calculation (Figure 14). Both these methods showed that HPH cell disruption resulted in a higher protein concentration than US treatment. HPH extracted a higher proportion of proteins (61.9 ± 3.9%) than US (44.8 ± 1.3%). In addition, the protein content of the HPH extract (26.1 ± 0.3% DW) was higher than those of the US extract (22.6 ± 0.3% DW) and the control extract (4.1 ± 0.4% DW). These findings highlight the superiority of HPH over US for protein extraction. Importantly, the conditions applied (two passes at 300 bars) were not very severe and were energy-efficient. Kinetic studies in the initial phase revealed that protein extraction with HPH was rapid and instantaneous, whereas protein extraction by US treatment required 25 min. These results are consistent with the findings of Safi et al. [29], who reported higher protein extraction yields with HPH than with US for various microalgae. HPH has also been shown to be more effective at extracting proteins from *P. kessleri* while consuming less energy, whereas US conditions are thought to cause protein degradation [19]. Furthermore, two separate studies conducted by Delran et al. [20,26] on the rupture of *Tetraselmis suecica* cells indicated that HPH at 400 bars allowed the extraction of 80% of the proteins, whereas US at 200 W power resulted in the extraction of 88% of the proteins after 60 min of treatment. The pressure range of 200–400 bars used in HPH corresponds to the threshold for cell disruption, with rupture of the membrane allowing the solvent access to proteins. Delran and coworkers also reported that the use of a power of 200 W resulted in changes in the shape of *Tetraselmis suecica* cells after 30 min of treatment, consistent with lower rates of cell lysis for US treatment than for HPH. This difference in protein extraction yields between the two techniques can be attributed to the extent of cell disruption. Based on the content, extraction yield, concentration, and protein release rate kinetics of the two techniques, HPH appears to be a more efficient method for the subsequent fractionation process.

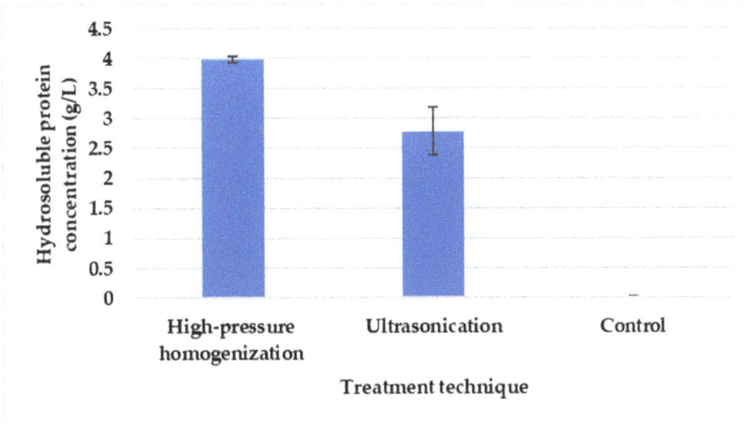

Figure 13. Differences in water-soluble protein concentrations between cell disruption techniques, as determined with the Bradford assay.

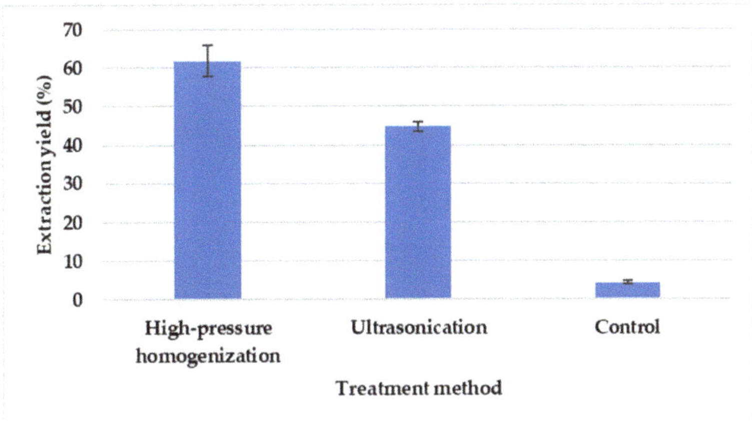

Figure 14. Differences in the total protein extraction yield according to the treatment technique, as analyzed by the Kjeldahl method.

3.3.4. Carbohydrates

The total carbohydrate content of the initial biomass was 7.3 ± 0.7% DW. The sugars identified in the initial biomass were mannitol, fucose, arabinose, galactose, glucose, mannose, and ribose. The monosaccharides identified are consistent with those detected by Pereira et al. [42]. Kermanshahi-Pour et al. [43] demonstrated that the structural monosaccharides present in the cell wall of *Tetraselmis suecica* are 3-deoxy-D-manno-oct-2-ulosonic acid (KDO), 3-deoxy-lyxo-2-heptulosaric acid (DHA), galacturonic acid, and galactose. The carbohydrate extraction yield did not differ significantly between the two techniques used ($p > 0.05$), with a value of 35% obtained for both of them (Figure 15). US (5.4 ± 0.7% DW) resulted in a slightly higher carbohydrate content than HPH (4.8 ± 0.6% DW), but the difference was almost negligible. These findings suggest that both methods are suitable for carbohydrate extraction. However, the kinetics of carbohydrate extraction remain an important factor for comparison. We found that carbohydrates were extracted significantly more rapidly with HPH than with US, for which 25 min of treatment were required. This is a crucial consideration, with HPH identified here as the superior technique for extracting carbohydrates from *Tetraselmis suecica*. In a study by Delran et al. [20,26], the carbohydrate extraction yields from *Tetraselmis suecica* were investigated in two separate experiments for HPH and US. US resulted in the extraction of 70% of carbohydrates at a power of 100–200 W after 60 min of treatment, whereas HPH released 60% of carbohydrates at a pressure of 400 bars. Zhang et al. [19] showed that carbohydrate extraction is correlated with energy consumption. For example, using US at a power of 400 W for 30 min allowed the extraction of 45% of carbohydrates from *P. Kessleri* with an energy consumption of 56 kJ/g DW, whereas HPH at a pressure of 400 bars enabled the extraction of 20% of carbohydrates with a consumption of 16 kJ/g DW.

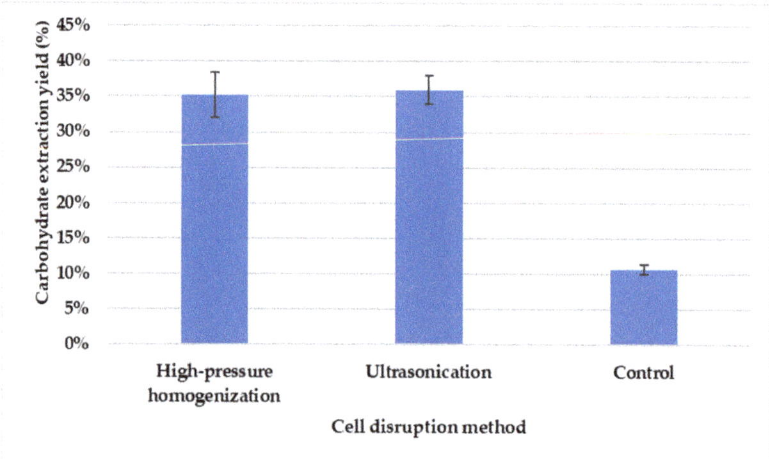

Figure 15. Comparison of carbohydrate extraction yields between cell disruption methods.

3.3.5. Uronic Acids

Total uronic acids in *Tetraselmis suecica* accounted for 17.1 ± 0.8% DW of the initial biomass. Uronic acids were extracted more efficiently by US, with an extraction yield of 24 ± 4% and a uronic acid content of 8.9 ± 1.8% DW. In comparison, HPH allowed the extraction of 20 ± 6% with a uronic acid content of 6.4 ± 2% DW. Both of these techniques outperformed the control, which released only 5% of uronic acids, with a content of 2.9 ± 0.2% DW (Figure 16). The difference in uronic acid extraction rates between HPH and US can be attributed to the impact of these molecules on cell structures. Galacturonic acid is located primarily in the cell wall in *Tetraselmis suecica*. US is a less severe cell disruption method than HPH and is not thought to destroy the cell wall of *Tetraselmis suecica*, instead creating pores via which uronic acids are released. Conversely, HPH is a more severe technique that fragments the cell wall, leading to uronic acid degradation. In terms of kinetics, HPH is considered superior to US because it leads to an immediate release of molecules, whereas a treatment time of 25 min is required for US.

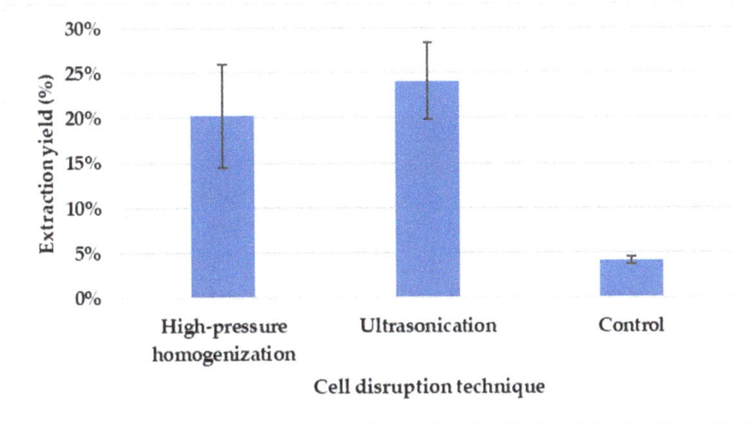

Figure 16. Comparison of the uronic acid extraction yield between disruption methods.

3.3.6. Pigments

The pigments detected and analyzed in *Tetraselmis suecica* were fucoxanthin, lutein, chlorophyll *b*, chlorophyll *a*, and *β*-carotene. The total pigment content of the initial biomass was 0.3 ± 0.07% DW. HPH was more efficient at pigment extraction, with an extraction yield of 25.7 ± 3.4%, than US, for which the extraction yield was much lower, at 11.2 ± 1.3%. This difference highlights the ability of HPH to facilitate the extraction of intracellular molecules more effectively than US, leading to greater cell lysis. The higher pigment content of the HPH extract (0.14 ± 0.02% DW) than of the US extract (0.07 ± 0.007% DW) suggests major alterations to the chloroplasts, releasing pigments from the thylakoids. Conversely, the low-intensity conditions used for US resulted in milder damage to the cells, impeding the passage of water across the phospholipid membrane of the thylakoid. The low extraction yields for pigments may reflect their hydrophobic nature, causing them to migrate toward the pellet. However, their presence in the aqueous phase was explained by Safi et al. [13], who suggested that pigments adsorb onto fine cell debris that does not settle in the pellet during centrifugation or are present in emulsions or attached to phospholipids. Similar findings were reported by Zhang et al. [19], who demonstrated that the extraction of pigments from *P. kessleri* was more effective with HPH than with US. Comparisons of the extraction yield and pigment content in *Tetraselmis suecica* extracts suggest that HPH is more effective for pigment extraction than US. The kinetics of pigment release were also better for HPH, which allowed faster extraction than US.

4. Conclusions

This study sheds light on the influence of cell disruption techniques and conditions on the extraction of biomolecules and the kinetics of their release behavior. A screening of extraction conditions based on mechanical cell lysis treatments revealed that both HPH and US were viable methods for the fractionation of *Tetraselmis suecica*. We evaluated and compared these rupture methods on the basis of the extraction yield, biomolecule content of the extract, and biomolecule release kinetics. Based on the parameters investigated, HPH is better for protein and pigment extraction, whereas US can release larger amounts of uronic acid. Interestingly, the two methods had similar extraction yields for carbohydrates. The release kinetics for all biomolecules showed that HPH was superior to US, resulting in the rapid release of metabolites. This study focused on the kinetics of molecule extraction, but additional studies are required to understand the morphological changes in cells over time. Such studies would provide detailed explanations for the phenomenon of biomolecule diffusion observed during cell disruption. The use of sequential cell disruption methods might increase the extraction yields for both the proteins and carbohydrates of *Tetraselmis suecica*. The initial focus could be on carbohydrate extraction, with a shift towards protein recovery in subsequent steps. Further studies are required for process optimization, and a techno-economic assessment is required to assess the feasibility of this approach. Our findings indicate that proteins are extracted alongside polysaccharides and other molecules, resulting in a lower protein content in the final extract. Additional separation techniques are required to purify target molecules by eliminating the molecules extracted with them.

Supplementary Materials: The following supporting information can be downloaded at: https://www.mdpi.com/article/10.3390/pr12061139/s1. Figure S1: Variation in total protein content with centrifugation delay time for different HPH conditions; Figure S2: Graphical solution for the determination of the proteins release rate constant k. "a" refers to the amplitude applied during US treatment; Figure S3: Variation of the fraction of released carbohydrates in the supernatant over time; Figure S4: Graphical solution for the determination of the proteins and carbohydrates release rate constant k during 10 min.

Author Contributions: Conceptualization, methodology, validation, investigation, H.R., H.T., A.I. and P.-Y.P.; formal analysis, H.R., J.P. and P.-Y.P.; resources, A.I. and P.-Y.P.; writing—original draft preparation, H.R.; writing—review and editing, H.T., A.I. and P.-Y.P.; visualization, H.R., J.P. and P.-Y.P.; supervision, H.T., A.I. and P.-Y.P.; project administration, funding acquisition, A.I. and P.-Y.P. All authors have read and agreed to the published version of the manuscript.

Funding: This research was supported by a SAFAR doctoral scholarship awarded to Mr. RIDA and cofinanced by the Lebanese University and the French Embassy in Lebanon (CAMPUS FRANCE).

Data Availability Statement: The original contributions presented in the study are included in the article; further inquiries can be directed to the corresponding author.

Conflicts of Interest: The authors declare no conflicts of interest.

References

1. Khanra, S.; Mondal, M.; Halder, G.; Tiwari, O.N.; Gayen, K.; Bhowmick, T.K. Downstream Processing of Microalgae for Pigments, Protein and Carbohydrate in Industrial Application: A Review. *Food Bioprod. Process.* **2018**, *110*, 60–84. [CrossRef]
2. Wicker, R.J.; Kwon, E.; Khan, E.; Kumar, V.; Bhatnagar, A. The Potential of Mixed-Species Biofilms to Address Remaining Challenges for Economically-Feasible Microalgal Biorefineries: A Review. *Chem. Eng. J.* **2023**, *451*, 138481. [CrossRef]
3. Show, P.-L.; Park, Y.-K.; Al-Zuhair, S.; Ashokkumar, V. Guest Editorial: Special Issue on "Microalgae Biorefinery: Current Bottlenecks, Challenges, and Future Directions". *Phytochem. Rev.* **2023**, *22*, 829–831. [CrossRef]
4. Nitsos, C.; Filali, R.; Taidi, B.; Lemaire, J. Current and Novel Approaches to Downstream Processing of Microalgae: A Review. *Biotechnol. Adv.* **2020**, *45*, 107650. [CrossRef]
5. Soto-Sierra, L.; Stoykova, P.; Nikolov, Z.L. Extraction and Fractionation of Microalgae-Based Protein Products. *Algal Res.* **2018**, *36*, 175–192. [CrossRef]
6. Patil, R.A.; Kausley, S.B.; Joshi, S.M.; Pandit, A.B. Process Intensification Applied to Microalgae-Based Processes and Products. In *Handbook of Microalgae-Based Processes and Products*; Elsevier: Amsterdam, The Netherlands, 2020; pp. 737–769, ISBN 978-0-12-818536-0.
7. Wang, Q.; Oshita, K.; Takaoka, M.; Shiota, K. Influence of Water Content and Cell Disruption on Lipid Extraction Using Subcritical Dimethyl Ether in Wet Microalgae. *Bioresour. Technol.* **2021**, *329*, 124892. [CrossRef] [PubMed]
8. Günerken, E.; D'Hondt, E.; Eppink, M.H.M.; Garcia-Gonzalez, L.; Elst, K.; Wijffels, R.H. Cell Disruption for Microalgae Biorefineries. *Biotechnol. Adv.* **2015**, *33*, 243–260. [CrossRef]
9. Alhattab, M.; Kermanshahi-Pour, A.; Brooks, M.S.-L. Microalgae Disruption Techniques for Product Recovery: Influence of Cell Wall Composition. *J. Appl. Phycol.* **2019**, *31*, 61–88. [CrossRef]
10. Ashok Kumar, N.; Sridhar, S.; Jayappriyan, K.R.; Raja, R. Chapter 33–Applications of Microalgae in Aquaculture Feed. In *Handbook of Food and Feed from Microalgae*; Jacob-Lopes, E., Queiroz, M.I., Maroneze, M.M., Zepka, L.Q., Eds.; Academic Press: Cambridge, MA, USA, 2023; pp. 421–433, ISBN 978-0-323-99196-4.
11. Fabregas, J.; Abalde, J.; Herrero, C.; Cabezas, B.; Veiga, M. Growth of the Marine Microalga *Tetraselmis suecica* in Batch Cultures with Different Salinities and Nutrient Concentrations. *Aquaculture* **1984**, *42*, 207–215. [CrossRef]
12. Jo, W.S.; Yang, K.M.; Park, H.S.; Kim, G.Y.; Nam, B.H.; Jeong, M.H.; Choi, Y.J. Effect of Microalgal Extracts of *Tetraselmis suecica* against UVB-Induced Photoaging in Human Skin Fibroblasts. *Toxicol. Res.* **2012**, *28*, 241–248. [CrossRef]
13. Safi, C.; Liu, D.Z.; Yap, B.H.J.; Martin, G.J.O.; Vaca-Garcia, C.; Pontalier, P.-Y. A Two-Stage Ultrafiltration Process for Separating Multiple Components of *Tetraselmis suecica* after Cell Disruption. *J. Appl. Phycol.* **2014**, *26*, 2379–2387. [CrossRef]
14. Corrêa, P.S.; Morais Júnior, W.G.; Martins, A.A.; Caetano, N.S.; Mata, T.M. Microalgae Biomolecules: Extraction, Separation and Purification Methods. *Processes* **2021**, *9*, 10. [CrossRef]

15. Gomes, T.A.; Zanette, C.M.; Spier, M.R. An Overview of Cell Disruption Methods for Intracellular Biomolecules Recovery. *Prep. Biochem. Biotechnol.* **2020**, *50*, 635–654. [CrossRef] [PubMed]
16. Gong, M.; Bassi, A. Carotenoids from Microalgae: A Review of Recent Developments. *Biotechnol. Adv.* **2016**, *34*, 1396–1412. [CrossRef]
17. Magpusao, J.; Giteru, S.; Oey, I.; Kebede, B. Effect of High Pressure Homogenization on Microstructural and Rheological Properties of *A. platensis*, *Isochrysis*, *Nannochloropsis* and *Tetraselmis* species. *Algal Res.* **2021**, *56*, 102327. [CrossRef]
18. Spiden, E.M.; Yap, B.H.J.; Hill, D.R.A.; Kentish, S.E.; Scales, P.J.; Martin, G.J.O. Quantitative Evaluation of the Ease of Rupture of Industrially Promising Microalgae by High Pressure Homogenization. *Bioresour. Technol.* **2013**, *140*, 165–171. [CrossRef]
19. Zhang, R.; Grimi, N.; Marchal, L.; Lebovka, N.; Vorobiev, E. Effect of Ultrasonication, High Pressure Homogenization and Their Combination on Efficiency of Extraction of Bio-Molecules from Microalgae Parachlorella Kessleri. *Algal Res.* **2019**, *40*, 101524. [CrossRef]
20. Delran, P.; Frances, C.; Guihéneuf, F.; Peydecastaing, J.; Pontalier, P.-Y.; Barthe, L. *Tetraselmis suecica* Biofilm Cell Destruction by High-Pressure Homogenization for Protein Extraction. *Bioresour. Technol. Rep.* **2023**, *21*, 101372. [CrossRef]
21. Wang, M.; Chen, S.; Zhou, W.; Yuan, W.; Wang, D. Algal Cell Lysis by Bacteria: A Review and Comparison to Conventional Methods. *Algal Res.* **2020**, *46*, 101794. [CrossRef]
22. Quesada-Salas, M.C.; Delfau-Bonnet, G.; Willig, G.; Préat, N.; Allais, F.; Ioannou, I. Optimization and Comparison of Three Cell Disruption Processes on Lipid Extraction from Microalgae. *Processes* **2021**, *9*, 369. [CrossRef]
23. Halim, R.; Rupasinghe, T.W.T.; Tull, D.L.; Webley, P.A. Mechanical Cell Disruption for Lipid Extraction from Microalgal Biomass. *Bioresour. Technol.* **2013**, *140*, 53–63. [CrossRef]
24. Natarajan, R.; Ang, W.M.R.; Chen, X.; Voigtmann, M.; Lau, R. Lipid Releasing Characteristics of Microalgae Species through Continuous Ultrasonication. *Bioresour. Technol.* **2014**, *158*, 7–11. [CrossRef]
25. Natarajan, R.; Chen, X.; Lau, R. Ultrasound Applications in Lipid Extractions from Microalgae. In *Production of Biofuels and Chemicals with Ultrasound*; Fang, Z., Smith, R.L., Jr., Qi, X., Eds.; Biofuels and Biorefineries; Springer: Dordrecht, The Netherlands, 2015; pp. 117–139, ISBN 978-94-017-9624-8.
26. Delran, P.; Frances, C.; Peydecastaing, J.; Pontalier, P.-Y.; Guihéneuf, F.; Barthe, L. Cell Destruction Level and Metabolites Green-Extraction of *Tetraselmis suecica* by Low and Intermediate Frequency Ultrasound. *Ultrason. Sonochem.* **2023**, *98*, 106492. [CrossRef]
27. Suarez Ruiz, C.A.; Emmery, D.P.; Wijffels, R.H.; Eppink, M.H.; van den Berg, C. Selective and Mild Fractionation of Microalgal Proteins and Pigments Using Aqueous Two-Phase Systems: Selective and Mild Fractionation of Microalgal. *J. Chem. Technol. Biotechnol.* **2018**, *93*, 2774–2783. [CrossRef] [PubMed]
28. Mear, H.; Gillon, P.; Gifuni, I.; Lavenant, L.; Poidevin, A.; Couallier, E. Extraction of Soluble Proteins by Bead Milling from *Tetraselmis chui* in Two Different Physiological States. *Algal Res.* **2023**, *74*, 103180. [CrossRef]
29. Safi, C.; Ursu, A.V.; Laroche, C.; Zebib, B.; Merah, O.; Pontalier, P.-Y.; Vaca-Garcia, C. Aqueous Extraction of Proteins from Microalgae: Effect of Different Cell Disruption Methods. *Algal Res.* **2014**, *3*, 61–65. [CrossRef]
30. Wang, Y.; Tibbetts, S.; McGinn, P. Microalgae as Sources of High-Quality Protein for Human Food and Protein Supplements. *Foods* **2021**, *10*, 3002. [CrossRef] [PubMed]
31. Bradford, M.M. A Rapid and Sensitive Method for the Quantitation of Microgram Quantities of Protein Utilizing the Principle of Protein-Dye Binding. *Anal. Biochem.* **1976**, *72*, 248–254. [CrossRef] [PubMed]
32. Van Wychen, S.; Laurens, L.M.L. *Determination of Total Carbohydrates in Algal Biomass, Laboratory Analytical Procedure (LAP)*; National Renewable Energy Laboratory (NREL): Golden, CO, USA, 2013.
33. Blumenkrantz, N.; Asboe-Hansen, G. New Method for Quantitative Determination of Uronic Acids. *Anal. Biochem.* **1973**, *54*, 484–489. [CrossRef]
34. Rowland, S.M.; Van Wychen, S.; Laurens, L.M.L. *Identification and Quantification of Photosynthetic Pigments in Algae (Laboratory Analytical Procedure (LAP))*; National Renewable Energy Laboratory (NREL): Golden, CO, USA, 2022.
35. Obeid, S. Ecodesign Process for Microalgae Fractionation: Use of Supercritical CO_2, Membrane Technology and Low Frequency Ultrasounds. Doctoral Dissertation, Institut National Polytechnique de Toulouse, Labège, France, 2018.
36. Liu, Y.; Liu, X.; Cui, Y.; Yuan, W. Ultrasound for Microalgal Cell Disruption and Product Extraction: A Review. *Ultrason. Sonochem.* **2022**, *87*, 106054. [CrossRef]
37. Suarez Garcia, E.; van Leeuwen, J.; Safi, C.; Sijtsma, L.; Eppink, M.H.M.; Wijffels, R.H.; van den Berg, C. Selective and Energy Efficient Extraction of Functional Proteins from Microalgae for Food Applications. *Bioresour. Technol.* **2018**, *268*, 197–203. [CrossRef]
38. Di Caprio, F.; Altimari, P.; Pagnanelli, F. Ultrasound-Assisted Extraction of Carbohydrates from Microalgae. *Chem. Eng. Trans.* **2021**, *86*, 25–30. [CrossRef]
39. Rida, H.; Peydecastaing, J.; Takache, H.; Ismail, A.; Pontalier, P.-Y. Concentration and Desalting of *Tetraselmis suecica* Crude Extract by Ultrafiltration. *Desalin. Water Treat.* **2024**, *317*, 100209. [CrossRef]
40. Safi, C.; Frances, C.; Ursu, A.V.; Laroche, C.; Pouzet, C.; Vaca-Garcia, C.; Pontalier, P.-Y. Understanding the Effect of Cell Disruption Methods on the Diffusion of Chlorella Vulgaris Proteins and Pigments in the Aqueous Phase. *Algal Res.* **2015**, *8*, 61–68. [CrossRef]

41. Liu, S.; Gifuni, I.; Mear, H.; Frappart, M.; Couallier, E. Recovery of Soluble Proteins from Chlorella Vulgaris by Bead-Milling and Microfiltration: Impact of the Concentration and the Physicochemical Conditions during the Cell Disruption on the Whole Process. *Process Biochem.* **2021**, *108*, 34–47. [CrossRef]
42. Pereira, H.; Silva, J.; Santos, T.; Gangadhar, K.N.; Raposo, A.; Nunes, C.; Coimbra, M.A.; Gouveia, L.; Barreira, L.; Varela, J. Nutritional Potential and Toxicological Evaluation of *Tetraselmis* sp. CTP4 Microalgal Biomass Produced in Industrial Photobioreactors. *Molecules* **2019**, *24*, 3192. [CrossRef]
43. Kermanshahi-Pour, A.; Sommer, T.J.; Anastas, P.T.; Zimmerman, J.B. Enzymatic and Acid Hydrolysis of *Tetraselmis suecica* for Polysaccharide Characterization. *Bioresour. Technol.* **2014**, *173*, 415–421. [CrossRef]

processes

MDPI

Review

A Comparative Study on Recent Developments for Individual Rare Earth Elements Separation

Subbu Venkata Satyasri Harsha Pathapati, Michael L. Free and Prashant K. Sarswat *

Department of Materials Science and Engineering, University of Utah, Salt Lake City, UT 84112, USA; u1398986@utah.edu (S.V.S.H.P.); michael.free@utah.edu (M.L.F.)
* Correspondence: prashant.sarswat@utah.edu; Tel.: +1-801-820-6919

Abstract: Facilitating the demands of modern society, namely, smartphones, televisions, electric vehicles, and high-stability aircraft structures, requires low-cost and high-performance materials and a corresponding change in the approach needed to design them. Rare earth elements (REEs) play a significant role in achieving these objectives by adding small amounts of these elements to alloys, thereby enhancing material properties. Despite being more abundant than precious metals, the 17 REEs exhibit subtle variations in their chemical and physical characteristics. Thus, their separation is still crucial for industrial applications. There is a corresponding need to develop more effective and efficient separation methods. Adding to the separation challenge is the complexity of the sources of REEs and related materials. Thus, large-scale production of REE materials is difficult. Current REE processing techniques can be categorized into pre-treatment, beneficiation, separation, and refining. Researchers have developed various technologies encompassing chemical, physical, and biological methods, focusing on economic and environmental considerations. However, not all these approaches can be scaled up for mass production. This article focuses on feasible strategies such as precipitation and crystallization, oxidation and reduction, ion exchange, adsorption, solvent extraction, and membrane separation. Further research into these traditional and modern methods can potentially revolutionize the separation dynamics of REEs.

Keywords: rare earth elements; precipitation; crystallization; oxidation; reduction; ion exchange; solvent extraction; membrane separation; adsorption

Citation: Pathapati, S.V.S.H.; Free, M.L.; Sarswat, P.K. A Comparative Study on Recent Developments for Individual Rare Earth Elements Separation. *Processes* **2023**, *11*, 2070. https://doi.org/10.3390/pr11072070

Academic Editor: Anna Wołowicz

Received: 6 June 2023
Revised: 4 July 2023
Accepted: 6 July 2023
Published: 11 July 2023

1. Introduction

Rare Earths Elements (REEs) are commonly associated with designing and developing sustainable technologies to meet human demands and standards of living. REEs are also recognized as critical minerals by some researchers and agencies. The REEs are the comprehensive series of elements in the IIIB group in the periodic table, with atomic numbers ranging from 57 to 71. These elements are also referred to as the lanthanide series, starting from Lanthanum to Lutetium. Further, Scandium (21 atomic number) and Yttrium (39 atomic number) are often considered to be part of the REE group [1,2]. The age-old classification of REE consists of the Ceric group (C.G.) with elements from La to Eu and the Yttric group (Y.G.) with elements from Gd to Lu including Y, the reason behind this is related to their solubility differences [3]. Additionally, Sc and Y are considered among the REEs because of their coexistence in common REE-bearing minerals and resembling properties [4]; in contrast, some authors argue not to include Sc with the rest due to ionic radii variation [5]. Later on, based on industrial applications, metallurgists classified the REEs/REMs into three groups: light/Ceric REEs from Lanthanum (La) to Neodymium (Nd), middle REEs from Samarium (Sm) to Gadolinium (Gd), and heavy REEs from Dysprosium (Dy) to Lutetium (Lu). Some of the literature uses terms like yttric REEs, which include Middle REEs, and heavy REEs, and yttrium is also considered a heavy REE because of its similar properties [6]. Figure 1 describes the chronological timeline of REE discovery.

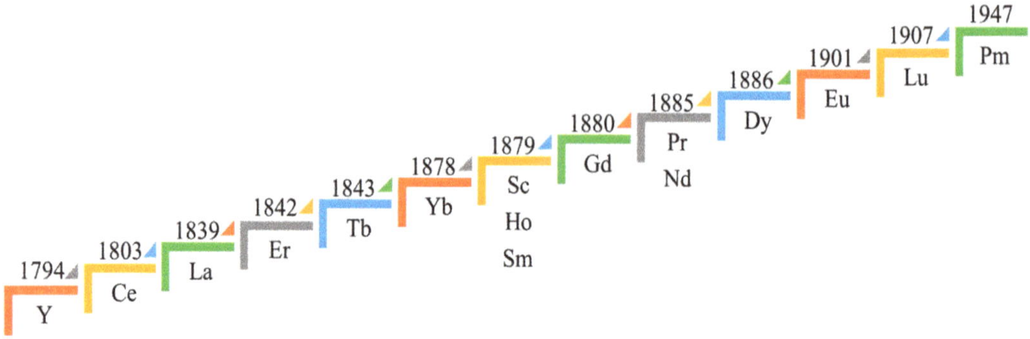

Figure 1. A Historical Path: Mapping the Discovery of REEs.

The British Geological Survey (BGS) (www.mineralsuk.com) has reported that due to the scarcity of the original minerals from which they are extracted, the designation "**rare earth elements**" has become misleading because they are relatively common in the globe's crust. REE materials are much more common in the earth's crust than many precious metals.

Also, almost all REEs can occur in the same deposits. When viewed from different perspectives, their separation and extraction are crucial at the industrial level. For instance, given individual elemental distribution, separation becomes challenging because of similarities in physical and chemical properties. The reason behind similar chemical properties in aqueous media is attributed to the continuous increase/decrease of electrons at 4f sublevels [6], resulting in a division of REEs into LREEs/MREEs/HREEs. The difficulty in separating individual REEs is related to the similarity in electronic structure that results from a slight change in ionic radii between adjacent rare earth elements (0.01–0.03 Å for adjacent atomic number elements). Thus, individual elemental separation by precipitation methods is challenging and complicated [7].

Regarding commercial extraction, the REE's content is often estimated and calculated based on REE oxides rather than elemental form. Jean-Claude G Bunzli stated that this is due to the electropositive nature of these ions to form oxides. Among REE series, the Oddo–Harkins rule (an odd–even strategy) is applicable where the odd atomic numbers have less availability in nature and vice versa [6].

1.1. Sources of REEs

The natural abundance (values averaged in μg/g except for Pm) of REEs in the earth's crust was in the order of **Ce > Nd > La > Y > Sc > Pr > Sm > Gd > Dy > Er > Yb > Eu > Ho > Tb > Lu > Tm** [4]. As indicated in the report from the United States Geological Survey (USGS)(https://www.usgs.gov/centers/national-minerals-information-center/rare-earths-statistics-and-information (accessed on 27 March 2023)), the most widely available mineral sources for REE are monazite, bastnaesite, and xenotime. The USGS report states that the minable and extractable quantity of world REE-oxides reserves worldwide is 130,000,000 metric tons, the mined production in 2021 was 290,000 metric tons, and in 2022 it was 300,000 metric tons. As shown in Figure 2, the statistical analysis of data on REE reserves from the USGS report for the year 2022 depicts China has the maximum amount of REE reserves and the United States of America has a lesser amount of REE reserves, but not the least. The data of other countries include the cumulative reserves for Canada, Greenland, South Africa, and Tanzania. Similarly, Figure 3 represents the worldwide production of REE oxides for 2022, which informs that China has the maximum share in the REE production market, thus emerging as the global REM supply chain controller. It is worth mentioning that in Figure 3, the data of other countries include the production details of both Burma and Madagascar.

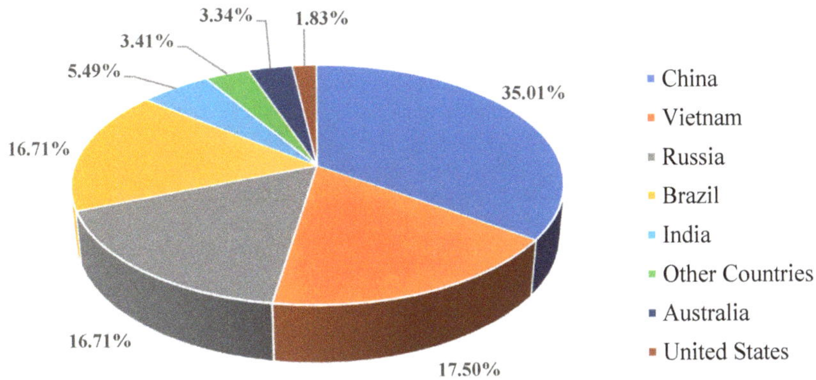

Figure 2. Economically mineable REE reserves across the world as of 2022. Data source: USGS report 2023.

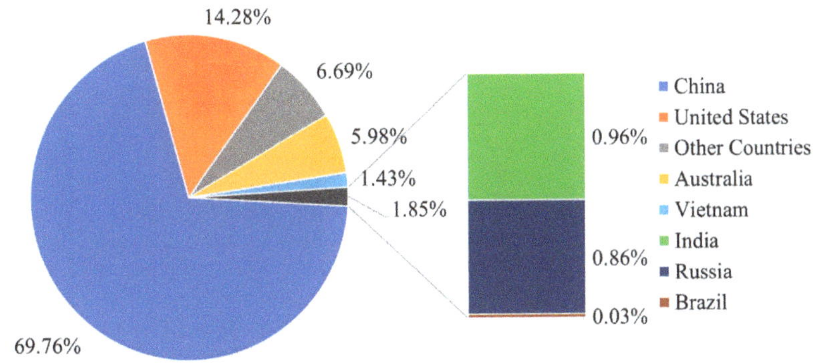

Figure 3. Production from REE reserves across the world as of 2022. Data source: USGS report 2023.

1.1.1. Primary Sources of REEs

According to the BGS report [8] from November 2011, it was stated that there are numerous mineral resources containing rare earth elements (REEs), numbering in the hundreds. These mineral resources are characterized into three groups, and the details of the most prominent resources are here.

Cerium-dominant resources: Aeschynite, Allanite, Bastnasite, Britholite, Fergusonite, Gadolinite, Monazite, Parisite, Huanghoite, Cebaite, Florencite, and Synchysite.

Yttrium-dominant resources: Euxenite, Kainosite, and Samarskite.

Other resources: Apatite, Brannerite, Eudialyte, Loparite, Xenotime, Yttrocerite, Knopite, Gittinsite, Iimoriite, Mosandrite, Pyrochlore, Rinkite, Steenstrupine, Zircon, Carbonatites, and Ion adsorption clays [9]. Economic limitations limit commercial extraction to a small group of resources. Among the global resources, the carbonatites (associated with Bastnasite and Monazite) are the more prominent and economically extractable deposits in the USA, Australia, China, and Mongolia regions, with REE content varying from 2–10% by weight [10]. The subsequent commercial extraction is possible through placers found in the ocean belts of India, Sri Lanka, Australia, and southern USA, where the deposit is widespread but with lower REE concentration (<0.5% by weight). Ion adsorption clays are another commercial source of extraction [10]. They are preconcentrated in some REE content, and the extraction process from these deposits has low levels of radioactivity in the resulting waste compared to other sources. However, studies across the globe are continuing concerning the extraction of low REE content.

1.1.2. Secondary Sources of REEs

Many countries and industries are performing metal-life cycle assessments to counteract the harmful environmental effects of traditional exploration and extraction of resources and to meet the global metal supply demand. They are accommodating the practice of implementing recycling left-to-recycle (LTR) material such as metal scraps, electronic waste, and phosphors. This practice benefits human, technological, and environmental sustainability by creating job opportunities, developing new technologies in place of traditional and outdated methods, and reducing scrap piling at dump yards, oceans, and landfills. Some such potential sources for rare earth extraction, according to [11–16], are listed below:

- Ash is one of the significant components in Coal Combustion and contributes to REEs.
- Dross and slags from metal processing and red mud from bauxite residue are a few sources of REEs.
- Tailings generated from the gold beneficiation process can be a source of LREEs (Ce, La, and Nd).
- Neodymium–Iron–Boron permanent magnets, phosphors, fluorescent lamps, Nickel–Metal-Hydride Batteries, motors, and fluid cracking catalysts contain some REEs, and recycling is the best way of recovering them.

According to the study on the criticality of raw materials and assessment [17], the European Union (E.U.) demonstrates a 4% of HREE recovery and 3% of LREE recovery through recycling these secondary resources.

1.2. Demand for REEs in the E.U.

The increase in population across the globe has increased the demand for utility vehicles, electronic gadgets, pharmaceuticals, fuel, and energy resources. Many countries have accommodated these demands with care and concern for the environment by adopting the transformation to using clean and green energy alternatives. This emerging transition once aroused the need for rare earth supply chains and production. As cited, 75% of REEs are utilized in industries of catalysts: 10% for ceramic, glass, and polishing industries; 10% for phosphor and other optical devices; and, finally, 5% for alloy design required for wind turbines, hard disk drives, and many others [2]. According to categorization by the E.U., Sc and LREEs (referring to La–Sm) are preferred for the sectors of catalysts, metallurgy, polishing, and magnets, and HREEs (referring to Eu–Lu and Y) are used in lasers (but not only limited to) and niche applications [17]. The E.U. Criticality Assessment report for 2023 states that compared to 2020, in the E.U., there has been a 50% increase in the economic importance of La, Pr, Nd, and Tb due to their usage in magnets and catalysts, and the comparison is presented in Figure 4.

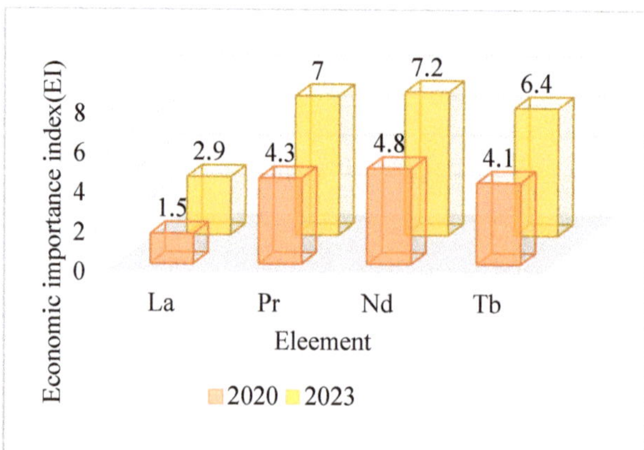

Figure 4. Comparison of rise in Economic Importance in the year 2023 compared to 2020.

Data in Figure 5 represent the statistical analysis of the economic importance of REEs for the E.U. as of the year 2023, and it is essential to note the data described here are on a point scale, but not on a % scale. Additionally, the plot indicates that within the European Union (EU), there is a dire need for Dy, Sm, Nd, Pr, and Tb, as these metals possess economic importance values exceeding 5.0. The supporting explanation for this emerging requirement can be understandable from Figure 6, where the orange bar on the chart explains the details of the REE metal applications for different industries as of 2023, and the blue bar represents the data for 2012. This analysis is performed from the E.U. study for Critical raw materials [17,18]. Here, the rise in demand is mainly due to a jump in the utilization of REE metals in magnets, which are used in mobile phones [19], and hard disk drives [4]. Further, the increased application of REEs for catalysts such as La can increase the recovery in oil refineries, and Ce-based catalysts can reduce the pollutants and toxic gases released from cars [18], and, additionally, the application of REEs in ceramics is increased due to the application of REE oxides of Pr, Nd, Y, and Ce in manufacturing tiles and electronic components such as capacitors [20].

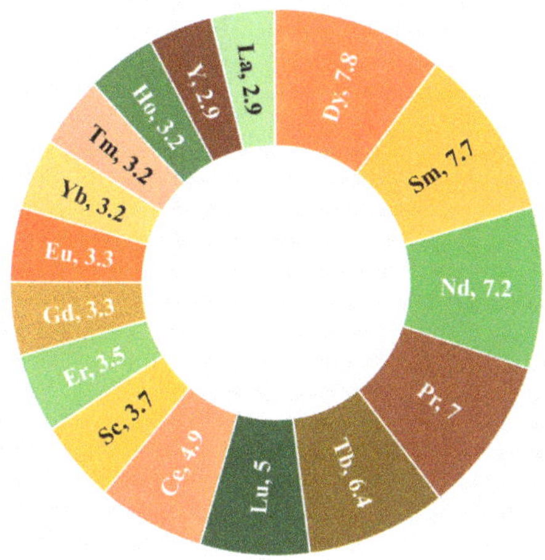

Figure 5. Statistical analysis of the economic importance of REEs for the E.U. in 2023.

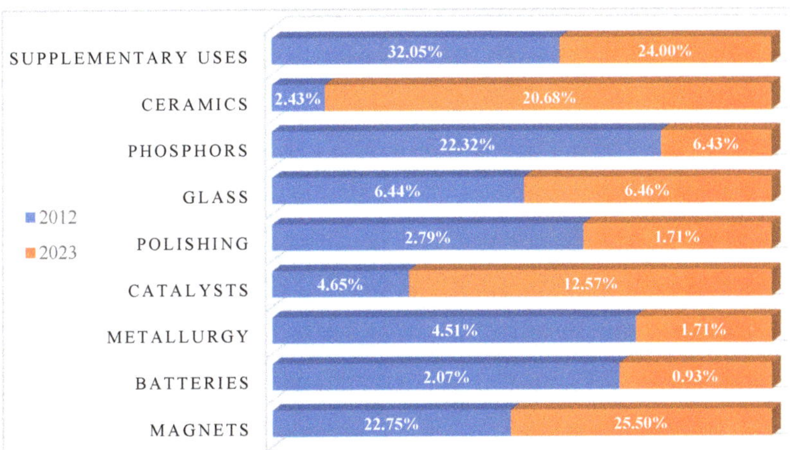

Figure 6. Comparison between applications of REE elements in various industrial sectors for 2012 and 2023.

Furthermore, Figures 2 and 3 illustrate that China holds a substantial market share in REE supply and demand. This same analysis is evident in the EU assessment report for the year 2023, which describes that among the global suppliers of REEs, China provides 100% of HREEs and 85% of LREEs. This analogy holds true for the EU as well. Thus, it may lead to the problem of a monopoly in the REE supply chain. To avoid these circumstances, the governments from almost all the countries globally had predominantly focused on demand-based REE reserve development and production. To support this, many researchers have invented different approaches for separating individual REEs, and their purpose for designing a system varies with the research interest. Most of the methods developed so far have been limited to laboratory demonstrations or research publications. Challenges such as economic and ecological concerns, time-consuming processes, and difficulties in scaling up for industrial purposes have hindered their implementation on a larger scale. Previous reviews on REE extraction and separation have often focused on promoting one or two main approaches while criticizing others, making it challenging to grasp a comprehensive understanding of the subject matter.

The critical distinction between previous reviews and this is that most of those existing reports revolve around some specific approaches for the separation of REEs. They all explained the merits, limitations, and developments that happened in a particular period. To the best of our knowledge, this article aims to bridge that gap by providing a clear and comprehensive explanation of those methods. This article will describe the mechanisms of these methods, using illustrations where necessary. By effectively comparing these approaches, this article intends to offer a comprehensive understanding and fill the void of proper explanations for each method. Thus, this article facilitates as a preliminary reference for any researcher interested in these separations.

2. Extraction of REE Minerals

Starting from the discovery of REE minerals, many scientists, industries, capitalists, and countries across the globe have invested vast amounts of knowledge, energy, time, and money in inventing and developing an Efficient-Easier-Economical-Environmentally safe (EEEE) strategy to extract these minerals for human sustainability and the circular economy [21]. Of these inventions, the most common technologies adopted by most processes from source to recycling are shown in Figure 7. As mentioned in Section 1.1.1, the concentration of REE varies by the type of deposit and, thus, the processing. In recent publications, some researchers have outlined the process of REE production from pre-concentration to refining [2]. A detailed description of each of these techniques is discussed in the subsequent sections.

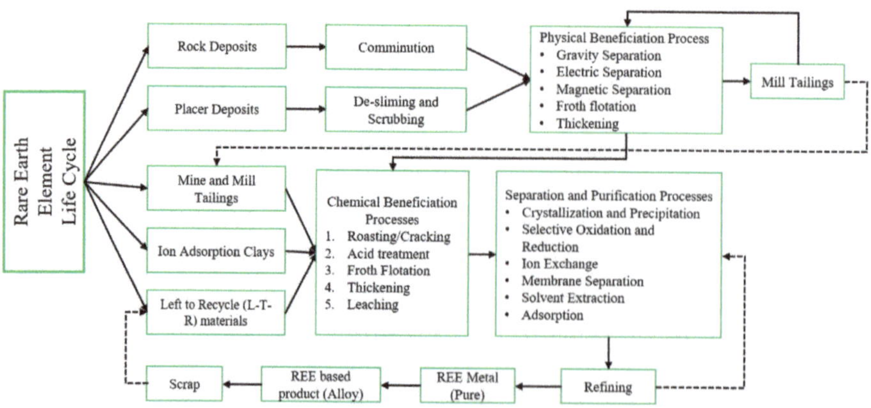

Figure 7. A Schematic explaining the REE lifecycle from Extraction to Utilization.

2.1. Enhancement of REE Content by Physical Beneficiation Process (PBP)

As shown in Figure 7, the rock deposits mined need to be processed to remove the valuable mineral locked in the host matrix. These are subjected to subsequent crushing and grinding operations where the valuable mineral content is liberated [22,23], whereas, in the case of placer deposits (in general beach sand minerals), they do not need any size reduction because of size distribution range; instead, they need to be de-slimed and attrition scrubbed to remove the clusters of fines accumulated. The processes and their fundamental principle of separation are discussed below [10,24–26].

- Gravity concentration processes usually separate the heavy and lighter minerals by exploiting differences in specific gravity (S.G.). The S.G. values for REE minerals are in the order of 4–5.
- Electrical separation processes operate by the charge transfer of the conducting material onto the separator's roll. The dominant mechanisms are charge transfer by ion bombardment, conductive induction, and triboelectric charging, usually seen in electrostatic separation. Monazite is non-conducting material, so it sticks to the roll of the separator.
- Magnetic separation uses the difference in magnetic susceptibility of the mineral particles in the presence of the applied magnetic field. In placer deposits, the difference is among the paramagnetic and ferromagnetic particles. The monazite is weakly magnetic and is separated using high-intensity magnetic separators.
- Froth Flotation is a physio-chemical separation process that operates based on the variations in the hydrophobicity differences of the minerals where the hydrophobic phase can be valuable/gangue minerals, and the hydrophobicity is induced by the addition of reagents termed as collectors and frothers. The most common collectors for REE-based mineral flotation are tall oils obtained from wood processing.
- Thickening is a dewatering process that reduces water content and increases the % of solids by weight in the slurry; thus, the product received is termed as concentrated and sent for further processing. If economically viable, the mill tailings obtained in PBS are sent for further mineral recovery.

2.2. Enhancement of REE Content by Chemical Beneficiation Process (CBP)

As shown in Figure 7, the ion adsorption clays, LTR material, and tailings (mine and mill) do not need a PBS route because they are already enriched in one or more REEs, so they are subjected to processing by chemical routes where the preliminary task is to break the bond between the coordinating elements of the REEs and waste (referred as gangue). To achieve this, the temperature of operations needs to be in the range of 573–1273 K. This process is technically termed roasting/cracking [22]; this operation produces oxides of REEs, and the purpose of this treatment is to achieve good coordination between water and mineral surface, which is needed for downstream processes [21,27]. This step is followed by acid treatment with concentrated HNO_3, HCl, and H_2SO_4 acids to produce a slurry that is further sent for flotation and thickening processes [27]. The final cake/concentrate from the thickening stage is sent for leaching operation. Solid particles are dissolved in the leaching circuit, leaving the undissolved particles as residue (to be disposed of). The pregnant leaching solution (PLS) obtained from the leaching circuit contains all the desired REEs associated with some transition elements. Mathematical models have been tabulated to determine the leaching rate and kinetics [15]. Recycling NdFeB permanent magnets involves using pyrometallurgical principles to assist selective dissolution in hydrometallurgical operations because, in pyrometallurgy, the REE is changed from one compound to another (often from a sulfide to an oxide by roasting) [28].

2.3. Technological Advances for Individual REE Separation

Sections 2.1 and 2.2 discussed the preliminary steps required for processing REE content from any source until it is obtained as a slurry (PLS). After PBP and CBP operations, the PLS is enriched with REE content in a solution. There can be some other multivalent ions (undesired metal ions) that need to be selectively separated to produce high-purity rare earth oxides of each element. Still, the approach from here is challenging because of the REEs' properties, as discussed earlier. In recent years some novel methods have been invented [29]. The University of Utah researchers developed the selective electrochemical recovery of REEs based on atomic mass where the oxidation state plays a significant role [30]; further, they discovered a novel approach to REE extraction, which was designed by introducing the concept of Bio oxidation and leaching followed by a subsequent stage of precipitation with subsequent acid production, which assists the leaching process; these are two of the few economic and environmentally viable approaches for recovering REEs from secondary sources, such as coal rejects [9,14,31–33]. Still, most producers rely on existing methods, either solid–liquid or liquid–liquid separation processes. In most systems with REE mixtures, it is possible to separate LREEs/MREEs and HREEs, but very challenging to separate REEs with adjacent atomic numbers. The detailed description of mechanisms for processes mentioned in Figure 7 is explained in upcoming sections. Before understanding the REE separation mechanisms, it is essential to understand the solution behavior of REEs.

i. **Solution behavior of REEs**

A few authors recently reported that coordination chemistry is crucial in segregating REEs [34]. The solid–liquid separation or liquid–liquid separation process involves the use of coordination chemistry. The larger charge densities of HREEs relative to LREEs increase electrostatic attraction, thus forming more stable complexes with ligands [34]. In the series of lanthanides, the unfilled 4f electron configuration leads to a predominant electrostatic interaction with the ligands compared to the 4d transition group of elements; the bonds formed during the interaction are relatively weaker [35]. For example, the properties of Sc are nearly equivalent to transition metals, while Y is more like lanthanides. Almost all the REEs are solely stable at +3 charge, but the same is not valid for Eu and Ce.

The polarizability of donor-atom (in other words, hard and soft acids and bases theory [36]) categorizes based on the ability of the external electronic field to cause ion distortion, where if ions are less influenced, they are hard and vice versa [34]. The hard acids (namely, the hydrated cations) interact with hard bases to form inner-sphere complexes, which form complexes into solids with low solubilities, thus assisting selectivity in precipitation.

The extent of complex formation depends on the anionic species present in the solution, and the degree to which the resulting complex can be measured by a constant of stability, the logarithm of the solubility product, logβ. The easier way of expressing this is:

$$\text{REE-Cation} + \text{Ligand} \Leftrightarrow \text{REE-complex}.$$

Also, the ligand denticity/tooth/coordination number plays a role in determining the stability of the REE complex. In aqueous chemistry, other factors that influence the logβ values are the temperature because the temperature change affects the reaction kinetics [26]; and the increase in hydrogen ion concentration increases the stability of REE complexes because of the increased tendency to deprotonate; therefore, large pKa values are often preferred for ligand design.

Additionally, it was noted that chelate stabilities also depend on the extent of the hydration layer of the REE ions in a solution [37]. Other than these factors, additional experimental variables include the strength of the ligand, ionic strength, and the hindrance caused by the steric bulk can also impact the logβ [35]. Anions of the soft bases such as nitrates, chlorides, and sulfates will only be able to produce outer-sphere complexes that are weaker in nature. Moreover, there are no such or lower logβ values of complex formation for ammonia and disulfide. Also, the interaction between these anions and a molecule is noteworthy only at high concentrations above 1 M; this is not true in the case of sulfate solutions, due to their higher logβ values; in addition, these sulfate solutions can form REE $(SO_4)_2^-$, which is a negatively charged complex at higher sulfate concentrations. In the presence of hard ligand groups, creating additional bonding with softer bases, such as amine groups, is feasible. The complexing stability usually increases from La to Eu or Gd. However, for HREEs like Gd, they reach a maximum and plateau [38]. Due to their large size, REE ions can form a limited number of complexes. Ligands with high chelation capacities and the ability to attract small-sized REE cations with increased charge density can form stable and water-soluble complexes. An overview of factors influencing ligand design is presented in Figure 8.

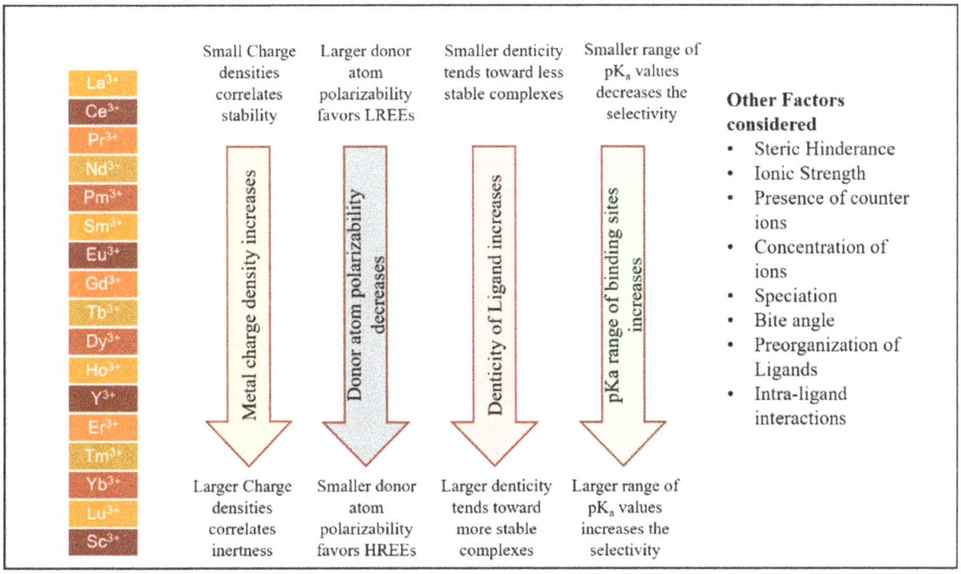

Figure 8. Design factors considered for ligand chemistry; the image is redrawn with permission from Hovey, Dittrich, and Allen 2023 [34], copyright 2022. Published by Elsevier B.V. on behalf of the Chinese Society of Rare Earths.

The underlying fundamentals for separating individual REEs follow the variations in the basicity resulting from reduced ionic radii from La to Lu [39]. This measure of the tendency to accept hydrogen ions from water impacts the solubility limit. Further, it influences the decomposition of the water molecule and complex formation. Most separation processes are designed by utilizing this concept. Moreover, in addition to these differences, a few more possibilities that influence separation are the valency states of REEs, which can be modified by oxidation or reduction [3]. Since the discovery of REEs, there have been attempts to separate each REE. From then until 1947, the methods of crystallization and precipitation, oxidation, and reduction seemed favorable [3].

ii. Separation by Crystallization and Precipitation Methods

The age-old traditional method for generating highly pure individual REEs is possible by the fractional crystallization (F.C.) approach. This method utilizes a mixture of different salts to produce REE oxides. Of these salts, nitrates based on ammonium are beneficial for LREEs (except Ce), nitrates of magnesium are helpful for MREEs (except Pm), and nitrates of manganese can separate the C.G. group of REEs, and for the Y.G. group, bromates/ethyl sulfates can be beneficial [3,40]. However, this approach is very challenging and not ideal for HREEs because of the large production cycle varying from 16 to 15,000 steps; historically, the crystallization process was performed by G. Urbain in 1907 for the separation of Yb and Lu [41], and Feit used a similar approach for the generation of Ho [3]. Moreover, salts are limited in tailoring the crystals needed for this approach [42].

Similarly, an alternative approach with extensive studies performed is the fractional precipitation (F.P.) like in F.C.; this approach generates precipitate REEs with salts of sulfates where the LREE group can form insoluble sulfates, followed by MREEs with partial solubility, and HREEs with high solubility [3]. In the case of Y.G., several precipitation techniques with hydroxides, chromates, and even nitrates were possible for preliminary Y enhancement and separation. However, the application of this method is limited to separating REEs into groups but not into individual ions [3,43].

Later on, in 2014, understanding the extensive efforts needed for F.C. separations, a size-selective crystallization (SSC) approach was tried to achieve REE separations. The duration of crystal synthesis is within a few hours, and it is environmentally friendly with a reduced use of reagents. The basis for this approach is a group of variations in a solubility product with an REE complex formation, thus resulting in the shape of a metal–organic framework (MOF) [42,44]. This approach relies on the coordination between the ionic radii and the prepared MOF to facilitate the crystallization of a specific target REE. If the ionic radii difference is significant, then a particular REE will concentrate on the specific type of crystal synthesized, which can be a suitable approach to large-scale production. However, this method fails when the difference is slight (in the range of 1–5 picometers), which is the case of adjacent REE separation leading to contamination in the range of 20–25% of undesired REEs in the final crystal [42].

A few more developments in 2017 were made to improve the separation by F.C., where crystals are formed by controlling the reaction kinetics and, later, selective separation by flotation. Still, this method did not explain the selectivity factor (S.F.) among adjacent REEs [1,45]. Later on, in 2019, a valence selective crystallization (VSC) approach was designed to selectively separate Ce from trivalent REEs because of its tetravalent nature; this work proved effective in the separation of Ce and other metal ions when there is a difference in the valence state of ion and without distinction in the valence state there is no chance of separation [1,45].

iii. Separation by Selective Oxidation and Reduction Method

As discussed above, among the REEs, Ce can exhibit a tetravalent state so that the separation can be even achieved by utilizing differences in the valence between other ions and Ce. Heat treatment of cerium ore in the presence of natural atmosphere/air at elevated temperatures of 650 °C can lead to oxidation as can treatment at lower temperatures (>100 °C) using hydroxides in the presence of chlorine [3]. Oxidation of Ce from a +3 to +4 state can be achieved using ozone. Leaching the ore in concentrated acids to dissolve +3 REEs, followed by precipitation of +4 Ce can facilitate separation from other REEs. Still, some Ce is left in the system, which can be further removed.

Among other lanthanides, Pr [46] and Tb [47] exhibit the tetravalent state and thus are less stable in aqueous media. These can be separated from the system by forming precipitated and separated oxides [3].

When the oxidation methods for REEs are applicable, there can be the possibility for an approach through reduction. Sm, Eu, and Yb separations can be performed after reducing them to a +2-valence state. The reduction can be performed electrolytically or by photolysis (referred as photochemical reduction using a laser), where a concentrated beam of consistent wavelength is directed to an aqueous solution. However, in the presence of calcium, reducing REE halides is not feasible, and thus they can be separated from the slag later on, if they are pyrometallurgically processed. But, Sm can be reduced to a +2 state electrolytically among a mixture of REE chlorides. Pre-concentration or enrichment by PBP/CBP forms the basis for any reduction method because of their low crystal abundance (values averaged in µg/g) in an ore body [3].

Due to the increasing demand for the supply of rare earth metals (REM) globally, REM producers employ the most versatile and widely applicable rare earth techniques for high-quality products. Thus, years after the initial separation work in the 1950s, the REM supply market changed tremendously. Before the 1960s, the ion exchange technology (solid–liquid separation strategy) was the only possible approach to separate REE. However, post-1960s, solvent extraction technology produced high-purity REEs for specified electronics and analytical measurements [48]. Other than these techniques, there are a few more methods: membrane separation and adsorption [1].

iv. Separation by Ion Exchange

The ion exchange techniques offer another approach to REE separation. The ion exchange (IX) process separates the desired ions from the solution/slurry with the help of either organic/inorganic materials [49]. Researchers have categorized IX media into natural, modified natural, and synthetic [50]. The additional classification is shown in Figure 9. As for inorganic materials, they can be clays and zeolites; the primary difference between them is layering for clays, and the porosity for zeolites [26]. Most of the ion exchange process uses resins with a porous network with embedded functional groups for enhancing selectivity. In most recent reviews, the list of commercial resins is tabulated [35].

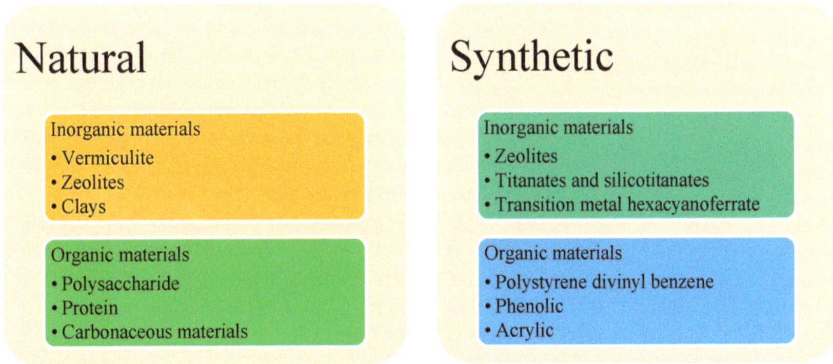

Figure 9. Classification of IX exchangers.

A critical feature of the material used in this IX process is its ability to stoichiometrically and reversibly swap the ions from its resin-ion functional group with other ions (REE) found in the solution in which it is submerged. The equation for ion exchange reaction is expressed in equation-1,

$$\text{REE ion} + \text{Resin-Ion-Functional group} \Leftrightarrow \text{Ion} + \text{Resin-REE-Functional group} \tag{1}$$

The selectivity index for the reaction in Equation (1) is defined as $K_{\text{Resin-Ion/REE ion}}$, and this ratio is explained in Equation (2) where

$$K_{\text{Resin-Ion/REE ion}} = \frac{\left((Conc\ of\ Resin\ Ion\ in\ resin\ phase)^{(Charge\ of\ REE\ ion)} * (Conc\ of\ REE\ ion\ in\ aqueous\ phase)^{(Charge\ of\ Resin\ ion)}\right)}{\left((Conc\ of\ REE\ Ion\ in\ resin\ phase)^{(Charge\ of\ Resin\ ion)} * (Conc\ of\ Resin\ ion\ in\ aqueous\ phase)^{(Charge\ of\ REE\ ion)}\right)} \tag{2}$$

The Extraction coefficient, $E_{\text{REE ion}}$, for the reaction in Equation (1) is explained using Equation (3) where

$$E_{\text{REE ion}} = \frac{(Conc\ of\ REE\ Ion\ in\ resin\ phase)^{(Charge\ of\ REE\ ion)}}{(Conc\ of\ REE\ ion\ in\ aqueous\ phase)^{(Charge\ of\ REE\ ion)}} \tag{3}$$

In addition to these equations, the IX data are correlated and analyzed at equilibrium using several models used to design the industrial systems. Some of the most widely used and applicable models are the Langmuir, Freundlich, and Tempkin models [51]. Generally, ion diffusion in the resin phase is the rate-controlling mechanism for the IX process. In some cases of low-concentrated solutions, the rate is often controlled by diffusion through the boundary layer due to a higher dilution ratio. Numerous models of the IX process are presented in the literature, but the most often used are Freundlich, Langmuir, and Tempkin models [51]. Also, based on experimental data, the pseudo-second-order kinetic model is most applicable to IX processes [35].

The most common IX resin matrix comprises crosslinked poly (styrene-co-divinylbenzene) with desired functional groups. The ion exchange rate is attributed to the type of active group, the extent of polymerization, the size of exchangers [1], and the classification of Exchangers is dependent on the charge of the functional group; thus, they are classified as strongly acidic cation exchangers with sulfonic acid functional groups and weakly acidic cation exchangers with carboxylic acid functional groups, strongly basic anion exchangers with quaternary ammonium groups and weakly basic anion exchangers with primary, secondary, and tertiary ammonium groups. The matrix of resin can be either microporous with permanent pores or gel where diffusion of ions happens through penetration in a concentrated solution, and the resulting diffusion coefficients (measured in m^2s^{-1}) are smaller than the former.

The process of IX is cyclic, with two primary steps of loading and elution. In this, the desired ions are extracted from PLS in the loading phase and are loaded into the resin phase, and the resin is removed from the liquid. Subsequently, an elution operation is performed to exchange the loaded ions into the eluent to obtain a purer form of the target ions. This operation can be performed either in batch or continuous process. The loading step can only be effective in separating the REE from other dissolved ions (like Cd, Mg, Fe, etc.) in the system but is difficult to use to fractionate into individual REE elements. Thus, an effective IX process needs a supporting selective elution that uses a complex solvent to desorb the loaded metals, and the degree of selectivity in the elution is a factor of $\log\beta$ and the type of complex solvent. The most commonly used solvent is EDTA, which typically elutes REEs in decreasing order of atomic number due to the increasing stability constants of the complexes from light to heavy REEs [51]. In recent studies, the separation of REE from sulfate solutions has been studied using conventional resins (CVRs), chelating resins (CRs), solvent-impregnated resins (SIRs), hybrid organic mineral materials (HOMSs), and ion-imprinted polymers (IIP).

CVRs with different types of functional groups showed a difference in adsorption behavior toward REEs; for instance, the strongly acidic cation type can load to higher capacities with greater affinity for LREEs than HREEs because of the small hydration layer surrounding the metal cation. But for strong basic anion types, there is a weaker affinity of REEs, other than Scandium, but further studies are needed to commercialize this type of resin. Also, for this resin type, the adsorption behavior depends on the kind of medium, like sulfate or chloride, for a given system. Strong acids can damage the resin, and weak acids can withstand the resin, thus limiting the choice of extractions. Elution operations from CVR depend more on concentrations, temperature (high temperatures are needed for REE compared to Sc [52]), and types of complexing agents like the EDTA, sodium carbonate, which are suitable for REE separation. EDTA is used for Nd separation from the bulk [35], and Sm/Eu and Eu/Tb can also be separated using EDTA, which has higher selectivity [53], but the only disadvantage is the solubility at a given pH, which needs to be addressed.

The CRs are another class of resins used to separate REEs from complex media, but these do not have good loading capacities compared to the CVRs (strongly acidic cation resins). Still, their selection depends on a trade-off between loading capacity and selectivity. As experiments with the iminodiacetic acid-based resin showed a good affinity for separating La, Sm, and Ho from the bulk or mixture of elements with REEs [35,54]. However, the CRs' loading capability varies with variations in sulfate concentration [35].

The HOMS are IX resins that combine inorganic supports and organic ion exchange groups. These materials offer non-swelling supports and outstanding operational stability by using inorganic matrices. At the same time, incorporating organic ion exchange groups provides good selectivity and excellent binding ability to metal ions. Silica is considered the most suitable inorganic matrix for producing organo-mineral compounds because it is affordable, readily available, and allows for precise control over structural parameters like grain sizes, average porosity, and specific surface area. However, inorganic supports such as titanium and zirconium oxides can also be used. There are three techniques for modifying inorganic support surfaces to create these resins: modification of the silica surface, alteration of inorganic matrices by polymeric functional groups, and polycondensation of applicable organosilicon modifiers or their co-condensation with a non-functional composite [35]. Various functional groups, such as N-methylimidazolium, iminodiacetic acid, and phosphonic and carboxylic acids, have been used to prepare these ion exchangers for complex REE separation mixtures.

The Ion imprinting method is another effective way to enhance the selectivity of metal ion separation. This method involves synthesizing functional monomers with chelating groups into a porous solid polymer particle, incorporating metal templates during polymerization to bind to metal cations selectively [35]. Numerous studies have demonstrated the potential of IIPs in selectively removing REEs from complex media. The selection of functional monomers is critical to preparing IIPs, with vinyl groups being the most suitable polymerizable functions for chelating ligands. Studies have shown that IIPs can achieve high sorption capacities and selectivity coefficients for targeted metals using a range of functional monomers such as methacrylic acid, ethylene glycol dimethyl acrylate, 4-vinyl pyridine-acetylacetone complexes/monomers, and N-methacryloylamido folic acid. Some studies have also explored using conductive polypyrrole and iminodiacetic acid (IDA) or sulfonic groups in developing IIPs. However, despite their potential, there are still uncertainties about isolating and separating IIPs from complex matrices. The SIRs are another class of IX resins where a desired solvent is impregnated into a resin matrix, thus enhancing the selectivity. Still, their application is limited by the leakage of organic solvents into the solution, thus leading to contamination. This problem can be addressed by adding a polymer layer to resin, but the merits and demerits of this approach are not yet published [1].

v. **Separation by Solvent Extraction**

After the development and implementation of solvent extraction (liquid–liquid separation strategy) in the 1960s, the commercial production of REMs received special attention using this method because it can handle vast volumes of solutions [22], consume less energy for operation, and increase the reaction rate by vigorous mixing for efficient separation of REEs [55]. This solvent extraction is a new approach in the field of extractive metallurgy, which is a subset of hydrometallurgy. The schematic in Figure 10 explains that the solute (target metal) is dissolved in the aqueous phase in a conventional solvent extraction. The target metal is distributed into the organic phase by vigorously mixing the organic phase with an aqueous phase [51]. The organic phase extracts the metal ion from the aqueous phase. However, the non-aqueous solvent extraction systems are less available in practical applications for rare earth systems and are primarily used for p-block and early d-block elemental separations [56]. In the traditional approach of conventional solvent extraction, the organic phase is filled with the utilization of numerous traditional organic solvents (TOS) (like kerosene, isooctane, and n-dodecane). These solvents are not eco-friendly, and the safe practices required for their handling and application can be tiresome due to their highly flammable nature, significant volatility, and potential risks to human and aquatic health [57]. Moreover, their utilization often involves saponification, which necessitates the use of acid and base [58]. A decade ago, the drafted term for this metallurgical process was ionometallurgy, in which the approach is similar to solvent extraction but with ionic solvents replacing the organic phase, which is typically made up of TOS [59]. Thus, the ionic solvents are broadly classified as ionic liquids, which consist of distinct cations and anions in the molecular structure [59]. Deep eutectic solvents are liquids that result from the eutectic mixtures of specific hydrogen bond acceptors and hydrogen bond donors [60]. The schematic classification of organic phases is explained in Figure 11.

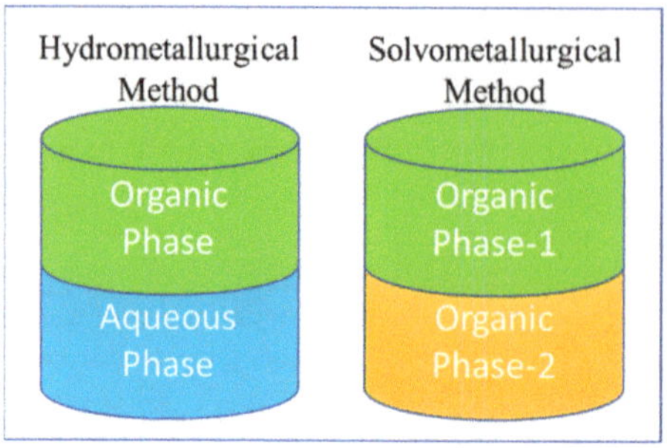

Figure 10. Schematic of solvent extraction in Hydrometallurgy and Solvometallurgy routes image is reprinted with permission from (Binnemans and Jones 2017 [56]), Copyright © 2017, The Author(s).

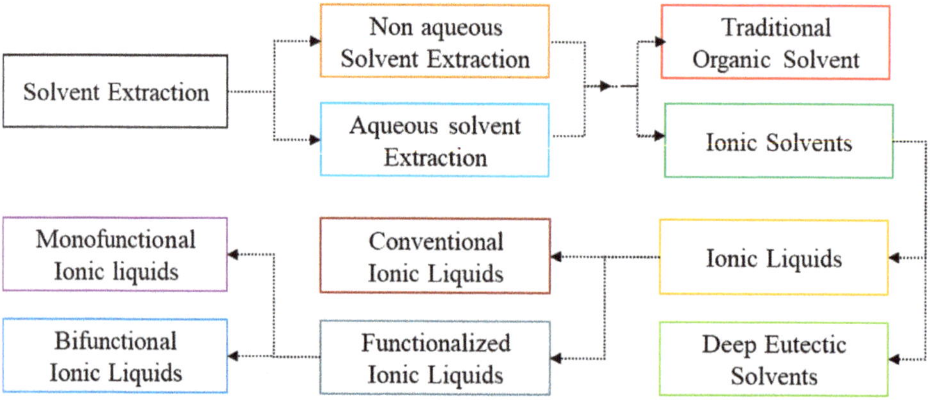

Figure 11. Schematic explaining the classification of solvents for the solvent extraction process.

Difference between the conventional method and the ionic liquid-assisted method of solvent extraction

a. Traditional solvent extraction system (Extractant based)

The separation in solvent extraction depends entirely on the efficiency of extracting a target metal from the aqueous phase to the organic phase; this transfer cannot be spontaneous due to differences in solvation energy. Adding an extractant is necessary to assist the rate of metal transfer. Therefore, a practical solvent extraction system can be expressed as

$$[M_{REE}{}^{3+}\text{-Acid}]_{\text{aqueous phase}}\text{-[Extractant-Organic phase]}_{\text{organic phase}}.$$

Based on the requirements for extractants, there are four categories of extractants [60–63], as discussed below:

- Acidic type (with cation as exchanger)
  - Carboxylic acid
    - Versatic acids (Versatic 10, Versatic 911);
    - Naphthenic acids.
  - Sulfonic acids (Dinonylnaphthalene sulfonic acid).
  - Phosphinic acids (P229, Cyanex-272).
  - Phosphonic acids (EHEHPA, HEHEHP, P507, and PC88A).
  - Phosphoric acids (D2EHPA).
  - Monothiophosphorous acids (Cyanex-302).
  - Dithiophosphorous acids (Cyanex-301).
- A basic type (with anion as exchanger)

  - Primary amines (Primene JMT, N1923).
  - Secondary amines (N-Methylaniline).
  - Tertiary amines (Alamines).
  - Quaternary amines (Aliquat 336, Adogen 464).
- Neutral type (solvation exchanger)
  - Phosphorous ester (TBP, DBBP).
  - Phosphine oxides (TOPO/Cyanex 923, Cyanex921, and CMPO).
- Chelating type
  - Beta diketones (LIX 54, HTTA).

Of these extractants for solvent extraction, the acidic type is most preferred, and the chelating ones are the least applicable [62,64]. However, their extraction mechanism is similar to the acidic ones. The extraction mechanism is a neutral exchange governed by the electroneutrality principle, where only neutral species are extracted. The application of this mechanism varies with the extractant type, and a detailed description is given in Table 1, and a pictorial representation of this mechanism is described in Figure 12 [60]. As in Figure 12, the red line explains that the hydrogen ion is released from the organic to the aqueous phase using an acidic extractant during the solvent extraction process. The black line indicates that the REE ion is captured from the aqueous to the organic phase. The same analogy applies to the basic extractant mechanism, where an anion is released in place of a hydrogen ion. For neutral extractants, additional cation/anion is not released into the aqueous phase [55]. In Table 1, M_{REE}^{3+} denotes an REE ion, E refers to the extractant, HE is the acidic extractant, H^+ is the hydrogen ion, BE refers to the basic extractant, and X refers to Halides.

Table 1. A detailed description of mechanisms applicable to traditional solvent extraction systems.

Underlying Mechanisms When Extractants and Traditional Organic Solvents Are Combined in the Organic Phase	Reactants		Products	
	Aqueous Phase	Organic Phase	Organic Phase	Aqueous Phase
Acidic Extractant	M_{REE}^{3+}	3HE	$M_{REE}E_3$	$3H^+$
Basic Extractant	$M_{REE}X_3$	3EB	$M_{REE}X_3E_3$	$3B^-$
Neutral Extractant	$M_{REE}^{3+} + 3X^-$	E	$M_{REE}E_3$	-

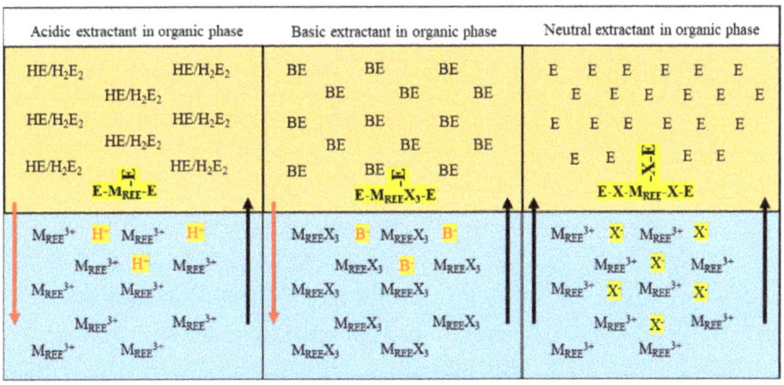

Figure 12. Pictorial representation of mechanisms applicable to traditional solvent extraction systems.

b. Modern Solvent Extraction system (Ionic Liquid Based)

1. Ionic liquids

   Ionic liquids (ILs) are the organic salts prepared by combining different cations (organic) and anions (can be organic/inorganic), forming a chemical structure, which generates designing a specific property for an ionic liquid [65]. Some of the dominating properties that widen the application of ionic liquids over TOS are the low melting points (<less than 100 °C), and hence these are referred to as room temperature ionic liquids (RTILs); the non-volatility due to negligible vapor pressure under room temperature conditions; and that almost all ionic liquids are non-flammable. Due to the wide versatility in changing the cation and anion combinations, the available number of solvents is in the order of 1,000,000. ILs have viscosity levels that range from 20 to 97,000 mPa. The low density of ionic liquids helps in more accessible phase separation. ILs have high thermal stability and high electrical conductivity values of nearly 120 mS cm^{-1}. An overview of promising high-potential IL utilization and potential future applications is reviewed and explained using Figure 13 [66].

Figure 13. Application potential of Ionic liquids in diverse fields of study.

2. Common Cations and Anions for ionic liquids [62]
    - Cations
        - Ammonium, Phosphonium, Imidazolium, Pyrrolindium, Pyridinium, Piperidinium, Morpholinium, Isoquinolinium, and Pyrazinium.
    - Anions
        - Inorganic Anions are Halides, Nitrate, Tetrafluoroborate, and Hexafluorophosphate.
        - Organic Anions are Trifluoromethanesulfonate, Bis(trifluoromethyl sulfonyl)imide, and Carboxylate.
3. Significance of properties on the selection of ionic liquids
    a. Viscosity
        - ✓ Some general parameters like temperature, water, and organic solvents modify viscosity in any ionic liquid.
        - ✓ Other than these subtle parameters, like the chemistry of ionic liquid, which is dependent on cationic and anion structures, impact the viscosity.
        - ✓ Cationic structure modifies the viscosity due to van der Waals interactions and ionic mobility [62].

- ✓ Anionic structure modifies viscosity due to hydrogen bonding and basicity [62].
- ✓ For a similar cation, the viscosity can vary from low to high by changing the anion structure from weak to complex [60].
- ✓ Viscosity and density are interrelated as these affect solute mass transport to the organic phase, impacting the extraction rate [67].
- ✓ Due to the addition of excess water, the Coulombic force of interaction between ions tries to reduce the viscosity [62,68].

b. Solubility

- ✓ The solubility of ionic liquids in water is a concern because of the cost involved. The solubility is dependent on the chemical structure of the ionic liquid and the charged species in the solution (predominantly the anions). The charge is related to the size of anions, and this is associated with the activity of the species in solution, which determines whether a species is favorable to interact with water or not [62,69].
- ✓ Ionic liquids made up of the anions Hexafluorophosphate and Bis (trifluoromethyl sulfonyl) imide have lower solubilities in water. Ionic liquids with halides and nitrates are readily soluble in water [62,70,71].

c. Hydrophobicity The hydrophobicity of ionic liquids is dependent on the alkyl chain length geometry of the cation and anion present in the structure, but having higher chains lengths of alkyl groups results in low values of density, more hydrophobicity, and high values of viscosity; these impacts may be different for different anion combinations [60,72].

4. Extraction mechanism of conventional (non-functional) ionic liquids in solvent extraction

➢ Organic phase: Conventional ionic liquids
In conventional ionic liquids, extraction is favored because they are hydrophobic and can rapidly extract the naturally hydrophobic compounds from the aqueous phase to the organic phase, which hastens the extraction process [63,73]. The extraction of cerium using $[C_8mim]PF_6$ is one such application, but it cannot fully extract the other REEs due to the non-coordination nature of ionic liquid with water.

➢ Organic phase: Mixture of extractants and conventional ionic liquids
As discussed previously, the mechanism of the extraction of REEs is only possible when the compounds are naturally hydrophobic, but such types of systems are not possible all the time; in such situations, extractants such as chelating agents will form a complex with metal ions of the aqueous phase, and these complexes are more soluble in ionic liquids. Other than this possibility, the extraction depends on the structure of the cations, and the anions play a significant role in leading to cation, anion, and neutral exchange mechanisms.
A detailed description and pictorial representation of these mechanisms are given in Figure 14 and Table 2, respectively. As in Figure 14, the red line explains the release of ionic liquid cation (for the cationic exchange mechanism), the mixture of ionic liquid anion and hydrogen ion (for the anionic exchange mechanism), and the mixture of ionic liquid cation and hydrogen ion (for anionic exchange mechanism with more hydrophilic anions) from the organic to the aqueous phase. The black line indicates the transfer of REE ions from the aqueous to the organic phase, except in the case of the neutral exchange mechanism, where both REE ions and hydrogen ions are transferred [63]. In Table 2, ${M_{REE}}^{3+}$ denotes REE ion,

E refers to the extractant, IL_C^+ is the cation of ionic liquid, HE is an acidic extractant, H^+ is the hydrogen ion, and IL_A^- is the anion of ionic liquid.

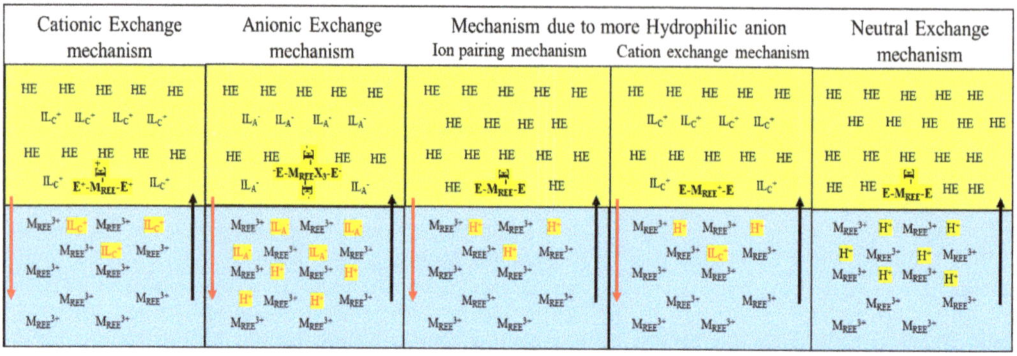

Figure 14. Pictorial representation of mechanisms applicable to conventional ionic liquid solvent extraction systems.

Table 2. Detailed description of mechanisms applicable to conventional ionic liquid solvent extraction systems.

Underlying Mechanisms When Extractants and Conventional Ionic Liquids Are Combined in the Organic Phase	Reactants		Products	
	Aqueous Phase	Organic Phase	Organic Phase	Aqueous Phase
Cation Exchange mechanism	M_{REE}^{3+}	$3E + 3IL_C^+$	$(M_{REE}E_3)^{3+}$	$3IL_C^+$
Anion Exchange mechanism	M_{REE}^{3+}	$4HE + IL_A^-$	$(M_{REE}E_4)^-$	$4H^+ + IL_A^-$
Anion Exchange mechanism with more hydrophilic anions (Ion pairing and Cation exchange)	M_{REE}^{3+}	3HE	$M_{REE}E_3$	$3H^+$
	M_{REE}^{3+}	$2HE + IL_C^+$	$(M_{REE}E_2)^+$	$2H^+ + IL_C^+$
Neutral exchange mechanism	M_{REE}^{3+}	HE	$(M_{REE}E_3)$	$3H^+$

➢ Cation exchange mechanism

From the available literature, many investigations have been conducted to demonstrate the effect of cation variation on the cation exchange mechanism on the extraction efficiency of the lanthanide series. Still, all of them depended on the cation alkyl chain length, where increased size resulted in hydrophobicity, thus resulting in a decrease in extraction efficiency and, thus, leading to a smaller loss of cations in the aqueous phase and vice versa [60,62,63]. So, an appropriate cost–benefit analysis should be performed before utilizing this approach when processing in large volumes.

➢ Anion exchange mechanism

Only a few experiments have been conducted to demonstrate the anion behavior in the extraction of REEs where the anion of the ligand forms an anionic-ligand rare earth complex (ALREC) with the REE, and this, when it interacts with the ionic liquid cation, forms a stable structure preventing the loss of the ionic liquid cation. According to the principle of electroneutrality for balancing the charge in both aqueous and anionic phases, the anionic component of the ionic liquid goes to the aqueous phase. After this, at the stripping stage, to prevent loss of ionic liquid, one possibility is to use the ionic liquid as a stripping solution for replacing ALREC with an anionic component of ionic liquid [60,62,63]. Some researchers tried another approach for the anionic exchange mechanism, whereby they modified the anionic part of an ionic liquid by applying the strategy of hydrophobicity as conducted in the cation exchange mechanism, but, here, the hydrophilicity of anion was improved to extract more ALREC. This modification resulted in a negative impact on the solvent extraction system where water tends to dissolve, leading to an aqueous–aqueous extraction system instead of organic–aqueous extraction system, thereby resulting in a decrease in the extraction of ALREC; the reason behind this impact is related to the more hydrophilic nature of anions of ionic liquid, which paves the way for ion-pairing and cation exchange mechanisms.

➢ Neutral exchange mechanism

To prevent the loss of either of the ions as described in the above two mechanisms neutral exchange mechanism works similarly to the extraction mechanism as in TOS, but modification is performed in the organic phase where it consists of acidic extractant and ionic liquid such that the protons exchange to the aqueous phase leaving behind the cationic and anionic components of the ionic liquids, thereby limiting the loss of ionic liquid, which also enables the potential for large scale industrial applications [60,62,63].

The extraction mechanism of conventional ionic liquids differs from TOS because the extraction mechanisms are more related to cation and anion interactions with the REEs; therefore, to consider ionic liquids as eco-friendly solvents, the loss of cation and anion should be prevented in the aqueous phase; at the same instance, there should not be any complications in operation as well as extraction efficiencies. Thus, the application of ionic liquids varies with the type of mechanism needed for a particular extraction process. One such approach is to modify the cation/anion structure of the ionic liquid with specialized functional groups, which have a higher tendency to attract REEs.

5. Functionalized ionic liquids (FILs) (referred to as task-specific ionic liquids in some of the literature)

In some solvent extractant systems, the targeted metal is associated with different types of ions, so the extraction of a specific metal can be challenging in such cases. Ionic liquids are modified with target-specific functional groups either at the cation or anion positions; these types of tailormade structures are referred to as functionalized ionic liquids [74]. The properties of ionic liquid, and the formation of the metal complex can be achieved within a functionalized ionic liquid. In most solvent extraction systems, ionic liquids with the functional group at the anionic site are utilized [75]. This approach has led to the development of different types of ionic liquids.

- Common functionalized cations and anions for ionic liquids [62]
  - Cations
    - Betainium, Trioctyl(2-ethoxy-2oxyethyl) ammonium, Tricapryl methylammonium, ammonium, Poly (ethylene glycol) functionalized imidazolium, and 1-Alkyl carboxylic acid 3-methyl imidazolium.
  - Anions
    - Di-(2-ethylhexyl) phosphate [P204], Di-(2-ethylhexyl) orthophosphinate [P507], Sec-octylphenoxyl acetate, Sec-non octylphenoxyl acetate, cyanex272, Dioctyldiglycolamate (DGA), and Dihexyldiglycolamate (DHDGA).
- Monofunctional ionic liquids

  The monofunctionalized ionic liquid structure is synthesized by placing the functional group of extractants either in the cation/anionic position of conventional ionic liquids, which results in two dominant benefits: One of them is the prevention of loss of the ionic liquid component to an aqueous phase, which correspondingly hinders the cation/anion exchange mechanism, thus supporting neutral extraction mechanism for the separation of REMs. Another benefit is that this single ionic liquid can perform the task of both diluent and extractant. Positioning the functional group at the anionic site is much easier than at the cationic site. Thus, most monofunctional ionic liquids are anionic-based. Some functional ionic liquids with carbamoylmethylphosphine oxide (CMPO) moieties showed reduced extraction due to high viscosity in such situations. It is noteworthy that a conventional ionic liquid can be used as a diluent to reduce the viscosity.

- Bifunctional ionic liquids

  Bifunctional ionic liquids are designed by attaching functional groups to cation and anion parts. The extraction can be achieved in the presence or absence of TOS, and the mechanism is related to the neutral extractant mechanism but in a much more complex way due to the selection of functional groups, thereby influencing interactions between the functional group and ionic liquid, and interactions between the ionic liquid and aqueous phase.

  Other functionalized ionic liquids discussed in the literature are acid-functionalized ionic liquids that give a mixed advantage in the form of liquid with an appreciable contact area [76]. Based on this, acid-functionalized ionic liquids are classified into carboxyl-functionalized ionic liquids and sulphonyl-functionalized ionic liquids. The selective exchange mechanism is similar to that of the neutral extractant mechanism where the protons (H^+) are released into aqueous media, thereby preventing the loss of cations and anions to the aqueous phase [76].

  A sulphonyl-functionalized ionic liquid was pioneered in solvent extraction for rare earth separation. According to the literature, acid–base-coupled ionic liquids are one other type of interesting ionic liquids for REE separations. Although most ionic liquids utilize an anionic-based exchange mechanism as part of the separation, this type of approach helps to combine the coordination ability and hydrophobic nature in a single ionic liquid [77]. On the whole, the ionic liquid application for solvent extraction enhances selectivity, and the type of ionic liquid application varies by the target metal required as a part of the extraction. Because these solvents are tailormade, it is easy to extract metal if the impact of the solution chemistry of the desired ions is understood and managed. The primary challenge lies in using these liquids appropriately. The selectivity enhancement of these liquids is not feasible below a certain feed concentration level. Thus, to extract them, a synergistic mixture of these liquids is required, which contributes to a factor of cost so there is a need to properly consider their application with an emphasis on not releasing some ecologic pollutant of either cations or anions.

vi. Separation Using Membranes

The membrane Separation Method (MSM) is an eco-friendly process that combines extraction and stripping without thermal heating. MSMs help separate organic mixtures, gases, and metal cations, including ferrous, non-ferrous, heavy, and precious metals. Liquid Membranes (LMs) have efficiently and rapidly separated metal ions. However, due to stability issues, research has shifted toward more stable Non-Liquid Membranes (NLM). MSMs are divided into liquid membranes (LMs) and non-liquid membranes (NLMs). LMs include several types, such as bulk liquid membranes (BLMs), Emulsification liquid membranes (ELMs), supported liquid membranes (SLMs), hollow-fiber-supported liquid membranes (HFSLMs), and electrostatic-pseudo-liquid membranes (ESPLMs). NLM comprises polymer inclusion membranes (PIMs) and ion-imprinted membranes (IIMs) [1]. HREEs, or heavy rare earth elements, are often preferred in membrane separation processes due to the absence of the issues typically associated with traditional extraction methods, such as the need for large amounts of organic solvents, supersaturated extraction, and poor phase separation. The membrane transport process involves a non-equilibrium mass transfer process across various types of membrane surfaces, including BLM, ELM, SLM, and HFSLM. In PIMs and IIMs, carriers are chemically or physically bound within the membrane matrix, making these membranes more stable. There are two types of mass transport in membrane separation: co-transport, which is suitable for neutral carriers like tributyl phosphate (TBP); and counter-transport, which is ideal for acidic carriers with cation exchange mechanisms like P507. The coupled transport process can be broken down into five possible steps [78].

1. Diffusion of REEs from the feed phase to the boundary layer of the feed-membrane side.
2. Carriers extract REEs at the interface.
3. Diffusion of REE complexes across the membrane.
4. Disintegration of REE complexes near the stripping interface of the membrane.
5. Diffusion of REEs from the stripping interface to the stripping phase and vice versa.

The use of BLMs was limited due to their small transport surface area.

ELMs have gained popularity for their rapid separation abilities, incorporating the benefits of solid-phase membranes and solvent extraction. The process of ELM involves carriers, surfactants, diluents, feed, and stripping solutions. The three primary stages include emulsion formation and target species permeation across the membrane, as well as emulsification of the emulsion after phase separation. Carriers, surfactants, and diluents are all crucial in the membrane separation process for rare earth ions, each with specific functions and properties. Despite their advantages, ELMs encounter instability problems such as membrane leakage and emulsion swelling that can impact transport efficiency.

Another effective method for the separation of REEs is the use of SLMs, which utilize a porous membrane that is infused with carrier-containing solvents. It separates the feed and stripping phases, has a higher interfacial area-to-volume ratio, and consumes fewer carriers (functional groups), thus providing an opportunity for the utilization of costly but effective extractants such as ionic liquids. Improved SLM strategies such as combined SLM and dispersion SLM (DSLM) have been developed to enhance stability and carrier renewal. However, challenges such as partial miscibility between the organic liquid membrane phase and aqueous phase, and blockage of carriers in the supporting membrane, must be addressed. Further advancements in supporting membrane structures and protection of the organic liquid membrane phase are required to fully realize the potential of SLMs for practical REE separation applications [78,79].

Hollow-fiber liquid membranes (HFLM) with micropores have a high surface area per volume and are promising for separating REEs. Polypropylene is commonly used in the matrix, and HFLMs can be operated in co-current or counter-current flow modes. HREEs permeate more quickly in the counter-current flow mode, and the presence of Na^+ accelerates the permeation rate. The stability of the HFLM process is good, and HFSLMs integrate extraction and stripping into one module [78,79]. Future research directions include developing highly selective and permeative membrane materials and inorganic membranes with thinner thicknesses and smaller pores.

The development of BLMs faced the challenge of a low interfacial area-to-volume ratio, while ELMs and SLMs encountered instability during long-term tests. To address these issues, ESPLMs with a high contact surface area per volume, low energy and reagent consumption, and minimal leakage and swelling rate were developed. A typical ESPLM consists of a baffle plate in the middle of the reaction tank that separates the organic phase, allowing it to move freely while the aqueous phases are isolated. The bottom of the tank contains two settler areas where the raffinate and concentrated solutions can settle independently without the high-voltage electrostatic field. ESPLMs rely on an electrical field to disintegrate the organic phase into tiny droplets, and the concentration gradient drives the transport of complex and carrier. ESPLMs have demonstrated high efficiency and selectivity for preconcentrating and separating REEs from dilute solutions but require careful maintenance of the electrode devices [78,79].

Polymeric membranes have become a preferred alternative to liquid membranes due to their stability and ease of synthesis. The PIM is the most commonly used polymeric membrane for water treatment, which includes a polymer matrix, plasticizer, and carrier. While there is limited literature on separating REEs using PIMs, various polymeric membranes have been developed and tested for separation, purification, or pre-concentration. However, there are challenges with the fouling of membranes, low flux, and limited thermal and chemical stability [1,78]. More research is needed to develop new synthetic polymers designed explicitly for REE separation and to improve the economy and sustainability of the membrane technique for industrial applications.

The molecular imprinting technique, developed in 1993, is widely applied for separating molecules or ions by creating cavities in polymers that match the size and configuration of the template. Since 1994, IIM/IIP has separated REEs through functional groups from polymerization monomers. Retarded and facilitated permeation mechanisms are two ways for an imprinted membrane, but the retarded mechanism is more common. Dual-template and interconnected 3D macro-porous structure methods improve the separation efficiency of adjacent REEs. However, higher demands for membrane applications have restricted the functional monomers and crosslinking agents for imprinted systems. Another developing direction is exploring functional monomers with a specific selectivity for target template REE ions [78,79].

vii. Separation by Adsorption

The adsorption-based separation for REEs is one of the most accessible and versatile methods, especially when traces of REEs are present in the source material. This method is an environmentally, gentle, and economically friendly approach. The operating principle mechanism behind adsorption is whereby the adsorbates (ions being adsorbed onto the surface) accumulate on adsorbent materials (the material on which ions are collected) with variations in surface area and porosity, thus forming a thin film on the adsorbent surface on the molecular scales [80]. The adsorption process has higher values of the selectivity index and enrichment indices. Talan et.al stated that on the surface of the adsorbent, specifically on the stern layer, the REE-complex formation occurs through covalent/ionic bonding, and the complex formation rate increases with the charge of the ion [81]. After the complex formation, the REE cation forms an outer sphere where the structure resembles a water molecule trapped between the REE cation and the adsorbent surface. There is another possibility of the REE cation remaining in the diffusion layer where charge cancellation happens with adjacent water molecules instead of complex formation, and the bonding here is purely ionic. The process is more susceptible to control parameters such as temperature, time, concentration of adsorbent, surface area, target ion concentration, etc. [7]. In extracting REEs, the adsorbent linked with the targeted ion selective functional group enhances the selectivity [1]. The process is exothermic; the continual mechanism of adsorption is the desorption process that focuses on removing adsorbed molecules from the surface. This adsorption process works in two ways: physisorption (adsorption through electrostatic forces and van der Waals interactions) and chemisorption (formation of irreversible chemical bonds, mainly the covalent bond between the adsorbent surface and the molecules) [80]. For any adsorption mechanism, we need to know the type of adsorbent, kinetics, isotherms, desorption efficiency, thermodynamics, and solution chemistry as prerequisites; the mathematical models for this are presented in [51,80,82]. In recent years, there has been widespread growth in the synthesis of these adsorbents, which are further classified and discussed in this section.

Natural and biodegradable adsorbents are cheaper, readily available, and exhibit considerable adsorption capacities. These materials also offer a typical advantage of being low cost. However, adsorption capacities in typical solutions with competitive ions are not as high as in media such as solvent extractants.

REE element production or separation using ion adsorption clays is another low-cost medium. REEs are naturally found on some clays and they can often be extracted from them due to their physisorption [7]. Thus, they can be easily separated. Some clays that are suitable for REE separation are kaolinite, halloysite, muscovite, montmorillonite, vermiculite, and illite [83,84]. The adsorption of REE on clay minerals is dependent on the surface morphology of these minerals. Natural and expanded vermiculite show good selectivity for REE extraction. Among kaolinite and halloysite, Eu adsorption was good on halloysite rather than Kaolinite due to the increased uptake capacity and surface area [85].

Zeolite is another material class that can be naturally available or synthetically manufactured. This material has good adsorption properties, especially for Nd and Gd, where a selective separation can be achieved with non-heat-treated zeolite-type adsorbent material for Nd and heat-treated adsorbent material for Gd [86]. Modifications to the surface of these materials can enhance the separation properties [1,7].

Chemical materials such as polysaccharides and carboxyl groups are found in some low-cost adsorbents, which are naturally available and biodegradable. Specifically, materials like cellulose, starch, gum arabic, chitosans, etc., function as binding sites for REE adsorption and surface modification of these adsorbents by chemical treatment can enhance the REE extraction and separation abilities. Selective separation of Dy from Nd, Pr, Tb, and Fe [87]; Gd from Dy, Tb, Pr, and Nd [88]; and Sc from Lanthanides [89] has been reported by using these kinds of materials.

Metal–organic frameworks (MOFs) are a well-known class of porous and crystalline materials with larger surface areas and adjustable pore properties. This type of material is formed by the chemical interaction of a single metal ion/cluster of ions with an organic linker [90]. As reported, a MOF-functionalized Cr-MIL-101 enhanced the separation of Gd from transition metals; also, for another compound, UiO-66, the adsorption rate of Gd increased due to the modification of structure with ionic solvents. Due to the poor mechanical properties of these types of materials, some literature used composite materials, which combine a MOF with natural adsorbents. Still, the literature did not explain the reusability of those materials. Similar to MOFs, a covalent organic framework has been found, but its application in the REE field is not yet known [1].

Using the impregnation of functional groups into adsorbents as discussed in ion-exchange technology, polymeric-based adsorbents have been developed that work on the ion-exchange mechanism [7,91]. Polymers can be either ion imprinted, natural polymers, or nanofibers. These nanostructures enhance REE adsorption due to a larger area-to-volume ratio, porosity, stability, and reduced density. In this sense, some geopolymers that help to separate HREE and LREE help to remove REE from the bulk and can facilitate subsequent separation [92].

Carbon-based adsorbents are also increasing due to the wide application of activated carbon in precious metals like gold [51]. This class of material can be graphene; the oxides of graphene with C, O, and H atoms in the structure; or carbon nanotubes, which include the single and multiwalled, activated carbon, fullerene, carbon dots, carbon nanofibers, and carbon black [18]. As reported, carbon-based nanomaterials showed good extraction values of REEs from harmful contaminants like Al, Fe, and Cu, but the selective fractionation of REE has not yet been achieved [7,18]. It is important to note that the term "harmful" indicates the negative impact these elements have on the separation process when present as contaminants.

Silica-based adsorbents are also booming for REE separation because of their increased surface area (KIT-6 silica), mechanical and thermal abilities, more readily available and easily manufactured materials, and more susceptibility for functionalization. Some functionalized silica adsorbents are selective for Nd separation from LREEs [93] and La separation from MREEs [94]. But due to the poor resistance of these materials to acidic environments and low selectivity compared to metal adsorbents, these materials are reinforced with carbon-based adsorbents to improve the acidic stability of the material [1]. Some carbon-silica-based composites showed selectivity for La and Sc separation from the rest of the REEs [7].

The magnetic-based adsorbents are developing and have improved properties compared to other adsorbents, and an easier separation from the REE stream; these adsorbents are quickly produced using co-precipitation and, similar to other adsorbents, surface modification helps to increase their selectivity for REE separation [1,7]. The principle of co-precipitation paved the way for the creation of nanocomposite adsorbents because they synthesize layered double hydroxides, but their potential application is yet to be researched [7].

The last class of adsorption materials discovered is hydrogel-based adsorbents, which show a promising application because of pore stability and functionalization; however, the penetration of REE ions and the stability of gel limits their use [1].

After these separations processes, the individual REEs are sent to an electrochemical refining process for metal recovery or recovery of REEs into oxides.

3. Holistic View of REE Separation: Perspectives and Conclusions

Information related to research and developments in REE separation is much needed because of the increased demand for the utilization of REEs in several sectors. Therefore, discovering a standard technology for base metals and other precious metals may not be suitable for REEs because of variations in sources of availability. However, for commercial operations, a solid knowledge of solution chemistry will be a guiding tool to decide which process will be technologically viable to meet business goals. Moreover, selecting a particular method depends on the operation's economic goals accordingly to a flowsheet designed for REE recovery streams. Of the methods explained in the previous sections, the merits and demerits of each approach are summarized.

a. The fractional crystallization (FC) approach has the potential for the selective separation of individual REEs, but the production cycle needed is tiresome. Additionally, the availability of salts required for this approach is limited, and the HREE products generated are less pure than any other method.
b. The fractional precipitation (FP) approach can resolve some of the problems caused by the fractional crystallization approach regarding salt availability. Still, this method is only suitable for the separation of REEs into groups as individual separation is challenging.
c. The size-selective crystallization (SSC) approach is the best suited when the REEs in aqueous media have a more significant ionic radii difference, but for the adjacent REEs with a smaller ionic radii difference, the contamination of unwanted REE cations into targeted REE crystals is the major setback.
d. The reaction-controlled crystallization (RCC) reported to be the best practical strategy for the Nd/Dy system but is impractical with the Nd/Sm system.
e. Valence selective crystallization (VSC) is a well-suited approach when dealing with systems that contain differences in valencies among metal cations, and thus it is limited to use for Ce.
f. The selective oxidation (SO) approach applies to Ce, Pr, and Tb where enhancement of their oxidation states to +4 can enhance their separation. The separation post enhancement of valence state differs for Ce compared to Pr and Tb.
g. The selective reduction (SR) approach can separate Sm, Eu, and Yb individually or in bulk from a mixture of REEs. The approach type varies depending on the interest of the operation needs and the desired product. However, this approach's detailed operation cost analysis is rarely discussed in the literature.
h. The ion exchange (IX) process is simple as it exchanges a counter ion for a targeted cation. This process is capable and can efficiently separate multiple elements in one go. However, the availability and development of appropriate resin material limit its use. Additionally, future research should address the issues of longer operation time, clogging of resin by unwanted metal, and loss of functional carriers to scale up this approach to industrial needs.
i. The membrane separation (MS) approach delimits the few concerns of IX as this is fast, versatile, and continuous. Further, this is eco-friendly and energy-saving because of the low operational equipment needed. Still, as in ion exchange, membrane

separation is also hindered due to the availability of suitable membrane material. The MS can potentially revolutionize the individual HREE separation from the bulk if the problem of suitable membrane material is resolved.

j. Solvent extraction (SX) is suitable for large-scale industrial operations as bulk volumes can be processed in a short span. Additionally, selectivity can be enhanced by modifying and tailoring the organic phase. The only demerit of this method is the loss and contamination of the organic phase, which contains both harmful and costlier species, thus leading to pollution of the aqueous phase and increasing the cost of operation. Therefore, this is a significant takeaway for further research in this direction.

k. Adsorption (AD) is a rapid and straightforward approach. This method needs the creation of a proper adsorbent that should meet the selectivity required for the targeted REE. In recent years researchers have synthesized many adsorbents; the hurdle in using them industrially is making adsorbents that have a high use of chemicals and more sophisticated equipment. Thus, there should be a trade-off between technology and the cost of operation. Moreover, desorption is the major challenge for natural clay-based adsorbents, and most of the available technology either uses highly concentrated acids or some less practical industrial procedures like microwave treatment or ultrasonicators. Thus, further research is needed to push these barriers for industrial application.

Of all the methods described in this study, adsorption and solvent extraction can ramp up the commercial production of REEs if few developments are performed. Additionally, it is necessary to know that other methods can be helpful and categorized as auxiliary tools to the above techniques. This hybrid application is again bound by the Efficient-Easier-Economical-Environmentally Safe (EEEE) strategy, where all these factors should be addressed commercially. However, relying entirely on auxiliary approaches for commercial REE production would require extensive research.

Author Contributions: Conceptualization, S.V.S.H.P. and P.K.S.; methodology, S.V.S.H.P. and P.K.S.; formal analysis, M.L.F. and P.K.S.; investigation, S.V.S.H.P.; resources, M.L.F. and P.K.S.; data curation, S.V.S.H.P.; writing—original draft preparation, S.V.S.H.P.; writing—review and editing, P.K.S. and M.L.F.; visualization, P.K.S. and M.L.F.; supervision, P.K.S. and M.L.F.; project administration, M.L.F. and P.K.S.; funding acquisition, M.L.F. and P.K.S. All authors have read and agreed to the published version of the manuscript.

Funding: This research received no external funding.

Data Availability Statement: The data presented in this study are available on request from the authors.

Conflicts of Interest: The authors declare no conflict of interest.

References

1. Chen, Z.; Li, Z.; Chen, J.; Kallem, P.; Banat, F.; Qiu, H. Recent advances in selective separation technologies of rare earth elements: A review. *J. Environ. Chem. Eng.* **2022**, *10*, 107104. [CrossRef]
2. Opare, E.O.; Struhs, E.; Mirkouei, A. A comparative state-of-technology review and future directions for rare earth element separation. *Renew. Sustain. Energy Rev.* **2021**, *143*, 110917. [CrossRef]
3. Gupta, C.K.; Krishnamurthy, N. Extractive metallurgy of rare earths. *Int. Mater. Rev.* **1992**, *37*, 197–248. [CrossRef]
4. Balaram, V. Rare earth elements: A review of applications, occurrence, exploration, analysis, recycling, and environmental impact. *Geosci. Front.* **2019**, *10*, 1285–1303. [CrossRef]
5. Galhardi, J.A.; Luko-Sulato, K.; Yabuki, L.N.; Santos, L.M.; da Silva, Y.J. *Rare Earth Elements and Radionuclides*; Elsevier Inc.: Amsterdam, The Netherlands, 2022. [CrossRef]
6. Bünzli, J.-C.G. Lanthanides. In *Kirk-Othmer Encyclopedia of Chemical Technology*; Wiley Blackwell: Hoboken, NJ, USA, 2013; pp. 1–43. Available online: https://www.researchgate.net/publication/257921324_Lanthanides (accessed on 27 March 2023).
7. Asadollahzadeh, M.; Torkaman, R.; Torab-Mostaedi, M. Extraction and Separation of Rare Earth Elements by Adsorption Approaches: Current Status and Future Trends. In *Separation and Purification Reviews*; Taylor and Francis Inc.: Abingdon, UK, 2020; pp. 1–28. [CrossRef]
8. Zhou, H.; Chen, X.; Fang, L.; Hu, C.; Wang, H. A new microwave dielectric ceramic for LTCC applications. *J. Mater. Sci. Mater. Electron.* **2010**, *21*, 849–853. Available online: http://www.mineralsuk.com/ (accessed on 6 March 2023). [CrossRef]

9. Sarswat, P.; Leake, M.; Allen, L.; Free, M.; Hu, X.; Kim, D.; Noble, A.; Luttrell, G. Efficient recovery of rare earth elements from coal based resources: A bioleaching approach. *Mater. Today Chem.* **2020**, *16*, 100246. [CrossRef]
10. McNulty, T.; Hazen, N.; Park, S. Processing the ores of rare-earth elements. *MRS Bull.* **2022**, *47*, 258–266. [CrossRef]
11. Gaustad, G.; Williams, E.; Leader, A. Rare earth metals from secondary sources: Review of potential supply from waste and byproducts. *Resour. Conserv. Recycl.* **2020**, *167*, 105213. [CrossRef]
12. Omodara, L.; Pitkäaho, S.; Turpeinen, E.-M.; Saavalainen, P.; Oravisjärvi, K.; Keiski, R.L. Recycling and substitution of light rare earth elements, cerium, lanthanum, neodymium, and praseodymium from end-of-life applications—A review. *J. Clean. Prod.* **2019**, *236*, 117573. [CrossRef]
13. Dutta, T.; Kim, K.-H.; Uchimiya, M.; Kwon, E.E.; Jeon, B.-H.; Deep, A.; Yun, S.-T. Global demand for rare earth resources and strategies for green mining. *Environ. Res.* **2016**, *150*, 182–190. [CrossRef]
14. Zhang, Z.; Allen, L.; Podder, P.; Free, M.L.; Sarswat, P.K. Recovery and Enhanced Upgrading of Rare Earth Elements from Coal-Based Resources: Bioleaching and Precipitation. *Minerals* **2021**, *11*, 484. [CrossRef]
15. Brahim, J.A.; Hak, S.A.; Achiou, B.; Boulif, R.; Beniazza, R.; Benhida, R. Kinetics and mechanisms of leaching of rare earth elements from secondary resources. *Miner. Eng.* **2022**, *177*, 107351. [CrossRef]
16. Echeverry-Vargas, L.; Ocampo-Carmona, L.M. Recovery of Rare Earth Elements from Mining Tailings: A Case Study for Generating Wealth from Waste. *Minerals* **2022**, *12*, 948. [CrossRef]
17. European Commission. Study on the Critical Raw Materials for the EU. 2023. Available online: https://single-market-economy.ec.europa.eu/publications/study-critical-raw-materials-eu-2023-final-report_en (accessed on 27 March 2023). [CrossRef]
18. Cardoso, C.E.D.; Almeida, J.C.; Lopes, C.B.; Trindade, T.; Vale, C.; Pereira, E. Recovery of Rare Earth Elements by Carbon-Based Nanomaterials—A Review. *Nanomaterials* **2019**, *9*, 814. [CrossRef] [PubMed]
19. Stein, R.T.; Kasper, A.C.; Veit, H.M. Recovery of Rare Earth Elements Present in Mobile Phone Magnets with the Use of Organic Acids. *Minerals* **2022**, *12*, 668. [CrossRef]
20. Yoldjian, G. The use of rare earths in ceramics. *J. Less Common Met.* **1985**, *111*, 17–22. [CrossRef]
21. Haque, N.; Hughes, A.; Lim, S.; Vernon, C. Rare Earth Elements: Overview of Mining, Mineralogy, Uses, Sustainability and Environmental Impact. *Resources* **2014**, *3*, 614–635. [CrossRef]
22. Castor, S.B.; Hedrick, J.B. Rare Earth Elements. *Encycl. Inorg. Bioinorg. Chem.* **2005**, 769–792. [CrossRef]
23. Suli, L.M.; Ibrahim, W.H.W.; Aziz, B.A.; Deraman, M.R.; Ismail, N.A. A Review of Rare Earth Mineral Processing Technology. *Chem. Eng. Res. Bull.* **2017**, *19*, 20. [CrossRef]
24. Wills, B.A.; Finch, J.A. *Wills' Mineral Processing Technology: An Introduction to the Practical Aspects of Ore Treatment and Mineral Recovery*; Butterworth-Heinemann: Oxford, UK, 2015.
25. Schulz, B.; Merker, G.; Gutzmer, J. Automated SEM Mineral Liberation Analysis (MLA) with Generically Labelled EDX Spectra in the Mineral Processing of Rare Earth Element Ores. *Minerals* **2019**, *9*, 527. [CrossRef]
26. Free, M.L. *Hydrometallurgy*; Springer International Publishing: Cham, Switzerland, 2022. [CrossRef]
27. Hidayah, N.N.; Abidin, S.Z. The evolution of mineral processing in extraction of rare earth elements using solid-liquid extraction over liquid-liquid extraction: A review. *Miner. Eng.* **2017**, *112*, 103–113. [CrossRef]
28. Yang, Y.; Walton, A.; Sheridan, R.; Güth, K.; Gauß, R.; Gutfleisch, O.; Buchert, M.; Steenari, B.-M.; Van Gerven, T.; Jones, P.T.; et al. REE Recovery from End-of-Life NdFeB Permanent Magnet Scrap: A Critical Review. *J. Sustain. Met.* **2017**, *3*, 122–149. [CrossRef]
29. Traore, M.; Gong, A.; Wang, Y.; Qiu, L.; Bai, Y.; Zhao, W.; Liu, Y.; Chen, Y.; Liu, Y.; Wu, H.; et al. Research progress of rare earth separation methods and technologies. *J. Rare Earths* **2023**, *41*, 182–189. [CrossRef]
30. Sarswat, P.K.; Free, M.L. Frequency and atomic mass based selective electrochemical recovery of rare earth metals and isotopes. *Electrochimica Acta* **2016**, *219*, 435–446. [CrossRef]
31. Sarswat, P.K.; Podder, P.; Zhang, Z.; Free, M.L. Autogenous acid production using a regulated bio-oxidation method for economical recovery of REEs and critical metals from coal-based resources. *Appl. Surf. Sci. Adv.* **2022**, *11*, 100283. [CrossRef]
32. Sarswat, P.K.; Zhang, Z.; Free, M.L. Rare Earth Elements Extraction from Coal Waste Using a Biooxidation Approach. In *Rare Metal Technology*; Springer International Publishing: Cham, Switzerland, 2021; pp. 211–216. [CrossRef]
33. Free, M.L.; Ilunga, J.K.; Podder, P.; Sarswat, P.K. Assessing and Improving Biooxidation for Acid Generation and Rare Earth Element Extraction. *Processes* **2023**, *11*, 1005. [CrossRef]
34. Hovey, J.L.; Dittrich, T.M.; Allen, M.J. Coordination chemistry of surface-associated ligands for solid–liquid adsorption of rare-earth elements. *J. Rare Earths* **2023**, *41*, 1–18. [CrossRef]
35. El Ouardi, Y.; Virolainen, S.; Mouele, E.S.M.; Laatikainen, M.; Repo, E.; Laatikainen, K. The recent progress of ion exchange for the separation of rare earths from secondary resources—A review. *Hydrometallurgy* **2023**, *218*, 106047. [CrossRef]
36. Pearson, R.G. Hard and soft acids and bases, HSAB, part I: Fundamental principles. *J. Chem. Educ.* **1968**, *45*, 581–587. [CrossRef]
37. Marcus, Y. Thermodynamics of solvation of ions. Part 5.—Gibbs free energy of hydration at 298.15 K. *J. Chem. Soc. Faraday Trans.* **1991**, *87*, 2995–2999. [CrossRef]
38. Moeller, T. *The Chemistry of the Lanthanides*; Elsevier: Amsterdam, The Netherlands, 1973. [CrossRef]
39. Wall, F. Rare Earth Elements. In *Encyclopedia of Geology*; Elsevier: Amsterdam, The Netherlands, 2021; pp. 680–693. [CrossRef]
40. James, C.; Grant, A.J. The separation of the rare earths giving the more soluble double sulfates from brazilian monazite sand. *J. Am. Chem. Soc.* **1916**, *38*, 41–47. [CrossRef]
41. Ishino, T. Separation of Rare Earth Elements. *J. Soc. Mater. Sci. Jpn.* **1963**, *12*, 559–571. [CrossRef]

42. Zhao, X.; Wong, M.; Mao, C.; Trieu, T.X.; Zhang, J.; Feng, P.; Bu, X. Size-Selective Crystallization of Homochiral Camphorate Metal–Organic Frameworks for Lanthanide Separation. *J. Am. Chem. Soc.* **2014**, *136*, 12572–12575. [CrossRef]
43. Moeller, T.; Kremers, H.E. Observations on Rare Earths Extraction of Ytterbium from Rare Earth Mixtures with Sodium Amalgam. *Ind. Eng. Chem. Anal. Ed.* **1945**, *17*, 798–800. [CrossRef]
44. Furukawa, H.; Cordova, K.E.; O'Keeffe, M.; Yaghi, O.M. The Chemistry and Applications of Metal-Organic Frameworks. *Science* **2013**, *341*, 1230444. [CrossRef]
45. Yin, X.; Wang, Y.; Bai, X.; Wang, Y.; Chen, L.; Xiao, C.; Diwu, J.; Du, S.; Chai, Z.; Albrecht-Schmitt, T.E.; et al. Rare earth separations by selective borate crystallization. *Nat. Commun.* **2017**, *8*, 14438. [CrossRef]
46. Willauer, A.R.; Palumbo, C.T.; Fadaei-Tirani, F.; Zivkovic, I.; Douair, I.; Maron, L.; Mazzanti, M. Accessing the +IV Oxidation State in Molecular Complexes of Praseodymium. *J. Am. Chem. Soc.* **2020**, *142*, 5538–5542. [CrossRef] [PubMed]
47. Palumbo, C.T.; Zivkovic, I.; Scopelliti, R.; Mazzanti, M. Molecular Complex of Tb in the +4 Oxidation State. *J. Am. Chem. Soc.* **2019**, *141*, 9827–9831. [CrossRef]
48. Jamrack, W. *Rare Metal Extraction by Chemical Engineering Techniques*; Elsevier: Amsterdam, The Netherlands, 1963. [CrossRef]
49. Chen, B.; He, M.; Zhang, H.; Jiang, Z.; Hu, B. Chromatographic Techniques for Rare Earth Elements Analysis. *Phys. Sci. Rev.* **2019**, *2*, 1–35. [CrossRef]
50. Kumar, S.; Jain, S. History, Introduction, and Kinetics of Ion Exchange Materials. *J. Chem.* **2013**, *2013*, 1–13. [CrossRef]
51. Free, M.L. *Hydrometallurgy, Fundemetals and Applications*; Springer Nature: Singapore, 2022.
52. Hérès, X.; Blet, V.; Di Natale, P.; Ouaattou, A.; Mazouz, H.; Dhiba, D.; Cuer, F. Selective Extraction of Rare Earth Elements from Phosphoric Acid by Ion Exchange Resins. *Metals* **2018**, *8*, 682. [CrossRef]
53. Mayer, S.W.; Freiling, E.C. Ion Exchange as a Separation Method. VI. Column Studies of the Relative Efficiencies of Various Complexing Agents for the Separation of Lighter Rare Earths. *J. Am. Chem. Soc.* **1953**, *75*, 5647–5649. [CrossRef]
54. Page, M.J.; Soldenhoff, K.; Ogden, M.D. Comparative study of the application of chelating resins for rare earth recovery. *Hydrometallurgy* **2017**, *169*, 275–281. [CrossRef]
55. Xie, F.; Zhang, T.A.; Dreisinger, D.; Doyle, F. A critical review on solvent extraction of rare earths from aqueous solutions. *Miner. Eng.* **2014**, *56*, 10–28. [CrossRef]
56. Binnemans, K.; Jones, P.T. Solvometallurgy: An Emerging Branch of Extractive Metallurgy. *J. Sustain. Met.* **2017**, *3*, 570–600. [CrossRef]
57. Joshi, D.R.; Adhikari, N. An Overview on Common Organic Solvents and Their Toxicity. *J. Pharm. Res. Int.* **2019**, *28*, 1–18. [CrossRef]
58. Sun, X.; Ji, Y.; Hu, F.; He, B.; Chen, J.; Li, D. The inner synergistic effect of bifunctional ionic liquid extractant for solvent extraction. *Talanta* **2010**, *81*, 1877–1883. [CrossRef]
59. Abbott, A.P.; Frisch, G.; Gurman, S.J.; Hillman, A.R.; Hartley, J.; Holyoak, F.; Ryder, K.S. Ionometallurgy: Designer redox properties for metal processing. *Chem. Commun.* **2011**, *47*, 10031–10033. [CrossRef] [PubMed]
60. Arrachart, G.; Couturier, J.; Dourdain, S.; Levard, C.; Pellet-Rostaing, S. Recovery of Rare Earth Elements (REEs) Using Ionic Solvents. *Processes* **2021**, *9*, 1202. [CrossRef]
61. Prodius, D.; Mudring, A.-V. Rare earth metal-containing ionic liquids. *Coord. Chem. Rev.* **2018**, *363*, 1–16. [CrossRef]
62. Trinh, H.B.; Lee, J.-C.; Lee, J. *Task-Specific Ionic Liquids for the Separation and Recovery of Rare Earth Elements*; INC: New York, NY, USA, 2022. [CrossRef]
63. Wang, K.; Adidharma, H.; Radosz, M.; Wan, P.; Xu, X.; Russell, C.K.; Tian, H.; Fan, M.; Yu, J. Recovery of rare earth elements with ionic liquids. *Green Chem.* **2017**, *19*, 4469–4493. [CrossRef]
64. Khodakarami, M.; Alagha, L. Separation and recovery of rare earth elements using novel ammonium-based task-specific ionic liquids with bidentate and tridentate O-donor functional groups. *Sep. Purif. Technol.* **2019**, *232*, 115952. [CrossRef]
65. Lei, Z.; Chen, B.; Koo, Y.-M.; MacFarlane, D.R. Introduction: Ionic Liquids. *Chem. Rev.* **2017**, *117*, 6633–6635. [CrossRef]
66. Kaur, G.; Kumar, H.; Singla, M. Diverse applications of ionic liquids: A comprehensive review. *J. Mol. Liq.* **2022**, *351*, 118556. [CrossRef]
67. Micheau, C.; Arrachart, G.; Turgis, R.; Lejeune, M.; Draye, M.; Michel, S.; Legeai, S.; Pellet-Rostaing, S. Ionic Liquids as Extraction Media in a Two-Step Eco-Friendly Process for Selective Tantalum Recovery. *ACS Sustain. Chem. Eng.* **2020**, *8*, 1954–1963. [CrossRef]
68. Seddon, K.R.; Stark, A.; Torres, M.-J. Influence of chloride, water, and organic solvents on the physical properties of ionic liquids. *Pure Appl. Chem.* **2000**, *72*, 2275–2287. [CrossRef]
69. Widegren, J.A.; Laesecke, A.; Magee, J.W. The effect of dissolved water on the viscosities of hydrophobic room-temperature ionic liquids. *Chem. Commun.* **2005**, *72*, 1610–1612. [CrossRef]
70. Huddleston, J.G.; Visser, A.E.; Reichert, W.M.; Willauer, H.D.; Broker, G.A.; Rogers, R.D. Characterization and comparison of hydrophilic and hydrophobic room temperature ionic liquids incorporating the imidazolium cation. *Green Chem.* **2001**, *3*, 156–164. [CrossRef]
71. Toh, S.L.I.; McFarlane, J.; Tsouris, C.; DePaoli, D.W.; Luo, H.; Dai, S. Room-Temperature Ionic Liquids in Liquid–Liquid Extraction: Effects of Solubility in Aqueous Solutions on Surface Properties. *Solvent Extr. Ion Exch.* **2006**, *24*, 33–56. [CrossRef]
72. Jacquemin, J.; Husson, P.; Padua, A.A.H.; Majer, V. Density and viscosity of several pure and water-saturated ionic liquids. *Green Chem.* **2006**, *8*, 172–180. [CrossRef]

73. Zuo, Y.; Liu, Y.; Chen, J.; Li, D.Q. The Separation of Cerium(IV) from Nitric Acid Solutions Containing Thorium(IV) and Lanthanides(III) Using Pure [C_8mim]PF_6 as Extracting Phase. *Ind. Eng. Chem. Res.* **2008**, *47*, 2349–2355. [CrossRef]
74. Hoogerstraete, T.V.; Onghena, B.; Binnemans, K. Homogeneous Liquid–Liquid Extraction of Rare Earths with the Betaine—Betainium Bis(trifluoromethylsulfonyl)imide Ionic Liquid System. *Int. J. Mol. Sci.* **2013**, *14*, 21353–21377. [CrossRef]
75. Rout, A.; Kotlarska, J.; Dehaen, W.; Binnemans, K. Liquid–liquid extraction of neodymium(iii) by dialkylphosphate ionic liquids from acidic medium: The importance of the ionic liquid cation. *Phys. Chem. Chem. Phys.* **2013**, *15*, 16533–16541. [CrossRef] [PubMed]
76. Dupont, D.; Raiguel, S.; Binnemans, K. Sulfonic acid functionalized ionic liquids for dissolution of metal oxides and solvent extraction of metal ions. *Chem. Commun.* **2015**, *51*, 9006–9009. [CrossRef] [PubMed]
77. Zhang, Y.; Guo, W.; Liu, D.; Xu, J. Rational design of novel carboxylic acid functionalized phosphonium based ionic liquids as high-performance extractants for rare earths. *J. Rare Earths* **2021**, *39*, 1435–1441. [CrossRef]
78. Chen, L.; Wu, Y.; Dong, H.; Meng, M.; Li, C.; Yan, Y.; Chen, J. An overview on membrane strategies for rare earths extraction and separation. *Sep. Purif. Technol.* **2018**, *197*, 70–85. [CrossRef]
79. Bartsch, R.A.; Way, J.D. *Chemical Separations with Liquid Membranes: An Overview*; American Chemical Society: Washington, DC, USA, 1996; Volume 642. [CrossRef]
80. Kecili, R.; Hussain, C.M. *Mechanism of Adsorption on Nanomaterials*; Elsevier Inc.: Amsterdam, The Netherlands, 2018. [CrossRef]
81. Talan, D.; Huang, Q. A review of environmental aspect of rare earth element extraction processes and solution purification techniques. *Miner. Eng.* **2022**, *179*, 107430. [CrossRef]
82. Anastopoulos, I.; Bhatnagar, A.; Lima, E.C. Adsorption of rare earth metals: A review of recent literature. *J. Mol. Liq.* **2016**, *221*, 954–962. [CrossRef]
83. Cheng, Y.-L. We are IntechOpen, the world's leading publisher of Open Access books Built by scientists, for scientists TOP 1%. *Intech* **2016**, *11*, 13. [CrossRef]
84. Brião, G.D.V.; da Silva, M.G.; Vieira, M.G.A. Expanded vermiculite as an alternative adsorbent for the dysprosium recovery. *J. Taiwan Inst. Chem. Eng.* **2021**, *127*, 228–235. [CrossRef]
85. Zhou, J.; Liu, H.; Liu, D.; Yuan, P.; Bu, H.; Du, P.; Fan, W.; Li, M. Sorption/desorption of Eu(III) on halloysite and kaolinite. *Appl. Clay Sci.* **2022**, *216*, 106356. [CrossRef]
86. Vasylechko, V.; Stechynska, E.; Stashkiv, O.; Gryshchouk, G.; Patsay, I. Sorption of Neodymium and Gadolinium on Transcarpathian Clinoptilolite. *Acta Phys. Pol. A* **2018**, *133*, 794–797. [CrossRef]
87. Liu, E.; Xu, X.; Zheng, X.; Zhang, F.; Liu, E.; Li, C. An ion imprinted macroporous chitosan membrane for efficiently selective adsorption of dysprosium. *Sep. Purif. Technol.* **2017**, *189*, 288–295. [CrossRef]
88. Zheng, X.; Liu, E.; Zhang, F.; Dai, J.; Yan, Y.; Li, C. Selective adsorption and separation of gadolinium with three-dimensionally interconnected macroporous imprinted chitosan films. *Cellulose* **2017**, *24*, 977–988. [CrossRef]
89. Ramasamy, D.L.; Porada, S.; Sillanpää, M. Marine algae: A promising resource for the selective recovery of scandium and rare earth elements from aqueous systems. *Chem. Eng. J.* **2019**, *371*, 759–768. [CrossRef]
90. Zhou, H.; Long, J.; Yaghi, O. Introduction to Metal—Organic Frameworks. *Chem. Rev.* **2012**, *112*, 673–674. [CrossRef]
91. Lee, Y.-R.; Yu, K.; Ravi, S.; Ahn, W.-S. Selective Adsorption of Rare Earth Elements over Functionalized Cr-MIL-101. *ACS Appl. Mater. Interfaces* **2018**, *10*, 23918–23927. [CrossRef]
92. Fiket, Ž.; Galović, A.; Medunić, G.; Turk, M.F.; Ivanić, M.; Dolenec, M.; Biljan, I.; Šoster, A.; Kniewald, G. Adsorption of Rare Earth Elements from Aqueous Solutions Using Geopolymers. *IECG 2018* **2018**, *2*, 567. [CrossRef]
93. Ravi, S.; Lee, Y.-R.; Yu, K.; Ahn, J.-W.; Ahn, W.-S. Benzene triamido-tetraphosphonic acid immobilized on mesoporous silica for adsorption of Nd3+ ions in aqueous solution. *Microporous Mesoporous Mater.* **2018**, *258*, 62–71. [CrossRef]
94. Liu, E.; Chen, L.; Dai, J.; Wang, Y.; Li, C.; Yan, Y. Fabrication of phosphate functionalized chiral nematic mesoporous silica films for the efficient and selective adsorption of lanthanum ions. *J. Mol. Liq.* **2019**, *277*, 786–793. [CrossRef]

processes MDPI

Review

Pulsed Electric Field Technology for Recovery of Proteins from Waste Plant Resources and Deformed Mushrooms: A Review

Ramya Ramaswamy [1,*], Sivaneasan Bala Krishnan [2] and Susanna Su Jan Leong [1]

[1] Food, Chemical and Biotechnology Cluster, Singapore Institute of Technology, Dover Drive, Singapore 138683, Singapore; susanna.leong@singaporetech.edu.sg
[2] Engineering Cluster, Singapore Institute of Technology, Dover Drive, Singapore 138683, Singapore; sivaneasan@singaporetech.edu.sg
* Correspondence: ramya.ramaswamy@singaporetech.edu.sg

Abstract: Proteins are complex molecules, which play a vital role in our body's function, the building of tissues, and the regulation of metabolic activity. They are crucial to children's growth and serve as a key component in the body's process of distributing oxygen. Proteins fuel the body by supplying the required nutrition and energy. Currently, there is an increasing demand for proteins on large scales with no detrimental effects. The adverse health effects of animal proteins have resulted in a growing preference for plant-based proteins, which offer a healthier daily dosage. Valuable proteins can be extracted from various parts of the plant, including stems, leaves, seeds, fruits, vegetables, and roots. Notably, protein extraction from waste plant and mushroom parts minimizes the product wastage and improves the overall production to support economic sustainability. There are several protein extraction techniques available, where the replacement of non-thermal methods with thermal ones is promising nowadays due to the appreciable retainment of protein quality. Pulsed Electric Field (PEF) technology is one of the most efficient non-thermal tools used to assist with extracting these proteins at the minimum processing time and energy consumption when compared with thermal techniques. It relies on the application of a high-voltage pulse between two electrodes to treat samples inside the treatment chamber. While electrode shapes and treatment chamber designs primarily govern the electric field's application, optimizing process parameters such as electric field strength, pulse width, number of pulses, and pulse waveshape assists in obtaining a desirable enhancement in the protein yield. The primary objective of this review is to explain the PEF-assisted protein extraction process applicable to waste plant parts and deformed mushrooms. While PEF is not a novel concept, utilizing it as a pre-extraction treatment to the aforementioned waste resources would aid in improving the production of value-added protein products economically. So far, PEF has shown immense promise in assisting with protein extraction studies, but requires further research in order to establish this area for large-scale industrial applications.

Keywords: pulsed electric field; protein extraction; waste plant resources; deformed mushroom parts

Citation: Ramaswamy, R.; Krishnan, S.B.; Leong, S.S.J. Pulsed Electric Field Technology for Recovery of Proteins from Waste Plant Resources and Deformed Mushrooms: A Review. *Processes* **2024**, *12*, 342. https://doi.org/10.3390/pr12020342

Academic Editors: Prashant K. Sarswat and Yanlin Zhang

Received: 20 January 2024
Revised: 2 February 2024
Accepted: 2 February 2024
Published: 6 February 2024

1. Introduction

Based on the Food and Agriculture Organization, around one third of the food produced for human consumption is wasted globally, i.e., 1.3 billion tons per year [1]. According to the United Nations Environment Programme's Food Waste Index Report, 17 percent of our food ends up being wasted in retail and by consumers, particularly in households [2]. For instance, India is among the countries that remains high in the production of fruits and vegetables. However, it suffers from 40% wastage of total food production. In that, fruits and vegetables contribute to 18% wastage of total food production [3]. These wastes comprise leaves, peels, stems, roots, and seeds, which have been found to have high nutritional value and contain a substantial amount of valuable bio-compounds, including proteins [4]. Upon extraction, the byproducts of these wastes can be efficiently utilized in food processing and pharma industries to produce nutritional and pharmaceutical supplements.

Several research studies have observed the significance of the protein requirement to the body [5,6]. This requirement is increasing globally due to the high population growth and the awareness of protein's positive impact on the body. Research studies indicate the Recommended Dietary Allowance of protein to be 0.8 g per kg body weight (BW) per day since insufficient protein intake can lead to anemia, physical weakness, stunting, edema, vascular dysfunction, and impaired immunity [7]. Dietary recommendations from other countries, including the joint FAO/WHO committee, do not significantly deviate from the US recommendations for macronutrient intake [8,9]. In the case of insufficient protein intake, protein supplements can provide dietary compensation. Hence, it is necessary to consume a protein-rich diet every day since protein's nutritional benefits are decided by both the quality and quantity. It has been reported that sufficient consumption of plant proteins may reduce the risk of all-cause and cardiovascular disease mortality and may increase longevity [10]. Thus, proteins play a significant role in providing nutritional value to foods and act as thickening, stabilizing, emulsifying, foaming, gelling, and binding agents.

Over the past few decades, a consistent increase in the demand for proteins has been observed due to multiple health benefits. With a global population of 7.3 billion, the demand for protein consumption has increased to approximately 202 million tons [11]. This demand may exceed the protein supply in future decades if the available resources are not efficiently utilized. This has led to the search for plant-based protein resources since the perception of the negative health effects of animal proteins has escalated. This extensive search is carried out with the objective of producing value-added products from the plant protein isolates in order to meet the ever-increasing protein demand. In order to improve the socioeconomic viability and to minimize product wastage, research is currently in the phase of extracting proteins from waste plant resources such as vegetable peels, fruit peels, seeds, stems, and roots from the processing industries. When compared with animal-derived waste, plant-derived waste makes up 63% of the overall food supply chain [12]. Globally, huge quantities of mushroom and plant pieces are discarded from packaging industries since they do not meet the required standards for size and shape regulations. These rejected or wasted plant parts, which are excellent sources of proteins and other necessary bioactive compounds, will typically be disposed of or utilized using conventional practices [13]. Here, valorization can play a major role in the efficient utilization of these resources to produce value-added products and minimize the overall wastage. Figure 1 illustrates the valorization of waste plant resources [14].

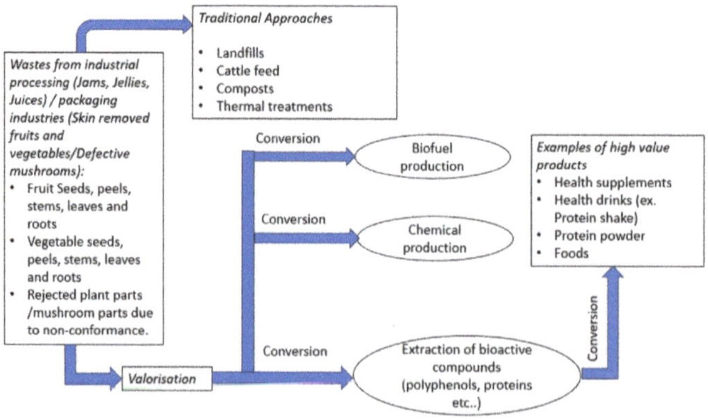

Figure 1. Valorization of waste plant/mushroom resources (reproduced with permission from Elsevier, license no. 5670740416741).

Plant waste valorization is the process of transforming plant wastes into higher-value products by converting them into chemicals or fuels or extracting them to produce nutrient-rich byproducts or supplements. The significant advantages of plant wastes include (i) ease of access, (ii) sustainability, and (iii) biodegradability [15]. Pertaining to the global protein requirement, efficient protein extraction techniques are prioritized to meet the rising demand for high-quality proteins. The food industry has constantly explored pathways to the extraction of plant-based proteins using several conventional processing methods, such as Soxhlet extraction, thermosonication, heat reflux, maceration, bead milling, mechanical grinding, ultrasonic treatment, osmotic treatments, and acid and alkaline pretreatments [16–19], which involve solvent usage, chemicals, heat, and the heavy mechanical disruption of plant cells, resulting in a requirement for additional processing due to purification and the low extraction efficiency. These poor results depend on the size of the testing particle, the usage of solvents, the pH value, the extraction time, and the temperature, which eventually degrade the quality and affect the functional properties of extracted proteins. Emerging non-thermal methods are gaining recognition due to their eco-friendly nature, consumption of less energy, and generation of minimal to no thermal effects on the extracted bio-compounds. These methods do not have an environmental impact and preserve the overall quality and nutrition of the resulting proteins. The application of a Pulsed Electric Field (PEF) is supposed to be a good alternative to thermal treatments in order to overcome the undesirable degradation in the protein quality. This non-thermal technique has been used for several years to inactivate food-borne pathogens in order to extend the shelf life and preserve the quality of foods [20–22] throughout their storage time [23,24]. When compared with other thermal food processing methods, PEF preserves the nutritive value of foods with the minimum degree of damage [25–27]. Initial PEF studies were reviewed by multiple authors [28–36], which aided in the development of PEF into having a global reach. The gradual scaling-up of lab-scale to industrial systems has led to developments in the design of high-voltage pulse generators that provide higher-end performance to suit PEF industrial processing [37,38]. Also, the optimization of electrode shapes and chamber designs have overcome the existing drawbacks of electric field enhancement and dielectric breakdown in foods [39]. Due to the myriad advantages of PEF, recent research has focused on using the technique in protein extraction procedures, where the finally derived protein isolates would be used for producing value-added products including health supplements and nutraceutical and pharmaceutical products. Though PEF enables physiological, functional, and enzymatic changes in proteins, the area still requires further insights into the factors causing these changes in order to promote them for industrial processing.

This review discusses protein extraction from waste plant resources and deformed mushrooms using PEF treatments. This paper covers waste resources, protein extraction methods, the Pulsed Electric Field technique, PEF-assisted extraction of proteins, optimization of the PEF-assisted extraction process, industrial processing, existing challenges, and future directions.

2. Waste Resources

2.1. Plant Wastes

Protein is the most significant macronutrient and is essential to our daily survival needs. Though animal proteins, including meat, fish, milk, and eggs, are rich in essential amino acids, minerals, and vitamins, they use extensive land and feed resources. Also, animal proteins cause adverse health effects that nowadays necessitate low-cost alternative protein sources [40]. Plant protein sources have several advantages, such as a lower cost of production, ease of access, minimal maintenance, and better environmental sustainability. Developing countries rely on plant proteins as they are more economical than animal proteins [9]. Specifically, when the proteins are extracted from waste plant resources, such as seeds, stems, peels, and roots, the extraction of proteins can be made more economical. Vegetable and fruit processing industries generate a huge amount of plant waste (about 90 to 92%) [41]. This leads to a high percentage of wastage at a global level. Plant wastes can be recycled for value-added purposes, thereby reducing environmental pollution and enabling sustainable green development. Some examples of nutrient-rich industrial plant wastes include papaya seeds, sesame press cake, olive pomace, flaxseed hulls, tomato residues, and brewer spent grains. These wastes will be dumped in the garbage, landfilled, or used as cattle feed. Selva Ganesh et al. reviewed the downsides of excessive vegetable and fruit waste and discussed the emerging opportunities to generate value-added products from these wastes [42]. They have also indicated the ever-increasing level of global food wastage and have recommended that we evaluate efficient waste management techniques to valorize and minimize wastage in the long term [42]. In this regard, the accurate monitoring of plant waste is necessary in order to minimize overall wastage. Although plant wastes are produced in all stages of the fruit and vegetable manufacturing chain, recovery and recycling are essential factors to be considered to improve sustainability. The recycling of plant wastes for the recovery of bioactive compounds to produce value-added products minimizes the environmental stress in the global scenario. Table 1 refers to examples of protein-rich plant wastes that can be used for ongoing protein extraction studies.

Table 1. Protein contents of plant wastes.

S.No	Plant Wastes	Protein (%) *	Protein (g/100 g) **	Ref.
1	Sesame Press Cake	35–50	X	[43]
2	Papaya Seeds	27.3–28.3	X	[44]
3	Olive pomace	4.51 (Crude)	X	[45]
4	Flaxseed hulls	17.45–19.14	X	[46]
5	Tomato wastes (i) Pomace (ii) Peels (iii) Seeds	 15.1–22.7 5.7–20 16.6–39.3	 X X X	[47]
6	Brewer's spent grains	26–30	X	[48]
7	Cabbage leaf residues	17.0–17.2 (Crude protein on DM*)	X	[49]
8	Potato peels	13.02 (dry powder)	X	[50]
9	Apple pomace	4	X	[51]
10	Orange peel waste	7.7	X	[52]
11	Rapeseed meal	36	X	[53]
12	Pumpkin peel	1.65	X	[54]
13	Ripe Banana peel	5.5–7.87 (Crude)	X	[55]
14	Pomegranate peel	4.90–8.97	X	[56]
15	Inodorus melon peel	34.90	X	[57]
16	Sweet potato peel (i) Primary peel (ii) Blanched peel	 6.40–6.49 8.11–8.20	 X X	[58]
17	Pumpkin seeds	~35 (Crude)	X	[59]
18	Wheat Bran	13.2–18.4	X	[60]
19	Rice bran	10–16	X	[61]
20	Outer scales of onion	2.64	X	[62]
21	Almond Husk	X	3.27 (Crude)	[63]
22	Brewer dry grain	X	19.96 (Crude)	[63]

* Representation of protein quantity in %. ** Representation of protein quantity in g/100 g. X, not applicable. DM*, Dry Matter.

For a protein source to be considered valuable, it must have substantial protein content with an essential amino acid composition [64]. From Table 1, plant wastes such as sesame cake, papaya seeds, brewer spent grains, rapeseed meal, and inodorus melon peel contain a higher percentage of proteins than other resources that, if extracted, would be beneficial in producing value-added products in an economical manner. Effective valorization of these plant wastes can improve food safety, security, and sustainability in food production.

2.2. Deformed Mushroom Wastes

Mushrooms have the potential to expel large quantities of proteins when extracted. Mushrooms belong to the fungi family, which has a 'plant-like' structure consisting of stems and heads lacking chlorophyll. Edible mushrooms are low-calorie, no-fat, and no-cholesterol foods that are rich in protein, fibers, and antioxidants. The protein quality of mushrooms has been found to be superior to that of plant sources. The nutritional value of edible mushrooms is primarily due to their protein content, consisting of all essential amino acids required to meet each day's dietary requirements, and they are more economical than plant and animal sources [65]. Generally, edible mushrooms have a protein level ranging from 6.60 to 36.87 g/100 g dry weight, with an average of 23.80 g/100 g [66]. The amino acid composition of the proteins in mushrooms is equivalent to that of animal proteins and, hence, they are considered to be more valuable than plant proteins [67,68]. However, different mushroom species will differ in terms of the overall amino acid composition [69]. Based on earlier studies, the protein digestibility of mushrooms ranges from 60% to 70% [70], which can be further improved by eliminating the components that prevent proteins from being hydrolyzed in the gastrointestinal system [70]. Additionally, mushrooms have a significant role in preventing and suppressing disease-causing factors [71].

Mushrooms are consumed widely due to their excellent organoleptic properties and are considered to be novel protein alternatives [65]. Due to their flavor-enhancing and nutritional characteristics, the demand for mushroom consumption is increasing every year. Mushrooms possess antiviral, anti-inflammatory, antifungal, antithrombotic, and antimutagenic functional properties, which are used extensively in foods and medicinal formulations [72–75]. Around 2000 species of mushrooms exist in our environment but only a few species can be used for edible purposes [76]. The most widely grown edible mushrooms are *Agaricus bisporus*, *Lentinula edodes*, and *Pleurotus* species, which are produced primarily in China [77]. The consumption of edible mushrooms on a per-person basis increased by approximately fourfold between 1997 and 2013 [78]. In addition, it is anticipated that the global production of mushrooms will increase to 20.84 million tons by 2026 [79]. On the other hand, the edible mushroom industry's accelerating growth and increasing production has led in parallel to a regrettable level of wastage (60 million tons of mushrooms per year) [80].

In the overall wastage, large quantities of misshapen parts are produced, whose size and shape do not comply with commercial standards. Since these misshapen mushroom parts are still edible, they are used as cattle feed to overcome wastage. It is noted that these misshapen parts are rich sources of carbohydrates, proteins, vitamins, minerals, and healthy fats. Unfortunately, there are few studies on mushroom-extracted bio-valuable compounds [80].

It is a known fact that proteins play a significant role in providing adequate nutrition to the body, and mushrooms are considered to be highly valuable sources of proteins in the human diet. Hence, in order to not neglect these valuable nutritional sources and to overcome food and nutrient wastage, extraction is a wise approach to addressing food sustainability and nutritional quality issues. Due to the extensive demand for alternate protein sources these days, rejected mushroom parts have been found to be an economic resource to boost protein production. Thus, the functional properties of rejected mushroom parts can be improved by extracting valuable proteins from them and processing them to produce food, medicinal, and health supplements. Numerous studies have projected the health benefits of mushrooms due to their medicinal properties. Table 2 lists the commercially cultivated mushroom varieties that have been recommended for protein extraction studies. When these mushrooms are transported to packaging and processing industries, large quantities of rejected waste would be generated, which are not only edible but also have numerous medicinal uses.

Table 2. Properties, health benefits, and protein content of commercially cultivated mushrooms.

Mushroom Species	Protein Content (%)	Nutritional Properties	Medicinal Properties	Reference
Agaricus Bisporus	29.64–39.84	• Rich in carbohydrates, amino acids, fats, and minerals	• Used in the treatment of cancer, bacterial and fungal infections, diabetes, heart disorders, and skin problems	[65,81]
Pleurotus: *Pleurotus ostreatus* *Pleurotus eryngii* *Pleurotus sajor-caju* *Pleurotus giganteus*	 7 11 37.4 17.7	• Rich in proteins and fiber, low in carbohydrates, fats, and sodium	• Used to treat atherosclerosis. Other properties: • Antimicrobial • Antiviral • Anticancer • Antioxidant • Hypolipidemic • Hypocholesterolemic • Antihyperglycemic • Immunomodulatory effects	[82–84]
Morchella esculenta	32.7 (DW)*	• Rich in protein, fiber, and necessary vitamins and minerals. Low in calories and fat	• Antioxidative • Immune protective • Anti-inflammatory • Anti-colon cancer effects	[85,86]
Lentinula edodes	15–23	• Low lipid contents • Rich in protein, vitamins, and minerals. • High concentrations of calcium, iron, phosphorus, potassium, zinc, manganese, phenolic compounds, and glucans	Medicinally used for the following diseases • Depressed immune function (including AIDS) • Cancer • Environmental allergies • Fungal infection • Frequent flu and colds • Bronchial inflammation • Heart disease • Hyperlipidemia (including high blood cholesterol) • Hypertension • Infectious diseases • Diabetes • Hepatitis • Regulating urinary inconsistencies.	[87–89]
Auricularia heimuer	10.62	• Plentiful in several amino acids, especially leucine and lysine. • High in protein, trace element, vitamin, and carbohydrate content • Low in fat	• Anticancer • Detoxifying Anticoagulant • Hypoglycemic • Cholesterol-lowering capacity	[90]

Table 2. *Cont.*

Mushroom Species	Protein Content (%)	Nutritional Properties	Medicinal Properties	Reference
Volvariella bombycina Fruit Body Mycelia	 28.30 25.50	• Rich in proteins, carbohydrates, vitamins, and minerals	• Resistance to ovarian, lung, breast, and prostate cancer	[91]
Enoki mushrooms: *Flammulina filiformis*, *Flammulina velutipes*	5	• Rich in carbohydrates, proteins, unsaturated fatty acids, many valuable micronutrients, and dietary fiber • Highly comparable to vegetables	• Anti-inflammatory activities	[92,93]
Tremella fuciformis	9.63	• Contains large quantities of vitamin D, proteins, vitamins, minerals, immune-boosting polysaccharides, trace minerals, and carbohydrates • Low in fat	• Used medicinally for cough, sore throat, stethalgia, constipation, and paramenia	[94–96]
Hypsizygus tessellatus,	33.9	• Rich in proteins, fiber, carbohydrates, potassium, calcium and phosphorous	• Has organoleptic and pharmacological properties, • Anticancer properties	[97,98]
Stropharia rugosoannulata,	30–50	• Rich in proteins, peptides, and amino acids	Essential Properties: • Antidiabetic • Antimicrobial • Antioxidant • Antiproliferative • Antitumor • Immunomodulatory and osteoclast-formation-suppressing properties	[99,100]
Cyclocybe aegerita	37.6	• Rich in carbohydrates, ash, and proteins	Essential Properties: • Antitumor, antioxidant, and antifungal • Hypocholesterolemic and hypolipidemic effects	[101–103]

Table 2. *Cont.*

Mushroom Species	Protein Content (%)	Nutritional Properties	Medicinal Properties	Reference
Hericium erinaceus	Dried powder, 20; Mycelia, 42	• Enriched protein and dietary fiber content and low in calories and fat	Essential Properties: • Antibiotic • Anticarcinogenic • Antidiabetic • Antifatigue • Antihypertensive • Anti-hyperlipodemic • Antisenescence • Cardioprotective • Hepatoprotective • Nephroprotective • Neuroprotective properties • Improvements in anxiety, cognitive function, and depression.	[104,105]
Phallus Indusiatus	4.813	• Contains very high water, crude fiber, protein, fat, carbohydrate, ash, and vitamin C content.	• Improves cardiovascular health • Reduces the risk of cancer • Promotes immune function • Prevents viruses, bacteria, and fungi • Reduces inflammation • Combats allergies • Helps to balance blood sugar levels • Supports the body's detoxification mechanism	[106]

DW*, Dry Weight.

Mushrooms are a great source of nutritionally valuable compounds, including proteins, lipids, polysaccharides, polyphenols, micronutrients, and vitamins. Rosello et al. describe the potential use of some novel non-conventional methods, including enzyme-assisted extraction, pulsed electric field, ultrasound, microwaves, and subcritical and supercritical fluid extraction, for the recovery of valuable compounds from mushrooms [107]. PEF technology is one such essential tool that can improve the functionality, extractability, and recovery of nutritionally valuable compounds as well as the bioavailability of micronutrients and components in a diverse variety of foods. To exploit mushroom extracts for various applications in food, pharmaceutical, and nutraceutical industries, they need to be characterized for their bioactive compounds through several extraction processes. The residues from the protein extraction can be further valorized into valuable products based on their potential, such as prebiotics, dietary fiber, and compost. Several research studies on mushrooms have provided a sustainable solution in the valorization of mushroom byproducts to produce valuable functional ingredients with health benefits, especially for the gastrointestinal system [108,109].

3. Protein Extraction Methods

An extraction process is described as the recovery of valuable bioactive compounds from plant cells by means of a mass transport phenomenon, where the intracellular components of the cell are transmitted to the solvent [110–112]. The stages of an extraction process typically include penetration of the solvent into the plant matrix, dissolution of solutes in the solvent, diffusion of the solute out of the matrix, and collection of extracted solutes [113]. Since these extracted solutes are bioactive compounds for which a high degree of purity is preferred, it is necessary to find efficient extraction techniques that are quick, easy, safe, and reliable, provide better yields, and, most importantly, require a minimal number of purification steps. Existing conventional protein extraction methods, including maceration, percolation, heat reflux extraction, and Soxhlet extraction, are widely adopted due to their simple operational techniques. However, these methods use large volumes of solvents and heat, have a long processing time, and are inefficient [114]. Under these methods, increasing the temperature increases the solubility and diffusion, leading to increased extraction. However, the huge rise in the temperature leads to the extraction of other undesirable impurities, which require additional processing for purification. While we discuss a few examples, alkaline extraction is an inexpensive chemical extraction method in which protein dissolution occurs in an alkaline medium, with the disadvantage of protein denaturation under highly alkaline conditions during the extraction procedure [115]. Also, the increase in temperature and extraction time induces thermal precipitation and denaturation effects that may degrade the quality of proteins [115]. Enzyme-assisted extraction is a biochemical method that relies on the use of enzymatic breaking of the rigid cell wall to boost the protein yield. Though the enzyme usage maintains the optimal pH in order to minimize protein denaturation, the process is an expensive one for industrial applications and the enzymes cannot achieve complete cell breakdown [116].

Since the primary goal is to achieve an enhanced yield in a shorter extraction time without affecting the protein quality to a high degree, the selection of the extraction technique is essential. As discussed earlier, conventional extraction methods have a long extraction time, have a high solvent requirement, cause increased damage to cells, and have low extraction efficiency. Modern extraction methods, including ultrasound-assisted extraction, high-voltage electrical-discharge-assisted extraction, high hydrostatic pressure extraction, and pulsed electric field-assisted extraction, are crucial in preserving a sustainable food chain by the recovery of essential bioactive compounds from plant wastes. These ecofriendly methods have several advantages, such as a minimal temperature increase, less solvent usage, a short extraction time, a quicker process, less damage to the plant matrix, and a higher quality of extracted products [117–120]. This can be made possible by optimizing the extraction parameters throughout the experimental approach. These physical methods of cell degradation provide enhanced protein yields with minimal or no usage of solvents and heat with no environmental impacts. Although research on these techniques is promising, large-scale applications would require a long period of time [121]. In the following section, we discuss the Pulsed Electric Field technique, which works under the phenomenon of electroporation. This method has shown favorable extraction qualities with minimal damage to the plant cell matrix.

4. PEF Technique

4.1. General Features

PEF is a non-thermal technique that uses ultrashort electric field pulses in various applications such as food processing, tissue ablation, electrochemotherapy, and electro-extraction processes [20,122–124]. In food applications, PEF has been used in microbial inactivation studies, the extraction of bio-compounds, assisting with the drying process, and oil extraction [125–128]. While applications in microbial inactivation were initially invented and earned a greater amount of potential in research, recent research has focused on using the method for the extraction of bioactive compounds from the plant cell matrix [129,130]. When compared with thermal methods, the qualities of plant-protein-based foods were minimally affected after the PEF treatment and were maintained throughout the storage period [24]. The PEF method involves subjecting the plant cells to intense high-voltage pulses at an electric field strength range between 0.5 and 50 kV/cm with a pulse duration of milliseconds, microseconds, or nanoseconds in samples positioned between two electrodes to inactivate microorganisms in foods and to improve the rate of mass transfer for extracting intracellular bioactive compounds [131,132]. Basically, the cell membrane is made up of a lipid bilayer with embedded protein channels that play a critical role in separating the intracellular organelles of the cell from the surrounding environment and modulate the transfer of ions in and out of the cell to maintain the cell balance. The application of high-voltage electric field pulses destabilizes the lipid bilayer by increasing the cell wall permeability, providing adequate contact between the solvent (plain water in this case) and the intracellular compounds. Hence, while the plant cell matrix receives the electric field pulses, the permeability of the cell wall would increase, resulting in the induction of a transmembrane voltage across the cell membrane, the intensity of which depends on the intensity of the electric field, the pulse profile, the nature of cells, and the cell orientation with reference to the direction of the electric field vector [133]. In general, increasing the transmembrane potential increases the electromechanical compression of the cell membrane, leading to electropermeabilization or electroporation [134–136]. As a consequence of electroporation, plant tissues will be softened, enabling them to be well equipped for the subsequent extraction of bio-compounds in a shorter processing time. Thus, electroporation has a wide range of potential applications, from changing the structure and function of plant cells to causing permanent cell death. Figure 2 illustrates the potential uses of PEF in food applications.

Beneficial features of the PEF method rely on not using harsh chemicals, heavy solvents, acids, alkalis, or high-frequency mechanical effects on cells. It merely focuses on optimizing ultrashort electric field pulses to enhance the electroporation effect on plant cells through mass transport phenomena for efficient protein extraction [112,126]. When compared with thermal methods, it is a promising method for preserving the quality of proteins throughout the storage period [24]. Additionally, using plain water as a solvent minimizes environmental impacts. The interest in using PEF-assisted extraction has increased significantly since it has developed into an environmentally friendly and green extraction technique [137,138] that can contribute to sustainable development in future food systems. During the PEF bio-extraction procedure, electroporation selectively obtains the valuable bio-compounds from the plant cells by optimizing the intensity of the electric field strength and pulse parameters [139]. Thus, PEF-assisted extraction is advantageous in providing high protein yields when compared with other conventional techniques [130,140]. Moreover, combining other extraction processes with PEF has been reported to increase the protein yield [141,142] when compared with PEF standalone treatments. In addition to the enhanced yield, this method also preserves the quality attributes and nutrition of extracted bio-compounds [143,144]. For instance, PEF enhanced the functional and emulsifying properties of rice bran protein by about 20.29–22.64% and 3.3–12.0%, respectively. Additionally, the foaming capacity and foam stability were increased by 1.8 to 2.9-fold, respectively. Moreover, the in vitro digestibility of proteins was also enhanced [145].

Optimizing process parameters plays a significant role in changing the structure and function of proteins. Some of the structural changes, such as protein unfolding, dissociation, protein denaturation, and reaggregation, highly depend on optimizing the intensity of electroporation, which in turn depends on the electric field strength. When PEF effects on proteins are analyzed, the PEF-induced electric field stress causes the electrostatic interactions to be disrupted inside the polypeptide chain of the protein, leading to the deformation of secondary structures [146,147]. Multi-protein food systems have more complex interactions in an electric field, which can impact the stability of the systems [148]. For instance, with an increase in the electric field strength and number of pulses, the rheological properties of soy milk, such as the viscosity, increase with the change in consistency [149]. Enhancing the surface hydrophobicity of the protein structure enhances the flavor of foods. In this regard, increasing the electric field strength and number of pulses resulted in protein unfolding and the destruction of hydrophobic interactions, leading to structural modifications that increased the surface hydrophobicity in the case of whey protein isolate [150].

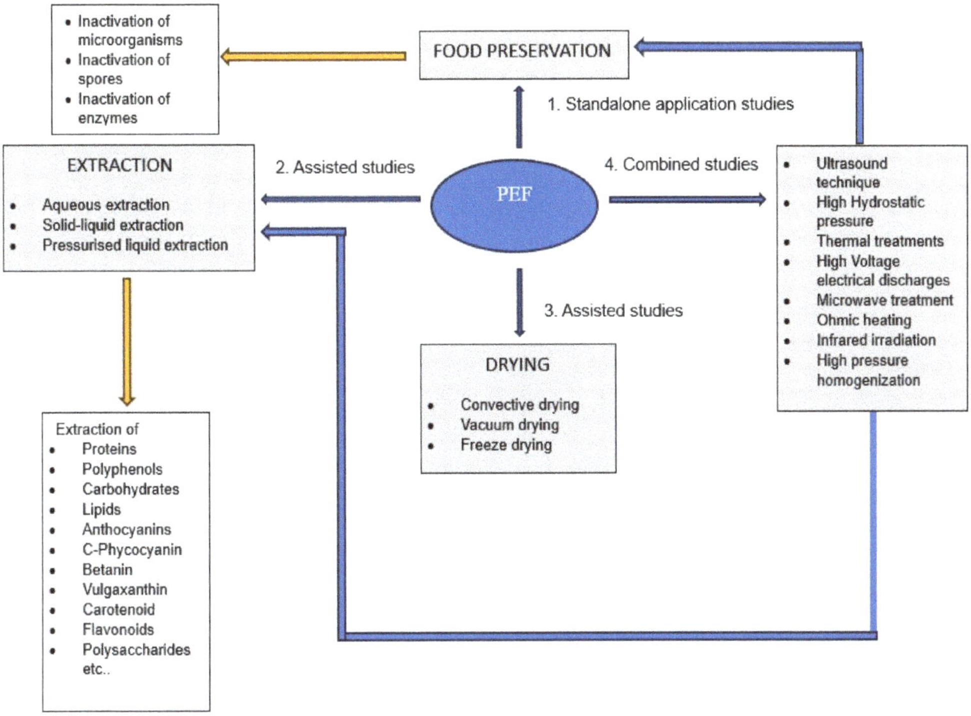

Figure 2. Potential uses of PEF in food applications.

4.2. PEF Setup

The working principle of PEF relies on the application of high-voltage pulses to samples placed between two electrodes, where one electrode will be charged and the other will be at the ground potential. A PEF prototype comprises a high-voltage pulse generator unit, a treatment chamber with two electrodes, a peristaltic pump for circulating samples, an oscilloscope for displaying pulses, and a control unit [22]. The untreated sample will be transferred to the PEF treatment chamber before it is subjected to the required high-voltage pulse. Samples will be tested in batches in the case of static treatment chambers or will be pumped into the chamber for continuous testing purposes. The pulse-forming network (not shown) plays a significant role in delivering the required pulse waveshape to the testing medium placed inside the PEF chamber. The pulses can be delivered in the form of square, exponential, or oscillatory waveshapes. The treatment chamber and the electrodes must be carefully designed since they govern the electric field distribution inside the chamber. Figure 3 represents a simplified sketch of a Pulsed Electric Field setup.

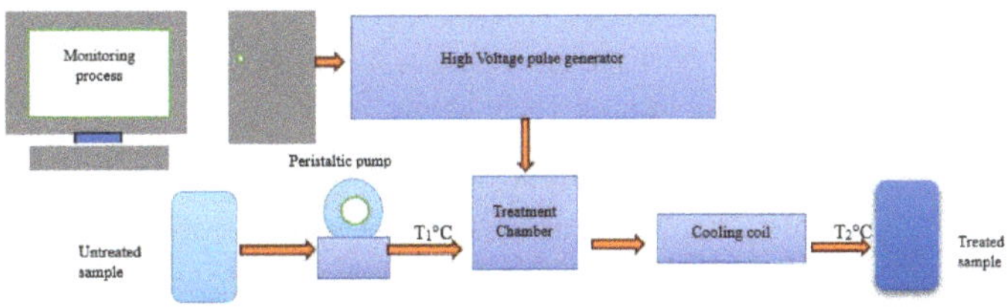

Figure 3. Pulsed Electric Field setup.

There are three basic electrode configurations, including parallel plate, coaxial, and collinear electrodes [151], the shapes of which can be optimized further to suit either batch or continuous modes of operation [152–155]. Among the designs, the parallel plate configuration was the earliest-invented design of high importance, with a simple construction that delivers a uniform electric field inside the treatment gap [156]. While applying the electric field pulses to the testing medium, the electrodes act as a mediating factor between the high-voltage pulse generator and the treatment gap. Though the treatment chamber and electrode designs play the primary role in providing the optimum electric field distribution, process parameters such as electric field intensity, pulse waveshape, pulse duration, and pulse frequency should be adjusted to produce beneficial impacts on the treated samples [157,158]. The impact of the treatment also depends on the properties of the testing medium, such as the electrical conductivity, pH, concentration, and permittivity [111,159]. Hence, to obtain the best treatment outcomes, the researcher has to optimize the combination of all the aforementioned parameters.

High-voltage pulse generators deliver the desired electric pulses repetitively to treatment chambers through a pulse-forming network (PFN) and high-voltage switches. These generators provide the required waveshape, polarity, and amplitude of the pulses based on the construction of the PFN. While square and exponential pulses were frequently used in earlier studies [160,161], the square waveshape has been shown to be more efficient in treating the samples due to the longer retainment of the peak voltage during the treatment [161]. Exposing the cells to oscillatory decay pulses were found to be less effective due to the shorter cell exposure time. To minimize undesirable electrochemical reactions and associated thermal effects, bipolar pulses are generally preferred over monopolar pulses [160,162,163] due to the alternate stress effects on the cell membrane.

During the PEF method's application, the sample will be treated either in batch or in continuous mode inside the treatment chamber [164]. Insulating spacers will be positioned between two electrode edges to facilitate the application of high-voltage pulses only in the treatment gap [112]. In earlier studies, batch testing modes were preferred for small-scale research and continuous modes were used for industrial testing purposes. Nowadays, lab-scale studies use both modes of operation. Batch operation involves testing the samples in batches using a standard treatment chamber. In this mode, there will be no continuous flow of the testing medium and the samples will be subjected to the static application of an electric field in the treatment gap for a specific treatment time. This mode of operation is slow to process due to the manpower involved in placing the testing medium inside the treatment chamber, transferring the medium, and collecting it after the treatment. Using a batch treatment chamber, the extraction of essential bio-compounds, including proteins, has been observed in several plant cells and waste plant tissues [4,165]. The sample to be treated will be placed either in small cuvettes or in a customized chamber, depending on the testing volume requirements. The product will be mixed with mild solvents or tap water in a prescribed ratio before it is subjected to the required high-voltage pulses [143].

During continuous operation, the peristaltic pump circulates the untreated samples and delivers them to the treatment chamber for the application of high-voltage pulses. When compared with batch systems, continuous systems are attractive due to their increased capacity to treat foods continuously in a shorter treatment time. Due to their high efficiency [166], continuous systems are preferred for large-scale industrial testing purposes not only in PEF-assisted extraction but also in other areas such as freezing, drying, cutting, and cold pasteurization [111,167]. In continuous mode, the electrodes and the chamber design will be constructed to facilitate the continuous flow of the testing medium inside the chamber at a constant flow rate per second/minute/hour depending on the volume of the testing sample. Tamborrino et al. used a collinear tubular electrode configuration for a continuous PEF-assisted olive oil extraction process, which resulted in an increase in the process efficiency and quantity [168]. One minor drawback of continuous chambers lies in the difficulty of having the product flow uniformly experience the electric field and subsequent electroporation effects. This has been observed in both microbial inactivation and bio-compound extraction studies [168,169]. The parallel plate and coaxial electrodes were generally used for both batch and continuous treatment processes [154,155,164,170–172] whereas collinear designs were allocated for the continuous treatment process in the majority of studies [168,173]. Despite fringing effects [174], the parallel plate design provides a more homogeneous electric field in the treatment area, whereas the coaxial shape ensures a uniform flow of the sample that is suitable for industrial-scale applications. Though all three electrode shapes are used for continuous PEF operations [175], the collinear design facilitates the easy flow of product through the chamber. The efficiency of both the batch and continuous testing modes is dependent on several factors, such as the chamber design, electrode design, electric field strength, number of pulses, pulse width, pulse waveshape, and properties of the testing medium such as the electrical conductivity, pH, permittivity, and concentration of the sample.

4.3. Mechanism of Extraction

PEF assists in the efficient extraction of intracellular proteins by means of a mass transport phenomenon through electroporation. Basically, the cell plasma membrane is considered to be a barrier, which protects the inside of the cell from dynamic environmental conditions. It is a semipermeable membrane that regulates the passage of nutrients and the ionic flow in and out of the cell, thereby creating an osmotic balance between the cell and its surroundings. This membrane acts as an insulator made up of lipid molecules consisting of polar (i.e., hydrophilic) and non-polar (i.e., hydrophobic) sections that together form a bilayer structure. When this membrane is subjected to high electric stress, the permeability and conductance of these membranes increase, leading to cell disruption that results in electropermeabilization or electroporation [176]. Electroporation is currently gaining recognition due to its cost-effective, time-saving, and chemical-free application in several areas, such as food processing, electrochemotherapy, tissue ablation, and bio-compound extraction protocols. Fundamentally, it is the process of inducing pore formation in the biological cell membrane, the radius of which depends on the intensity of the electric field strength [177]. The capacity of electroporation to cause the mass transport phenomenon in the plant matrix depends on the electric field strength, pulse number, pulse frequency, and total treatment duration [178]. The rationale behind producing electrical pores is to extract valuable bio-compounds from the cell wall matrix. Due to the increase in the electric field, transmembrane potential (tmp) will be induced across the cell membrane that enables the membrane to experience electromechanical stress leading to the creation of pores. The pores will continue to increase depending on the supercritical electric field Ec [177]. Hence, the electroporation mechanism involves a significant number of stages, including application of ultrashort electric field pulses, lipid bilayer charging, local structural rearrangements within the cell, perforation of the cell membrane, and increased ionic and molecular transport through the pores [134]. The structural rearrangements indicate the permeation of the lipid bilayer leading to the formation of hydrophilic pores for an increase in the transmembrane potential value [134]. After being exposed to an electric field, cells cannot regain their original semipermeable properties completely [178].

When the electric field E is increased such that E > Ec, structural changes, cell rupture, and an increase in the cell membrane permeability will take place, leading to pore formation. Based on the increase in the electric field E, a larger number of pores will be formed, resulting in the leakage of intracellular biomolecules, including proteins, lipids, and carbohydrates. After the pulse's application and the subsequent release of intracellular components, the resealing of pores will occur, and the cell will return back to its original state in the case of reversible electroporation (RE). The life span of pores ranges from milliseconds to minutes after the pulse's application [176]. When the increase in the electric field E is very high such that E >> Ec, which progresses beyond the radius of the pores, permanent cell damage will occur, leading to irreversible electroporation (IRE) [179,180]. Therefore, the electroporation parameters should be optimized based on the treatment requirements and cell types. Figure 4 illustrates the electroporation stages in a plant cell.

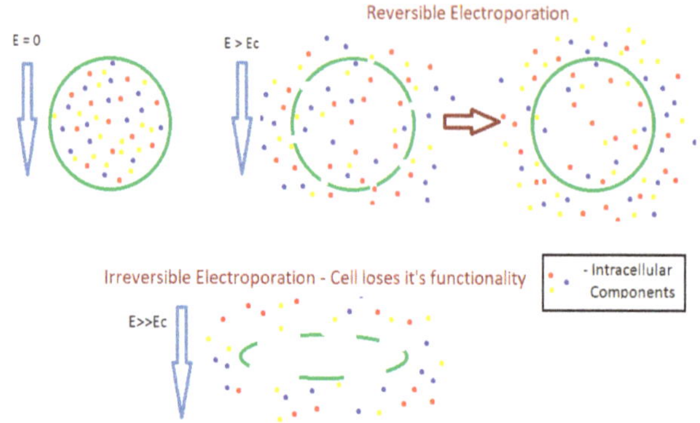

Figure 4. Electroporation stages in a plant cell.

5. PEF-Assisted Extraction of Proteins

In PEF-assisted extraction, the intensity of the electric field can significantly influence electroporation to soften the plant tissues and facilitate a better protein yield from the cell matrix. Hence, PEF-induced permeability changes in the cell membrane depend on the electric field intensity, extraction time, pH, specific energy, temperature, and properties of the cell matrix, especially the conductivity [181]. These electroporation effects can be observed on the cell surface using scanning electron microscopy (SEM), which can help the researcher to make analogous studies with other treatment effects on the cells [182]. Figure 5 illustrates the PEF-assisted protein extraction process.

Over the past few years, studies on PEF-assisted protein extraction have escalated exponentially due to the potential of this treatment to provide better extraction yields with quality retainment. This technique to extract proteins through the mass transport mechanism works by the diffusion of the solvent into the intracellular medium and the dissolution of the solute out of the cell matrix. Due to its beneficial advantages, several studies have incorporated PEF into protein extraction research [4,114,119,129,159,183–185]. Moreover, thermolabile compounds are well preserved using PEF when compared with existing thermal methods [18,112]. Table 3 presents several remarkable studies on PEF-assisted protein extraction from waste plant parts and mushrooms. These studies represent the protein yield through process parameter optimization.

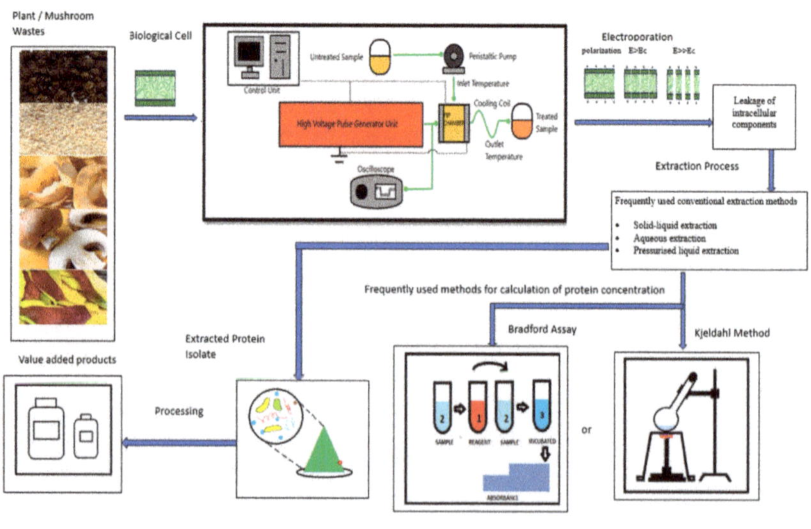

Figure 5. PEF-assisted protein extraction process.

Table 3. PEF-assisted protein extraction studies from waste plant parts and mushrooms.

Protein Source	Process Parameters	Outcome	Essential Findings	Reference
Agaricus Bisporus	38.4 kV/cm, Pulse width of 2 µS at both 400 Hz and 800 Hz frequencies Inlet Temperature: 20 °C	Protein extraction of 40.8%	* Significant increase in protein yield when combined with mild heating * Protein concentration increased with an increase in the electric field, temperature, and treatment time	[119]
Agaricus Bisporus	2.5 kV/cm, Specific energy of 50 kJ/kg, and 6 h of extraction time	Maximum protein recovery of 140.72 +/− 15.14 mg BSA/g DM*	* PEF increased the recovery of proteins	[143]
Papaya peels	13.3 kV/cm, 400 pulses, 40 kV applied voltage, treatment time of 2720 secs.	Protein concentration of 20 mg/L	* Two stages of PEF + supplementary aqueous extraction resulted in good recovery of bioactive compounds including proteins	[165]
Rapeseed stems and leaves	20 kV/cm 200 pulses, 40 kV applied voltage, pulse duration of 10 µs	Protein extraction of up to 80%	* PEF treatment increased the yield of proteins. * Protein contents changed during plant development	[4]
Rapeseed green biomass (stems)	8 kV/cm, Pulse duration of 10 µS, pressure of 10 bar, maximum applied voltage of 40 kV, frequency of 0.5 Hz, 200 pulses	0.14 g BSA/100 g DM*	* Protein content increased by about 2 times compared with the untreated sample and 50 pulse applications	[186]
Sesame cake	13.3 kV/cm Pulse duration of 10 µs Frequency of 0.5 Hz (2 s between pulses or discharges) 0–700 pulses Temperature of 40 °C Variable treatment time between 1 and 7 ms Energy inputs between 0 and 291 kJ/kg	Between 374 and 2001 mg BSA/100 g DM*.	* Protein contents increased with energy inputs until 83 kJ/kg was reached. * PEF improved the protein extraction yield.	[183]
Tomato wastes	5 kV/cm Treatment time of 1.5 ms 200 pulses	145.1 mg/100 g	* Protein concentration increased with increasing the electric field strength and time	[158]
Rice bran	2.3 kV/cm Treatment time of 25 min 250 pulses per minute	20.71–22.8%	* Improved the extraction yield of rice bran protein * Improved oil's holding and emulsifying properties * Increased foaming ability and stability * Enhanced in vitro digestibility	[145]

Table 3. *Cont.*

Protein Source	Process Parameters	Outcome	Essential Findings	Reference
Brewers spent grains (BSG) Light Dark	2.8 kV/cm 3000 pulses 20 µs pulse-width	20.31 ± 0.01 (%DWe*) 23.78 ± 0.16 (%DWe*)	* Increased yield of proteins from dark BSG compared with light BSG * Light BSG exhibited higher antimicrobial activity than untreated ones	[187]
Nettle leaves	3 kV/cm Specific energy of 10–24 kJ/kg Temperature between 70 and 78°C Extraction time of 5 min	>60%	PEF specific energy input, extraction temperature, and particle size had high impacts on solid liquid extraction kinetics	[188]

DM*, Dry Matter; DWe*, Dry weight extract.

The integration of PEF into the protein extraction process in mushrooms and waste plant parts is evident from earlier studies [4,119,143,165]. In addition to protein extraction, PEF–US combination techniques increased the drying rates, decreased the drying time, provided better mass transfer and rehydration capacities, changed the structure of the cell tissue, and showed high antioxidant activity in shiitake mushrooms [189]. In 2014, Boussetta et al. studied the impact of pulsed electric field-assisted extraction of polyphenols from waste flaxseed hulls. Their findings highlighted a research gap in protein extraction from flaxseed hulls [190].

PEF has also been proven to be effective in bio-compound extraction from seeds. Sesame seeds are extensively used in cooking to prepare tasty dishes and act as a flavoring agent in many recipes. These seeds are highly nutritious, are a good source of vitamins, healthy fats, protein, minerals, fiber, and antioxidants, and contain other beneficial characteristics for health. These seeds contain about 20% protein [191]. After oil extraction, the remaining byproducts of sesame seeds are still high in nutritional value. Sarkis et al. carried out extraction studies on sesame press cake for the protein and polyphenol content. A significant increase in protein extraction was observed at an optimum temperature of 40 °C, an electric field strength of 13.3 kV/cm, and a pulse duration of 10 µS [183]. The PEF method's application increased the electrical conductivity and cell permeability of the residual seed cake, causing minimal cell damage. This enabled an easier extraction in a shorter time [183].

Papaya seeds are rich in isothiocyanates, phenolic compounds, and proteins. Parniakov et al. observed an increased degree of disintegration in papaya seeds at 13.3 kV/cm and 2000 pulses when coupled with supplementary aqueous extraction (SAE). The authors recovered a significant amount of protein after three hours of PEF-assisted extraction. Upon the PEF method's application, the degree of extraction of ionic components facilitated the release of embedded proteins from the cell wall matrix during SAE [159]. Therefore, PEF and supplementary aqueous extraction together produced a higher protein yield, opening the door for industrial processing [159]. Yu et al. studied the PEF-assisted extraction of proteins from rapeseed stems and leaves, which primarily depended on the electric field strength and growth phase of the plant. The specific parameters that caused tissue damage were observed at 800 V/cm and a protein extraction yield of up to 80% was achieved at 20 kV/cm [4]. Kamboj et al. observed an increase in the protein yield of up to 10.58% on perilla seed meal at processing parameters of a 4 kV/cm electric field and a 120 µS pulse width [192]. Though PEF-assisted protein extraction studies have been performed on mushrooms and plant wastes, these edible products are currently in the amateur stage of study and require further exploration to improve the productivity economically.

6. Optimization of the PEF-Assisted Extraction Process

The optimization of PEF-assisted extraction requires the execution of systematic procedures to enhance the recovery of proteins from the cell matrices. The primary factors that influence the PEF performance in the extraction process include the treatment chamber design and the electrode shape, which govern the electric field distribution inside the chamber. These factors must be carefully designed in order to provide the optimal treatment to the testing samples. Other process parameters that need to be adjusted include the electric field strength, number of pulses, pulse width, pulse waveshape, specific energy input, and temperature of the untreated sample, as these factors optimize the electroporation effects on the cells to be treated [193], facilitating the mass transport phenomenon to assist with extraction. If these process parameters are increased, the ability of electroporation to puncture the plant cells would increase, enhancing the overall extraction process. Also, a decrease in impedance would be noted on the cell tissues upon increasing the electric field strength [194]. However, increasing the process parameters by many folds beyond the extractable threshold of the cell matrix and a prolonged extraction time would cause a significant reduction in protein extraction levels. On the other hand, the electroporation effects also depend on the characteristics of the cell, such as the cell wall rigidity, cell arrangement, structure, water content, and usage of solvents. Hence, the process parameter optimization may be variable depending on each cell matrix.

Electric field strength is one of the most significant parameters to be optimized in order to obtain an enhanced protein extraction rate. While PEF technology relies on the application of an electric field in a lower range from 0.1 to 0.3 kV/cm and in a higher range from 20 kV/cm to 80 kV/cm [131,195], reducing the pulse duration with a higher electric field would aid in providing a higher extraction yield in a short processing time [126]. In dairy and plant proteins, the intensity of the electric field can modify the structure and techno-functional characteristics of proteins. In addition, a higher electric field strength can yield better emulsifying properties of proteins and protein–polysaccharide complexes [196]. Xue and Farid analyzed the impact of PEF-assisted protein extraction on white button mushrooms (*Agaricus Bisporus*) and obtained 49% $\pm$ 0.09 protein at a 38.4 kV/cm field intensity for 272 μs of treatment time and a treatment temperature of 85 °C. In addition, reducing the pulse width to 136 μs while keeping constant the electric field value retained the pink-to-reddish color similar to fresh extracts [119]. Although a high electric field intensifies the electroporation effect, care must be taken not to increase the electric field exponentially as it may synergistically increase the product temperature, resulting in a loss of nutrients in the extracted outcome. So far, more research findings have been reported on the fact that the electric field strength induces higher protein yields in microalgal cells [124,188,197] than in plant resources and mushrooms. The aim of this article is to open up new avenues for more PEF-assisted protein extraction research on waste plant tissues and mushrooms.

The PEF treatment time can be calculated by multiplying the pulse duration by the number of pulses. Modulating either the pulse duration or the pulse number would modify the treatment time. The extraction time and temperature are the essential parameters to be optimized during the extraction process. An exponential increase in either of these parameters would degrade the quality of the testing medium. For instance, Boussetta and Segovia reported an increase in the yield of valuable compounds in grape seeds and borage leaves at a lower temperature and shorter extraction time than the control samples [137,198]. Hence, these two factors must be regulated appropriately to preserve nutritional and sensory aspects in the samples.

The efficiency of PEF technology also depends on the conductivity of the sample [20]. In microbial inactivation studies, high-conductivity foods created a peak electric field across the treatment chamber, which intervened into the electric field's application during the treatment, resulting in a low-efficiency outcome. Hence, it is always preferred to choose low-conductivity foods to obtain optimum results in electroporating microorganisms. In plant-based bio-extraction studies, samples with high conductivity increased the ionic strength of the medium, which hastened the cell membrane charging, resulting in enhanced protein extraction levels. Hence, according to Buchmann, a highly conductive extracellular medium would enable the diffusion of proteins through the permeable membrane, resulting in an enhanced protein yield [124].

The application of ultrashort pulses (microseconds or nanoseconds) is considered to be one of the major advantages of this treatment that renders it non-thermal and consumes less energy when compared with thermal methods. This feature would assist in lowering the processing cost, though the initial cost of the equipment is high [199]. Hence, a suitable combination of optimized pulse parameters would provide effective extraction results. However, the process parameters must be selected based on the plant tissues in order to achieve high protein yields. Generally, the application of high-voltage pulses may not have an impact on intracellular organelles due to the presence of the rigid cell wall. However, if the pulse duration is made to be ultrashort, i.e., nanoseconds, sub-nanoseconds, or picoseconds, the electric shots will directly charge intracellular organelles in the cytoplasm without charging the outer cell membrane [200]. Hence, the cell wall characteristics play a significant role in the extractability of proteins [201]. However, when a long pulse duration is applied, it may lead to an undesirable increase in the temperature, resulting in adverse electrode reactions and thereby degrading the quality of the testing medium [202]. Hence, the application of a large number of ultrashort pulses would help in minimizing the temperature increase and associated detrimental effects [203].

The use of solvent at a suitable concentration has been highly beneficial in achieving the mass transport phenomenon in extraction studies [204]. By using PEF, the amount of energy and organic solvent needed to extract the desired compounds would be minimized [137,201]. Water is recognized as a green solvent liquid. However, the low solubility of valuable compounds in water makes it less preferable [205,206]. While the majority of extraction methods use ethanol, methanol, aqueous NaOH, KOH, $NaHCO_3$, or NaCl as the solvent in a high concentration, PEF-assisted extraction has used different solvent types, such as water alone and a mixture of low-concentration organic solvents with water, for research. For instance, Min Wang [207] reported a higher protein yield from *Chlorella* using water alone as a solvent. This study gave the best extraction efficiency at an extraction time of 180 min at 3 kV/cm, 44 pulses, and 99 kJ/kg energy. Barba et al. studied PEF-assisted extraction of valuable bio-compounds from stevia leaves using water alone as a solvent, which facilitated an increase in the extraction of soluble matter (~33%), an increase in the activity of phenolic compounds (80%), chlorogenic acid (93%), caffeic acids (55%), ferulic acid (90%), and protocatechuic acids (45%), and increased the conductivity by 25% [111,208]. Parniakov et al. [209] compared one-stage and two-stage extraction procedures using the microalgae *Nannochloropsis* spp. In two-stage procedures, the combination of less-concentrated organic solvents and water obtained high non-degraded protein levels at 20 kV/cm for 400 pulses with a one-minute pause after each application of 200 pulses for better temperature control to 20 °C. The one-stage procedure resulted in the undesirable degradation of proteins due to the high concentration of solvents used. Serena Carpentieri et al. studied the PEF-assisted extractability of aroma and bioactive compounds from rehydrated plant materials. Here, green solvents such as a water–ethanol mixture and propylene glycol were used in solid–liquid extraction after the PEF application. PEF improved the solvent's penetration into the plant cell's cytoplasm and the diffusion rate of solubilized compounds during solid–liquid extraction, thereby enhancing the overall extractability rate [210].

7. Industrial Processing

In the 1960s, several possible applications were identified using electroporation in food and bioprocessing areas [211]. A PEF commercial installation for fruit juice processing was first achieved in 2006 in the USA followed by a line of installations in Europe in 2009, which was extended to applications for the pretreatment of vegetables in 2010 [211]. A PEF system has recently been installed in several countries, including Italy (Amica Chips), Canada

(Potatopro), and New Zealand (McCain Foods), from machine manufacturers including Pulsemaster from The Netherlands and Elea Technology GmbH from Germany, allowing these countries to manufacture their own food products. For industrial applications, sustainable production is essential to feasible applicability in PEF-assisted protein extraction research, which requires further insights to become a commercially mature technique. Substantial factors to be considered for this analysis include reliability, economic evaluation, environmental impact, social impact, and relevant contribution to society. Therefore, it is essential to analyze all these factors for the successful integration of PEF into the protein extraction process in order to evaluate its efficiency. Based on earlier studies, PEF has been proven to be highly energy efficient in terms of applying ultrashort pulses when compared with other thermal processing methods. Due to the low energy consumption, the temperature of the PEF-treated sample would be lower when compared with conventional thermal techniques [212]. Hence, the overall processing cost would be lower, even though the initial cost of the machine is high [199].

Currently, there is high demand for green extraction processes that replace the non-ecofriendly conventional chemical and heavy solvent usage techniques. The objective of green techniques is to minimize the interference with solvent usage and enhance the qualitative and quantitative characteristics of the target extractable compounds. The PEF technology started with testing small quantities in batches using simple treatment chamber designs and electrodes at laboratory scales and has grown into a sustainable and commercially established technique in microbial inactivation studies. Later, many other divisions of application emerged, including solid and liquid food processing, spore inactivation, enzyme inactivation, dehydration, freezing, and medical science areas [213]. Moreover, the increase in demand led to an increase in the flexibility of PEF systems appropriate for each testing sample [112,214,215]. Over the past few years, PEF has been found to be effective in bio-compound extraction studies, but they are still in a preliminary stage [126]. Hence, further research is required in extraction studies to streamline the energy efficiency and safety processes suitable for industrial applications.

Due to the growing consumer demand for nutritional, safe, and high-quality food products, PEF technology has upscaled systems for industrial production. Section 4 discusses the PEF layout in detail. There are many significant parameters involved in the successful operation of a PEF system at the commercial scale. An effective treatment depends on optimizing these process parameters based on the volume of the sample treated at a larger scale. Arshad et al. have discussed the interdependency of these process parameters and the difficulty of modulating these parameters in larger-scale operations [216].

PEF systems deliver high-voltage pulses with high functional capabilities that meet large-scale industrial standards [217]. Stefan Toepfl installed high-power semiconductors that facilitated the evolution of industrial-scale PEF systems with a high degree of competence [218]. Research on the scaling-up of the PEF extraction process can be understood from several studies [219,220]. Toepfl's analysis on upscaling from the lab scale to the industrial scale was achieved through an appropriate transfer of processing conditions suitable for larger scales. In this regard, the major processing parameters, such as the electric field strength, specific energy input, and temperature, were optimized, where a specific energy input of 5–10 kJ/kg was required for the permeabilization of plant tissue [38].

8. Existing Challenges for PEF Systems

At present, the occurrence of electrochemical reactions is a significant concern for pulsed electric field treatment in the food industry [221,222]. The main consequences of electrochemical reactions include corrosive effects, electrode fouling, migration of metallic particles from electrodes, chemical changes, and water electrolysis in the food matrix [222,223]. So far, research has been performed to minimize the electrolysis by (i) using bipolar square wave pulses [162,224], (ii) reducing the pulse duration [220,225], (iii) maintaining the minimum peak voltage [225], and (iv) coating the electrode [155]. Moreover, the availability of commercially established PEF units is limited worldwide, which requires further research to overcome the existing limitations [199]. Johan Morren et al. studied the electrolysis that occurs during PEF treatment, the effect of which could be minimized in the treated food by using small pulses [221]. If the electrodes are corroded, the PEF-treated samples would be contaminated with metal particles [221].

Another drawback is the occurrence of gas bubbles during the PEF treatment that affect the uniformity of the electric field's application in the treatment area, resulting in operational issues [196]. Moreover, the majority of food samples are homogeneous in nature. The PEF's application in heterogeneous foods leads to dielectric ruptures due to air and gas bubbles in the food product that occur due to the electrical discharge between the two electrodes inside the treatment chamber. This would create sparks, thereby increasing the probability of explosions [226].

The major disadvantage of a pulsed electric field is the extensive initial cost of the machine, which requires significant financial investment [30,227,228]. However, since the operational cost is very low [216,228], many industries and research institutes propose to buy the equipment, thereby evaluating PEF as a long-term profitable food processing business that has the potential to provide high research integrity.

Moreover, during the PEF's application, varying the process parameters would cause electric-field-induced temperature changes in the food samples based on the conductivity value of the food. High-conductivity foods will have a high ionic concentration that would produce a peak electric field across the treatment chamber, thereby inducing heat energy. Hence, the temperature should be monitored and controlled to minimize the degradation of the thermolabile compounds in the plant cells [113]. For this purpose, mild treatments or circulating cooling water would aid in minimizing the associated temperature increase in the system, which would also help to preserve the quality attributes of the tested food [167,229,230].

The extraction of proteins from the plant matrix involves process parameter optimization, which poses a greater challenge to protein extraction research teams. Varying degrees of optimization would be a requirement to suit each type of plant matrix to obtain enhanced protein levels. In addition, promoting this research area on a larger scale requires established research reports pertaining to specific doses of the PEF to each protein source. Hence, further research is still required on molecular perception to find out the changes that occur in the protein levels from the plant matrix for each processing condition. In addition to protein yield, other factors such as the initial cost of equipment, maintenance and operating costs, sustainability, and the feasibility of PEF application under varying environmental conditions should be analyzed to upscale this technology to industrial processing. However, studies involving all of the above factors are still limited and we need deeper insights for further progress.

9. Conclusions and Future Directions

Protein is a mandatory requirement for the body's everyday functioning. The protein requirement varies based on age and conditions in the human body. When it comes to plant-based dietary proteins, the extraction of proteins from plant wastes using green techniques favors both the economy and the environment, promoting sustainable production. From a global perspective, valorizing raw plant wastes and rejected mushroom parts would help to reduce the amount of fruit, vegetable and mushroom wastage and provide value-added products. In this regard, the PEF-assisted protein extraction method has the potential to exponentially grow in the future to meet the ever-escalating protein demand since this technique facilitates a high protein yield in a non-thermal manner that preserves the bioactive and associated thermolabile compounds. However, further studies are required in order to optimize process parameters and scale up the technology for industrial applications. PEF is considered to be a green technique, due to which its utilization will increase in the following years. As a limited number of studies are available on PEF-assisted protein extraction, emerging unknown aspects need to be examined in the future. In particular, the impact of different electric field strength values must be studied to find out the optimum value for enhanced protein yield at a controlled temperature, since temperature is one of the most important parameters to monitor. Other process parameters should also be optimized with the electric field until we obtain clear conclusions, which should be evaluated multiple times to obtain valid and established reports. These studies would facilitate the technique for industrial applications.

Recently, combinations of techniques with PEF have been found to be effective in obtaining enhanced protein levels when compared with PEF standalone treatments [231,232]. However, the additional method should not use severe processing conditions to degrade the proteins, which may in turn require further processing steps including purification. Zhou et al. studied protein extraction with a combination of PEF and pressurized liquid extraction (PLE) treatments compared with a PEF pretreatment with conventional extraction methods. The enhanced extraction was due to the initial cell breakdown effect caused by PEF followed by the easier diffusion of the solvent into the cell during PLE to achieve a high level of protein recovery from *Spirulina* [233]. Carullo et al. showed improved recovery of bioactive compounds including proteins from a combination of high-shear homogenization (HSH) and PEF treatments on *A. platensis* suspensions due to synergistic effects. The disaggregating effect of *A. platensis trichome* clusters due to the HSH pretreatment enhanced the cell membrane permeabilization, followed by the PEF treatment, thus intensifying the extractability of valuable compounds [142]. The experiment was maintained at a feasible temperature in order to achieve non-thermal effects in the samples. Therefore, the integration of two techniques would aid in enhancing the overall protein levels in the samples. However, further studies should be conducted in order to select an appropriate green technology to combine with PEF to obtain high protein yields, also facilitating the combination for larger scales. Moreover, the cost effectiveness and sustainability should also be considered in order to promote combined techniques to industrial levels.

The challenges and future developments that should be addressed are as follows:

- The protein extraction research on PEF process parameter optimization should be broadened to promote the applications in a large-scale manner.
- Deeper insights should be gained into PEF-assisted protein extraction in non-conforming botanical parts such as rejected stems, leaves, roots, flowers, fruits, and other green waste resources from industrial processing.
- Research on combining other green techniques with PEF to obtain the optimum protein yield should be extended.
- The research on electrolysis is underdeveloped in extraction studies and can be further improved to minimize the effects in the final byproducts.
- Advanced simulation studies should be conducted to find out the protein's structural and functional changes due to electric stress, which would improve our understanding of the interaction between the protein molecule and electric field and consequently be beneficial for future experimental research.

Funding: This work is supported by Singapore Institute of Technology (SIT) under the Food Extraction Programme.

Data Availability Statement: No new data were created or analyzed in this study. Data sharing is not applicable to this article.

Conflicts of Interest: The authors declare no conflicts of interest.

References

1. Global Food Losses and Food Waste. Available online: https://www.fao.org (accessed on 22 November 2023).
2. UNEP Food Waste Index Report 2021 | UNEP—UN Environment Programme. Available online: https://www.unep.org (accessed on 22 November 2023).
3. NPCS Team. *India Emerging Business Opportunities: Cold Chain Sector (Why to Invest, Project Potential, Core Financials, Market Size & Industry Analysis)*; NIIR Project Consultancy Services: New Delhi, India, 2014; pp. 11–12.
4. Yu, X.; Bals, O.; Grimi, N.; Vorobiev, E. A new way for the oil plant biomass valorization: Polyphenols and proteins extraction from rapeseed stems and leaves assisted by pulsed electric fields. *Ind. Crops Prod.* **2015**, *74*, 309–318. [CrossRef]

5. Lonnie, M.; Hooker, E.; Brunstrom, J.M.; Corfe, B.M.; Green, M.A.; Watson, A.W.; Williams, E.A.; Stevenson, E.J.; Penson, S.; Johnstone, A.M. Protein for Life: Review of Optimal Protein Intake, Sustainable Dietary Sources and the Effect on Appetite in Ageing Adults. *Nutrients* **2018**, *10*, 360. [CrossRef] [PubMed]
6. Pikosky, M.A.; Ragalie-Carr, J.; Miller, G.D. Recognizing the importance of protein quality in an era of food systems transformation. *Front. Sustain. Food Syst.* **2022**, *6*, 1–7. [CrossRef]
7. Wolfe, R.R.; Cifelli, A.M.; Kostas, G.; Kim, I.-Y. Optimizing Protein Intake in Adults: Interpretation and Application of the Recommended Dietary Allowance Compared with the Acceptable Macronutrient Distribution Range. *Adv. Nutr.* **2017**, *8*, 266–275. [CrossRef] [PubMed]
8. FAO; WHO; UNU. *Technical Report Series 935. Protein and Amino Acid Requirements in Human Nutrition*; WHO Press: Geneva, Switzerland, 2007; pp. 1–265.
9. Institute of Medicine (IOM). *Dietary Reference Intakes for Energy, Carbohydrates, Fiber, Fat, Fatty Acids, Cholesterol, Proteins, and Amino Acids*; The National Academies Press: Washington, DC, USA, 2005.
10. Naghshi, S.; Sadeghi, O.; Willett, W.C.; Esmaillzadeh, A. Dietary intake of total, animal, and plant proteins and risk of all cause, cardiovascular, and cancer mortality: Systematic review and dose-response meta-analysis of prospective cohort studies. *BMJ* **2020**, *370*, m2412. [CrossRef] [PubMed]
11. Henchion, M.; Hayes, M.; Mullen, A.M.; Fenelon, M.; Tiwari, B. Future Protein Supply and Demand: Strategies and Factors Influencing a Sustainable Equilibrium. *Foods* **2017**, *6*, 53. [CrossRef] [PubMed]
12. Jin, Q.; Yang, L.; Poe, N.; Huang, H. Integrated processing of plant-derived waste to produce value-added products based on the biorefinery concept. *Trends Food Sci. Technol.* **2018**, *74*, 119–131. [CrossRef]
13. Basri, M.S.M.; Shah, N.N.A.K.; Sulaiman, A.; Tawakkal, I.S.M.A.; Nor, M.Z.M.; Ariffin, S.H.; Ghani, N.H.A.; Salleh, F.S.M. Progress in the Valorization of Fruit and Vegetable Wastes: Active Packaging, Biocomposites, By-Products, and Innovative Technologies Used for Bioactive Compound Extraction. *Polymers* **2021**, *13*, 3503. [CrossRef]
14. Esparza, I.; Jiménez-Moreno, N.; Bimbela, F.; Ancín-Azpilicueta, C.; Gandía, L.M. Fruit and vegetable waste management: Conventional and emerging approaches. *J. Environ. Manag.* **2020**, *265*, 110510. [CrossRef]
15. Rana, A.K. Green approaches in the valorization of plant wastes: Recent insights and future directions. *Curr. Opin. Green Sustain. Chem.* **2022**, *38*, 1–9. [CrossRef]
16. Barbarino, E.; Lourenço, S.O. An evaluation of methods for extraction and quantification of protein from marine macro- and microalgae. *J. Appl. Phycol.* **2005**, *17*, 447–460. [CrossRef]
17. Fleurence, J.; Le Coeur, C.; Mabeau, S.; Maurice, M.; Landrein, A. Comparison of different extractive procedures for proteins from the edible seaweeds *Ulva rigida* and *Ulva rotundata*. *J. Appl. Phycol.* **1995**, *7*, 577–582. [CrossRef]
18. Urango, A.C.M.; Strieder, M.M.; Silva, E.K.; Meireles, M.A.A. Thermosonication Process Design for Recovering Bioactive Compounds from Fennel: A Comparative Study with Conventional Extraction Techniques. *Appl. Sci.* **2021**, *11*, 12104. [CrossRef]
19. Harnedy, P.A.; FitzGerald, R.J. Extraction of protein from the macroalga Palmaria palmata. *LWT—Food Sci. Technol.* **2013**, *51*, 375–382. [CrossRef]
20. Nowosad, K.; Sujka, M.; Pankiewicz, U.; Kowalski, R. The application of PEF technology in food processing and human nutrition. *J. Food Sci. Technol.* **2021**, *58*, 397–411. [CrossRef] [PubMed]
21. Lammerskitten, A.; Mykhailyk, V.; Wiktor, A.; Toepfl, S.; Nowacka, M.; Bialik, M.; Czyżewski, J.; Witrowa-Rajchert, D.; Parniakov, O. Impact of pulsed electric fields on physical properties of freeze-dried apple tissue. *Innov. Food Sci. Emerg. Technol.* **2019**, *57*, 1–7. [CrossRef]
22. Min, S.; Evrendilek, G.A.; Zhang, H.Q. Pulsed Electric Fields: Processing System, Microbial and Enzyme Inhibition, and shelf life extension of Foods. *IEEE Trans. Plasma Sci.* **2007**, *35*, 59–73. [CrossRef]
23. Dziadek, K.; Kopeć, A.; Dróżdż, T.; Kiełbasa, P.; Ostafin, M.; Bulski, K.; Oziembłowski, M. Effect of pulsed electric field treatment on shelf life and nutritional value of apple juice. *J. Food Sci. Technol.* **2019**, *56*, 1184–1191. [CrossRef] [PubMed]
24. Odriozola-Serrano, I.; Aguiló-Aguayo, I.; Soliva-Fortuny, R.; Martín-Belloso, O. Pulsed electric fields processing effects on quality and health-related constituents of plant-based foods. *Trends Food Sci. Technol.* **2013**, *29*, 98–107. [CrossRef]
25. Kantala, C.; Supasin, S.; Intra, P.; Rattanadecho, P. Evaluation of Pulsed Electric Field and Conventional Thermal Processing for Microbial Inactivation in Thai Orange Juice. *Foods* **2022**, *11*, 1102. [CrossRef]
26. Barsotti, L.; Dumay, E.; Mu, T.H.; Diaz, M.D.F.; Cheftel, J.C. Effects of high voltage electric pulses on protein-based food constituents and structures. *Trends Food Sci. Technol.* **2001**, *12*, 136–144. [CrossRef]
27. Zulueta, A.; Esteve, M.J.; Frasquet, I.; Frígola, A. Fatty acid profile changes during orange juice-milk beverage processing by high-pulsed electric field. *Eur. J. Lipid Sci. Technol.* **2007**, *109*, 25–31. [CrossRef]
28. Wouters, P.C.; Smelt, J.P. Inactivation of microorganisms with pulsed electric fields: Potential for food preservation. *Food Biotechnol.* **1997**, *11*, 193–229. [CrossRef]
29. Barsotti, L.; Cheftel, J.C. Food processing by pulsed electric fields. *II. Biological aspects. Food Rev. Int.* **1999**, *15*, 181–213. [CrossRef]
30. Jeyamkondan, S.; Jayas, D.S.; Holley, R.A. Pulsed electric field processing of foods: A review. *J. Food Prot.* **1999**, *62*, 1088–1096. [CrossRef]
31. Zhang, Q.; Barbosa-Cánovas, G.V.; Swanson, B.G. Engineering aspects of pulsed electric field pasteurization. *J. Food Eng.* **1995**, *25*, 261–281. [CrossRef]
32. Ho, S.; Mittal, G.S. High voltage pulsed electrical field for liquid food pasteurization. *Food Rev. Int.* **2000**, *16*, 395–434. [CrossRef]

33. Dunn, J.E.; Pearlman, J.S.; La Costa, R. Methods and Apparatus for Extending the Shelf Life of Fluid Food Products. U.S. Patent 4,838,154, 22 September 1987.
34. Dunn, J. Pulsed electric field processing: An overview. In *Pulsed Electric Fields in Food Processing, Fundamental Aspects and Applications*; Barbosa-Cánovas, G., Zhang, Q.H., Eds.; Technomic Press: Lancaster, PA, USA, 2001; p. 130.
35. Sampedro, F.; Rodrigo, D.; Martínez, A.; Barbosa-Cánovas, G.V.; Rodrigo, M. Review: Application of Pulsed Electric Fields in Egg and Egg Derivatives. *Food Sci. Technol. Int.* **2006**, *12*, 397–405. [CrossRef]
36. Góngora-Nieto, M.M.; Sepúlveda, D.R.; Pedrow, P.; Barbosa-Cánovas, G.V.; Swanson, B.G. Food Processing by Pulsed Electric Fields: Treatment Delivery, Inactivation Level, and Regulatory Aspects. *LWT—Food Sci. Technol.* **2002**, *35*, 375–388. [CrossRef]
37. Puértolas, E.; Luengo, E.; Álvarez, I.; Raso, J. Improving Mass Transfer to Soften Tissues by Pulsed Electric Fields: Fundamentals and Applications. *Annu. Rev. Food Sci. Technol.* **2012**, *3*, 263–282. [CrossRef]
38. Toepfl, S. Pulsed Electric Field food treatment—Scale up from lab to industrial scale. *Procedia Food Sci.* **2011**, *1*, 776–779. [CrossRef]
39. Góngora-Nieto, M.M.; Pedrow, P.D.; Swanson, B.G.; Barbosa-Cánovas, G.V. Impact of air bubbles in a dielectric liquid when subjected to high field strengths. *Innov. Food Sci. Emerg. Technol.* **2003**, *4*, 57–67. [CrossRef]
40. Mariotti, F. Animal and Plant Protein Sources and Cardiometabolic Health. *Adv. Nutr.* **2019**, *10*, S351–S366. [CrossRef] [PubMed]
41. Rifna, E.J.; Misra, N.N.; Dwivedi, M. Recent advances in extraction technologies for recovery of bioactive compounds derived from fruit and vegetable waste peels: A review. *Crit. Rev. Food Sci. Nutr.* **2023**, *63*, 719–752. [CrossRef]
42. Ganesh, K.S.; Sridhar, A.; Vishali, S. Utilization of fruit and vegetable waste to produce value-added products: Conventional utilization and emerging opportunities-A review. *Chemosphere* **2022**, *287*, 132221. [CrossRef]
43. Mathews, A.D.; Tangirala, A.S.; Kumar, S.; Anandharaj, A.; Rawson, A. Extraction and modification of protein from sesame oil cake by the application of emerging technologies. *Food Chem. Adv.* **2023**, *2*, 100326. [CrossRef]
44. Kumoro, A.C.; Alhanif, M.; Wardhani, D.H. A Critical Review on Tropical Fruits Seeds as Prospective Sources of Nutritional and Bioactive Compounds for Functional Foods Development: A Case of Indonesian Exotic Fruits. *Int. J. Food Sci.* **2020**, *2020*, 4051475. [CrossRef] [PubMed]
45. Ruschioni, S.; Loreto, N.; Foligni, R.; Mannozzi, C.; Raffaelli, N.; Zamporlini, F.; Pasquini, M.; Roncolini, A.; Cardinali, F.; Osimani, A.; et al. Addition of Olive Pomace to Feeding Substrate Affects Growth Performance and Nutritional Value of Mealworm (*Tenebrio Molitor* L.) Larvae. *Foods* **2020**, *9*, 317. [CrossRef]
46. Herchi, W.; Bahashwan, S.; Sakouhi, F.; Boukhchina, S. Influence of harvest year in the physicochemical properties and antioxidant activity of flaxseed hull oils from Tunisia. *Food Sci. Technol.* **2015**, *35*, 175–182. [CrossRef]
47. Liadakis, G.; Kekes, T.; Frakolaki, G.; Giannou, V.; Tzia, C. *Ingredients for Food Products in Mejdi Jeguirim. Tomato Processing By-Products*; Zorpas, A., Ed.; Academic Press: Cambridge, MA, USA, 2022; pp. 117–148.
48. Wen, C.; Zhang, J.; Duan, Y.; Zhang, H.; Ma, H. A Mini-Review on Brewer's Spent Grain Protein: Isolation, Physicochemical Properties, Application of Protein, and Functional Properties of Hydrolysates. *J. Food Sci.* **2019**, *84*, 3330–3340. [CrossRef]
49. Mustafa, A.F.; Baurhoo, B. Evaluation of dried vegetables residues for poultry: II. Effects of feeding cabbage leaf residues on broiler performance, ileal digestibility and total tract nutrient digestibility. *Poult. Sci.* **2017**, *96*, 681–686. [CrossRef]
50. Kaur, M.; Gautam, A.; Kaur, H. Nutritional, techno-functional, structural, and rheological properties of potato peel powder: A valuable biowaste being potential source of dietary fibre and antioxidants in cookie formulation. *J. Food Process. Preserv.* **2022**, *46*, 1–13. [CrossRef]
51. Lobo, M.G.; Dorta, E. Utilization and management of horticultural waste in Elhadi. In *Postharvest Technology of Perishable Horticultural Commodities*; Yahia, M., Ed.; Woodhead Publishing: Cambridge, UK, 2019; pp. 639–666.
52. Wikandari, R.; Nguyen, H.; Millati, R.; Niklasson, C.; Taherzadeh, M.J. Improvement of biogas production from orange peel waste by leaching of limonene. *Biomed. Res. Int.* **2015**, 494182. [CrossRef] [PubMed]
53. Cunha, T.J. *Value of Feeds for Horses in Horse Feeding and Nutrition*, 2nd ed.; Academic Press: San Diego, CA, USA, 2012; pp. 233–273.
54. Hussain, A.; Kausar, T.; Sehar, S.; Sarwar, A.; Ashraf, A.H.; Jamil, M.A.; Noreen, S.; Rafique, A.; Iftikhar, K.; Quddoos, M.Y.; et al. A Comprehensive review of functional ingredients, especially bioactive compounds present in pumpkin peel, flesh and seeds, and their health benefits. *Food Chem. Adv.* **2022**, *1*, 100067. [CrossRef]
55. Zaini, H.M.; Roslan, J.; Saallah, S.; Munsu, E.; Sulaiman, N.S.; Pindi, W. Banana peels as a bioactive ingredient and its potential application in the food industry. *J. Funct. Foods* **2022**, *92*, 105054. [CrossRef]
56. Ain, H.B.U.; Tufail, T.; Bashir, S.; Ijaz, N.; Hussain, M.; Ikram, A.; Farooq, M.A.; Saewan, S.A. Nutritional importance and industrial uses of pomegranate peel: A critical review. *Food Sci. Nutr.* **2023**, *11*, 2589–2598. [CrossRef] [PubMed]
57. Gómez-García, R.; Campos, D.A.; Oliveira, A.; Aguilar, C.N.; Madureira, A.R.; Pintado, M. A chemical valorisation of melon peels towards functional food ingredients: Bioactives profile and antioxidant properties. *Food Chem.* **2021**, *335*, 127579. [CrossRef] [PubMed]
58. Maloney, K.P.; Truong, V.D.; Allen, J.C. Susceptibility of sweet potato (*Ipomoea batatas*) peel proteins to digestive enzymes. *Food Sci. Nutr.* **2014**, *2*, 351–360. [CrossRef] [PubMed]
59. Dotto, J.M.; Chacha, J.S. The potential of pumpkin seeds as a functional food ingredient: A review. *Sci. Afr.* **2020**, *10*, e00575. [CrossRef]
60. Apprich, S.; Tirpanalan, O.; Hell, J.; Reisinger, M.; Böhmdorfer, S.; Siebenhandl-Ehn, S.; Novalin, S.; Kneifel, W. Wheat bran-based biorefinery 2: Valorization of products. *LWT—Food Sci. Technol.* **2014**, *56*, 222–231. [CrossRef]

61. Kalpanadevi, C.; Muthukumar, S.P.; Govindaraju, K.; Subramanian, R. Rice bran protein: An alternative plant-based protein to ameliorate protein malnourishment. *J. Cereal Sci.* **2021**, *97*, 103154. [CrossRef]
62. Kumar, M.; Barbhai, M.D.; Hasan, M.; Punia, S.; Dhumal, S.; Radha; Rais, N.; Chandran, D.; Pandiselvam, R.; Kothakota, A.; et al. Onion (*Allium cepa* L.) peels: A review on bioactive compounds and biomedical activities. *Biomed. Pharmacother.* **2022**, *146*, 112498. [CrossRef]
63. Kamal, H.; Le, C.F.; Salter, A.M.; Ali, A. Extraction of protein from food waste: An overview of current status and opportunities. *Compr. Rev. Food Sci. Food Saf.* **2021**, *20*, 2455–2475. [CrossRef]
64. Kumar, M.; Tomar, M.; Punia, S.; Dhakane-Lad, J.; Dhumal, S.; Changan, S.; Senapathy, M.; Berwal, M.K.; Sampathrajan, V.; Sayed, A.A.; et al. Plant-based proteins and their multifaceted industrial applications. *LWT* **2022**, *154*, 112620. [CrossRef]
65. Ayimbila, F.; Keawsompong, S. Nutritional Quality and Biological Application of Mushroom Protein as a Novel Protein Alternative. *Curr. Nutr. Rep.* **2023**, *12*, 290–307. [CrossRef]
66. Zhou, J.; Chen, M.; Wu, S.; Liao, X.; Wang, J.; Wu, Q.; Zhuang, M.; Ding, Y. A review on mushroom-derived bioactive peptides: Preparation and biological activities. *Food Res. Int.* **2020**, *134*, 109230. [CrossRef]
67. Meenu, M.; Xu, B. Application of vibrational spectroscopy for classification, authentication and quality analysis of mushroom: A concise review. *Food Chem.* **2019**, *289*, 545–557. [CrossRef] [PubMed]
68. Wang, X.M.; Zhang, J.; Wu, L.H.; Zhao, Y.L.; Li, T.; Li, J.Q.; Wang, Y.Z.; Liu, H.G. A mini-review of chemical composition and nutritional value of edible wild-grown mushroom from China. *Food Chem.* **2014**, *151*, 279–285. [CrossRef] [PubMed]
69. Elhusseiny, S.M.; El-Mahdy, T.S.; Awad, M.F.; Elleboudy, N.S.; Farag, M.M.S.; Aboshanab, K.M.; Yassien, M.A. Antiviral, Cytotoxic, and Antioxidant Activities of Three Edible Agaricomycetes Mushrooms: Pleurotus columbinus, Pleurotus sajor-caju, and Agaricus bisporus. *J. Fungi—Open Access Mycol. J.* **2021**, *7*, 645. [CrossRef]
70. González, A.; Nobre, C.; Simões, L.S.; Cruz, M.; Loredo, A.; Rodríguez-Jasso, R.M.; Contreras, J.; Texeira, J.; Belmares, R. Evaluation of functional and nutritional potential of a protein concentrate from *Pleurotus ostreatus* mushroom. *Food Chem.* **2021**, *346*, 128884. [CrossRef]
71. Miles, P.G.; Chang, S.-T. *Mushrooms: Cultivation, Nutritional Value, Medicinal Effect, and Environmental Impact*, 2nd ed.; CRC Press: Boca Raton, FL, USA, 2004. [CrossRef]
72. Rangsinth, P.; Sillapachaiyaporn, C.; Nilkhet, S.; Tencomnao, T.; Ung, A.T.; Chuchawankul, S. Mushroom-derived bioactive compounds potentially serve as the inhibitors of SARS-CoV-2 main protease: An in silico approach. *J. Tradit. Complement. Med.* **2021**, *11*, 158–172. [CrossRef] [PubMed]
73. Yang, M.; Belwal, T.; Devkota, H.P.; Li, L.; Luo, Z. Trends of utilizing mushroom polysaccharides (MPs) as potent nutraceutical components in food and medicine: A comprehensive review. *Trends Food Sci. Technol.* **2019**, *92*, 94–110. [CrossRef]
74. Lu, X.; Brennan, M.A.; Serventi, L.; Liu, J.; Guan, W.; Brennan, C.S. Addition of mushroom powder to pasta enhances the antioxidant content and modulates the predictive glycaemic response of pasta. *Food Chem.* **2018**, *264*, 199–209. [CrossRef]
75. Venturella, G.; Ferraro, V.; Cirlincione, F.; Gargano, M.L. Medicinal Mushrooms: Bioactive Compounds, Use, and Clinical Trials. *Int. J. Mol. Sci.* **2021**, *22*, 634. [CrossRef]
76. Das, A.K.; Nanda, P.K.; Dandapat, P.; Bandyopadhyay, S.; Gullón, P.; Sivaraman, G.K.; McClements, D.J.; Gullón, B.; Lorenzo, J.M. Edible Mushrooms as Functional Ingredients for Development of Healthier and More Sustainable Muscle Foods: A Flexitarian Approach. *Molecules* **2021**, *26*, 2463. [CrossRef]
77. Jacinto-Azevedo, B.; Valderrama, N.; Henríquez, K.; Aranda, M.; Aqueveque, P. Nutritional value and biological properties of Chilean wild and commercial edible mushrooms. *Food Chem.* **2021**, *356*, 129651. [CrossRef] [PubMed]
78. Ghose, A.; Mitra, S. Spent waste from edible mushrooms offers innovative strategies for the remediation of persistent organic micropollutants: A review. *Environ. Pollut.* **2022**, *305*, 119285. [CrossRef] [PubMed]
79. Atallah, E.; Zeaiter, J.; Ahmad, M.N.; Leahy, J.J.; Kwapinski, W. Hydrothermal carbonization of spent mushroom compost waste compared against torrefaction and pyrolysis. *Fuel Process. Technol.* **2021**, *216*, 106795. [CrossRef]
80. Leong, Y.K.; Ma, T.-W.; Chang, J.-S.; Yang, F.-C. Recent advances and future directions on the valorization of spent mushroom substrate (SMS): A review. *Bioresour. Technol.* **2021**, *344*, 126157. [CrossRef] [PubMed]
81. Usman, M.; Murtaza, G.; Ditta, A. Nutritional, Medicinal, and Cosmetic Value of Bioactive Compounds in Button Mushroom (*Agaricus bisporus*): A Review. *Appl. Sci.* **2021**, *11*, 5943. [CrossRef]
82. Carrasco-González, J.A.; Serna-Saldívar, S.O.; Gutiérrez-Uribe, J.A. Nutritional composition and nutraceutical properties of the *Pleurotus* fruiting bodies: Potential use as food ingredient. *J. Food Compos. Anal.* **2017**, *58*, 69–81. [CrossRef]
83. Abidin, M.H.Z.; Abdullah, N.; Abidin, N.Z. Therapeutic properties of *Pleurotus* species (oyster mushrooms) for atherosclerosis: A review. *Int. J. Food Prop.* **2017**, *20*, 1251–1261. [CrossRef]
84. Valverde, M.E.; Hernández-Pérez, T.; Paredes-López, O.; Hernández-Pérez, T.; Paredes-López, O. Edible Mushrooms: Improving Human Health and Promoting Quality Life. *Int. J. Microbiol.* **2015**, *2015*, 376387. [CrossRef]
85. Li, Y.; Chen, H.; Zhang, X. Cultivation, nutritional value, bioactive compounds of morels, and their health benefits: A systematic review. *Front. Nutr.* **2023**, *10*, 1159029. [CrossRef]
86. Zhang, Q.; Wu, C.; Fan, G.; Li, T.; Sun, Y. Improvement of antioxidant activity of *Morchella esculenta* protein hydrolysate by optimized glycosylation reaction. *CyTA—J. Food* **2018**, *16*, 238–246. [CrossRef]
87. Das, S.; Prakash, B. Edible mushrooms: Nutritional composition and medicinal benefits for improvement in quality life. In *Research and Technological Advances in Food Science*; Prakash, B., Ed.; Academic Press: Cambridge, MA, USA, 2022; pp. 269–300.

88. Spim, S.R.V.; Castanho, N.R.C.M.; Pistila, A.M.H.; Jozala, A.F.; Oliveira Júnior, J.M.; Grotto, D. Lentinula edodes mushroom as an ingredient to enhance the nutritional and functional properties of cereal bars. *J. Food Sci. Technol.* **2021**, *58*, 1349–1357. [CrossRef]
89. Bisen, P.S.; Baghel, R.K.; Sanodiya, B.S.; Thakur, G.S.; Prasad, G.B. Lentinus edodes: A macrofungus with pharmacological activities. *Curr. Med. Chem.* **2010**, *17*, 2419–2430. [CrossRef]
90. Sun, X.; Yang, C.; Ma, Y.; Zhang, J.; Wang, L. Research progress of Auricularia heimuer on cultivation physiology and molecular biology. *Front. Microbiol.* **2022**, *13*, 148249. [CrossRef]
91. Hrudayanath, T.; Sameer, K.S. Diversity, nutritional composition and medicinal potential of Indian mushrooms: A review. *Afr. J. Biotechnol.* **2014**, *13*, 523–545. [CrossRef]
92. Tang, C.; Hoo, P.C.; Tan, L.T.; Pusparajah, P.; Khan, T.M.; Lee, L.H.; Goh, B.H.; Chan, K.G. Golden Needle Mushroom: A Culinary Medicine with Evidenced-Based Biological Activities and Health Promoting Properties. *Front. Pharmacol.* **2016**, *7*, 474. [CrossRef] [PubMed]
93. Chen, J.; Li, J.M.; Tang, Y.J.; Ma, K.; Li, B.; Zeng, X.; Liu, X.-B.; Li, Y.; Yang, Z.-L.; Xu, W.-N.; et al. Genome-wide analysis and prediction of genes involved in the biosynthesis of polysaccharides and bioactive secondary metabolites in high-temperature-tolerant wild *Flammulina filiformis*. *BMC Genom.* **2020**, *21*, 719. [CrossRef] [PubMed]
94. Zhang, L.; Chen, J.; Xu, F.; Han, R.; Quan, M.; Wang, L. Effect of Tremella fuciformis on dough structure and rheology, noodle flavor, and quality characteristics. *LWT* **2022**, *172*, 114180. [CrossRef]
95. Shahrajabian, M.H.; Sun, W.; Cheng, Q.; Khoshkharam, M. Exploring the quality of foods from ancient China based on traditional Chinese medicine. In *Functional Foods and Nutraceuticals in Metabolic and Non-Communicable Diseases*; Singh, R.B., Watanabe, S., Isaza, A.A., Eds.; Academic Press: Cambridge, MA, USA, 2022; pp. 87–105. [CrossRef]
96. Rajarathnam, S.; Shashirekha, M.N. Mushrooms and Truffles | Use of Wild Mushrooms. In *Encyclopaedia of Food Sciences and Nutrition*, 2nd ed.; Academic Press: Cambridge, MA, USA, 2003; pp. 4048–4054. [CrossRef]
97. Mikola, E.; Geösel, A.; Stefanovits-Bányai, É.; Fodor, M. Quantitative determination of macro components and classification of some cultivated mushrooms using near-infrared spectroscopy. *J. Food Process Preserv.* **2020**, *44*, e14540. [CrossRef]
98. Chauhan, G.; Prasad, S.; Himanshi, R.; Satyawati, S. Nutritional profiling and value addition of products from *Hypsizygus tessellatus*. *Food Biol.* **2017**, *6*, 1–6. [CrossRef]
99. Li, W.; Chen, W.; Ma, H.; Wang, J.; Li, Z.; Wang, Q.; Zhang, Z.; Wu, D.; Zhang, J.; Yang, Y. Study on the relationship between structure and taste activity of the umami peptide of *Stropharia rugosoannulata* prepared by ultrasound. *Ultrason. Sonochem.* **2022**, *90*, 106206. [CrossRef] [PubMed]
100. Hu, Y.; Kakumyan, P.; Bandara, A.; Peter, M. The nutrition, cultivation and biotechnology of *Stropharia rugosoannulata*. *Fungal Biotec.* **2021**, *1*, 13–25. [CrossRef]
101. Petrović, J.; Glamočlija, J.; Stojković, D.; Ćirić, A.; Barros, L.; Ferreira, I.C.F.R.; Soković, M. Nutritional value, chemical composition, antioxidant activity and enrichment of cream cheese with chestnut mushroom *Agrocybe aegerita* (Brig.) Sing. *J. Food Sci. Technol.* **2015**, *52*, 6711–6718. [CrossRef] [PubMed]
102. Wang, M.; Gu, B.; Huang, J.; Jiang, S.; Chen, Y.; Yin, Y.; Pan, Y.; Yu, G.; Li, Y.; Wong, B.H.C.; et al. Transcriptome and Proteome Exploration to Provide a Resource for the Study of *Agrocybe aegerita*. *PLoS ONE* **2013**, *8*, e56686. [CrossRef]
103. Bandura, I.; Kulyk, A.; Makohon, S.; Khareba, O.; Khareba, V. Influence of the substrate composition on the yield and nutritional value of the fruiting bodies of the edible mushrooms *Pleurotus citrinopileatus* and *Cyclocybe aegerita*. *Plant Var. Stud. Prot.* **2021**, *17*, 130–138. [CrossRef]
104. Friedman, M. Chemistry, Nutrition, and Health-Promoting Properties of *Hericium erinaceus* (Lion's Mane) Mushroom Fruiting Bodies and Mycelia and Their Bioactive Compounds. *J. Agric. Food Chem.* **2015**, *63*, 7108–7123. [CrossRef] [PubMed]
105. Chaiyasut, C.; Sivamaruthi, B.S. Anti-hyperglycemic property of *Hericium erinaceus*—A mini review. *Asian Pac. J. Trop. Biomed.* **2017**, *7*, 1036–1040. [CrossRef]
106. Sitinjak, R. The Nutritional Content of the Mushroom Phallus indusiatus Vent., which Grows in the Cocoa Plantation, Gaperta-Ujung, Medan. *Der Pharma Chem.* **2017**, *9*, 44–47.
107. Roselló-Soto, E.; Barba, F.J.; Parniakov, O.; Galanakis, C.M.; Lebovka, N.; Grimi, N.; Vorobiev, E. High Voltage Electrical Discharges, Pulsed Electric Field, and Ultrasound Assisted Extraction of Protein and Phenolic Compounds from Olive Kernel. *Food Bioprocess. Technol.* **2015**, *8*, 885–894. [CrossRef]
108. Antunes, F.; Marçal, S.; Taofiq, O.; Morais, A.M.M.B.; Freitas, A.C.; Ferreira, I.C.F.R.; Pintado, M. Valorization of Mushroom By-Products as a Source of Value-Added Compounds and Potential Applications. *Molecules* **2020**, *25*, 2672. [CrossRef] [PubMed]
109. Jayachandran, M.; Xiao, J.; Xu, B.A. Critical Review on Health Promoting Benefits of Edible Mushrooms through Gut Microbiota. *Int. J. Mol. Sci.* **2017**, *18*, 1934. [CrossRef] [PubMed]
110. Welti-Chanes, J.; Vergara-Balderas, F.; Bermúdez-Aguirre, D. Transport phenomena in food engineering: Basic concepts and advances. *J. Food Eng.* **2005**, *67*, 113–128. [CrossRef]
111. Barba, F.J.; Parniakov, O.; Pereira, S.A.; Wiktor, A.; Grimi, N.; Boussetta, N.; Saraiva, J.A.; Raso, J.; Martin-Belloso, O.; Witrowa-Rajchert, D.; et al. Current applications and new opportunities for the use of pulsed electric fields in food science and industry. *Food Res. Int.* **2015**, *77*, 773–798. [CrossRef]
112. Barba, F.J.; Zhu, Z.; Koubaa, M.; Sant'Ana, A.S.; Orlien, V. Green alternative methods for the extraction of antioxidant bioactive compounds from winery wastes and by-products: A review. *Trends Food Sci. Technol.* **2016**, *49*, 96–109. [CrossRef]

113. Zhang, Q.W.; Lin, L.G.; Ye, W.C. Techniques for extraction and isolation of natural products: A comprehensive review. *Chin. Med.* **2018**, *13*, 20. [CrossRef]
114. Chua, L.S. A review on plant-based rutin extraction methods and its pharmacological activities. *J. Ethnopharmacol.* **2013**, *150*, 805–817. [CrossRef]
115. Chandran, A.S.; Suri, S.; Choudhary, P. Sustainable plant protein: An up-to-date overview of sources, extraction techniques and utilization. *Sustain. Food Technol.* **2023**, *1*, 466–483. [CrossRef]
116. Nadar, S.S.; Rao, P.; Rathod, V.K. Enzyme assisted extraction of biomolecules as an approach to novel extraction technology: A review. *Food Res. Int.* **2018**, *108*, 309–330. [CrossRef]
117. Chakka, A.K.; Sriraksha, M.S.; Ravishankar, C.N. Sustainability of emerging green non-thermal technologies in the food industry with food safety perspective: A review. *LWT* **2021**, *151*, 112140. [CrossRef]
118. Soquetta, M.B.; Terra, L.d.M.; Bastos, C.P. Green technologies for the extraction of bioactive compounds in fruits and vegetables. *CyTA—J. Food* **2018**, *16*, 400–412. [CrossRef]
119. Xue, D.; Farid, M.F. Pulsed Electric Field Extraction of Valuable Compounds from White Button Mushroom (Agaricus Bisporus). *Innov. Food Sci. Emerg. Technol.* **2015**, *29*, 178–186. [CrossRef]
120. Jadhav, H.B.; Annapure, U.S.; Deshmukh, R.R. Non-thermal Technologies for Food Processing. *Front. Nutr.* **2021**, *8*, 657090. [CrossRef] [PubMed]
121. Galanakis, C.M.; Schieber, A. Editorial. *Food Res. Int.* **2014**, *65*, 299–300. [CrossRef]
122. Reddy, V.Y.; Koruth, J.; Jais, P.; Petru, J.; Timko, F.; Skalsky, I.; Hebeler, R.; Labrousse, L.; Barandon, L.; Kralovec, S.; et al. Ablation of Atrial Fibrillation with Pulsed Electric Fields: An Ultra-Rapid, Tissue-Selective Modality for Cardiac Ablation. *JACC Clin. Electrophysiol.* **2018**, *4*, 987–995. [CrossRef]
123. Nuccitelli, R. Application of Pulsed Electric Fields to Cancer Therapy. *Bioelectricity* **2019**, *1*, 30–34. [CrossRef]
124. Buchmann, L.; Brändle, I.; Haberkorn, I.; Hiestand, M.; Mathys, A. Pulsed electric field based cyclic protein extraction of microalgae towards closed-loop biorefinery concepts. *Bioresour. Technol.* **2019**, *291*, 121870. [CrossRef]
125. Rios-Corripio, G.; Morales-de la Peña, M.; Welti-Chanes, J.; Ángel Guerrero-Beltrán, J. Pulsed electric field processing of a pomegranate (*Punica granatum* L.) fermented beverage. *Innov. Food Sci. Emerg. Technol.* **2022**, *79*, 103045. [CrossRef]
126. Naliyadhara, N.; Kumar, A.; Girisa, S.; Daimary, U.D.; Hegde, M.; Kunnumakkara, A.B. Pulsed electric field (PEF): Avant-garde extraction escalation technology in food industry. *Trends Food Sci. Technol.* **2022**, *122*, 238–255. [CrossRef]
127. Punthi, F.; Yudhistira, B.; Gavahian, M.; Chang, C.-K.; Cheng, K.-C.; Hou, C.-Y.; Hsieh, C.-W. Pulsed electric field-assisted drying: A review of its underlying mechanisms, applications, and role in fresh produce plant-based food preservation. *Compr. Rev. Food Sci. Food Saf.* **2022**, *21*, 5109–5130. [CrossRef]
128. Navarro, A.; Ruiz-Méndez, M.-V.; Sanz, C.; Martínez, M.; Rego, D.; Pérez, A.G. Application of Pulsed Electric Fields to Pilot and Industrial Scale Virgin Olive Oil Extraction: Impact on Organoleptic and Functional Quality. *Foods* **2022**, *11*, 2022. [CrossRef] [PubMed]
129. Parniakov, O.; Barba, F.J.; Grimi, N.; Lebovka, N.; Vorobiev, E. Extraction assisted by pulsed electric energy as a potential tool for green and sustainable recovery of nutritionally valuable compounds from mango peels. *Food Chem.* **2016**, *192*, 842–848. [CrossRef] [PubMed]
130. Käferböck, A.; Smetana, S.; de Vos, R.; Schwarz, C.; Toepfl, S.; Parniakov, O. Sustainable extraction of valuable components from Spirulina assisted by pulsed electric fields technology. *Algal Res.* **2020**, *48*, 101914. [CrossRef]
131. Kumari, B.; Tiwari, B.K.; Hossain, M.B.; Brunton, N.P.; Rai, D.K. Recent Advances on Application of Ultrasound and Pulsed Electric Field Technologies in the Extraction of Bioactives from Agro-Industrial By-products. *Food Bioprocess. Technol.* **2018**, *11*, 223–241. [CrossRef]
132. Saldaña, G.; Puértolas, E.; Álvarez, I.; Meneses, N.; Knorr, D.; Raso, J. Evaluation of a static treatment chamber to investigate kinetics of microbial inactivation by pulsed electric fields at different temperatures at quasi-isothermal conditions. *J. Food Eng.* **2010**, *100*, 349–356. [CrossRef]
133. Golberg, A.; Sack, M.; Teissie, J.; Pataro, G.; Pliquett, U.; Saulis, G.; Stefan, T.; Miklavcic, D.; Vorobiev, E.; Frey, W. Energy-efficient biomass processing with pulsed electric fields for bioeconomy and sustainable development. *Biotechnol. Biofuels* **2016**, *9*, 1–22. [CrossRef]
134. Weaver, J.C. Electroporation of cells and tissues. *IEEE Trans. Plasma Sci.* **2000**, *28*, 24–33. [CrossRef]
135. Wang, M.-S.; Wang, L.-H.; Bekhit, A.E.-D.A.; Yang, J.; Hou, Z.-P.; Wang, Y.-Z.; Dai, Q.-Z.; Zeng, X.-A. A review of sublethal effects of pulsed electric field on cells in food processing. *J. Food Eng.* **2018**, *223*, 32–41. [CrossRef]
136. Napotnik, T.B.; Polajžer, T.; Miklavčič, D. Cell death due to electroporation—A review. *Bioelectrochemistry* **2021**, *141*, 107871. [CrossRef]
137. Boussetta, N.; Lesaint, O.; Vorobiev, E. A study of mechanisms involved during the extraction of polyphenols from grape seeds by pulsed electrical discharges. *Innov. Food Sci. Emerg. Technol.* **2013**, *19*, 124–132. [CrossRef]
138. Redondo, D.; Venturini, M.E.; Luengo, E.; Raso, J.; Arias, E. Pulsed electric fields as a green technology for the extraction of bioactive compounds from thinned peach by-products. *Innov. Food Sci. Emerg. Technol.* **2018**, *45*, 335–343. [CrossRef]
139. Sánchez-Vega, R.; Elez-Martínez, P.; Martín-Belloso, O. Influence of high-intensity pulsed electric field processing parameters on antioxidant compounds of broccoli juice. *Innov. Food Sci. Emerg. Technol.* **2015**, *29*, 70–77. [CrossRef]

140. Carullo, D.; Abera, B.D.; Casazza, A.A.; Donsì, F.; Perego, P.; Ferrari, G.; Pataro, G. Effect of pulsed electric fields and high pressure homogenization on the aqueous extraction of intracellular compounds from the microalgae *Chlorella vulgaris*. *Algal Res.* **2018**, *31*, 60–69. [CrossRef]
141. Postma, P.R.; Pataro, G.; Capitoli, M.; Barbosa, M.J.; Wijffels, R.H.; Eppink, M.H.M.; Olivieri, G.; Ferrari, G. Selective extraction of intracellular components from the microalga Chlorella vulgaris by combined pulsed electric field–temperature treatment. *Bioresour. Technol.* **2016**, *203*, 80–88. [CrossRef] [PubMed]
142. Carullo, D.; Donsì, F.; Ferrari, G.; Pataro, G. Extraction improvement of water-soluble compounds from Arthrospira platensis through the combination of high-shear homogenization and pulsed electric fields. *Algal Res.* **2021**, *57*, 102341. [CrossRef]
143. Calleja-Gómez, M.; Castagnini, J.M.; Carbó, E.; Ferrer, E.; Berrada, H.; Barba, F.J. Evaluation of Pulsed Electric Field-Assisted Extraction on the Microstructure and Recovery of Nutrients and Bioactive Compounds from Mushroom (*Agaricus bisporus*). *Separations* **2022**, *9*, 302. [CrossRef]
144. Moreira, S.A.; Alexandre, E.M.; Pintado, M.; Saraiva, J.A. Effect of emergent non-thermal extraction technologies on bioactive individual compounds profile from different plant materials. *Food Res. Int.* **2019**, *115*, 177–190. [CrossRef]
145. Thongkong, S.; Klangpetch, W.; Unban, K.; Tangjaidee, P.; Phimolsiripol, Y.; Rachtanapun, P.; Jantanasakulwong, K.; Schönlechner, R.; Thipchai, P.; Phongthai, S. Impacts of Electroextraction Using the Pulsed Electric Field on Properties of Rice Bran Protein. *Foods* **2023**, *12*, 835. [CrossRef]
146. Zhao, W.; Yang, R. Pulsed Electric Field Induced Aggregation of Food Proteins: Ovalbumin and Bovine Serum Albumin. *Food Bioprocess. Technol.* **2012**, *5*, 1706–1714. [CrossRef]
147. Liu, Y.Y.; Zeng, X.A.; Deng, Z.; Yu, S.J.; Yamasaki, S. Effect of pulsed electric field on the secondary structure and thermal properties of soy protein isolate. *Eur. Food Res. Technol.* **2011**, *233*, 841–850. [CrossRef]
148. Wu, L.; Zhao, W.; Yang, R.; Chen, X. Effects of pulsed electric fields processing on stability of egg white proteins. *J. Food Eng.* **2014**, *139*, 13–18. [CrossRef]
149. Xiang, B.Y.; Simpson, M.V.; Ngadi, M.O.; Simpson, B.K. Effect of pulsed electric field on the rheological and colour properties of soy milk. *Int. J. Food Sci. Nutr.* **2011**, *62*, 787–793. [CrossRef] [PubMed]
150. Xiang, B.Y.; Ngadi, M.O.; Ochoa-Martinez, L.A.; Simpson, M.V. Pulsed Electric Field-Induced Structural Modification of Whey Protein Isolate. *Food Bioprocess Technol.* **2011**, *4*, 1341–1348. [CrossRef]
151. Cullen, P.J.; Tiwari, B.K.; Valdramidis, V. *Novel Thermal and Non-Thermal Technologies for Fluid Foods*; Academic Press: Cambridge, MA, USA, 2011; pp. 65–66. [CrossRef]
152. El-Hag, A.H.; Rodriguez Gonzalez, O.; Jayaram, S.H.; Griffiths, M.W. A Performance Study of a Multilevel Electrode Treatment Chamber for Food Processing. *IEEE Trans. Ind. Appl.* **2013**, *49*, 1091–1097. [CrossRef]
153. Huang, K.; Wang, J. Designs of Pulsed Electric Fields Treatment Chambers for Liquid Foods Pasteurization Process: A Review. *J. Food Eng.* **2009**, *95*, 227–239. [CrossRef]
154. Qin, S.; Timoshkin, I.V.; Maclean, M.; Wilson, M.P.; Given, M.J.; Wang, T.; Anderson, J.G.; Macgregor, S.J. Pulsed electric field treatment of saccharomyces cerevisiae using different waveforms. *IEEE Trans. Dielectr. Electr. Insul.* **2015**, *22*, 1841–1848. [CrossRef]
155. Qin, S.; Timoshkin, I.V.; Maclean, M.; Wilson, M.P.; Given, M.J.; Wang, T.; Anderson, J.G.; Macgregor, S.J. TiO_2-Coated Electrodes for Pulsed Electric Field Treatment of Microorganisms. *IEEE Trans. Plasma Sci.* **2016**, *44*, 2121–2128. [CrossRef]
156. Achayuthakan, P.; Wongsagonsup, R.; Sriprablom, J.; Suphantharika, M.; Intra, P. Effect of Pulsed Electric Field Treatment on the Protein, Digestibility, and Physicochemical Properties of Starch Granules in Wheat Flour. *Polymers* **2023**, *15*, 4087. [CrossRef]
157. Raso, J.; Frey, W.; Ferrari, G.; Pataro, G.; Knorr, D.; Teissie, J.; Miklavčič, D. Recommendations guidelines on the key information to be reported in studies of application of PEF technology in food and biotechnological processes. *Innov. Food Sci. Emerg. Technol.* **2016**, *37*, 312–321. [CrossRef]
158. Andreou, V.; Dimopoulos, G.; Dermesonlouoglou, E.; Taoukis, P. Application of pulsed electric fields to improve product yield and waste valorization in industrial tomato processing. *J. Food Eng.* **2020**, *270*, 109778. [CrossRef]
159. Parniakov, O.; Roselló-Soto, E.; Barba, F.J.; Grimi, N.; Lebovka, N.; Vorobiev, E. New approaches for the effective valorization of papaya seeds: Extraction of proteins, phenolic compounds, carbohydrates, and isothiocyanates assisted by pulsed electric energy. *Food Res. Int.* **2015**, *77*, 711–717. [CrossRef]
160. Qin, B.L.; Zhang, Q.; Canovas, G.V.B. Inactivation of microorganisms by Pulsed Electric Fields of different voltage waveforms. *IEEE Trans. Dielectr. Electr. Insul.* **1994**, *1*, 1047–1057. [CrossRef]
161. Canovas, G.V.B.; Tapia, M.S.; Cano, M.P. *Novel Food Processing Technologies*; CRC Press: Boca Raton, FL, USA, 2004.
162. Brito, P.S.; Canacsinh, H.; Mendes, J.P.; Redondo, L.M.; Pereira, M.T. Comparison between monopolar and bipolar microsecond range Pulsed Electric Fields in enhancement of apple juice extraction. *IEEE Trans. Plasma Sci.* **2012**, *40*, 2348–2354. [CrossRef]
163. Sobrino-López, Á.; Raybaudi-Massilia, R.; Martín-Belloso, O. High intensity pulsed electric field variables affecting Staphylococcus aureus inoculated in milk. *J. Dairy Sci.* **2006**, *89*, 3739–3748. [CrossRef] [PubMed]
164. Qin, B.L.; Barbosa-Canovas, G.V.B.; Swanson, B.G.; Pedrow, P.D.; Olsen, R.G. Inactivating microorganisms using a Pulsed Electric Field continuous treatment system. *IEEE Trans. Ind. Appl.* **1998**, *34*, 43–50. [CrossRef]
165. Parniakov, O.; Barba, F.J.; Grimi, N.; Lebovka, N.; Vorobiev, E. Impact of pulsed electric fields and high voltage electrical discharges on extraction of high-added value compounds from papaya peels. *Food Res. Int.* **2014**, *65*, 337–343. [CrossRef]

166. Qin, B.; Zhang, Q.; Barbosa-Canovas, G.V.; Swanson, B.G.; Pedrow, P.D. Pulsed electric field treatment chamber design for liquid food pasteurization using a finite element method. *Trans. ASAE* **1995**, *38*, 557–565. [CrossRef]
167. Ghoshal, G. Comprehensive review on pulsed electric field in food preservation: Gaps in current studies for potential future research. *Heliyon* **2023**, *9*, e17532. [CrossRef]
168. Tamborrino, A.; Mescia, L.; Taticchi, A.; Berardi, A.; Lamacchia, C.M.; Leone, A.; Servili, M. Continuous pulsed electric field pilot plant for olive oil extraction process. *Innov. Food Sci. Emerg. Technol.* **2022**, *82*, 103192. [CrossRef]
169. Schottroff, F.; Knappert, J.; Eppmann, P.; Krottenthaler, A.; Horneber, T.; McHardy, C.; Rauh, C.; Jaeger, H. Development of a Continuous Pulsed Electric Field (PEF) Vortex-Flow Chamber for Improved Treatment Homogeneity Based on Hydrodynamic Optimization. *Front. Bioeng. Biotechnol.* **2020**, *8*, 340. [CrossRef] [PubMed]
170. Arshad, R.N.; Abdul Malek, Z.; Munir, A.; Ahmad, M.H.; Nawawi, Z.; Sidik, M.A.B.; Jumani, T.A.; Khan, I.; Alotabi, H.; Khan, A. An improved electroporator with continuous liquid flow and double-exponential waveform for liquid food pasteurization. *IEEE Access.* **2021**, *9*, 147732–147742. [CrossRef]
171. Seshakamal, J.H.; Kameswara, L.R. Electric Field Fluid Treatment Chamber. CA Patent CA2560858C, 22 September 2006.
172. Rego, D.; Costa, L.; Pereira, M.T.; Redondo, L.M. Cell Membrane Permeabilization Studies of *Chlorella* sp. by Pulsed Electric Fields. *IEEE Trans. Plasma Sci.* **2015**, *43*, 3483–3488. [CrossRef]
173. Evrendilek, G.A.; Li, S.; Dantzer, W.R.; Zhang, Q.H. Pulsed Electric Field Processing of Beer: Microbial, Sensory, and Quality Analyses. *J. Food Sci.* **2004**, *69*, M228–M232. [CrossRef]
174. Donsì, G.; Ferrari, G.; Pataro, G. Inactivation kinetics of Saccharomyces cerevisiae by pulsed electric fields in a batch treatment chamber: The effect of electric field unevenness and initial cell concentration. *J. Food Eng.* **2007**, *78*, 784–792. [CrossRef]
175. Toepfl, S.; Volker, H.; Dietrich, K. Applications of pulsed electric fields technology for the food industry. In *Pulsed Electric Fields Technology for the Food Industry*; Springer: Berlin/Heidelberg, Germany, 2006; pp. 197–221.
176. Kotnik, T.; Kramar, P.; Pucihar, G.; Miklavcic, D.; Tarek, M. Cell membrane electroporation-Part 1: The phenomenon. *IEEE Electr. Insul. Mag.* **2012**, *28*, 14–23. [CrossRef]
177. Sugar, I.P.; Neumann, E. Stochastic model for electric field-induced membrane pores electroporation. *Biophys. Chem.* **1984**, *19*, 211–225. [CrossRef]
178. Kandušer, M.; Belič, A.; Čorović, S.; Škrjanc, I. Modular Serial Flow Through device for pulsed electric field treatment of the liquid samples. *Sci. Rep.* **2017**, *7*, 8115. [CrossRef]
179. Sale, A.J.H.; Hamilton, W.A. Effect of high electric fields on microorganisms. I. Killing of bacteria and yeast. *Biochim. Biophys. Acta.* **1967**, *148*, 781–788. [CrossRef]
180. Saulis, G. Electroporation of Cell Membranes: The Fundamental Effects of Pulsed Electric Fields in Food Processing. *Food Eng. Reviews.* **2010**, *2*, 52–73. [CrossRef]
181. Martí-Quijal, F.J.; Ramon-Mascarell, F.; Pallarés, N.; Ferrer, E.; Berrada, H.; Phimolsiripol, Y.; Barba, F.J. Extraction of Antioxidant Compounds and Pigments from Spirulina (*Arthrospira platensis*) Assisted by Pulsed Electric Fields and the Binary Mixture of Organic Solvents and Water. *Appl. Sci.* **2021**, *11*, 7629. [CrossRef]
182. Vernon-Parry, K.D. Scanning Electron Microscopy: An Introduction. *III-Vs Rev.* **2000**, *13*, 40–44. [CrossRef]
183. Sarkis, J.R.; Boussetta, N.; Blouet, C.; Tessaro, I.C.; Marczak, L.D.F.; Vorobiev, E. Effect of pulsed electric fields and high voltage electrical discharges on polyphenol and protein extraction from sesame cake. *Innov. Food Sci. Emerg. Technol.* **2015**, *29*, 170–177. [CrossRef]
184. Jaeschke, D.P.; Mercali, G.D.; Marczak, L.D.F.; Müller, G.; Frey, W.; Gusbeth, C. Extraction of valuable compounds from Arthrospira platensis using pulsed electric field treatment. *Bioresour. Technol.* **2019**, *283*, 207–212. [CrossRef] [PubMed]
185. Gateau, H.; Blanckaert, V.; Veidl, B.; Burlet-Schiltz, O.; Pichereaux, C.; Gargaros, A.; Marchand, J.; Schoefs, B. Application of pulsed electric fields for the biocompatible extraction of proteins from the microalga *Haematococcus pluvialis. Bioelectrochemistry* **2021**, *137*, 107588. [CrossRef] [PubMed]
186. Yu, X.; Gouyo, T.; Grimi, N.; Bals, O.; Vorobiev, E. Pulsed electric field pretreatment of rapeseed green biomass (stems) to enhance pressing and extractives recovery. *Bioresour. Technol.* **2016**, *199*, 194–201. [CrossRef] [PubMed]
187. Kumari, B.; Tiwari, B.K.; Walsh, D.; Griffin, T.P.; Islam, N.; Lyng, J.G.; Brunton, N.P.; Rai, D.K. Impact of pulsed electric field pre-treatment on nutritional and polyphenolic contents and bioactivities of light and dark brewer's spent grains. *Innov. Food Sci. Emerg. Technol.* **2019**, *54*, 200–210. [CrossRef]
188. Kronbauer, M.; Shorstkii, I.; Silva, S.; Toepfl, S.; Lammerskitten, A.; Siemer, C. Pulsed electric field assisted extraction of soluble proteins from nettle leaves (*Urtica dioica* L.): Kinetics and optimization using temperature and specific energy. *Sustain. Food Technol.* **2023**, *1*, 886–895. [CrossRef]
189. Li, X.; Li, J.; Wang, R.; Rahaman, A.; Zeng, S.X.-A.; Brennan, C.S. Combined effects of pulsed electric field and ultrasound pretreatments on mass transfer and quality of mushrooms. *LWT* **2021**, *150*, 112008. [CrossRef]
190. Boussetta, N.; Soichi, E.; Lanoisellé, J.L.; Vorobiev, E. Valorization of oilseed residues: Extraction of polyphenols from flaxseed hulls by pulsed electric fields. *Ind. Crops Prod.* **2014**, *52*, 347–353. [CrossRef]
191. Martinchik, A.N. Nutritional value of sesame seeds. *Vopr. Pitan.* **2011**, *80*, 41–43. [PubMed]
192. Kamboj, A.; Chopra, R.; Singh, R.; Saxena, V.; Prassana Kumar, G.V. Effect of pulsed electric field parameters on the alkaline extraction of valuable compounds from perilla seed meal and optimization by central composite design approach. *Appl. Food Res.* **2022**, *2*, 100240. [CrossRef]

193. Pappas, V.M.; Lakka, A.; Palaiogiannis, D.; Athanasiadis, V.; Bozinou, E.; Ntourtoglou, G.; Makris, D.P.; Dourtoglou, V.G.; Lalas, S.I. Optimization of Pulsed Electric Field as Standalone "Green" Extraction Procedure for the Recovery of High Value-Added Compounds from Fresh Olive Leaves. *Antioxidants* **2021**, *10*, 1554. [CrossRef]
194. Genovese, J.; Kranjc, M.; Serša, I.; Petracci, M.; Rocculi, P.; Miklavčič, D.; Mahnič-Kalamiza, S. PEF-treated plant and animal tissues: Insights by approaching with different electroporation assessment methods. *Innov. Food Sci. Emerg. Technol.* **2021**, *74*, 102872. [CrossRef]
195. Saldaña, M.D.A.; Silva, E.K.; Olmos Cornejo, J.E.; Orozco Lopez, C.L. Comprehensive foodomics. In *Green Processes in Foodomics: Biorefineries in the Food Industry*; Elsevier: Amsterdam, The Netherlands, 2021; pp. 808–824. [CrossRef]
196. Taha, A.; Casanova, F.; Šimonis, P.; Stankevič, V.; Gomaa, M.A.E.; Stirkė, A. Pulsed Electric Field: Fundamentals and Effects on the Structural and Techno-Functional Properties of Dairy and Plant Proteins. *Foods* **2022**, *11*, 1556. [CrossRef] [PubMed]
197. Coustets, M.; Joubert-Durigneux, V.; Hérault, J.; Schoefs, B.; Blanckaert, V.; Garnier, J.-P.; Teissié, J. Optimization of protein electroextraction from microalgae by a flow process. *Bioelectrochemistry* **2015**, *103*, 74–81. [CrossRef] [PubMed]
198. Segovia, F.J.; Luengo, E.; Corral-Pérez, J.J.; Raso, J.; Almajano, M.P. Improvements in the aqueous extraction of polyphenols from borage (*Borago officinalis* L.) leaves by pulsed electric fields: Pulsed electric fields (PEF) applications. *Ind. Crops Prod.* **2015**, *65*, 390–396. [CrossRef]
199. Priyadarshini, A.; Rajauria, G.; O'Donnell, C.P.; Tiwari, B.K. Emerging food processing technologies and factors impacting their industrial adoption. *Crit. Rev. Food Sci. Nutr.* **2019**, *59*, 3082–3101. [CrossRef]
200. Schoenbach, K.H.; Joshi, R.P.; Stark, R.H.; Dobbs, F.C.; Beebe, S.J. Bacterial decontamination of liquids with pulsed electric fields. *IEEE Trans. Dielectr. Electr. Insul.* **2000**, *7*, 637–645. [CrossRef]
201. Safi, C.; Charton, M.; Pignolet, O.; Silvestre, F.; Vaca-Garcia, C.; Pontalier, P.-Y. Influence of microalgae cell wall characteristics on protein extractability and determination of nitrogen-to-protein conversion factors. *J. Appl. Phycol.* **2013**, *25*, 523–529. [CrossRef]
202. Gad, A.; Jayaram, S.H.; Pritzker, M. Performance of electrode materials during food processing by Pulsed Electric Fields. *IEEE Trans. Plasma Sci.* **2014**, *42*, 3161–3166. [CrossRef]
203. Moonesan, M.S.; Jayaram, S.H. Effect of pulsewidth on medium temperature rise and microbial inactivation under Pulsed Electric Field food treatment. *IEEE Trans. Ind. Appl.* **2013**, *49*, 1767–1772. [CrossRef]
204. Joana Gil-Chávez, G.; Villa, J.A.; Fernando Ayala-Zavala, J.; Basilio Heredia, J.; Sepulveda, D.; Yahia, E.M.; González-Aguilar, G.A. Technologies for Extraction and Production of Bioactive Compounds to be Used as Nutraceuticals and Food Ingredients: An Overview. *Compr. Rev. Food Sci. Food Saf.* **2013**, *12*, 5–23. [CrossRef]
205. Carpentieri, S.; Jambrak, A.R.; Ferrari, G.; Pataro, G. Pulsed Electric Field-Assisted Extraction of Aroma and Bioactive Compounds from Aromatic Plants and Food By-Products. *Front. Nutr.* **2022**, *8*, 792203. [CrossRef] [PubMed]
206. Lajoie, L.; Fabiano-Tixier, A.-S.; Chemat, F. Water as Green Solvent: Methods of Solubilisation and Extraction of Natural Products—Past, Present and Future Solutions. *Pharmaceuticals* **2022**, *15*, 1507. [CrossRef] [PubMed]
207. Wang, M.; Zhou, J.; Castagnini, J.M.; Berrada, H.; Barba, F.J. Pulsed electric field (PEF) recovery of biomolecules from Chlorella: Extract efficiency, nutrient relative value, and algae morphology analysis. *Food Chem.* **2023**, *404*, 134615. [CrossRef]
208. Ranjha, M.M.A.N.; Kanwal, R.; Shafique, B.; Arshad, R.N.; Irfan, S.; Kieliszek, M.; Kowalczewski, P.Ł.; Irfan, M.; Khalid, M.Z.; Roobab, U.; et al. A Critical Review on Pulsed Electric Field: A Novel Technology for the Extraction of Phytoconstituents. *Molecules* **2021**, *26*, 4893. [CrossRef]
209. Parniakov, O.; Barba, F.J.; Grimi, N.; Marchal, L.; Jubeau, S.; Lebovka, N.; Vorobiev, E. Pulsed electric field and pH assisted selective extraction of intracellular components from microalgae *Nannochloropsis*. *Algal Res.* **2015**, *8*, 128–134. [CrossRef]
210. Carpentieri, S.; Mazza, L.; Nutrizio, M.; Jambrak, A.R.; Ferrari, G.; Pataro, G. Pulsed electric fields- and ultrasound-assisted green extraction of valuable compounds from *Origanum vulgare* L. and *Thymus serpyllum* L. *Int. J. Food Sci. Technol.* **2021**, *56*, 4834–4842. [CrossRef]
211. Toepfl, S. Pulsed electric field food processing -industrial equipment design and commercial applications. *Stewart Postharvest Rev.* **2012**, *8*, 1–7. [CrossRef]
212. Timmermans, R.A.H.; Mastwijk, H.C.; Berendsen, L.B.J.M.; Nederhoff, A.L.; Matser, A.M.; Van Boekel, M.A.J.S.; Nierop Groot, M.N. Moderate intensity Pulsed Electric Fields (PEF) as alternative mild preservation technology for fruit juice. *Int. J. Food Microbiol.* **2019**, *298*, 63–73. [CrossRef]
213. Li, X.; Farid, M. A review on recent development in non-conventional food sterilization technologies. *J. Food Eng.* **2016**, *182*, 33–45. [CrossRef]
214. Kempkes, M. Industrial pulsed electric field systems. In *Handbook of Electroporation*; Miklavcic, D., Ed.; Springer: Berlin/Heidelberg, Germany, 2020; pp. 1–22. [CrossRef]
215. Bhat, Z.F.; Morton, J.D.; Mason, S.L.; Bekhit, A.E.-D.A. Current and future prospects for the use of pulsed electric field in the meat industry. *Crit. Rev. Food Sci. Nutr.* **2019**, *59*, 1660–1674. [CrossRef] [PubMed]
216. Arshad, R.N.; Abdul-Malek, Z.; Munir, A.; Buntat, Z.; Ahmad, M.H.; Jusoh, Y.M.M.; Bekhit, A.E.-D.; Roobab, U.; Manzoor, M.F.; Aadil, R.M. Electrical systems for pulsed electric field applications in the food industry: An engineering perspective. *Trends Food Sci. Technol.* **2020**, *104*, 1–13. [CrossRef]
217. Puértolas, E.; Barba, F. Electrotechnologies applied to valorization of by-products from food industry: Main findings, energy and economic cost of their industrialization. *Food Bioprod. Process.* **2016**, *100*, 172–184. [CrossRef]

218. Toepfl, S.; Kinsella, J.; Parniakov, O. *Industrial Scale Equipment, Patents, and Commercial Applications in Pulsed Electric Fields to Obtain Healthier and Sustainable Food for Tomorrow*; Academic Press: Cambridge, MA, USA, 2020. [CrossRef]
219. Jaeger, H.; Balasa, A.; Knorr, D. Food industry applications for pulsed electric fields. In *Electrotechnologies for Extraction from Food Plants and Biomaterials*; Food Engineering Series; Springer: New York, NY, USA, 2009; pp. 181–216. [CrossRef]
220. Mohammed, M.; Eiss, A.A. Pulsed electric fields for food processing technology. In *Structure and Function of Food Engineering*; Eissa, A.A., Ed.; Intech: Rijeka, Croatia, 2012; pp. 275–306. [CrossRef]
221. Morren, J.; Roodenburg, B.; de Haan, S.W.H. Electrochemical reactions and electrode corrosion in pulsed electric field (PEF) treatment chambers. *Innov. Food Sci. Emerg. Technol.* **2003**, *4*, 285–295. [CrossRef]
222. Pataro, G.; Barca, G.M.J.; Donsì, G.; Ferrari, G. On the modeling of electrochemical phenomena at the electrode-solution interface in a PEF treatment chamber: Methodological approach to describe the phenomenon of metal release. *J. Food Eng.* **2015**, *165*, 34–44. [CrossRef]
223. Samaranayake, C.P.; Sastry, S.K.; Zhang, H. Pulsed Ohmic Heating–A Novel Technique for Minimization of Electrochemical Reactions During Processing. *J. Food Sci.* **2005**, *70*, e460–e465. [CrossRef]
224. Kotnik, T.; Miklavčič, D.; Mir, L.M. Cell membrane electropermeabilization by symmetrical bipolar rectangular pulses: Part II. Reduced electrolytic contamination. *Bioelectrochemistry* **2001**, *54*, 91–95. [CrossRef]
225. Gad, A.; Jayaram, S.H. Effect of electric pulse parameters on releasing metallic particles from stainless steel electrodes during PEF processing of milk. *IEEE Trans. Ind. Appl.* **2013**, *50*, 1402–1409. [CrossRef]
226. Bocker, R.; Silva, E.K. Pulsed electric field assisted extraction of natural food pigments and colorings from plant matrices. *Food Chem. X* **2022**, *15*, 100398. [CrossRef] [PubMed]
227. Picart-Palmade, L.; Cunault, C.; Chevalier-Lucia, D.; Belleville, M.-P.; Marchesseau, S. Potentialities and Limits of Some Non-thermal Technologies to Improve Sustainability of Food Processing. *Front. Nutr.* **2019**, *5*, 130. [CrossRef] [PubMed]
228. Martínez, J.M.; Delso, C.; Álvarez, I.; Raso, J. Pulsed Electric Field-assisted extraction of valuable compounds from microorganisms. *Compr. Rev. Food Sci. Food Saf.* **2020**, *19*, 530–552. [CrossRef] [PubMed]
229. Schottroff, F.; Gratz, M.; Krottenthaler, A.; Johnson, N.B.; Bédard, M.F.; Jaeger, H. Pulsed electric field preservation of liquid whey protein formulations—Influence of process parameters, pH, and protein content on the inactivation of *Listeria innocua* and the retention of bioactive ingredients. *J. Food Eng.* **2019**, *243*, 142–152. [CrossRef]
230. Martínez-Beamonte, R.; Ripalda, M.; Herrero-Continente, T.; Barranquero, C.; Dávalos, A.; López de las Hazas, M.C.; Álvarez-Lanzarote, I.; Sánchez-Gimeno, A.C.; Raso, J.; Arnal, C.; et al. Pulsed electric field increases the extraction yield of extra virgin olive oil without loss of its biological properties. *Front. Nutr.* **2022**, *9*, 1065543. [CrossRef] [PubMed]
231. Manzoor, M.F.; Zeng, X.A.; Rahaman, A.; Siddeeg, A.; Aadil, R.M.; Ahmed, Z.; Li, J.; Niu, D. Combined impact of pulsed electric field and ultrasound on bioactive compounds and FT-IR analysis of almond extract. *J. Food Sci. Technol.* **2019**, *56*, 2355–2364. [CrossRef]
232. Steinbruch, E.; Wise, J.; Levkov, K.; Chemodanov, A.; Israel, A.; Livney, Y.D.; Golberg, A. Enzymatic cell wall degradation combined with pulsed electric fields increases yields of water-soluble-protein extraction from the green marine macroalga *Ulva* sp. *Innov. Food Sci. Emerg. Technol.* **2023**, *84*, 103231. [CrossRef]
233. Zhou, J.; Wang, M.; Berrada, H.; Zhu, Z.; Grimi, N.; Barba, F.J. Pulsed electric fields (PEF), pressurized liquid extraction (PLE) and combined PEF + PLE process evaluation: Effects on Spirulina microstructure, biomolecules recovery and Triple TOF-LC-MS-MS polyphenol composition. *Innov. Food Sci. Emerg. Technol.* **2022**, *77*, 102989. [CrossRef]

processes

Review

Hollow Fiber Membrane Modification by Interfacial Polymerization for Organic Solvent Nanofiltration

Abdulaziz Y. Alammar [1,*], Seung-Hak Choi [1] and Maria Giovanna Buonomenna [2,*]

1 Oil & Gas Treatment Division, Research & Development Center, Saudi Aramco, Dhahran 34472, Saudi Arabia; seunghak.choi@aramco.com
2 Ordine Regionale dei Chimici e Fisici della Campania, 80138 Naples, Italy
* Correspondence: abdulaziz.ammar.1@aramco.com (A.Y.A.); mg.buonomenna@chimici.it (M.G.B.)

Abstract: Hollow fiber (HF) organic solvent nanofiltration (OSN) membranes have recently attracted significant interest in the field of membrane technology. Their popularity stems from comparative advantages, such as high packing density, fouling resistance, and easier scalability for larger applications, unlike flat-sheet/spiral-wound OSN membranes, which may present challenges in these aspects. The combination of interfacial polymerization (IP) and HF configuration has opened up new opportunities for developing advanced membranes with enhanced separation performance that can be tailored for various OSN applications. The objective of this review is to discuss the latest advancements in developing thin film composite (TFC) HF membranes, with a focus on the IP method. Novel materials and processes are discussed in detail, emphasizing the fabrication of greener, interfacially polymerized HF OSN membranes. In addition, the commercial viability and limitations of TFC HF membranes are highlighted, providing perspectives on future research directions.

Keywords: hollow fiber; interfacial polymerization; thin film composite (TFC) membrane; organic solvent nanofiltration

Citation: Alammar, A.Y.; Choi, S.-H.; Buonomenna, M.G. Hollow Fiber Membrane Modification by Interfacial Polymerization for Organic Solvent Nanofiltration. *Processes* **2024**, *12*, 563. https://doi.org/10.3390/pr12030563

Academic Editors: Yanlin Zhang and Prashant K. Sarswat

Received: 14 February 2024
Revised: 7 March 2024
Accepted: 8 March 2024
Published: 13 March 2024

1. Introduction: Setting the Scene

Organic solvents are critical components in a wide range of industries, such as agriculture, pharmaceuticals, oil, petrochemicals, electronics, cosmetics, and food and beverage production [1]. They are predominantly used across all stages of a product's lifecycle due to their ability to dissolve, extract, and facilitate chemical reactions. In these processes, solvents are often required for purification. Particularly in the pharmaceutical and fine chemicals industry, a significant volume of organic solvents is utilized, sometimes constituting as much as 90% of the total chemical consumption on a mass basis [2,3]. The extensive use of organic solvents underscores the necessity for efficient separation methods. However, the standard methods for chemical separations in the industry are distillation, evaporation, and liquid-liquid extraction. Notably, membrane-based processes, such as organic solvent nanofiltration (OSN), also known as organophilic nanofiltration (oNF) or solvent-resistant nanofiltration (SRNF), can save up to 90% of the energy consumed by chemical separations [3,4].

Membrane separation operations, such as reverse osmosis (RO), microfiltration (MF), ultrafiltration (UF) and nanofiltration (NF), electrodialysis (ED), dialysis (D), pervaporation (PV), gas separation (GS), membrane distillation (MD), electro-osmosis (EO), and forwarded osmosis (FO) (Table 1), are largely used in many different applications [5]. The OSN process, as a pressure-driven membrane process (Table 1), requires an external hydraulic pressure to make the separation among the components of the feed. Depending on the operating pressure, two different types of processes can be distinguished, i.e., OSN, which uses relatively low pressures (<30 bar) for the separation of solutes with molecular weights in the range of 150–1000 g mol^{-1}, and organic solvent reverse osmosis (OSRO), where the required operating pressure to separate solutes with molecular weights <150 gmol^{-1}

is higher than 30 bar [6,7]. Moreover, organic solvent forward osmosis (OSFO) (Table 1) has been demonstrated recently [8] as a promising technology for solvent recovery, as it possesses a reasonable solvent flux, low reverse solute flux, and requires no external pressure. Indeed, the solvent, due to its chemical potential gradient between the two membrane sides, spontaneously flows from the feed solution to the draw solution, thus concentrating the feed solution.

Table 1. Classification of membrane processes according to their driving forces. Adapted from [5] under CCBY-4.0.

Driving Force	Pressure Difference (DP)	Osmotic Pressure Difference ($\Delta\pi$) [Solvent Chemical Potential Gradient ($\Delta\mu$)]	Concentration Difference (ΔC)	Temperature Difference (ΔT)	Electrical Potential Difference (Δf)
Phenomenological equation	Darcy's law	Fick's law	Fick's law	Fourier's law	Ohm's law
Membrane operations	MF UF NF OSN OSRO RO	FO OSFO	GS PV D RO	MD	ED EO

The OSN process consumes less energy, by an order of magnitude, compared to the traditional thermal separations, owing to its operation at lower temperatures [9,10]; hence, there is no need for phase change, such as vaporization and/or condensation steps, to perform the separation [11]. Additionally, OSN membranes provide more selective separation, being particularly beneficial for complex mixtures and facilities with space constraints due to their smaller footprint, along with preserving the quality of heat-sensitive molecules.

There are primarily three configurations for OSN membranes: flat-sheet (FS), hollow fiber (HF), and tubular configurations, as shown in Figure 1 [12]. FS configuration is the simplest form of an OSN membrane, often used in lab-scale experiments and small-scale applications [13]. The popularity of FS OSN membranes in academia can be attributed to being simpler to fabricate, handle, and characterize in lab settings compared to the other configurations. Flat sheets of membrane material can be stacked/layered in a plate-and-frame module or wound around a central tube, resembling a spiral-wound (SW) module, to maximize the filtration area. On the other hand, a hollow fiber (HF) configuration consists of thin, hollow strands of membrane material with an internal diameter typically less than 1 mm [14]. A similar configuration is a tubular configuration, which is larger in diameter compared to the HF configuration (i.e., internal diameter > 1 mm) [15]. They are often used in microfiltration (MF) and ultrafiltration (UF) applications, but have also been used recently to develop ceramic membranes with MoS_2 selective layers [16] and polymeric selective layers [17] for OSN applications. While FS OSN membranes are more reported in the literature, there has been recently a substantial research interest in HF OSN membranes due to their self-supporting structure and high surface area-to-volume ratio, enabling dense packaging in a single module and allowing ease of scalability and separation efficiency in processing large volumes of solvent [18,19].

HF OSN membranes can be classified based on their structure/morphology into integrally skinned asymmetric (ISA) and composite membranes. ISA OSN membranes are typically fabricated using phase inversion and crosslinking methods to be stable in a range of organic solvents. They are characterized by a monolithic structure, consisting of a dense top layer (selective layer) and a porous sublayer (support layer), which are made from the same material. ISA OSN membranes are widely studied in academia and commercialized in SW module configuration from, e.g., Evonik (Essen, Germany), GMT

Membrantechnik (Rheinfelden, Germany) (the parent company is Borsig GmbH, Berlin, Germany) and SolSep BV (Apeldoorn, The Netherlands). However, crosslinking the ISA HF membranes, to enable adequate chemical stability and molecular sieving in the nanoscale range, increases the solvent transport resistance through its thick structure and reduces the flux/permeance. On the other hand, composite OSN membranes can be prepared by coextrusion, dip coating, spray coating, grafting, introducing nanomaterials and interfacial polymerization. Even though in co-extrusion processes, a coating layer forms in situ onto a support membrane in a one-step process, its major challenge concerns the delamination of the two layers because the dope solution undergoes phase separation at both the inner and outer surfaces. Instead, the IP process allows the preparation of thin film composites (TFCs) with independently optimized layered structures, although commonly resulting in better adhesion between the support and the top layer. A thin selective layer is made of one material at the interface between two immiscible liquids, supported by a porous layer made of another material (relatively cost-effective), for mechanical support. Because of this, IP is a major breakthrough compared to the above mentioned processes.

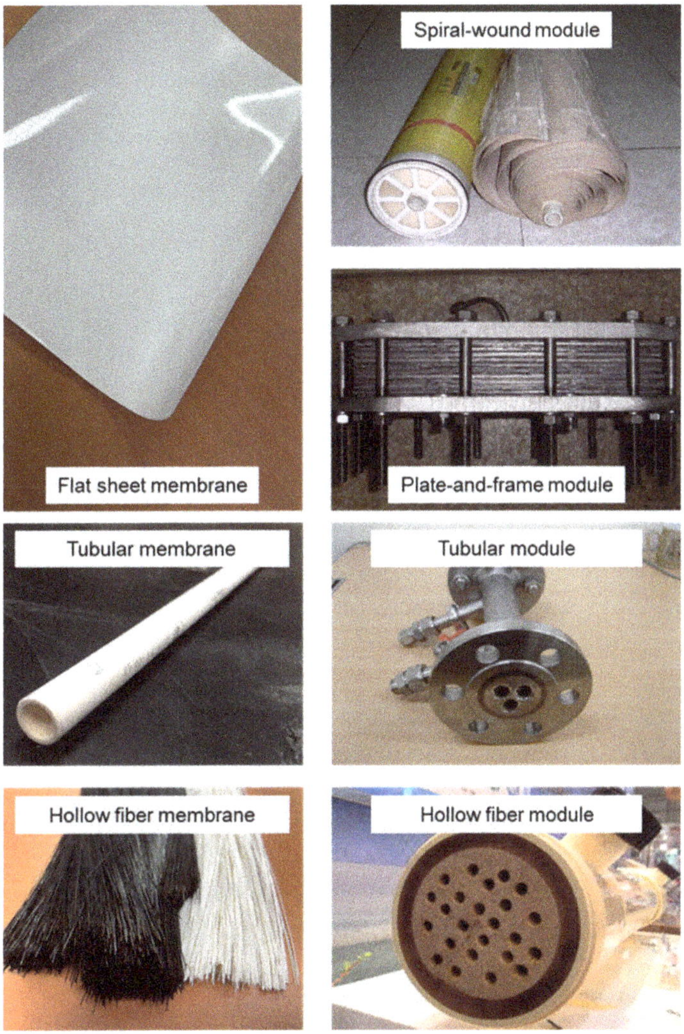

Figure 1. Classification of membrane geometry types and corresponding modules.

Numerous review articles have been published on the latest material developments, fabrication methods and post-treatment methods, with a focus on FS OSN membranes [13,20–23] and their industrial applications [24,25]. Additionally, there are extensive reviews covering HF NF membranes for aqueous solutions and gas separation, which also briefly explore OSN as a potential application [14,26,27]. Fewer review articles have been published on HF OSN membranes in comparison, which mainly discuss in detail crosslinking and its methods [10], as well as polybenzimidazole (PBI) HF membranes [19]. One cannot find a recent review specifically addressing HF OSN membranes, including their material development, fabrication, and treatment methods [12]. This review aims to highlight the current research advancements in the fabrication of TFC HF membranes, with a focus on interfacial polymerization and the use of bio-based materials. It also aims to discuss the challenges encountered in OSN technology, its commercial viability, and future perspectives.

2. Separation Performance of HF OSN Membranes

2.1. Separation Mechanism: The Swelling Challenge

While both OSN and NF membranes for aqueous solutions share the fundamental principle of size-exclusion separation, the specifics of their separation mechanisms differ significantly, due to the distinct properties of organic solvents and aqueous solutions [28]. The pore size of OSN membranes is typically tailored to the size and shape of the solutes, as the size exclusion is the dominant separation factor [11]. However, the separation mechanism can involve a combination of size exclusion, solubility-diffusion, and possibly adsorptive and electrostatic interactions between charged solutes and the membrane selective layer. The solvent/membrane interactions are more pronounced in the separation performance of polymeric OSN membranes than the solute/solvent interactions [23]. The solubility and swelling behavior of polymers in solvents are often predicted using the Hildebrandt and Hansen solubility parameters [29,30]. In general, membranes are less chemically stable in solvents with solubility parameters similar to the polymer, leading to solvation and possible swelling or dissolution. Conversely, membranes show better chemical stability in non-solvents, which have dissimilar solubility parameters and interact less with the polymer. Indeed, OSN membranes often suffer severe swelling and solvation when being exposed to harsh organic solvents, which cause alterations of membrane characteristics and filtration performance [31–33]. Ebert and Cuperus [34] reported that polymeric chains in dense polymeric membranes move further apart during swelling, thus increasing the free volume; the membrane becomes more open (Figure 2a). On the other hand, when a porous membrane swells, the pores become narrower. The membrane becomes less open, which results in higher rejections and the deterioration of flux.

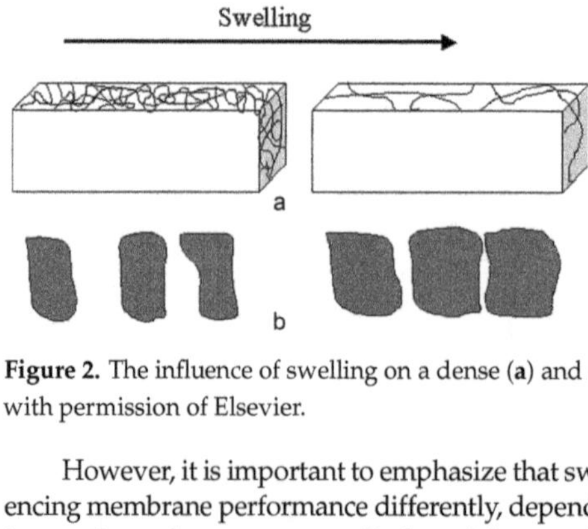

Figure 2. The influence of swelling on a dense (**a**) and a porous (**b**) membrane. Reproduced from [30] with permission of Elsevier.

However, it is important to emphasize that swelling can work in different manners, influencing membrane performance differently, depending on factors such as solvent/membrane interactions, the pressure applied, and the crosslinking of the membrane material. These issues are well described by Tarleton et al. [35], who tested cellulose fiber/PAN/PDMS composite membranes (Figure 3).

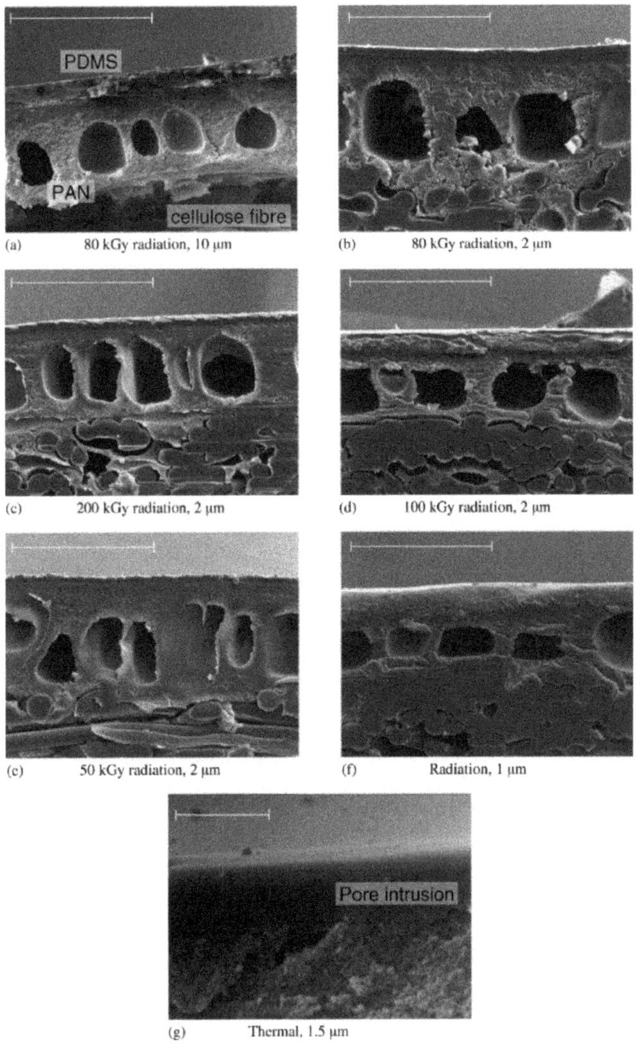

Figure 3. Micrographs of the membranes tested by Tarleton et al. in [35]. Membranes (**a**,**b**,**f**) were each exposed to a "standard" radiation dose of 80 kGy during manufacture (1 kGy = 1 kJ/kg), whilst three other specially prepared samples, designated as membranes (**c**–**e**), were, respectively, treated with radiation doses of 200, 100, and 50 kGy. The dimension on a sub-legend refers to the nominal PDMS thickness in the unswollen state. In the membrane (**g**) the selective PDMS layer was formed by phase-inversion and subsequently crosslinked using a purely thermal technique. The size bar represents a length of 60 μm on (**a**–**f**) and 2 μm on (**g**). Reproduced from [35] with permission of Elsevier.

Tarleton et al. [35] assessed the swelling of four membranes, as reported in Figure 3, across the pressure range of 0–2000 kPa (Figure 4). All of the tested membranes showed a reduction in swelling with increased pressure. This reduction, attributed to membrane compaction, reduces the phenomenon, even though in different ways, dependent upon factors such as the starting thickness of PDMS and the degree of crosslinking.

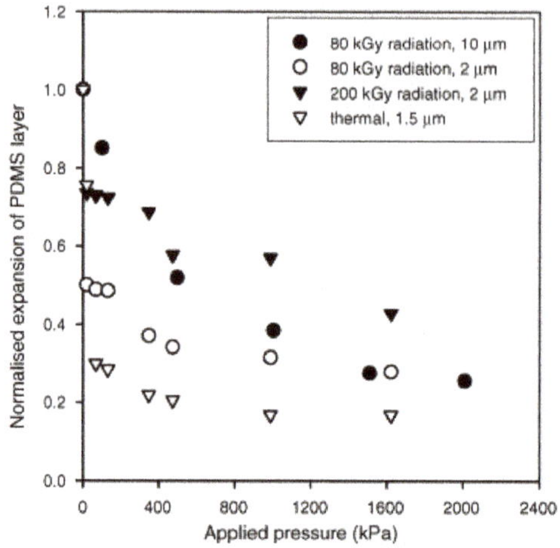

Figure 4. Swelling of the PDMS layer on four composite PAN/PDMS membranes with heptane solvent and a range of applied pressures. Reproduced from [35] with permission of Elsevier.

For membrane materials that are different from rubbery polymers such as PDMS, but are instead based on glassy polymers with high glass transition temperatures, such as polybenzimidazole (PBI) and polyimides (PI), which have high strength, the challenge is their brittleness when being fabricated into self-support hollow fiber membranes [36,37]. The mechanical strength of hollow fiber membranes under high pressure operations has been increased by various cross-linking modifications and fabrication approaches [38–44]. Among cross-linking modifications, one by Chen et al. [44] is worthy of mention, being a green approach: it doesn't use toxic organic solvents such as hexane, but rather a potassium persulphate ($K_2S_2O_8$) water solution at room temperature, as discussed below, in Section 4.5, "Alternative fabrication methods". The approach was successfully employed to modify PBI hollow fibers with superior separation performance for OSN applications [45]. Regarding fabrication approaches, braid-reinforced hollow fibers offer a solution to increase mechanical strength and extend the useful lifetime of hollow fiber membranes [46,47]. Zhao et al. [48] reported braid-reinforced PBI hollow fibers (Figure 5) as solvent-resistant membranes for OSN applications with the aid of the aforementioned green cross-linking method based on the ($K_2S_2O_8$) water solution [10,44].

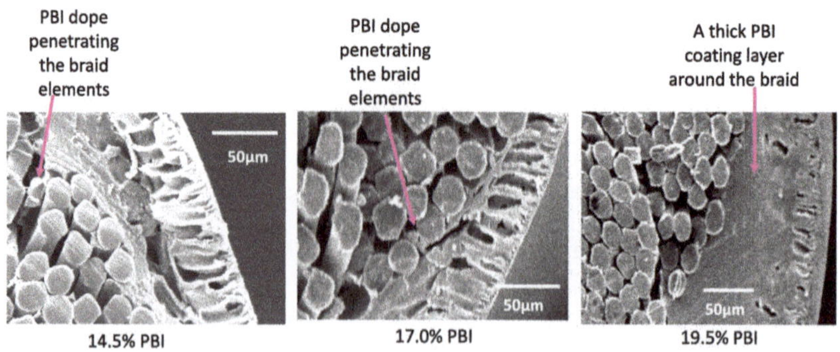

Figure 5. FESEM images of cross-lined braid-reinforced PBI hollow fibers with different dope concentrations. Reproduced from [48] with permission of Elsevier.

2.2. Performance Characterization

The performance of OSN membranes (or OSN process) can be evaluated in terms of productivity and separation efficiency. Productivity indicates how fast the separation can be achieved and expressed as flux or permeance. The permeance can be calculated using Equation (1), in which the permeate volume (V) is divided by the membrane effective area (A), time (t), and the transmembrane pressure (ΔP). Separation efficiency indicates how well separation can be achieved and is commonly denoted as the rejection or removal efficiency as percentage of the solutes rejected by the membrane. Rejection can be calculated using Equation (2) based on permeate concentration (C_p) and feed concentration (C_f).

$$Permeance\left(L\,m^{-2}h^{-1}bar^{-1}\right) = \frac{V}{A \times t \times \Delta P} \quad (1)$$

$$Rejection(\%) = \left(1 - \frac{C_p}{C_f}\right) \times 100 \quad (2)$$

The molecular weight cut-off (MWCO) is defined as the molecular weight of a solute to be rejected at least 90% by the membrane, as shown in Figure 6a, which is often used as a metric to estimate the molecular sieving property of the HF membrane. Various types of dyes, food additives and APIs with different molecular weights (e.g., 100–1000 g mol^{-1}) are used as "markers" or "solutes" to determine the MWCO of HF OSN membranes. However, the types of solutes and solvents used for performance screening in the OSN membrane literature are not standardized. However, it is recommended that solutes with similar physicochemical properties and different molecular sizes be used to determine the MWCO, such as a homologous series of alkenes, polystyrene oligomers, and poly(propylene glycols) (PPGs) [28,49]. Moreover, regarding solvent selection, it is recommended to test at least one solvent from each class, such as apolar solvents (hexane, heptane, toluene), polar protic solvents (isopropanol, methanol, ethanol), mild polar aprotic solvents (acetone, acetonitrile, ethyl acetate), and harsh polar aprotic solvents (DMSO: dimethyl sulfoxide, DMF: *N,N*-dimethylformamide, THF: tetrahydrofuran) [28]. However, alkenes have limited solubility in polar solvents, especially those with higher molecular weights [50]. PPGs are also utilized but pose detection challenges, and they are not as well established in the literature compared to other markers [51]. On the contrary, polystyrene oligomers are commonly used due to their solubility in many organic solvents and commercial availability, although lower molecular weight variants can be expensive [52].

For the OSN process, there are mainly two process configurations (dead-end and cross-flow configurations) to test the separation performance of the HF OSN membranes. The cross-flow configuration is more relevant industrially and the most reported in academia [53], as depicted in Figure 6b, since it minimizes concentration polarization on the membrane surface, allowing continuous operation and steady-state conditions to be reached.

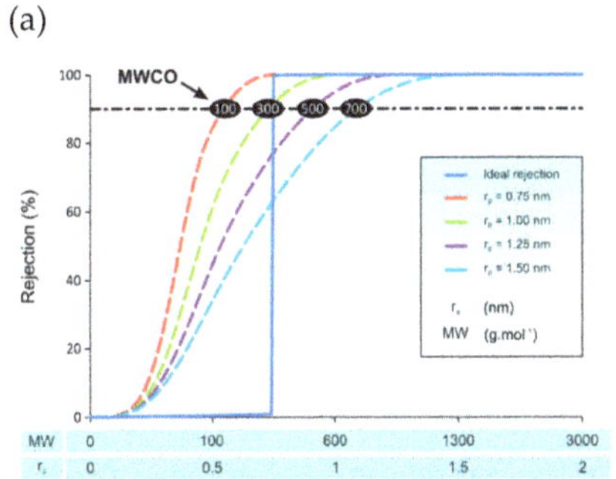

Figure 6. (**a**) Simulated MWCO curves of different membrane (dash lines) are compared to the desired sharp MWCO curve of 100% rejection (solid line). Reprinted with permission from Ref. [11]. Copyright 2014, American Chemical Society. (**b**) Schematic illustration of laboratory scale cross-flow filtration set-up. Reproduced under CCBY 4.0 from Ref. [54].

3. Fabrication of Interfacially Polymerized HF OSN Membranes

3.1. Interfacial Polymerization Fundamentals

The IP process begins with the fabrication of the porous support, which will provide the mechanical stability of the TFC membrane, followed by the dispersion of two immiscible monomer phases, often referred to as the aqueous and organic phases. These monomers possess complementary reactive groups that can undergo a rapid and selective reaction when brought into contact at the interface, as shown in Figure 7. One of the significant advantages of IP is its ability to independently optimize the support layer to be mechanically robust and highly permeable, along with tailoring the selectivity and thickness of the top layer. The selective layer can be optimized by adjusting the reaction conditions, such as concentration and type of monomers, reaction time, and temperature, and by modifying the support layer. Common reactive groups of monomers include amines, phenols, and acid chlorides. A comprehensive review of the IP method and advancements in monomers is conducted by Li et al. [23].

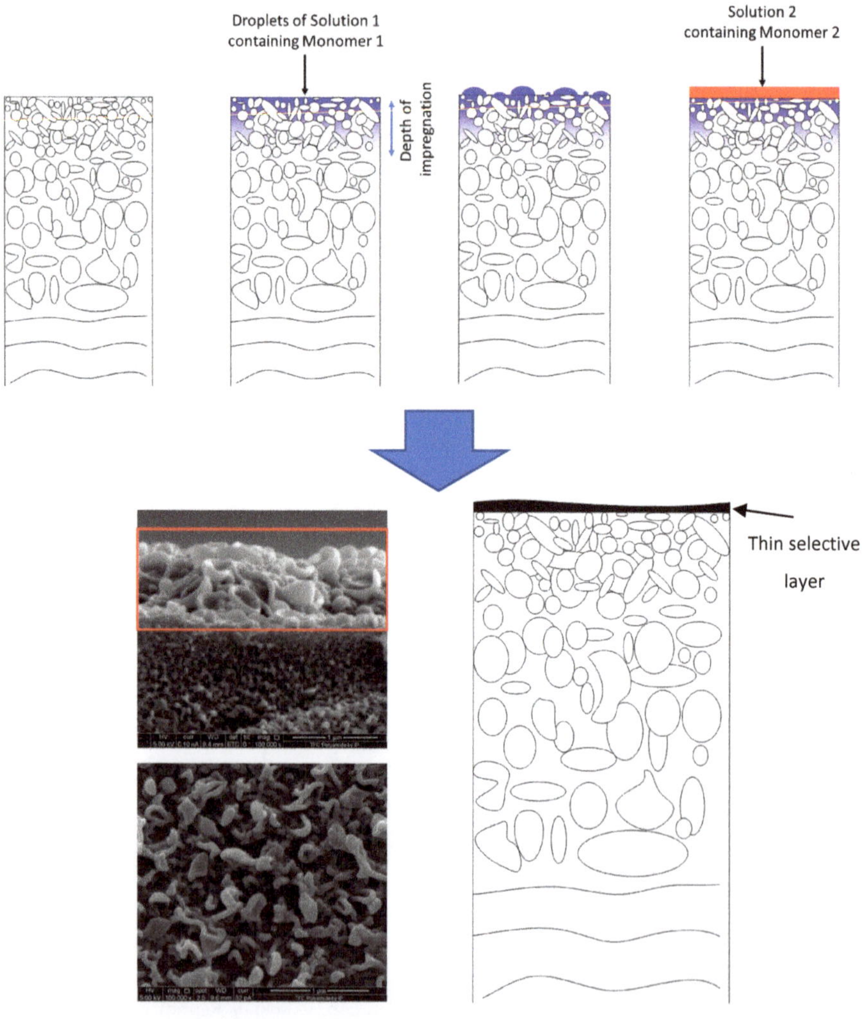

Figure 7. Schematic representation of conventional interfacial polymerization to form TFC membranes.

The IP process for fabricating the TFC HF membrane is described in Figure 8. First, the HF support is stored in a water bath (Bath 1). The porous support then contacts the 1st solution, which contains reactive monomer 1. Depending on the contacting time and the viscosity and surface tension of the solution, the penetration depth of the solution into the porous support can vary. It needs to be controlled well to manage the thickness of the selective layer. The saturation degree can be controlled by increasing or decreasing the immersion time of the support in the reactive solution 1. More specifically, the immersion time is controlled by adjusting the motor speed M1 and the number of loops in Bath 2. Depending on the speed of M1, the tension controller (R1) controls the motor speed of the support bobbin (M2). Removing the excess reactive solution from the membrane surface is one of the most challenging steps, and it can be achieved by an air blower (AB). At a fixed coating speed, adjusting the air flow rate and pressure can completely remove any excess. The saturated porous support with solution 1 then comes into contact with the 2nd solution, which contains reactive monomer 2. Then, the reaction between monomers 1 and 2 occurs at the pores' openings, starting to form the selective layer. Herein, the reaction rate of the monomers is controlled by kinetics and then slowed down by diffusion [55]. The selective layer could hinder reactive monomer diffusion and inhibit further growth, namely the self-sealing process [56]. The resultant fibers pass through this drying tower (DT), where the solvent is dried, and the selective layer is thermally crosslinked. The drying tower consists of several heating guns and temperature sensors to monitor and control the drying temperature. Besides the heating gun, infrared (IR) and/or ultraviolet (UV) light can be employed. The tower length and the temperature profile can determine the drying speed. Finally, the TFC HF membrane is collected on the bobbin in Bath 4. The collecting speed is controlled by a second tension controller (R2) and bobbin-connected motor (M3).

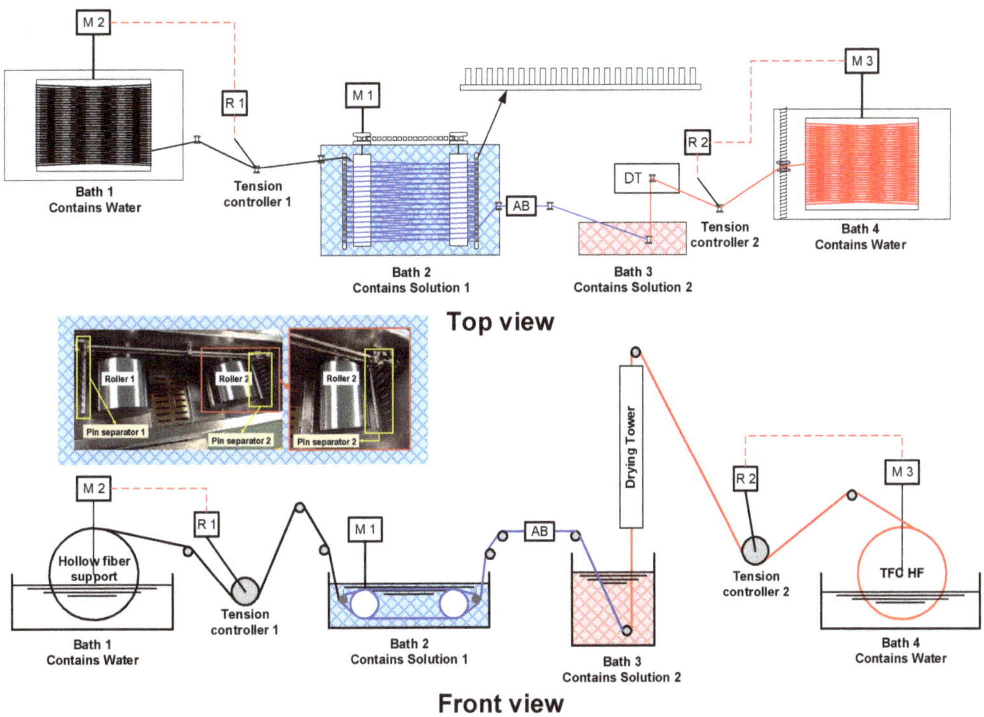

Figure 8. Schematic illustration of a continuous interfacial polymerization procedure for fabricating TFC HF membranes.

3.2. Porous Support

To develop the optimal structure for an interfacial polymerized OSN HF membrane, the design should include a porous support that is both solvent-resistant and mechanically robust, capable of withstanding high pressures while maintaining high permeance. Additionally, it should feature a selective layer with nanofiltration-sized pores to ensure efficient solute rejection. This can be achieved by initially fabricating a UF membrane to act as the support layer, and subsequently applying a thin polyamide selective layer. The optimum porous support (including both ISA and composite morphologies) should have the following features, as mentioned by Chung's group: (i) a no-defect and uniform top layer, (ii) optimized pore size with a narrow distribution, (iii) minimum filtration resistance (i.e., high solvent permeance), and (iv) adequate mechanical stability under high pressures [57]. Additionally, the porous support must be chemically stable in harsh organic solvents, and this is usually achieved by integrally crosslinking the membrane matrix (refer to Section 3.4.1 for a discussion on this matter).

Polymeric membranes, such as polyimide (PI) [57–61], polyacrylonitrile (PAN) [62], polyvinylidene fluoride (PVDF) [63], polypropylene (PP) [32], and polytetrafluoroethylene (PTFE) [64] are widely used to fabricate the porous support due to their cost-effectiveness and ease of synthesis. Phase inversion/separation is still the most common method to fabricate these polymers onto HF porous support. Phase inversion is simply described as a process where a solvent–polymer solution, often referred to as a dope solution, solidifies from a liquid phase to a solid phase. This phase transformation can be achieved by various techniques, including thermally induced phase separation (TIPS), non-solvent induced phase separation (NIPS), adsorption of non-solvent vapor (VIPS), and evaporation-induced phase separation (SIPS) [65]. NIPS is the most reported technique for the fabrication of ISA HF porous support via liquid-liquid demixing, usually using water and mild or harsh polar aprotic solvents. The fabrication process can be explained thermodynamically by the ternary phase system (i.e., a polymer–solvent–non-solvent system) [66]. The final morphology of the HF membrane can be controlled by several parameters, such as the concentration of dope solution and bore fluid, dope flow rate, take-up speed, air gap, temperature (i.e., temperature of coagulation bath and spinneret), and type of spinneret (double or triple spinneret). Single-layer HF membranes (one type of polymer) can be fabricated using a double spinneret, whereas double-layer HF membranes (two different polymers) can be fabricated using a triple spinneret, as shown in Figure 9a,b respectively.

A single-layer HF membrane via NIPS is the preferred option due to its simplicity. However, optimizing the fabrication parameters (e.g., pore-forming additives, concentration, and viscosity of the dope solution, etc.) to control the ISA structure is challenging. Delaying the demixing leads to a tighter structure (low permeance), while instantaneous demixing leads to a higher number of macrovoids (high permeance) with a broader pore size distribution (unfavorable for depositing the selective layer via IP). Additionally, high mechanical strength is critical for the TFC HF membrane to operate under high pressures, which is challenging for the HF support layer with the ISA morphology. These parameters, which are related to the shear force/stress exerted in the spinneret, such as air gap length and take-up speed, were extensively investigated by Chung's group [67–69] and highlighted by Wang et al. in a recent publication [70]. One way to improve mechanical strength involves braiding the fibers, similar to the concept of non-woven support, which is usually used in FS configurations [71,72]. Zhao et al. [48] as mentioned above, developed a braided PBI HF membrane with high mechanical strength (i.e., withstanding more than 10 bar), lower swelling ratio, and improved separation performance compared to the self-support PBI HF membrane in organic solvents (methanol performance of 3.6 L m^{-2} h^{-1} bar^{-1} and 90% rejection of rose Bengal, RB).

Dual layer HF membranes incorporate a similar benefit to those of TFC HF membranes. This involves utilizing a more economical polymer for the support layer to provide mechanical strength while using less of the more costly polymer for the selective layer. It is worth highlighting that dual layer HF membrane can also be referred to, in the literature, as the coextrusion of one polymer type using a triple spinneret with different fabrication parameters/conditions. This can solve one of the main limitations of dual layer HF membranes, which is the poor adhesion/delamination between two polymers or layers [73,74]. Sun et al. [57] fabricated a dual layer PI HF support utilizing less non-solvent to slow down the coagulation of the outer surface layer while the inner surface layer undergoes fast coagulation (i.e., forming highly porous morphology), followed by chemical crosslinking with 1,6-Hexanediamine (HDA) to enhance the chemical resistance of the membrane. Specifically, the addition of polyvinylpyrrolidone (PVP) to the outer dope solution was investigated for its ability to slow precipitation and, consequently, achieve a narrower pore size distribution suitable for the IP of the polyamide selective layer (methanol permeance of 0.9 L m^{-2} h^{-1} bar^{-1} and 99.3% rejection of Remazol brilliant blue, RBB) [57]. Similarly, HF OSN membranes can be fabricated in one step, combining precipitation and crosslinking steps simultaneously. This method, referred to as "Chemistry in a spinneret", follows the work of Kopeć et al. [75] and Dutczak et al. [76], aiming to optimize the fabrication steps of a crosslinked polyimide support (i.e., polyethylenimine (PEI) was used as a crosslinking agent). A novel crosslinked HF polyelectrolyte membrane was developed via combining ionic crosslinking between PEI and polystyrene sulfonate (PSS), covalent crosslinking using glutaraldehyde (GA), and phase inversion in the spinneret, as depicted in Figure 9c [77]. Additionally, Sun et al. [78] applied the same concept by combining crosslinking with phase inversion using a PBI dope solution as the outer surface layer (selective layer) and feeding a polyimide dope solution to the inner surface, due to its imide groups, to be further crosslinked with PEI. The optimal membranes showed more than 99% rejection toward methylene blue (MB) in methanol and acetonitrile [78].

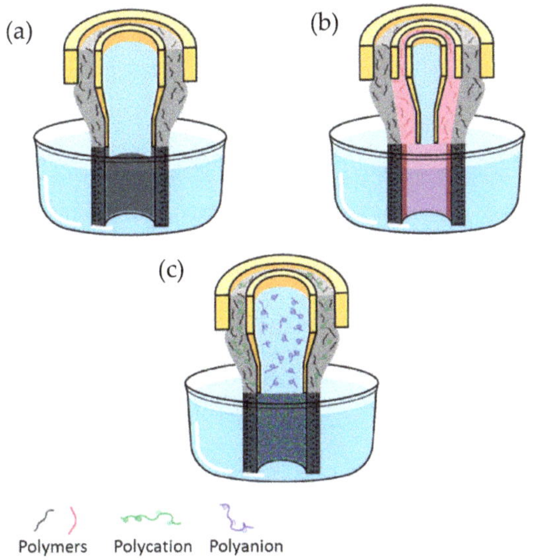

Figure 9. Schematic representation of the phase inversion-based methods for HF membranes: (**a**) single layer, (**b**) double layer, (**c**) chemistry in the spinneret. The fabrication methods can be performed on the shell or the lumen side of the fiber. These methods can be applied to fabricate the support layer of TFC HF membranes for OSN applications. Reproduced from Ref. [27] under CCBY 4.0.

Another limitation of the ISA porous support is structural relaxation, often associated with physical aging, which occurs as the polymer system moves towards thermodynamic equilibrium. When polymers are processed or used at temperatures that are different from their production temperature, their molecular chains may not be in equilibrium. Over time, these chains rearrange to minimize the system's free energy, leading to changes in various physical properties of the polymer, such as density, free volume, and mechanical properties. On the other hand, membrane compaction refers to a phenomenon where the physical structure of a membrane undergoes densification, often because of prolonged exposure to pressure, temperature, and harsh chemical environments. In the long run, pressure can cause the polymer chains in the membrane to pack more closely together, reducing the membrane's adequate pore size and porosity and, hence, leading to a gradual loss of permeance. Physical aging and compaction challenges are more pronounced in ISA membranes, where larger pores and macrovoids are present, compared to TFC membranes [79]. Incorporating TiO_2 nanoparticles into the crosslinked polyimide matrix (mixed matrix HF membranes, MMMs) has been demonstrated to enhance compaction resistance and the hydrophilicity of the resultant membranes by decreasing porosity (eliminating the macrovoids in the membrane sublayer) without affecting its separation performance (i.e., permeance and rejection of styrene oligomers in DMF) [79]. In addition, single-layer PI HF membranes compromising amine-functionalized multi-walled carbon nanotubes (NH_2-MWCNTs) were fabricated using NIPS by Farahani and colleagues [80]. The MMMs, crosslinked with HDA for 15 h at room temperature, showed higher mechanical strength and separation performance of tetracycline, L-αlecithin, and BINAP-Ru(II) complexes compared to pristine polyimide HF membranes in polar solvents (e.g., acetone, methanol, isopropanol, and ethanol) [80].

Instead of using polymers as support membranes, ceramics have also been investigated as support layers, owing to several advantages, such as high thermal stability, chemical resistance, and superior mechanical strength for OSN [81]. A multi-layer ceramic UF membrane (in a tubular configuration), with intermediate layers, was successfully demonstrated as a support membrane to deposit a polyamide selective layer via IP [81]. Furthermore, ceramic HF membranes are often fabricated by spinning out precursors via the NIPS approach followed by sintering. In recent work, high-performing TFC HF membranes (MWCO of 340 g mol^{-1}) were fabricated via direct IP of PEI/PIP selective layers on a ceramic HF membrane without an intermediate layer and found to be stable in aqueous solutions at high temperatures (i.e., 80 °C) [82]. Further, a hybrid of PAN and silicon nitride (Si_3N_4) ceramic HF membranes have been successfully applied as support layers in the IP fabrication of polyamide selective layers [62]. However, ceramic HF membranes are cost-prohibitive, fragile, and nonflexible regarding pore size and surface properties [23].

Other methods include grafting treatment; although it can be directly used in the spinning step, its utilization in modifying the HF support layer is rarely reported. This method includes applying UV/photo irradiation, e-beam, and plasm on the outside layer of the membrane, allowing the tuning of its properties, such as hydrophilicity and surface charge [83,84]. Notably, plasma grafting has been demonstrated to enhance the mechanical and chemical stability of the membrane (i.e., PDMS membrane) in isopropanol, methanol, and diethyl ether. These enhancements are ideal for modifying the UF membrane as a support layer [85]. However, the main drawback of these methods is their limitation to the outside-in geometry only, limiting the available options to be applied for modifying the support layer for IP fabrication.

3.3. Selective Layer

In aqueous solution applications, the state-of-the-art selective layer of TFC HF membranes (NF and RO) is a polyamide layer formed by the IP reaction between diamine and diacyl chloride [86]. It has also been adapted to fabricate TFC HF membranes for OSN applications. The IP reaction typically takes place on the inner surface (lumen side) of the HF support, as shown in Figure 10a, and it can also be performed on the outside surface (shell side), depending on the specific application. The latter type allows for the outside-in operation mode, which is generally more commercially applicable due to the larger effective membrane area and reduced pressure drop.

It is worth mentioning that the work of Kosaraju and colleagues was among the first to pioneer the fabrication of TFC HF membranes specifically for OSN [32,86]. In their work, pre-wetted PP microporous support, with acetone, was oxidized using a chromic acid solution to have better adhesion with the selective layer. Following this, a solution containing PEI and isophthaloyl dichloride (IPD) was interfacially polymerized into the inner side of the PP fibers [32]. The resulting membranes were thermally treated, and the SEM of the membrane cross-section with a thickness of ~100 nm is shown in Figure 10b. They demonstrated a methanol permeance of 1.47 L m^{-2} h^{-1} bar^{-1} but exhibited a low rejection of 88% for brilliant blue R (BBR) [32]. To improve rejection and solvent permeance, a less dense polyamide layer with reduced crosslinking can be created, using the branched PEI as the aqueous phase monomer [32]. This has been shown to increase solvent permeance (e.g., acetone permeance of 11.6 L m^{-2} h^{-1} bar^{-1}) [58]. In addition, incorporating a small quantity of piperazine (PIP) can enhance crosslinking density, improving solute rejection (i.e., 91.8% rejection of acid fuchsin) without significantly affecting solvent permeance [58]. However, the developed TFC HF membranes were unstable in harsh polar aprotic solvents such as DMF. In addition, the IP reaction between *m*-phenylenediamine (MPD) and trimesoyl chloride (TMC) has a high degree of crosslinking and results in a very selective layer suitable for separation in organic solvents. Nevertheless, it typically results in low solvent permeance [57]. On this basis, Goh et al. developed a defects-free MPD–TMC layer on crosslinked PI support for the concentration of APIs [59]. The optimal membrane, achieved by applying solvent activation with DMF as the post-treatment step as discussed below has a low MWCO of 269 g mol^{-1} and high acetone permeance of 24.2 L m^{-2} h^{-1} bar^{-1}. They also demonstrated the scalability of the fabrication method of the HF module from a 5-piece module to a 100-piece module with an active membrane area of 724 cm^2. Despite these promising results, controlling the thickness of the polyamide layer by IP is a significant challenge for scale-up fabrication. In particular, interfacial reactions occur at the boundary between the aqueous and organic phases; the reaction occurs rapidly, releasing a substantial amount of heat that can cause local increases in temperature. In this context, the locally heated areas facing the hexane layer become less dense and rise, while the cooler, denser areas sink. This creates a convective cell pattern (Rayleigh–Bénard convection) driven by buoyancy [87]. An innovative approach is to utilize a sacrificial support or intermediate layer to overlay the thin selective layer. Using the sacrificial layer approach allows precise control over the thickness of the selective layer and minimizes its defects, such as forming pinholes or cracks. The final step involves removing the sacrificial nanostrand layer. This is typically done using a solvent or etching process that dissolves or lifts away the nanostrands without damaging the thin selective layer. Karan et al. [88] fabricated an ultrathin polyamide layer with less than 10 nm thickness on cadmium hydroxide nanostrands (sacrificial layer), deposited on a crosslinked PI or alumina support. However, this technique of fabricating the ultrathin polyamide layer may face challenges in scaling up, particularly for HF support membranes. Interestingly, polymers of intrinsic microporosity (PIMs) have garnered interest due to their high free volume, porosity, and pore tunability. They are used for developing the selective layer for TFC membranes in OSN applications, mostly applied by coating methods such as dip coating and spray coating [89].

PIM-1 have been successfully fabricated in FS configuration by dip coating on a PAN support [90], followed by solvent vapor annealing (SVA) to control solvent evaporation [89]. Notably, Park et al. [91] fabricated HF TFC membranes via dip coating onto a crosslinked PI, followed by intermolecular crosslinking. The resultant membranes exhibited a MWCO of 952 g mol^{-1} and a high ethanol permeance of 10.9 L m^{-2} h^{-1} bar^{-1}. Although solvent permeance and selective layer thickness have an inverse correlation, decreasing the thickness of the selective layer does not always translate to high solvent permeance [92]. Livingston's group demonstrated that reducing the thickness of a selective layer containing linear polymers (<140 nm) resulted in reduced permeance, which was attributed to their structural relaxation toward the equilibrium state [93]. In contrast, network polymers such as the highly crosslinked MPD–TMC polyamide can be fabricated successfully with a thickness of less than 100 nm.

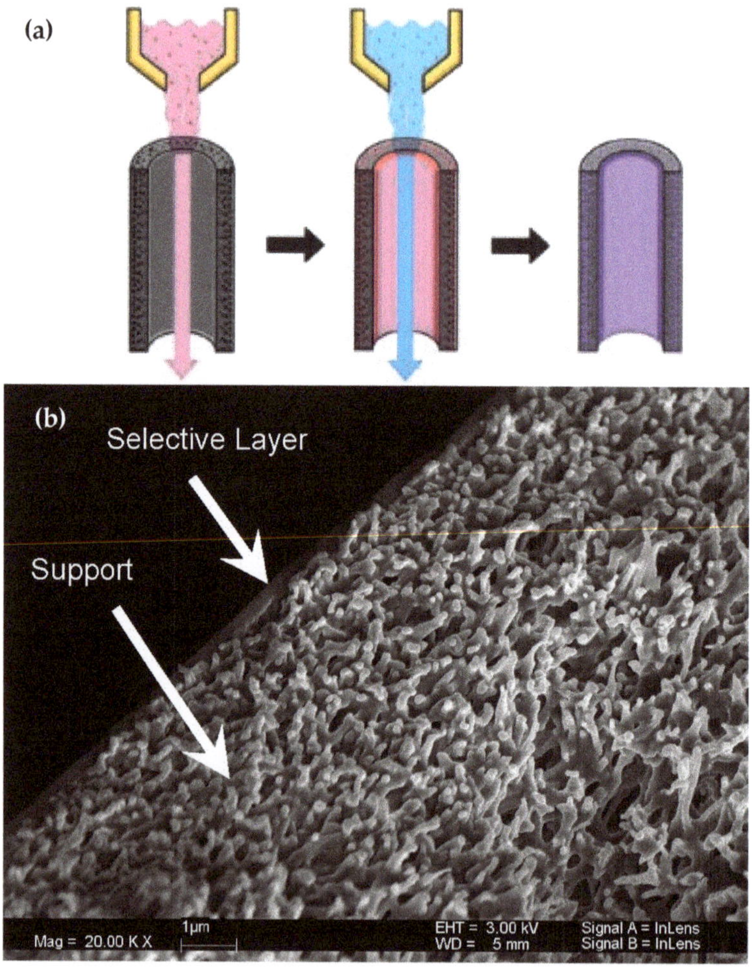

Figure 10. Schematic illustration of interfacial polymerization (**a**) fabrication method with inner selective layer. Reproduced under CCBY 4.0 from Ref. [27]. Copyright 2022, Elsevier; (**b**) SEM image of the PEI–IPD selective layer on porous PP support. Reprinted with permission from Ref. [32]. Copyright 2008, Elsevier.

Various fabrication techniques are employed to enhance the TFC HF membranes, such as incorporating additional nanomaterials into the selective layer, referred to as thin film nanocomposite (TFN) [93]. While these methods effectively modify membrane characteristics, such as permeance and chemical resistance, the compatibility of swelling properties across different layers is crucial for preventing delamination. Su et al. [60] developed a TFN HF membrane by incorporating graphene oxide (GO) nanosheets into the selective layer of TFC HF membrane, which is fabricated via IP of MPD and triethanolamine (TEA) with the addition of surfactants such as sodium dodecyl sulfate (SDS) to uniformly depositing the MPD aqueous solution on the HF support. The addition of GO improved ethanol permeance by approx. 66% while maintaining solute rejection. Further, Nunes' group reported spray coating for depositing a thin layer of GO on a crosslinked polyetherimide HF membrane (i.e., HDA was used as the crosslinker) via an airbrush, resulting in a MWCO of 973 g mol^{-1} and acetone permeance of 4 L m^{-2} h^{-1} bar^{-1} [94]. Recently, covalent organic frameworks (COFs) have gained significant attention for their potential use in membrane technology. COFs comprise hydrogen, boron, carbon, nitrogen, and oxygen, linked by strong covalent bonds. This structure endows COFs with several unique properties for OSN, such as high surface area, thermal stability, and well-defined pores (1–5 nm channels) [95,96]. Gao et al. [61] reported that COFs could be uniformly disturbed on the inner surface of crosslinked PI HF support, using a tris(4-aminophenyl)amine (TAPA) aqueous solution on the shell side and a benzene-1,3,5-tricarboxaldehyde (BTCA) organic solution on the lumen side to react at the interface and grow the COFs. The resultant membranes showed Janus-like properties (i.e., having hydrophilic pores to allow polar solvents to pass through the crosslinked PI support and having hydrophobic cavities of imine-linked COFs for non-polar solvent transport), leading to a high isopropanol permeance of ~62 L m^{-2} h^{-1} bar^{-1} (MWCO~784 g mol^{-1}). The separation performance of these membranes is the best reported yet compared to other HF OSN membranes (including TFC and TFN HF membranes). However, a primary limitation of incorporating nanomaterials is the poor adhesion and aggregation of the discrete additives within the polymer matrix. Such issues can lead to the development of non-selective voids, ultimately diminishing the membrane's separation efficiency and overall performance, leading to high cost and the possibility of leaching from the membrane during operation [97].

3.4. Post-Treatment Methods

3.4.1. Crosslinking

Crosslinking is a post-treatment method, after HF spinning, commonly employed to enhance the rigidity of porous support membranes (e.g., PI, PAN, polyetherimide, etc.), thereby reducing swelling and improving their resistance to organic solvents, especially against strong polar aprotic solvents such as DMF (Figure 11). Crosslinking involves creating bonds between polymer chains, which can be either covalent or ionic. However, most research and applications in OSN predominantly utilize covalent crosslinking, which includes chemical, thermal, or photo-induced processes. Among them, chemical crosslinking has become the go-to option for modifying the porous support for OSN. This method has also effectively reduced the pore size, enhanced the mechanical strength, and improved the molecular sieving properties of HF OSN membranes.

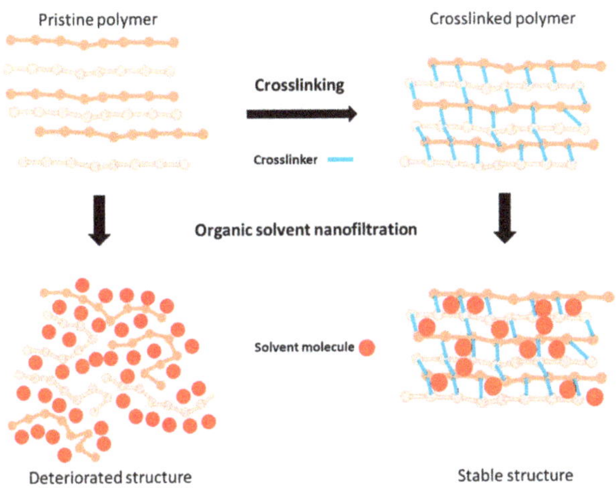

Figure 11. Schematic representation of crosslinking treatment method. Reprinted with permission from Ref. [10]. Copyright 2019, American Chemical Society.

Diamines (e.g., aliphatic, aromatic, and polymeric diamines) are widely used as crosslinking agents owing to their compatibility and reactive nature with PI and polyamide-imide (PAI) [98]. The existence of amine groups can interact with the imide groups of PI, leading to the opening of the rings and the formation of an integrated structure with improved chemical stability in the most common polar aprotic solvents, such as dichloromethane (DCM), *N*-methyl-pyrrolidone (NMP), THF, and DMF [99,100]. Lim et al. [101] developed PAI HF membranes for low-pressure OSN applications (i.e., the operating pressure is under 2 bar) using 3-aminopropyl trimethoxysilane (APTMS) as a crosslinking agent. The best-performing membranes achieved an RB rejection of over 97% and permeance of 0.9 L m^{-2} h^{-1} bar^{-1} and 6.4 L m^{-2} h^{-1} bar^{-1} for isopropanol and DMF, respectively [101]. Furthermore, Tham et al. [102] fabricated OSN HF membranes by chemical crosslinking a PAN UF membrane with cost-effective hydrazine monohydrate at 70 °C. Crosslinking time, dope solution composition, dope and bore fluid flow rate, and take-up speed were adjusted to minimize the macrovoids in the ISA membrane's spongy structure, achieving nanofiltration performance in ethanol without the need for coating or interfacial polymerization (with a permeance of 2.32 L m^{-2} h^{-1} bar^{-1} and 99.9% rejection of Remazol brilliant blue R) [102]. On the other hand, diisocyanates (i.e., 1,6-Hexamethylene Diisocyanate, HMDI) are also reported to crosslink the polyamide layer and PI support layer, achieving a low MWCO of 260 g mol^{-1} [103]. However, the permeance was very low at 1.2 L m^{-2} h^{-1} bar^{-1} and 0.1 L m^{-2} h^{-1} bar^{-1} for acetone and NMP, respectively (the membrane swelled in NMP over time) [103].

Another approach to achieving high solute rejection was demonstrated by Wang et al. [104]. This method involved the dual modification of α,α′-dibromo-p-xylene (DBX), followed by impregnation with 4-sulfocalix [8]arene (SCA8), which was demonstrated to enhance the mechanical strength and molecular sieving properties of PBI HF membranes. The resultant membranes have a low MWCO of 217 g mol^{-1} but with a dramatically reduced acetone permeance of 0.35 L m^{-2} h^{-1} bar^{-1} [104]. Comprehensive reviews specialized in crosslinking PBI and PI have been published elsewhere [19,98]. Interestingly, Loh et al. [105] showed that adding large organic acids such as malic acid into polyaniline (PANi) before spinning and removing the acid after spinning can introduce molecular sieving in the nanofiltration range with a MWCO of ~350 g mol^{-1} at 20–30 bar in acetone (permeance of 1.5 L m^{-2} h^{-1} bar^{-1}), making the acid doping and de-doping method suitable for fabricating OSN HF membranes in ISA morphology. Achieving the optimal crosslinking level is crucial, as over crosslinking can lead to a sharp drop in solvent permeance. In contrast, insufficient crosslinking may not provide the necessary enhancements in mechanical strength and solvent resistance. The key takeaway is that crosslinking is inevitable as an additional step in fabricating TFC HF membranes for OSN, as it is a prerequisite for improving the chemical resistance of the membrane's porous support when exposed to various organic solvents.

3.4.2. Solvent Activation

Livingston's research group has pioneered a promising approach that leverages a simple solvent treatment using strong polar aprotic solvents such as DMF to enhance membrane performance (it is also known as solvent activation or solvent annealing) [106,107]. The solvent activation technique showed applicability for OSN and aqueous solution separations, such as desalination, with benzyl alcohol acting as a solvent activator [108]. In addition, a range of common solvents, such as methanol, ethanol, isopropanol, and DMSO, have demonstrated efficacy in solvent activation, attributed to their ability to swell and induce structural deformation in the polyamide selective layer, coupled with a consequent alteration in its microstructural arrangement, thereby facilitating enhanced solvent permeance [107].

The extent of swelling depends on the type of selective/support layer, activation duration, and the solubility parameters of the solvents (i.e., solvency power). Studies have revealed that solvent-induced changes in the performance of polyamide-based TFC membranes can be predicted using Hansen solubility parameters [108,109]. For example, the crosslinked PI HF support membranes (reported by Wang's group in the previous paragraph) were fabricated to be thinner to increase the packaging density of the fibers, and the spinning parameters were optimized to obtain a tighter selective layer. This enhanced the selectivity of the resultant TFC HF membranes and hence lowered the permeance. The MPD-based TFC membranes were activated with DMF to increase the acetone permeance by almost 5-fold to 24.3 L m^{-2} h^{-1} bar^{-1} while maintaining its solute rejection (MWCO of 269 g mol^{-1} in acetone). This can be attributed to the swelling of weaker molecular chains of the polyamide layer and the crosslinked PI support layer due to similar solubility parameters with DMF. The aliphatic part of the polyamide layer tends to stretch, enabling the entry of DMF into the membrane matrix, while the aromatic part of the crosslinked network prevents it from dissolving, as represented in Figure 12 [59,110]. Therefore, semi-aromatic polyamide structures containing aromatic rings and hexagonal structures showed minimal change with solvent activation [88,110].

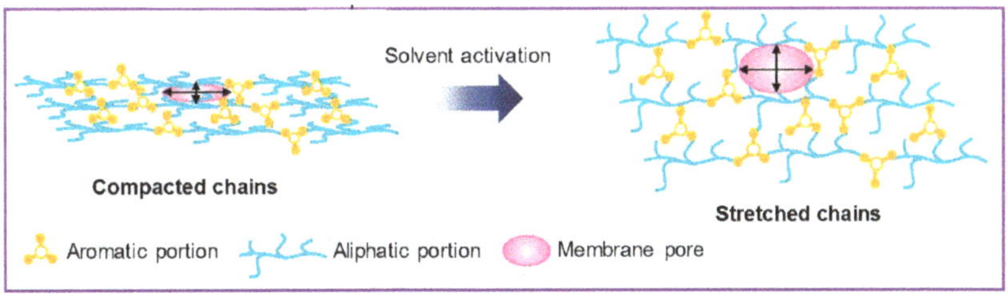

Figure 12. Schematic illustration of the solvent activation of PEI–TMC TFC membrane. Reprinted with permission from [110]. Copyright 2020, American Chemical Society.

Vankelecom and co-author [109] conducted a study on the effect of solvent activation on PI-based ISA membranes. They noted that solvent activation could lead to the reorganization and densification of polymer chains in the membrane rather than a looser structure [109]. They found that solvent activation with DMF transforms the crosslinked PI membrane from UF to OSN membrane with a MWCO of approx. 300–500 g mol^{-1} and higher ethanol permeance (up to 400%) compared to commercial Duramem membranes with similar MWCOs [109]. The high densification of the top selective layer is possibly due to the favorable interactions (e.g., π-π interactions and hydrogen bonding) between the flexible PI chains and the additional crosslinking of the PI matrix with the remaining crosslinking agent (i.e., crosslinking the unreacted PI chains with HDA). In this case, the densification depends on the degree of crosslinking, as a highly crosslinked PI will probably not be affected, and the duration of solvent activation.

3.4.3. Thermal Annealing

Thermal annealing (referred to as thermal crosslinking or thermal treatment) of TFC OSN membranes can significantly influence the properties of each layer, depending on the temperature and duration, by rearranging the polymer chains and adjusting the crosslinking density, which affects the membrane's solvent resistance and separation characteristics. This technique can achieve the desired selectivity and permeance by reducing defects in the selective layer, leading to improved rejection of solutes [111]. The support layer of TFC membranes also benefits from thermal annealing by strengthening the mechanical integrity and chemical resistance. For example, thermal annealing is an effective method to crosslink a PANi HF membrane (i.e., at a temperature of 180 for 1 h °C) to make it stable in an organic solvent environment [105]. In addition, Lai et al. [112] deposited a polypyrrole (PPy) selective layer onto a poly(p-phenylene terephthamide) (PPTA) braided support via chemical vapor deposition (CVD), which involves applying a thermal annealing step with an oxidative initiator ($FeCl_3$) to achieve the reaction. Furthermore, a thermal annealing step can be used to accelerate the crosslinking reaction (i.e., crosslinking a PAN membrane with hydrazine monohydrate at 70 °C) [102,113].

The advantages and disadvantages of the fabrication methods of HF OSN membranes discussed in this section are summarized in Table 2, including the IP fabrication method and the post-treatment methods.

Table 2. Advantages and disadvantages of the fabrication and post-treatment methods for HF OSN membranes.

Method	Advantage	Disadvantages
Single-layer phase inversion	• Simple one-step fabrication process • Cost-effective	• Lack of selectivity and chemical resistance • Balancing the trade-off between permeance and selectivity is challenging • The possibility of defects is high
Dual-layer phase inversion	• Simple one-step fabrication process • Versatility in polymer selection • Less of the expensive polymers used for the selective layer is needed	• Poor adhesion between the support and selective layers • Thicker selective layer compared to TFC membranes • Difficult to control the resultant properties
Interfacial Polymerization (TFC)	• High selectivity and permeance • Adequate adhesion between the support layer and selective layer • Thin selective layer	• Complex fabrication process • Difficulty in reproducibility • Limitation on the monomers selection that can be used for interfacial polymerization
Incorporating fillers (MMM, TFN)	• Modifications can be made to the spinning line • High selectivity and permeance • Improved mechanical and chemical stability	• Achieving a uniform dispersion of fillers is challenging • The possibility of interfacial and non-selective voids is high • Difficulty in scaling up
Post-treatment methods		
Crosslinking	• Enhanced mechanical and chemical stability • Enhanced selectivity	• Complexity in fabrication and the generation of chemical waste • Low permeance • Leads to brittleness
Solvent activation and Acid/alkali doping	• Simple treatment step • Cost-effective • Enhanced permeance or selectivity	• High consumption of organic solvents typically involving harsh polar aprotic solvents
Thermal annealing	• Enhanced mechanical and chemical stability • Cost-effective • Enhanced selectivity • Minimize defects	• Complexity in fabrication • Low permeance • Leads to brittleness

Table 3 presents the separation performance of the HF OSN membranes reported in the literature. It is worth highlighting that the "screening" refers to the filtration durations of less than one day, which is different from the long-term static immersion test for characterizing the chemical resistance of the membranes. Long-term cross-flow filtration (>1 day) is highlighted herein as more industrially viable. Relative to ISA HF OSN membranes (single or double layer), TFC HF membranes exhibited superior solvent permeance and comparable rejection. The median permeance of TFC HF membranes is 5.8 L m^{-2} h^{-1} bar^{-1}, 132% higher than the recommended solvent permeance of approx. 2.5 L m^{-2} h^{-1} bar^{-1} highlighted by Livingston's group. However, the median molecular weight of the solute is approx. 627 g mol^{-1} (with an average of ~693 g mol^{-1}), which is higher than the average molecular weight of APIs (i.e., 300 g mol^{-1}) in the pharmaceutical industry [107,114]. Notably, among these composite membranes, those with the COF selective layer demonstrated remarkably high solvent permeance, surpassing others by an order of magnitude [61].

Table 3. Summary of OSN HF membranes reported in the literature.

Selective Layer	Support Layer	Membrane Active Area (cm^2)	Membrane Type	Filtration Type	Solute	Solute MW ($g\ mol^{-1}$)	Rejection (%)	Solvent	Permeance ($L\ m^{-2}\ h^{-1}\ bar^{-1}$)	Pressure (bar)	Operation (days)	Ref.
PANi	PANi	~79	Single layer	-	Oligostyrene	400	~93	Acetone	1.50	6	Screening	[105]
PEI/IPD	PP	9	TFC	Inside-out	SO BBR	351 826	43 88	Methanol	1.47	4.13	Screening	[32]
PBI	Crosslinked PI	50	Dual layer	Outside-in	MB	319.85	99.17 99.55	Methanol Acetonitrile	2.60 1.58	1	Screening	[78]
MPD–TMC	Crosslinked dual layer PI	~45	TFC	Outside-in	RBB	626.54	99.3	Methanol	0.90	16	Screening	[57]
PMDA–ODA PI/TiO_2	PMDA-ODA PI/TiO_2	20	MMM	Outside-in	FG	808	90	DMF	2.5	10	~2	[115]
PAI	PAI	~16	Single layer	Outside-in	RB	1017.64	97.3 98.6	Isopropanol DMF	6.4 0.9	2	7 3	[101]
PAN	PAN	19	Single layer	Outside-in	RBB	626.54	99.9	Ethanol	2.32	2.8	Screening	[102]
PI/NH_2–MWCNTs	PI	-	MMM	Outside-in	MB BBR RBB TC	319.85 826 626.54 444.4	99.8 99.9 99.9 97.4	Isopropanol Acetone Methanol Ethanol	0.53 4.31 2.26 1.17	5	2.5	[80]
Polyamide 6	Polyamide 6	2.59	Single layer	Inside-out	Vitamin B_{12}	1355	98	Methanol	0.21	3	Screening	[116]
Cellulose	Cellulose	-	Single layer	Outside-in	CR BBR	696 826	>90 ~40	Ethanol	6	0.2	1	[117]
MPD–TMC/TiO_2@rGO	Si_3N_4/PAN	22	TFN	Outside-in	BTB RB	624 1017.64	95 98	Ethanol	~4.6 ~3.3	8	Screening	[62]
PIM–1/AlO_x	Crosslinked PI	-	TFN	Outside-in	RB	1017.64	86	Ethanol	~1.5	0.69	Screening	[118]
PBI	PBI	20.10	Single layer	Outside-in	MB RB	319.85 1017.64	~99 ~99	Acetone Methanol	~4 ~1	5	Screening 1	[46]
GO	Crosslinked polyetherimide	4.73	TFN	Outside-in	RB	1017.64	91	Acetone	4	1	Screening	[94]
PI/FTB	Crosslinked polyetherimide	-	Dual layer	Outside-in	TC	444.4	>99	Methanol	3.7	-	Screening	[46]
PEI/PIP–TMC	Crosslinked PI	~32	TFC	Inside-out	AF RB	585 1017.64	90.3 99.9	Acetone Isopropanol	10.8 4.5	2	3 Screening	[58]
PAN	PAN	23	Single layer	Outside-in	EB MB	960.81 319.85	100 71.5	Methanol	1.28	3	7 Screening	[113]
PI	PI	-	Single layer	Outside-in	RDB	479	96.5	Ethanol	0.45	10	Screening	[119]
PI	Modified PET braids with alkaline treatment	2000	Composite	Outside-in	TC	444.4	99.2	DMF Methanol	2.03 4.94	6	Screening	[71]

Table 3. *Cont.*

Selective Layer	Support Layer	Membrane Active Area (cm^2)	Membrane Type	Filtration Type	Solute	Solute MW ($g\ mol^{-1}$)	Rejection (%)	Solvent	Permeance ($L\ m^{-2}\ h^{-1}\ bar^{-1}$)	Pressure (bar)	Operation (days)	Ref.
MPD–TMC	Crosslinked PI	724	TFC	Inside-out	MR MR Levofloxacin MO AF RB	269 269 361 327 585 1017.84	90 90.1 98.2 98.6 99.4 99.8	Acetone	23.5 24.2	2	7 Screening	[59]
Polyamide 6	Polyamide 6	60	MMM	Inside-out	Vitamin B_{12}	1355	>99	Methanol	0.1	8	Screening	[120]
PAI	PAI	~9	Single layer	Outside-in	MB VbB	319.85 506.1	99.7 99.9	Methanol	1 1.87	2	Screening	[121]
PBI	PBI	2200	Single layer	Outside-in	MO RBB TG	327.33 626.54 885	77.5 99.5 99	Acetone	1.85 1.95 0.35	10	Screening 30	[104]
PI	PI	-	Single layer	Outside-in	RDB	479.01	98	Ethanol	1.83	10	Screening	[70]
PPy	Braid reinforced PPTA	~25	Composite	Outside-in	MB Eosin Y CR	319.85 647.89 696.7	84.5 93.4 99.3 99.5	DMAc Ethanol	1.04 1.10 1.17 1.64	6	Screening	[112]
MPD–TMC/GO	Crosslinked PI	0.17	TFC	Outside-in	RDB RB	479.01 1017.84	~100 99	Ethanol Methanol	1.80 5.80	5	Screening	[59]
BTCA–TAPA/COF	Crosslinked PI	0.44	TFC	Inside-out	L-α-lecithin FG RB	758 808 1017.84	~100 98.9 99.5 ~99 92 99.1	Hexane Acetone Isopropanol Ethanol Isopropanol	266.27 395.21 61.68 ~50 98.44 61.68	1	Screening 7.29 Screening	[61]
PBI	PET	544	Composite	Outside-in	RB	1017.84	99.5	Methanol	3.60 1.37	10	Screening ~3	[48]
PEI/SMA	PVDF	-	Composite	Outside-in	MO CV	327.3 407	99.9 ~98	Ethanol	6.4 1.90 0.70	1	Screening 1	[63]
PEI–TMC PEI/PIP–TMC	PTFE modified with PDA coating	6	TFC	Inside-out	AF MO	585 327.3	94.8 93 99.9 95.4	DMF DMF	3.70 2.44 1.79	2	Screening 3 Screening	[64]

SO: safranin O; BBR: brilliant blue R; MB: methylene blue, RBB: Remazol brilliant blue; FG: fast green; RB: rose Bengal; TC: tetracycline; CR: Congo red; BTB: bromothymol blue; AF: acid fuchsin; EB: Evans blue; RDB: rhodamine B; MR: methyl red; MO: methyl orange; VbB: Victoria blue B; TG: triglycerides; CV: crystal violet. PANi: polyaniline; PEI: polyethyleneimine; PP: polypropylene; IPD: isophthaloyl dichloride; PBI: polybenzimidazole; PI: polyimide; MPD: M-phenylenediamine; TMC: trimesoyl chloride; PMDA-ODA: poly(4,4′-oxydiphenylene pyromellitimide); PAI: polyamide-imide; PAN: polyacrylonitrile; PIM: polymer of intrinsic microporosity; GO: graphene oxide; FTB: 4-Fluoro-2-(trifluoromethyl) benzylamine; PPy: polypyrrole; PPTA: poly(p-phenylene terephthamide); BTCA: benzene-1,3,5-tricarboxaldehyde; TAPA: tris(4-aminophenyl)amine; SMA: styrene maleic anhydride copolymer; *TEA*: triethylamine; EGDGE: ethylene glycol diglycidyl ether; TAEA: tris(2-aminoethyl)amine; SCA8: 4-sulfocalix [8]arene; DMAc: N,N-dimethylacetamide.

3.5. Module Configuration

An HF module generally consists of four parts: the membranes themselves, membrane housing, potting, and end caps, as schematized in Figure 13A. Commercial modules are commonly around 1.5 m length. Instead, in academia, modules of 10–30 cm length are typically used [122]. As emphasized in the context of other membrane processes [123], it is important to realize that research on small, lab-scale hollow fibers cannot be simply translated into the performance of full-scale membrane modules due to membrane length being an important factor in recovery and retention [123–126]. Another important issue in module design is the operation mode (Figure 13B). Most of the studies regarding HFs for OSN applications are focus on inside-out HFs, where the selective layer is formed via interfacial polymerization onto the interior layer of the fiber. However, there is the possibility of using outside-in HFs with the selective layer applied to the outer layer of the fiber.

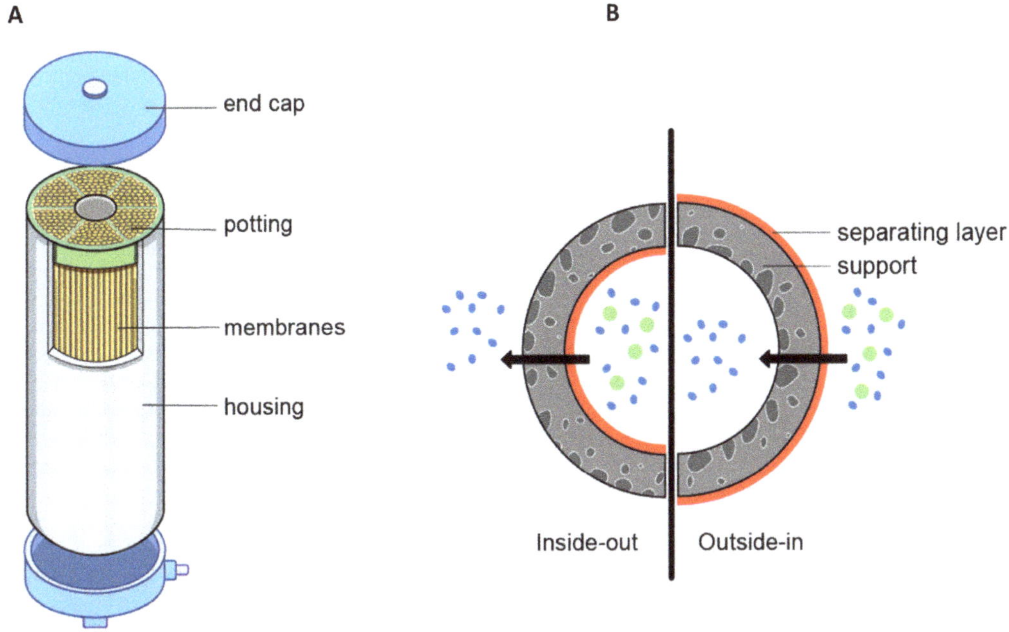

Figure 13. (**A**) Schematic representation of a hollow fiber module. (**B**) Inside-out vs. outside-in filtration. Reproduced from [27] with permission of Elsevier.

Both modes have their specific advantages: inside-out membranes are less prone to abrasive wear [112] due to reduced fiber movement and concentration polarization [68,126]; an outside-in configuration yields a higher specific surface area [27].

4. Greener Interfacially Polymerized HF OSN Membranes

Despite the widespread application of OSN technology, most commercially available OSN membranes are currently manufactured from fossil-based polymers, such as polybenzimidazole, polysulfone, and polyamide using environmentally harmful solvents. Nanofiltration membranes fabricated from these conventional polymers undergo a linear lifecycle (Figure 14), necessitating disposal in a landfill or environmental setting (approximately 79% of polymers) or incineration (approximately 12% of polymers) at the end of their lifecycle, which in turn leads to substantial carbon dioxide emissions [121].

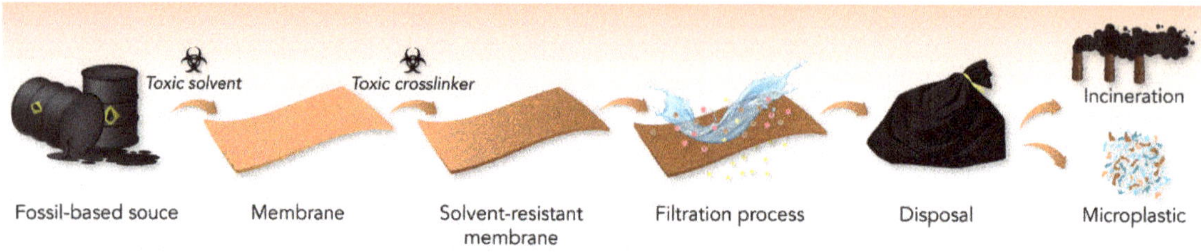

Figure 14. Schematic of conventional OSN membranes lifecycle. Reproduced under CCBY 4.0 from Ref. [127].

The overall research landscape in the domain of interfacially polymerized TFC HF membranes for OSN is particularly limited, especially in aspects concerning environmentally friendly, or 'green', fabrication methods. In this section, we will explore studies that focus on the green fabrication of TFC membranes, which are mainly in flat sheet configurations. The aim is to draw inspiration from these studies and explore how their environmentally conscious approaches can be adapted and applied to the fabrication of TFC membranes in HF configuration.

4.1. Greener Solvents

The fabrication of OSN membranes involves using hazardous polar aprotic solvents, such as NMP, DMAc, and DMF for the fabrication of the support layer, which is prepared by non-solvent-induced phase separation and apolar aprotic solvents such as n-hexane, used in the interfacial polymerization process. All of these solvents pose significant challenges in terms of handling and disposal. They are often released into the environment, causing severe harm to aquatic and terrestrial ecosystems. In particular, n-hexane, a fossil-derived solvent, is toxic to aquatic life and potentially damaging to fertility [127]. Green alternatives to these solvents would be those derived from renewable biomass sources, and they should not only be less toxic and safer to handle, thereby reducing health and safety risks, but also match or surpass the solubilization power of their fossil-based counterparts, contributing to the conservation of fossil energy resources while offering a lower ecological footprint. There has been a notable shift towards developing bio-based solvents in recent years. In a recent study, Lin et al. [128] proposed cyclopentyl methyl ether (CPME) and 2-methyltetrahydrofuran (2-MeTHF) as alternative solvents to n-hexane for interfacial polymerization. CPME is a fossil-fuel-derived solvent that can be potentially produced from bio-derived adipic acid and furfurals [129], and 2-MeTHF [130] could be synthesized from various biomasses. These solvents are promising alternatives to n-hexane for interfacial polymerization because they have low miscibility with water and can dissolve chemicals used in polyamide synthesis, such as TMC [129]. Lin et al. [128] produced sustainable TFC membranes with better performances: a TFC membrane fabricated via CPME demonstrated a higher NaCl rejection (97.8%), while the same membrane fabricated using n-hexane only presented 92.4% rejection; a TFC membrane fabricated with 2-MeTHF showed an ethanol permeance of 9.87 L m^{-2} h^{-1} bar^{-1} and 97.1% RB rejection, 3.7-fold higher than the TFC fabricated using n-hexane. This demonstrated the feasibility and advantages of replacing toxic and hazardous solvents that have long been the standard solvents used in membrane fabrication with benign alternatives [130]. Regarding the preparation of the membrane support layer, several green solvents are gaining attraction in OSN membrane fabrication, such as methyl 5-(dimethylamino)-2-methyl-5-oxopentanoate (PolarClean) [131,132], decanoic acid [133], γ-valerolactone (GVL) [134], MeSesamol [135], TamiSolve (trade name for *N*,*N*-Dimethyl Ethanolamine, produced by Eastman) [136,137], eucalyptol [43], *p*-cymene [138], dihydrolevoglucosenone (Cyrene), [139] and dimethyl isosorbide (DMI) [140,141]. Figure 15 presents the pricing of common organic solvents alongside a guide to their sustainability. However, it is important to note that these prices may not accurately reflect the actual costs, as they tend to be lower at industrial production scales. Interestingly, the cost of more sustainable solvents is competitive with that of hazardous solvents. Despite this, transitioning away from harsh polar aprotic solvents (e.g., NMP, DMAc, and DMF) is still a significant challenge. For instance, Cyrene's cost is at least four times higher than that of DMF, which is the most economical option among the solvents compared. Notably, Isobar G is a greener option to replace hexane in fabricating TFC membranes [24,142]. There is an anticipation that the future will bring more effective, sustainable, and commercially viable bio-based solvents, offering a practical and environmentally friendly alternative to conventional polar aprotic solvents.

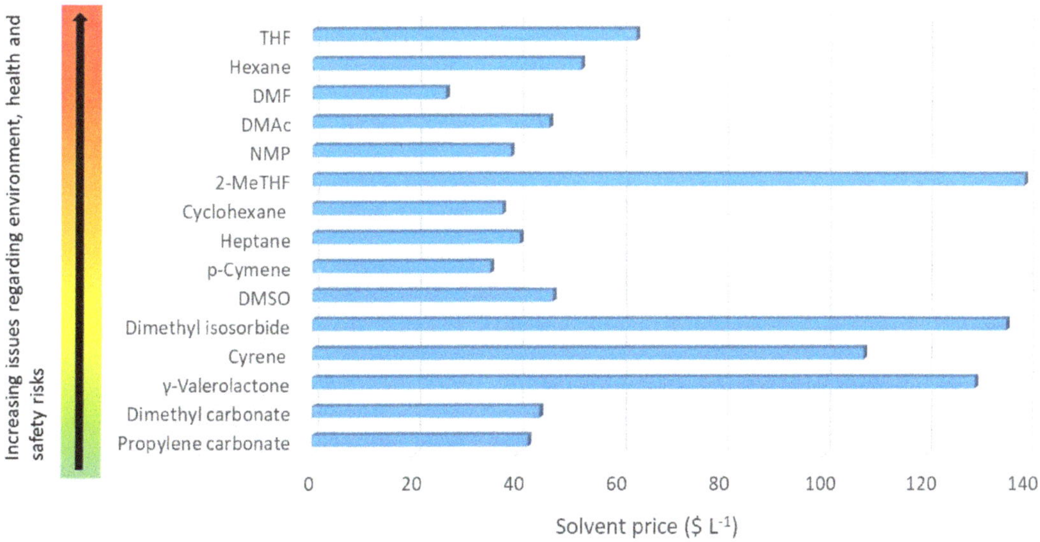

Figure 15. Price of common organic solvents (in USD per liter, 2023) used for the fabrication of OSN membranes. The prices were based on purity of ≥99% (i.e., ReagentPlus and extra pure grades, according to the manufacturer). The prices were obtained from the databases of Sigma Aldrich (St. Louis, MO, USA) and Thermo Fisher Scientific (Waltham, MA, USA) (accessed on 15 December 2023). On the left panel, solvents were classified based on GlaxoSmithKline (GSK) (Tokyo, Japan) solvent sustainability guide [143].

4.2. Biopolymers

While significant progress has been made in developing green solvents for membrane fabrication, exploring biopolymers as substitutes for fossil-derived polymers in OSN membrane preparation is still nascent [144]. This is due to biopolymers' poor mechanical and chemical stability compared to conventional fossil-derived polymers. Furthermore, crosslinking and modifying biopolymers will affect their biodegradability [145]. However, recent research has begun exploring various renewable polymers for OSN membrane fabrication. Examples of such polymers include polylactic acid (PLA) [138], cellulose and its derivatives [117,139,146,147], bamboo biomass [148], date seed biomass [149], chitosan [139,150], and alginate [151].

Among these biopolymers, cellulose is the most abundant and contributes to approximately 1.5 trillion tons of the world's annual biomass production, being used in paper production, textiles, and construction materials [152]. Its robustness against harsh polar aprotic solvents is attributed to its strong inter- and intramolecular hydrogen bonds [153]. Due to this, cellulose has limited solubility in common organic solvents, posing a challenge for OSN membrane fabrication. Recent research has explored the use of cellulose in ISA OSN membranes [154,155] and as support for TFC OSN membranes [145,146], mostly in FS configurations. Membrane fabrication involves using ionic liquids with DMSO or acetone as a cosolvent. Falca et al. [117] developed cellulose HF OSN membranes at 90 °C spinneret temperature using different ionic liquids, such as 1-ethyl-3-metthylimidazolium acetate ([Emim][OAc]), 1-ethyl-3-methyimidazolium diethyl phosphate ([Emim][DEP]), and 1,3-dimethylimidazolium dimethyl phosphate ([Emim][DMP]). However, certain challenges currently hinder the broader industrial adoption of ionic liquids in the membrane fabrication process. A primary concern is their cost, as they are still relatively expensive compared to more conventional solvents, especially when using large amounts, such as in membrane fabrication [156].

Another approach is the regeneration of cellulose fully from cellulose acetate (CA) in a dilute alkaline solution. CA membranes were prepared using green solvents (e.g., PolarClean, DMSO/acetone, triethyl phosphate, and methyl lactate), followed by thermal annealing in water at 100 °C and deacetylation of the resultant membranes using a NaOH aqueous solution [127]. Abdellah et al. [140] investigated the degree of acetylation (i.e., 1.2–39.3%) on the solvent-resistance and separation performance of the OSN membrane, prepared via NIPS using Cyrene as a green solvent. Decreasing the degree of acetylation improved chemical resistance and separation performance (i.e., MWCO of 325 g mol^{-1}). This improvement makes it a candidate for use as a support layer in the fabrication of TFC HF membranes [140].

Chitosan, derived from the deacetylation of chitin, ranks as the second most abundant polysaccharide after cellulose and shows significant promise for liquid-phase separations. This is mainly due to its solubility in mildly acidic media and the presence of amino groups [157]. Despite these benefits, chitosan's main limitation for OSN membrane fabrication lies in its hygroscopic nature, high tendency to degrade over time, poor mechanical strength, and insolubility in most common solvents. Thus, chitosan has been investigated recently as a filler, either in combination with cellulose to fabricate ISA OSN membranes [149], or interfacially polymerized with plant-based 2,5-furandicarboxaldehyde (FDA) to produce TFC OSN membranes [140].

4.3. Recyclable Polymers

Recycling stands out as a promising sustainable option compared to other methods, including landfill use and incineration. The European Union (EU) has set ambitious goals to ensure that by the year 2040, all plastic packaging is either completely recyclable or recoverable [158]. Transforming waste materials into valuable resources will help to reduce the environmental footprint across various industries. An illustrative example of this is the work conducted by Nunes's group on recycling used hydrophobic polyethylene terephthalate (PET) plastic bottles into porous supports fabricated via NIPS for OSN [159]. The recycled PET supports can be effectively utilized in the IP process to develop TFC OSN with thin selective layers (~30 nm) [136,139]. It is noteworthy that many sustainable polymers discussed in academic research have not undergone a comprehensive lifecycle assessment. Such an assessment is essential to quantify their environmental impact and draw comparisons with existing fossil-fuel-based polymers. For example, specific polymers that are derived from fossil fuels, such as polycaprolactone (PCL) and polybutylene adipate terephthalate (PBAT), have biodegradable properties. On the other hand, some polymers derived from natural resources (sugarcane and corn), such as bio-based polyethylene (bio-PE) and bio-polycarbonate (bio-PC), are not biodegradable [160]. In a recent study, Hardian et al. [161] fabricated freestanding recyclable nanofiltration membranes from novel polyester P $(BIL)_{ROP}$ and poly(cyclic olefin) $P(BIL)_{ROMP}$ polymers using γ-butyrolactone as the green solvent (Figure 16). Noteworthy, both polymers can be obtained from a single hybrid bicyclic lactone-olefin (BIL) monomer, a hybridization of γ-butyrolactone and cyclohexene constituents. Both γ-butyrolactone and cyclohexene can be derived from natural products and biologically synthesized, classifying them as renewable sources [161].

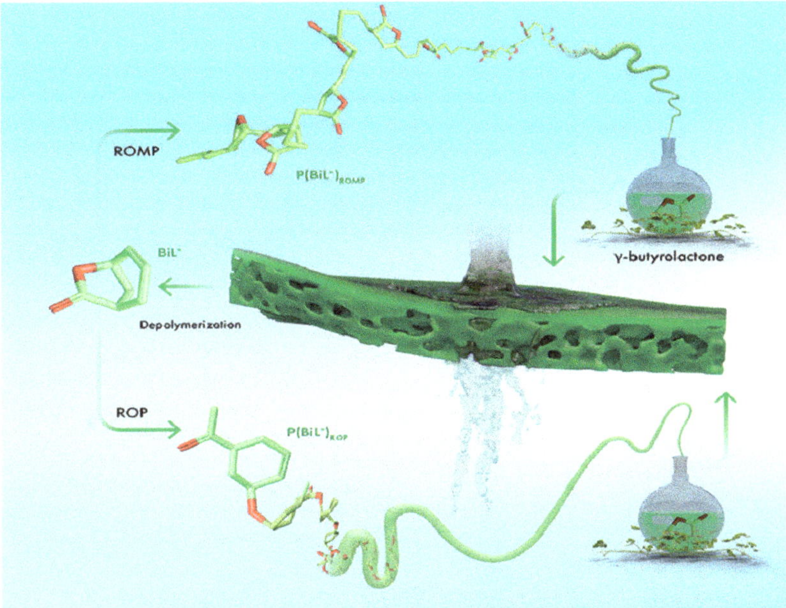

Figure 16. Sustainable lifecycle of recyclable "one monomer–two polymers–one monomer" membranes for organic solvent nanofiltration (OSN): ring-opening polymerization (ROP) and ring-opening metathesis polymerization (ROMP) of hybrid lactone-olefin monomer that yielded P(BiL=)ROP and P(BiL=)ROMP polymer structures; both polymers can be fabricated into OSN membranes using a γ-butyrolactone green solvent. Membrane fabrication: the dope solution containing polymer and green solvent is cast, and the formed film is immersed in a water coagulation bath, followed by annealing. Reproduced from [161] under CCBY 4.0.

4.4. Bio-Based Monomers

Several research groups have investigated the utilization of bio-based monomers in developing interfacially polymerized TFC OSN membranes. One of the key challenges lies in fabricating hydrophobic separation materials exclusively from sustainable resources. This is particularly complex when it comes to most bio-based materials, as they are inherently hydrophilic. However, there has been success in fabricating hydrophobic TFC membranes through IP using plant-derived monomers in the FS configuration. For example, Park et al. [139] have pioneered the use of priamine and tannic acid (TA) to develop a thin selective layer (~30 nm) on recycled PET for OSN, utilizing *p*-cymene and water for the organic phase and the aqueous phase, respectively. The resulting TFC membranes demonstrated a low MWCO of 235 g mol^{-1} and high acetone permeance of 13.7 L m^{-2} h^{-1} bar^{-1} [139]. Further, a polyamide selective layer was established via IP from priamine and genipin monomers using eucalyptol solvent in the organic phase on an electrospun nanofibrous PLA (the support layer was thermally annealed to enhance its mechanical properties) [138]. Citric acid was used to adjust the pH and accelerate the IP process. A gelatin intermediate layer was applied to enhance the compatibility between the highly porous PLA support and the priamine–genipin layer. In particular, narrowing the pore size of its loose structure and increasing the hydrophilicity of the hydrophobic PLA support (i.e., at least CA ~128°C) can facilitate the uptake of the genipin aqueous solution. The optimized membranes showed a high acetone permeance of 10 L m^{-2} h^{-1} bar^{-1} with a low MWCO of 281 g mol $^{-1}$ [138].

Another green TFC membrane fabrication approach involved using chitosan sourced from shrimp shells and a biomass-derived FDA monomer atop a recycled PET support [141]. Additionally, the green solvent Tamisolve was utilized in the solvent activation process post-fabrication to precisely control the pore size of the selective layer. The best-performing membrane demonstrated an acetone permeance of 12 L m^{-2} h^{-1} bar^{-1} (MWCO~317 g mol^{-1}) [136]. Moreover, cyclodextrins have gained tremendous attention in nanofiltration for precise molecular sieving, owing to their rigid structure and sub-nanometer cavities with lipophilic interiors. They are categorized in the industry as α-, β-, and γ-cyclodextrins, based on their six, seven, and eight glucopyranose units, respectively [162].

Figure 17 presents the prices of common monomers, as reported in the literature, in line with the discussion on green solvents from the previous section. It is important to note that these prices may not reflect the exact market values, as they typically decrease when produced on a larger scale. However, for the purpose of lab-scale testing, this pricing information is appropriate for conducting comparative analyses on a uniform mass basis sourced from standard manufacturers, such as Sigma Aldrich. Notably, it is evident from the data that bio-based monomers are highly cost-effective, being significantly more affordable than some commonly used monomers, such as trimethylsilyl chloride.

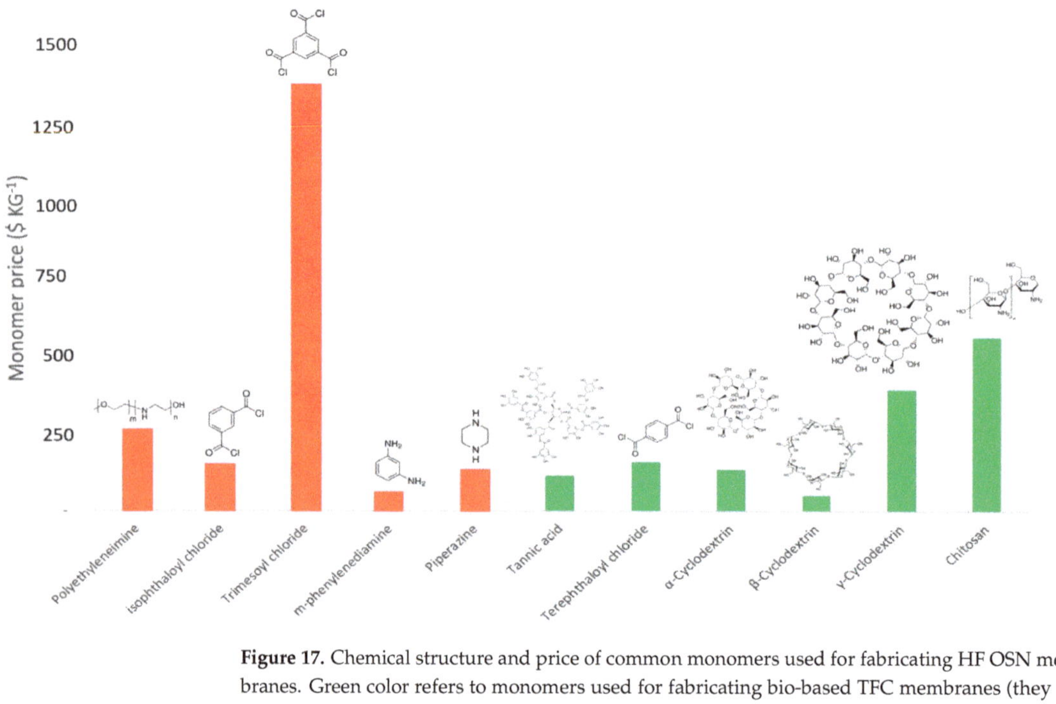

Figure 17. Chemical structure and price of common monomers used for fabricating HF OSN membranes. Green color refers to monomers used for fabricating bio-based TFC membranes (they are reported by the studies shown in Table 4); red color refer to the fossil-based monomers. The prices were based on mass basis of 1 KG and obtained from the databases of Sigma Aldrich [163] and Thermo Fisher Scientific [164].

Table 4 summarizes the performance of the green TFC membranes. Generally, their rejection rates are comparable to those of the ISA and TFC HF membranes listed in Table 3. However, it is noteworthy that the median permeance of these green TFC membranes is considerably lower compared to the TFC HF membranes, which have a reported median permeance of 5.8 L m^{-2} h^{-1} bar^{-1}, whereas the former exhibit a lower median permeance of 2.5 L m^{-2} h^{-1} bar^{-1}. This finding underscores the potential of exploring bio-based monomers further for the development of more environmentally friendly TFC HF membranes.

Table 4. Summary of OSN TFC membranes using bio-based monomers via IP, as reported in the literature.

Selective Layer	Support Layer	Membrane Active Area (cm^2)	Solute	Solute MW ($g\ mol^{-1}$)	Rejection (%)	Solvent	Permeance ($L\ m^{-2}\ h^{-1}\ bar^{-1}$)	Pressure (bar)	Operation (Days)	Ref.
Chitosan and FDA	Recycled PET	52	MO RB	327.33 1017.64	91.6 ~100	Acetone	11.9	10	14	[136]
Priamine–genipin	PLA with gelatin as the interlayer layer	28	Oleuropein AF	540.51 585.54	99 100	Acetone	9.9	30	7	[138]
Priamine–TA	Recycled PET	52	Methyl styrene dimer	235	~90 ~93 ~75	Acetone Heptane Toluene	13.5 ~2.5 ~3.5	-	7	[139]
Quercetin–TPC	Cellulose	12.6	SO MO TB AB BB Reactive black Reactive green	248.28 327.33 466.59 595.50 792.84 991.80 1418.93	58 90 94 96 97 98 100	DMF	2.6 2.4 2.7 2.6 2.4 2.6 2.8	10	Screening	[146]
TA–TPC	Crosslinked PAN	12.6	CR BB RB	696.66 825.97 1017.64	88 95 93	NMP	0.09 0.08 0.08	20	Screening	[165]
Morin hydrate–TPC	Crosslinked PAN	12.6	Solvent blue 35 Polystyrene CR RBB BB RB	350 500 590.71 626 825.97 1017.64	62 93 89 94 96 97	NMP	0.1 0.06 0.5 0.2 0.3 0.2	20	Screening	[166]
Catechin–TPC	Cellulose	12.6	AB	617	92	DMF	1.4	10	Screening	[147]
Cyclodextrin–TMC	Crosslinked PI	-	RBB	626	98.8	Ethanol	2.9	10	Screening	[167]

MO: methyl orange; RB: rose Bengal; AF: acid fuchsin; SO: Sudan orange; TB: thymol blue; AB: amido-black; BB: brilliant blue; CR: Congo red; RBB: Remazol brilliant blue; FDA: 2,5-furandicarboxaldehyde; PET: polyethylene terephthalate; PLA: polylactic acid; TA: tannic acid; TPC: terephthaloyl chloride; PAN: polyacronitrile; TMC: trimesoyl chloride; PI: polyimide.

4.5. Alternative Fabrication Methods

There are various approaches to enhance the greenness of TFC OSN membranes other than the selection of materials and solvents for the fabrication. For instance, Chen et al. [44] developed a crosslinked PBI membrane in a FS configuration, using potassium persulfate ($K_2S_2O_8$) as a green crosslinker in an aqueous solution at room temperature, resulting in superior chemical stability in harsh solvents such as DMAc and DMF. Similarly, Cheng's group employed the same chemistry to fabricate a crosslinked PBI HF membrane using $K_2S_2O_8$ at a temperature of 35 °C over a 14 h reaction time [46]. The resultant membranes showed improvement in molecular separation (i.e., more than 99% rejection of rose Bengal, RB in acetone) and chemical stability in strong polar aprotic solvents (e.g., DMSO, NMP, and DMF). However, the main drawback of the green crosslinking is the long reaction time (i.e., a crosslinking time of at least 14 h). Alternatively, acid doping (e.g., using a 2 wt% H_2SO_4 solution) has been employed to crosslink PBI HF, as it allows the acid molecules to protonate the fibers, forming hydrogen bonds with the PBI backbone to obtain a crosslinked structure with improved chemical resistance, mechanical strength, and selectivity (reducing MWCO from 2000 g mol^{-1} to 500 g mol^{-1}) [38].

The development of a spray-coating method for producing TFC membranes has led to several environmental benefits. This technique notably decreases the overall carbon footprint, along with reducing the consumption of polymers and solvents. Additionally, it streamlines the production process by curtailing the time required for coagulation and washing. Kim et al. [168] reported that the fabrication of sulfonated poly(arylene ether sulfone) (SPAES) membranes via simultaneous spray coating, thermal crosslinking, and drying procedures is more feasible economically than the other conventional methods, such as fabricating ISA PI membranes via phase inversion, followed by crosslinking and conditioning. They demonstrated that the fabrication procedure of spray coating to produce a SPAES membrane resulted in eliminating wastewater generation from the phase inversion/washing and solvent consumption in the crosslinking and conditioning steps and, therefore, reducing the overall chemical consumption substantially to only 0.29 kg m^{-2} [168]. Additionally, the processing time was around 0.5 day, which is much lower when compared to the 7 days and the 3–5 days required for the conventional fabrication procedure of PI and PEEK membranes, respectively. Interestingly, the spray coating approach has been exploited recently for developing advanced interfacially polymerized TFC membranes. In a procedure reported recently by Dedvukaj et al. [169], the porous support layer is cast in the liquid state (i.e., containing one monomer) and solidified after spray-coating the other monomer to initiate the IP. Then, the film undergoes a final solidification process in a non-solvent bath to complete the membrane formation [170]. Spray-assisted IP is more time-efficient, utilizes fewer chemicals, and generates less waste than the traditional IP method. Lin et al. [168] showed that the technique can be applied to both the porous support and the selective layers for TFC membrane fabrication, reducing the process time from 3–4 days to 1 day and 40 min.

5. Challenges and Commercial Viability

The practicality of OSN was first demonstrated by operating an industrial plant with a capacity of up to 72,000 barrels per day. This plant employed OSN technology pioneered by ExxonMobil to recover organic solvents from lube oil. This innovative approach shifted from the traditional thermal distillation methods, leading to a remarkable reduction in energy and capital costs by 20% and 60%, respectively [171,172]. Both in academia and commercially, OSN technology has been demonstrated to be highly effective in various applications, including the extraction of APIs, solvent recovery from the residual liquids left after crystallization, and the separation processes in lube oil production (e.g., solvent recovery and dewaxing) [24]. Additionally, OSN has been utilized in biorefineries, the isolation of natural products, organocatalysis, and in processes that involve solvent exchange, among other applications [17,24].

There is no commercially available HF OSN membrane, although there are several advantages to providing IP TFC in HF configuration compared to ISA membranes in FS configuration (in SW modules) for OSN applications. However, scaling up and commercialization present some technical challenges, which are the following: (1) The fabrication process for TFC HF membranes is more complex, as it involves multiple steps, including the preparation of a porous support layer, deposition of a thin selective layer, and subsequent modification steps for both layers to be chemically stable, especially in polar aprotic solvents. (2) Achieving a uniform and defect-free selective layer on the outer surface of the HF membrane is particularly challenging, with a high occurrence of defects. (3) Ensuring a proper attachment between the support and selective layers is essential for the durability of TFC HF membranes, especially for long-term operation in harsh organic solvent environments. The long-term filtration test (operation time > 1 day) is rarely reported in the literature (Table 3). (4) HF modules are typically not suited for high operating pressures compared to SW modules, limiting the available options for deployment. (5) There is a tremendous gap between academia and the industry when it comes to the HF size tested for OSN. In academic settings, the median active area of HF membranes tested is approx. 21 cm^2, as shown in Table 3. This is considerably smaller when compared to the standard active area of industrial HF modules, which is typically at least 40 m^2 [27]. (6) There is lack in research regarding the green fabrication of interfacially polymerized TFC HF membranes for OSN applications, which is particularly important under the increasingly stringent environmental regulations on the generated waste from membrane suppliers and the global shift toward achieving the United Nations' Sustainable Development Goals (SDGs). On the other hand, it is important to note that there are additional challenges and considerations that, while not specific to TFC HF membranes, are relevant and noteworthy in the broader context of research in the OSN field. For example, the scope of solutes currently tested in OSN is quite narrow, primarily focusing on APIs, dyes, and oligomers. This limited range only represents a fraction of the potential molecules available, which may impede the expansion of OSN applications to enhance the commercial viability of OSN membranes [173]. Further, currently, the OSN field lacks a universally accepted standard for the systematic testing of OSN membranes. Despite numerous proposals and a variety of solutes suggested to be an ideal testing protocol, no single protocol has gained universal acceptance [17,174]. Furthermore, moving away from the traditional trial-and-error methodology has shown a growing interest in leveraging automation and artificial intelligence to streamline and enhance the development process in OSN membrane technology (Figure 18).

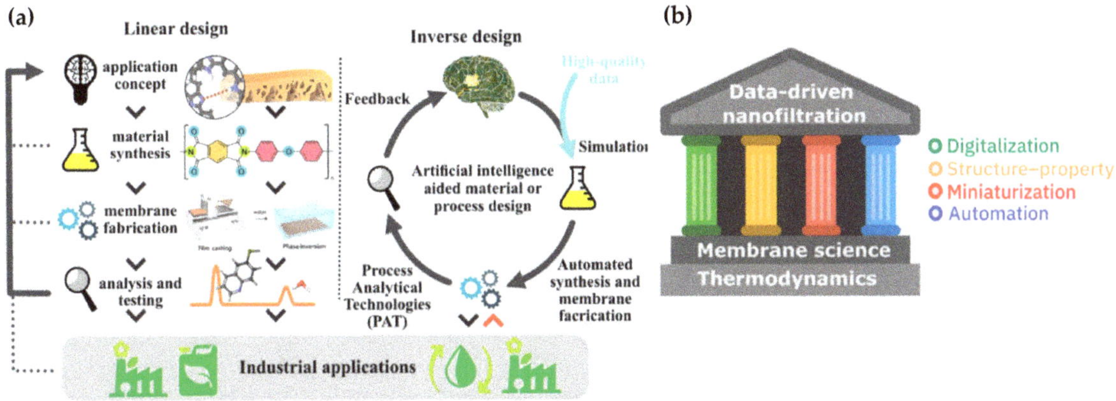

Figure 18. (**a**) The comparison between linear design and inverse design for developing nanofiltration membranes from conceptualization to industrial application. (**b**) The four pillars of developing membrane materials and processes, based on available data. Reproduced from [175] under CCBY 4.0.

6. Conclusions and Future Perspectives

To develop an optimal structure for an interfacial polymerized OSN HF membrane, the design should include a porous support that is both solvent-resistant and mechanically robust, capable of withstanding high pressures while maintaining high permeance. The selective layer should feature more precise pore size distributions (300 g mol^{-1}) to ensure efficient solute rejection for most applications in the pharmaceutical and fine chemical industry (where OSN has the highest demand) and to enable previously impossible separations, while maintaining a solvent permeance of at least 2.5 L m^{-2} h^{-1} bar^{-1}. Interfacial polymerization has significantly elevated the performance of TFC HF membranes, rendering them more suitable for challenging separations in organic solvent environments. Bio-based monomers and renewable polymers have not only provided environmentally friendly alternatives but also bring unique properties that could enhance the performance of OSN membranes. Academia often investigates novel monomers, polymers, and advanced fabrication methods in OSN membrane development. While this continues to yield more findings, translating these developments into practical, large-scale applications is a complex process that is currently scarce in the literature, especially for interfacially polymerized TFC HF membranes. Moreover, for sustainable polymers to be viable alternatives to conventional ones, they must demonstrate comparable or superior performance. Key attributes include thermal stability, sufficient mechanical strength, and the ability to be produced on an industrial scale to meet the need for temperature and solvent-resistant membranes. Adopting a comprehensive cradle-to-grave approach is essential for confirming the sustainability of membrane technology. This approach should encompass all aspects, including the materials used for membranes, module construction, preparation methodologies, operational processes, and the recyclability of membranes post-use. However, challenges persist in the fabrication process, particularly in ensuring defect-free layers and long-term durability in harsh solvent environments. Despite these efforts, the limited scope of solutes tested in OSN research could restrict these membranes' broader application and commercial viability. The future of OSN membrane technology lies in addressing these fabrication challenges, expanding the range of solutes for testing, and integrating greener, more sustainable practices, potentially aided by automation and artificial intelligence innovations. Collaboration with pioneering industries is crucial to identify and implement HF OSN membranes successfully, either alone or in combination with other technologies.

Author Contributions: A.Y.A. conceived the original concept and designed the work, made substantial contribution to the reference acquisition, critical analysis/interpretation of data for the work and to the writing. M.G.B. made substantial contributions to the reference acquisition, interpretation of data, revising the work. The manuscript was written through contribution of all authors. All authors have read and agreed to the published version of the manuscript.

Funding: This research received no external funding.

Data Availability Statement: No new data were created or analyzed in this study. Data sharing is not applicable to this article.

Conflicts of Interest: Authors A.Y.A. and S.-H.C. are employed by the company Saudi Aramco. The remaining author declares that the research was conducted in the absence of any commercial or financial relationships that could be construed as a potential conflict of interest.

References

1. Cseri, L.; Razali, M.; Pogany, P.; Szekely, G. Organic Solvents in Sustainable Synthesis and Engineering. In *Green Chemistry: An Inclusive Approach*; Elsevier: Amsterdam, The Netherlands, 2018.
2. Constable, D.J.C.; Jimenez-Gonzalez, C.; Henderson, R.K. Perspective on Solvent Use in the Pharmaceutical Industry. *Org. Process Res. Dev.* **2007**, *11*, 133–137. [CrossRef]
3. Buonomenna, M.G.; Bae, J. Organic solvent nanofiltration in pharmaceutical industry. *Sep. Purif. Rev.* **2015**, *44*, 157–182. [CrossRef]
4. Sholl, D.S.; Lively, R.P. Seven Chemical Separations to Change the World. *Nature* **2016**, *532*, 435–437. [CrossRef] [PubMed]

5. Buonomenna, M.G.; Golemme, G.; Perrotta, E. Membrane operations for industrial applications. In *Advances in Chemical Engineering*; Nawaz, Z., Naveed, S., Eds.; IntechOpen: Rijeka, Croatia, 2011.
6. Liu, C.; Dong, G.Y.; Tsuru, T.; Matsuyama, H. Organic solvent reverse osmosis membranesfor organic liquid mixture separation: A review. *J. Membr. Sci.* **2021**, *620*, 118882. [CrossRef]
7. Chau, J.; Sirkar, K.K. Organic solvent mixture separation during reverse osmosis and nanofiltration by a perfluorodioxole copolymer membrane. *J. Membr. Sci.* **2021**, *618*, 118663. [CrossRef]
8. Cui, Y.; Chung, T.S. Pharmaceutical concentration using organic solvent forward osmosis for solvent recovery. *Nat. Commun.* **2018**, *9*, 1426. [CrossRef]
9. Vandezande, P.; Gevers, L.E.M.; Vankelecom, I.F.J. Solvent Resistant Nanofiltration: Separating on a Molecular Level. *Chem. Soc. Rev.* **2008**, *37*, 365–405. [CrossRef]
10. Asadi Tashvigh, A.; Feng, Y.; Weber, M.; Maletzko, C.; Chung, T.S. 110th Anniversary: Selection of Cross-Linkers and Cross-Linking Procedures for the Fabrication of Solvent-Resistant Nanofiltration Membranes: A Review. *Ind. Eng. Chem. Res.* **2019**, *58*, 10678–10691. [CrossRef]
11. Marchetti, P.; Jimenez Solomon, M.F.; Szekely, G.; Livingston, A.G. Molecular Separation with Organic Solvent Nanofiltration: A Critical Review. *Chem. Rev.* **2014**, *114*, 10735–10806. [CrossRef]
12. Liu, L.; Liu, S.; Wang, E.; Su, B. Hollow Fiber Membrane for Organic Solvent Nanofiltration: A Mini Review. *Membranes* **2022**, *12*, 995. [CrossRef]
13. Shi, G.M.; Feng, Y.; Li, B.; Tham, H.M.; Lai, J.Y.; Chung, T.S. Recent Progress of Organic Solvent Nanofiltration Membranes. *Prog. Polym. Sci.* **2021**, *123*, 101470. [CrossRef]
14. Lau, H.S.; Lau, S.K.; Soh, L.S.; Hong, S.U.; Gok, X.Y.; Yi, S.; Yong, W.F. State-of-the-Art Organic-and Inorganic-Based Hollow Fiber Membranes in Liquid and Gas Applications: Looking Back and Beyond. *Membranes* **2022**, *12*, 539. [CrossRef]
15. Scholz, M.; Wessling, M.; Balster, J. Chapter 5. Design of Membrane Modules for Gas Separations. In *Membrane Engineering for the Treatment of Gases*; Drioli, E., Barbieri, G., Eds.; RSC: London, UK, 2011.
16. Liu, T.; Qin, Z.; Liu, Q.; Li, X.; Liu, Y.; An, Q.F.; Guo, H. In Situ Growth of a Tubular MoS2membrane on a Ceramic Tube with Improved Organic Solvent Nanofiltration Performance. *Mater. Chem. Front.* **2021**, *5*, 3184–3191. [CrossRef]
17. Merlet, R.; Winnubst, L.; Nijmeijer, A.; Amirilargani, M.; Sudhölter, E.J.R.; de Smet, L.C.P.M.; Salvador Cob, S.; Vandezande, P.; Dorbec, M.; Sluijter, S.; et al. Comparing the Performance of Organic Solvent Nanofiltration Membranes in Non-Polar Solvents. *Chem. Ing. Tech.* **2021**, *93*, 1389–1395. [CrossRef]
18. Mulder, M. *Basic Principles of Membrane Technology*; Springer: Berlin/Heidelberg, Germany, 1991.
19. Wang, K.Y.; Weber, M.; Chung, T.S. Polybenzimidazoles (PBIs) and State-of-the-Art PBI Hollow Fiber Membranes for Water, Organic Solvent and Gas Separations: A Review. *J. Mater. Chem. A Mater.* **2022**, *10*, 8687–8718. [CrossRef]
20. Hoek, E.M.V.; Tarabara, V.V. *Encyclopedia of Membrane Science and Technology*; Wiley: Hoboken, NJ, USA, 2013.
21. Tul Muntha, S.; Kausar, A.; Siddiq, M. Advances in Polymeric Nanofiltration Membrane: A Review. *Polym. Plast. Technol. Eng.* **2017**, *56*, 841–856. [CrossRef]
22. Lim, S.K.; Goh, K.; Bae, T.H.; Wang, R. Polymer-Based Membranes for Solvent-Resistant Nanofiltration: A Review. *Chin. J. Chem. Eng.* **2017**, *25*, 1653–1675. [CrossRef]
23. Li, Y.; Guo, Z.; Li, S.; Van der Bruggen, B. Interfacially Polymerized Thin-Film Composite Membranes for Organic Solvent Nanofiltration. *Adv. Mater. Interfaces* **2021**, *8*, 2001671. [CrossRef]
24. Szekely, G.; Jimenez-Solomon, M.F.; Marchetti, P.; Kim, J.F.; Livingston, A.G. Sustainability Assessment of Organic Solvent Nanofiltration: From Fabrication to Application. *Green Chem.* **2014**, *16*, 4440–4473. [CrossRef]
25. Peeva, L.G.; Marchetti, P.; Livingston, A.G. Nanofiltration Operations in Nonaqueous Systems. In *Comprehensive Membrane Science and Engineering*, 2nd ed.; Elsevier: Amsterdam, The Netherlands, 2018; Volume 2.
26. Sewerin, T.; Elshof, M.G.; Matencio, S.; Boerrigter, M.; Yu, J.; de Grooth, J. Advances and Applications of Hollow Fiber Nanofiltration Membranes: A Review. *Membranes* **2021**, *11*, 890. [CrossRef] [PubMed]
27. Jonkers, W.A.; Cornelissen, E.R.; de Vos, W.M. Hollow Fiber Nanofiltration: From Lab-Scale Research to Full-Scale Applications. *J. Membr. Sci.* **2023**, *669*, 121234. [CrossRef]
28. Marchetti, P.; Peeva, L.; Livingston, A. The Selectivity Challenge in Organic Solvent Nanofiltration: Membrane and Process Solutions. *Annu. Rev. Chem. Biomol. Eng.* **2017**, *8*, 473–497. [CrossRef] [PubMed]
29. Gilbert, M. Relation of Structure to Chemical Properties. In *Brydson's Plastics Materials*, 8th ed.; Elsevier: Amsterdam, The Netherlands, 2017.
30. Buonomenna, M.G.; Golemme, G.; Jansen, J.C.; Choi, S.-H. Asymmetric PEEKWC membranes for treatment of organic solvent solutions. *J. Membr. Sci.* **2011**, *368*, 144–149. [CrossRef]
31. Xu, Y.; You, F.; Sun, H.; Shao, L. Realizing mussel-inspired polydopamine selective layer with strong solvent resistance in nanofiltration toward sustainable reclamation. *ACS Sustain. Chem. Eng.* **2017**, *5*, 5520–5528. [CrossRef]
32. Kosaraju, P.B.; Sirkar, K.K. Interfacially polymerized thin film composite membranes on microporous polypropylene supports for solvent-resistant nanofiltration. *J. Membr. Sci.* **2008**, *321*, 155–161. [CrossRef]

33. Sun, S.P.; Chung, T.S.; Lu, K.J.; Chan, S.Y. Enhancement of flux and solvent stability of Matrimid®thin-film composite membranes for organic solvent nanofiltration. *AIChE J.* **2014**, *60*, 3623–3633. [CrossRef]
34. Ebert, K.; Cuperus, F.P. Solvent resistant nanofiltration in edible oil processing. *Membr. Technol.* **1999**, *107*, 5–8.
35. Tarleton, E.S.; Robinson, J.P.; Salman, M. Solvent-induced swelling of membranes—Measurements and influence in nanofiltration. *J. Membr. Sci.* **2006**, *280*, 442–451. [CrossRef]
36. Wang, K.Y.; Chung, T.S.; Rajagopalan, R. Dehydration of tetrafluoropropanol (TFP) by pervaporation via novel PBI/BTDA-TDI/MDI co-polyimide (P84) dual-layer hollow fiber membranes. *J. Membr. Sci.* **2007**, *287*, 60–66. [CrossRef]
37. Wang, Y.; Gruender, M.; Chung, T.S. Pervaporation dehydration of ethylene glycol through polybenzimidazole (PBI)-based membranes. 1. Membrane fabrication. *J. Membr. Sci.* **2010**, *363*, 149–159. [CrossRef]
38. Tashvigh, A.A.; Chung, T.S. Robust polybenzimidazole (PBI) hollow fiber membranes for organic solvent nanofiltration. *J. Membr. Sci.* **2019**, *572*, 580–587. [CrossRef]
39. Farahani, M.H.D.A.; Chung, T.S. A novel crosslinking technique towards the fabrication of high-flux polybenzimidazole (PBI) membranes for organic solvent nanofiltration (OSN). *Sep. Purif. Technol.* **2019**, *209*, 182–192. [CrossRef]
40. Xing, D.Y.; Chan, S.Y.; Chung, T.S. The ionic liquid [EMIM] OAc as a solvent to fabricate stable polybenzimidazole membranes for organic solvent nanofiltration. *Green Chem.* **2014**, *16*, 1383–1392. [CrossRef]
41. Valtcheva, I.B.; Kumbharkar, S.C.; Kim, J.F.; Bhole, Y.; Livingston, A.G. Beyond polyimide: Crosslinked polybenzimidazole membranes for organic solvent nanofiltration (OSN) in harsh environments. *J. Membr. Sci.* **2014**, *457*, 62–72. [CrossRef]
42. Chen, D.; Yu, S.; Yang, M.; Li, D.; Li, X. Solvent resistant nanofiltration membranes based on crosslinked polybenzimidazole. *RSC Adv.* **2016**, *6*, 16925–16932. [CrossRef]
43. Tashvigh, A.A.; Chung, T.S. Facile fabrication of solvent resistant thin film composite membranes by interfacial crosslinking reaction between polyethylenimine and dibromo-p-xylene on polybenzimidazole substrates. *J. Membr. Sci.* **2018**, *560*, 115–124. [CrossRef]
44. Chen, D.; Yan, C.; Li, X.; Liu, L.; Wu, D.; Li, X. A highly stable PBI solvent resistant nanofiltration membrane prepared via versatile and simple crosslinking process. *Sep. Purif. Technol.* **2019**, *224*, 15–22. [CrossRef]
45. Zhao, B.; Shi, G.M.; Wang, K.Y.; Lai, J.Y.; Chung, T.S. Employing a green cross-linking method to fabricate polybenzimidazole (PBI) hollow fiber membranes for organic solvent nanofiltration (OSN). *Sep. Purif. Technol.* **2021**, *255*, 117702. [CrossRef]
46. Quan, Q.; Xiao, C.; Liu, H.; Huang, Q.; Zhao, W.; Hu, X.; Huan, G. Preparation and characterization of braided tube reinforced polyacrylonitrile hollow fiber membranes. *J. Appl. Polym. Sci.* **2015**, *132*, 41795. [CrossRef]
47. Fan, Z.; Xiao, C.; Liu, H.; Huang, Q.; Zhao, J. Structure design and performance study on braid-reinforced cellulose acetate hollow fiber membranes. *J. Membr. Sci.* **2015**, *486*, 248–256. [CrossRef]
48. Zhao, B.; Min Shi, G.; Lai, J.H.; Chung, T.S. Braid-reinforced polybenzimidazole (PBI) hollow fiber membranes for organic solvent nanofiltration (OSN). *Sep. Purif. Technol.* **2022**, *290*, 120811. [CrossRef]
49. Böcking, A.; Koleva, V.; Wind, J.; Thiermeyer, Y.; Blumenschein, S.; Goebel, R.; Skiborowski, M.; Wessling, M. Can the Variance in Membrane Performance Influence the Design of Organic Solvent Nanofiltration Processes? *J. Membr. Sci.* **2019**, *575*, 217–228. [CrossRef]
50. White, L.S. Transport Properties of a Polyimide Solvent Resistant Nanofiltration Membrane. *J. Membr. Sci.* **2002**, *205*, 191–202. [CrossRef]
51. Low, Z.X.; Patterson, D.A.; Patterson, D.A. Molecular Weight Cut-off Determination of Organic Solvent Nanofiltration Membranes Using Poly(Propylene Glycol). *J. Membr. Sci.* **2017**, *526*, 221–228.
52. See Toh, Y.H.; Loh, X.X.; Li, K.; Bismarck, A.; Livingston, A.G. In Search of a Standard Method for the Characterisation of Organic Solvent Nanofiltration Membranes. *J. Membr. Sci.* **2007**, *291*, 120–125. [CrossRef]
53. Silva, P.; Han, S.; Livingston, A.G. Solvent Transport in Organic Solvent Nanofiltration Membranes. *J. Membr. Sci.* **2005**, *262*, 49–59. [CrossRef]
54. Ohkame, T.; Minegishi, K.; Sugihara, H.; Nakagawa, K.; Shintani, T.; Matsuyama, H.; Yoshioka, T. Hollow-Fiber Ro Membranes Fabricated via Adsorption of Low-Charge Poly(Vinyl Alcohol) Copolymers. *Membranes* **2021**, *11*, 981. [CrossRef]
55. Zhang, F.; Fan, J.B.; Wang, S. Interfacial Polymerization: From Chemistry to Functional Materials. *Angew. Chem. Int. Ed.* **2020**, *59*, 21840–21856. [CrossRef]
56. Pabby, A.K.; Sastre, A.M. Hollow Fiber Membrane-Based Separation Technology: Performance and Design Perspectives. In *Solvent Extraction and Liquid Membranes: Fundamentals and Applications in New Materials*; CRC Press: Boca Raton, FL, USA, 2008.
57. Sun, S.P.; Chan, S.Y.; Chung, T.S. A Slow-Fast Phase Separation (SFPS) Process to Fabricate Dual-Layer Hollow Fiber Substrates for Thin-Film Composite (TFC) Organic Solvent Nanofiltration (OSN) Membranes. *Chem. Eng. Sci.* **2015**, *129*, 232–242. [CrossRef]
58. Goh, K.S.; Chong, J.Y.; Chen, Y.; Fang, W.; Bae, T.H.; Wang, R. Thin-Film Composite Hollow Fibre Membrane for Low Pressure Organic Solvent Nanofiltration. *J. Membr. Sci.* **2020**, *597*, 117760. [CrossRef]
59. Goh, K.S.; Chen, Y.; Chong, J.Y.; Bae, T.H.; Wang, R. Thin Film Composite Hollow Fibre Membrane for Pharmaceutical Concentration and Solvent Recovery. *J. Membr. Sci.* **2021**, *621*, 119008. [CrossRef]

60. Su, J.; Lv, X.; Li, S.; Jiang, Y.; Liu, S.; Zhang, X.; Li, H.; Su, B. High Separation Performance Thin Film Composite and Thin Film Nanocomposite Hollow Fiber Membranes via Interfacial Polymerization for Organic Solvent Nanofiltration. *Sep. Purif. Technol.* **2022**, *278*, 119567. [CrossRef]
61. Gao, Z.F.; Liu, J.; Chung, T.S. Rapid In-Situ Growth of Covalent Organic Frameworks on Hollow Fiber Substrates with Janus-like Characteristics for Efficient Organic Solvent Nanofiltration. *Sep. Purif. Technol.* **2022**, *294*, 121166. [CrossRef]
62. Abadikhah, H.; Kalali, E.N.; Behzadi, S.; Khan, S.A.; Xu, X.; Shabestari, M.E.; Agathopoulos, S. High Flux Thin Film Nanocomposite Membrane Incorporated with Functionalized TiO_2@reduced Graphene Oxide Nanohybrids for Organic Solvent Nanofiltration. *Chem. Eng. Sci.* **2019**, *204*, 99–109. [CrossRef]
63. Song, C.; Tang, S.; Yue, S.; Cui, Z.; Du, X.; Jiang, T.; He, B.; Li, J. Design of Microstructure for Hollow Fiber Loose Nanofiltration Separation Layer and Its Compactness-Tailoring Mechanism. *J. Hazard. Mater.* **2022**, *421*, 126800. [CrossRef] [PubMed]
64. Francis, V.N.; Chong, J.Y.; Yang, G.; Che, L.; Wang, R. Robust Polyamide-PTFE Hollow Fibre Membranes for Harsh Organic Solvent Nanofiltration. *Chem. Eng. J.* **2023**, *452*, 139333. [CrossRef]
65. Feng, W.; Li, J.; Fang, C.; Zhang, L.; Zhu, L. Controllable Thermal Annealing of Polyimide Membranes for Highly-Precise Organic Solvent Nanofiltration. *J. Membr. Sci.* **2022**, *643*, 120013. [CrossRef]
66. Baker, R.W. Membranes and Modules. In *Membrane Technology and Applications*; Wiley: Hoboken, NJ, USA, 2012.
67. Chung, T.S.; Hu, X. Effect of Air-Gap Distance on the Morphology and Thermal Properties of Polyethersulfone Hollow Fibers. *J. Appl. Polym. Sci.* **1997**, *66*, 1067–1077. [CrossRef]
68. Chung, T.S.; Teoh, S.K.; Lau, W.W.Y.; Srinivasan, M.P. Effect of Shear Stress within the Spinneret on Hollow Fiber Membrane Morphology and Separation Performance. *Ind. Eng. Chem. Res.* **1998**, *37*, 3930–3938. [CrossRef]
69. Qin, J.J.; Wang, R.; Chung, T.S. Investigation of Shear Stress Effect within a Spinneret on Flux, Separation and Thermomechanical Properties of Hollow Fiber Ultrafiltration Membranes. *J. Membr. Sci.* **2000**, *175*, 197–213. [CrossRef]
70. Wang, Z.Y.; Li, S.; Xu, S.; Tian, L.; Su, B.; Han, L.; Mandal, B. Fundamental Understanding on the Preparation Conditions of High-Performance Polyimide-Based Hollow Fiber Membranes for Organic Solvent Nanofiltration (OSN). *Sep. Purif. Technol.* **2021**, *254*, 117600. [CrossRef]
71. Wang, Z.Y.; Feng, R.; Wang, W.J.; Sun, Y.X.; Tao, S.N.; Wang, Y.M.; Chen, Y.F.; Fu, Z.J.; Cao, X.L.; Sun, S.P.; et al. Robust Braid Reinforced Hollow Fiber Membranes for Organic Solvent Nanofiltration (OSN). *Adv. Membr.* **2021**, *1*, 100007. [CrossRef]
72. Xia, L.; Ren, J.; McCutcheon, J.R. Braid-Reinforced Thin Film Composite Hollow Fiber Nanofiltration Membranes. *J. Membr. Sci.* **2019**, *585*, 109–114. [CrossRef]
73. He, T.; Mulder, M.H.V.; Strathmann, H.; Wessling, M. Preparation of Composite Hollow Fiber Membranes: Co-Extrusion of Hydrophilic Coatings onto Porous Hydrophobic Support Structures. *J. Membr. Sci.* **2002**, *207*, 143–156. [CrossRef]
74. Li, X.M.; Ji, Y.; Yin, Y.; Zhang, Y.Y.; Wang, Y.; He, T. Origin of Delamination/Adhesion in Polyetherimide/Polysulfone Co-Cast Membranes. *J. Membr. Sci.* **2010**, *352*, 173–179. [CrossRef]
75. Kopeć, K.K.; Dutczak, S.M.; Wessling, M.; Stamatialis, D.F. Chemistry in a Spinneret-On the Interplay of Crosslinking and Phase Inversion during Spinning of Novel Hollow Fiber Membranes. *J. Membr. Sci.* **2011**, *369*, 308–318. [CrossRef]
76. Dutczak, S.M.; Tanardi, C.R.; Kopeć, K.K.; Wessling, M.; Stamatialis, D. "Chemistry in a Spinneret" to Fabricate Hollow Fibers for Organic Solvent Filtration. *Sep. Purif. Technol.* **2012**, *86*, 183–189. [CrossRef]
77. Emonds, S.; Roth, H.; Wessling, M. Chemistry in a Spinneret—Formation of Hollow Fiber Membranes with a Cross-Linked Polyelectrolyte Separation Layer. *J. Membr. Sci.* **2020**, *612*, 118325. [CrossRef]
78. Sun, S.P.; Chan, S.Y.; Xing, W.; Wang, Y.; Chung, T.S. Facile Synthesis of Dual-Layer Organic Solvent Nanofiltration (OSN) Hollow Fiber Membranes. *ACS Sustain. Chem. Eng.* **2015**, *3*, 3019–3023. [CrossRef]
79. Soroko, I.; Livingston, A. Impact of TiO_2 Nanoparticles on Morphology and Performance of Crosslinked Polyimide Organic Solvent Nanofiltration (OSN) Membranes. *J. Membr. Sci.* **2009**, *343*, 189–198. [CrossRef]
80. Davood Abadi Farahani, M.H.; Chung, T.S. Solvent Resistant Hollow Fiber Membranes Comprising P84 Polyimide and Amine-Functionalized Carbon Nanotubes with Potential Applications in Pharmaceutical, Food, and Petrochemical Industries. *Chem. Eng. J.* **2018**, *345*, 174–185. [CrossRef]
81. Xia, L.; Ren, J.; Weyd, M.; McCutcheon, J.R. Ceramic-Supported Thin Film Composite Membrane for Organic Solvent Nanofiltration. *J. Membr. Sci.* **2018**, *563*, 857–863. [CrossRef]
82. Huang, W.Z.; Lin, F.; Lee, S.L.; Tao, F.T.; Tung, K.L. Fabrication of Microporous Polyamide Selective Layer on Macroporous Ceramic Hollow Fibers via Direct Interfacial Polymerization for Nanofiltration Applications. *J. Membr. Sci.* **2022**, *658*, 120710. [CrossRef]
83. Wang, X.L.; Wei, J.F.; Dai, Z.; Zhao, K.Y.; Zhang, H. Preparation and Characterization of Negatively Charged Hollow Fiber Nanofiltration Membrane by Plasma-Induced Graft Polymerization. *Desalination* **2012**, *286*, 138–144. [CrossRef]
84. Xu, H.M.; Wei, J.F.; Wang, X.L. Nanofiltration Hollow Fiber Membranes with High Charge Density Prepared by Simultaneous Electron Beam Radiation-Induced Graft Polymerization for Removal of Cr(VI). *Desalination* **2014**, *346*, 122–130. [CrossRef]
85. Aerts, S.; Vanhulsel, A.; Buekenhoudt, A.; Weyten, H.; Kuypers, S.; Chen, H.; Bryjak, M.; Gevers, L.E.M.; Vankelecom, I.F.J.; Jacobs, P.A. Plasma-Treated PDMS-Membranes in Solvent Resistant Nanofiltration: Characterization and Study of Transport Mechanism. *J. Membr. Sci.* **2006**, *275*, 212–219. [CrossRef]

86. Korikov, A.P.; Kosaraju, P.B.; Sirkar, K.K. Interfacially Polymerized Hydrophilic Microporous Thin Film Composite Membranes on Porous Polypropylene Hollow Fibers and Flat Films. *J. Membr. Sci.* **2006**, *279*, 588–600. [CrossRef]
87. Vanhook, S.J.; Schatz, M.F.; Swift, J.B.; McCormick, W.D.; Swinney, H.L. Long-Wavelength Surface-Tension-Driven Bénard Convection: Experiment and Theory. *J. Fluid Mech.* **1997**, *345*, 45–78. [CrossRef]
88. Karan, S.; Jiang, Z.; Livingston, A.G. Sub-10 Nm Polyamide Nanofilms with Ultrafast Solvent Transport for Molecular Separation. *Science* **2015**, *348*, 1347–1351. [CrossRef]
89. Li, J.; Zhang, M.; Feng, W.; Zhu, L.; Zhang, L. PIM-1 Pore-Filled Thin Film Composite Membranes for Tunable Organic Solvent Nanofiltration. *J. Membr. Sci.* **2020**, *601*, 117951. [CrossRef]
90. Fritsch, D.; Merten, P.; Heinrich, K.; Lazar, M.; Priske, M. High Performance Organic Solvent Nanofiltration Membranes: Development and Thorough Testing of Thin Film Composite Membranes Made of Polymers of Intrinsic Microporosity (PIMs). *J. Membr. Sci.* **2012**, *401–402*, 222–231. [CrossRef]
91. Yang, E.; Liang, Y.; Yanar, N.; Kim, M.; Park, H.; Choi, H. Intermolecular Cross-Linked Polymer of Intrinsic Microporosity-1 (PIM-1)-Based Thin-Film Composite Hollow Fiber Membrane for Organic Solvent Nanofiltration. *J. Membr. Sci.* **2023**, *671*, 121370. [CrossRef]
92. Gorgojo, P.; Karan, S.; Wong, H.C.; Jimenez-Solomon, M.F.; Cabral, J.T.; Livingston, A.G. Ultrathin Polymer Films with Intrinsic Microporosity: Anomalous Solvent Permeation and High Flux Membranes. *Adv. Funct. Mater.* **2014**, *24*, 4729–4737. [CrossRef]
93. Sorribas, S.; Gorgojo, P.; Téllez, C.; Coronas, J.; Livingston, A.G. High Flux Thin Film Nanocomposite Membranes Based on Metal-Organic Frameworks for Organic Solvent Nanofiltration. *J. Am. Chem. Soc.* **2013**, *135*, 15201–15208. [CrossRef] [PubMed]
94. Mahalingam, D.K.; Falca, G.; Upadhya, L.; Abou-Hamad, E.; Batra, N.; Wang, S.; Musteata, V.; da Costa, P.M.; Nunes, S.P. Spray-Coated Graphene Oxide Hollow Fibers for Nanofiltration. *J. Membr. Sci.* **2020**, *606*, 118006. [CrossRef]
95. Lu, H.; Wang, C.; Chen, J.; Ge, R.; Leng, W.; Dong, B.; Huang, J.; Gao, Y. A Novel 3D Covalent Organic Framework Membrane Grown on a Porous α-Al_2O_3 Substrate under Solvothermal Conditions. *Chem. Commun.* **2015**, *51*, 15562–15565. [CrossRef] [PubMed]
96. Shi, X.; Xiao, A.; Zhang, C.; Wang, Y. Growing Covalent Organic Frameworks on Porous Substrates for Molecule-Sieving Membranes with Pores Tunable from Ultra- to Nanofiltration. *J. Membr. Sci.* **2019**, *576*, 116–122. [CrossRef]
97. Zhao, D.L.; Japip, S.; Zhang, Y.; Weber, M.; Maletzko, C.; Chung, T.S. Emerging Thin-Film Nanocomposite (TFN) Membranes for Reverse Osmosis: A Review. *Water Res.* **2020**, *173*, 115557. [CrossRef]
98. Vanherck, K.; Koeckelberghs, G.; Vankelecom, I.F.J. Crosslinking Polyimides for Membrane Applications: A Review. *Prog. Polym. Sci.* **2013**, *38*, 874–896. [CrossRef]
99. Vanherck, K.; Cano-Odena, A.; Koeckelberghs, G.; Dedroog, T.; Vankelecom, I. A Simplified Diamine Crosslinking Method for PI Nanofiltration Membranes. *J. Membr. Sci.* **2010**, *353*, 135–143. [CrossRef]
100. Guo, X.; Liu, D.; Han, T.; Huang, H.; Yang, Q.; Zhong, C. Preparation of Thin Film Nanocomposite Membranes with Surface Modified MOF for High Flux Organic Solvent Nanofiltration. *AIChE J.* **2017**, *63*, 1303–1312. [CrossRef]
101. Lim, S.K.; Setiawan, L.; Bae, T.H.; Wang, R. Polyamide-Imide Hollow Fiber Membranes Crosslinked with Amine-Appended Inorganic Networks for Application in Solvent-Resistant Nanofiltration under Low Operating Pressure. *J. Membr. Sci.* **2016**, *501*, 152–160. [CrossRef]
102. Tham, H.M.M.; Wang, K.Y.; Hua, D.; Japip, S.; Chung, T.S. From Ultrafiltration to Nanofiltration: Hydrazine Cross-Linked Polyacrylonitrile Hollow Fiber Membranes for Organic Solvent Nanofiltration. *J. Membr. Sci.* **2017**, *542*, 289–299. [CrossRef]
103. Dutczak, S.M.; Cuperus, F.P.; Wessling, M.; Stamatialis, D.F. New Crosslinking Method of Polyamide-Imide Membranes for Potential Application in Harsh Polar Aprotic Solvents. *Sep. Purif. Technol.* **2013**, *102*, 142–146. [CrossRef]
104. Wang, K.Y.; Li, B.; Chung, T.S. 3D-Macrocycles Impregnated Polybenzimidazole Hollow Fiber Membranes with Excellent Organic Solvent Resistance for Industrial Solvent Recovery. *J. Membr. Sci.* **2021**, *638*, 119678. [CrossRef]
105. Loh, X.X.; Sairam, M.; Steinke, J.H.G.; Livingston, A.G.; Bismarck, A.; Li, K. Polyaniline Hollow Fibres for Organic Solvent Nanofiltration. *Chem. Commun.* **2008**, *47*, 6324–6326. [CrossRef]
106. See Toh, Y.H.; Lim, F.W.; Livingston, A.G. Polymeric Membranes for Nanofiltration in Polar Aprotic Solvents. *J. Membr. Sci.* **2007**, *301*, 3–10. [CrossRef]
107. Jimenez Solomon, M.F.; Bhole, Y.; Livingston, A.G. High Flux Membranes for Organic Solvent Nanofiltration (OSN)-Interfacial Polymerization with Solvent Activation. *J. Membr. Sci.* **2012**, *423–424*, 371–382. [CrossRef]
108. Shin, M.G.; Park, S.H.; Kwon, S.J.; Kwon, H.E.; Park, J.B.; Lee, J.H. Facile Performance Enhancement of Reverse Osmosis Membranes via Solvent Activation with Benzyl Alcohol. *J. Membr. Sci.* **2019**, *578*, 220–229. [CrossRef]
109. Mariën, H.; Vankelecom, I.F.J. Transformation of Cross-Linked Polyimide UF Membranes into Highly Permeable SRNF Membranes via Solvent Annealing. *J. Membr. Sci.* **2017**, *541*, 205–213. [CrossRef]
110. Li, Y.; Zhu, J.; Li, S.; Guo, Z.; Van Der Bruggen, B. Flexible Aliphatic-Aromatic Polyamide Thin Film Composite Membrane for Highly Efficient Organic Solvent Nanofiltration. *ACS Appl. Mater. Interfaces* **2020**, *12*, 31962–31974. [CrossRef]
111. Hołda, A.K.; Vankelecom, I.F.J. Understanding and Guiding the Phase Inversion Process for Synthesis of Solvent Resistant Nanofiltration Membranes. *J. Appl. Polym. Sci.* **2015**, *132*, 42130. [CrossRef]

112. Lai, X.; Wang, C.; Wang, L.; Xiao, C. A Novel PPTA/PPy Composite Organic Solvent Nanofiltration (OSN) Membrane Prepared by Chemical Vapor Deposition for Organic Dye Wastewater Treatment. *J. Water Process Eng.* **2022**, *45*, 102533. [CrossRef]
113. Tham, H.M.; Chung, T.S. One-Step Cross-Linking and Tannic Acid Modification of Polyacrylonitrile Hollow Fibers for Organic Solvent Nanofiltration. *J. Membr. Sci.* **2020**, *610*, 118294. [CrossRef]
114. Grodowska, K.; Parczewski, A. Organic Solvents in the Pharmaceutical Industry. *Acta Pol. Pharm. Drug Res.* **2010**, *67*, 3–12.
115. Li, Y.; Cao, B.; Li, P. Effects of Dope Compositions on Morphologies and Separation Performances of PMDA-ODA Polyimide Hollow Fiber Membranes in Aqueous and Organic Solvent Systems. *Appl. Surf. Sci.* **2019**, *473*, 1038–1048. [CrossRef]
116. Jeon, S.; Nishitani, A.; Cheng, L.; Fang, L.F.; Kato, N.; Shintani, T.; Matsuyama, H. One-Step Fabrication of Polyamide 6 Hollow Fibre Membrane Using Non-Toxic Diluents for Organic Solvent Nanofiltration. *RSC Adv.* **2018**, *8*, 19879–19882. [CrossRef]
117. Falca, G.; Musteata, V.E.; Behzad, A.R.; Chisca, S.; Nunes, S.P. Cellulose Hollow Fibers for Organic Resistant Nanofiltration. *J. Membr. Sci.* **2019**, *586*, 151–161. [CrossRef]
118. McGuinness, E.K.; Zhang, F.; Ma, Y.; Lively, R.P.; Losego, M.D. Vapor Phase Infiltration of Metal Oxides into Nanoporous Polymers for Organic Solvent Separation Membranes. *Chem. Mater.* **2019**, *31*, 5509–5518. [CrossRef]
119. Wang, Z.Y.; Fu, Z.J.; Shao, D.D.; Lu, M.J.; Xia, Q.C.; Xiao, H.F.; Su, B.W.; Sun, S.P. Bridging the Miscibility Gap to Fabricate Delamination-Free Dual-Layer Nanofiltration Membranes via Incorporating Fluoro Substituted Aromatic Amine. *J. Membr. Sci.* **2020**, *610*, 118270. [CrossRef]
120. Xu, S.; Wang, Z.Y.; Li, S.; Tian, L.; Su, B. Fabrication of Polyimide-Based Hollow Fiber Membrane by Synergetic Covalent-Crosslinking Strategy for Organic Solvent Nanofiltration (OSN) Application. *Sep. Purif. Technol.* **2020**, *241*, 116751. [CrossRef]
121. Kato, N.; Gonzales, R.R.; Nishitani, A.; Negi, Y.; Ono, T.; Matsuyama, H. Single-Step Preparation of Nanocomposite Polyamide 6 Hollow Fiber Membrane with Integrally Skinned Asymmetric Structure for Organic Solvent Nanofiltration. *Colloids Surf. A Physicochem. Eng. Asp.* **2021**, *620*, 126538. [CrossRef]
122. Junker, M.A.; de Vos, W.M.; Lammertink, R.G.H.; de Grooth, J. Bridging the gap between lab-scale and commercial dimensions of hollow fiber nanofiltration membranes. *J. Membr. Sci.* **2021**, *624*, 119100. [CrossRef]
123. Buonomenna, M.G. *Nano-Enhanced and Nanostructured Polymer-Based Membranes for Energy Applications*; Series in Composites Science and Engineering; Woodhead Publishing: Sawston, UK, 2022.
124. Labban, O.; Chong, T.H.; Lienhard, J.H. Design and modeling of novel low-pressure nanofiltration hollow fiber modules for water softening and desalination pretreatment. *Desalination* **2018**, *439*, 58–72. [CrossRef]
125. Jährig, J.; Vredenbregt, L.; Wicke, D.; Miehe, U.; Sperlich, A. Capillary nanofiltration under anoxic conditions as post-treatment after bank filtration. *Water* **2018**, *10*, 1599. [CrossRef]
126. Geyer, R.; Jambeck, J.R.; Law, K.L. Production, Use, and Fate of All Plastics Ever Made. *Sci. Adv.* **2017**, *3*, e1700782. [CrossRef]
127. Nguyen Thi, H.Y.; Kim, S.; Duy Nguyen, B.T.; Lim, D.; Kumar, S.; Lee, H.; Szekely, G.; Kim, J.F. Closing the Sustainable Life Cycle Loop of Membrane Technology via a Cellulose Biomass Platform. *ACS Sustain. Chem. Eng.* **2022**, *10*, 2532–2544. [CrossRef]
128. Lin, S.; Semiao, A.C.; Zhang, Y.; Lu, S.; Hon Lau, C. Interfacial polymerization using biobased solvents and their application as desalination and organic solvent nanofiltration membranes. *J. Membr. Sci.* **2024**, *692*, 122281. [CrossRef]
129. de Gonzalo, G.; Alcantara, A.R.; Dominguez de Maria, P. Cyclopentyl Methyl Ether (CPME): A Versatile Eco-Friendly Solvent for Applications in Biotechnology and Biorefineries. *ChemSusChem* **2019**, *12*, 2083–2097. [CrossRef] [PubMed]
130. Khoo, H.H.; Wong, L.L.; Tan, J.; Isoni, V.; Sharratt, P. Synthesis of 2-methyl tetrahydrofuran from various lignocellulosic feedstocks: Sustainability assessment via Lca. *Resour. Conserv. Recycl.* **2015**, *95*, 174–182. [CrossRef]
131. Cseri, L.; Szekely, G. Towards Cleaner PolarClean: Efficient Synthesis and Extended Applications of the Polar Aprotic Solvent Methyl 5-(Dimethylamino)-2-Methyl-5-Oxopentanoate. *Green Chem.* **2019**, *21*, 4178–4188. [CrossRef]
132. Hardian, R.; Cywar, R.M.; Chen, E.Y.X.; Szekely, G. Sustainable Nanofiltration Membranes Based on Biosourced Fully Recyclable Polyesters and Green Solvents. *J. Membr. Sci. Lett.* **2022**, *2*, 100016. [CrossRef]
133. Ong, C.; Falca, G.; Huang, T.; Liu, J.; Manchanda, P.; Chisca, S.; Nunes, S.P. Green Synthesis of Thin-Film Composite Membranes for Organic Solvent Nanofiltration. *ACS Sustain. Chem. Eng.* **2020**, *8*, 11541–11548. [CrossRef]
134. Rasool, M.A.; Vankelecom, I.F.J. Use of γ-Valerolactone and Glycerol Derivatives as Bio-Based Renewable Solvents for Membrane Preparation. *Green Chem.* **2019**, *21*, 1054–1064. [CrossRef]
135. Dargo, G.; Kis, D.; Gede, M.; Kumar, S.; Kupai, J.; Szekely, G. MeSesamol, a Bio-Based and Versatile Polar Aprotic Solvent for Organic Synthesis and Depolymerization. *Chem. Eng. J.* **2023**, *471*, 144365. [CrossRef]
136. Park, S.H.; Yang, C.; Ayaril, N.; Szekely, G. Solvent-Resistant Thin-Film Composite Membranes from Biomass-Derived Building Blocks: Chitosan and 2,5-Furandicarboxaldehyde. *ACS Sustain. Chem. Eng.* **2022**, *10*, 998–1007. [CrossRef]
137. Alqadhi, N.; Abdellah, M.H.; Nematulloev, S.; Mohammed, O.F.; Abdulhamid, M.A.; Szekely, G. Solution-Processable Poly(Ether-Ether-Ketone) Membranes for Organic Solvent Nanofiltration: From Dye Separation to Pharmaceutical Purification. *Sep. Purif. Technol.* **2024**, *328*, 125072. [CrossRef]
138. Yang, C.; Topuz, F.; Park, S.H.; Szekely, G. Biobased Thin-Film Composite Membranes Comprising Priamine-Genipin Selective Layer on Nanofibrous Biodegradable Polylactic Acid Support for Oil and Solvent-Resistant Nanofiltration. *Green Chem.* **2022**, *24*, 5291–5303. [CrossRef]

139. Park, S.H.; Alammar, A.; Fulop, Z.; Pulido, B.A.; Nunes, S.P.; Szekely, G. Hydrophobic Thin Film Composite Nanofiltration Membranes Derived Solely from Sustainable Sources. *Green Chem.* **2021**, *23*, 1175–1184. [CrossRef]
140. Abdellah, M.H.; Oviedo, C.; Szekely, G. Controlling the Degree of Acetylation in Cellulose-Based Nanofiltration Membranes for Enhanced Solvent Resistance. *J. Membr. Sci.* **2023**, *687*, 122040. [CrossRef]
141. Zheng, D.; Hua, D.; Hong, Y.; Ibrahim, A.R.; Yao, A.; Pan, J.; Zhan, G. Functions of Ionic Liquids in Preparing Membranes for Liquid Separations: A Review. *Membranes* **2020**, *10*, 395. [CrossRef] [PubMed]
142. Falca, G.; Musteata, V.E.; Chisca, S.; Hedhili, M.N.; Ong, C.; Nunes, S.P. Naturally Extracted Hydrophobic Solvent and Self-Assembly in Interfacial Polymerization. *ACS Appl. Mater. Interfaces* **2021**, *13*, 44824–44832. [CrossRef] [PubMed]
143. Alder, C.M.; Hayler, J.D.; Henderson, R.K.; Redman, A.M.; Shukla, L.; Shuster, L.E.; Sneddon, H.F. Updating and Further Expanding GSK's Solvent Sustainability Guide. *Green Chem.* **2016**, *18*, 3879–3890. [CrossRef]
144. Xie, W.; Li, T.; Tiraferri, A.; Drioli, E.; Figoli, A.; Crittenden, J.C.; Liu, B. Toward the Next Generation of Sustainable Membranes from Green Chemistry Principles. *ACS Sustain. Chem. Eng.* **2021**, *9*, 50–75. [CrossRef]
145. Nunes, S.P.; Culfaz-Emecen, P.Z.; Ramon, G.Z.; Visser, T.; Koops, G.H.; Jin, W.; Ulbricht, M. Thinking the Future of Membranes: Perspectives for Advanced and New Membrane Materials and Manufacturing Processes. *J. Membr. Sci.* **2020**, *598*, 117761. [CrossRef]
146. Abdellah, M.H.; Pérez-Manríquez, L.; Puspasari, T.; Scholes, C.A.; Kentish, S.E.; Peinemann, K.V. Effective Interfacially Polymerized Polyester Solvent Resistant Nanofiltration Membrane from Bioderived Materials. *Adv. Sustain. Syst.* **2018**, *2*, 1800043. [CrossRef]
147. Abdellah, M.H.; Pérez-Manríquez, L.; Puspasari, T.; Scholes, C.A.; Kentish, S.E.; Peinemann, K.V. A Catechin/Cellulose Composite Membrane for Organic Solvent Nanofiltration. *J. Membr. Sci.* **2018**, *567*, 139–145. [CrossRef]
148. Le Phuong, H.A.; Izzati Ayob, N.A.; Blanford, C.F.; Mohammad Rawi, N.F.; Szekely, G. Nonwoven Membrane Supports from Renewable Resources: Bamboo Fiber Reinforced Poly(Lactic Acid) Composites. *ACS Sustain. Chem. Eng.* **2019**, *7*, 11885–11893. [CrossRef]
149. Alammar, A.; Hardian, R.; Szekely, G. Upcycling Agricultural Waste into Membranes: From Date Seed Biomass to Oil and Solvent-Resistant Nanofiltration. *Green Chem.* **2022**, *24*, 365–374. [CrossRef]
150. Hardian, R.; Alammar, A.; Holtzl, T.; Szekely, G. Fabrication of Sustainable Organic Solvent Nanofiltration Membranes Using Cellulose–Chitosan Biopolymer Blends. *J. Membr. Sci.* **2022**, *658*, 120743. [CrossRef]
151. Aburabie, J.H.; Puspasari, T.; Peinemann, K.V. Alginate-Based Membranes: Paving the Way for Green Organic Solvent Nanofiltration. *J. Membr. Sci.* **2020**, *596*, 117615. [CrossRef]
152. George, N.; Varghese, V.; Kavitha, O. Earth's versatile gift: Cellulose. In *Handbook of Biomass*; Springer: Singapore, 2023.
153. Durmaz, E.N.; Zeynep Çulfaz-Emecen, P. Cellulose-Based Membranes via Phase Inversion Using [EMIM]OAc-DMSO Mixtures as Solvent. *Chem. Eng. Sci.* **2018**, *178*, 93–103. [CrossRef]
154. Sukma, F.M.; Çulfaz-Emecen, P.Z. Cellulose Membranes for Organic Solvent Nanofiltration. *J. Membr. Sci.* **2018**, *545*, 329–336. [CrossRef]
155. Konca, K.; Çulfaz-Emecen, P.Z. Effect of Carboxylic Acid Crosslinking of Cellulose Membranes on Nanofiltration Performance in Ethanol and Dimethylsulfoxide. *J. Membr. Sci.* **2019**, *587*, 117175. [CrossRef]
156. Nasirpour, N.; Mohammadpourfard, M.; Zeinali Heris, S. Ionic Liquids: Promising Compounds for Sustainable Chemical Processes and Applications. *Chem. Eng. Res. Des.* **2020**, *160*, 264–300. [CrossRef]
157. Qiu, W.Z.; Zhong, Q.Z.; Du, Y.; Lv, Y.; Xu, Z.K. Enzyme-Triggered Coatings of Tea Catechins/Chitosan for Nanofiltration Membranes with High Performance. *Green Chem.* **2016**, *18*, 6205–6208. [CrossRef]
158. Biron, M. A Practical Guide to Plastics Sustainability. In *Concept, Solutions, and Implementation*; Elsevier: Amsterdam, The Netherlands, 2020.
159. Pulido, B.A.; Habboub, O.S.; Aristizabal, S.L.; Szekely, G.; Nunes, S.P. Recycled Poly(Ethylene Terephthalate) for High Temperature Solvent Resistant Membranes. *ACS Appl. Polym. Mater.* **2019**, *1*, 2379–2387. [CrossRef]
160. Zhu, Y.; Romain, C.; Williams, C.K. Sustainable Polymers from Renewable Resources. *Nature* **2016**, *540*, 354–362. [CrossRef]
161. Hardian, R.; Ghaffar, A.; Shi, C.; Chen, E.Y.-X.; Szekely, G. Chemically recyclable nanofiltration membranes fabricated from two circular polymer classes of the same monomer origin. *J. Membr. Sci.* **2024**, *4*, 100067. [CrossRef]
162. Szejtli, J. Introduction and General Overview of Cyclodextrin Chemistry. *Chem. Rev.* **1998**, *98*, 1743–1754. [CrossRef]
163. Available online: https://www.sigmaaldrich.com/SA/en/products (accessed on 15 December 2023).
164. Available online: https://www.thermofisher.com/it/en/home.html (accessed on 15 December 2023).
165. Pérez-Manríquez, L.; Neelakanda, P.; Peinemann, K.V. Tannin-Based Thin-Film Composite Membranes for Solvent Nanofiltration. *J. Membr. Sci.* **2017**, *541*, 137–142. [CrossRef]
166. Pérez-Manríquez, L.; Neelakanda, P.; Peinemann, K.V. Morin-Based Nanofiltration Membranes for Organic Solvent Separation Processes. *J. Membr. Sci.* **2018**, *554*, 1–5. [CrossRef]
167. Liu, J.; Hua, D.; Zhang, Y.; Japip, S.; Chung, T.S. Precise Molecular Sieving Architectures with Janus Pathways for Both Polar and Nonpolar Molecules. *Adv. Mater.* **2018**, *30*, 1705933. [CrossRef] [PubMed]

168. Kim, J.H.; Cook, M.; Park, S.H.; Moon, S.J.; Kim, J.F.; Livingston, A.G.; Lee, Y.M. A Compact and Scalable Fabrication Method for Robust Thin Film Composite Membranes. *Green Chem.* **2018**, *20*, 1887–1898. [CrossRef]
169. Dedvukaj, A.; Raemdonck, S.; Vankelecom, I.F.J. Spray Coating as a Novel, Versatile Tool to Prepare Membranes via IP. *J. Membr. Sci.* **2023**, *685*, 121905. [CrossRef]
170. Lin, S.; Zhang, Y.; Shao, L.; Lau, C.H. Spray-Assisted Assembly of Thin-Film Composite Membranes in One Process. *Adv. Membr.* **2024**, *4*, 100080. [CrossRef]
171. Gould, R.M.; White, L.S.; Wildemuth, C.R. Membrane Separation in Solvent Lube Dewaxing. *Environ. Prog.* **2001**, *20*, 12–16. [CrossRef]
172. Lively, R.P.; Sholl, D.S. From Water to Organics in Membrane Separations. *Nat. Mater.* **2017**, *16*, 276–279. [CrossRef]
173. Hu, J.; Kim, C.; Halasz, P.; Kim, J.F.; Kim, J.; Szekely, G. Artificial Intelligence for Performance Prediction of Organic Solvent Nanofiltration Membranes. *J. Membr. Sci.* **2021**, *619*, 118513. [CrossRef]
174. Le Phuong, H.A.; Blanford, C.F.; Szekely, G. Reporting the Unreported: The Reliability and Comparability of the Literature on Organic Solvent Nanofiltration. *Green Chem.* **2020**, *22*, 3397–3409. [CrossRef]
175. Ignacz, G.; Beke, A.K.; Szekely, G. Data-Driven Future for Nanofiltration: Escaping Linearity. *J. Membr. Sci. Lett.* **2023**, *3*, 100040. [CrossRef]

processes

MDPI

Review

The Use of Immobilised Enzymes for Lipid and Dairy Processing and Their Waste Products: A Review of Current Progress

Fawzih Alzahrani [1,2], Taiwo O. Akanbi [1,*], Christopher J. Scarlett [1] and Alberta N. A. Aryee [3]

1 School of Environmental and Life Sciences, College of Engineering, Science and Environment, The University of Newcastle, Brush Road, Ourimbah, NSW 2258, Australia; fawzih.alzahrani@uon.edu.au (F.A.); c.scarlett@newcastle.edu.au (C.J.S.)
2 Department of Biology, University of Ha'il, Ha'il 55476, Saudi Arabia
3 Food Science & Biotechnology Program, Department of Human Ecology, College Agriculture, Science and Technology, Delaware State University, 1200 N DuPont Highway, Dover, DE 19901, USA; aaryee@desu.edu
* Correspondence: taiwo.akanbi@newcastle.edu.au; Tel.: +61-243484117

Abstract: The use of edible oils and fats in dairy products is becoming increasingly important in the food industry because of their complementary functional properties. Most of these products are produced using food-grade enzymes as processing aids because processes involving enzymes are considered mild and environmentally friendly for regulatory purposes. The poor stability and recovery of enzymes in their native state limit their performance, and to enhance their activity, stability, and reusability, enzymes are often immobilised—a process that involves attaching them to a solid support. Additionally, immobilisation enables enzymes to selectively target specific substrates or products, making them highly efficient. These features have led to the increased use of immobilised enzymes in dairy and lipid processing and enzymes have been used to produce a broad range of products such as whey protein concentrates and isolates, peptide–lipid conjugates, lipid concentrates, structured lipids, and human milk fat substitutes. Therefore, this article reviews the current progress on different enzyme preparations and their use in lipid and dairy processing. It also summarises opportunities in enzyme-catalysed valorisation of dairy and lipid waste streams with the ultimate goals of sustainable food production and reductions in waste.

Keywords: enzymes; immobilisation; activity; applications; processing

Citation: Alzahrani, F.; Akanbi, T.O.; Scarlett, C.J.; Aryee, A.N.A. The Use of Immobilised Enzymes for Lipid and Dairy Processing and Their Waste Products: A Review of Current Progress. *Processes* **2024**, *12*, 634. https://doi.org/10.3390/pr12040634

Academic Editors: Yanlin Zhang and Prashant K. Sarswat

Received: 22 February 2024
Revised: 18 March 2024
Accepted: 18 March 2024
Published: 22 March 2024

1. Introduction

Enzymes are proteins that act as biological catalysts in living organisms, including plants, animals, and microorganisms [1]. They play a crucial role in accelerating reactions, enabling the various biochemical processes necessary for life. Enzymes are highly specific in their actions, catalysing a particular reaction or a group of related reactions [2]. They function by lowering the activation energy required for a reaction to occur, making it easier and faster for the reaction to take place [3]. Enzymes are used extensively in industrial processes, such as food and beverage production, pharmaceuticals, and bioremediation [4]. However, to ensure their efficient and cost-effective use, enzymes often need to be stabilised for repeated reuse. This necessity underscores the significance of enzyme immobilisation, a vital process for enhancing their stability and repeated reuse in industrial applications.

Immobilisation is a technique used in biotechnology to attach enzymes onto a support material or carrier, allowing them to function in a variety of different applications [5]. Immobilised enzymes have become an important tool in many industries due to their unique advantages over free enzymes, including increased stability, reusability, and selectivity (Figure 1). Immobilisation protects enzymes from degradation, extreme temperatures, and other harsh conditions, thereby extending their functional life and efficiency [5–8].

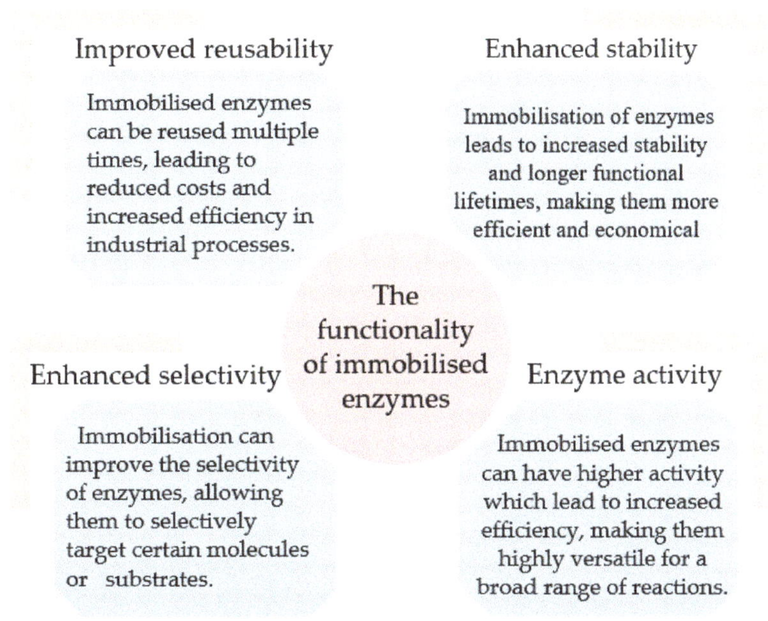

Figure 1. The functionality of immobilised enzymes.

Enzymes play a major role in lipid and dairy processing and offer various benefits over chemical processes. Enzymes have been used for milk modifications to produce several dairy-based products such as whey protein isolate (WPI), whey protein concentrate (WPC), hydrolysed infant formula, and milk fat [9–12]. For instance, whey was obtained via enzymatic coagulation of cow milk by rennet [13] and whey protein hydrolysates were produced by the hydrolysis of cow and goat milk using different proteases including alcalase, neutrase, trypsin, bromelain, pepsin, papain, and proteinase [14–16]. Enzymes have also been used to produce bioactive peptides from milk proteins and dairy by-products [17,18], as well as structured and specialty lipids with functional properties via processes such as hydrolysis, esterification, and interesterification [19–21]. This suggests that food-grade enzymes and appropriate immobilisation supports are needed for these applications. An ideal support material or matrix for immobilisation must be robust, inert, stable, and affordable [22,23].

Several studies have been carried out on the fortification of dairy products with functional lipids, resulting in improved health and sensory qualities. These functional lipids are mostly added in pure forms, and they include omega-3, -6, and -9 polyunsaturated fatty acids (PUFAs) such as eicosapentaenoic (EPA), docosahexaenoic (DHA), *alpha*-linolenic (ALA), linoleic (LA), and oleic (OA) acids. A recent study found that yoghurt fortified with PUFAs showed neuroprotective effects against aluminium-chloride-mediated neurotoxicity and enhanced brain dopamine and serotonin [24]. Also, yoghurts fortified with EPA and DHA showed high bioaccessibility values as a result of efficient intestinal absorption [25]. Other dairy products fortified with omega-3 fatty acids [26–28] and other functional lipids have been reported (Table 1). Since pure forms of these PUFAs are often required, they must be extracted or concentrated from oils. Industrial production of PUFA concentrates is achieved by environmentally unfriendly methods such as fractional distillation and urea complexation techniques, which lead to partial oxidation and polymerisation of these lipids [29]. The use of enzymes for concentrating PUFAs has been reported as the preferred method as they are milder and environmentally friendly [21,30]. In addition, enzymes are used as food processing aids and are regarded as natural for regulatory purposes [31,32].

Table 1. Dairy product fortification with functional lipids.

Functional Lipids	Fortified Dairy Product	References
Docosahexaenoic acid (DHA)	DHA-fortified yoghurt and infant formula	[33,34]
Docosapentaenoic acid (DPA)	DPA-fortified milk	[35]
Eicosapentaenoic acid (EPA)	Fermented milk fortified with EPA	[36]
Stearidonic acid (SDA)	SDA-fortified yoghurt	[37]
Arachidonic acid (ARA)	ARA-fortified infant formula powders	[34]
Linoleic acid (LA)	Butter fortified with LA	[26]
Alpha-linolenic acid (ALA)	Yoghurt fortified with linolenic acid	[38]
Oleic acid (OA)	White soft cheese fortified with oleic acid	[39]
Flaxseed oil (rich in ALA)	Cheese fortified with flaxseed oil	[40]
Encapsulated echium oil (rich in SDA)	Yoghurt fortified with encapsulated echium oil	[37]

Therefore, this review focuses on different enzyme immobilisation techniques using various supports, their use in lipid and dairy processing, and the valorisation of their waste products.

2. Enzyme Immobilisation Methods

Enzyme immobilisation is a process of attaching enzymes to a solid support, resulting in an enhanced activity, stability, reusability, and efficiency [41,42]. Immobilisation can also significantly reduce the cost and loss of enzymes during reactions as immobilised enzymes are easier to recover than non-immobilised ones. There are several methods of enzyme immobilisation, including adsorption, entrapment, covalent bonding, and crosslinking. The choice of an immobilisation method depends on many factors, such as the specific characteristics of the enzyme and its applications [43].

Enzyme immobilisation can be achieved through physical and chemical methods, and the strength of enzyme attachment to the support differs (Figure 2). For instance, in the physical method, enzymes can be adsorbed on the surface of the support (adsorption) or entrapped within the support matrix (entrapment) without forming any chemical bonds with the support [44]. Meanwhile, the chemical method involves the formation of covalent (chemical) bonds between the enzyme and the support material. They can also be chemically bonded to the functional groups on the support material via crosslinkers [45,46].

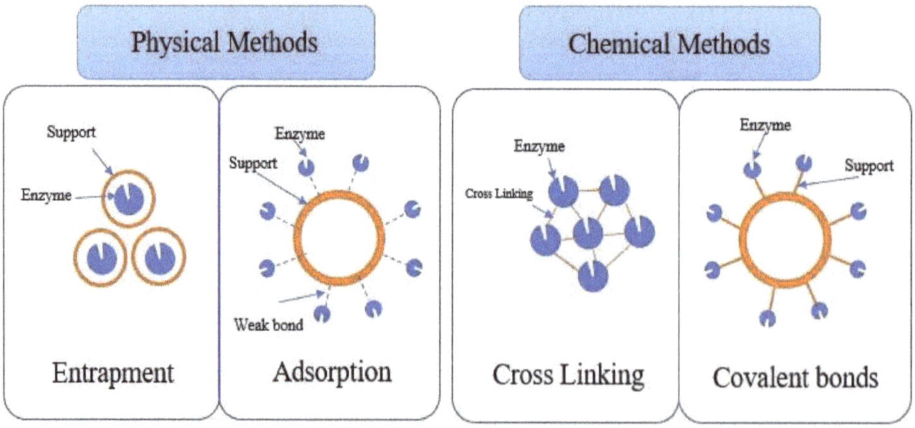

Figure 2. Enzyme immobilisation methods.

2.1. Physical Methods of Enzyme Immobilisation

2.1.1. Adsorption

The adsorption method involves attaching enzymes to a solid support through weak interactions such as van der Waals forces, hydrogen bonding, and hydrophobic interactions [47]. With this method, enzymes attach to the surface of the support material without forming strong covalent bonds. The adsorption method is simple and no chemical modification of the enzyme or the support material is required. It has been reported that enzymes immobilised by this interfacial adsorption method often have a high activity because they are attached to the support in an open conformation [48].

A previous study found that *Candida antacrtica* lipase A immobilised on octadecyl resin via physical adsorption showed a higher immobilisation yield and enzyme activity than when immobilised covalently [41]. However, the immobilised enzyme could be reused for five cycles before a sharp decline in activity [41]. This is the major drawback of immobilisation by adsorption, as enzymes can easily leach from the support due to the weak bonding, thus reducing the stability and reusability of the immobilised enzymes. Despite this drawback, the method is still being used in some applications where a simple immobilisation method like adsorption is desired.

2.1.2. Entrapment

Immobilisation by entrapment involves trapping the enzyme within a porous support material that allows free diffusion of the substrates and product but not the enzyme [44]. The support material must have the appropriate pore sizes for the enzyme to be entrapped. A highly porous material or a material with too large pores will cause enzyme leaching and a low immobilisation efficiency and activity [49,50]. To prevent leaching, the enzyme is mixed with a polymer, which forms a complex structure within the material, thus trapping the enzyme [49]. It is important to note that the entrapped enzyme is not attached to the support, but the support acts as a physical barrier that prevents the enzyme from leaching.

A β-fructofuranosidase enzyme from *Aspergillus japonicus* was immobilised by entrapment on Amberlite IRA 900 resin (a food-grade resin) for the synthesis of fructooligosaccharide (FOS) [51]. The authors found that the immobilised enzyme had a higher FOS yield, activity, and reusability in up to 12 batch cycles [51]. Several studies have shown that enzymes immobilised by entrapment often have a high degree of reusability, which is desired in industrial applications [44,49–51]. However, others have reported that the process of entrapment can cause enzyme rigidity and inactivation [52,53].

2.2. Chemical Methods of Enzyme Immobilisation

2.2.1. Covalent Bond

Covalent bond formation is widely used for enzyme immobilisation as it ensures a strong and permanent attachment of the enzyme to the support material, thus preventing enzyme leaching [44]. This method greatly relies on the amino or carboxyl functional group of the enzymes and the reactive group of the support through various types of linkages and interactions [54]. The most common active groups on the support include glutaraldehyde, epoxide, glyoxyl, and vinyl sulfone [55,56]. The covalent binding method is irreversible, and enzymes immobilised this way are always stable as they do not detach from the support during use. Therefore, this is the ideal immobilisation method when the complete removal of the enzyme from the final product is desired [57]. Covalent bonding also improves the selectivity, reusability, thermostability, and organic solvent tolerance of enzymes, making it highly versatile in many applications [54,55,58].

A commercial alkaline protease from *Bacillus licheniformis* protease immobilised covalently on chitosan modified by glutaraldehyde and ethylenediamine showed higher enzyme activity at 70 °C compared to the value exhibited at 60 °C by the free enzyme. The immobilised enzyme was also highly effective in hydrolysing casein, haemoglobin, soy proteins, and gelatin compared to the free enzyme [59]. Protease (Novo-Pro D) immobilised on agarose activated with a glyoxyl group was used to hydrolyse casein ten times without any decrease in activity [60]. Others have also reported multiple reuses of enzymes immobilised covalently [16,61,62].

2.2.2. Crosslinking

The crosslinking method involves forming a covalent bond between enzymes and further attaching them to the support material [44]. Generally, this immobilisation technique is straightforward, and it involves two steps. First, enzymes are precipitated using precipitants such as ammonium sulfate, ethanol, tert-butanol, acetone, acetonitrile, glycol, and isopropanol [2,63]. Afterwards, the precipitated enzymes are crosslinked to form a crosslinked enzyme aggregate or crystal (CLEA or CLEC, respectively). Examples of crosslinkers include glutaraldehyde, benzoquinone, dextran-polyaldehyde, polyethyleneimine, and carbodiimide [44]. However, glutaraldehyde is the most reported crosslinker. The crosslinking process using glutaraldehyde ensures the stable bond of the enzyme to the support, allowing for efficient immobilisation [6]. Crosslinking is a well-established immobilisation technique, and the immobilised enzymes are always stable and highly reusable. The only drawback with this method is that enzymes can be easily denatured during precipitation as some are not solvent-tolerant [64]. Consequently, choosing a precipitant that will not denature the enzyme is essential.

A recent study found that immobilisation of lipase from *Pseudomonas stutzeri* via crosslinking resulted in a more robust and highly stable enzyme that is more active in deep eutectic solvents (DESs) than the free enzyme that was not active in these solvents [65]. In another study, invertase was immobilised on chitosan using glutaraldehyde as the crosslinker [66]. The results showed that the enzyme had higher pH stability, thermal stability, and reusability, retaining 70% of its activity after nine batches of reuse. It was highly effective in producing bioethanol [66].

3. Immobilised Enzymes in Dairy Processing

Enzymes have been widely used as processing aids during the production of dairy products including WPC, WPI, infant formula, and milk fat (Table 2). The use of immobilised enzymes in the production of these products is discussed in this section.

Table 2. Enzymes in dairy product manufacturing.

Product	Enzyme	References
Whey protein concentrate	Alcalase	[60,67]
	β-galactosidase	[68,69]
	Proteinase	[70,71]
	Aspartic protease	[72,73]
	Neutrase	[30,74]
Whey protein isolate	Alcalase	[75,76]
	Trypsin	[77,78]
	Flavourzyme	[79,80]
Peptide	Ficin	[81,82]
	Rennin (Chymosin)	[83,84]
	Alcalase	[67,85]
Infant formula	Lipase	[86,87]
	Alcalase	[88,89]
	Trypsin	[90,91]
	Flavourzyme	[92,93]
Milk fat	Lipase	[94,95]

3.1. Whey Protein Concentrate

WPC is a popular supplement and ingredient in various food products due to its high protein content (66.5–95.8%) and nutritional value [96]. Traditional methods of producing WPC involve the use of various filtration and separation processes to concentrate the protein content from whey, a by-product of cheese making [97]. Enzyme immobilisation techniques have provided opportunities for more efficient and environmentally friendly methods of producing WPC. Immobilised enzymes in the production of whey protein concentrate can potentially lower production costs in the long run due to enzyme reusability. Additionally, the process can be tailored to achieve specific protein concentrations and meet the desired product specifications [68].

The most used enzymes to produce WPC are lactase and β-galactosidase, which are mixed with the whey to hydrolyse the lactose [98]. The enzymes break down the lactose into glucose and galactose while leaving the proteins intact. Studies describing WPC production via hydrolysis using immobilised lactase and β-galactosidase have been reported [68,69,99]. After the hydrolysis, the solid supports with the immobilised enzymes are separated from the liquid phase via centrifugation. The liquid phase containing the whey proteins is then concentrated using various methods such as evaporation, ultrafiltration, or spray drying to produce whey protein concentrate. During concentration, the liquid is reduced in volume, resulting in a higher protein content [42,98].

Immobilised enzymes have emerged as a promising solution to produce WPC, offering a multitude of advantages over free enzymes. β-galactosidase from *Kluyveromyces lactis* was immobilised on calcium alginate spheres by entrapment and used for the production of WPC from bovine, goat, and buffalo whey [68]. The authors found that the storage and thermal stability of the enzyme increased after immobilisation and it showed a high operational stability, which allowed multiple reuses for more than 10 cycles [68]. In another study, *Kluyveromyces lactis* β-galactosidase immobilised on chitosan activated with glutaraldehyde, hydrolysed 43% of milk lactose compared to the 38% hydrolysed by the free (non-immobilised) enzyme [100]. These enzymes experience enhanced stability when immobilised, enabling multiple reuses and specificity, leading to targeted hydrolysis of lactose while preserving whey proteins and minimising by-product formation [68,98,100].

3.2. *Whey Protein Isolate*

Whey protein isolate (WPI) is a highly purified form of whey protein, recognised for its exceptional nutritional profile and effectiveness in supporting muscle recovery and growth [101]. Derived from milk during the cheese-making process, WPI undergoes further processing to remove most of the fats and carbohydrates, resulting in a product that contains an impressive concentration of protein, typically over 90% [102]. This purity not only makes it an ideal choice for athletes but also makes it a suitable option for individuals who may be lactose intolerant, as the lactose content is significantly reduced, as well as its rapid absorption and rich essential amino acid content [103].

WPI production using enzymes involves a specialised process known as enzymatic hydrolysis or proteolysis. This process is used to break down the whey protein into its constituent peptides and amino acids, followed by further purification to isolate the whey protein in its purest form [104]. However, using immobilised enzymes is a highly efficient and advanced process, with these methods carefully chosen to ensure the highest quality of the final product, with minimal impact on the nutritional value and bioactivity of the whey proteins [88].

One of the key benefits of using immobilised enzymes for WPI production is the control over the enzymatic reactions. The immobilised enzymes function under controlled conditions, such as temperature and pH, leading to consistent and predictable outcomes. This control results in a purer form of WPI with a higher protein content, fewer impurities, and a more refined amino acid profile. Moreover, immobilised enzymes have an extended operational life compared to free enzymes, reducing the need for frequent enzyme replacements and making the production process more cost-effective and sustainable [105]. Additionally, the immobilised enzymes can be easily recovered and reused, reducing the overall demand for enzymes. As consumer demand for high-quality protein supplements continues to grow, WPI production using immobilised enzymes offers a promising solution to meet these needs efficiently and sustainably [59].

3.3. Bioactive Peptide Synthesis

Proteases have been used in dairy processing to hydrolyse milk proteins into smaller peptides and free amino acids [81] to improve the digestibility of proteins and absorption in humans [82] and to introduce the desired sensory characteristics [83]. For example, in cheese making, proteases were used to break down casein into peptides, which enhances the flavour and texture of the cheese during the production process [83]. Apart from improved sensory characteristics, several studies have shown that milk-derived peptides can possess a broad range of functional properties, such as antioxidant and antimicrobial properties. For instance, bioactive peptides with antioxidant and antimicrobial properties have been produced from goat milk via hydrolysis using proteases derived from *Aspergillus oryzae* and *Aspergillus flavipes* [106].

Peptide synthesis plays a significant role in dairy processing, particularly in the production of dairy products with improved functional and nutritional properties. Bioactive peptides are important in the process of cheese making as they selectively impede the activity of proteolytic enzymes found in lactic acid bacteria. As a result, these peptides have a direct impact on the quality of the cheese produced [107]. Immobilising enzymes on a support material enhances their stability and reusability, allowing for efficient and controlled enzymatic reactions. In dairy processing, immobilised enzymes are utilised to selectively cleave or modify milk proteins, leading to the generation of bioactive peptides with desirable functionalities, such as antioxidant and antimicrobial activity. This enzymatic approach offers advantages in terms of specificity, efficiency, and the ability to generate peptides with tailored properties, contributing to the development of innovative dairy products with enhanced health benefits and sensory attributes [108].

3.4. Hydrolysed Infant Formula

Hydrolysed milk is commonly used for children genetically predisposed to a cow's milk allergy [109]. A recent study found that partially hydrolysed milk can reduce the skin barrier function in infants and reduce the risk of atopic dermatitis [110]. Lactose intolerance has been widely reported to be common among infants and children [111]. While this condition may be transient in some infants, it could be severe in others [112]. So, the intake of lactose-free milk has been suggested as a possible solution as it does not adversely affect normal growth in term infants [113,114]. Immobilised enzymes can be used to produce some of these specialty milk-based foods [82].

Lactase can catalyse the hydrolysis of lactose and is used to produce lactose-reduced and lactose-free dairy foods. For instance, lactase immobilised on chitosan was used to produce low-dose-lactose milk [115]. The immobilised enzyme was used for ten successive hydrolysis cycles, making the process highly economical with higher performance than using the free enzyme [115]. Several other studies have shown that immobilised enzymes can be used to hydrolyse lactose to produce a broad range of lactose-free dairy-based foods for infants [116,117].

A recent study used immobilised lipase to hydrolyse lipids in infant formula to understand if this can improve intestinal structures in preterm infants [118]. Another study found an enhanced absorption of long-chain PUFAs following consumption of functional milk formula partially hydrolysed with immobilised lipase [119]. The authors found that this process helped improve lipid absorption, tissue accretion, and the intestinal structure. These findings indicate that food-grade enzymes immobilised on different support materials can be used to produce hydrolysed infant formula.

3.5. Butter, Cheese, and Related Milk Fat Products

Milk fat can be used to make a broad range of products including butter, cheese, milk powder, and cream. It has been reported that by 2025, about 35 million tonnes of milk fat will be produced globally and enzymes will play a major role in this production [120].

Cheese making involves the coagulation of milk via proteolysis or acidification reactions. The enzymatic coagulation process involves rennet that is often isolated and purified from calf stomachs [121]. However, plant and microbial (fungi) sources have also been reported and have gained wide acceptance because of the growing demand for cheese and the reduction in animal rennet production [122].

Rennet obtained from *Rhizomucor miehei* was immobilised on alginate-chitosan nanoparticles for cheese making [122]. The immobilised rennet can be recovered after use, unlike the free form that remains in the curd, which may accelerate the degradation of caseins, thereby influencing the flavour and texture of the finished cheese [122]. An immobilised chymosin, a major protease in rennet, was used for the production of rennet-free Gouda cheese [123]. A recent study also found that chymosin immobilised on eggshell membranes was more effective than the free enzyme in cheese making [124].

4. Immobilised Enzymes for Lipid Processing and Modification

Lipids are a diverse group of biomolecules that include fats, oils, waxes, and other related compounds. One of the key enzymes involved in lipid processing is lipase. Lipases are a class of enzymes that catalyse the hydrolysis, esterification, and interesterification of lipids [8,19,125,126]. Immobilised lipases play crucial roles in lipid modification for a broad range of applications in the food industry (Table 3).

Table 3. Immobilised lipases for lipid hydrolysis, esterification, and interesterification.

	Enzyme	Support Material	Substrate	Reference
Hydrolysis	*Candida antarctica* lipase	Octadecyl, epoxy, and amino resins	Anchovy oil	[41]
	Mucor racemosus lipase	Alginate-gelatin	Olive oil	[127]
	Escherichia coli lipolytic enzymes	Calcium alginate beads	Avocado oil	[128]
	Rhizomucor miehei	Silica nanoparticles	Fish oil	[30]
	Rhizomucor miehei lipase	Magnetic nanoparticles	Babassu oil	[129]
	Candida antarctica lipase	Magnetic carrier	Rice bran oil	[130]
	Thermomyces lanuginosus lipase	Magnetic nanoparticles	Soybean oil	[131]
Esterification	*Thermomyces lanuginosus* lipase	Silica	Soybean oil	[132]
	Candida rugosa lipase	Amino resin	Fish oil	[133]
	Rhizomucor miehei lipase	Duolite A568	Coconut oil	[134]
	Candida antarctica lipase	Acrylic resin	Argan oil	[135]
	Rhizomucor miehei lipase	Magnetic nanoparticles	Coffee grounds oil	[136]
	Candida antarctica lipase	Acrylic resin	Olive oil	[19]
	Thermomyces lanuginose lipase	Succinilated celluloses	Sunflower oil	[137]
	Candida antarctica lipase	Acrylic resin	Palm oil	[138]
	Thermomyces lanuginose lipase	Biogenic nanoparticles	Pomace olive oil	[139]
Interesterification	*Bacillus licheniformis* lipase	Magnetic mesoporous biochar	Perilla seed oil	[140]
	Thermomyces lanuginose lipase	Silica gel	Soybean oil	[141]
	Candida antarctica lipase	Acrylic resin	Sunflower oil	[142]
	Thermomyces lanuginosus lipase	Silica	Kernel oil	[143]

4.1. Production of Omega-3 Fatty Acid Concentrates for Dairy Product Fortification

Hydrolysis is the primary function of lipases and involves the breakdown of triacylglycerols into free fatty acids. Immobilised lipases have been widely reported for the hydrolysis of oil to produce concentrates of specific fatty acids. Lipases have positional or fatty acid specificities, which allow them to hydrolyse some fatty acids selectively from the glycerol backbone of oils [41,144]. For instance, in fish and thraustochytrid oils, studies have shown that DHA is predominantly located at the central position (*sn*-2) on the glycerol backbone [145,146]. So, to produce concentrates of DHA from these oils, positional selective lipases that discriminate against fatty acids in the *sn*-2 position would be ideal.

Immobilised lipases with positional selectivity have been used to produce concentrates of omega-3 EPA and DHA via hydrolysis. Recently, lipase from *Alcaligenes* species immobilised on Accurel MP1000 was used to concentrate 70% EPA from microalga oil. Highly pure forms of EPA like this are being used for cheese and yoghurt fortification [33,39] and in nutrition drinks [147]. Also, *Candida antarctica* lipase A (CAL-A) immobilised on octadecyl-activated resin was used for concentrating 89% DHA by hydrolysing saturated fatty acids from thraustochytrid oil. The high level of saturated fatty acids in this oil makes it an excellent oil for DHA concentration [41]. This shows that lipases can be used to produce high-quality DHA concentrate for food fortification, such as infant formula and other dairy-based products such as butter, cheese, and ice cream [24,26–28,148].

4.2. Production of Flavour Esters, Structured Lipids, and Human Milk Fat Substitutes

Lipases also possess esterification activities, enabling them to catalyse the synthesis of esters from fatty acids. Lipases are commonly used to produce flavour esters and a broad range of food additives [32]. In a recent study, the rancid and tart off-flavour of goat milk was removed via the esterification of goat milk fat using immobilised *Thermomyces lanuginosus* lipase. Esters with sweet, floral, and fruity flavours were produced and the unpleasant odour in the original goat milk was decreased by 75% [149].

Interesterification is a process in which fatty acids are rearranged within triglyceride molecules, resulting in the modification of the physical and chemical properties of lipids [150]. Lipases are widely used for interesterification reactions in the production of structured lipids, which are tailored fats with specific fatty acid compositions [138,151]. These modified lipids offer improved nutritional profiles, enhanced functional properties, and better stability compared to natural fats and oils [150]. Interesterification is commonly applied in the food industry to produce margarine, shortenings, and other specialty fats [152].

Human breast milk fat consists of saturated, monounsaturated, and long-chain polyunsaturated fatty acids that are beneficial for the growth and development of infants. The distribution of these fatty acids on the glycerol backbone of human breast milk fat determines their bioavailability when consumed by infants [153,154]. Several studies have shown that enzymes such as lipases can be used to produce human milk fat (HMF) substitutes (HMFSs) [155–157]. HMFSs are synthesised to mimic HMF which has palmitic and oleic acids with a unique positional distribution. HMFs are generally triacylglycerols (TAGs) that contain mainly palmitic acid esterified at the *sn*-2 position on the glycerol backbone, while oleic acids and other unsaturated fats are located at *sn*-1 and *sn*-3 positions [158]. Since breast milk fat is rich in long-chain PUFAs such as EPA, DHA, and ARA, which are precursors of prostaglandins and eicosanoids that play regulatory roles [159], lipases can be used to produce HMFSs that are rich in these fatty acids because of their positional selectivity [41,138].

Immobilised lipases play a crucial role in the esterification (conjugation) of fatty acids with other non-lipid compounds. The conjugation of oils, lipids, or fatty acids with other compounds such as polyphenols has been reported, resulting in the formation of phenolipids [5–8]. These phenolipids possess unique properties that combine the beneficial attributes of both polyphenols and lipids [160]. They can exhibit an enhanced antioxidant activity, improved solubility in lipophilic environments, and increased bioavailability. Phenolipids have found applications in various industries, including pharmaceuticals, cosmetics, and food [160,161].

5. Immobilised Enzymes for Bioconversion of Dairy and Lipid Waste

5.1. Enzymes for Dairy Waste Processing

The utilisation of immobilised enzymes for the treatment of dairy waste has gained considerable attention in recent years due to their potential to enhance waste management practices and contribute to sustainability in the dairy industry. They can also be used under harsher reaction conditions, such as non-neutral pH and higher temperatures, which are typical of dairy processing waste streams [162,163]. The immobilisation material or matrix for enzymes used in dairy waste processing must be robust, inert, and stable (Table 4). This can reduce the cost and increase the sustainability of dairy waste processing [164]. Because enzymes are more specific, efficient, and environmentally friendly than traditional chemical processes, they can selectively break down complex organic molecules in dairy waste into simpler compounds that can be further processed into valuable products [165].

Lactose hydrolysis, achieved through immobilised lactase enzymes, stands out as a prominent application. By converting lactose into its constituent sugars, immobilised lactase facilitates the production of lactose-free dairy products and reduces the environmental impact of dairy waste streams [166]. Furthermore, immobilised enzymes like lipases hold promise in modifying the lipid profile of dairy waste, enabling the development of novel value-added products or facilitating lipid breakdown for more efficient waste treatment [167]. The interaction between immobilised enzymes and dairy waste treatment not only addresses economic and environmental concerns but also fosters improvement in dairy by-product utilisation.

The efficacy of immobilised enzymes in dairy waste treatment is underscored by the diverse array of solid supports that can be employed to immobilise these biocatalysts. Calcium alginate, chitosan beads, and agarose gels are examples of widely used matrices that offer favourable environments for enzyme activity while allowing efficient separation and recovery [168]. These immobilisation matrices facilitate targeted enzymatic hydrolysis processes, such as lactose breakdown and lipid modification, which are crucial for value creation and waste reduction [169].

The use of immobilised enzymes in dairy waste processing has been widely reported. For instance, dairy by-products are used as the main ingredient for the production of kashk, an Iranian dairy product produced enzymatically [170]. A highly potent protease isolated from *Bacillus cereus* was immobilised on magnetic nanoparticles and used to treat wastewater from yoghurt production with a protein recovery yield of 44% [171]. Another study proposed a β-galactosidase-enzyme-mediated process for valorisation of lactose-rich whey [172]. The authors described how β-galactosidase immobilised on polystyrene nanofibers was used to convert lactose into galactooligosaccharides (GOSs) in batch and repeated batch systems [172].

Therefore, the use of immobilised enzymes in dairy waste treatment aligns with the broader movement toward sustainable and circular practices in the food industry.

Table 4. Enzyme immobilisation support for dairy waste processing.

Type of Support	Enzyme	Application in Dairy Waste Processing	Reference
Calcium alginate	Lactase	Hydrolysis of lactose in dairy waste for lactose-intolerant products.	[173]
Agarose gel	Peptidase	Production of bioactive peptides from dairy by-products.	[170]
Modified glass beads	β-galasctosidase	Extraction of bioactive compounds from dairy waste for use as functional food ingredients.	[174]
Zeolite nanoparticles	Laccase	Removal of harmful pollutants from dairy waste to protect the environment.	[175]
Magnetic nanoparticles	Protease	Removal of pathogens from dairy waste streams.	[176]
Methacrylatic monoliths	β-galactosidase	Prebiotic production from a waste product of the dairy industry.	[177]
Magnetic nanoparticles	Lipase	Applied to produce biodiesel from dairy waste.	[167]
Magnetite nanoparticles	protease	Hydrolysis of protein from dairy wastewater to produce whey protein isolate.	[171]

By converting waste materials into useful products or reducing the environmental burden of waste disposal, immobilised enzyme technologies contribute to the optimisation of resource utilisation and promote a greener dairy sector. However, challenges and considerations persist in the application of immobilised enzymes for dairy waste treatment. The choice of immobilisation method, enzyme selection, and support material must be carefully tailored to the specific characteristics of the waste stream and the desired treatment outcomes. Factors such as enzyme loading, operational stability, and economic feasibility require a thorough assessment to ensure practical implementation [178].

5.2. Enzymes for Lipid Waste Processing

The application of immobilised enzymes in treating lipid-rich waste presents a promising avenue for enhancing the management of industrial wastes and by-products. Immobilised enzymes offer distinct advantages over free enzymes, making them well suited to address the challenges associated with lipid-rich waste streams. Lipase, a widely studied enzyme, stands out as a key biocatalyst in this context (Table 5). Immobilised lipases can hydrolyse the triglycerides present in the lipid-rich waste, breaking them down into valuable fatty acids and glycerol [179]. This process not only facilitates the conversion of lipid waste into potentially valuable materials for various applications but also reduces the environmental burden of untreated lipid wastes [180].

By converting lipid waste into valuable products, such as fatty acids that can be used as biofuels or biochemicals, immobilised enzyme approaches contribute to the optimisation of resource utilisation and the reduction in environmental impacts. Nevertheless, challenges remain in the widespread adoption of immobilised enzyme-based strategies for lipid-rich waste treatment [181]. Ongoing research focuses on refining immobilisation techniques, exploring alternative enzyme sources, and assessing the environmental implications of immobilised-enzyme-based processes [182]. As industries seek to address the environmental and economic concerns associated with lipid waste, the integration of immobilised enzymes holds potential for advancing sustainable practices and resource-efficient waste management [183].

Several studies have shown that lipases can be used for lipid waste processing. Immobilised lipases have been used for the remediation of hazardous pollutants from lipid-rich waste via hydrolysis of fats and oils [176,182]. A new bioremediation method was developed for the removal of oils in wastewater using *Acinetobacter haemolyticus* lipase. The lipase, which was immobilised on an eggshell membrane, was highly effective in hydrolysing olive, canola, cottonseed, palm, and peanut oils from wastewater while retaining its activity after 19 reuses [184]. *Aspergillus niger* lipase immobilised on the microporous acrylic resin was used for the deacidification of high-acid soy sauce residue oil [185]. This enzyme exhibited a high deacidification activity, thermostability, and reusability. It was also able to convert the free fatty acid in soy source residue oil to diacylglycerol-enriched oil that could be used for other food applications [185].

Lipid waste from different food processes has been widely used to produce biodiesel. Biodiesel is usually prepared using edible vegetable oils via transesterification, which can be catalysed by lipases [186]. Interestingly, old or used cooking oil can be a good source of oil for biodiesel production, and several studies have shown that spent frying or cooking oils could be used to make biodiesel [187–191].

Therefore, lipases are versatile in the bioprocessing of lipid waste for value addition, which is consistent with a circular economy, in which food processing wastes are kept in use while minimising negative environmental impacts and footprints.

Table 5. Enzyme immobilisation support for lipid waste processing.

Type of Support	Enzyme	Application in Lipid-Rich Waste Processing	Reference
Eggshell membrane	Lipase	Hydrolysis of lipids in waste with a high oleic acid composition	[184]
Chitosan beads	Lipase	Modification of lipid profiles in waste for value-added products	[192]
Silica cellulose	Lipase	Remediation of hazardous pollutants from lipid-rich waste	[176]
Calcium alginate beads	Lipase	Breakdown of fats and oils in lipid waste	[180]
Magnetic nanoparticles	Lipase	Conversion of lipid waste into biodiesel or bioproducts	[193]
Microporous acrylic resins	Lipase	Biotransformation of lipid waste components into valuable compounds	[185]
Kieselguhr (diatomaceous earth)	Lipase	Production of lipolytic enzymes for lipid waste treatment	[194]
Rice straw	Lipase	Breakdown of waste oil	[195]

6. Conclusions and Future Research Direction

Enzyme use in dairy and lipid processing is becoming increasingly attractive due to numerous advantages. Enzymes offer greater advantages over the use of chemicals because reactions involving enzymes can be performed under milder conditions, such as low temperatures and near-neutral pH, and products are considered natural for regulatory purposes. Also, because the reaction conditions are mild, the product quality is enhanced and fewer side products are formed, making this an environmentally friendly technique. In addition, since enzymes are highly specific in their actions, they can be used to produce a broad range of dairy and lipid-based products with a high purity and functionality. In this review, the numerous applications of immobilised enzymes for lipid and dairy processing have been discussed. The selection of an appropriate immobilisation method and support material was also addressed, as this is crucial to maintaining the enzyme's stability and activity. Despite the numerous advantages of enzyme immobilisation, there is still a low uptake of this technology in several dairy and lipid processing applications.

In any event, enzyme immobilisation remains a promising technology with significant potential for industrial applications, as it offers a range of potential benefits in various applications in lipid and dairy processing. The prospects of immobilised enzymes for industrial-scale applications hold significant promise as they can be applied to a broad range of catalytic applications in dairy and lipid processing through an enhanced operational efficiency, environmental sustainability, and economic viability. To utilise them at an industrial scale, several challenges must be addressed. Further research is needed to optimise immobilisation techniques, ensuring that enzyme activity, stability, and selectivity are not compromised during the immobilisation process. Also, more food-grade immobilisation supports must be investigated to avoid contamination when used for dairy and lipid processing. Further studies should focus on the development of scalable and cost-effective immobilisation methods for viable large-scale industrial use.

Funding: This research was funded by a PhD scholarship received from the University of Ha'il, Kingdom of Saudi Arabia.

Data Availability Statement: No new data were created or analyzed in this study. Data sharing is not applicable to this article.

Acknowledgments: Fawzih Alzahrani gratefully acknowledges the support from the University of Ha'il, Kingdom of Saudi Arabia, and the University of Newcastle, Australia.

Conflicts of Interest: The authors declare no conflicts of interest.

Abbreviations

PUFA	Polyunsaturated fatty acid
EPA	Eicosapentaenoic acid
WPI	Whey protein isolate
WPC	Whey protein concentrate
DHA	Docosahexaenoic acid
ALA	Alpha-linolenic acid
LA	Linoleic acid
OA	Oleic acid
FOS	Fructooligosaccharide
CLEA	Crosslinked enzyme aggregate
CLEC	Crosslinked enzyme crystal
CAL-A	*Candida antarctica* lipase A
HMF	Human milk fat
TAGs	Triacylglycerols
ARA	Arachidonic acid

References

1. Akanbi, T.O.; Ji, D.; Agyei, D. Revisiting the scope and applications of food enzymes from extremophiles. *J. Food Biochem.* **2020**, *44*, e13475. [CrossRef]
2. Chen, N.; Chang, B.; Shi, N.; Yan, W.; Lu, F.; Liu, F. Cross-linked enzyme aggregates immobilization: Preparation, characterization, and applications. *Crit. Rev. Biotechnol.* **2023**, *43*, 369–383. [CrossRef]
3. Zhang, R.; Yan, X.; Fan, K. Nanozymes inspired by natural enzymes. *Acc. Mater. Res.* **2021**, *2*, 534–547. [CrossRef]
4. Singh, R.; Singh, A.; Sachan, S. Enzymes used in the food industry: Friends or foes? In *Enzymes in Food Biotechnology*; Elsevier: Amsterdam, The Netherlands, 2019; pp. 827–843.
5. Silva, J.M.F.; Dos Santos, K.P.; Dos Santos, E.S.; Rios, N.S.; Gonçalves, L.R.B. Immobilization of *Thermomyces lanuginosus* lipase on a new hydrophobic support (Streamline phenyl™): Strategies to improve stability and reusability. *Enzym. Microb. Technol.* **2023**, *163*, 110166. [CrossRef]
6. Abellanas-Perez, P.; Carballares, D.; Fernandez-Lafuente, R.; Rocha-Martin, J. Glutaraldehyde modification of lipases immobilized on octyl agarose beads: Roles of the support enzyme loading and chemical amination of the enzyme on the final enzyme features. *Int. J. Biol. Macromol.* **2023**, *248*, 125853. [CrossRef]
7. Basso, A.; Serban, S. Industrial applications of immobilized enzymes—A review. *Mol. Catal.* **2019**, *479*, 110607. [CrossRef]
8. Akanbi, T.O.; Marshall, S.N.; Barrow, C.J. Polydatin-fatty acid conjugates are effective antioxidants for stabilizing omega 3-containing bulk fish oil and fish oil emulsions. *Food Chem.* **2019**, *301*, 125297. [CrossRef]
9. Eberhardt, A.; López, E.C.; Marino, F.; Mammarella, E.J.; Manzo, R.M.; Sihufe, G.A. Whey protein hydrolysis with microbial proteases: Determination of kinetic parameters and bioactive properties for different reaction conditions. *Int. J. Dairy Technol.* **2021**, *74*, 489–504. [CrossRef]
10. Amigo-Benavent, M.; FitzGerald, R.J. Impact of thermal inactivation conditions on the residual proteolytic activity and the viscosity properties of whey protein concentrate enzymatic hydrolysates. *Food Hydrocoll.* **2022**, *124*, 107333. [CrossRef]
11. Mazorra-Manzano, M.A.; Mora-Cortes, W.G.; Leandro-Roldan, M.M.; González-Velázquez, D.A.; Torres-Llanez, M.J.; Ramírez-Suarez, J.C.; González-Córdova, A.F.; Vallejo-Córdoba, B. Production of whey protein hydrolysates with angiotensin-converting enzyme-inhibitory activity using three new sources of plant proteases. *Biocatal. Agric. Biotechnol.* **2020**, *28*, 101724. [CrossRef]
12. Lermen, A.M.; Clerici, N.J.; Borchartt Maciel, D.; Daroit, D.J. Characterization and application of a crude bacterial protease to produce antioxidant hydrolysates from whey protein. *Prep. Biochem. Biotechnol.* **2023**, *53*, 12–21. [CrossRef]
13. León-López, A.; Pérez-Marroquín, X.A.; Estrada-Fernández, A.G.; Campos-Lozada, G.; Morales-Peñaloza, A.; Campos-Montiel, R.G.; Aguirre-Álvarez, G. Milk whey hydrolysates as high value-added natural polymers: Functional properties and applications. *Polymers* **2022**, *14*, 1258. [CrossRef]
14. Du, X.; Jing, H.; Wang, L.; Huang, X.; Wang, X.; Wang, H. Characterization of structure, physicochemical properties, and hypoglycemic activity of goat milk whey protein hydrolysate processed with different proteases. *LWT* **2022**, *159*, 113257. [CrossRef]
15. Wróblewska, B.; Karamać, M.; Amarowicz, R.; Szymkiewicz, A.; Troszyńska, A.; Kubicka, E. Immunoreactive properties of peptide fractions of cow whey milk proteins after enzymatic hydrolysis. *Int. J. Food Sci. Technol.* **2004**, *39*, 839–850. [CrossRef]
16. Oliveira, J.P.; Gonçalves, L.R.; Amorim, K.P.; Pinheiro, B.B.; Ramos, M.V.; Souza, P.F.; Oliveira, J.S.; Freitas, D.C.; Freitas, C.D. Immobilization and characterization of latex cysteine peptidases on different supports and application for cow's milk protein hydrolysis. *Process Biochem.* **2022**, *117*, 180–190. [CrossRef]
17. Monari, S.; Ferri, M.; Russo, C.; Prandi, B.; Tedeschi, T.; Bellucci, P.; Zambrini, A.V.; Donati, E.; Tassoni, A. Enzymatic production of bioactive peptides from scotta, an exhausted by-product of ricotta cheese processing. *PLoS ONE* **2019**, *14*, e0226834. [CrossRef]

18. Kleekayai, T.; Cermeño, M.; FitzGerald, R.J. The production of bioactive peptides from milk proteins. In *Agents of Change: Enzymes in Milk and Dairy Products*; Springer: Berlin/Heidelberg, Germany, 2021; pp. 447–497.
19. Gunathilake, T.; Akanbi, T.O.; Barrow, C.J. Lipase-produced omega-3 acylglycerols for the fortification and stabilization of extra virgin olive oil using hydroxytyrosyl palmitate. *Future Foods* **2021**, *4*, 100045. [CrossRef]
20. Aryee, A.N.; Akanbi, T.O.; Nwachukwu, I.D.; Gunathilake, T. Perspectives on preserving lipid quality and strategies for value enhancement. *Curr. Opin. Food Sci.* **2022**, *44*, 100802. [CrossRef]
21. Xia, Q.; Akanbi, T.O.; Wang, B.; Li, R.; Liu, S.; Barrow, C.J. Investigation of enhanced oxidation stability of microencapsulated enzymatically produced tuna oil concentrates using complex coacervation. *Food Funct.* **2020**, *11*, 10748–10757. [CrossRef]
22. Gan, J.; Bagheri, A.R.; Aramesh, N.; Gul, I.; Franco, M.; Almulaiky, Y.Q.; Bilal, M. Covalent organic frameworks as emerging host platforms for enzyme immobilization and robust biocatalysis–A review. *Int. J. Biol. Macromol.* **2021**, *167*, 502–515. [CrossRef]
23. Taheri-Kafrani, A.; Kharazmi, S.; Nasrollahzadeh, M.; Soozanipour, A.; Ejeian, F.; Etedali, P.; Mansouri-Tehrani, H.-A.; Razmjou, A.; Yek, S.M.-G.; Varma, R.S. Recent developments in enzyme immobilization technology for high-throughput processing in food industries. *Crit. Rev. Food Sci. Nutr.* **2021**, *61*, 3160–3196. [CrossRef]
24. Ramadan, M.M.; El-Said, M.M.; El-Messery, T.M.; Mohamed, R.S. Development of flavored yoghurt fortified with microcapsules of triple omega 3-6-9 for preventing neurotoxicity induced by aluminum chloride in rats. *J. Food Process. Preserv.* **2021**, *45*, e15759. [CrossRef]
25. Guadarrama-Flores, B.; Matencio, A.; Navarro-Orcajada, S.; Martínez-Lede, I.; Conesa, I.; Vidal-Sánchez, F.J.; García-Carmona, F.; López-Nicolás, J.M. Development if healthy milk and yogurt products for reducing metabolic diseases using cyclodextrin and omega-3 fatty acids from fish oil. *Food Funct.* **2022**, *13*, 5528–5535. [CrossRef] [PubMed]
26. Pandule, V.S.; Sharma, M.; HC, D. Omega-3 fatty acid-fortified butter: Preparation and characterisation of textural, sensory, thermal and physico-chemical properties. *Int. J. Dairy Technol.* **2021**, *74*, 181–191. [CrossRef]
27. Muñoz-Tébar, N.; De la Vara, J.; de Elguea-Culebras, G.O.; Cano, E.; Molina, A.; Carmona, M.; Berruga, M.I. Enrichment of sheep cheese with chia (*Salvia hispanica* L.) oil as a source of omega-3. *LWT* **2019**, *108*, 407–415. [CrossRef]
28. Gowda, A.; Sharma, V.; Goyal, A.; Singh, A.; Arora, S. Process optimization and oxidative stability of omega-3 ice cream fortified with flaxseed oil microcapsules. *J. Food Sci. Technol.* **2018**, *55*, 1705–1715. [CrossRef] [PubMed]
29. Akanbi, T.O.; Adcock, J.L.; Barrow, C.J. Selective concentration of EPA and DHA using *Thermomyces lanuginosus* lipase is due to fatty acid selectivity and not regioselectivity. *Food Chem.* **2013**, *138*, 615–620. [CrossRef] [PubMed]
30. Zhang, H.; Secundo, F.; Sun, J.; Mao, X. Advances in enzyme biocatalysis for the preparation of functional lipids. *Biotechnol. Adv.* **2022**, *61*, 108036. [CrossRef] [PubMed]
31. Gunathilake, T.; Akanbi, T.O.; Bucher, T.; Barrow, C.J. Enzymes in nutrition, baby foods, and food safety. In *Value-Addition in Food Products and Processing through Enzyme Technology*; Elsevier: Amsterdam, The Netherlands, 2022; pp. 153–161.
32. Gunathilake, T.; Akanbi, T.O.; Van Vuong, Q.; Scarlett, C.J.; Barrow, C.J. Enzyme technology in the production of flavors and food additives. In *Value-Addition in Food Products and Processing through Enzyme Technology*; Elsevier: Amsterdam, The Netherlands, 2022; pp. 45–55.
33. Gumus, C.E.; Gharibzahedi, S.M.T. Yogurts supplemented with lipid emulsions rich in omega-3 fatty acids: New insights into the fortification, microencapsulation, quality properties, and health-promoting effects. *Trends Food Sci. Technol.* **2021**, *110*, 267–279. [CrossRef]
34. Jia, H.-X.; Qu, Y.-Z.; Chen, W.-L.; Su, M.-Y. Effect of high DHA and ARA fortification on lipid oxidation of infant formula powder. *CyTA-J. Food* **2024**, *22*, 2300812. [CrossRef]
35. Madrigal, C.; Soto-Méndez, M.J.; Leis, R.; Hernández-Ruiz, Á.; Valero, T.; Lara Villoslada, F.; Martínez de Victoria, E.; Moreno, J.M.; Ortega, R.M.; Ruiz-López, M.D. Dietary intake, nutritional adequacy and food sources of total fat and fatty acids, and relationships with personal and family factors in Spanish children aged one to <10 Years: Results of the EsNuPI study. *Nutrients* **2020**, *12*, 2467.
36. Moghadam, F.V.; Pourahmad, R.; Mortazavi, A.; Davoodi, D.; Azizinezhad, R. Use of fish oil nanoencapsulated with gum arabic carrier in low fat probiotic fermented milk. *Food Sci. Anim. Resour.* **2019**, *39*, 309. [CrossRef]
37. Zahran, H.; Mabrouk, A.M.; Salama, H.H. Evaluation of yoghurt fortified with encapsulated echium oil rich in stearidonic acid as a low-fat dairy food. *Egypt. J. Chem.* **2022**, *65*, 29–41. [CrossRef]
38. Zahed, O.; Khosravi-Darani, K.; Mortazavian, A.M.; Mohammadi, A. Bacterial conjugated linoleic acid bio-fortification of synbiotic yogurts using *Propionibacterium freudenreichii* as adjunct culture. *Ital. J. Food Sci.* **2021**, *33*, 1–11. [CrossRef]
39. Villamil, R.-A.; Guzmán, M.-P.; Ojeda-Arredondo, M.; Cortés, L.Y.; Archila, E.G.; Giraldo, A.; Mondragón, A.-I. Cheese fortification through the incorporation of UFA-rich sources: A review of recent (2010–2020) evidence. *Heliyon* **2021**, *7*, e05785. [CrossRef] [PubMed]
40. Lehaçani, S.M.; Al-Abdullah, B. Characterisation of soft white cheese fortified with flaxseed oil to enhance its quality, lipid profile and health benefits. *Riv. Ital. Delle Sostanze Grasse* **2023**, *100*, 165–175.
41. Akanbi, T.O.; Barrow, C.J. *Candida antarctica* lipase A effectively concentrates DHA from fish and thraustochytrid oils. *Food Chem.* **2017**, *229*, 509–516. [CrossRef] [PubMed]
42. Maghraby, Y.R.; El-Shabasy, R.M.; Ibrahim, A.H.; Azzazy, H.M.E.-S. Enzyme immobilization technologies and industrial applications. *ACS Omega* **2023**, *8*, 5184–5196. [CrossRef]

43. Mohammadi, Z.B.; Zhang, F.; Kharazmi, M.S.; Jafari, S.M. Nano-biocatalysts for food applications; immobilized enzymes within different nanostructures. *Crit. Rev. Food Sci. Nutr.* **2022**, *63*, 11351–11369. [CrossRef] [PubMed]
44. Imam, H.T.; Marr, P.C.; Marr, A.C. Enzyme entrapment, biocatalyst immobilization without covalent attachment. *Green Chem.* **2021**, *23*, 4980–5005. [CrossRef]
45. Sicard, C. In Situ Enzyme Immobilization by Covalent Organic Frameworks. *Angew. Chem.* **2023**, *135*, e202213405. [CrossRef]
46. Guisan, J.M.; Fernandez-Lorente, G.; Rocha-Martin, J.; Moreno-Gamero, D. Enzyme immobilization strategies for the design of robust and efficient biocatalysts. *Curr. Opin. Green Sustain. Chem.* **2022**, *35*, 100593. [CrossRef]
47. Du, X.; Wang, L.; Li, Y.; Wu, J.; Chen, G.; Liang, H.; Gao, D. Cooperative adsorption-degradation of benzo[a]pyrene by extracellular enzymes of white rot fungi immobilized on macroporous adsorption resin. *Int. Biodeterior. Biodegrad.* **2023**, *178*, 105564. [CrossRef]
48. Binhayeeding, N.; Yunu, T.; Pichid, N.; Klomklao, S.; Sangkharak, K. Immobilisation of *Candida rugosa* lipase on polyhydroxybutyrate via a combination of adsorption and cross-linking agents to enhance acylglycerol production. *Process Biochem.* **2020**, *95*, 174–185. [CrossRef]
49. Cao, Y.-P.; Zhi, G.-Y.; Han, L.; Chen, Q.; Zhang, D.-H. Biosynthesis of benzyl cinnamate using an efficient immobilized lipase entrapped in nano-molecular cages. *Food Chem.* **2021**, *364*, 130428. [CrossRef]
50. Vetrano, A.; Gabriele, F.; Germani, R.; Spreti, N. Characterization of lipase from *Candida rugosa* entrapped in alginate beads to enhance its thermal stability and recyclability. *New J. Chem.* **2022**, *46*, 10037–10047. [CrossRef]
51. Bedzo, O.K.; Trollope, K.; Gottumukkala, L.D.; Coetzee, G.; Görgens, J.F. Amberlite IRA 900 versus calcium alginate in immobilization of a novel, engineered β-fructofuranosidase for short-chain fructooligosaccharide synthesis from sucrose. *Biotechnol. Prog.* **2019**, *35*, e2797. [CrossRef]
52. Fopase, R.; Paramasivam, S.; Kale, P.; Paramasivan, B. Strategies, challenges and opportunities of enzyme immobilization on porous silicon for biosensing applications. *J. Environ. Chem. Eng.* **2020**, *8*, 104266. [CrossRef]
53. Nawawi, N.N.; Hashim, Z.; Rahman, R.A.; Murad, A.M.A.; Bakar, F.D.A.; Illias, R.M. Entrapment of porous cross-linked enzyme aggregates of maltogenic amylase from *Bacillus lehensis* G1 into calcium alginate for maltooligosaccharides synthesis. *Int. J. Biol. Macromol.* **2020**, *150*, 80–89. [CrossRef]
54. Thangaraj, B.; Solomon, P.R. Immobilization of lipases–A review. Part I: Enzyme immobilization. *Chem. Biochem. Eng. Rev.* **2019**, *6*, 157–166. [CrossRef]
55. Rodrigues, R.C.; Berenguer-Murcia, Á.; Carballares, D.; Morellon-Sterling, R.; Fernandez-Lafuente, R. Stabilization of enzymes via immobilization: Multipoint covalent attachment and other stabilization strategies. *Biotechnol. Adv.* **2021**, *52*, 107821. [CrossRef] [PubMed]
56. Wahab, R.A.; Elias, N.; Abdullah, F.; Ghoshal, S.K. On the taught new tricks of enzymes immobilization: An all-inclusive overview. *React. Funct. Polym.* **2020**, *152*, 104613. [CrossRef]
57. Hassan, M.E.; Yang, Q.; Xiao, Z.; Liu, L.; Wang, N.; Cui, X.; Yang, L. Impact of immobilization technology in industrial and pharmaceutical applications. *3 Biotech* **2019**, *9*, 440. [CrossRef] [PubMed]
58. Xing, C.; Mei, P.; Mu, Z.; Li, B.; Feng, X.; Zhang, Y.; Wang, B. Enhancing enzyme activity by the modulation of covalent interactions in the confined channels of covalent organic frameworks. *Angew. Chem.* **2022**, *134*, e202201378. [CrossRef]
59. Ramalho, E.X.; de Castro, R.J.S. Covalent bonding immobilization of a *Bacillus licheniformis* protease on chitosan and its application in protein hydrolysis. *Biocatal. Agric. Biotechnol.* **2023**, *50*, 102713. [CrossRef]
60. Lopes, L.A.; Novelli, P.K.; Fernandez-Lafuente, R.; Tardioli, P.W.; Giordano, R.L.C. Glyoxyl-activated agarose as support for covalently link Novo-Pro D: Biocatalysts performance in the hydrolysis of casein. *Catalysts* **2020**, *10*, 466. [CrossRef]
61. Kızıldağ, S.; Işık, C.; Teke, M. Milk lactose removal by β-galactosidase immobilized on eggshell membrane. *Eur. Food Res. Technol.* **2023**, *249*, 2125–2136. [CrossRef]
62. Rios, N.S.; Mendez-Sanchez, C.; Arana-Peña, S.; Rueda, N.; Ortiz, C.; Gonçalves, L.R.; Fernandez-Lafuente, R. Immobilization of lipase from *Pseudomonas fluorescens* on glyoxyl-octyl-agarose beads: Improved stability and reusability. *Biochim. Biophys. Acta (BBA)-Proteins Proteom.* **2019**, *1867*, 741–747. [CrossRef]
63. Rojas, M.J.; Amaral-Fonseca, M.; Zanin, G.M.; Fernandez-Lafuente, R.; Giordano, R.d.L.C.; Tardioli, P.W. Preparation of crosslinked enzyme aggregates of a thermostable cyclodextrin glucosyltransferase from *Thermoanaerobacter* sp. critical effect of the crosslinking agent. *Catalysts* **2019**, *9*, 120. [CrossRef]
64. Reis, C.L.B.; Sousa, E.Y.A.d.; Serpa, J.d.F.; Oliveira, R.C.; Santos, J.C.S.d. Design of immobilized enzyme biocatalysts: Drawbacks and opportunities. *Química Nova* **2019**, *42*, 768–783. [CrossRef]
65. Guajardo, N.; Ahumada, K.; de María, P.D. Immobilization of pseudomonas stutzeri lipase through cross-linking aggregates (CLEA) for reactions in deep eutectic solvents. *J. Biotechnol.* **2021**, *337*, 18–23. [CrossRef]
66. Malhotra, I.; Basir, S.F. Application of invertase immobilized on chitosan using glutaraldehyde or tris (Hydroxymethyl) phosphine as cross-linking agent to produce bioethanol. *Appl. Biochem. Biotechnol.* **2020**, *191*, 838–851. [CrossRef]
67. Tacias-Pascacio, V.G.; Morellon-Sterling, R.; Siar, E.-H.; Tavano, O.; Berenguer-Murcia, A.; Fernandez-Lafuente, R. Use of Alcalase in the production of bioactive peptides: A review. *Int. J. Biol. Macromol.* **2020**, *165*, 2143–2196. [CrossRef]
68. Argenta, A.B.; Nogueira, A.; Scheer, A.d.P. Hydrolysis of whey lactose: Kluyveromyces lactis β-galactosidase immobilisation and integrated process hydrolysis-ultrafiltration. *Int. Dairy J.* **2021**, *117*, 105007. [CrossRef]

69. Hackenhaar, C.R.; Spolidoro, L.S.; Flores, E.E.E.; Klein, M.P.; Hertz, P.F. Batch synthesis of galactooligosaccharides from co-products of milk processing using immobilized β-galactosidase from *Bacillus circulans*. *Biocatal. Agric. Biotechnol.* **2021**, *36*, 102136. [CrossRef]
70. Guo, Y.; Jiang, X.; Xiong, B.; Zhang, T.; Zeng, X.; Wu, Z.; Sun, Y.; Pan, D. Production and transepithelial transportation of angiotensin-I-converting enzyme (ACE)-inhibitory peptides from whey protein hydrolyzed by immobilized *Lactobacillus helveticus* proteinase. *J. Dairy Sci.* **2019**, *102*, 961–975. [CrossRef]
71. Bustamante, S.Z.; Gonzalez, J.G.; Sforza, S.; Tedeschi, T. Bioactivity and peptide profile of whey protein hydrolysates obtained from Colombian double-cream cheese production and their products after gastrointestinal digestion. *LWT* **2021**, *145*, 111334. [CrossRef]
72. Rocha, G.F.; Cotabarren, J.; Obregón, W.D.; Fernández, G.; Rosso, A.M.; Parisi, M.G. Milk-clotting and hydrolytic activities of an aspartic protease from *Salpichroa origanifolia* fruits on individual caseins. *Int. J. Biol. Macromol.* **2021**, *192*, 931–938. [CrossRef]
73. Balan, A.; Murthy, V.V.; Kadeppagari, R.K. Immobilized enzymes for bioconversion of waste to wealth. In *Biotechnology for Zero Waste: Emerging Waste Management Techniques*; Wiley: Hoboken, NJ, USA, 2022; pp. 33–46.
74. Wafaa, N.; Elbarbary, H.; Ibrahim, E.; Mohamed, H.; Jenssen, H. Effect of enzyme type and hydrolysis time on antibacterial and antioxidant activity of whey protein hydrolysates. *Iraqi J. Agric. Sci.* **2022**, *53*, 1340–1357.
75. Zhu, H.; Zhang, Y.; Yang, T.; Zheng, D.; Liu, X.; Zhang, J.; Zheng, M. Preparation of immobilized Alcalase based on metal affinity for efficient production of bioactive peptides. *LWT* **2022**, *162*, 113505. [CrossRef]
76. da Cruz, C.Z.P.; de Mendonça, R.J.; Guimaraes, L.H.S.; dos Santos Ramos, M.A.; Garrido, S.S.; de Paula, A.V.; Monti, R.; Massolini, G. Assessment of the bioactive potential of cheese whey protein hydrolysates using immobilized alcalase. *Food Bioprocess Technol.* **2020**, *13*, 2120–2130. [CrossRef]
77. Mao, Y.; Krischke, M.; Hengst, C.; Kulozik, U. Influence of salts on hydrolysis of β-lactoglobulin by free and immobilised trypsin. *Int. Dairy J.* **2019**, *93*, 106–115. [CrossRef]
78. Xu, L.; Gong, Y.; Gern, J.E.; Lucey, J.A. Influence of whey protein hydrolysis in combination with dextran glycation on immunoglobulin E binding capacity with blood sera obtained from patients with a cow milk protein allergy. *J. Dairy Sci.* **2020**, *103*, 1141–1150. [CrossRef]
79. Salvado, A.F.; Leitão, J.H.; Fonseca, L.P. Enzymatic production of bioactive peptides from whey proteins: Their active role and potential health benefits. In *Enzymes for Solving Humankind's Problems: Natural and Artificial Systems in Health, Agriculture, Environment and Energy*; Springer: Berlin/Heidelberg, Germany, 2021; pp. 473–506.
80. Kazemi-Taskooh, Z.; Varidi, M. Food-based iron delivery systems: A review. *Trends Food Sci. Technol.* **2021**, *116*, 75–89. [CrossRef]
81. Siar, E.-H.; Morellon-Sterling, R.; Zidoune, M.N.; Fernandez-Lafuente, R. Use of glyoxyl-agarose immobilized ficin extract in milk coagulation: Unexpected importance of the ficin loading on the biocatalysts. *Int. J. Biol. Macromol.* **2020**, *144*, 419–426. [CrossRef] [PubMed]
82. Morellon-Sterling, R.; El-Siar, H.; Tavano, O.L.; Berenguer-Murcia, Á.; Fernández-Lafuente, R. Ficin: A protease extract with relevance in biotechnology and biocatalysis. *Int. J. Biol. Macromol.* **2020**, *162*, 394–404. [CrossRef] [PubMed]
83. Souza, T.S.P.d.; de Andrade, C.J.; Koblitz, M.G.B.; Fai, A.E.C. Microbial peptidase in food processing: Current state of the art and future trends. *Catal. Lett.* **2023**, *153*, 114–137. [CrossRef]
84. Zeng, J.; Zou, J.; Zhao, J.; Lin, K.; Zhang, L.; Yi, H.; Gong, P. Chymosin pretreatment accelerated papain catalysed hydrolysis for decreasing casein antigenicity by exposing the cleavage site at tyrosine residues. *Food Chem.* **2023**, *404*, 134777. [CrossRef]
85. Sun, S.; Gao, Y.; Chen, J.; Liu, R. Identification and release kinetics of peptides from tilapia skin collagen during alcalase hydrolysis. *Food Chem.* **2022**, *378*, 132089. [CrossRef] [PubMed]
86. Xie, D.; Chen, Y.; Yu, J.; Yang, Z.; Wang, X.; Wang, X. Progress in enrichment of n-3 polyunsaturated fatty acid: A review. *Crit. Rev. Food Sci. Nutr.* **2022**, *63*, 11310–11326. [CrossRef]
87. Ghide, M.K.; Li, K.; Wang, J.; Abdulmalek, S.A.; Yan, Y. Immobilization of Rhizomucor miehei lipase on magnetic multiwalled carbon nanotubes towards the synthesis of structured lipids rich in sn-2 palmitic acid and sn-1,3 oleic acid (OPO) for infant formula use. *Food Chem.* **2022**, *390*, 133171. [CrossRef]
88. Li, H.; Yang, J.; Qin, A.; Yang, F.; Liu, D.; Li, H.; Yu, J. Milk protein hydrolysates obtained with immobilized alcalase and neutrase on magnetite nanoparticles: Characterization and antigenicity study. *J. Food Sci.* **2022**, *87*, 3107–3116. [CrossRef]
89. Abd El-Salam, M.H.; El-Shibiny, S. Reduction of milk protein antigenicity by enzymatic hydrolysis and fermentation. A review. *Food Rev. Int.* **2021**, *37*, 276–295. [CrossRef]
90. Aggarwal, S.; Ikram, S. Zinc oxide nanoparticles-impregnated chitosan surfaces for covalent immobilization of trypsin: Stability & kinetic studies. *Int. J. Biol. Macromol.* **2022**, *207*, 205–221.
91. Rivera del Rio, A.; Boom, R.M.; Janssen, A.E. Effect of fractionation and processing conditions on the digestibility of plant proteins as food ingredients. *Foods* **2022**, *11*, 870. [CrossRef]
92. Katić, K.; Banjanac, K.; Simović, M.; Ćorović, M.; Milivojević, A.; Marinković, A.; Bezbradica, D. Development of protease nanobiocatalysts and their application in hydrolysis of sunflower meal protein isolate. *Int. J. Food Sci. Technol.* **2021**, *56*, 4287–4297. [CrossRef]
93. Li, J.; Hartinger, C.; Zhu, F. Physicochemical properties of infant formula model emulsions stabilised by different whey protein hydrolysates and characteristics of interfacial peptides. *Food Hydrocoll.* **2023**, *145*, 109035. [CrossRef]

94. Budžaki, S.; Velić, N.; Ostojčić, M.; Stjepanović, M.; Rajs, B.B.; Šereš, Z.; Maravić, N.; Stanojev, J.; Hessel, V.; Strelec, I. Waste management in the agri-food industry: The conversion of eggshells, spent coffee grounds, and brown onion skins into carriers for Lipase immobilization. *Foods* **2022**, *11*, 409. [CrossRef] [PubMed]
95. Jasińska, K.; Zieniuk, B.; Jankiewicz, U.; Fabiszewska, A. Bio-Based Materials versus Synthetic Polymers as a Support in Lipase Immobilization: Impact on Versatile Enzyme Activity. *Catalysts* **2023**, *13*, 395. [CrossRef]
96. Pellegrino, L.; Hogenboom, J.A.; Rosi, V.; Sindaco, M.; Gerna, S.; D'Incecco, P. Focus on the Protein Fraction of Sports Nutrition Supplements. *Molecules* **2022**, *27*, 3487. [CrossRef] [PubMed]
97. Pires, A.F.; Marnotes, N.G.; Rubio, O.D.; Garcia, A.C.; Pereira, C.D. Dairy by-products: A review on the valorization of whey and second cheese whey. *Foods* **2021**, *10*, 1067. [CrossRef]
98. Bella, K.; Pilli, S.; Venkateswara Rao, P.; Tyagi, R. Bio-conversion of whey lactose using enzymatic hydrolysis with β-galactosidase: An experimental and kinetic study. *Environ. Technol.* **2022**, *45*, 1234–1247. [CrossRef]
99. Shafi, A.; Ahmed, F.; Husain, Q. β-Galactosidase mediated synthesized nanosupport for the immobilization of same enzyme: Its stability and application in the hydrolysis of lactose. *Int. J. Biol. Macromol.* **2021**, *184*, 57–67. [CrossRef]
100. de Freitas, M.d.F.M.; Hortêncio, L.C.; de Albuquerque, T.L.; Rocha, M.V.P.; Gonçalves, L.R.B. Simultaneous hydrolysis of cheese whey and lactulose production catalyzed by β-galactosidase from Kluyveromyces lactis NRRL Y1564. *Bioprocess Biosyst. Eng.* **2020**, *43*, 711–722. [CrossRef] [PubMed]
101. Kelkar, N. Clinical applications of whey protein. In *Nutrition Science, Marketing Nutrition, Health Claims, and Public Policy*; Elsevier: Amsterdam, The Netherlands, 2023; pp. 13–22.
102. Srikanth, D.; Gopi, D.; Sunil, C.; Michael, K.; Rawson, A. Proteins as Fat Replacers in the Food Industry. In *Fat Mimetics Food Applications*; Wiley: Hoboken, NJ, USA, 2023; pp. 133–154.
103. Sharma, K.; Saini, R.; Patil, S. A Systematic Review on the Emerging Role of Protein Powder. *Int. J. Res. Eng. Sci.* **2022**, *10*, 88–90.
104. Mora, L.; Toldrá, F. Advanced enzymatic hydrolysis of food proteins for the production of bioactive peptides. *Curr. Opin. Food Sci.* **2022**, *49*, 100973. [CrossRef]
105. Yu, A.E. Continuous hydrolysis of milk proteins in membrane reactors of various configurations. *Foods Raw Mater.* **2021**, *9*, 271–281.
106. Zanutto-Elgui, M.R.; Vieira, J.C.S.; do Prado, D.Z.; Buzalaf, M.A.R.; de Magalhães Padilha, P.; de Oliveira, D.E.; Fleuri, L.F. Production of milk peptides with antimicrobial and antioxidant properties through fungal proteases. *Food Chem.* **2019**, *278*, 823–831. [CrossRef]
107. Smacchi, E.; Gobbetti, M. Bioactive peptides in dairy products: Synthesis and interaction with proteolytic enzymes. *Food Microbiol.* **2000**, *17*, 129–141. [CrossRef]
108. Punia, H.; Tokas, J.; Malik, A.; Sangwan, S.; Baloda, S.; Singh, N.; Singh, S.; Bhuker, A.; Singh, P.; Yashveer, S. Identification and detection of bioactive peptides in milk and dairy products: Remarks about agro-foods. *Molecules* **2020**, *25*, 3328. [CrossRef] [PubMed]
109. van Esch, B.C.; Knipping, K.; Jeurink, P.; van der Heide, S.; Dubois, A.E.; Willemsen, L.E.; Garssen, J.; Knippels, L.M. In vivo and in vitro evaluation of the residual allergenicity of partially hydrolysed infant formulas. *Toxicol. Lett.* **2011**, *201*, 264–269. [CrossRef]
110. Holvoet, S.; Nutten, S.; Dupuis, L.; Donnicola, D.; Bourdeau, T.; Hughes-Formella, B.; Simon, D.; Simon, H.-U.; Carvalho, R.S.; Spergel, J.M. Partially hydrolysed whey-based infant formula improves skin barrier function. *Nutrients* **2021**, *13*, 3113. [CrossRef]
111. Heine, R.G.; AlRefaee, F.; Bachina, P.; De Leon, J.C.; Geng, L.; Gong, S.; Madrazo, J.A.; Ngamphaiboon, J.; Ong, C.; Rogacion, J.M. Lactose intolerance and gastrointestinal cow's milk allergy in infants and children–common misconceptions revisited. *World Allergy Organ. J.* **2017**, *10*, 1–8. [CrossRef]
112. del Carmen Tocaa, M.; Fernándezb, A.; Orsic, M.; Tabaccod, O.; Vinderolae, G. Lactose intolerance: Myths and facts. An update. *Arch. Argent. Pediatr.* **2022**, *120*, 59–66.
113. Lasekan, J.B.; Jacobs, J.; Reisinger, K.S.; Montalto, M.B.; Frantz, M.P.; Blatter, M.M. Lactose-free milk protein-based infant formula: Impact on growth and gastrointestinal tolerance in infants. *Clin. Pediatr.* **2011**, *50*, 330–337. [CrossRef] [PubMed]
114. Dipasquale, V.; Serra, G.; Corsello, G.; Romano, C. Standard and specialized infant formulas in Europe: Making, marketing, and health outcomes. *Nutr. Clin. Pract.* **2020**, *35*, 273–281. [CrossRef] [PubMed]
115. Wolf, M.; Belfiore, L.A.; Tambourgi, E.B.; Paulino, A.T. Production of low-dosage lactose milk using lactase immobilised in hydrogel. *Int. Dairy J.* **2019**, *92*, 77–83. [CrossRef]
116. Schulz, P.; Rizvi, S.S. Hydrolysis of lactose in milk: Current status and future products. *Food Rev. Int.* **2021**, *39*, 2875–2894. [CrossRef]
117. Selvarajan, E.; Nivetha, A.; Subathra Devi, C.; Mohanasrinivasan, V. Nanoimmobilization of β-galactosidase for lactose-free product development. In *Nanoscience and Biotechnology for Environmental Applications*; Springer: Berlin/Heidelberg, Germany, 2019; pp. 199–223.
118. Zaworski, K.; Woliński, J.; Słupecka-Ziemilska, M.; Pierzynowski, S.; Pierzynowska, K. Pre-digestion of the lipids in infant formula affects gut maturation of the preterm pig. *PLoS ONE* **2022**, *17*, e0265144. [CrossRef] [PubMed]

119. Goncharova, K.; Kirko, S.; Grujic, D.; Kardas, M.; Grochowska-Niedworok, E.; Prykhodko, O.; Woliński, J.; Ushakova, G.; Lozinska, L.; Pierzynowski, S.G. Enhanced absorption of long-chain polyunsaturated fatty acids following consumption of functional milk formula, pre-digested with immobilized lipase ex vivo, in an exocrine pancreatic insufficient (EPI) pig model. *J. Funct. Foods* **2017**, *34*, 422–430. [CrossRef]
120. Mohan, M.S.; O'Callaghan, T.F.; Kelly, P.; Hogan, S.A. Milk fat: Opportunities, challenges and innovation. *Crit. Rev. Food Sci. Nutr.* **2021**, *61*, 2411–2443. [CrossRef]
121. Fox, P.F.; Guinee, T.P.; Cogan, T.M.; McSweeney, P.L.; Fox, P.F.; Guinee, T.P.; Cogan, T.M.; McSweeney, P.L. Enzymatic coagulation of milk. In *Fundamentals of Cheese Science*; Springer: Berlin/Heidelberg, Germany, 2017; pp. 185–229.
122. Hosseini, S.; Varidi, M. Optimization of microbial rennet encapsulation in alginate–chitosan nanoparticles. *Food Chem.* **2021**, *352*, 129325. [CrossRef] [PubMed]
123. Jeong, S.-H.; Yun, S.-E.; Mun, S.-P. The role of immobilized rennet on carbon cloth in flavor development during ripening of Gouda cheese. *Food Sci. Biotechnol.* **2016**, *25*, 1561–1567. [CrossRef] [PubMed]
124. Kessi, E.; Arias, J. Recombinant Camel Chymosin Effectively Acts on Milk Coagulation after Immobilization on Eggshell Membranes. *ACS Food Sci. Technol.* **2022**, *2*, 808–814. [CrossRef]
125. He, W.S.; Sun, Y.; Li, Z.; Yang, H.; Li, J.; Wang, Q.; Tan, C.; Zou, B. Enhanced antioxidant capacity of lipoic acid in different food systems through lipase-mediated esterification with phytosterols. *J. Sci. Food Agric.* **2022**, *102*, 7115–7125. [CrossRef]
126. Mustafa, A.; Ramadan, R.; Niikura, F.; Inayat, A.; Hafez, H. Highly selective synthesis of glyceryl monostearate via lipase catalyzed esterification of triple pressed stearic acid and glycerin. *Sustain. Energy Technol. Assess.* **2023**, *57*, 103200. [CrossRef]
127. Abdel-Mageed, H.M.; Nada, D.; Radwan, R.A.; Mohamed, S.A.; Gohary, N.A.E. Optimization of catalytic properties of *Mucor racemosus* lipase through immobilization in a biocompatible alginate gelatin hydrogel matrix for free fatty acid production: A sustainable robust biocatalyst for ultrasound-assisted olive oil hydrolysis. *3 Biotech* **2022**, *12*, 285. [CrossRef] [PubMed]
128. Savickaite, A.; Sadauskas, M.; Gudiukaite, R. Immobilized GDEst-95, GDEst-lip and GD-95RM lipolytic enzymes for continuous flow hydrolysis and transesterification reactions. *Int. J. Biol. Macromol.* **2021**, *173*, 421–434. [CrossRef] [PubMed]
129. Moreira, K.d.S.; de Oliveira, A.L.; Júnior, L.S.d.M.; Monteiro, R.R.; da Rocha, T.N.; Menezes, F.L.; Fechine, L.M.; Denardin, J.C.; Michea, S.; Freire, R.M. Lipase from *Rhizomucor miehei* immobilized on magnetic nanoparticles: Performance in fatty acid ethyl ester (FAEE) optimized production by the Taguchi method. *Front. Bioeng. Biotechnol.* **2020**, *8*, 693. [CrossRef]
130. Dong, T.; Zhou, X.; Dai, Y.; Yang, X.; Zhang, W.; Yu, D.; Liu, T. Application of magnetic immobilized enzyme of nano dialdehyde starch in deacidification of rice bran oil. *Enzym. Microb. Technol.* **2022**, *161*, 110116. [CrossRef]
131. Bilal, M.; Fernandes, C.D.; Mehmood, T.; Nadeem, F.; Tabassam, Q.; Ferreira, L.F.R. Immobilized lipases-based nano-biocatalytic systems—A versatile platform with incredible biotechnological potential. *Int. J. Biol. Macromol.* **2021**, *175*, 108–122. [CrossRef] [PubMed]
132. Sabi, G.J.; Gama, R.S.; Fernandez-Lafuente, R.; Cancino-Bernardi, J.; Mendes, A.A. Decyl esters production from soybean-based oils catalyzed by lipase immobilized on differently functionalized rice husk silica and their characterization as potential biolubricants. *Enzym. Microb. Technol.* **2022**, *157*, 110019. [CrossRef] [PubMed]
133. Iuliano, M.; Ponticorvo, E.; Cirillo, C.; Castaldo, R.; De Pasquale, S.; Gentile, G.; Sarno, M. Wax esters from waste fish oil catalysed by immobilized *Candida rugosa* lipase. *Process Biochem.* **2023**, *130*, 386–400. [CrossRef]
134. Ungcharoenwiwat, P.; Aran, H. Enzymatic synthesis of coconut oil based wax esters by immobilized lipase EQ3 and commercial lipozyme RMIM. *Electron. J. Biotechnol.* **2020**, *47*, 10–16. [CrossRef]
135. Simões, T.; Ferreira, J.; Lemos, M.F.; Augusto, A.; Félix, R.; Silva, S.F.; Ferreira-Dias, S.; Tecelão, C. Argan oil as a rich source of linoleic fatty acid for dietetic structured lipids production. *Life* **2021**, *11*, 1114. [CrossRef]
136. Mota, D.A.; Rajan, D.; Heinzl, G.C.; Osorio, N.M.; Gominho, J.; Krause, L.C.; Soares, C.M.; Nampoothiri, K.M.; Sukumaran, R.K.; Ferreira-Dias, S. Production of low-calorie structured lipids from spent coffee grounds or olive pomace crude oils catalyzed by immobilized lipase in magnetic nanoparticles. *Bioresour. Technol.* **2020**, *307*, 123223. [CrossRef]
137. Brisola, J.; Andrade, G.J.S.; Oliveira, S.A.d.; Viana, R.M.R.; Tischer, P.C.d.S.F.; Tischer, C.A. Covalent Immobilization of Lipase on Bacterial Cellulose Membrane and Nanocellulose. *Mater. Res.* **2022**, *25*, e20210350. [CrossRef]
138. Xia, Q.; Akanbi, T.O.; Li, R.; Wang, B.; Yang, W.; Barrow, C.J. Lipase-catalysed synthesis of palm oil-omega-3 structured lipids. *Food Funct.* **2019**, *10*, 3142–3149. [CrossRef]
139. Fotiadou, R.; Lefas, D.; Vougiouklaki, D.; Tsakni, A.; Houhoula, D.; Stamatis, H. Enzymatic Modification of Pomace Olive Oil with Natural Antioxidants: Effect on Oxidative Stability. *Biomolecules* **2023**, *13*, 1034. [CrossRef]
140. Zhao, J.; Ma, M.; Yan, X.; Zhang, G.; Xia, J.; Zeng, Z.; Yu, P.; Deng, Q.; Gong, D. Green synthesis of polydopamine functionalized magnetic mesoporous biochar for lipase immobilization and its application in interesterification for novel structured lipids production. *Food Chem.* **2022**, *379*, 132148. [CrossRef]
141. Zhang, Z.; Zhang, S.; Lee, W.J.; Lai, O.M.; Tan, C.P.; Wang, Y. Production of Structured triacylglycerol via enzymatic interesterification of medium-chain triacylglycerol and soybean oil using a pilot-scale solvent-free packed bed reactor. *J. Am. Oil Chem. Soc.* **2020**, *97*, 271–280. [CrossRef]
142. Hwang, J.; Aum, J.; Lee, S.J.; Mun, J.M.; Kim, S.W.; Chung, M.-Y.; Kim, I.-H.; Kim, B.H. Immobilized *Candida antarctica* lipase B as an sn-1,3 regiospecific biocatalyst for the interesterification of triacylglycerols with fatty acid ethyl esters. *Food Sci. Biotechnol.* **2023**, *33*, 159–170. [CrossRef]

143. Miotti Jr, R.H.; Cortez, D.V.; De Castro, H.F. Transesterification of palm kernel oil with ethanol catalyzed by a combination of immobilized lipases with different specificities in continuous two-stage packed-bed reactor. *Fuel* **2022**, *310*, 122343. [CrossRef]
144. Xia, Q.; Akanbi, T.O.; Wang, B.; Li, R.; Yang, W.; Barrow, C.J. Investigating the mechanism for the enhanced oxidation stability of microencapsulated omega-3 concentrates. *Mar. Drugs* **2019**, *17*, 143. [CrossRef]
145. Zhu, H.; Wang, X.; Zhang, W.; Zhang, Y.; Zhang, S.; Pang, X.; Lu, J.; Lv, J. Dietary *Schizochytrium microalgae* affect the fatty acid profile of goat milk: Quantification of docosahexaenoic Acid (DHA) and its distribution at Sn-2 position. *Foods* **2022**, *11*, 2087. [CrossRef] [PubMed]
146. Dovale-Rosabal, G.; Rodríguez, A.; Espinosa, A.; Barriga, A.; Aubourg, S.P. Synthesis of EPA-and DHA-enriched structured acylglycerols at the sn-2 position starting from commercial salmon oil by enzymatic lipase catalysis under supercritical conditions. *Molecules* **2021**, *26*, 3094. [CrossRef] [PubMed]
147. Khoonin, W.; Shantavasinkul, P.C.; Santivarangkna, C.; Trachootham, D. Loss of Eicosapentaenoic Acid (EPA) after Retort Sterilization of the EPA-BCAA Fortified Complete Nutrition Drink. *Foods* **2022**, *11*, 2023. [CrossRef] [PubMed]
148. Akanbi, T.O.; Sinclair, A.J.; Barrow, C.J. Pancreatic lipase selectively hydrolyses DPA over EPA and DHA due to location of double bonds in the fatty acid rather than regioselectivity. *Food Chem.* **2014**, *160*, 61–66. [CrossRef] [PubMed]
149. Chen, X.; Wang, J.; Stevenson, R.J.; Ang, X.; Peng, Y.; Quek, S.Y. Lipase-catalyzed modification of milk fat: A promising way to alter flavor notes of goat milk products. *LWT* **2021**, *145*, 111286. [CrossRef]
150. Singh, P.K.; Chopra, R.; Garg, M.; Dhiman, A.; Dhyani, A. Enzymatic Interesterification of Vegetable Oil: A Review on Physicochemical and Functional Properties, and Its Health Effects. *J. Oleo Sci.* **2022**, *71*, 1697–1709. [CrossRef] [PubMed]
151. Akanbi, T.O.; Barrow, C.J. Lipase-catalysed incorporation of EPA into emu oil: Formation and characterisation of new structured lipids. *J. Funct. Foods* **2015**, *19*, 801–809. [CrossRef]
152. Nagpal, T.; Sahu, J.K.; Khare, S.K.; Bashir, K.; Jan, K. Trans fatty acids in food: A review on dietary intake, health impact, regulations and alternatives. *J. Food Sci.* **2021**, *86*, 5159–5174. [CrossRef]
153. Chen, Q.; Xie, Q.; Jiang, C.; Evivie, S.E.; Cao, T.; Wang, Z.; Zhao, L.; Liang, S.; Li, B.; Huo, G. Infant formula supplemented with 1, 3-olein-2-palmitin regulated the immunity, gut microbiota, and metabolites of mice colonized by feces from healthy infants. *J. Dairy Sci.* **2022**, *105*, 6405–6421. [CrossRef] [PubMed]
154. Zhang, Z.; Wang, Y.; Li, Y.; Lin, X.; Hong, Z.; Huang, J.; Zhang, X.; Yang, Y.; Su, Y. Effects of Sn-2-palmitate-enriched formula feeding on infants' growth, stool characteristics, stool fatty acid soap contents and bone mineral content: A systematic review and meta-analysis of randomized controlled trials. *Crit. Rev. Food Sci. Nutr.* **2023**, *63*, 10256–10266. [CrossRef] [PubMed]
155. Cao, X.; Pan, Y.; Qiao, M.; Yuan, Y. Synthesis of human milk fat substitutes based on enzymatic preparation of low erucic acid acyl-donors from rapeseed oil. *Food Chem.* **2022**, *387*, 132907. [PubMed]
156. Li, Y.; Zhang, Y.; Zhou, Y.; Zhang, Y.; Zheng, M. A novel and controllable method for simultaneous preparation of human milk fat substitutes (OPL, OPO and LPL): Two-step enzymatic ethanolysis-esterification strategy. *Food Res. Int.* **2023**, *163*, 112168. [CrossRef] [PubMed]
157. Wang, X.; Huang, Z.; Hua, L.; Zou, F.; Cheng, X.; Wang, X. Preparation of human milk fat substitutes similar to human milk fat by enzymatic acidolysis and physical blending. *LWT* **2021**, *140*, 110818. [CrossRef]
158. Soares, C.M.; Barbosa, M.S.; Santos, S.B.; Mattedi, S.; Lima, Á.S.; Pereira, M.M.; Tecelão, C.; Ferreira-Dias, S. Production of Human Milk Fat Substitutes by Lipase-Catalyzed Acidolysis: Immobilization, Synthesis, Molecular Docking and Optimization Studies. *Catalysts* **2023**, *13*, 825. [CrossRef]
159. Bryś, J.; Górska, A.; Ostrowska-Ligęza, E.; Wirkowska-Wojdyła, M.; Bryś, A.; Brzezińska, R.; Dolatowska-Żebrowska, K.; Małajowicz, J.; Ziarno, M.; Obranović, M. Human milk fat substitutes from lard and hemp seed oil mixtures. *Appl. Sci.* **2021**, *11*, 7014. [CrossRef]
160. Arzola-Rodríguez, S.I.; Muñoz-Castellanos, L.-N.; López-Camarillo, C.; Salas, E. Phenolipids, amphipilic phenolic antioxidants with modified properties and their spectrum of applications in development: A review. *Biomolecules* **2022**, *12*, 1897. [CrossRef]
161. Akanbi, T.O.; Barrow, C.J. Lipase-produced hydroxytyrosyl eicosapentaenoate is an excellent antioxidant for the stabilization of omega-3 bulk oils, emulsions and microcapsules. *Molecules* **2018**, *23*, 275. [CrossRef] [PubMed]
162. Ahmad, T.; Aadil, R.M.; Ahmed, H.; Rahman, U.U.; Soares, B.C.V.; Souza, S.L.Q.; Pimentel, T.C.; Scudino, H.; Guimaraes, J.T.; Esmerino, E.A.; et al. Treatment and utilization of dairy industrial waste: A review. *Trends Food Sci. Technol.* **2019**, *88*, 361–372. [CrossRef]
163. Awasthi, M.K.; Paul, A.; Kumar, V.; Sar, T.; Kumar, D.; Sarsaiya, S.; Liu, H.; Zhang, Z.; Binod, P.; Sindhu, R. Recent trends and developments on integrated biochemical conversion process for valorization of dairy waste to value added bioproducts: A review. *Bioresour. Technol.* **2022**, *344*, 126193.
164. Zanuso, E.; Gomes, D.G.; Ruiz, H.A.; Teixeira, J.A.; Domingues, L. Enzyme immobilization as a strategy towards efficient and sustainable lignocellulosic biomass conversion into chemicals and biofuels: Current status and perspectives. *Sustain. Energy Fuels* **2021**, *5*, 4233–4247. [CrossRef]
165. Bilal, M.; Iqbal, H.M. Sustainable bioconversion of food waste into high-value products by immobilized enzymes to meet bio-economy challenges and opportunities–A review. *Food Res. Int.* **2019**, *123*, 226–240. [CrossRef] [PubMed]
166. Charles, M. *Applications of Lactase and Immobilized Lactase*; CRC Press: Boca Raton, FL, USA, 2019.
167. Sarnoab, M.; Iuliano, M. Biodiesel production from dairy waste scum by using an efficient nano-biocatalyst. *Chem. Eng.* **2020**, *79*, 181–186.

168. Mehrotra, T.; Dev, S.; Banerjee, A.; Chatterjee, A.; Singh, R.; Aggarwal, S. Use of immobilized bacteria for environmental bioremediation: A review. *J. Environ. Chem. Eng.* **2021**, *9*, 105920. [CrossRef]
169. Cieh, N.L.; Mokhtar, M.N.; Baharuddin, A.S.; Mohammed, M.A.P.; Wakisaka, M. Progress on lipase immobilization technology in edible oil and fat modifications. *Food Rev. Int.* **2023**, *40*, 457–503. [CrossRef]
170. Pourjoula, M.; Picariello, G.; Garro, G.; D'Auria, G.; Nitride, C.; Ghaisari, A.R.; Ferranti, P. The protein and peptide fractions of kashk, a traditional Middle East fermented dairy product. *Food Res. Int.* **2020**, *132*, 109107. [CrossRef]
171. Hashemabadi, M.; Badoei-Dalfard, A. Fabrication of magnetic CLEA-protease nanocomposite: High progression in biotechnology and protein waste management. *Catal. Lett.* **2019**, *149*, 1753–1764. [CrossRef]
172. Misson, M.; Saallah, S.; Zhang, H. Nanofiber-immobilized β-galactosidase for dairy waste conversion into galacto-oligosaccharides. In *Advances in Waste Processing Technology*; Springer: Berlin/Heidelberg, Germany, 2020; pp. 37–48.
173. Wang, Z.; Qi, J.; Goddard, J.M. Concentrated sugar solutions protect lactase from thermal inactivation. *Int. Dairy J.* **2021**, *123*, 105168. [CrossRef]
174. Ng, H.S.; Kee, P.E.; Yim, H.S.; Chen, P.-T.; Wei, Y.-H.; Lan, J.C.-W. Recent advances on the sustainable approaches for conversion and reutilization of food wastes to valuable bioproducts. *Bioresour. Technol.* **2020**, *302*, 122889. [CrossRef]
175. Ekeoma, B.C.; Ekeoma, L.N.; Yusuf, M.; Haruna, A.; Ikeogu, C.K.; Merican, Z.M.A.; Kamyab, H.; Pham, C.Q.; Vo, D.-V.N.; Chelliapan, S. Recent advances in the biocatalytic mitigation of emerging pollutants: A comprehensive review. *J. Biotechnol.* **2023**, *369*, 14–34. [CrossRef] [PubMed]
176. Yaashikaa, P.; Devi, M.K.; Kumar, P.S. Advances in the application of immobilized enzyme for the remediation of hazardous pollutant: A review. *Chemosphere* **2022**, *299*, 134390. [CrossRef] [PubMed]
177. Pottratz, I.; Schmidt, C.; Müller, I.; Hamel, C. Immobilization of β-Galactosidase on Monolithic Discs for the Production of Prebiotics Galacto-oligosaccharides. *Chem. Ing. Tech.* **2021**, *93*, 838–843. [CrossRef]
178. Zdarta, J.; Meyer, A.S.; Jesionowski, T.; Pinelo, M. Multi-faceted strategy based on enzyme immobilization with reactant adsorption and membrane technology for biocatalytic removal of pollutants: A critical review. *Biotechnol. Adv.* **2019**, *37*, 107401. [CrossRef]
179. Zieniuk, B.; Białecka-Florjańczyk, E.; Wierzchowska, K.; Fabiszewska, A. Recent advances in the enzymatic synthesis of lipophilic antioxidant and antimicrobial compounds. *World J. Microbiol. Biotechnol.* **2022**, *38*, 11. [CrossRef] [PubMed]
180. Nimkande, V.D.; Bafana, A. A review on the utility of microbial lipases in wastewater treatment. *J. Water Process Eng.* **2022**, *46*, 102591. [CrossRef]
181. Rather, A.H.; Khan, R.S.; Wani, T.U.; Beigh, M.A.; Sheikh, F.A. Overview on immobilization of enzymes on synthetic polymeric nanofibers fabricated by electrospinning. *Biotechnol. Bioeng.* **2022**, *119*, 9–33. [CrossRef] [PubMed]
182. Cárdenas-Alcaide, M.F.; Godínez-Alemán, J.A.; González-González, R.B.; Iqbal, H.M.; Parra-Saldívar, R. Environmental impact and mitigation of micro (nano) plastics pollution using green catalytic tools and green analytical methods. *Green Anal. Chem.* **2022**, *3*, 100031. [CrossRef]
183. Chakraborty, D.; Chatterjee, S.; Althuri, A.; Mohan, S.V. Sustainable enzymatic treatment of organic waste in a framework of circular economy. *Bioresour. Technol.* **2022**, *370*, 128487. [CrossRef] [PubMed]
184. Işık, C.; Saraç, N.; Teke, M.; Uğur, A. A new bioremediation method for removal of wastewater containing oils with high oleic acid composition: *Acinetobacter haemolyticus* lipase immobilized on eggshell membrane with improved stabilities. *New J. Chem.* **2021**, *45*, 1984–1992. [CrossRef]
185. Feng, K.; Huang, Z.; Peng, B.; Dai, W.; Li, Y.; Zhu, X.; Chen, Y.; Tong, X.; Lan, Y.; Cao, Y. Immobilization of *Aspergillus niger* lipase onto a novel macroporous acrylic resin: Stable and recyclable biocatalysis for deacidification of high-acid soy sauce residue oil. *Bioresour. Technol.* **2020**, *298*, 122553. [CrossRef] [PubMed]
186. Guo, J.; Sun, S.; Liu, J. Conversion of waste frying palm oil into biodiesel using free lipase A from *Candida antarctica* as a novel catalyst. *Fuel* **2020**, *267*, 117323. [CrossRef]
187. Altinok, F.; Albayrak, S.; Arslan, N.P.; Taskin, M.; Aygun, E.; Sisecioglu, M.; Adiguzel, A. Application of *Anoxybacillus gonensins* UF7 lipase as a catalyst for biodiesel production from waste frying oils. *Fuel* **2023**, *334*, 126672. [CrossRef]
188. Nezhad, M.K.; Aghaei, H. Tosylated cloisite as a new heterofunctional carrier for covalent immobilization of lipase and its utilization for production of biodiesel from waste frying oil. *Renew. Energy* **2021**, *164*, 876–888. [CrossRef]
189. Patchimpet, J.; Simpson, B.K.; Sangkharak, K.; Klomklao, S. Optimization of process variables for the production of biodiesel by transesterification of used cooking oil using lipase from Nile tilapia viscera. *Renew. Energy* **2020**, *153*, 861–869. [CrossRef]
190. Parandi, E.; Safaripour, M.; Abdellattif, M.H.; Saidi, M.; Bozorgian, A.; Nodeh, H.R.; Rezania, S. Biodiesel production from waste cooking oil using a novel biocatalyst of lipase enzyme immobilized magnetic nanocomposite. *Fuel* **2022**, *313*, 123057. [CrossRef]
191. Parandi, E.; Safaripour, M.; Mosleh, N.; Saidi, M.; Nodeh, H.R.; Oryani, B.; Rezania, S. Lipase enzyme immobilized over magnetic titanium graphene oxide as catalyst for biodiesel synthesis from waste cooking oil. *Biomass Bioenergy* **2023**, *173*, 106794. [CrossRef]
192. Sharma, T.; Xia, C.; Sharma, A.; Raizada, P.; Singh, P.; Sharma, S.; Sharma, P.; Kumar, S.; Lam, S.; Nadda, A.K. Mechano-chemical and biological energetics of immobilized enzymes onto functionalized polymers and their applications. *Bioengineered* **2022**, *13*, 10518–10539.
193. Almeida, F.L.; Travalia, B.M.; Goncalves, I.S.; Forte, M.B.S. Biodiesel production by lipase-catalyzed reactions: Bibliometric analysis and study of trends. *Biofuels Bioprod. Biorefining* **2021**, *15*, 1141–1159. [CrossRef]

194. Skliar, V.; Krusir, G.; Zakharchuk, V. Investigation of enzymatic hydrolysis of fatty waste processing of sunflower seeds by immobilized lipase. *Grain Prod. Mix. Fodder's* **2020**, *20*, 5–11. [CrossRef]
195. Kumari, M.; Chattopadhyay, S. The evaluation of the performance of rice husk and rice straw as potential matrix to obtain the best lipase immobilized system: Creating wealth from wastes. *Prep. Biochem. Biotechnol.* **2023**, *53*, 763–772. [CrossRef]

processes

MDPI

Review

Retaining Resveratrol Content in Berries and Berry Products with Agricultural and Processing Techniques: Review

Audrone Ispiryan [1,*], Ingrida Kraujutiene [2] and Jonas Viskelis [3]

[1] Bioeconomy Research Institute, Agriculture Academy, Vytautas Magnus University, Studentu Str. 11, LT-53361 Akademija, Lithuania
[2] Department of Food and Agrotechnology, Kauno Kolegija Higher Education Institution, Pramones pr. 20, LT-50468 Kaunas, Lithuania; ingrida.kraujutiene@go.kauko.lt
[3] Institute of Horticulture, Lithuanian Research Centre for Agriculture and Forestry, Kauno Str. 30, Kaunas District, LT-54333 Babtai, Lithuania; jonas.viskelis@lammc.lt
* Correspondence: audrone.ispiryan@vdu.lt; Tel.: +370-672-90015

Abstract: Resveratrol is a natural compound that can be found in red wine, grapes, and berries. It has attracted attention due to its potential health benefits. The aim of this review was to align ways of retaining resveratrol contents in berries and products made of berries, and to show which agricultural and processing techniques can maximize the content in the berries and their products and how this can be achieved. The scientific literature has revealed that resveratrol concentration in berries and berry-derived products varies significantly depending on the source and the processing techniques applied. Resveratrol content can range from 0.03–0.06 mg/kg in blueberries to 5–10 mg/kg in grape skins. Agricultural techniques such as controlled water stress (e.g., increasing resveratrol in grapes to 8.3–11.5 mg/kg), optimal sun exposure (e.g., enhancing blueberries to 1.5–2.1 mg/kg), balanced nutrient management, and selecting high-resveratrol cultivars (e.g., up to 15 mg/kg in certain grapes) can substantially increase resveratrol content. Processing methods like cold pressing, centrifugation, ultrafiltration, and freeze-drying are effective in preserving resveratrol levels, while traditional pasteurization tends to reduce its concentration. For instance, high-temperature short-time pasteurization can reduce resveratrol in juice from 1.5 mg/kg to 0.8 mg/kg, whereas cold pressing retains more resveratrol (1.5 mg/kg to 1.4 mg/kg). By optimizing these agricultural and processing techniques, manufacturers can enhance the resveratrol content in berry-derived products, meeting the growing consumer demand for health-enhancing natural products and supporting a healthier society. This approach aligns with the commitment to overcoming the technical challenges associated with resveratrol use, ensuring its potential is fully realized in both health-related and non-health-related applications.

Keywords: resveratrol; berries; berry products; natural products; influence of processes

Citation: Ispiryan, A.; Kraujutiene, I.; Viskelis, J. Retaining Resveratrol Content in Berries and Berry Products with Agricultural and Processing Techniques: Review. *Processes* **2024**, *12*, 1216. https://doi.org/10.3390/pr12061216

Academic Editors: Yanlin Zhang and Prashant K. Sarswat

Received: 30 April 2024
Revised: 9 June 2024
Accepted: 10 June 2024
Published: 13 June 2024

1. Introduction

Resveratrol (3,5,4′-trihydroxystilbene) is a phytoalexin belonging to the class of polyphenolic compounds called stilbene [1], isolated from grape berries, skin, and seeds for the first time in 1976 [2]. Resveratrol exists in trans- and cis- forms. Its molecular formula is $C_{14}H_{12}O_3$, with a weight of 228.24 g/mol, and the chemical structures of its trans- and cis-isomers are shown in Figure 1. These structures show that resveratrol contains three hydroxyl groups (-OH) bonded to the two benzene rings, which contribute to its antioxidant properties.

Resveratrol and its human metabolites are widely discussed in scientific research. It has low bioavailability and an extensive metabolism [3,4]. Recently, Springer, M, and Moko, S., discussed human metabolism and biotransformation of resveratrol, and reported molecular mechanisms of action [5]. Scientists have identified 4-hydroxydibenzyl as a novel gut microbiota-derived metabolite of resveratrol in humans [6–9], and the positive effects of resveratrol metabolites on human health have been discussed [10–13].

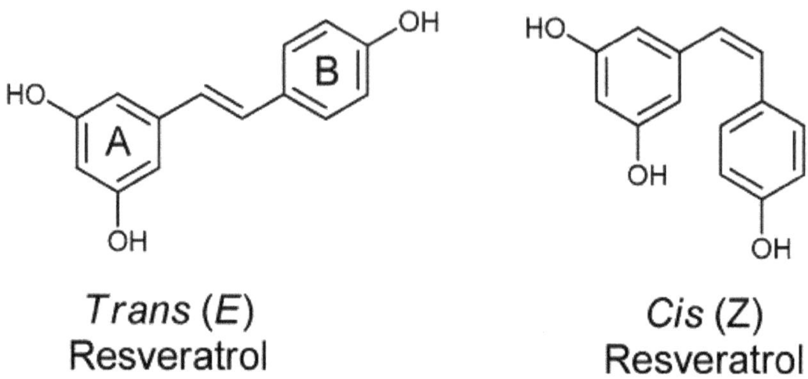

Figure 1. Chemical structures of trans-resveratrol and its cis-isomer.

Scientists revealed how specific resveratrol metabolites improve insulin resistance, highlighting their role in inhibiting cancer cell progression and enhancing the effectiveness of chemotherapy [14–16]. Resveratrol's antioxidant [17,18], anti-inflammatory [19,20], antibacterial [21], cardioprotective [22,23], anti-cancer [24,25], and potential anti-aging and neuroprotective [26] properties are widely discussed in recent scientific research. The field continues to evolve as researchers uncover new insights into resveratrol's metabolism, biological activity, and therapeutic potential for human health [27–29]. The typical dietary intake of resveratrol from these sources is relatively low, usually less than 1 mg per day, significantly lower than the doses commonly used in research studies. Consequently, the actual impact of dietary resveratrol remains unclear, largely due to the minimal amounts typically consumed through fresh berries [30–32].

Berries are an important natural source of resveratrol, particularly grapes [33,34] and strawberries. Additionally, red wine is produced by the fermentation of grapes and different types of wines contain different amounts of resveratrol [35,36]. Resveratrol synthesis in berries can be affected by various factors which increase resveratrol accumulation [37]. Some studies have delved into the impact of nitrogen supply on resveratrol concentrations in grape berries, alongside exploring methods for extracting resveratrol from blueberries.

This polyphenolic compound can also be found in various parts of the berry plant, notably in grapes, which have been extensively studied. Resveratrol is predominantly found in the skins of grape berries [38,39]. This is particularly significant as the skin is used extensively in the production of red wine, where resveratrol is extracted during the fermentation process [40–43]. Leaves also contain resveratrol, although the concentration can vary widely depending on the berry variety and the environmental factors affecting the plant [44,45]. Some studies have shown that resveratrol is present in the stems and vines of berry plants, suggesting that almost all parts of the plant can contribute to resveratrol content in various products derived from the vine [46].

The presence of resveratrol in these parts of the plant underscores its role in plant defense and its potential health benefits when consumed in products like wines, juice [47], or dried berries [48]. This distribution also highlights the importance of agricultural and processing techniques that can maximize the content of resveratrol from these parts for use in health supplements and functional foods. To maximize or maintain resveratrol content in fruits and wine, employing specific processing techniques is crucial. For fruits, gentle handling, cold storage, controlled atmosphere storage, minimal processing, and vacuum packaging are effective strategies. For wine, selecting high-resveratrol grape varieties, increasing skin contact time, optimizing fermentation temperatures, using antioxidants, controlled micro-oxygenation, proper aging conditions, stabilization techniques, and protective packaging can help preserve resveratrol levels. It can be concluded that avoiding high heat and minimizing light exposure are general best practices. These processing techniques collectively ensure that the beneficial properties of resveratrol are retained in both fruits and wine.

The Preferred Reporting Items for Systematic Reviews and Meta-Analyses (PRISMA) flowchart was used to systematically review and synthesize the relevant literature on resveratrol in berries and berry products. Records were identified through searches in multiple academic databases, including Google Scholar, Pubmed, MDPI, ScienceDirect, PubMed, Web of Science, Scopus, Food Science Source, European Patent Office, Agricola, and reputable websites. Keywords such as "Resveratrol", "Berries", "Berries' products", and "Impact of the processes" were used to locate relevant studies. The titles and abstracts of the identified records were screened to determine their relevance. The studies that met all the inclusion criteria were included in the final review and analysis. Inclusion criteria were as follows: (1) studies discussing resveratrol metabolism and biotransformation; (2) research on the health effects of resveratrol; (3) articles detailing agricultural techniques affecting resveratrol content; (4) studies on processing techniques influencing resveratrol in berries and products; (5) articles published in peer-reviewed journals; (6) studies providing quantitative data on resveratrol concentrations. Exclusion criteria were as follows: (1) studies not focused on resveratrol; (2) articles without substantial data on metabolism or health effects; (3) research not detailing agricultural or processing techniques; (4) duplicates and inaccessible full-text articles; (5) non-peer-reviewed articles, opinion pieces, and editorials; (6) studies published in languages other than English (unless translations were available). To complete this review, relevant articles were collected from the academic databases mentioned above. This flowchart in Figure 2 and the summary outline the process of identifying, screening, assessing, and including articles for the review, ensuring clarity and transparency in the research methodology. This methodical approach allowed the researchers to transparently report how they arrived at the final set of studies included in their review, providing a clear and concise summary of the literature selection process.

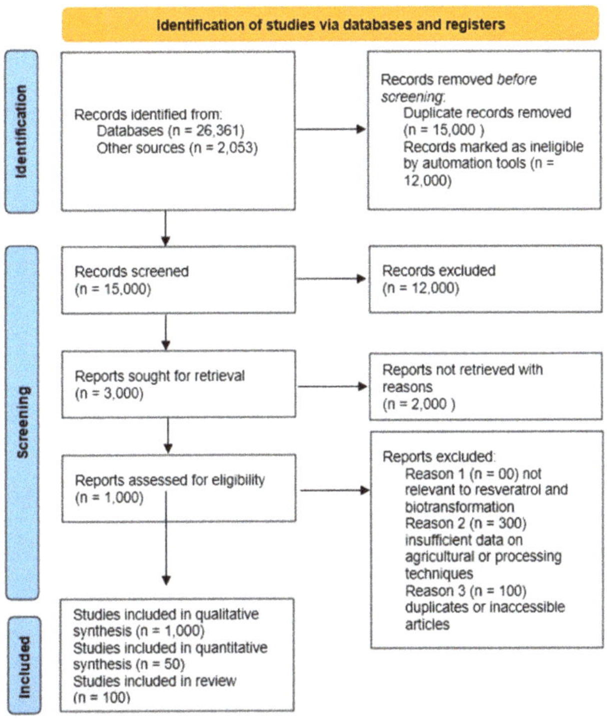

Figure 2. PRISMA flowchart on resveratrol in berries and berry products.

Relevant information was extracted from the selected articles and organized systematically. We analyzed the collected data, identified existing trends and gaps, compared and contrasted the findings of different studies, and critically evaluated the strengths and weaknesses of each piece of evidence. On this basis, we created the conceptual structure of our review article. We categorized our findings and they are presented in the tables.

Table data visualization software (PRISMA 2020) was used to create interactive and visually appealing dashboards and reports. It connects to various data sources and enables users to explore, analyze, and present data in a user-friendly manner. Therefore, it can be used in business intelligence, data analytics, and decision-making processes. Our aim was to follow this methodology to reveal how to retain resveratrol contents in berries and products made of berries, and to show how agricultural and processing techniques can maximize resveratrol content in the berries and berry products.

2. Agricultural Technologies

Resveratrol is predominantly associated with grapes, particularly red and purple grapes, as they are one of the richest sources of this compound, which is widely recognized due to its presence in red wine [49–51]. Notably, the Pinot noir grape variety contains the highest content of resveratrol, with concentrations reaching 5–10 mg/kg in the grape skins [52–55]. Blueberries contain resveratrol, though in smaller amounts than grapes. Both wild and cultivated blueberries have resveratrol, with variations depending on the specific type and growing conditions [56]. The concentration of resveratrol in blueberries, bilberries, strawberries, cranberries, mulberries, lingonberries, and raspberries can vary significantly depending on the specific type of berry and the environmental conditions in which they are cultivated [57–60]. Agricultural techniques play a critical role in influencing the production and retention of resveratrol in fresh berries, leaves, and stems [61,62]. From the agricultural methods found in the research, we identified the following, presented in Table 1.

Table 1. Agricultural technologies influencing resveratrol content in berries.

Agricultural Techniques	Description	Advantages	Disadvantages	Resveratrol Content (mg/kg)	Refs.
Water Stress	Controlled water stress, especially deficit irrigation, can significantly increase resveratrol concentration in berries. Limiting water at critical growth stages can induce a stress response in plants, leading to the enhanced synthesis of resveratrol as part of the plant's defense mechanism.	Enhances resveratrol production as a stress response; can also conserve water usage, making it environmentally sustainable.	Risk of under-irrigation leading to stress that could negatively impact overall plant health and yield if not carefully managed.	Grapes: 8.3–11.5 mg/kg	[63–66]
Sun Exposure	Ensuring optimal sun exposure can stimulate resveratrol production, as this is part of the plant's natural response to ultraviolet (UV) radiation. However, excessive sun exposure can be detrimental, so balance is key.	Proper sunlight exposure can naturally increase resveratrol production through UV-induced stress responses.	Requires careful management to avoid overexposure, which can lead to sunburn or heat stress, adversely affecting the plants.	Blueberries: 1.5–2.1 mg/kg	[67–71]
Nutrient Management	Adequate nutrition, especially phosphorus and nitrogen management, influences the general health of the plants and their ability to produce phytochemicals like resveratrol. However, excessive use of nitrogen can sometimes decrease phenolic compounds, so careful management is necessary.	Healthy soil supports overall plant growth and optimizes conditions for resveratrol production.	Balancing soil nutrients precisely to avoid over- or under-fertilization can be challenging.	Strawberries: 0.3–0.5 mg/kg	[72,73]

Table 1. *Cont.*

Agricultural Techniques	Description	Advantages	Disadvantages	Resveratrol Content (mg/kg)	Refs.
Organic Farming Practices	Some studies suggest that organic farming practices, which avoid synthetic pesticides and fertilizers, might enhance the level of resveratrol. This could be due to the increased reliance of the plant on its own defense mechanisms to combat pests and diseases.	Organic practices enrich soil biodiversity, enhancing plant resilience.	Organic fertilization often requires more labor and can be more costly than conventional methods.	Raspberries: 0.8–1.2 mg/kg	[74–76]
Cultivar Selection	Choosing berry varieties that are naturally high in resveratrol can be an effective way to ensure higher levels in the final produce. Genetic variability significantly affects resveratrol content.	Ensures a genetic advantage in producing more of the compound, potentially reducing the need for intensive agricultural inputs.	Limited availability of high-resveratrol cultivars and possible higher costs associated with obtaining these specialized varieties.	Grapes: Up to 15 mg/kg in certain cultivars	[77–79]
Biotic Stress	Introducing controlled amounts of biotic stress, such as exposure to certain pathogens or insects, can stimulate the production of resveratrol as a defensive phytoalexin. However, this needs to be carefully managed to avoid significant damage to crops.	Reduces reliance on chemical pesticides, promoting natural plant defense mechanisms	Can be labor-intensive and requires a deep understanding of pest and predator relationships. Initial setup and maintenance costs can be high.	Grapes: 9–12 mg/kg when exposed to Botrytis cinerea	[80–82]
Use of Fungicides	Certain fungicides might induce resveratrol synthesis as a stress response, although the reliance on chemical treatments can have other environmental or health implications.	The same as for biotic stress	The same as for biotic stress	Grapes: Variable, can induce significant increases	[83,84]
Harvest Time	The stage of maturity at harvest can impact the resveratrol content in berries. Generally, berries accumulate resveratrol as they mature, so timing the harvest to coincide with peak resveratrol levels can optimize resveratrol content in the harvested fruit.	Harvesting at peak maturity when resveratrol levels are highest ensures the maximum content is retained in the harvested berries.	Narrow harvest windows require precise timing and may conflict with market prices or labor availability.	Grapes: 10–15 mg/kg at peak maturity	[85–87]
Pruning and Canopy Management	Proper pruning and canopy management can affect the light and air penetration within the plant foliage, which influences overall plant health and its metabolic processes, including resveratrol production.	Proper handling and storage maintain resveratrol levels post-harvest, preserving the quality and extending the shelf life of berries.	Requires investment in storage facilities. Mishandling can quickly degrade quality, despite optimal pre-harvest conditions.	Grapes: Improved resveratrol with balanced canopy	[88–90]

The qualitative analysis of agricultural techniques reveals that various methods can significantly influence the resveratrol content in berries. The enhancement of resveratrol primarily occurs through the plant's natural stress responses, whether induced by controlled water stress [91], optimal sun exposure [92], or biotic stress [93] from pathogens or insects. Organic farming practices and cultivar selection also play crucial roles by leveraging natural plant defense mechanisms and genetic advantages, respectively. Organic farming practices, for instance, emphasize soil biodiversity and plant resilience, leading to potentially higher resveratrol levels. Cultivar selection ensures that genetically predisposed plants produce more resveratrol, reducing the need for intensive agricultural inputs. Biotic and abiotic stresses such as controlled pathogen exposure and optimal sun exposure further stimulate resveratrol synthesis as part of the plant's defense strategy. However, these methods come with challenges, such as the need for careful management to avoid adverse effects on plant health and yield. Techniques like water stress and biotic stress must be precisely controlled to achieve the beneficial stress response without causing harm to the plant. Similarly, managing sun exposure requires ensuring there is enough UV radiation to stimulate resveratrol production without causing heat stress or sunburn.

The stability and concentration of resveratrol in agricultural produce are primary indicators of the effectiveness of agricultural techniques. Studies show that water stress can increase resveratrol concentration in grapes from 8.3 to 11.5 mg/kg. Blueberries show resveratrol contents of 1.5 to 2.1 mg/kg under proper sun exposure. Proper nutrient balance is crucial, as excessive nitrogen can reduce phenolic compounds. Strawberries exhibit resveratrol levels of 0.3 to 0.5 mg/kg with balanced nutrients. Organic farming is labor-intensive and costlier but leads to resveratrol levels of 0.8 to 1.2 mg/kg in raspberries [94]. To optimize resveratrol content in berries, a combination of these agricultural techniques should be carefully implemented, considering the specific operational conditions required for each method. By carefully weighing the advantages and obstacles linked with each technique, farmers can effectively boost resveratrol levels in berries, thereby augmenting their nutritional and economic value [95].

3. Processing Technologies

The preservation of berries provides a multitude of benefits that extend beyond simply preserving their freshness. It significantly enhances their practical utility, safety, nutritional profile, and economic value. It can sometimes increase the concentration of certain nutrients and improve their bioavailability, making it easier for the body to absorb essential vitamins and antioxidants. Processed berry products offer convenience, providing quick and easy ways to incorporate berries into diverse dietary habits, from jams and juices to frozen berries that retain much of the nutritional value of fresh berries [96–101].

The discovery of resveratrol immediately attracted the attention of scientists and manufacturers. This is shown not only by the research, but also by the number of patented products, production methods, and technologies. The European patent office contains 26,361 patents related to resveratrol. The first patented product was unregistered in 1985, which included resveratrol and aimed to improve liver health. A patent search with the term "resveratrol and berries" led to 2053 results. Resveratrol is commonly included in the composition of various patented products, particularly in sectors like pharmaceuticals, nutraceuticals, cosmetics, and the food and beverage industries. There are patents focused on chemical derivatives of resveratrol that enhance its efficacy, stability, or bioavailability for therapeutic purposes, and formulations that use resveratrol as a key ingredient in health supplements aimed at providing anti-aging, anti-inflammatory, or antioxidant benefits, as it is a molecule with anti-inflammatory and anti-cancer activities [102–105]. Some of them focus on methods for infusing resveratrol into alcoholic beverages to provide health benefits alongside traditional consumption, and on standardizing the bioactivity and content of resveratrol in other products to ensure consistent health benefits [106–111].

These patents demonstrate the diverse applications of resveratrol across different industries, leveraging its health-promoting properties to enhance products ranging from dietary supplements to skincare items and beyond, and reflect the broad interest and ongoing innovation in utilizing resveratrol for various health-related applications, from enhanced synthesis methods to novel therapeutic derivatives. Each patent represents a step towards better understanding and leveraging resveratrol's potential benefits.

At present, berries can be preserved using traditional or very innovative techniques. In this area of product production, as many production technologies have recently been invented, e.g., ultrasonic homogenization, the need for resveratrol in berry products remains underestimated [112]. Based on the largest supply of berry products on the market, we selected the most-used processing techniques and describe their effects in Table 2.

Table 2. Processing methods influencing resveratrol content in berry products.

Processing Technology and Product Form	Technical Characteristics	Resveratrol Content before Processing (mg/kg)	Resveratrol Content after Processing (mg/kg)	Influence on Resveratrol	Refs.
Pasteurization: High Temperature, Short Time *Juice*	Quickly heats the juice to a high temperature for a brief period to kill pathogens and deactivate enzymes.	1.5	0.8	Heat involved in pasteurization can reduce the levels of resveratrol due to thermal degradation.	[113,114]
Pasteurization: Low Temperature Long Time *Juice*	Uses lower temperatures but for longer durations, also aiming to minimize microbial growth and enzyme activity.	1.5	0.9	Heat involved in pasteurization can reduce the levels of resveratrol due to thermal degradation.	[115–119]
Cold Pressing *Juice*	Utilizes a hydraulic press to extract juice under cold conditions, preserving more nutrients and flavors.	1.5	1.4	Retains more resveratrol because it avoids heat	[120,121]
Centrifugation *Juice*	Separates juice from solids quickly using centrifugal force	1.5	1.4	Preserves resveratrol content because it inactivates microbes and enzymes at cold or room temperatures without the use of heat.	[122]
Ultrafiltration *Juice*	Uses semi-permeable membranes to clarify juice and remove particles without heating	1.5	1.4	Preserves resveratrol content well because it inactivates microbes and enzymes at cold or room temperatures without the use of heat.	[123]
Reverse Osmosis *Juice*	Concentrates juice by removing water through a semi-permeable membrane under pressure.	1.5	Data unavailable	The impact of the process has been little studied; no reliable information has been found	[124,125]
Flash Pasteurization *Juice*	Rapidly heats and cools juice to kill microbes while preserving quality.	1.5	Data unavailable	The impact of the process has been little studied; no reliable information has been found	[126]
Fermentation *Wine*	Uses carefully controlled temperatures and yeast strains to optimize the fermentation process. Converts malic acid into lactic acid, softening the wine's taste.	1.5	1.8	Fermentation can actually increase the availability of resveratrol in wine, as the process involves the conversion of compounds in grape skins into more bioavailable forms.	[127–134]
Oak Aging *Wine*	Wine is aged in oak barrels, which contributes to flavor complexity through slow oxygenation and infusion of wood compounds	1.8	Data unavailable	The impact of the process has been little studied; no reliable information has been found	[135]

Table 2. *Cont.*

Processing Technology and Product Form	Technical Characteristics	Resveratrol Content before Processing (mg/kg)	Resveratrol Content after Processing (mg/kg)	Influence on Resveratrol	Refs.
Oven Drying *Dried berries*	Utilizes conventional ovens or specialized drying ovens where temperature and airflow are controlled to ensure even drying.	1.5	Depends on temperature during processing	Depends on temperature during processing	[136,137]
Dehydrator Drying *Dried berries*	Uses electric dehydrators equipped with heating elements, fans, and air vents to provide a controlled environment for moisture removal.	1.5	1.2	This method is more consistent than oven-drying and is excellent for producing high-quality dried fruits.	[138]
Microwave Drying *Dried berries*	Employs microwave energy to rapidly heat and evaporate water from the fruit.	1.5	Data unavailable	The impact of the process has been little studied; no reliable information has been found	[139,140]
Freeze Drying (Lyophilization) *Dried berries*	Involves freezing the fruit and then reducing the surrounding pressure to allow the frozen water in the fruit to directly transition from ice to vapor.	1.5	1.4	Freeze-drying is excellent for preserving resveratrol as it removes water at low temperatures, thus avoiding the heat that causes degradation. This method best retains the biochemical structure of the berry, including resveratrol.	[141,142]
Spray Drying *Powdered extracts*	Converts liquid fruit extracts into a powder by spraying the liquid into a hot drying chamber where the moisture quickly evaporates, leaving behind fine particles of dry powder.	1.5	Data unavailable	The impact of the process has been little studied; no reliable information has been found	[143]

While the specific methods and technologies may vary significantly depending on the product type—juice, wine, or dried fruits—the overarching principles of food processing are fundamentally aligned. They focus on enhancing product appeal, ensuring safety, preserving nutritional and sensory qualities, and achieving economic benefits. As the food processing industry continues to evolve, these principles guide the development of innovative technologies and processing techniques that meet the growing demands of consumers for high-quality, safe, and flavorful food products. The processing methods and technologies discussed in the context of juices, wines, and dried products are deeply interconnected with the preservation and enhancement of resveratrol content in berry-derived products. Here is how these processes specifically relate to optimizing resveratrol levels.

Scientific research shows that resveratrol is sensitive to thermal degradation. Processes that use lower temperatures (e.g., cold pressing or freeze-drying) are more likely to preserve resveratrol content in berry products. In contrast, the high temperatures used in pasteurization or conventional drying may reduce resveratrol levels. Another aspect is time management: extended exposure to processing conditions can degrade resveratrol. Minimizing the process time, such as with flash pasteurization or quick drying methods, helps to preserve its content. It is worth noting that resveratrol oxidation can occur if berry products are exposed to air during processing. Techniques that limit oxygen exposure, such as vacuum-packing in dried products or inert gas-flushing during wine bottling, help maintain resveratrol integrity.

Ensuring the stability of berry products through pH control and microbial safety measures also indirectly benefits resveratrol preservation by preventing the conditions that might lead to its breakdown. Technological parameters as pH management are important in preserving berries. Maintaining an optimal pH during fermentation (wine making) not only influences the growth of yeast and bacteria but also affects the stability of resveratrol. A stable pH can help preserve resveratrol throughout the fermentation process. In drying processes, controlling humidity is crucial not only for moisture removal but also for preventing the hydrolytic degradation of sensitive compounds like resveratrol [144,145]. Utilizing advanced technologies such as high-pressure processing (HPP) or pulsed electric fields (PEF) can enhance the extraction and retention of resveratrol during juice processing by disrupting cell walls without using significant heat, thus releasing more resveratrol while maintaining its structure [146–150].

In summary, it can be concluded that the relationship between processing methods, common goals, and technological parameters is critical to how resveratrol is preserved and enhanced in berry products. By understanding and optimizing these aspects, producers can maximize the health benefits of resveratrol in their berry-derived products, making them more appealing to health-conscious consumers. These strategies highlight the importance of selecting appropriate processing methods to retain bioactive compounds, ultimately influencing the nutritional quality and market value of the final products.

4. Challenges and Future Development

The review of the relevant literature highlighted that retaining resveratrol in fresh berries involves navigating several challenges, such as environmental variability, the resource-intensive nature of optimal agricultural practices, genetic differences among berry varieties, and post-harvest degradation. Future strategies to overcome these issues include adopting precision agriculture to finetune growing conditions, developing genetically enhanced varieties with higher natural resveratrol production, improving post-harvest and storage technologies, and enhancing consumer awareness of resveratrol's benefits. Additionally, integrated supply chain management and supportive policies from governmental bodies can further aid in maintaining the high resveratrol content from the farm to the consumer. These efforts will require coordinated actions across multiple sectors, from agriculture to education and policy-making, to ensure berries retain their maximum health benefits.

Increasing the amount of resveratrol in processed berry products involves several techniques that focus on both maximizing the extraction from the berries and preserving the compound during processing. Our scientific literature analysis showed these to be the most effective strategies for the preservation of resveratrol in berry products. First of all, it should be mentioned that employing specific extraction methods can maximize the amount of resveratrol extracted from berries. Techniques like cold pressing, enzymatic extraction, or the use of solvents that effectively dissolve resveratrol can increase its concentration in the final product. Scientists note that it is useful to add antioxidants such as ascorbic acid (vitamin C) or other stabilizers during processing to protect resveratrol from oxidative degradation. These additives can help to maintain the stability and bioactivity of resveratrol throughout the shelf life of the product. Encapsulating resveratrol during processing can protect it from degradation due to environmental factors and enhance its absorption when consumed. Techniques like nanoencapsulation or liposomal encapsulation are effective in stabilizing resveratrol in food matrices [151–153].

To maximize resveratrol levels in products, innovative preservation techniques such as pulse electric fields or ultrasound treatments can be implemented. These methods disrupt the cell structures in berries, potentially increasing resveratrol release and preservation during processing [154]. Additionally, formulating products with resveratrol combined with other functional ingredients, like vitamin E, flavonoids from various plant sources, and omega-3 fatty acids, can stabilize resveratrol and enhance its bioavailability and efficacy, thanks to their synergistic antioxidant properties. It is essential to ensure that any method that is used complies with local and international food safety regulations [155–157]. Transparent labeling regarding the enhanced resveratrol content can inform consumers, empowering them to make health-conscious choices. Thus, consumers should be educated about resveratrol and its health benefits, as well as the available methods to preserve it in products [158,159].

Fermentation is the most significant and widely studied method at present. Fermenting berries can naturally increase the concentration of resveratrol. The fermentation process, especially when conducted under controlled conditions, can enhance the biosynthesis of resveratrol by activating certain pathways in the berries [160–162]. The authors would like to acknowledge that many scientists point out that more in-depth research is needed in this area.

Further investigation into the molecular mechanisms by which resveratrol exerts its effects can help to understand how it influences aging, cancer, cardiovascular health, and neurodegenerative diseases. Pinpointing specific pathways can lead to targeted therapeutic strategies and optimized dosages. The literature analysis shows that, despite the promising preclinical studies, there is a need for more comprehensive and large-scale clinical trials to confirm resveratrol's benefits in humans [163–166]. These studies should focus on specific health outcomes, optimal dosages, and long-term effects to validate its efficacy and safety. Developing new formulations that enhance the bioavailability and stability of resveratrol is crucial for maximizing its therapeutic potential. The research could focus on nanotechnology-based delivery systems, like liposomes or nanoparticles, to improve cellular uptake and targeted delivery. Also, investigating the synergistic effects of resveratrol with other compounds could lead to more effective therapies. Combining resveratrol with other nutraceuticals or pharmaceuticals might enhance its efficacy or mitigate the side effects. Scientists should stay updated on ongoing research in food science that could offer new insights or technologies for enhancing resveratrol content. Investing in research and development can lead to innovative solutions that keep your products at the forefront of health trends.

5. Conclusions

This paper reviewed and discussed the agricultural and processing techniques that influence the content of resveratrol in berries and berry products, and possible solutions for retention. The most significant conclusions are as follows:

Resveratrol is a versatile and valuable phytochemical with a wide range of sources and applications. Its benefits are leveraged in health and wellness industries, and ongoing research continues to uncover new potentials and refine existing technologies to harness its properties more effectively.

Various agricultural methods significantly influence the resveratrol content in berries. Resveratrol, which is primarily found in grapes and other berries, is affected by factors such as berry variety, soil quality, climate conditions, and agricultural techniques. Implementing effective agricultural techniques involves balancing the benefits and challenges. For instance, while water stress and biotic stress enhance resveratrol, they must be precisely managed to avoid harming the plant. Techniques like organic farming and cultivar selection offer sustainable benefits but come with higher costs and labor requirements.

Processing methods also significantly affect resveratrol content in berry products. Techniques that minimize heat and oxidative stress are more likely to preserve resveratrol. High-temperature processes like pasteurization reduce resveratrol levels. Techniques must be optimized to balance microbial safety, nutritional preservation, and resveratrol retention. Optimizing resveratrol content in berries involves a combination of precise agricultural practices, innovative processing techniques, and advanced research [167–169]. Balancing the benefits and challenges of these methods can enhance the nutritional and economic value of berry products, contributing to a healthier society.

Author Contributions: Conceptualization, A.I.; methodology, A.I., J.V. and I.K.; validation, A.I.; formal analysis and investigation, A.I.; data curation, A.I.; resources, A.I.; writing—original draft, A.I.; writing—review and editing, A.I., J.V. and I.K.; visualization, A.I.; supervision, A.I., J.V. and I.K.; project administration, A.I.; funding acquisition, I.K.; software, J.V. All authors have read and agreed to the published version of the manuscript.

Funding: This research received no external funding.

Data Availability Statement: Not applicable.

Conflicts of Interest: The authors declare no conflicts of interest.

References

1. Catalogna, G.; Moraca, F.; D'Antona, L.; Dattilo, V.; Perrotti, G.; Lupia, A.; Costa, G.; Ortuso, F.; Iuliano, R.; Trapasso, F.; et al. Review about the multi-target profile of resveratrol and its implication in the SGK1 inhibition. *Eur. J. Med. Chem.* **2019**, *183*, 111675. [CrossRef] [PubMed]
2. Wu, C.F.; Yang, J.Y.; Wang, F.; Wang, X.X. Resveratrol: Botanical origin, pharmacological activity and applications. *Chin. J. Nat. Med.* **2013**, *11*, 1–15. [CrossRef]
3. Wenzel, E.; Somoza, V. Metabolism and bioavailability of trans-resveratrol. *Mol. Nutr. Food Res.* **2005**, *49*, 472–481. [CrossRef] [PubMed]
4. Chen, K.H.; Cheng, M.L.; Jing, Y.H.; Chiu, D.T.; Shiao, M.S.; Chen, J.K. Resveratrol ameliorates metabolic disorders and muscle wasting in streptozotocin-induced diabetic rats. *Am. J. Physiol. Endocrinol. Metab.* **2011**, *301*, E853–E863. [CrossRef] [PubMed]
5. Springer, M.; Moco, S. Resveratrol and Its Human Metabolites-Effects on Metabolic Health and Obesity. *Nutrients* **2019**, *11*, 143. [CrossRef] [PubMed]
6. Iglesias-Aguirre, C.E.; Vallejo, F.; Beltran, D.; Berna, J.; Puigcerver, J.; Alajarin, M.; Selma, M.V.; Espin, J.C. 4-Hydroxydibenzyl: A novel metabolite from the human gut microbiota after consuming resveratrol. *Food Funct.* **2022**, *13*, 7487–7493. [CrossRef]
7. Bode, L.M.; Bunzel, D.; Huch, M.; Cho, G.S.; Ruhland, D.; Bunzel, M.; Bub, A.; Franz, C.; Kulling, S.E. In vivo and in vitrometabolism of trans-resveratrol by human gut microbiota. *Am. J. Clin. Nutr.* **2013**, *97*, 295–309. [CrossRef] [PubMed]
8. Wang, P.; Gao, J.; Ke, W.; Wang, J.; Li, D.; Liu, R.; Jia, Y.; Wang, X.; Chen, X.; Chen, F.; et al. Resveratrol reduces obesity in high-fat diet-fed mice via modulating the composition and metabolic function of the gut microbiota. Free Radic. *Biol. Med.* **2020**, *156*, 83–98. [CrossRef] [PubMed]
9. Berman, A.Y.; Motechin, R.A.; Wiesenfeld, M.Y.; Holz, M.K. The Therapeutic Potential of Resveratrol: A Review of Clinical Trials. *NPJ Precis. Oncol.* **2017**, *1*, 35. [CrossRef] [PubMed]
10. Pannu, N.; Archana Bhatnagar, A. Resveratrol: From enhanced biosynthesis and bioavailability to multitargeting chronic diseases. *Biomed. Pharmacother.* **2019**, *109*, 2237–2251. [CrossRef]
11. Baur, J.A.; Pearson, K.J.; Price, N.L.; Jamieson, H.A.; Lerin, C.; Kalra, A.; Prabhu, V.V.; Allard, J.S.; Lopez-Lluch, G.; Lewis, K.; et al. Resveratrol Improves Health and Survival of Mice on a High-Calorie Diet. *Nature* **2006**, *444*, 337–342. [CrossRef] [PubMed]
12. Tiao, M.-M.; Lin, Y.-J.; Yu, H.-R.; Sheen, J.-M.; Lin, I.-C.; Lai, Y.-J.; Tain, Y.-L.; Huang, L.-T.; Tsai, C.-C. Resveratrol Ameliorates Maternal and Post-Weaning High-Fat Diet-Induced Nonalcoholic Fatty Liver Disease via Renin-Angiotensin System. *Lipids Health Dis.* **2018**, *17*, 178. [CrossRef] [PubMed]
13. Xu, K.; Liu, S.; Zhao, X.; Zhang, X.; Fu, X.; Zhou, Y.; Xu, K.; Miao, L.; Li, Z.; Li, Y.; et al. Treating Hyperuricemia Related Non-Alcoholic Fatty Liver Disease in Rats with Resveratrol. *Biomed. Pharmacother.* **2019**, *110*, 844–849. [CrossRef] [PubMed]
14. Nutakul, W.; Sobers, H.S.; Qiu, P.; Dong, P.; Decker, E.A.; McClements, D.J.; Xiao, H. Inhibitory Effects of Resveratrol and Pterostilbene on Human Colon Cancer Cells: A Side-by-Side Comparison. *J. Agric. Food Chem.* **2011**, *59*, 10964–10970. [CrossRef] [PubMed]
15. Jiang, Z.; Chen, K.; Cheng, L.; Yan, B.; Qian, W.; Cao, J.; Li, J.; Wu, E.; Ma, Q.; Yang, W. Resveratrol and cancer treatment: Updates. *Ann. N. Y. Acad. Sci.* **2017**, *1403*, 59–69. [CrossRef] [PubMed]
16. Aggarwal, B.B.; Takada, Y.; Oommen, O.V. From chemoprevention to chemotherapy: Common targets and common goals. *Expert Opin. Investig. Drugs* **2004**, *13*, 1327–1338. [CrossRef] [PubMed]
17. Orallo, F. Comparative studies of the antioxidant effects of Cisand trans-resveratrol. *Curr. Med. Chem.* **2006**, *13*, 87–98. [CrossRef] [PubMed]
18. Iuga, C.; Alvarez-Idaboy, J.R.; Russo, N. Antioxidant activity of trans-resveratrol toward hydroxyl and hydroperoxyl radicals: A quantum chemical and computational kinetics study. *J. Org. Chem.* **2012**, *77*, 3868–3877. [CrossRef] [PubMed]
19. Das, S.; Das, D.K. Anti-inflammatory responses of resveratrol. *Inflamm. Allergy-Drug Targets (Former. Curr. Drug Targets-Inflamm. Allergy)* **2007**, *6*, 168–173. [CrossRef]
20. Timmers, S.; Auwerx, J.; Schrauwen, P. The journey of resveratrol from yeast to human. *Aging* **2012**, *4*, 146–153. [CrossRef]
21. Vestergaard, M.; Ingmer, H. Antibacterial and antifungal properties of resveratrol. *Int. J. Antimicrob. Agents* **2019**, *53*, 716–723. [CrossRef] [PubMed]
22. Hao, H.D.; He, L.R. Mechanisms of Cardiovascular Protection by Resveratrol. *J. Med. Food* **2004**, *7*, 290–298. [CrossRef] [PubMed]

23. Tomé-Carneiro, J.; Gonzálvez, M.; Larrosa, M.; Yáñez-Gascón, M.J.; García-Almagro, F.J.; Ruiz-Ros, J.A.; Conesa, M.T.G.; Tomás-Barberán, F.A.; Espín, J.C. One-year consumption of a grape nutraceutical containing resveratrol improves the inflammatory and fibrinolytic status of patients in primary prevention of cardiovascular disease. *Am. J. Cardiol.* **2012**, *110*, 356–363. [CrossRef] [PubMed]
24. Vitrac, X.; Desmoulière, A.; Brouillaud, B.; Krisa, S.; Deffieux, G.; Barthe, N.; Rosenbaum, J.; Mérillon, J.-M. Distribution of [14C]-trans-resveratrol, a cancer chemopreventive polyphenol, in mouse tissues after oral administration. *Life Sci.* **2003**, *72*, 2219–2233. [CrossRef] [PubMed]
25. Pozo-Guisado, E.; Merino, L.M.; Mulero-Navarro, S.; Lorenzo-Benayas, M.J.; Centeno, F.; Alvarez-Barrientos, A.; Salguero, P.M.F. Resveratrol-induced apoptosis inMCF-7 human breast cancer cells involves a caspase-independent mechanism with downregulation of Bcl-2 andNF-κB. *Int. J. Cancer* **2005**, *115*, 74–84. [CrossRef]
26. Rocha-González, H.I.; Ambriz-Tututi, M.; Granados-Soto, V. Resveratrol: A Natural Compound with Pharmacological Potential in Neurodegenerative Diseases. *CNS Neurosci. Ther.* **2008**, *14*, 234–247. [CrossRef] [PubMed]
27. Patel, K.R.; Scott, E.; Brown, V.A.; Gescher, A.J.; Steward, W.P.; Brown, K. Clinical trials of resveratrol. *Ann. N. Y. Acad. Sci.* **2011**, *1215*, 161–169. [CrossRef] [PubMed]
28. Öztürk, E.; Arslan, A.K.K.; Yerer, M.B.; Bishayee, A. Resveratrol and diabetes: A critical review of clinical studies. *Biomed. Pharmacother.* **2017**, *95*, 230–234. [CrossRef] [PubMed]
29. Kasprzak-Drozd, K.; Niziński, P.; Kasprzak, P.; Kondracka, A.; Oniszczuk, T.; Rusinek, A.; Oniszczuk, A. Does Resveratrol Improve Metabolic Dysfunction-Associated Steatotic Liver Disease (MASLD)? *Int. J. Mol. Sci.* **2024**, *25*, 3746. [CrossRef]
30. Cho, Y.J.; Kim, J.E.; Chun, H.S.; Kim, C.T.; Kim, S.S.; Kim, C.J. Contents of resveratrol in different parts of grapes. *Korean J. Food Sci. Technol.* **2003**, *35*, 306–308.
31. Vincenzi, S.; Tomasi, D.; Gaiotti, F.; Lovat, L.; Giacosa, S.; Torchio, F.; Segade, S.R.; Rolle, L. Comparative study of the resveratrol content of twenty-one Italian red grape varieties. *S. Afr. J. Enol. Vitic.* **2013**, *34*, 30–35. [CrossRef]
32. Romero-Pérez, A.I.; Ibern-Gómez, M.; Lamuela-Raventós, R.M.; De la Torre-Boronat, M.C. Piceid, the major resveratrol derivative in grape juices. *J. Agric. Food Chem.* **1999**, *47*, 1533–1536. [CrossRef]
33. Romero-Pérez, A.I.; Lamuela-Raventós, R.M.; Waterhouse, A.L.; De La Torre-Boronat, M. Levels of cis-and trans-resveratrol and their glucosides in white and rose Vitis vinifera wines from Spain. *J. Agric. Food Chem.* **1996**, *44*, 2124–2128. [CrossRef]
34. Ferrandino, A.; Lovisolo, C. Abiotic stress effects on grapevine (*Vitis vinifera* L.): Focus on abscisic acid-mediated consequences on secondary metabolism and berry quality. *Environ. Exp. Bot.* **2014**, *103*, 138–147. [CrossRef]
35. Wang, S.Y.; Chen, C.-T.; Wang, C.Y.; Chen, P. Resveratrol Content in Strawberry Fruit Is Affected by Preharvest Conditions. *J. Agric. Food Chem.* **2007**, *55*, 8269–8274. [CrossRef]
36. Geana, E.I.; Dinca, O.R.; Ionete, R.E.; Artem, V.; Niculescu, V.C. Monitoring trans-resveratrol in grape berry skins during ripening and in corresponding wines by HPLC. *Food Technol. Biotechnol.* **2015**, *53*, 73–80. [CrossRef]
37. Bartolomé, B.; Nuñez, V.; Monagas, M.; Gómez-Cordovés, C. In vitro antioxidant activity of red grape skins. *Eur. Food Res. Technol.* **2004**, *218*, 173–177. [CrossRef]
38. Storchi, P.; Puccioni, S.; Santini, C. 10—The case of resveratrol-enriched wines: Is it possible to create a new niche or to revitalize mature products? In *Case Studies in the Wine Industry*; Woodhead Publishing: Sawston, UK, 2019.
39. Tøíska, J.; Houška, M. Physical methods of resveratrol induction in grapes and grape products—A review. *Czech. J. Food Sci.* **2012**, *30*, 489–502. [CrossRef]
40. Kriel, C.; Pott, R.W.M. Quantification of Resveratrol in Different Parts of Solid Pinotage Winery Waste: Investigating the Variance between Consecutive Harvests. *S. Afr. J. Enol. Vitic.* **2020**, *41*, 121–127. [CrossRef]
41. Bábíková, P.; Vrchotová, N.; Tříska, J.; Kyseláková, M. Content of trans -resveratrol in leaves and berries of in-terspecific grapevine (*Vitis* sp.) varieties. *Czech J. Food Sci.* **2008**, *26*, 13–17. [CrossRef]
42. Goufo, P.; Singh, R.K.; Cortez, I. A reference listof phenolic compounds (including stilbenes) in grapevine (*Vitis vinifera* L.) roots, woods, canes, stems and leaves. *Antioxidants* **2020**, *9*, 398. [CrossRef]
43. Tian, B.; Liu, J. Resveratrol: A Review of Plant Sources, Synthesis, Stability, Modification and Food Application. *J. Sci. Food Agric.* **2020**, *100*, 1392–1404. [CrossRef]
44. Wang, Y.; Catana, F.; Yang, Y.; Roderick, R.; van Breemen, R.B. An LC-MS method for analyzing total resveratrol in grape juice, cranberry juice, and in wine. *J. Agric. Food Chem.* **2002**, *50*, 431–435. [CrossRef]
45. Chedea, V.S.; Vicaş, S.I.; Sticozzi, C.; Pessina, F.; Frosini, M.; Maioli, E.; Valacchi, G. Resveratrol: From diet to topical usage. *Food Funct.* **2017**, *8*, 3879–3892. [CrossRef]
46. De Lorgeril, M.; Salen, P.; Martin, J.L.; Boucher, F.; Paillard, F.; de Leiris, J. Wine drinking and risks of cardiovascular complications after recent acute myocardial infarction. *Circulation* **2002**, *106*, 1465–1469. [CrossRef]
47. Briggs, N.C.; Levine, R.S.; Bobo, L.D.; Haliburton, W.P.; Brann, E.A.; Hennekens, C.H. Wine drinking and risk of non- Hodgkin's lymphoma among men in the United States: A population-based case-control study. *Am. J. Epidemiol.* **2002**, *156*, 454–462. [CrossRef]
48. Platz, E.A.; Leitzmann, M.F.; Rimm, E.B.; Willett, W.C.; Giovannucci, E. Alcohol intake, drinking patterns, and risk of prostate cancer in a large prospective cohort study. *Am. J. Epidemiol.* **2004**, *159*, 444–453. [CrossRef]
49. Nayak, B.S.; Ramdath, D.D.; Marshall, J.R.; Isitor, G.N.; Eversley, M.; Xue, S.; Shi, J. Wound-healing activity of the skin of the common grape (Vitis Vinifera) variant, Cabernet Sauvignon. *Phytother. Res.* **2010**, *24*, 1151–1157. [CrossRef]

50. Iacopini, P.; Baldi, M.; Storchi, P.; Sebastiani, L. Catechin, epicatechin, quercetin, rutin and resveratrol in red grape: Content, in vitro antioxidant activity and interactions. *J. Food Compos. Anal.* **2008**, *21*, 589–598. [CrossRef]
51. Vanyo, H.; Yoncheva, T.; Dimitrov, D. Study of Resveratrol Content in Grapes and Wine of the Varieties Storgozia, Kaylashki Rubin, Trapezitsa, Rubin, Bouquet and Pinot Noir. 2017, pp. 300–311. Available online: https://www.researchgate.net/publication/319489770_Study_of_resveratrol_content_in_grapes_and_wine_of_the_varieties_Storgozia_Kaylashki_Rubin_Trapezitsa_Rubin_Bouquet_and_Pinot_Noir (accessed on 15 April 2024).
52. Weiskirchen, S.; Weiskirchen, R. Resveratrol: How Much Wine Do You Have to Drink to Stay Healthy? *Adv. Nutr.* **2016**, *7*, 706–718. [CrossRef]
53. Detenchuk, E.A.; Trebše, P.; Marjanović, A.; Kosyakov, D.S.; Ul'yanovskii, N.V.; Kralj, M.B.; Lebedev, A.T. Transformation of resveratrol under disinfectionconditions. *Chemosphere* **2020**, *260*, 127557. [CrossRef]
54. Gomes, J.S.; Rita, M.S.; Altomare, J.M.C.; Raposo, N.R.B.; Brandao, M.A.F.; Polonini, H.C. Resveratrol, Blueberry, Cranberry and Raspberry Exhibit Protection Effect to Photodamaged Hair. *J. Dermatol. Res. Ther.* **2018**, *4*, 61. [CrossRef]
55. Borowska, E.J.; Mazur, B.; Kopciuch, R.G.; Buszewski, B. Polyphenol, Anthocyanin and Resveratrol Mass Fractions and Antioxidant Properties of Cranberry Cultivars. *Food Technol. Biotechnol.* **2009**, *47*, 56.
56. Lyons, M.M.; Yu, C.; Toma, R.B.; Cho, S.Y.; Reiboldt, W.; Lee, J.; van Breemen, R.B. Resveratrol in raw and baked blue-berries and bilberries. *J. Agric. Food Chem.* **2003**, *51*, 5867–5870. [CrossRef]
57. Ma, Y.Q.; Han, C.R.; Liu, Y. Extraction of Resveratrol from Wild Blueberry (*Vaccinium uliginosum* L). *Adv. Mater. Res.* **2011**, *183–185*, 2032–2036. [CrossRef]
58. Xu, Y.; Fang, M.; Li, X.; Wang, D.; Yu, L.; Ma, F.; Jiang, J.; Zhang, L.; Li, P. Contributions of Common Foods to Resveratrol Intake in the Chinese Diet. *Foods* **2024**, *13*, 1267. [CrossRef]
59. Vang, O. What is new for resveratrol? Is a new set of recommendations necessary? *Ann. N. Y. Acad. Sci.* **2013**, *1290*, 1–11. [CrossRef]
60. Khattar, S.; Khan, S.A.; Zaidi, S.A.A.; Darvishikolour, M.; Farooq, U.; Naseef, P.P.; Kurunian, M.S.; Khan, M.Z.; Shamim, A.; Khan, M.M.U.; et al. Resveratrol from Dietary Supplement to a Drug Candidate: An Assessment of Potential. *Pharmaceuticals* **2022**, *15*, 957. [CrossRef] [PubMed] [PubMed Central]
61. Ubalde, J.M.; Sort, X.; Zayas, A.; Poch, R.M. Effects of soil and climatic conditions on grape ripening and wine quality of cabernet sauvignon. *J. Wine Res.* **2010**, *21*, 1–17. [CrossRef]
62. Rätsep, R.; Karp, K.; Maante-Kuljus, M.; Aluvee, A.; Bhat, R. Polyphenols and Resveratrol from Discarded Leaf Biomass of Grapevine (*Vitis* sp.): Effect of Cultivar and Viticultural Practices in Estonia. *Agriculture* **2020**, *10*, 393. [CrossRef]
63. Souid, I.; Toumi, I.; Hermosín-Gutiérrez, I.; Nasri, S.; Mliki, A.; Ghorbel, A. The effect of salt stress on resveratrol and piceid accumulation in two *Vitis vinifera* L. cultivars. *Physiol. Mol. Biol. Plants* **2019**, *25*, 625–635. [CrossRef] [PubMed] [PubMed Central]
64. Ojeda, H. Influence of pre-and postveraison water deficit on synthesis and concentration of skin phenolic compounds during berry growth of Vitis vinifera cv. *Shiraz. Am. J. Enol. Vitic.* **2002**, *53*, 261–267.
65. Villangó, S. The effect of postveraison water deficit on the phenolic composition and concentration of the Kékfrankos (*Vitis vinifera* L.) berry. *Sci. Hortic.* **2016**, *209*, 113–116. [CrossRef]
66. Chaves, M.M. Deficit irrigation in grapevine improves water-use efficiency while controlling vigour and production quality. *Ann. Appl. Biol.* **2007**, *150*, 237–252. [CrossRef]
67. Yin, X.; Singer, S.D.; Qiao, H.; Liu, Y.; Jiao, C.; Wang, H.; Li, Z.; Fei, Z.; Wang, Y.; Fan, C.; et al. Insights into the mechanisms underlying ultraviolet-c induced resveratrol metabolism in grapevine (*V. amurensis* rupr.) cv. "Tonghua-3". *Front. Plant Sci.* **2016**, *7*, 530. [CrossRef]
68. Wang, J.-F.; Ma, L.; Xi, H.-F.; Wang, L.-J.; Li, S.-H. Resveratrol synthesis under natural conditions and after UV-C irradiation in berry skin is associated with berry development stages in 'Beihong' (*V. vinifera* × *V. amurensis*). *Food Chem.* **2015**, *168*, 430–438. [CrossRef]
69. Crupi, P.; Pichierri, A.; Basile, T.; Antonacci, D. Postharvest stilbenes and flavonoids enrichment of table grape cv Redglobe (*Vitis vinifera* L.) as affected by interactive UV-C exposure and storage conditions. *Food Chem.* **2013**, *141*, 802–808. [CrossRef]
70. Wang, W.; Tang, K.; Yang, H.-R.; Wen, P.-F.; Zhang, P.; Wang, H.-L.; Huang, W.-D. Distribution of resveratrol and stilbene synthase in young grape plants (*Vitis vinifera* L. cv. Cabernet Sauvignon) and the effect of UV-C on its accumulation. *Plant Physiol. Biochem.* **2010**, *48*, 142–152. [CrossRef]
71. Cantos, E.; García-Viguera, C.; de Pascual-Teresa, S.; Tomás-Berberán, F.A. Effect of postharvest ultraviolet irradiation on resveratrol and other phenolics of cv. Napoleon table grapes. *J. Agric. Food Chem.* **2000**, *48*, 4606–4612. [CrossRef]
72. Braga, A.; Oliveira, J.; Silva, R.; Ferreira, P.; Rocha, I.; Kallscheuer, N.; Marienhagen, J.; Faria, N. Impact of the cultivation strategy on resveratrol production from glucose in engineered *Corynebacterium glutamicum*. *J. Biotechnol.* **2018**, *265*, 70–75. [CrossRef]
73. Versari, A.; Parpinello, G.P.; Tornielli, G.B.; Ferrarini, R.; Giulivo, C. Stilbene compounds and stilbene synthase expression during ripening, wilting, and UV treatment in grape cv. Corvina. *J. Agric. Food Chem.* **2001**, *49*, 5531–5536. [CrossRef]
74. De Souza, C.R.; Da Mota, R.V.; Dias, F.A.N.; De Melo, E.T.; Pimentel, R.M.D.A.; De Souza, L.C.; Regina, M.D.A. Physiological and agronomical responses of Syrah grapevine under protected cultivation. *Bragantia* **2015**, *74*, 270–278. [CrossRef]
75. Wang, S.Y.; Lin, H.S. Compost as a soil supplement increases the level of antioxidant compounds and oxygen radical absorbance capacity in strawberries. *J. Agric. Food Chem.* **2003**, *51*, 6844–6850. [CrossRef]

76. Carbone, F.; Preuss, A.; de Vos, R.C.; D'Amico, E.; Perrotta, G.; Bovy, A.G.; Martens, S.; Rosati, C. Developmental, genetic and environmental factors affect the expression of flavonoid genes, enzymes and metabolites in strawberry fruits. *Plant Cell Environ.* **2009**, *32*, 1117–1131. [CrossRef]
77. Lubin, B.-C.R.; Inbar, N.; Pinkus, A.; Stanevsky, M.; Cohen, J.; Rahimi, O.; Anker, Y.; Shoseyov, O.; Drori, E. Ecogeographic Conditions Dramatically Affect Trans-Resveratrol and Other Major Phenolics' Levels in Wine at a Semi-Arid Area. *Plants* **2022**, *11*, 629. [CrossRef]
78. Keylor, M.H.; Matsuura, B.S.; Stephenson, C.R.J. Chemistry and biology of resveratrol-derived natural products. *Chem. Rev.* **2015**, *115*, 8976–9027. [CrossRef]
79. Yang, B.; Zheng, J.; Laaksonen, O.; Tahvonen, R.; Kallio, H. Effects of latitude and weather conditions on phenolic compounds in currant (*Ribes* spp.) cultivars. *J. Agric. Food Chem.* **2013**, *61*, 3517–3532. [CrossRef]
80. Fernández-González, M.; Rodríguez-Rajo, F.J.; Escuredo, O.; Aira, M.J. Influence of Thermal Requirement in the Aerobiological and Phenological Behavior of Two Grapevine Varieties. *Aerobiologia* **2013**, *29*, 523–535. [CrossRef]
81. Dobrei, A.; Dobrei, A.G.; Nistor, E.; Poșta, G.; Mălăescu, M.; Balint, M. Characterization of Grape and Wine Quality Influenced by Terroir in Different Ecosystems from Romania Cultivated with Feteasca Neagra. *Sci. Pap. Ser. B Hortic.* **2018**, *62*, 247–253.
82. Cardozo, L.F.; Pedruzzi, L.M.; Stenvinkel, P.; Stockler-Pinto, M.B.; Daleprane, J.B.; Leite, M., Jr.; Mafra, D. Nutritional strategies to modulate inflammation and oxidative stress pathways via activation of the master antioxidant switch Nrf2. *Biochimie* **2013**, *95*, 1525–1533. [CrossRef]
83. Jiménez, J.B.; Orea, J.M.; Montero, C.; Ureña, Á.G.; Navas, E.; Slowing, K.; Gómez-Serranillos, M.P.; Carretero, E.; De Martinis, D. Resveratrol treatment controls microbial flora, prolongs shelf life, and preserves nutritional quality of fruit. *J. Agric. Food Chem.* **2005**, *53*, 1526–1530. [CrossRef]
84. Zhang, W.; Li, X.; Zheng, J.; Wang, G.; Sun, C.; Ferguson, I.B.; Chen, K. Bioactive components and antioxidant capacity of Chinese bayberry (*Myrica rubra* Sieb. and Zucc.) fruit in relation to fruit maturity and postharvest storage. *Eur. Food Res. Technol.* **2008**, *227*, 1091–1097. [CrossRef]
85. Jeandet, P.; Bessis, R.; Gautheron, B. The production of resveratrol (3,5,4′-trihydroxystilbene) by grape berries in different developmental stages. *Am. J. Enol. Vitic.* **1991**, *42*, 41–46. [CrossRef]
86. Tomás-Barberán, F.; Espín, J.C.; Cantos Villar, E. Postharvest Treatment of Fruits and Vegetables Using Ultraviolet Irradiation Pulses. Patent WO/2002/085137, 31 October 2002.
87. Vazquez-Cruz, M.A.; Jiménez-Garcia, S.N.; Torres-Pacheco, I.; Guzman-Maldonado, S.H.; Guevara-Gonzalez, R.G.; Miranda-Lopez, R. Effect of maturity stage an storage on flavor compounds and sensory description of berrycactus (*Myrtillocactus geometrizans*). *J. Food Sci.* **2012**, *77*, C366–C373.
88. Acquadro, S.; Appleton, S.; Marengo, A.; Bicchi, C.; Sgorbini, B.; Mandrone, M.; Gai, F.; Peiretti, P.G.; Cagliero, C.; Rubiolo, P. Grapevine green pruning residues as a promising and sustainable source of bioactive phenolic compounds. *Molecules* **2020**, *25*, 464. [CrossRef]
89. Sarig, P.; Zahavi, T.; Zutkhi, Y.; Yannai, S.; Lisker, N.; Ben-Arie, R. Ozone for control of post-harvest decay of table grapes caused by Rhizopus stolonifer. *Physiol. Mol. Plant Pathol.* **1996**, *48*, 403–415. [CrossRef]
90. Srivastava, A.; Akoh, C.C.; Yi, W.; Fischer, J.; Krewer, G. Effect of storage conditions on the biological activity of phenolics compounds of blueberry extract packed in glass bottles. *J. Agric. Food Chem.* **2007**, *55*, 2705–2713. [CrossRef]
91. Sun, Y.; Xi, B.; Dai, H. Effects of Water Stress on Resveratrol Accumulation and Synthesis in 'Cabernet Sauvignon' Grape Berries. *Agronomy* **2023**, *13*, 633. [CrossRef]
92. Bergqvist, J.; Dokoozlian, N.; Ebisuda, N. Sunlight Exposure and Temperature Effects on Berry Growth and Composition of Cabernet Sauvignon and Grenache in the Central San Joaquin Valley of California. *Am. J. Enol. Vitic.* **2001**, *52*, 1–7. [CrossRef]
93. Tariq, A.; Ambreen, A. *Plant Phenolics Production: A Strategy for Biotic Stress Management*; Springer: Singapore, 2024. [CrossRef]
94. Murtić, S.; Đurić, M.; Fazlić, J.; Behmen, F.; Haseljić, S.; Šerbo, A.; Berbić, H. Organic versus conventional 'Willamette' raspberry: Yield, bioactive compounds and antioxidant properties. *Hort. Sci.* **2023**, *50*, 275–282. [CrossRef]
95. Delmas, D.; Aires, V.; Limagne, E.; Dutartre, P.; Mazué, F.; Ghiringhelli, F. Transport, stability, and biological activity of resveratrol. *Ann. N. Y. Acad. Sci.* **2011**, *1215*, 48–59. [CrossRef] [PubMed]
96. Nunes, S.; Danesi, F.; Del Rio, D.; Silva, P. Resveratrol and inflammatory bowel disease: The evidence so far. *Nutr. Res. Rev.* **2017**, *31*, 85–97. [CrossRef]
97. Jimenez-Garcia, S.N.; Guevara-Gonzalez, R.G.; Miranda-Lopez, R.; Feregrino-Perez, A.A.; Torres-Pacheco, I.; Vazquez-Cruz, M.A. Functional Properties and Quality Characteristics of Bioactive Compounds in Berries: Biochemistry, Biotechnology, and Genomics. *Food Res. Int.* **2013**, *54*, 1195–1207. [CrossRef]
98. Klopotek, Y.; Otto, K.; Böhm, V. Processing strawberries to different products alters contents of vitamin C, total phenolics, total anthocyanins, and antioxidant capacity. *J. Agric. Food Chem.* **2005**, *53*, 5640–5646. [CrossRef]
99. Kopjar, M.; Tiban, N.N.; Pilizota, V.; Babic, J. Stability of anthocyanins, phenols and free radical scavenging activity through sugar addition during frozen storage of blackberries. *J. Food Process. Preserv.* **2009**, *33*, 1–11. [CrossRef]
100. Oke, M.; Jacob, J.K.; Paliyath, G. Biochemistry of Fruit Processing. In *Food Biochemistry and Food Processing*, 2nd ed.; John Wiley & Sons, Inc.: Hoboken, NJ, USA, 2012; pp. 554–568. [CrossRef]

101. Singla, R.K.; Dubey, A.K.; Ameen, S.M.; Montalto, S.; Parisi, S. Melanoidins and Browning Reactions in Processed Foods. Quantitative Determinations, Colour Measurement, and Sensorial Assessment. In *Analytical Methods for the Assessment of Maillard Reactions in Foods*; Springer: Cham, Switzerland, 2018; pp. 47–54. [CrossRef]
102. Meng, T.; Xiao, D.; Muhammed, A.; Deng, J.; Chen, L.; He, J. Anti-Inflammatory Action and Mechanisms of Resveratrol. *Molecules* **2021**, *26*, 229. [CrossRef] [PubMed] [PubMed Central]
103. Lastra, C.A.; Villegas, I. Resveratrol as an anti-inflammatory and anti-aging agent: Mechanisms and clinical implications. *Mol. Nutr. Food Res.* **2005**, *49*, 405–430. [CrossRef] [PubMed]
104. Singh, A.; Yau, Y.F.; Leung, K.S.; El-Nezami, H.; Lee, J.C.-Y. Interaction of Polyphenols as Antioxidant and Anti-Inflammatory Compounds in Brain–Liver–Gut Axis. *Antioxidants* **2020**, *9*, 669. [CrossRef] [PubMed]
105. Savouret, J.F.; Quesne, M. Resveratrol and cancer: A review. *Biomed. Pharmacother.* **2002**, *56*, 84–87. [CrossRef]
106. Salla, M.; Karaki, N.; Kaderi, B.; Ayoub, A.; Younes, S.; Chahla, M.; Baksh, S.; El Khatib, S. Enhancing the Bioavailability of Resveratrol: Combine It, Derivatize It, or Encapsulate It? *Pharmaceutics* **2024**, *16*, 569. [CrossRef]
107. De Vries, K.; Strydom, M.; Steenkamp, V. Bioavailability of resveratrol: Possibilities for enhancement. *J. Herb. Med.* **2018**, *11*, 71–77. [CrossRef]
108. Zhang, T.; Salmeron, I.; Perez-Vega, S.; Morales-Oyervides, L.; Montañez, J. Editorial: The role of food processing in the production and bioavailability of bio-compounds critical for boosting the immune system. *Front. Nutr.* **2022**, *9*, 1011211. [CrossRef] [PubMed]
109. Baur, J.A.; Sinclair, D.A. Therapeutic potential of resveratrol: The in vivo evidence. *Nat. Rev. Drug Discov.* **2006**, *5*, 493–506. [CrossRef] [PubMed]
110. Gescher, A.J.; Steward, W.P. Relationship between mechanisms, bioavailability, and preclinical chemopreventive efficacy of resveratrol: A conundrum. Cancer Epidemiol. *Biomark. Prev.* **2003**, *12*, 953–957.
111. Walle, T.; Hsieh, F.; DeLegge, M.H.; Oatis, J.E., Jr.; Walle, U.K. High absorption but very low bioavailability of oral resveratrol in humans. *Drug Metab. Dispos.* **2004**, *32*, 1377–1382. [CrossRef] [PubMed]
112. Cho, Y.-J.; Hong, J.-Y.; Chun, H.S.; Lee, S.K.; Min, H.-Y. Ultrasonication-assisted extraction of resveratrol from grapes. *J. Food Eng.* **2006**, *77*, 725–730. [CrossRef]
113. Cory, H.; Passarelli, S.; Szeto, J.; Tamez, M.; Mattei, J. The Role of Polyphenols in Human Health and Food Systems: A Mini-Review. *Front. Nutr.* **2018**, *5*, 87. [CrossRef]
114. Ifie, I.; Marshall, L.J.; Yildiz, F. Food processing and its impact on phenolic constituents in food. *Cogent Food Agric.* **2018**, *4*, 1507782. [CrossRef]
115. Ribas-Agustí, A.; Martín-Belloso, O.; Soliva-Fortuny, R.; Elez-Martínez, P. Food processing strategies to enhance phenolic compounds bioaccessibility and bioavailability in plant-based foods. *Crit. Rev. Food Sci. Nutr.* **2018**, *58*, 2531–2548. [CrossRef]
116. Hollands, W.; Brett, G.M.; Radreau, P.; Saha, S.; Teucher, B.; Bennett, R.N.; Kroon, P.A. Processing blackcurrants dramatically reduces the content and does not enhance the urinary yield of anthocyanins in human subjects. *Food Chem.* **2008**, *108*, 869–878. [CrossRef]
117. Minatel, I.O.; Borges, C.V.; Ferreira, M.I.; Gomez, H.A.G.; Chen, C.-Y.O.; Lima, G.P.P. Phenolic Compounds: Functional Properties, Impact of Processing and Bioavailability. In *Phenolic Compounds—Biological Activity*; InTech: London, UK, 2017. [CrossRef]
118. Musilova, J.; Lidikova, J.; Vollmannova, A.; Frankova, H.; Urminska, D.; Bojnanska, T.; Toth, T. Influence of Heat Treatments on the Content of Bioactive Substances and Antioxidant Properties of Sweet Potato (*Ipomoea batatas* L.) Tubers. *J. Food Qual.* **2020**, *2020*, 8856260. [CrossRef]
119. Lupu, M.; Puchianu, G.; Padureanu, V.; Canja, C.; Maier, A. The influence of ultrasound and pasteurisation on microbiological properties of grape juice. In Proceedings of the 18th International Multidisciplinary Scientific GeoConference SGEM, Albena, Bulgaria, 30 June–9 July 2018; pp. 7234–7728.
120. Howard, L.R.; Prior, R.L.; Liyanage, R.; Lay, J.O. Processing and storage effect on berry polyphenols: Challenges and implications for bioactive properties. *J. Agric. Food Chem.* **2012**, *60*, 6678–6693. [CrossRef] [PubMed]
121. Verbeyst, L.; Crombruggen, K.V.; van der Plancken, I.; Hendrickx, M.; van Loey, A. Anthocyanin degradation kinetics during thermal and high pressure treatments of raspberries. *J. Food Eng.* **2011**, *105*, 513–521. [CrossRef]
122. Genova, G.; Tosetti, R.; Tonutti, P. Berry ripening, pre-processing and thermal treatments affect the phenolic composition and antioxidant capacity of grape (*Vitis vinifera* L.) juice. *J. Sci. Food Agric.* **2015**, *96*, 664–671. [CrossRef] [PubMed]
123. Pap, N.; Mahosenaho, M.; Pongrácz, E.; Mikkonen, H.; Jaakkola, M.; Virtanen, V.; Myllykoski, L.; Horváth-Hovorka, Z.; Hodúr, C.; Vatai, G. Effect of ultrafiltration on anthocyanin and flavonol content of black currant juice (*Ribes nigrum* L.). *Food Bioprocess Technol.* **2012**, *5*, 921–928. [CrossRef]
124. Kårlund, A.; Moor, U.; Sandell, M.; Karjalainen, R.O. The Impact of Harvesting, Storage and Processing Factors on Health-Promoting Phytochemicals in Berries and Fruits. *Processes* **2014**, *2*, 596–624. [CrossRef]
125. Sarkis, J.R.; Jaeschke, D.P.; Tessaro, I.C.; Marczak, L.D.F. Effects of ohmic and conventional heating on anthocyanin degradation during the processing of blueberry pulp. *LWT Food Sci. Technol.* **2013**, *51*, 79–85. [CrossRef]
126. Hellström, J.; Mattila, P.; Karjalainen, R. Stability of anthocyanins in berry juices stored at different temperatures. *J. Food Compos. Anal.* **2013**, *31*, 12–19. [CrossRef]
127. Rinaldi, M.; Caligiani, A.; Borgese, R.; Palla, G.; Barbanti, D.; Massini, R. The effect of fruit processing and enzymatic treatments on pomegranate juice composition, antioxidant activity and polyphenols content. *LWT Food Sci. Technol.* **2013**, *53*, 355–359. [CrossRef]

128. Clodoveo, M.L.; Dipalmo, T.; Rizzello, C.G.; Corbo, F.; Crupi, P. Emerging technology to develop novel red winemaking practices: An overview. *Innov. Food Sci. Emerg. Technol.* **2016**, *38*, 41–56. [CrossRef]
129. Yang, K.-R.; Yu, H.-C.; Huang, C.-Y.; Kuo, J.-M.; Chang, C.; Shieh, C.-J.; Kuo, C.-H. Bioprocessed Production of Resveratrol-Enriched Rice Wine: Simultaneous Rice Wine Fermentation, Extraction, and Transformation of Piceid to Resveratrol from Polygonum cuspidatum Roots. *Foods* **2019**, *8*, 258. [CrossRef] [PubMed]
130. Karmakar, M.; Ray, R.R. A statistical approach for optimization of simultaneous production of β-glucosidase and endoglucanase by Rhizopus oryzae from solid-state fermentation of water hyacinth using central composite design. *Biotechnol. Res. Int.* **2011**, *2011*, 574983. [CrossRef] [PubMed]
131. Ma, T.; Wu, W.; Tu, Y.; Zhang, N.; Diao, Q. Resveratrol affects in vitro rumen fermentation, methane production and prokaryotic community composition in a time-and diet-specific manner. *Microb. Biotechnol.* **2020**, *13*, 1118–1131. [CrossRef] [PubMed]
132. Deveci, G.; Çelik, E.; Ağagündüz, D.; Bartkiene, E.; Rocha, J.M.F.; Özogul, F. Certain Fermented Foods and Their Possible Health Effects with a Focus on Bioactive Compounds and Microorganisms. *Fermentation* **2023**, *9*, 923. [CrossRef]
133. Ray, R.; Joshi, V. Fermented Foods: Past, Present and Future. In *Microorganisms and Fermentation of Traditional Foods*; CRC Press: Boca Raton, FL, USA, 2014.
134. Annunziata, G.; Arnone, A.; Ciampaglia, R.; Tenore, G.C.; Novellino, E. Fermentation of foods and beverages as a tool for increasing availability of bioactive compounds. *Focus on Short-Chain Fatty Acids. Foods* **2020**, *9*, 999. [PubMed]
135. Monagas, M.; Bartolome, B.; Gomez-Cordoves, C. Evolution of the polyphenols in red wine from *Vitis vinifera* L. during aging in the bottle. II. Non-anthocyanin phenolic compounds. *Eur. Food Res. Technol.* **2005**, *220*, 331–334. [CrossRef]
136. Sontakke, M.S.; Salve, S.P. Solar Drying Technologies: A Review. *Int. J. Eng. Sci.* **2015**, *4*, 29–35.
137. Abhay, S.; Hii, C.; Law, C.; Suzannah, S.; Djaeni, M. Effect of hot-air drying temperature on the polyphenol content and the sensory properties of cocoa beans. *Int. Food Res. J.* **2016**, *23*, 1479–1484.
138. Wojdyło, A.; Lech, K.; Nowicka, P.; Hernandez, F.; Figiel, A.; Carbonell-Barrachina, A.A. Influence of different drying techniques on phenolic compounds, antioxidant capacity and colour of *Ziziphus jujube* mill. Fruits. *Molecules* **2019**, *24*, 2361. [CrossRef]
139. Paniwnyk, L. Applications of ultrasound in processing of liquid foods: A review. *Ultrason. Sonochem.* **2017**, *38*, 794–806. [CrossRef]
140. Dzah, C.S.; Duan, Y.; Zhang, H.; Wen, C.; Zhang, J.; Chen, G.; Ma, H. The effects of ultrasound assisted extraction on yield, antioxidant, anticancer and antimicrobial activity of polyphenol extracts: A review. *Food Biosci.* **2020**, *35*, 100547. [CrossRef]
141. Almeida, H.; Ferreira, B.; Fernandes-Lopes, C.; Araújo, F.; Bonifácio, M.J.; Vasconcelos, T.; Sarmento, B. Third-Generation Solid Dispersion Through Lyophilization Enhanced Oral Bioavailability of Resveratrol. *ACS Pharmacol. Transl. Sci.* **2024**, *7*, 888–898. [CrossRef] [PubMed]
142. Flink, J.M. *An Introduction to Freeze Drying*; Knudsen, H., Ed.; Bogtryk Offset: Valby, Denmark, 1983.
143. Bustos, M.C.; Rocha-Parra, D.; Sampedro, I.; De Pascual-Teresa, S.; León, A.E. The Influence of Different Air-Drying Conditions on Bioactive Compounds and Antioxidant Activity of Berries. *J. Agric. Food Chem.* **2018**, *66*, 2714–2723. [CrossRef] [PubMed]
144. Robinson, K.; Mock, C.; Liang, D. Pre-formulation studies of resveratrol. *Drug Dev. Ind. Pharm.* **2015**, *41*, 1464–1469. [CrossRef] [PubMed] [PubMed Central]
145. Zupančič, Š.; Lavrič, Z.; Kristl, J. Stability and solubility of trans-resveratrol are strongly influenced by pH and temperature. *Eur. J. Pharm. Biopharm.* **2015**, *93*, 196–204. [CrossRef] [PubMed]
146. Casazza, A.; Aliakbarian, B.; Montoya, E.; Perego, P. T-resveratrol recovery from grape skins using high pressure and temperature extraction. *J. Biotechnol.* **2010**, *150*, 333. [CrossRef]
147. Kaushik, N.; Kaur, B.P.; Rao, P.S.; Mishra, H. Effect of high pressure processing on color, biochemical and microbiological characteristics of mango pulp (*Mangifera indica* cv. Amrapali). *Innov. Food Sci. Emerg. Technol.* **2014**, *22*, 40–50. [CrossRef]
148. Yi, J.; Kebede, B.T.; Dang, D.N.H.; Buvé, C.; Grauwet, T.; Van Loey, A.; Hu, X.; Hendrickx, M. Quality change during high pressure processing and thermal processing of cloudy apple juice. *LWT* **2017**, *75*, 85–92. [CrossRef]
149. Marszałek, K.; Woźniak, Ł.; Kruszewski, B.; Skąpska, S. The Effect of High Pressure Techniques on the Stability of Anthocyanins in Fruit and Vegetables. *Int. J. Mol. Sci.* **2017**, *18*, 277. [CrossRef] [PubMed] [PubMed Central]
150. Chatzimitakos, T.; Athanasiadis, V.; Kalompatsios, D.; Mantiniotou, M.; Bozinou, E.; Lalas, S.I. Pulsed Electric Field Applications for the Extraction of Bioactive Compounds from Food Waste and By-Products: A Critical Review. *Biomass* **2023**, *3*, 367–401. [CrossRef]
151. Marinheiro, D.; Ferreira, B.J.M.L.; Oskoei, P.; Oliveira, H.; Daniel-Da-Silva, A.L. Encapsulation and Enhanced Release of Resveratrol from Mesoporous Silica Nanoparticles for Melanoma Therapy. *Materials* **2021**, *14*, 1382. [CrossRef] [PubMed]
152. Balanč, B.; Trifkovic, K.; Pravilovic, R.; Djordjevic, V.; Markovic, S.; Nedovic, V.; Bugarski, B. Encapsulation of resveratrol in spherical particles of food grade hydrogels. *Food Feed. Res.* **2017**, *44*, 23–29. [CrossRef]
153. Davidov-Pardo, G.; McClements, D.J. Resveratrol encapsulation: Designing delivery systems to overcome solubility, stability and bioavailability issues. *Trends Food Sci. Technol.* **2014**, *38*, 88–103. [CrossRef]
154. Syed, Q.A.; Syed, A.I.Q.A.; Rahman, S.A.U.U.; Shukat, R. Pulsed Electric Field Technology in Food Preservation: A Review. *J. Nutr. Health Food Eng.* **2017**, *6*, 168–172. [CrossRef]
155. Serini, S.; Cassano, R.; Corsetto, P.A.; Rizzo, A.M.; Calviello, G.; Trombino, S. Omega-3 PUFA Loaded in Resveratrol-Based Solid Lipid Nanoparticles: Physicochemical Properties and Antineoplastic Activities in Human Colorectal Cancer Cells In Vitro. *Int. J. Mol. Sci.* **2018**, *19*, 586. [CrossRef] [PubMed]

156. Santaniello, M.; Giannini, G. A Versatile and Stable Mixture of Fish Oil and Resveratrol in a Powder Formulation. *Nutr. Food Technol.* **2016**, *2*, 1–5. Available online: https://www.researchgate.net/publication/303880940_A_Versatile_and_Stable_Mixture_of_Fish_Oil_and_Resveratrol_in_a_Powder_Formulation (accessed on 20 April 2024).
157. Arab-Tehrany, E.; Jacquot, M.; Gaiani, C.; Imran, M.; Desobry, S.; Linder, M. Beneficial effects and oxidative stability of omega-3 long-chain polyunsaturated fatty acids. *Trends Food Sci. Technol.* **2011**, *25*, 24–33. [CrossRef]
158. Freeman, A. Transparency for Food Consumers: Nutrition Labeling and Food Oppression. *Am. J. Law Med.* **2015**, *41*, 315. [CrossRef] [PubMed]
159. Reed, G.; Phartiyal, P. Transparency in Food Labeling: Food Labels Inform Consumer Choices—And Industry Pushes Back. 2016. Available online: https://www.researchgate.net/publication/307908554_Transparency_in_Food_Labeling_Food_Labels_Inform_Consumer_Choices-and_Industry_Pushes_Back?channel=doi&linkId=57d1801508ae0c0081e02516&showFulltext=true (accessed on 20 April 2024). [CrossRef]
160. Surguladze, M.A.; Bezhuashvili, M.G. Impact of wine technology on the variability of resveratrol and piceids in Saperavi (*Vitis vinifera* L.). *Ann. Agrar. Sci.* **2017**, *15*, 137–140. [CrossRef]
161. Atanacković, M.; Petrović, A.; Jović, S.; Gojković- Bukarica, L.; Bursać, M.; Cvejić, J. Influence of winemaking techniques on the resveratrol content, total phenolic content and antioxidant potential of red wines. *Food Chem.* **2012**, *131*, 513–518. [CrossRef]
162. Cvejic, J.M.; Djekic, S.V.; Petrovic, A.V.; Atanackovic, M.T.; Jovic, S.M.; Brceski, I.D.; Gojkovic-Bukarica, L.C. Determination of trans- and cis-Resveratrol in Serbian Commercial Wines. *J. Chromatogr. Sci.* **2010**, *48*, 229–234. [CrossRef] [PubMed]
163. Muñoz, O.; Muñoz, R.; Bustamante, S. Pharmacological Properties of Resveratrol. A Pre-Clinical and Clinical Review. *Biochem. Pharmacol.* **2015**, *4*, 184. [CrossRef]
164. Singh, A.P.; Singh, R.; Verma, S.S.; Rai, V.; Kaschula, C.H.; Maiti, P.; Gupta, S.C. Health benefits of resveratrol: Evidence from clinical studies. *Med. Res. Rev.* **2019**, *39*, 1851–1891. [CrossRef] [PubMed]
165. Athar, M.; Back, J.H.; Tang, X.; Kim, K.H.; Kopelovich, L.; Bickers, D.R.; Kim, A.L. Resveratrol: A review of preclinical studies for human cancer prevention. *Toxicol. Appl. Pharmacol.* **2007**, *224*, 274–283. [CrossRef] [PubMed]
166. Samuel, V.P.; Gupta, G.; Dahiya, R.; Jain, D.A.; Mishra, A.; Dua, K. Current Update on Preclinical and Clinical Studies of Resveratrol, a Naturally Occurring Phenolic Compound. *Crit. Rev. Eukaryot. Gene Expr.* **2019**, *29*, 529–537. [CrossRef] [PubMed]
167. Skroza, D.; Generalić Mekinić, I.; Svilović, S.; Šimat, V.; Katalinić, V. Investigation of the potential synergistic effect of resveratrol with other phenolic compounds: A case of binary phenolic mixtures. *J. Food Compos. Anal.* **2015**, *38*, 13–18. [CrossRef]
168. Boța, M.; Vlaia, L.; Jîjie, A.-R.; Marcovici, I.; Crişan, F.; Oancea, C.; Dehelean, C.A.; Mateescu, T.; Moacă, E.-A. Exploring Synergistic Interactions between Natural Compounds and Conventional Chemotherapeutic Drugs in Preclinical Models of Lung Cancer. *Pharmaceuticals* **2024**, *17*, 598. [CrossRef]
169. Prakash, V.; Bose, C.; Sunilkumar, D.; Cherian, R.M.; Thomas, S.S.; Nair, B.G. Resveratrol as a Promising Nutraceutical: Implications in Gut Microbiota Modulation, Inflammatory Disorders, and Colorectal Cancer. *Int. J. Mol. Sci.* **2024**, *25*, 3370. [CrossRef] [PubMed] [PubMed Central]

MDPI AG
Grosspeteranlage 5
4052 Basel
Switzerland
Tel.: +41 61 683 77 34

Processes Editorial Office
E-mail: processes@mdpi.com
www.mdpi.com/journal/processes

www.ingramcontent.com/pod-product-compliance
Lightning Source LLC
Chambersburg PA
CBHW061505220926
51850CB00004B/172